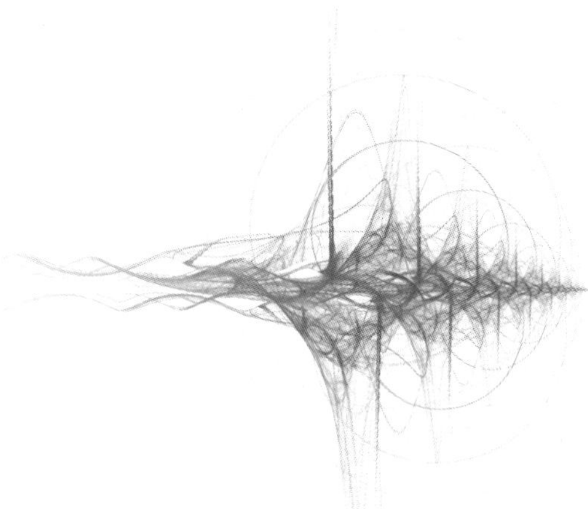

STOCHASTIC PROCESSES, FINANCE and CONTROL

A Festschrift in Honor of Robert J. Elliott

Advances in Statistics, Probability and Actuarial Science
(ISSN: 2010-1317)

Published

Vol. 1 Stochastic Processes, Finance and Control:
A Festschrift in Honor of Robert J Elliott
edited by Samuel N Cohen, Dilip Madan, Tak Kuen Siu and Hailiang Yang

Advances in Statistics, Probability and Actuarial Science – Vol. 1

STOCHASTIC PROCESSES, FINANCE and CONTROL

A Festschrift in Honor of Robert J. Elliott

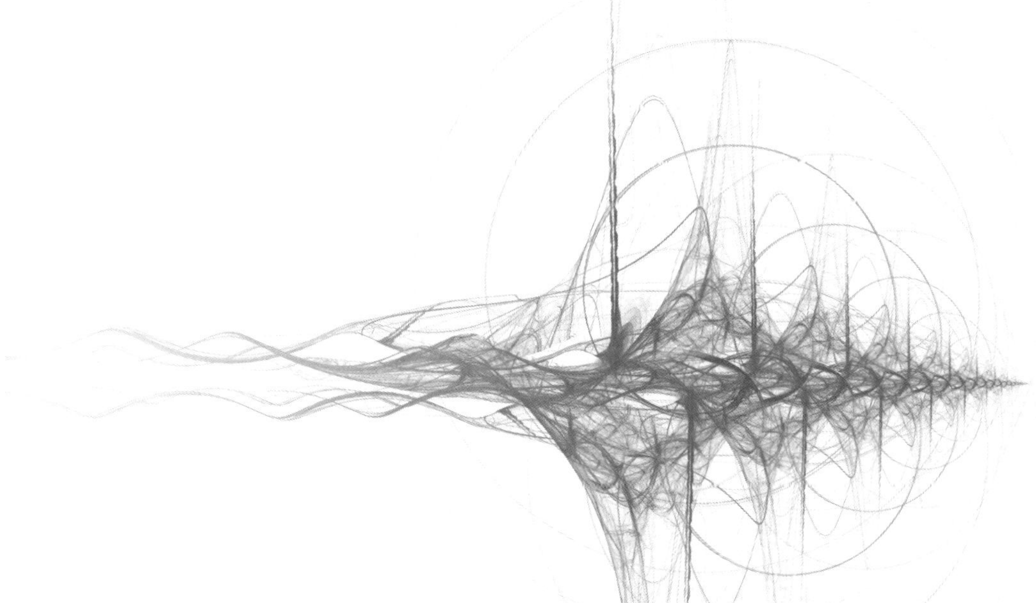

Editors

Samuel N. Cohen (University of Oxford, UK)

Dilip Madan (University of Maryland, USA)

Tak Kuen Siu (Macquarie University, Australia)

Hailiang Yang (The University of Hong Kong, Hong Kong)

World Scientific

NEW JERSEY • LONDON • SINGAPORE • BEIJING • SHANGHAI • HONG KONG • TAIPEI • CHENNAI

Published by

World Scientific Publishing Co. Pte. Ltd.
5 Toh Tuck Link, Singapore 596224
USA office: 27 Warren Street, Suite 401-402, Hackensack, NJ 07601
UK office: 57 Shelton Street, Covent Garden, London WC2H 9HE

British Library Cataloguing-in-Publication Data
A catalogue record for this book is available from the British Library.

Advances in Statistics, Probability and Actuarial Science — Vol. 1
STOCHASTIC PROCESSES, FINANCE AND CONTROL
A Festschrift in Honor of Robert J Elliott

ISBN 978-981-4383-30-1

Printed in Singapore by Mainland Press Pte Ltd.

Preface

This Festschrift is dedicated to Robert James Elliott to celebrate his seventieth birthday. It is also a tribute of our admiration for Robert's many significant and continuing contributions to stochastic processes, filtering, control and their applications in finance and electrical engineering through his distinguished and long academic career (almost 50 years!). We, as Robert's collaborators and (grand-)students, have greatly benefited from Robert's friendship, mentoring and enthusiasm. Robert is not only a distinguished scholar and a great teacher, but also a very nice person. This view is shared by Robert's many students and collaborators, as well as all other people who know Robert. Indeed, Robert has nurtured several generations of academics who are now renowned scholars in their fields.

Robert was born in Derbyshire in the English Midlands in 1940, one year after the outbreak of World War II. Robert received his BA and MA at Oxford University in the early 1960s. He received his PhD at Cambridge University under the supervision of John Williamson in 1965, thus making him a mathematical descendant of G.H. Hardy and Sir Isaac Newton. He received his DSc from Cambridge University in 1983. Robert began his academic career in the United Kingdom as a Lecturer at the University of Newcastle-upon-Tyne. He then held academic positions in various places including, in chronological order, Yale, Oxford, Warwick and Northwestern. He took up a Professorship in Pure Mathematics at the University of Hull in the United Kingdom in 1973 when he was only 33.

Following this, he held various senior academic positions, including the G.F. Grant Professorship of Mathematics and Head of Department at the University of Hull, Professor of Mathematics and A.F. Collins Professor of Finance at the University of Alberta, and Royal Bank of Canada Professor of Finance at the University of Calgary. He is currently an Australian Professorial Fellow at the University of Adelaide. Robert holds a number of Adjunct Professorships in places such as the Australian National University, the University of Calgary and McMaster University. Robert was also a Visiting Professor at Brown University, the University of Maryland, University of Kentucky, University of Bonn, University of Konstanz, the Technical University of Denmark, University of Western Australia, Université d'Evry and Université de Paris IX.

Looking at Robert's impressive CV, you can find everything that should be used to describe a distinguished world-leading scholar. Robert has authored over 450 research papers and nine monographs across different fields including abstract harmonic analysis, hypoelliptic operators, differential games, stochastic processes, calculus, filtering, control and mathematical finance. Many of his papers and monographs are highly influential and have become classics. He is the author of the monograph "Stochastic Calculus and Applications", which was the first book to introduce the French general theory of stochastic processes to the English speaking world and is still a standard reference for Stochastic Calculus. This monograph was translated into Russian. It has influenced several generations of outstanding researchers in Stochastic Calculus and related fields, and has been highly cited, not only in the world of Stochastics, but also in disciplines such as finance, economics and electrical engineering. In particular, this monograph has been cited in many path-breaking works in (quantitative) finance. He authored, (with Lahkdar Aggoun and John Moore), the monograph "Hidden Markov Models" which is frequently referenced in the engineering literature and has been reprinted twice. His monograph (with P. Ekkehard Kopp) "Mathematics of Financial Markets" has been adopted in major programs in Financial Mathematics at top universities worldwide. This monograph has been reprinted three times. The Hungarian edition was published in 2000 and the second edition was published in September 2004. An edition in China was published in 2010.

It is difficult to summarize Robert's many distinguished and outstanding contributions to such a wide range of areas. Following Churchill's dictum, "Continuous effort—not strength or intelligence—is the key to unlocking our potential", we shall try.

Robert started working on functional analysis, in particular abstract harmonic analysis, in the 1960s. His PhD thesis concerned "Spectral Synthesis". Robert's work on abstract harmonic analysis was awarded a Senior Mathematics Prize by Oxford University. Returning from Yale to Oxford, Robert talked with Sir Michael Atiyah and started working on pseudo-differential operators and hypo-elliptic operators in the late 1960s.

After taking up a position at University of Warwick, Robert's interests moved to control theory and differential games (i.e., control with competing controllers) in the 1970s. He wrote (with Nigel Kalton) a Memoir of the American Mathematical Society, where they proved a dynamic version of von-Neumann's saddle point theorem. This theorem is known as the Elliott–Kalton theorem and is highly influential in differential games. In particular, it has been cited in a number of works on differential games by Wendell H. Fleming and others. Robert's work on differential games resulted in an invitation from Avner Friedman to visit Northwestern University in the mid 1970s, where Robert's interest in stochastic processes began.

With his mastery of the general theory of stochastic processes, in particular semimartingale theory, Robert published a number of fundamental and influential results in stochastic processes, calculus and control in the late 1970s. A notable contribution was his beautiful proof of a stochastic minimum principle using the unique decomposition of special semi-martingales. This work was published in the Bulletin of the American Mathematical Society in 1976 and has influenced several generations of researchers in stochastic optimal control theory. Robert also made many fundamental contributions to the study of discontinuous processes and their stochastic calculus in the late 1970s. In the 1980s and the early 1990s, Robert continued to make contributions to stochastic calculus and control theory. Armed with powerful mathematical concepts and tools such as stochastic flows, the general differentiation rule and martingale representations, Robert developed many novel results for stochastic calculus and control, in particular, stochastic minimum principles for both completely observed and partially observed stochastic systems.

Robert also developed an interest in the Stochastic Calculus of Variation (Malliavin calculus). He wrote a number of papers (with Michael Kohlmann), on the topic. They adopted the concept of stochastic flows and applied iteratively a martingale representation to establish a Chaos expansion. Using the Chaos expansion, they established an integration-by-parts formula (the heart of Malliavin calculus), and proved the smoothness of densities of solutions of stochastic differential equations with and without jumps. The novelty was that they established these major results in Malliavin calculus without using heavy machinery from infinite-dimensional calculus. Robert's work (with Michael Kohlmann) was cited in the classic monograph "Stochastic Analysis" written by Paul Malliavin.

Robert focused on filtering, in particular filtering hidden Markov models, in the 1990s. He had previously derived many beautiful filtering equations using the unique decomposition of special semimartingales and a reference probability approach in the early 1980s (see Chapter 18 of the monograph "Stochastic Calculus and Applications"). Robert pioneered a canonical representation of the space state of a finite state Markov chain using a set of standard unit vectors in the Euclidean space, which facilitates the use of techniques from linear algebra. Using the canonical representation of the state space, the semimartingale dynamics for the Markov chain can be written down. The equation is a fundamental result which plays a key role in deriving filtering equations such as analogs of Zakai's stochastic differential equations. In recent years, these semimartingale dynamics have been widely used in mathematical finance, in particular regime-switching finance and asset pricing.

With his mastery of the change of measure techniques, Robert derived many novel and practically useful results for filtering and control of hidden Markov models. These results have been widely used in electrical engineering, in particular in signal processing and communication systems, and are highly regarded in these fields. Robert has also worked with many world-leading researchers in these fields, and a number of them have contributed to this volume. In addition, Robert's work on filtering and control of hidden Markov models also inspired some work by leading economists. In particular, Robert's monograph (with Lahkdar Aggoun and John Moore) on "Hidden Markov Models" has been cited in works by Thomas Sargent, a Nobel Laureate in Economics in 2011.

Robert's interest in mathematical finance began in the late 1980s. He has made significant contributions to many areas of mathematical finance, including option pricing and hedging, asset allocation, risk measures, financial signal processing, credit risk analysis, stochastic interest rate models, stochastic volatility, and many others. Recently, he has worked with backward stochastic differential equations and nonlinear expectations in finance, as well as regime-switching models. Robert also has interests in actuarial mathematics.

Robert has contributed to many major professional activities of the international academic community. He has been invited by the Royal Swedish Academy to nominate for the Nobel Prize in Economics each year since 2002. He has served as a member and Chair of selection committees for the National Sciences and Engineering Research Council of Canada, and the Social Sciences and Humanities Research Council of Canada. He is an

Associate Editor of Mathematical Finance, Stochastics, Stochastic Analysis and Applications and the Canadian Applied Mathematics Quarterly.

We wish to express our admiration to Robert for his achievements and hope he will enjoy many more years of good health and happiness.

Sam Cohen
Dilip Madan
Tak Kuen Siu
Hailiang Yang
Editors

Robert James Elliott

Contents

PART 1
Stochastic Analysis

Chapter 1

On the connection between discrete and continuous Wick calculus with an application to the fractional Black-Scholes model

Christian Bender and Peter Parczewski

Department of Mathematics, Saarland University,
PO Box 151150, D-66041 Saarbrücken, Germany
bender@math.uni-sb.de, parcz@math.uni-sb.de

Dedicated to Robert J. Elliott

In this chapter we discuss some questions related to the discrete Wick calculus which are motivated by a joint work of Robert Elliott with the first author [3]. One question is to what extent the connection between discrete and continuous Wick calculus can be made rigorous in terms of weak convergence results. Another question is how to sample from difference equations which involve discrete Wick products. The results are applied to financial models driven by a fractional Brownian motion.

Contents

1.1. Introduction

During a research stay of the first author at Calgary in 2002, Robert Elliott suggested to study arbitrage issues in a discrete version of the fractional Black-Scholes model. The fractional Black-Scholes model can be considered as a generalization of the classical Black-Scholes model where the driving Brownian motion is replaced by a fractional Brownian motion. In the context of stochastic integration with respect to a fractional Brownian motion in terms of Wick calculus, this model was discussed by Hu and Øksendal [18] and Elliott and van der Hoek [10]. In order to avoid the technicalities of the continuous Wick calculus, in the discrete version of Bender and Elliott [3] the fractional Brownian motion is replaced by a disturbed binary random walk approximation due to Sottinen [31] and the continuous Wick product is replaced by a discrete analogue acting on binary random variables. This discrete Wick product was introduced by Holden et al. [15, 16] and motivated by formal analogies in terms of the chaos decomposition. Bender and Elliott [3] showed that the discrete version of the fractional Black-Scholes model, defined via discrete Wick calculus, admits arbitrage opportunities provided the number of time steps is sufficiently large. Indeed, an arbitrage can be obtained, because the stock price will move upwards once more, if it has increased for sufficiently many previous steps, due to the memory effects of the random walk approximation to the fractional Brownian motion.

Two questions concerning this discrete model are posed in Remark 4.1 of [3], which are also of interest from the more general point of view of approximating stochastic differential equations driven by a fractional Brownian motion by discrete Wick difference equations:

(1) Does this discrete version of the fractional Black-Scholes model weakly converge to its continuous counterpart? Based on some heuristic arguments, it was conjectured that the answer is yes.
(2) How can one draw sample paths of this discrete version of the fractional Black-Scholes model? Here the main difficulty is that the discrete Wick product cannot be evaluated pathwise, but may, in principle, depend on all other paths.

The present paper is devoted to answering these two questions. Concerning the first question, a proof of weak convergence of this discrete fractional Black-Scholes model was recently provided by Bender and Parczewski [4] in a more general study of linear difference equations involving discrete Wick products. It is based on so-called Wick-analytic functions, and we will here present the main ideas of the proof without going into the more technical estimates. Concerning the second question, we suggest an algorithm for generating sample paths of a discrete Wick difference equation. The main idea is to replace the discrete Wick products by an ordinary product plus a trace term which involves a discrete Malliavin derivative. This approach can be seen as a discrete analogue of a formula by Alòs et al. [1] for rewriting a Wick-Itô integral with respect to a fractional Brownian motion as a pathwise integral plus a trace term. The numerical experiments presented at the end of the paper are, to the best of our knowledge, the first ones performed for discrete Wick difference equations.

We also present a parallel introduction to some aspects of the continuous and discrete Wick calculus in Section 1.2. Our approach for motivating the discrete Wick product is different from the one usually provided in the literature. Instead of exploiting analogies in terms of the chaos decomposition as in [13, 15, 16], our starting point for the introduction of the continuous and discrete Wick product is a functional equation for Wick exponentials. These Wick exponentials can be easily interpreted as stochastic Doléans-Dade exponentials. The defining property of the Wick products is then to enforce that the Wick product of two Wick exponentials equals the Wick exponential of the sum of the exponents. We do not merely give an introduction to discrete Wick exponentials, discrete Wick products, and discrete Wick power series by formal analogies to their continuous counterparts, but also provide weak convergence results for these objects.

The paper is organized as follows: In Section 1.2 we introduce some basic objects of the continuous and discrete Wick calculus, namely (discrete) Wiener integrals and Wick exponentials, continuous and discrete Wick products, and continuous and discrete Wick powers and Wick power series. The main result of this section is Theorem 1.3 which yields weak convergence of discrete Wick power series to their continuous counterparts under rather weak assumptions. Section 1.3 is devoted to the fractional Black-Scholes model. We first briefly discuss the model and its discrete version based on a discrete Wick difference equation due to [3]. Then we explain how the results in Section 1.2 can be applied to prove weak convergence of the discrete version of the fractional Black-Scholes model to the continuous model in the Skorokhod space, following the arguments of [4] In Section 1.4, we finally illustrate the weak convergence result by some numerical experiments. Recalling that the discrete Wick product cannot be calculated path by path, we also discuss how much information regarding the whole set of sample paths is required to generate a single path of the discrete fractional Black-Scholes model.

1.2. Continuous and discrete Wick calculus

In this section we give an introduction to the Wick product in a Brownian motion setting and a discrete analogue based on the binary random walk approximation of Brownian motion. In physics, the Wick product first appears in the study of Heisenberg's S-matrix [32]. Its related but different probabilistic counterpart, which we consider throughout this paper, was first defined by Hida and Ikeda [14]. We refer to [12] for a comparison of the physical and probabilistic Wick products. Based on a formal analogy in terms of the chaos decomposition, Holden et al. [15, 16] introduced the notion of the discrete Wick product. We will here avoid the technicalities related to the chaos decomposition and, instead, motivate continuous and discrete Wick products in terms of a functional identity for Wick exponentials. The definitions related to discrete and continuous Wick calculus are not only justified by formal analogies, but we also provide some convergence results in the weak sense.

1.2.1. *Wiener integrals and Wick exponentials*

We suppose that $(\Omega, \mathcal{F}, P)$ is a probability space which carries a Brownian motion $(B_t)_{t\in[0,1]}$ on the interval $[0,1]$. We shall assume that $\mathcal{F}$ is the σ-field generated by the Brownian motion. For every $f \in L^2([0,1])$, we denote by $I(f) := \int_0^1 f(s)dB_s$ the Wiener integral of f with respect to the Brownian motion B. Recall that the Wiener integral $I(f)$ is a centered Gaussian random variable and the covariance of two Wiener integrals is given by $E[I(f)I(g)] = \int_0^1 f(s)g(s)ds$.

In view of Donsker's theorem a discrete counterpart of the Brownian motion is given by a binary random walk

$$B_l^n = \frac{1}{\sqrt{n}} \sum_{i=1}^{l} \xi_i^n, \quad l = 0, \ldots, n,$$

for every $n \in \mathbb{N}$. Here, $(\xi_1^n, \ldots, \xi_n^n)$ is an n-tuple of independent symmetric Bernoulli random variables with $P_n(\xi_i^n = 1) = P_n(\xi_i^n = -1) = \frac{1}{2}$, living on a probability space $(\Omega_n, \mathcal{F}_n, P_n)$. Moreover, we assume that $\mathcal{F}_n$ is the σ-field generated by the binary trials $(\xi_1^n, \ldots, \xi_n^n)$. For a vector

$f^n = (f_i^n, \ldots, f_n^n) \in \mathbb{R}^n$, the *discrete Wiener integral* is then given by

$$I^n(f^n) := \frac{1}{\sqrt{n}} \sum_{i=1}^{n} f_i^n \xi_i^n.$$

We first characterize weak convergence of these discrete Wiener integrals to the continuous Wiener integrals. The following result can be derived from a nonclassical central limit theorem e.g. in the formulation of [11]. We here give a direct proof which only relies on computations of some higher order moments of discrete Wiener integrals.

Theorem 1.1. *For all m-tuples of Wiener integrals $(I(f_1), \ldots, I(f_m))$, respectively discrete Wiener integrals $(I^n(f_1^n), \ldots, I^n(f_m^n))$, the following three assertions are equivalent as n tends to infinity:*

(a) The vector of discrete Wiener integrals weakly converges to the vector of continuous Wiener integrals:

$$(I^n(f_1^n), \ldots, I^n(f_m^n)) \xrightarrow{d} (I(f_1), \ldots, I(f_m)).$$

(b) All mixed moments converge: For all $l_1, \ldots, l_k \in \{1, \ldots, m\}$,

$$\boldsymbol{E}\left[I^n(f_{l_1}^n) \cdots I^n(f_{l_k}^n)\right] \to \boldsymbol{E}\left[I(f_{l_1}) \cdots I(f_{l_k})\right].$$

(c) The condition of infinite smallness,

$$\lim_{n\to\infty} \max_{l=1,\ldots,m} \max_{i=1,\ldots,n} \frac{1}{\sqrt{n}} |f_{l,i}^n| = 0,$$

is satisfied and the covariance matrix converges, i.e.

$$\forall k, l \in \{1, \ldots, m\} \quad \frac{1}{n} \sum_{i=1}^{n} (f_{k,i}^n)(f_{l,i}^n) \to \int_0^1 f_k(u) f_l(u) du.$$

Proof. By the Cramér-Wold device (Billingsley [7, Theorem 7.7]) it is no loss of generality to consider the one-dimensional case. Hence we shall assume $m = 1$ and write f^n instead of f_1^n and f instead of f_1.
c) ⇒ a): If f is equivalent to 0, the claim is trivial. Otherwise the assertion easily follows by checking the Lindeberg condition and applying the classical central limit theorem.
a) ⇒ b): By the de la Vallée-Poussin criterion for uniform integrability, it suffices to show that, for every $k \in \mathbb{N}$,

$$\sup_{n\in\mathbb{N}} \mathbf{E}[|I^n(f^n)|^{2k}] < \infty. \tag{1.1}$$

Exploiting the fact that $(2k-1)!! := \frac{(2k)!}{2^k k!}$ gives the number of partitions of a set with $2k$ elements into pairs, we obtain

$$\begin{aligned}
\mathbf{E}\left[(I^n(f^n))^{2k}\right] &= n^{-k}\mathbf{E}\left[\sum_{i_1,\dots,i_{2k}=1}^{n} f_{i_1}^n \cdots f_{i_{2k}}^n \xi_{i_1}^n \cdots \xi_{i_{2k}}^n\right] \\
&= n^{-k}\left(\sum_{\substack{i_1,\dots,i_{2k}=1 \\ \{i_1,\dots,i_{2k}\} \text{ is a multiset of pairs}}}^{n} f_{i_1}^n \cdots f_{i_{2k}}^n\right) \\
&\le (2k-1)!! \left(\frac{1}{n}\sum_{i=1}^{n}(f_i^n)^2\right)^k \\
&= (2k-1)!!\,\mathbf{E}\left[(I^n(f^n))^2\right]^k. \qquad (1.2)
\end{aligned}$$

Hence, it remains to show (1.1) for $k = 1$. By the assumption of weak convergence we can choose K and n_0 sufficiently large such that $P((I^n(f^n))^2 > K) \le \frac{1}{2\sqrt{3}}$ for $n \ge n_0$. Thus we obtain, by the Cauchy-Schwarz inequality and (1.2) with $k = 2$, for $n \ge n_0$,

$$\begin{aligned}
\mathbf{E}\big[(I^n(f^n))^2\big] &= \mathbf{E}\big[(I^n(f^n))^2 \mathbf{1}_{\{(I^n(f^n))^2 \le K\}}\big] + \mathbf{E}\big[(I^n(f^n))^2 \mathbf{1}_{\{(I^n(f^n))^2 > K\}}\big] \\
&\le \mathbf{E}\left[(I^n(f^n))^2 \wedge K\right] + \left(\mathbf{E}\left[(I^n(f^n))^4\right]\right)^{\frac{1}{2}} P((I^n(f^n))^2 > K) \\
&\le \mathbf{E}\left[(I^n(f^n))^2 \wedge K\right] + \frac{1}{2}\mathbf{E}\left[(I^n(f^n))^2\right].
\end{aligned}$$

Since $x^2 \wedge K$ is a continuous and bounded function and $I^n(f^n)$ weakly converges, $\mathbf{E}\left[(I^n(f^n))^2 \wedge K\right]$ is convergent and hence $\sup_n \mathbf{E}\left[(I^n(f^n))^2\right]$ is finite.

b) $\Rightarrow$ *c)*. The convergence of the variance is an immediate consequence of b). In order to prove the infinite smallness condition, we first observe that

$$\begin{aligned}
\mathbf{E}\left[(I^n(f^n))^4\right] &= n^{-2}\left(\sum_{\substack{i_1,\dots,i_4=1 \\ \{i_1,\dots,i_4\} \text{ is a multiset of pairs}}}^{n} f_{i_1}^n \cdots f_{i_4}^n\right) \\
&= 3\left(\frac{1}{n}\sum_{i=1}^{n}(f_i^n)^2\right)^2 - 2\sum_{i=1}^{n}\left(\frac{1}{\sqrt{n}}f_i^n\right)^4. \qquad (1.3)
\end{aligned}$$

Here, the left hand side and the first term on the right hand side in (1.3) both converge to $\mathbf{E}\left[I(f)^4\right] = 3\left(\int_0^1 (f(u))^2 du\right)^2$ by b). Thus we have that $0 \le \sum_{i=1}^{n}(\frac{1}{\sqrt{n}}f_i^n)^4 \to 0$, which implies $\max_{i \le n} \frac{1}{\sqrt{n}}|f_i^n| \to 0$. □

We next introduce the *Wick exponential* of the Wiener integral $I(f)$ as

$$\exp^{\diamond}(I(f)) := \exp\left(I(f) - \frac{1}{2}\int_0^1 f^2(s)ds\right), \tag{1.4}$$

which plays an important role in white noise analysis, see e.g. [17, 21]. It is a renormalization of the ordinary exponential of $I(f)$ in the sense that

$$\exp^{\diamond}(I(f)) = \frac{\exp(I(f))}{\mathbf{E}[\exp(I(f))]}.$$

From the point of view of stochastic calculus we note that the dynamic version of the Wick exponential, $X_t = \exp^{\diamond}(I(f\mathbf{1}_{[0,t]}))$, is the usual stochastic exponential and satisfies the Doléans-Dade equation

$$dX_t = f(t)X_t dB_t, \quad X_0 = 1.$$

Hence, a natural candidate for the discrete Wick exponential of the discrete Wiener integral $I^n(f^n)$ is X_n^n, where X_i^n, $i = 0, \dots, n$, solves the stochastic difference equation

$$X_i^n = X_{i-1}^n\left(1 + f_i^n \frac{1}{\sqrt{n}}\xi_i\right), \quad X_0^n = 1.$$

We thus define the *discrete Wick exponential* as

$$\exp^{\diamond_n}(I^n(f^n)) := \prod_{i=0}^{n}\left(1 + \frac{1}{\sqrt{n}} f_i^n \xi_i^n\right). \tag{1.5}$$

It corresponds to the terminal value of a discrete exponential martingale and was studied in [13] in the context of discrete Wick calculus. We shall next see that weak convergence of the discrete Wiener integrals easily extends to weak convergence of the discrete Wick exponentials.

Theorem 1.2. *Suppose we have*

$$(I^n(f_1^n), \dots, I^n(f_m^n)) \xrightarrow{d} (I(f_1), \dots, I(f_m)).$$

Then

$$(\exp^{\diamond_n}(I^n(f_1^n)), \dots, \exp^{\diamond_n}(I^n(f_m^n))) \xrightarrow{d} (\exp^{\diamond}(I(f_1)), \dots, \exp^{\diamond}(I(f_m)))$$

as n tends to infinity.

Proof. The proof relies on a logarithmic transform and is analogous to establishing weak convergence of the Cox-Ross-Rubinstein model to the Black-Scholes model. We fix some arbitrary $a_1, \dots, a_m \in \mathbb{R}$. As

$$\lim_{n\to\infty} \max_{l\leq m} \max_{i\leq n} \frac{1}{\sqrt{n}} |f_{l,i}^n| = 0$$

by Theorem 1.1, we can apply the logarithm for sufficiently large n and obtain by a Taylor expansion

$$\sum_{l=1}^{m} a_l \log \exp^{\diamond_n} (I^n(f_l^n)) = \sum_{l=1}^{m} a_l \sum_{i=1}^{n} \log\left(1 + \frac{1}{\sqrt{n}} f_{l,i}^n \xi_i^n\right)$$
$$= \sum_{l=1}^{m} a_l \sum_{i=1}^{n} \left(\frac{1}{\sqrt{n}} f_{l,i}^n \xi_i^n - \frac{1}{2n}(f_{l,i}^n \xi_i^n)^2\right) + \mathcal{O}\left(n^{-\frac{1}{2}} \sup_j |f_{l,j}^n| \sum_{i=1}^{n} \frac{1}{n}(f_{l,i}^n)^2\right)$$
$$= \sum_{l=1}^{m} a_l \left(I^n(f_l^n) - \frac{1}{2n}\sum_{i=1}^{n}(f_{l,i}^n)^2\right) + \mathcal{O}\left(n^{-\frac{1}{2}} \sup_j |f_{l,j}^n| \sum_{i=1}^{n} \frac{1}{n}(f_{l,i}^n)^2\right).$$

By Theorem 1.1 and Slutsky's Theorem we obtain

$$\sum_{l=1}^{m} a_l \log \exp^{\diamond_n} (I^n(f_l^n)) \xrightarrow{d} \sum_{l=1}^{m} a_l \left(I(f_l) - \frac{1}{2}\mathbf{E}\left[(I(f_l))^2\right]\right).$$

Thanks to the the Cramér-Wold device and the continuous mapping theorem we obtain the assertion. □

1.2.2. *Wick product*

Having the Wick exponential at our disposal, we can motivate the introduction of the Wick product. The Wick product is often defined in terms of the chaos decomposition of (generalized) random variables, see e.g. [17]. We here take a functional identity for Wick exponentials as starting point (cf. also [9]), which then leads to the S-transform characterization of the Wick product which was already noted in [23] and studied in detail in [20, 21].

Suppose $I(f)$ and $I(g)$ are two Wiener integrals. Multiplying the corresponding Wick exponentials, one easily obtains

$$\exp^{\diamond}(I(f)) \cdot \exp^{\diamond}(I(g)) = \exp\left(\int_0^1 f(s)g(s)ds\right) \cdot \exp^{\diamond}(I(f+g)). \quad (1.6)$$

Hence, if the Wiener integrals $I(f)$ and $I(g)$ are correlated, the product of their renormalized exponentials is itself not renormalized. In order to enforce the renormalization, the defining functional equation for the (continuous) Wick product is

$$\exp^{\diamond}(I(f)) \diamond \exp^{\diamond}(I(g)) = \exp^{\diamond}(I(f+g)). \tag{1.7}$$

Analogously we wish to define for two, possibly correlated, discrete Wiener integrals $I^n(f^n)$, $I^n(g^n)$ a discrete Wick product of the corresponding discrete Wick exponentials via the property

$$\exp^{\diamond_n}(I^n(f^n)) \diamond_n \exp^{\diamond_n}(I^n(g^n)) = \exp^{\diamond_n}(I^n(f^n + g^n)). \tag{1.8}$$

The defining identities for the continuous and discrete Wick product raise two questions, which we briefly discuss before we can further study their relation:

(1) Continuous case: How can one reasonably extend the Wick product defined by (1.7) to a bilinear operator on a sufficiently large domain $\mathcal{D} \subset L^2(\Omega, \mathcal{F}, P) \times L^2(\Omega, \mathcal{F}, P)$?
(2) Discrete case: It is easy to construct a basis of $L^2(\Omega_n, \mathcal{F}_n, P_n)$ consisting of 2^n discrete Wick exponentials, but there are infinitely many discrete Wick exponentials. So here the question is: Is a bilinear extension of (1.8) on $L^2(\Omega_n, \mathcal{F}_n, P_n) \times L^2(\Omega_n, \mathcal{F}_n, P_n)$ well-defined?

We first explain how (1.7) can be used as a starting point to extend the (continuous) Wick product to a closed operator on a dense subset of $L^2(\Omega, \mathcal{F}, P) \times L^2(\Omega, \mathcal{F}, P)$. To this end we denote $X = \exp^{\diamond}(I(f))$ and $Y = \exp^{\diamond}(I(g))$. By (1.7), (1.6) and the renormalization property of the Wick exponential, we get for every $h \in L^2([0,1])$

$$\begin{aligned}
\mathbf{E}[(X \diamond Y)\exp^{\diamond}(I(h))] &= \mathbf{E}[\exp^{\diamond}(I(f+g))\exp^{\diamond}(I(h))] \\
&= \exp\left(\int_0^1 (f(s)+g(s))h(s)ds\right) \\
&= \exp\left(\int_0^1 f(s)h(s)ds\right)\exp\left(\int_0^1 g(s)h(s)ds\right) \\
&= \mathbf{E}[(X\exp^{\diamond}(I(h))] \cdot \mathbf{E}[Y\exp^{\diamond}(I(h))].
\end{aligned}$$

One can use this property to extend the Wick product beyond Wick exponentials as stated in the following proposition. The definition of the Wick product below just rephrases the S-transform characterization of the Wick product in the present L^2-setting, cf. e.g. Definition 16.25 in [20] or Definition 8.11 in [21].

Proposition 1.1. Define the *Wick product* by

$$\mathcal{D} := \big\{(X,Y) \in L^2(\Omega) \times L^2(\Omega) : \ \exists Z_{X,Y} \in L^2(\Omega)\ \forall h \in L^2(\mathbb{R})$$
$$\mathbf{E}\left[Z_{X,Y} \exp^{\diamond}(I(h))\right] = \mathbf{E}\left[X \exp^{\diamond}(I(h))\right] \mathbf{E}\left[Y \exp^{\diamond}(I(h))\right]\big\}$$
$$\diamond : \mathcal{D} \to L^2(\Omega), \quad (X,Y) \mapsto Z_{X,Y}.$$

Then, $\mathcal{D}$ is a dense subset of $L^2(\Omega) \times L^2(\Omega)$, the Wick product $\diamond$ is well-defined (i.e $Z_{X,Y}$ is uniquely determined), and the following properties hold:
(a) For every $f, g \in L^2([0,1])$, $(\exp^{\diamond}(I(f)), \exp^{\diamond}(I(g)) \in \mathcal{D}$ and (1.7) holds.
(b) The Wick product is bilinear.
(c) The Wick product is closed, i.e. if $(X_k, Y_k)_{k\in\mathbb{N}} \subset \mathcal{D}$, $(X_k, Y_k) \to (X,Y)$ in $L^2(\Omega) \times L^2(\Omega)$ and $X_k \diamond Y_k \to Z$ in $L^2(\Omega)$, then $(X,Y) \in \mathcal{D}$ and $X \diamond Y = Z$.

The proof of these well-known properties easily follows from the fact that the Wick exponentials form a total subset of $L^2(\Omega)$, see [20, Cor. 3.40]. Several sufficient conditions for a pair (X,Y) to belong to $\mathcal{D}$ can be found in Hu and Yan [19]. We emphasize that $\mathcal{D}$ is a true subset of $L^2(\Omega) \times L^2(\Omega)$. For instance, the Wick product of B_1 and $\operatorname{sgn}(B_1)$ does not exist in $L^2(\Omega)$. The Wick product can be further extended to a continuous bilinear operator acting on the space of Hida distributions (generalized random variables), see e.g. the monographs of Kuo [21] and Holden et al. [17].

We now turn to the question concerning the well-posedness of a discrete Wick product satisfying (1.8). To this end we first fix some notation. Recall that

$$\mathcal{F}_n := \sigma\left(\xi_1^n, \xi_2^n, \ldots, \xi_n^n\right)$$

is the σ-field generated by the binary trials, and hence $L^2(\Omega_n, \mathcal{F}_n, P_n)$ is a 2^n-dimensional vector space. A canonical orthonormal basis of $L^2(\Omega_n, \mathcal{F}_n, P_n)$ consists of

$$\Xi_A^n := \prod_{i\in A} \xi_i^n, \quad A \subset \{1, \ldots, n\}.$$

Every $X \in L^2(\Omega_n, \mathcal{F}_n, P_n)$ has a unique expansion in terms of this basis, which is called the *Walsh decomposition* of X,

$$X = \sum_{A\subset\{1,\ldots,n\}} X_A^n \Xi_A^n,$$

where $X_A \in \mathbb{R}$. It follows immediately that expectation and L^2-inner product can be computed in terms of the Walsh decomposition via $E[X] = X_\emptyset$

and

$$\mathbf{E}[XY] = \sum_{A\subset\{1,\ldots,n\}} X_A^n Y_A^n, \tag{1.9}$$

where Y_A^n are the coefficients of the Walsh decomposition of Y.

The next lemma clarifies how a discrete Wick product in the sense of (1.8) acts on the canonical basis of $L^2(\Omega_n, \mathcal{F}_n, P_n)$. In particular, it shows that such discrete Wick product must coincide with the one introduced in [15, 16] in terms of the Walsh decomposition.

Lemma 1.1. *Suppose there is an operator* $\diamond_n : L^2(\Omega_n, \mathcal{F}_n, P_n) \times L^2(\Omega_n, \mathcal{F}_n, P_n) \to L^2(\Omega_n, \mathcal{F}_n, P_n)$ *such that* $(L^2(\Omega_n, \mathcal{F}_n, P_n), +, \diamond_n)$ *is a commutative ring and (1.8) holds. Then, for every* $A, B \subset \{1, \ldots, n\}$,

$$\prod_{i\in A}\xi_i^n \diamond_n \prod_{i\in B}\xi_i^n := \begin{cases} \prod_{i\in A\cup B}\xi_i^n & \text{if } A\cap B=\emptyset \\ 0 & \text{otherwise}\end{cases}. \tag{1.10}$$

Proof. We first note the elementary facts that Ξ_A^n can be expressed as a linear combination of $\{\exp^{\diamond_n}(I^n(\mathbf{1}_C)),\ C \subset A\}$, and, that for disjoint sets $A, B \subset \{1, \ldots, n\}$,

$$\exp^{\diamond_n}(I^n(\mathbf{1}_A)) \diamond_n \exp^{\diamond_n}(I^n(\mathbf{1}_B)) = \exp^{\diamond_n}(I^n(\mathbf{1}_A)) \cdot \exp^{\diamond_n}(I^n(\mathbf{1}_B))$$

by (1.8). Combining these two facts one immediately gets (1.10) when A and B are disjoint. We now suppose that $j \in A \cap B$. Then,

$$\begin{aligned}\Xi_A^n &= \Xi_{A\setminus\{j\}}^n \exp^{\diamond_n}(I^n(\mathbf{1}_{\{j\}})) - \Xi_{A\setminus\{j\}}^n \\ &= \Xi_{A\setminus\{j\}}^n \diamond_n \exp^{\diamond_n}(I^n(\mathbf{1}_{\{j\}})) - \Xi_{A\setminus\{j\}}^n = \Xi_{A\setminus\{j\}}^n \diamond_n \xi_j^n\end{aligned}$$

as $A \setminus \{j\}$ and $\{j\}$ are disjoint. Hence,

$$\Xi_A^n \diamond_n \Xi_B^n = (\Xi_{A\setminus\{j\}}^n \diamond_n \Xi_{B\setminus\{j\}}^n) \diamond_n (\xi_j^n \diamond_n \xi_j^n) = 0,$$

because, due to (1.8) and the definition of the discrete Wick exponential,

$$\begin{aligned}\xi_j^n \diamond_n \xi_j^n &= \exp^{\diamond_n}(I^n(\mathbf{1}_{\{j\}})) \diamond_n \exp^{\diamond_n}(I^n(\mathbf{1}_{\{j\}})) - 2\xi_j^n - 1 \\ &= \exp^{\diamond_n}(I^n(2\cdot\mathbf{1}_{\{j\}})) - 2\xi_j^n - 1 - 0.\end{aligned}$$

□

The previous lemma motivates to define the *discrete Wick product* of two random variables $X, Y \in L^2(\Omega_n, \mathcal{F}_n, P_n)$ in terms of their Walsh decomposition,

$$X = \sum_{A\subset\{1,\ldots,n\}} X_A^n \Xi_A^n, \quad Y - \sum_{B\subset\{1,\ldots,n\}} Y_B^n \Xi_B^n,$$

as

$$X \diamond_n Y = \sum_{C \subset \{1,\dots,n\}} \left(\sum \{ X_A^n Y_B^n : A \cap B = \emptyset, A \cup B = C \} \right) \Xi_C^n. \quad (1.11)$$

This definition was first suggested and motivated by some analogies in terms of the chaos decomposition in [15, 16]. Note that relation (1.10), which is the core of the definition of the discrete Wick product, may roughly be interpreted as follows: The discrete Wick product kills all terms which include the interaction of one of the noise sources ξ_j^n with itself.

It remains to verify that the discrete Wick product indeed satisfies (1.8), which was the starting point of our motivation.

Lemma 1.2. *Equation (1.8) holds for the discrete Wick product defined in (1.11).*

Proof. We observe that the Walsh decomposition of a discrete Wick exponential is given by

$$\exp^{\diamond_n}(I^n(f^n)) = \sum_{A \subset \{1,\dots,n\}} n^{-|A|/2} \prod_{i \in A} f_i^n \xi_i^n = \sum_{A \subset \{1,\dots,n\}} n^{-|A|/2} f_A^n \Xi_A^n, \quad (1.12)$$

where $f_A^n := \prod_{i \in A} f_i^n$. Thus, given two discrete Wiener integrals $I^n(f^n)$ and $I^n(g^n)$, we have

$$\begin{aligned} &\exp^{\diamond_n}(I^n(f^n)) \diamond_n \exp^{\diamond_n}(I^n(g^n)) \\ &= \left(\sum_{B \subset \{1,\dots,n\}} n^{-|B|/2} f_B^n \Xi_B^n \right) \diamond_n \left(\sum_{C \subset \{1,\dots,n\}} n^{-|C|/2} g_C^n \Xi_C^n \right) \\ &= \sum_{A \subset \{1,\dots,n\}} n^{-|A|/2} \left(\sum_{\substack{B \subset \{1,\dots,n\} \\ C \subset \{1,\dots,n\} \\ A = B \cup C, B \cap C = \emptyset}} f_B^n g_C^n \right) \Xi_A^n. \end{aligned}$$

We have to show that the right hand side coincides with the Walsh decomposition of $\exp^{\diamond_n}(I^n(f^n + g^n))$, which, due to (1.12), is given by

$$\sum_{A \subset \{1,\dots,n\}} n^{-|A|/2} \prod_{i \in A} (f_i^n + g_i^n) \Xi_A^n.$$

This follows from

$$\prod_{i\in A}(f_i^n + g_i^n) = \sum_{B\subset A}\prod_{i\in B} f_i^n \prod_{i\in A\setminus B} g_i^n = \sum_{\substack{B\subset\{1,\ldots,n\}\\ C\subset\{1,\ldots,n\}\\ A=B\cup C,\, B\cap C=\emptyset}} f_B^n g_C^n.$$

□

We close this subsection with two important observations:

(i) The definitions of the discrete and continuous Wick products are not pathwise definitions and, in general, the calculation of a realization of a Wick product requires information about many other realizations. For the discrete Wick product we illustrate this feature in more detail in Section 1.4, where we suggest an algorithm for simulating discrete Wick difference equations. In the continuous case the Wick product can alternatively be defined in terms of the Hermite expansion of the random variables, see [17]. This definition illustrates that the Wick product behaves like a convolution in the ω-variable. Hence the Wick product is not a local operator, i.e. $(X \diamond Y)$ cannot be calculated along a Brownian path using only sample paths in a small neighborhood. For a specific counterexample we refer to [20, Example 16.37].

(ii) The distribution of a discrete Wick product $X \diamond_n Y$ is not determined by the joint distribution of (X, Y). In order to illustrate this, we present a simple example from [16]. For $n \geq 2$, the random vectors (ξ_1^n, ξ_2^n) and $(\xi_1^n, \xi_1^n\xi_2^n)$ have the same distribution, but the discrete Wick products fulfill

$$\xi_1^n \diamond_n \xi_2^n = \xi_1^n \xi_2^n, \quad \xi_1^n \diamond_n (\xi_1^n \xi_2^n) = 0.$$

In the continuous case, the situation is even more intricate. The random vectors $(B_{1/2}, \operatorname{sgn}(B_1 - B_{1/2}))$ and $(B_{1/2}, \operatorname{sgn}(B_{1/2}))$ have the same distribution. It is easily seen from the definition of the continuous Wick product that,

$$B_{1/2} \diamond \operatorname{sgn}(B_1 - B_{1/2}) = B_{1/2} \cdot \operatorname{sgn}(B_1 - B_{1/2}).$$

But, $B_{1/2} \diamond \operatorname{sgn}(B_{1/2})$ does not even exist in $L^2(\Omega)$.

By this second observation, it is immediate that weak convergence of (X^n, Y^n) to (X, Y) cannot generally imply weak convergence of $X_n \diamond_n Y_n$ to $X \diamond Y$. In the next subsection we will present a class of important examples, for which weak convergence of the Wick products is inherited from weak convergence of the random vectors, namely Wick powers and Wick-analytic functions of (discrete) Wiener integrals.

1.2.3. Wick powers and Wick-analytic functions

Wick powers of a Wiener integral $I(f)$, respectively *discrete Wick powers* of a discrete Wiener integral $I^n(f^n)$, are defined recursively by

$$I(f)^{\diamond k} := I(f)^{\diamond(k-1)} \diamond I(f), \quad I(f)^{\diamond 0} = 1, \quad k \in \mathbb{N}$$

and

$$I^n(f^n)^{\diamond_n k} := I^n(f^n)^{\diamond_n (k-1)} \diamond_n I^n(f^n), \quad I^n(f^n)^{\diamond_n 0} = 1, \quad k \in \mathbb{N}.$$

We now illustrate how the defining property (1.7) and the closedness of the Wick product can be applied to calculate the continuous Wick powers recursively. In this way we derive the well-known representation of Wick powers in terms of Hermite polynomials and the Hermite recursion formula for Wick powers. One can first observe that

$$I(f)^{\diamond k} = \frac{d^k}{dw^k} \exp^{\diamond}(wI(f))\big|_{w=0}. \tag{1.13}$$

Indeed, this is easily seen for $k = 0$ and $k = 1$ and can thus be deduced inductively by

$$\begin{aligned} I(f)^{\diamond(k+1)} &= \frac{d^k}{dw^k} \exp^{\diamond}(wI(f))\bigg|_{w=0} \diamond \frac{d}{dx} \exp^{\diamond}(xI(f))\bigg|_{x=0} \\ &= \frac{d^k}{dw^k}\frac{d}{dx} \exp^{\diamond}((w+x)I(f))\bigg|_{\substack{w=0,\\ x=0}} = \frac{d^{k+1}}{dz^{k+1}} \exp^{\diamond}(zI(f))\bigg|_{z=0}, \end{aligned}$$

making use of (1.7) and the closedness of the Wick product. Thanks to (1.13) and applying the Leibniz rule, one then obtains the recursion formula

$$\begin{aligned} &I(f)^{\diamond k+1} \\ &= \frac{d^k}{dw^k}\left(\left(I(f) - w\mathbf{E}\left[|I(f)|^2\right]\right)\exp\left(wI(f) - \frac{1}{2}\mathbf{E}\left[|wI(f)|^2\right]\right)\right)\bigg|_{w=0} \\ &= \left(I(f) - w\mathbf{E}\left[|I(f)|^2\right]\right)\frac{d^k}{dw^k}\exp\left(wI(f) - \frac{1}{2}\mathbf{E}\left[|wI(f)|^2\right]\right)\bigg|_{w=0} \\ &\quad + k(-\mathbf{E}\left[|I(f)|^2\right])\frac{d^{k-1}}{dw^{k-1}}\exp\left(wI(f) - \frac{1}{2}\mathbf{E}\left[|wI(f)|^2\right]\right)\bigg|_{w=0} \\ &= I(f)I(f)^{\diamond k} - k\mathbf{E}\left[|I(f)|^2\right]I(f)^{\diamond k-1}. \end{aligned} \tag{1.14}$$

We now recall that the same recursion formula is satisfied by the *Hermite polynomials of degree* $k \in \mathbb{N}$ *with parameter* σ^2,

$$h^k_{\sigma^2}(x) := (-\sigma^2)^k \exp\left(\frac{x^2}{2\sigma^2}\right)\frac{d^k}{dx^k}\exp\left(\frac{-x^2}{2\sigma^2}\right),$$

when $\sigma^2 = \mathbf{E}\left[|I(f)|^2\right]$, see e.g. [21]. This implies the representation of the Wick powers in terms of Hermite polynomials as

$$I(f)^{\diamond k} = h^k_{\mathbf{E}[|I(f)|^2]}(I(f)). \tag{1.15}$$

Taking the generating function property of the Hermite polynomials into account one can finally express the Wick exponential as a Wick power series

$$\exp^{\diamond}(I(f)) = \sum_{k=0}^{\infty} \frac{1}{k!} I(f)^{\diamond k}, \tag{1.16}$$

which a posteriori justifies the name Wick exponential.

Remark 1.1. (i) If, alternatively, the Wick product is defined in terms of the chaos decomposition as in [17], then (1.16) is usually taken as definition of the Wick exponential and identity (1.4) is shown.
(ii) One can derive from (1.15) that $I(f)^{\diamond k}$ is the orthogonal projection of $I(f)^k$ on the kth Wiener chaos. More generally, the Wick product of k Gaussian random variables can equivalently be defined as the orthogonal projection of the ordinary product of these random variables on the kth Wiener chaos. An alternative approach to the Wick product based on this observation can be be found in [20].

The discrete Wick exponential can also be expressed as a power series in terms of discrete Wick powers. Given a discrete Wiener integral $I^n(f^n) = \frac{1}{\sqrt{n}} \sum_{i=1}^n f_i^n \xi_i^n$, we can apply the definition of the discrete Wick product in (1.11) in order to compute the Walsh decomposition of the discrete Wick powers as

$$\frac{1}{k!} I^n(f^n)^{\diamond_n k} = \frac{1}{k!} \left(\frac{1}{\sqrt{n}} \sum_{i=1}^n f_i^n \xi_i^n \right)^{\diamond_n k} = n^{-k/2} \sum_{\substack{A \subset \{1,\ldots,n\} \\ |A|=k}} f_A^n \Xi_A^n. \tag{1.17}$$

Note that $I^n(f^n)^{\diamond_n k} = 0$ for all $k > n$. By summing up this expression and thanks to (1.12) we obtain

$$\exp^{\diamond_n}(I^n(f^n)) = \sum_{k=0}^{n} \frac{1}{k!} I^n(f^n)^{\diamond_n k} = \sum_{k=0}^{\infty} \frac{1}{k!} I^n(f^n)^{\diamond_n k}. \tag{1.18}$$

We can now reinterpret Theorem 1.2, which asserts weak convergence of the discrete Wick exponential to the continuous Wick exponential. Although this result was formulated and proved without Wick products, it actually is a first example where a discrete Wick power series converges to its continuous counterpart.

In the remainder of this section we discuss the convergence behaviour of Wick power series in more general situations. To this end let us first fix some notations. Given an analytic function $F(x) = \sum_{k=0}^{\infty} \frac{a_k}{k!} x^k$, we define its *Wick versions* as

$$F^{\diamond}(x) = \sum_{k=0}^{\infty} \frac{a_k}{k!} x^{\diamond k}, \quad F^{\diamond_n}(x) = \sum_{k=0}^{n} \frac{a_{n,k}}{k!} x^{\diamond_n k},$$

and the associated partial sums as

$$F^{N,\diamond}(x) = \sum_{k=0}^{N} \frac{a_k}{k!} x^{\diamond k}, \quad F^{N,\diamond_n}(x) = \sum_{k=0}^{N\wedge n} \frac{a_{n,k}}{k!} x^{\diamond_n k}.$$

We recall that Wick products cannot be evaluated pointwise. Hence, the above Wick versions are to be understood as formal expressions, which only make sense when a random variable is plugged into the x-variable. We shall prove the following result, which generalizes Theorem 5 in [4].

Theorem 1.3. *Assume that* $(I^n(f_1^n), \dots, I^n(f_m^n)) \xrightarrow{d} (I(f_1), \dots, I(f_m))$. *Additionally suppose that the coefficients in the Wick-analytic functionals satisfy*

(1) $\lim_{n\to\infty} a_{n,k} = a_k$ *exists for all* $k \in \mathbb{N}$.
(2) *For every* $l = 1, \dots, m$,

$$\sum_{k=0}^{\infty} \frac{1}{k!} \sup_{n\in\mathbb{N}} \left\{ a_{n,k}^2 \left(\frac{1}{n} \sum_{i=1}^{n} (f_{l,i}^n)^2 \right)^k \right\} < \infty.$$

Then it holds that

$$(F^{\diamond_n}(I^n(f_1^n)), \dots, F^{\diamond_n}(I^n(f_m^n))) \xrightarrow{d} (F^{\diamond}(I(f_1)), \dots, F^{\diamond}(I(f_m))),$$

as n tends to infinity.

Remark 1.2. (i) A sufficient condition for (2), assuming the other conditions of the previous theorem, is:

(2') There exists a $C \in \mathbb{R}_+$, so that $|a_{n,k}| \le C^k$ for all $n, k \in \mathbb{N}$.

This immediately follows from the boundedness of $\frac{1}{n}\sum_{i=1}^{n} (f_{l,i}^n)^2$ under the assumptions of the theorem.
(ii) Thanks to Theorem 1.1, Condition (2) implies that

$$\sum_{k=0}^{\infty} \frac{a_k^2}{k!} \left(\int_0^1 f_l^2(s)ds \right)^k < \infty, \quad l = 1, \dots, m,$$

which is equivalent to the existence of $F^\diamond(I(f_l))$, $l = 1,\dots,m$, in $L^2(\Omega,\mathcal{F},P)$, see e.g. Hu and Yan [19].

In order to prove this theorem, we require some preparations. The first one is an analogue of the Hermite recursion (1.14) for discrete Wick powers.

Proposition 1.2. Given a discrete Wiener integral $I^n(f^n) = \frac{1}{\sqrt{n}}\sum_{i=1}^n f_i^n \xi_i^n$, it holds for every $N \geq 1$ that

$$\begin{aligned}&I^n(f^n)^{\diamond_n(N+1)}\\ &= I^n(f^n)(I^n(f^n)^{\diamond_n N}) - N\mathbf{E}\left[I^n(f^n)^2\right] I^n(f^n)^{\diamond_n(N-1)} + R(I^n(f^n),N),\end{aligned}$$

with remainder term

$$R(I^n(f^n),N) = N! n^{-(N+1)/2} \sum_{\substack{A\subset\{1,\dots,n\}\\ |A|=N-1}} f_A^n \Xi_A^n \left(\sum_{i\in A}(f_i^n)^2\right).$$

Moreover,

$$\mathbf{E}\left[(R(I^n(f^n),N))^2\right] \leq N^3\mathbf{E}\left[\left(I^n(f^n)^{\diamond_n(N-1)}\right)^2\right]\left(\max_{i\leq n}\frac{1}{\sqrt{n}}|f_i^n|\right)^2. \tag{1.19}$$

Proof. The proof is similar to the one of Proposition 5 in [4]. We briefly sketch the argument. One first writes

$$\begin{aligned}I^n(f^n)^{\diamond_n(N+1)} &= I^n(f^n)^{\diamond_n N} \diamond_n I^n(f^n)\\ &= \left(I^n(f^n)^{\diamond_n N}\right) I^n(f^n) - \sum_{\substack{A\subset\{1,\dots,n\}\\ |A|=N}} \sum_{i\in A} N! n^{-\frac{N+1}{2}} f_i^n f_A^n \Xi_A^n \xi_i^n,\end{aligned}$$

making use of the Walsh decomposition of the discrete Wick powers in (1.17) and the definition of the discrete Wick product. Rearranging terms, the recursion formula easily follows. Noticing that the remainder term is given in terms of the Walsh decomposition, the L^2 estimate can be obtained by (1.9):

$$\begin{aligned}\mathbf{E}\left[(R(I^n(f^n),N))^2\right] &= N!^2 n^{-(N+1)} \sum_{\substack{A\subset\{1,\dots,n\}\\ |A|=N-1}} (f_A^n)^2 \left(\sum_{i\in A}(f_i^n)\right)^2\\ &\leq N^3\mathbf{E}\left[\left(I^n(f^n)^{\diamond_n(N-1)}\right)^2\right]\left(\max_{i\leq n}\frac{1}{\sqrt{n}}|f_i^n|\right)^2. \qquad \square\end{aligned}$$

If $I^n(f^n)$ weakly converges to $I(f)$, we observe by Theorem 1.1 that the remainder term in the discrete Hermite expansion converges to zero, provided that the L^2-norm of the discrete Wick powers does not explode. The next Proposition shows that the L^2-norm of the discrete Wick power indeed converges to the L^2-norm of the continuous Wick power.

Proposition 1.3. Suppose we have $I^n(f^n) \xrightarrow{d} I(f)$. Then it holds for all $N \in \mathbb{N}$ that

$$\mathbf{E}\left[\left(I^n(f^n)^{\diamond_n N}\right)^2\right] \to \mathbf{E}\left[\left(I(f)^{\diamond N}\right)^2\right] \tag{1.20}$$

as n tends to infinity.

Proof. Since

$$I^n(f^n)^{\diamond_n N} = \left(\frac{1}{\sqrt{n}}\sum_{i=1}^{n} f_i^n \xi_i^n\right)^{\diamond_n N} = n^{-\frac{N}{2}} N! \sum_{\substack{A\subset\{1,\dots,n\}\\ |A|=N}} \prod_{i\in A} f_i^n \xi_i^n,$$

we have by (1.9) that

$$\mathbf{E}\left[\left(I^n(f^n)^{\diamond_n N}\right)^2\right] = n^{-N} N!^2 \sum_{\substack{A\subset\{1,\dots,n\}\\ |A|=N}} \prod_{i\in A} (f_i^n)^2. \tag{1.21}$$

Hence,

$$\mathbf{E}\left[\left(I^n(f^n)^{\diamond_n N}\right)^2\right] = n^{-N} N! \left(\sum_{i_1,\dots,i_N=1}^{n} \prod_{k=1}^{N} (f_{i_k}^n)^2 - \sum_{\substack{i_1,\dots,i_N=1\\ \exists(l,m):i_l=i_m}}^{n} \prod_{k=1}^{N} (f_{i_k}^n)^2 \right).$$

Concerning the first term we get by Theorem 1.1

$$n^{-N} N! \sum_{i_1,\dots,i_N=1}^{n} \prod_{k=1}^{N} (f_{i_k}^n)^2 = N! \left(\frac{1}{n}\sum_{i=1}^{n} (f_i^n)^2\right)^N$$

$$\to N! \left(\int_0^1 (f(s))^2 ds\right)^N = \mathbf{E}\left[\left(I(f)^{\diamond N}\right)^2\right],$$

where the L^2-norm of the continuous Wick power can be directly calculated thanks to the representation in terms of Hermite polynomials in (1.14).

Hence it remains to show that

$$\alpha_N := n^{-N} N! \sum_{\substack{i_1,\dots,i_N=1\\ \exists(l,m):i_l=i_m}}^{n} \prod_{k=1}^{N} (f_{i_k}^n)^2 \to 0, \quad \text{as } n \to \infty.$$

This term can be estimated as follows:

$$0 \le \alpha_N \le n^{-2}\left(\sum_{i=1}^{n}(f_i^n)^4\right) N!N^2 n^{-(N-2)} \sum_{i_1,\dots,i_{N-2}=1}^{n} \prod_{k=1}^{N-2}(f_{i_k}^n)^2$$

$$\le \left(\max_{i\le n}\frac{1}{\sqrt{n}}|f_i^n|\right)^2 \left(\frac{1}{n}\sum_{i=1}^{n}(f_i^n)^2\right) N!N^2 \left(\frac{1}{n}\sum_{i=1}^{n}(f_i^n)^2\right)^{N-2}$$

$$\le \sup_n \left(\frac{1}{n}\sum_{i=1}^{n}(f_i^n)^2\right)^{N-1} N!N^2 \left(\max_{i\le n}\frac{1}{\sqrt{n}}|f_i^n|\right)^2,$$

which tends to zero as n tends to infinity due to Theorem 1.1. □

We are now in the position to prove Theorem 1.3.

Proof. [Proof of Theorem 1.3] By Billingsley [7, Theorem 4.2] it is sufficient to show that the following three conditions are satisfied:

$$\forall N \in \mathbb{N} \ \left(F^{N,\diamond_n}(I^n(f_1^n)),\dots,F^{N,\diamond_n}(I^n(f_m^n))\right)$$
$$\xrightarrow{d} \left(F^{N,\diamond}(I(f_1)),\dots,F^{N,\diamond}(I(f_m))\right), \tag{1.22}$$

$$\forall l \le m \quad \lim_{N\to\infty}\limsup_{n\to\infty}\mathbf{E}\left[\left(F^{\diamond_n}(I^n(f_l^n)) - F^{N,\diamond_n}(I^n(f_l^n))\right)^2\right] = 0, \tag{1.23}$$

$$\forall l \le m \quad \lim_{N\to\infty}\mathbf{E}\left[\left(F^{\diamond}(I(f_l)) - F^{N,\diamond}(I(f_l))\right)^2\right] = 0. \tag{1.24}$$

Condition (1.24) is a direct consequence of Remark 1.2 (ii). Concerning condition (1.23) we get by the orthogonality of the different Wick powers

$$\limsup_{n\to\infty}\mathbf{E}\left[\left(F^{\diamond_n}(I^n(f_l^n)) - F^{N,\diamond_n}(I^n(f_l^n))\right)^2\right]$$

$$= \limsup_{n\to\infty}\sum_{k=N+1}^{\infty}\frac{a_{n,k}^2}{(k!)^2}\mathbf{E}\left[|(I^n(f_l^n))^{\diamond_n k}|^2\right]$$

$$= \sum_{k=N+1}^{\infty}\frac{a_k^2}{(k!)^2}\mathbf{E}\left[|(I(f_l))^{\diamond k}|^2\right] = \mathbf{E}\left[\left(F^{\diamond}(I(f_l)) - F^{N,\diamond}(I(f_l))\right)^2\right],$$

making use of Proposition 1.3 and assumption (2) in Theorem 1.3 for introducing the limit superior under the series. Hence, (1.23) follows from (1.24).

Condition (1.22) is a consequence of the generalized continuous mapping theorem (Billingsley [7, Theorem 5.5]), once we have established the weak convergence of the Wick powers

$$\left(1, I^n(f_1^n),\dots,I^n(f_1^n)^{\diamond_n N},\dots,I^n(f_m^n),\dots,I^n(f_m^n)^{\diamond_n N}\right)$$
$$\xrightarrow{d} \left(1, I(f_1),\dots,I(f_1)^{\diamond N},\dots,I(f_m),\dots,I(f_m)^{\diamond N}\right) \tag{1.25}$$

for all $N \in \mathbb{N}$. This will be done by induction on N, exploiting the continuous Hermite recursion (1.14) and its discrete counterpart in Proposition 1.2. By Theorem 1.1 we obtain (1.25) for $N = 1$. Now suppose that (1.25) is proved for some $N \geq 1$. We fix some arbitrary $r_j^l \in \mathbb{R}$ for $j = 0, \ldots, N+1$ and $l = 1, \ldots, m$. By the discrete Hermite recursion in Proposition 1.2 we get

$$\sum_{j=0}^{N+1} \left(\sum_{l=1}^{m} r_j^l I^n(f_l^n)^{\diamond_n j} \right) = \sum_{j=0}^{N} \left(\sum_{l=1}^{m} r_j^l I^n(f_l^n)^{\diamond_n j} \right)$$
$$+ \sum_{l=1}^{m} r_{N+1}^l \left(I^n(f_l^n) \left(I^n(f_l^n)^{\diamond_n N} \right) - N\mathbf{E}\left[I^n(f_l^n)^2 \right] \left(I^n(f_l^n)^{\diamond_n (N-1)} \right) \right)$$
$$+ \sum_{l=1}^{m} r_{N+1}^l \left(R(I^n(f_l^n), N) \right).$$

Due to Propositions 1.2 and 1.3 and Theorem 1.1, the sum with the remainder terms $R(I^n(f_l^n), N)$ weakly converges to zero. Thus, by Slutsky's theorem, the generalized continuous mapping theorem (Billingsley [7, Theorem 5.5]) and the induction hypothesis, we obtain that

$$\sum_{j=0}^{N+1} \left(\sum_{l=1}^{m} r_j^l I^n(f_l^n)^{\diamond_n j} \right)$$
$$\xrightarrow{d} \sum_{j=0}^{N} \left(\sum_{l=1}^{m} r_j^l I(f_l)^{\diamond j} \right)$$
$$+ \sum_{l=1}^{m} r_{N+1}^l \left(I(f_l) \left(I(f_l)^{\diamond N} \right) - N\mathbf{E}\left[I(f_l)^2 \right] \left(I(f_l)^{\diamond (N-1)} \right) \right)$$
$$= \sum_{j=0}^{N+1} \left(\sum_{l=1}^{m} r_j^l I(f_l)^{\diamond j} \right),$$

where the last identity follows from the Hermite recursion (1.14) for the continuous Wick powers. Hence, by the Cramér-Wold device, we obtain the assertion (1.25) for $N + 1$, and the proof is finished. □

1.3. Application to the fractional Black-Scholes model

The fractional Black-Scholes model is a modification of the classical Black-Scholes model, where the driving Brownian motion is replaced by a fractional Brownian motion. In the context of the continuous Wick calculus, the fractional Black-Scholes model was studied by Hu and Øksendal [18] and Elliott and van der Hoek [10]. In [3], Bender and Elliott introduced a discrete version of the fractional Black-Scholes model based on a difference equation involving discrete Wick products. Following the lines of [4], we will now explain how the convergence results obtained in the previous section can be applied in order to prove weak convergence of the discrete fractional Black-Scholes model to the fractional Black-Scholes model. To this end we first introduce the general setting.

1.3.1. *The fractional Black-Scholes model*

Recall that a *fractional Brownian motion* $(B_t^H)_{t\in[0,1]}$ on the unit time interval with Hurst parameter $H \in (0,1)$ is a continuous zero mean Gaussian process with stationary increments and covariance function

$$\mathbf{E}[B_t^H B_s^H] = \frac{1}{2}\left(t^{2H} + s^{2H} - |t-s|^{2H}\right), \quad s,t \in [0,1].$$

The process $B^{\frac{1}{2}}$ is a standard Brownian motion, but a fractional Brownian motion is not a semimartingale for $H \neq \frac{1}{2}$. We restrict ourselves to the case of Hurst parameter $H \in (\frac{1}{2}, 1)$.

Based on works by Molchan and Golosov [25, 26], a fractional Brownian motion can be represented by

$$B_t^H = \int_0^t z_H(t,s)dB_s, \quad t \in [0,1]. \tag{1.26}$$

Here B is a standard Brownian motion and, for $H > \frac{1}{2}$, the deterministic kernel takes the form

$$z_H(t,s) = \mathbf{1}_{\{t\geq s\}} c_H (H - \frac{1}{2}) s^{\frac{1}{2}-H} \int_s^t u^{H-\frac{1}{2}}(u-s)^{H-\frac{3}{2}} du \tag{1.27}$$

with the constant

$$c_H = \sqrt{\frac{2H\Gamma(\frac{3}{2}-H)}{\Gamma(H+\frac{1}{2})\Gamma(2-2H)}},$$

where Γ is the Gamma function (Norros et al. [28, 5.2] or Nualart [29, section 5.1.3]). For an introduction to some elementary properties of fractional Brownian motion we refer to Nualart [29, chapter 5] or Mishura [24].

The *fractional Black-Scholes model* with Hurst parameter $H > 1/2$ consists of two assets, a riskless bond with interest rate r, $A_t := e^{rt}$ and a risky stock of the form

$$S_t := s_0\, e^{\mu t + \sigma B_t^H - \sigma^2 t^{2H}/2} = s_0\, e^{\mu t} \exp^\diamond(\sigma B_t^H) = s_0\, e^{\mu t} \exp^\diamond(\sigma I(z_H(t,\cdot)).$$

Here, $s_0 > 0$ is the initial price of the stock, $\sigma > 0$ can be interpreted as volatility and $\mu \in \mathbb{R}$ is the average growth rate of the stock. As the stock is not a semimartingale, there cannot exist an equivalent local martingale measure. As a consequence, arbitrage can be obtained when dynamic readjustment of the portfolio is admissible, see e.g. [5] for some examples and a survey on arbitrage issues related to fractional Brownian motion. By the fractional Girsanov theorem [28, Theorem 4.1], [24, Theorem 2.8.1] one can however construct an equivalent measure Q such that the discounted fractional Black-Scholes model has dynamics

$$\frac{S_t}{A_t} = s_0 \, e^{\sigma \tilde{B}_t^H - \sigma^2 t^{2H}/2},$$

where $\tilde{B}^H$ is a fractional Brownian motion under Q. It is straightforward that S_t/A_t has martingale marginals under Q and, hence, call option prices calculated as expectations of discounted call option payoffs under Q do not admit static arbitrage. In particular Q is an average risk-neutral measure in the sense that

$$\mathbf{E}^Q \left[\frac{S_t}{A_t} \right] = \frac{S_0}{A_0}$$

for every $t \in [0, 1]$.

Analogously to the case of the classical Black-Scholes model, the dynamics of the fractional Black-Scholes model can be described by the SDE

$$dS_t = \mu S_t dt + \sigma S_t d^{\diamond} B_t^H, \quad S_0 = s_0, \tag{1.28}$$

see e.g. [10]. Here, the diamond denotes that the stochastic integral is to be understood in the fractional Wick-Itô sense. The fractional Wick-Itô integral, introduced by Duncan et al. [9], is an extension of the Itô integral beyond the semimartingale framework. There are several approaches to the fractional Wick-Itô integral. Essentially these approaches are via white noise theory as in Elliott and von der Hoek [10] and Hu and Øksendal [18], by Malliavin calculus in Alòs et al. [1], or by an S-transform approach in Bender [2]. For a sufficiently good process $(X_s)_{s \in [0,t]}$, the fractional Wick-Itô integral with respect to fractional Brownian motion $(B_s^H)_{[0,t]}$ can be defined by Wick-Riemann sums (cf. Duncan et al. [9] or Mishura [24, Theorem 2.3.10]). Suppose $\pi_n = \{0 = t_0 < t_1 < \cdots < t_n = t\}$ with $\max_{t_i \in \pi_n} |t_i - t_{i-1}| \to 0$ for $n \to \infty$, then

$$\int_0^t X_s d^{\diamond} B_s^H := \lim_{n \to \infty} \sum_{t_i \in \pi_n} X_{t_{i-1}} \diamond \left(B_{t_i}^H - B_{t_{i-1}}^H \right),$$

if the Wick products and the $L^2(\Omega)$-limit exist (independently of the choice of the partitions). We note that the fractional Wick-Itô integral has zero mean, which is a crucial property for an additive noise. For more information on Wick-Itô integral with respect to fractional Brownian motion we refer to Mishura [24, chapter 2].

1.3.2. *A discrete version of the fractional Black-Scholes market*

The dynamic description of the fractional Black-Scholes model in Eq. (1.28) and the definition of the fractional Wick-Itô integral in terms of Wick-Riemann sums suggest to define a discrete version by a discrete Wick difference equation, where the fractional Brownian motion is replaced by an approximation based on binary trials. Such definition of a discrete fractional Black-Scholes model was proposed by Bender and Elliott [3].

The starting point is Sottinen's [31] approximation of a fractional Brownian motion by a disturbed binary random walk. Recalling that a fractional Brownian motion B_t^H can be represented as a Wiener integral of the kernel $z_H(t,\cdot)$ in (1.27) with respect to a Brownian motion, the idea of Sottinen [31] is to consider a discrete Wiener integral of a suitable approximation of the kernel as discrete analogue of a fractional Brownian motion. Precisely,

$$B_t^{H,n} := \sum_{i=1}^{\lfloor nt \rfloor} b_{t,i}^n \frac{1}{\sqrt{n}} \xi_i^n, \tag{1.29}$$

where

$$b_{t,i}^n := n \int_{\frac{i-1}{n}}^{\frac{i}{n}} z_H(\frac{\lfloor nt \rfloor}{n}, s) ds$$

(suppressing the dependence of $H > 1/2$, which we think of as fixed from now on).

It is shown in [31, Theorem 1] that $B^{H,n}$ weakly converges to a fractional Brownian motion $(B_t^H)_{t\in[0,1]}$ in the Skorokhod space $D([0,1],\mathbb{R})$.

We can now define a discrete version of the fractional Black-Scholes model by

$$S_i^n = \left(1 + \frac{\mu}{n}\right) S_{i-1}^n + \sigma S_{i-1}^n \diamond_n \left(B_{\frac{i}{n}}^{H,n} - B_{\frac{i-1}{n}}^{H,n}\right), \; S_0^n = s_0, \tag{1.30}$$

$i = 1, \ldots, n$, which is the natural discrete counterpart of the SDE (1.28). Bender and Elliott [3] show that this discrete stock price will move upwards once more, if it has increased for sufficiently many previous steps. Hence it admits a simple buy-and-hold arbitrage.

Before discussing weak convergence of this discrete model to the fractional Black-Scholes model, it is instructive to compare the situation with the case of approximating a standard Black-Scholes model by a Cox-Ross-Rubinstein model.

Example 1.1. (i) The stock price in the Cox-Ross-Rubinstein model is given by

$$X_i^n = \left(1 + \frac{\mu}{n}\right) X_{i-1}^n + \sigma X_{i-1}^n \frac{1}{\sqrt{n}} \xi_i^n, \qquad X_0^n = x_0, \qquad i = 1, \dots, n.$$

Denoting the canonical binary random walk approximation of a Brownian motion by

$$B_t^{1/2} = \frac{1}{\sqrt{n}} \sum_{i=1}^{\lfloor nt \rfloor} \xi_i^n,$$

we can rewrite X^n as

$$X_i^n = \left(1 + \frac{\mu}{n}\right) X_{i-1}^n + \sigma X_{i-1}^n \diamond_n \left(B_{\frac{i}{n}}^{1/2,n} - B_{\frac{i-1}{n}}^{1/2,n}\right), \; X_0^n = x_0,$$

$i = 1, \dots, n$. This shows that the Cox-Ross-Rubinstein model naturally appears in Eq. (1.30), if we formally set $H = 1/2$. The solution of the difference equation for the Cox-Ross-Rubinstein model is

$$X_i^n = \prod_{j=1}^{i} \left(1 + \frac{\mu}{n} + \sigma \frac{1}{\sqrt{n}} \xi_j^n\right). \tag{1.31}$$

If we assume for simplicity that $\mu = 0$ and $\sigma = 1$, then X_j^n is just the discrete Wick exponential of $B_{j/n}^{1/2}$. Hence, in this situation the discrete Wick exponential solves the (discrete Wick) difference equation. Passing to the logarithm as in the proof of Theorem 1.2 is then the standard way to prove the convergence of the finite-dimensional distributions of the Cox-Ross-Rubinstein model to the Black-Scholes model. The weak convergence of the processes follows by Donsker's theorem and by applying the exponential as a continuous mapping on the Skorokhod space.

(ii) By part (i) of this example, it is tempting to hope that the discrete Wick difference equation (1.30) is also solved by the Wick exponential of $B_{j/n}^H$, if $\mu = 0$ and $\sigma = 1$. However, due to the memory of the perturbed random walk approximation of the fractional Brownian motion, one obtains by elementary calculations e.g.

$$\exp^{\diamond_n}(B^H_{2/n}) - S^n_2 = \frac{1}{n}(b^n_{2/n,1} - b^n_{1/n,1})b^n_{2/n,2}\Xi^n_{\{1,2\}}.$$

Thus, in order to discuss convergence of the discrete model S^n to the fractional Black-Scholes model, Theorem 1.2 cannot be applied directly.

1.3.3. *Weak convergence to the fractional Black-Scholes model*

Weak convergence of the discrete version of the fractional Black-Scholes model (1.30) to its continuous counterpart was conjectured by Bender and Elliott [3, Remark 4.1]. It was recently proved by Bender and Parczewski [4, Theorem 3] in the following form.

Theorem 1.4. *Suppose $\mu, s_0 \in \mathbb{R}$, $\sigma > 0$. Then $\widetilde{S}^n_t := S^n_{\lfloor nt \rfloor}$, the piecewise constant interpolation of the discrete Wick difference equation (1.30), converges weakly to the stock S in the fractional Black-Scholes model, i.e. the solution of the SDE (1.28), in the Skorokhod space $D([0,1],\mathbb{R})$.*

We now sketch the main arguments of the proof of Theorem 1.4. We will first consider the case where $\mu = 0$ and $\sigma = 1$. Hence, the stock S_t of the fractional Black-Scholes model coincides with the Wick exponential $\exp^\diamond(B^H_t)$. Recall that the Wick exponential has the series representation

$$\exp^\diamond\left(B^H_t\right) = \sum_{k=0}^{\infty} \frac{1}{k!}\left(B^H_t\right)^{\diamond k}. \tag{1.32}$$

Writing $(B^H_t)^{\diamond k} = h^k_{t^{2H}}(B^H_t)$ in terms of Hermite polynomials as in (1.15) and applying the fractional Itô formula (cf. [2, Theorem 5.3] or [6, Theorem 3.7.2]), we have

$$d(B^H_t)^{\diamond k} = k(B^H_t)^{\diamond k-1} d^\diamond B^H_t, \qquad (B^H_0)^{\diamond k} = \mathbf{1}_{\{k=0\}}. \tag{1.33}$$

Summing up this identity (1.33), we observe that the Wick exponential $\exp^\diamond(B^H_t)$ indeed solves the fractional Doléans-Dade equation,

$$dS_t = S_t d^\diamond B^H_t, \qquad S_0 = 1. \tag{1.34}$$

The discrete counterpart of the recursive system of SDEs (1.33) is the following system of discrete Wick difference equations,

$$U^{k,n}_i = U^{k,n}_{i-1} + kU^{k-1,n}_{i-1} \diamond_n \left(B^{H,n}_{\frac{i}{n}} - B^{H,n}_{\frac{i-1}{n}}\right), \; U^{0,n}_i = 1, \; U^{k,n}_0 = \mathbf{1}_{\{k=0\}}, \tag{1.35}$$

for all $i = 0, \ldots, n$ and $k \in \mathbb{N}$. As in Example 1.1 (ii), it is straightforward to check that the discrete Wick powers of $B^{H,n}_{i/n}$ do not solve this system of difference equations. However, by summing up the equation (1.35) over k we observe that

$$\left(\sum_{k=0}^{n} \frac{1}{k!} U_i^{k,n} \right)_{i=0,\ldots,n}$$

solves the difference equation

$$S_i^n = S_{i-1}^n + S_{i-1}^n \diamond_n \left(B^{H,n}_{\frac{i}{n}} - B^{H,n}_{\frac{i-1}{n}} \right), \; S_0^n = 1, \; i = 1, \ldots, n. \tag{1.36}$$

Denoting by $\widetilde{U}_t^{k,n} := U_{\lfloor nt \rfloor}^{k,n}$ the piecewise constant interpolation of (1.35), Theorem 1.4 for $\mu = 0$ and $\sigma = 1$ is hence a direct consequence of the following more general result (setting $a_{n,k} \equiv 1$).

Theorem 1.5 ([4], Theorem 2). *Suppose that the coefficients $a_{n,k}$ satisfy condition (1) in Theorem 1.3 and condition (2') in Remark 1.2. Then the sequence of processes $\sum_{k=0}^{n} \frac{a_{n,k}}{k!} \widetilde{U}^{k,n}$ converges weakly to the Wick power series $\sum_{k=0}^{\infty} \frac{a_k}{k!} \left(B^H\right)^{\diamond k}$ in the Skorokhod space $D([0,1], \mathbb{R})$.*

We now first show how Theorem 1.4 with a general choice of the parameter μ and σ can be obtained from Theorem 1.5 following the argumentation in [4]. Then we explain how Theorem 1.5 is related to Theorem 1.3 on the convergence of Wick-analytic functionals, which was proved in Section 2.

Proof. [Proof of Theorem 1.4] We define

$$a_{n,k} := \left(\frac{\sigma}{1 + \mu/n} \right)^k.$$

Then $a_{n,k} \to \sigma^k$ as n tends to infinity and assumption (2') in Remark 1.2 is satisfied. Hence, by Theorem 1.5,

$$\widetilde{V}^n := \sum_{k=0}^{n} \frac{a_{n,k}}{k!} \widetilde{U}^{k,n} \xrightarrow{d} \sum_{k=0}^{\infty} \frac{\sigma^k}{k!} \left(B^H\right)^{\diamond k} = \exp^{\diamond}(\sigma B^H).$$

By (1.35) we observe that $V_i^n := \widetilde{V}^n_{\frac{i}{n}}$ satisfies

$$V_i^n = V_{i-1}^n + \left(\frac{\sigma}{1 + \frac{\mu}{n}} \right) V_{i-1}^n \diamond_n \left(B^{H,n}_{\frac{i}{n}} - B^{H,n}_{\frac{i-1}{n}} \right), \; V_0^n = 1, \; i = 1, \ldots, n.$$

Consider now the piecewise constant function $\left(\widetilde{W}_t^n\right)_{t\in[0,1]}$ determined by $\widetilde{W}_t^n := W_{\lfloor nt \rfloor}^n$ and

$$W_i^n = \left(1 + \frac{\mu}{n}\right) W_{i-1}^n, \; W_0^n = s_0, \; i = 1, \ldots, n.$$

Obviously,

$$\left(\widetilde{W}_t^n\right)_{t\in[0,1]} \longrightarrow s_0 \left(\exp(\mu t)\right)_{t\in[0,1]} \tag{1.37}$$

holds in the sup-norm on $[0,1]$. Moreover,

$$\begin{aligned}
& V_i^n W_i^n \\
&= \left(1 + \frac{\mu}{n}\right) V_{i-1}^n W_{i-1}^n + \left[\left(\frac{\sigma}{1 + \frac{\mu}{n}}\right) V_{i-1}^n \diamond_n \left(B_{\frac{i}{n}}^{H,n} - B_{\frac{i-1}{n}}^{H,n}\right)\right] \left(1 + \frac{\mu}{n}\right) W_{i-1}^n \\
&= \left(1 + \frac{\mu}{n}\right) V_{i-1}^n W_{i-1}^n + \sigma V_{i-1}^n W_{i-1}^n \diamond_n \left(B_{\frac{i}{n}}^{H,n} - B_{\frac{i-1}{n}}^{H,n}\right),
\end{aligned}$$

$V_0^n W_0^n = s_0$, $i - 1, \ldots, n$. Hence, $S_i^n = V_i^n W_i^n$ is the solution of the discrete Wick difference equation (1.30). Multiplication with the deterministic function $s_0 \exp(\mu t)$ is continuous in the Skorokhod topology. Now, thanks to (1.37) and Theorem 4.1 in [7], we obtain

$$\left(\widetilde{S}_t^n\right)_{t\in[0,1]} = \left(\widetilde{V}_t^n \widetilde{W}_t^n\right)_{t\in[0,1]} \xrightarrow{d} s_0 \left(\exp(\mu t) \exp^{\diamond}\left(\sigma B_t^H\right)\right)_{t\in[0,1]}$$

in the Skorokhod space $D([0,1],\mathbb{R})$. Finally, by applying the fractional Itô formula, one easily observes that $s_0 \exp(\mu t) \exp^{\diamond}\left(\sigma B_t^H\right)$ solves the SDE (1.28), see e.g. [2, Corollary 5.5], and so the proof is complete. □

We close this section with a sketch of the proof of Theorem 1.5.

Proof. [Sketch of the proof of Theorem 1.5] As usual, weak convergence in the Skorokhod topology is established by showing convergence of the finite-dimensional distributions and tightness. We will here only discuss the convergence of the finite-dimensional distributions and refer to Section 6 in [4] for the rather technical estimates which yield tightness.

Recall from (1.29) that

$$B_t^{H,n} = \sum_{i=1}^{\lfloor nt \rfloor} b_{t,i}^n \frac{1}{\sqrt{n}} \xi_i^n,$$

where

$$b_{t,i}^n := n \int_{\frac{i-1}{n}}^{\frac{i}{n}} z_H\left(\frac{\lfloor nt \rfloor}{n}, s\right) ds.$$

By a result of Nieminen [27],

$$\frac{1}{n}\sum_{i=1}^{\lfloor nt\rfloor} b^n_{t,i} b^n_{s,i} \longrightarrow \int_0^1 z_H(t,u) z_H(s,u)du, \quad s,t \in [0,1].$$

Moreover, it is rather straightforward to estimate

$$\frac{1}{\sqrt{n}}|b^n_{t,i}| \leq 2c_H n^{-(1-H)}, \quad t \in [0,1],\ i \in \{1,\dots,n\},$$

see e.g. [4, Proposition 2]. Hence, by Theorems 1.1 and 1.3, we have that

$$\sum_{k=0}^{n} \frac{a_{n,k}}{k!} (B^{H,n})^{\diamond_n k} \overset{fd}{\longrightarrow} \sum_{k=0}^{\infty} \frac{a_k}{k!} \left(B^H\right)^{\diamond k} \quad (n \to \infty),$$

where $\overset{fd}{\to}$ denotes convergence of the finite-dimensional distributions. In order to get

$$\sum_{k=0}^{n} \frac{a_{n,k}}{k!} \widetilde{U}^{k,n} \overset{fd}{\longrightarrow} \sum_{k=0}^{\infty} \frac{a_k}{k!} \left(B^H\right)^{\diamond k} \quad (n \to \infty),$$

by Slutsky's theorem it is hence sufficient to show that, for every $t \in [0,1]$,

$$\lim_{n\to\infty} \mathbf{E}\left[\left|\sum_{k=0}^{n} \frac{a_{n,k}}{k!} \left(\widetilde{U}^{k,n}_t - (B^{H,n}_t)^{\diamond_n k}\right)\right|^2\right] = 0. \tag{1.38}$$

Making use of the Walsh decomposition of the discrete Wick powers in (1.17), one can inductively (over k) derive the Walsh decomposition of $U^{k,n}_t - (B^{H,n}_t)^{\diamond_n k}$. It is

$$\frac{1}{k!}\left(B^{H,n}_{\frac{l}{n}}\right)^{\diamond_n k} - \frac{1}{k!}U^{k,n}_l$$
$$= n^{-k/2} \sum_{\substack{C\subset\{1,\dots,l\}\\ |C|=k}} \left(\sum_{\substack{m:C\to\{1,\dots,l\}\\ \text{not injective}}} \prod_{p\in C} \left(b^n_{\frac{m(p)}{n},p} - b^n_{\frac{m(p)-1}{n},p}\right)\right) \Xi^n_C$$

for $l = 1, \dots, n$, with the convention $b^n_{l/n,i} = 0$ for $i > l$, see [4, Proposition 1]. By (1.9), we hence obtain

$$\mathbf{E}\left[\left|\sum_{k=0}^{n} \frac{a_{n,k}}{k!}\left(\widetilde{U}_t^{k,n} - (B_t^{H,n})^{\diamond_n k}\right)\right|^2\right]$$

$$= \sum_{k=0}^{\infty} \frac{a_{n,k}^2\, n^k}{(k!)^2} \sum_{\substack{C\subset\{1,\dots,\lfloor nt\rfloor\}\\ |C|=k}} \left(\sum_{\substack{m:C\to\{1,\dots,\lfloor nt\rfloor\}\\ \text{not injective}}} \prod_{p\in C} \left(b^n_{\frac{m(p)}{n},p} - b^n_{\frac{m(p)-1}{n},p}\right)\right)^2.$$

As shown in [4, Proposition 4], the right hand side can be estimated by Kn^{1-2H} for some constant K (independent of t and n), which yields (1.38) as $H > \frac{1}{2}$. □

1.4. Simulating the discrete version of the fractional Black-Scholes model

In this section we propose an algorithm for simulating sample paths of the discrete version of the fractional Black-Scholes model which is determined by the discrete Wick difference equation (1.30). In particular, we will discuss how much information about the whole family of sample paths is required, in order to sample a single paths of the model. Finally, we will then present some numerical examples which illustrate how the behaviour of the weak convergence, stated in Theorem 1.4, is influenced by the choice of the Hurst parameter.

1.4.1. *Discussion of the algorithm*

Our aim is to draw sample paths from the solution of the discrete Wick difference equation

$$S_i^n = \left(1 + \frac{\mu}{n}\right) S_{i-1}^n + \sigma S_{i-1}^n \diamond_n \left(B^{H,n}_{\frac{i}{n}} - B^{H,n}_{\frac{i-1}{n}}\right), \; S_0^n = s_0, \; i = 1, \dots, n.$$

The main difficulty is the presence of the discrete Wick product, which is defined in terms of the Walsh decomposition. Here we can exploit that one of the factors in the discrete Wick product is an increment of the 'discrete fractional Brownian motion'. This fact can be used to reformulate the discrete Wick product as an ordinary product plus a trace term which involves a discrete Malliavin derivative.

Suppose $X \in L^2(\Omega_n, \mathcal{F}_n, P_n)$ is represented in the form $X = f(\xi_1, \dots, \xi_n)$ for some deterministic function. We think of the *discrete*

Malliavin derivative as the derivative of X with respect to the increments $\frac{1}{\sqrt{n}}\xi_i$, $i = 1, \ldots, n$. This leads to the definition

$$D_i^n X := \frac{f(\xi_1^n, \ldots, \xi_{i-1}^n, 1, \xi_{i+1}^n, \ldots, \xi_n^n) - f(\xi_1^n, \ldots, \xi_{i-1}^n, -1, \xi_{i+1}^n, \ldots, \xi_n^n)}{2/\sqrt{n}} \tag{1.39}$$

for $i = 1, \ldots, n$, which is easily seen to be independent of the choice of f. Hence, this definition coincides with the one given by Leitz-Martini [22] or Privault [30]. The following lemma is a consequence of Proposition 5.5 in [22] adapted to our framework.

Lemma 1.3. *Suppose $X \in L^2(\Omega_n, \mathcal{F}_n, P_n)$ and $i \in \{1, \ldots, n\}$. Then,*

$$X \diamond_n \left(B_{\frac{i}{n}}^{H,n} - B_{\frac{i-1}{n}}^{H,n}\right) = X \left(B_{\frac{i}{n}}^{H,n} - B_{\frac{i-1}{n}}^{H,n}\right) - \frac{1}{n}\sum_{j=1}^{i} \left(b_{\frac{i}{n},j}^n - b_{\frac{i-1}{n},j}^n\right) D_j^n X.$$

Proof. By the bilinearity of the discrete Wick product and since

$$B_{\frac{i}{n}}^{H,n} - B_{\frac{i-1}{n}}^{H,n} = \frac{1}{\sqrt{n}}\sum_{j=1}^{i} \left(b_{\frac{i}{n},j}^n - b_{\frac{i-1}{n},j}^n\right) \xi_i^n,$$

it is sufficient to show

$$X \diamond_n \xi_i^n = X\xi_i^n - \frac{1}{\sqrt{n}} D_i^n X.$$

Taking the definition of the discrete Malliavin derivative (1.39) into account, $D_i^n X$ can be expressed in terms of the Walsh decomposition $X = \sum_{A \subset \{1,\ldots,n\}} X_A^n \Xi_A$ as

$$D_i^n X = \sqrt{n} \sum_{\substack{A \subset \{1,\ldots,n\} \\ i \in A}} X_A^n \Xi_{A \setminus \{i\}}^n.$$

As $\Xi_A^n \xi_i^n = \Xi_{A\setminus\{i\}}^n$ for $i \in A$, we get

$$X \diamond_n \xi_i^n = \sum_{A \subset \{1,\ldots,n\}} X_A^n (\Xi_A^n \diamond_n \xi_i^n) = \sum_{\substack{A \subset \{1,\ldots,n\} \\ i \notin A}} X_A^n \Xi_A^n \xi_i^n = X\xi_i^n - \frac{1}{\sqrt{n}} D_i^n X.$$

□

Remark 1.3. The term involving the discrete Malliavin derivative in Lemma 1.3 can be interpreted as a discrete analogue of the trace term which appears in continuous time, when a Wick-Itô integral is expressed in terms of a pathwise forward integral, see Proposition 3 in [1].

Applying Lemma 1.3, we can now reformulate the discrete Wick difference equation (1.30) into a 'Wick-free' version as

$$S_i^n = S_{i-1}^n \left(1 + \frac{\mu}{n} + \sigma \left(B^{H,n}_{\frac{i}{n}} - B^{H,n}_{\frac{i-1}{n}}\right)\right) - \sigma \frac{1}{n} \sum_{j=1}^{i-1} \left(b^n_{\frac{i}{n},j} - b^n_{\frac{i-1}{n},j}\right) D_j^n S_{i-1}^n. \tag{1.40}$$

(Here we sum up to $i-1$ since $D_i^n S_{i-1}^n = 0$, because S_{i-1}^n does not contain the binary trial ξ_i^n.)

By the definition of the discrete Malliavin derivative (1.39), Eq. (1.40) allows to construct recursively functions $s_i^n(x_1, \ldots, x_i)$ such that $S_i^n = s_i^n(\xi_1^n, \ldots, \xi_i^n)$. This is the algorithm which we implement for our numerical results presented below.

We emphasize that the construction of $s_i^n(\xi_1^n, \ldots, \xi_i^n)$ requires the evaluation of $s_{i-1}^n(\xi_1, , \ldots, \xi_{i-1}^n)$ and, additionally, for every $j = 1, \ldots, i-1$, the evaluation of $s_{i-1}^n(\xi_1^n, \ldots, \xi_{j-1}^n, 1, \xi_{j+1}^n, \ldots, \xi_{i-1}^n)$ and $s_{i-1}^n(\xi_1^n, \ldots, \xi_{j-1}^n, -1, \xi_{j+1}^n, \ldots, \xi_{i-1}^n)$, due to the presence of the discrete Malliavin derivative. Recursively, we observe that the construction of $s_i^n(\xi_1^n, \ldots, \xi_i^n)$ along a fixed path requires the evaluation of s_j^n, $j \leq i-1$, along many other paths. We shall now compute the number of such evaluations which to some extent illustrates the complexity of the algorithm. To this end we denote by $p_n(k)$ the number of different vectors $(x_1, \ldots, x_k) \in \{-1, 1\}^k$ at which s_k^n must be evaluated in order to generate a single path of $(s_i^n(\xi_1^n, \ldots, \xi_i^n),\ i = 1, \ldots, n)$. Moreover,

$$q_n := \sum_{k=1}^{n} p_n(k)$$

denotes the total number of different evaluations of $(s_j^n)_{j=1,\ldots,n}$ required to sample a single path.

Proposition 1.4. It holds that

$$p_n(k) = \sum_{i=0}^{k \wedge (n-k)} \binom{k}{i}, \tag{1.41}$$

$$q_n = F_{n+3} - 2 \sim \frac{\rho^3}{\sqrt{5}} \rho^n, \tag{1.42}$$

where F_n denote the Fibonacci numbers and $\rho = \frac{1+\sqrt{5}}{2} \approx 1.618$ is the golden ratio.

Proof. By (1.40) and the definition of the discrete Malliavin derivative (1.39) we have to vary up to $n-k$ of the entries $(\xi_1, \ldots, \xi_k)$ of $s_k^n(\xi_1^n, \ldots, \xi_k^n)$ in order to compute $s_i^n(\xi_1^n, \ldots, \xi_i^n)$ for $i = 1, \ldots, n$. This implies (1.41).

Summing $p_n(k)$ over k, we get

$$q_n = \sum_{k=1}^{n} \sum_{i=0}^{k \wedge (n-k)} \binom{k}{i},$$

which is actually a sum over a triangular array in Pascal's triangle. By the change of variables $l = k + j$ we obtain

$$\sum_{k=1}^{n} \sum_{i=0}^{k \wedge (n-k)} \binom{k}{i} = \sum_{l=1}^{n} \sum_{j=0}^{\lfloor \frac{l}{2} \rfloor} \binom{l-j}{j}, \tag{1.43}$$

where the inner sums on the right hand side are built across diagonals in Pascal's triangle, as illustrated in Figure 1.1 on the right hand side.

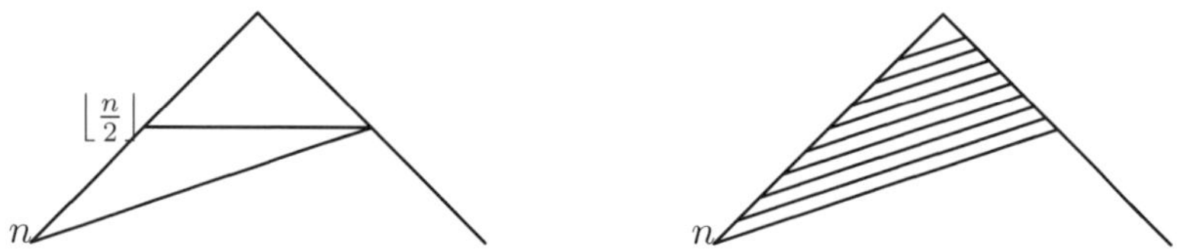

Fig. 1.1. Illustration of the change of variables in (1.43).

We will now use the following facts on Fibonacci numbers, which can be easily obtained by induction on k:

$$F_{k+1} = \sum_{j=0}^{\lfloor \frac{k}{2} \rfloor} \binom{k-j}{j}, \qquad \sum_{k=1}^{n} F_k = F_{n+2} - 1. \tag{1.44}$$

Gathering the above identities yields

$$q_n = \sum_{k=1}^{n} F_{k+1} = \sum_{k=2}^{n+1} F_k = F_{n+3} - 2.$$

We can finally apply the asymptotical behaviour of the Fibonacci numbers

$$F_n \sim \frac{1}{\sqrt{5}} \rho^n,$$

in order to get

$$q_n \sim \frac{\rho^{n+3}}{\sqrt{5}} - 2 \sim \frac{\rho^3}{\sqrt{5}} \rho^n.$$

□

Remark 1.4. We now compare the situation with the generation of a path in the Cox-Ross-Rubinstein model and reinterpret the results of Proposition 1.4 in terms of a binomial tree.

The stock price in the Cox-Ross-Rubinstein model in (1.31) forms a recombining tree in the sense that X_i^n only depends on how many binary trials ξ_j^n, $j = 1, \dots, i$, are in the up-state and how many are in the down-state. At the i-th time step this tree has $(i + 1)$ nodes and, hence, the whole tree has $(n+2)(n+1)/2$ nodes. Due to the explicit pathwise solution formula, the generation of a single fixed path of the Cox-Ross-Rubinstein model requires to know the value of one node per time step, and hence the value of n of the $(n + 2)(n + 1)/2$ nodes in the binomial tree.

In contrast, the binomial tree generated by the discrete version of the fractional Black-Scholes model in (1.30) is not recombining. Thus, at the i-th time step this tree has 2^i nodes and the total number of nodes in the tree is $2^{n+1} - 1$. The number $p_n(i)$ is just the number of nodes required at time i in order to generate a single path of the model. Thus, $q_n = F_{n+3} - 2$ is the total number of the $2^{n+1} - 1$ nodes in the tree which is required to generate one sample path. This illustrates that, due to the presence of the discrete Wick products, the value of the stock price along a single path is heavily influenced by the value of the stock price along the other paths.

1.4.2. *Numerical results*

We now illustrate the dependence of the weak convergence result in Theorem 1.4 on the Hurst parameter by some numerical experiments. The paths of the discrete version of the fractional Black-Scholes model (1.30) are generated as described in the previous subsection. We consider a discounted, average risk-neutral version of the discrete models (i.e $\mu = 0$) with volatility $\sigma = 0.2$ and initial price normalized to $s_0 = 1$. In order to make the dependence on $H > 1/2$ explicit, we write $S_i^{H,n}$ instead of S_i^n. Moreover $S_i^{1/2,n}$ denotes the Cox Ross Rubinstein model (1.31).

We first examine, how well the distribution of $S_n^{H,n}$ approximates the marginal distribution of the continuous time limit $\exp^\diamond(0.2\, B_1^H)$, which is a log-normal distribution with parameters $(\mu, \sigma) = (0, 0.2)$ independent of $H \geq 1/2$. Figure 1.2 shows the histograms of $S_n^{0.75,n}$ for $n = 5, 9, 12, 15$ and a histogram of the exact limiting distribution. For the histograms the positive real line is divided into $2^7 + 1 = 129$ intervals. Figures 1.3 and 1.4 contain the histograms for $S_n^{0.6,n}$ and $S_n^{1/2,n}$ for the same choices of n.

In the case of the Cox-Ross-Rubinstein model ($H = 1/2$), $S_n^{1/2,n}$ takes

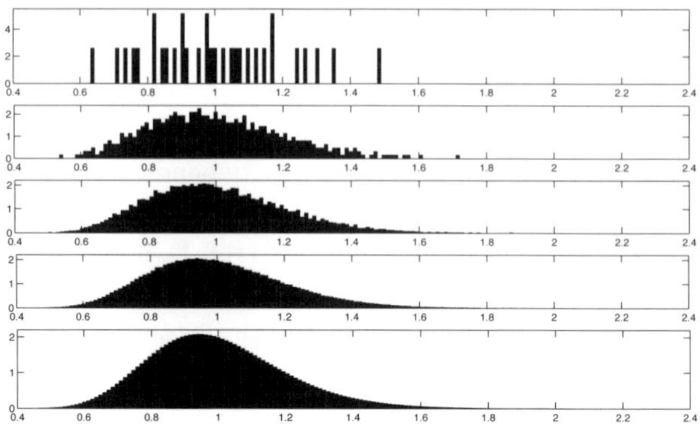

Fig. 1.2. Histograms of $S_n^{0.75,n}$ (1.30) for $n = 5, 9, 12, 15$ and of the exact lognormal distribution (from the top to the bottom).

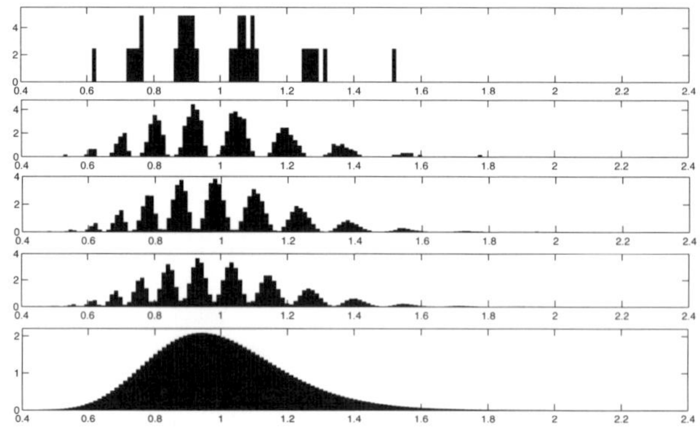

Fig. 1.3. Histograms of $S_n^{0.6,n}$ (1.30) for $n = 5, 9, 12, 15$ and of the exact lognormal distribution.

$(n+1)$ values only. Hence, the histograms show $(n+1)$ columns which slowly tend to the histogram of the log-normal distribution as n increases. As the Hurst parameter H and, hence the memory effect, increases, the probability mass of $S_n^{H,n}$ is spreading out. For $H = 0.6$, the peaks are still clearly visible, while for $H = 0.75$ and $n = 15$, the histogram is already very close to that of the continuous limiting distribution. The maximal absolute error between the cumulative distribution function of $S_{15}^{0.75,15}$ and the cumulative distribution function of a log-normal with parameters $(\mu, \sigma) = (0, 0.2)$ is as small as 0.015. In conclusion, we observe that the larger the Hurst parameter, the better $S_n^{H,n}$ approximates the marginal distribution of the

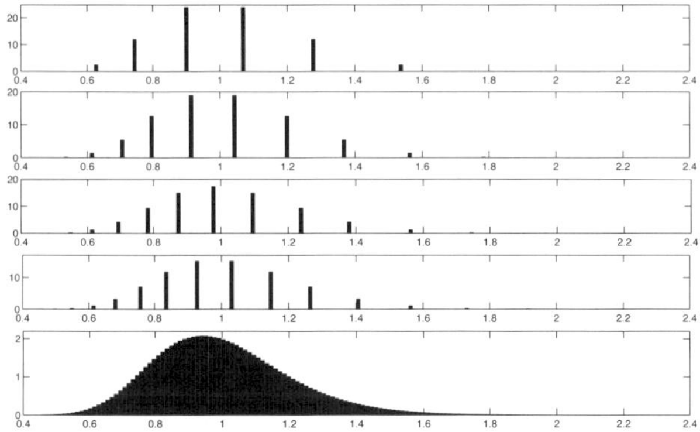

Fig. 1.4. Histograms of $S_n^{1/2,n}$ (1.30) for $n = 5, 9, 12, 15$ and of the exact lognormal distribution.

fractional Black-Scholes model.

We now compare the convergence behaviour of the average risk-neutral expectations of a call-payoff for different Hurst parameters. Figure 1.5 displays the absolute error

$$|E[(\exp^{\diamond}(0.2\, B_1^H) - K)_+] - E[(S_n^{H,n} - K)_+]|$$

for $H = 0.5, 0.6, 0.75$ (solid line, dashed line, dot-dashed line) as a function of the number of time steps n. We consider the at the money case ($K = 1$). For out-of-the-money options and in-the-money-options, the behaviour is qualitatively similar. As expected in view of the histograms, the error is strongly fluctuating for small Hurst parameters and decreases more regularly for $H = 0.75$. Moreover, for a small number of time steps the error is significantly smaller for the case of a large Hurst parameter $H = 0.75$.

An analysis of a log-log-plot indicates that the weak rate of convergence is about $2H$ (indeed a little bit faster in all cases), but, due the relatively small number of time steps, the analysis is not fully conclusive.

We finally illustrate the influence of the Hurst parameter on the path behaviour of $S_i^{H,n}$, $i = 0, \ldots, n$. To this end we draw a sample path of the random walk approximation $B^{1/2,n}$ of the Brownian motion and plot the random walk approximations of the fractional Brownian motion $B^{H,n}$ for $H = 0.65, 0.75, 0.85$, constructed as discrete Wiener integral driven by the path of $B^{1/2,n}$ in Figure 1.6. This figure also shows the corresponding paths of the discrete versions of the fractional Black-Scholes model $S^{H,n}$. We choose $n = 20$ and, hence, by Proposition 1.4, $F_{23} - 2 = 28,655$ nodes of the binomial tree (consisting of $2,097,151$ nodes) are required to construct

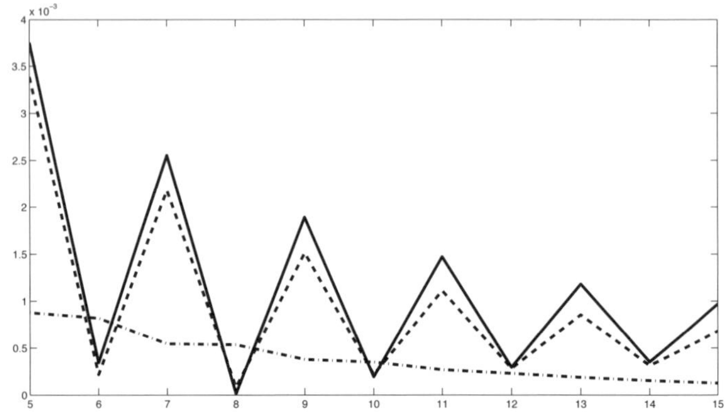

Fig. 1.5. Absolute error of the average risk-neutral expectations of an at-the-money call as a number of the time steps n for different Hurst parameters $H = 0.5, 0.6, 0.75$ (solid line, dashed line, dot-dashed line).

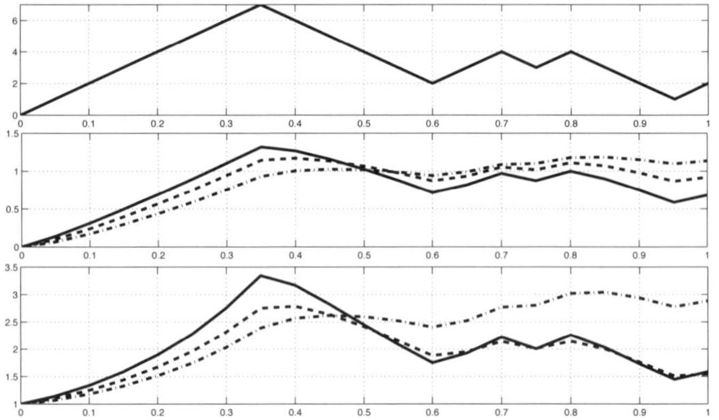

Fig. 1.6. From the top to the bottom: Random walk approximation of a Brownian motion; corresponding approximations of a fractional Brownian motion with Hurst parameters $H = 0.65$ (solid line), 0.75 (dashed line) and 0.85 (dot-dashed line); solution of (1.30) for $H = 0.65, 0.75, 0.85$.

a single path of $S^{H,n}$.

The path of the driving random walk (at the top of Figure 1.6) steeply increases until time 0.35 and, then, has a tendency to decrease. Due to the memory effect the steep increase of $B^{1/2,n}$ persists in the corresponding paths of $B^{H,n}$ and $S^{H,n}$. For the smaller Hurst parameter $H = 0.65$ the paths increase even more steeply at the beginning. Although the early increase of $B^{1/2,n}$ still influences the path behaviour of $B^{0.65,n}$ and $S^{0.65,n}$ at the later time steps, the paths have a tendency to fall in the long run, as does the path of the driving random walk. For the case of a large Hurst parameter $H = 0.85$ the steep increase is less pronounced at the beginning, but the tendency to increase persists up to time 1. These observations come, of course, at no surprise, because the corresponding dependency of the memory effect and the Hurst parameter is well known to be present in the continuous time limit.

References

[1] Alòs, E. and Mazet, O. and Nualart, D. Stochastic calculus with respect to Gaussian processes. *Ann. Probab.* **29** (2), 766-801 (2001).

[2] Bender, C. An S-transform approach to integration with respect to a fractional Brownian motion. *Bernoulli* **6**, 955-983 (2003).

[3] Bender, C. and Elliott, R. J. Arbitrage in a discrete version of the Wick-fractional Black-Scholes market. *Math. Oper. Res.* **29**, 935-945, (2004).

[4] Bender, C., Parczewski, P. Approximating a geometric fractional Brownian motion and related processes via discrete Wick calculus. *Bernoulli* **16** (2), 389-417 (2010).

[5] Bender, C. and Sottinen, T. and Valkeila, E. Fractional Processes as Models in Stochastic Finance. Book chapter in: Di Nunno, Oksendal (Eds.), *AMaMeF: Advanced Mathematical Methods for Finance*, Springer, 75-103 (2011).

[6] Biagini, F. and Hu, Y. and Øksendal, B. and Zhang, T. *Stochastic calculus for fractional Brownian motion and applications.* Probability and its Applications. London: Springer (2008).

[7] Billingsley, P. *Convergence of probability measures.* New York-London-Sydney-Toronto: John Wiley and Sons, Inc. (1968).

[8] Billingsley, P. *Probability and Measure. Third edition.* New York-London-Sydney-Toronto: John Wiley and Sons, Inc. (1995).

[9] Duncan, T. E. and Hu, Y. and Pasik-Duncan, B. Stochastic calculus for fractional Brownian motion. I: Theory. *SIAM J. Control Optimization* **38**, 582-612, (2000).

[10] Elliott, R. J. and van der Hoek, J. A general fractional white noise theory and applications to finance. *Math. Finance* **13** (2), 301-330, (2003).

[11] Formanov, Sh. K. On the SteinTikhomirov method and its applications in nonclassical limit theorems. *Discr. Math. Appl.* **17**, 13–23, (2007).

[12] Gjessing, H. and Holden, H. and Lindstrøm, T. and Øksendal, B. and Ubøe, J. and Zhang, T. The Wick product. In Niemi, Högnas, Shiryaev, Melnikov *Frontiers in Pure and Applied Probability. Vol. 1.* TVP Publishers, Moscow, 29-67, (1993).

[13] Gzyl, H. An exposé on discrete Wiener chaos expansions. *Bol. Asoc. Mat. Venez.* **13** (1), 3-27, (2006).

[14] Hida, T. and Ikeda, N. Analysis on Wiener space with reproducing kernel arising from multiple Wiener integrals. Proceedings of the fifth Berkeley Symposium on Mathematical Statistics and Probability, II. Californian University Press, Los Angeles, 117–143, (1965).

[15] Holden, H. and Lindstrøm, T. and Øksendal, B. and Ubøe, J. Discrete Wick calculus and stochastic functional equations *Potential Anal.* **1** (3), 291-306, (1992).

[16] Holden, H. and Lindstrøm, T. and Øksendal, B. and Ubøe, J. Discrete Wick products. Stochastic analysis and related topics (Oslo, 1992), Stochastics Monogr., **8**, Gordon and Breach, Montreux, 123-148, (1993).

[17] Holden H. and Øksendal, B. and Ubøe, J. and Zhang, T. *Stochastic partial differential equations. A modeling, white noise functional approach. Second Edition* New York: Springer (2010).

[18] Hu, Y. and Øksendal, B. Fractional white noise calculus and applications to finance. *Infin. Dimens. Anal. Quantum Probab. Relat. Top.* **6** (1), 1-32, (2003).

[19] Hu, Y. and Yan, J.A. Wick calculus for nonlinear Gaussian functionals. *Acta Math. Appl. Sin. Engl. Ser.* **25** (3), 399-414, (2009).

[20] Janson, S. *Gaussian Hilbert Spaces.*, Cambridge: Cambridge University Press (1997).

[21] Kuo, H.-H. *White noise distribution theory.* Probability and Stochastics Series. Boca Raton, FL: CRC Press (1996).

[22] Leitz-Martini, M. A discrete Clark-Ocone formula., Maphysto Research Report No 29 (2000).

[23] Meyer, P. A., Yan, J.-A. Distributions sur l'espace Wiener (suite). *Séminaire de Probabilités* **XXIII**, 382–392, (1989).

[24] Mishura, Y. *Stochastic calculus for fractional Brownian motion and related processes.* Lecture Notes in Mathematics 1929. Berlin: Springer (2008).

[25] Molchan, G. Gaussian processes with spectra which are asymptotically equivalent to a power of λ. *Theory Prob. Appl.* **14**, 530-532 ,(1969).

[26] Molchan, G. and Golosov, J. Gaussian stationary processes with asymptotic power spectrum. *Soviet. Math. Dokl.* **10**, 134-137 ,(1969).

[27] Nieminen, A. Fractional Brownian motion and martingale-differences. *Stat. Probab. Lett.* **70**, 1-10 (2004).

[28] Norros, I. and Valkeila, E. and Virtamo, J. An elementary approach to a Girsanov formula and other analytical results on fractional Brownian motions. *Bernoulli* **5**,571-587 ,(1999).

[29] Nualart, D. *The Malliavin calculus and related topics.* Second Edition. Probability and its Applications (New York). Springer (2006).

[30] Privault, N. *Stochastic Analysis in Discrete and Continuous Settings.* Lecture Notes in Mathematics 1982. Berlin: Springer (2009).

[31] Sottinen, T. Fractional Brownian motion, random walks and binary market models. *Finance and Stochastics.* **5**, 343-355 (2001).

[32] Wick, G. C. The evaluation of the collinear matrix. *Phys. Rev.* **80**, 278–272 (1950).

Chapter 2

Malliavin differentiability of a class of Feller-diffusions with relevance in Finance

Christian-Oliver Ewald[1], Yajun Xiao[2], Yang Zou[3] and Tak Kuen Siu[4]

[1] *Department of Economics, University of Glasgow, Glasgow, G12 8RT, United Kingdom*
christian.ewald@glasgow.ac.uk

[2] *Institute for Research in Economic Evolution, Platz der Alten Synagoge, KGII, 79098 Freiburg i. Br. Germany*
xiao.yajun@vwl.uni-freiburg.de

[3] *Mathematics Department, University of Kaiserslautern, D-67653 Kaiserslautern, Germany*
zou@mathematik.uni-kl.de

[4] *Department of Applied Finance and Actuarial Studies, Faculty of Business and Economics, Macquarie University, Sydney, NSW 2109, Australia*
ken.siu@mq.edu.au,   ktksiu2005@gmail.com

In this chapter we discuss the Malliavin differentiability of a particular class of Feller diffusions which we call δ-diffusions. This class is given by

$$d\nu_t = \kappa(\theta - \nu_t)dt + \eta\nu_t^\delta dW_t, \delta \in \left[\frac{1}{2}, 1\right]$$

and appears to be of relevance in Finance, in particular in the context of interest and foreign-exchange rate models, as well as stochastic volatility models. We extend the result obtained in Alos and Ewald (2008) [1] for $\delta = \frac{1}{2}$ and prove the Malliavin differentiability for all $\delta \in [\frac{1}{2}, 1]$

Contents

2.1. Introduction

Malliavin calculus has been extensively applied in Mathematical Finance with particular emphasize on portfolio optimization with anticipative information and the Monte Carlo computation of Greeks. In the case that the dynamics of the underlying assets in a market model are specified as the solutions of stochastic differential equations, coefficients are in general assumed to be regular in the sense that they are continuously differentiable with bounded derivatives. Only then do standard results such as Theorem 2.2.1 in Nualart (1995) [2] guarantee Malliavin differentiability. In all other cases it is, in general, not easy to prove Malliavin differentiability. Here we show that the class of processes specified by

$$d\nu_t = \kappa(\theta - \nu_t)dt + \eta\nu_t^\delta dW_t, \tag{2.1}$$

with $\delta \in [\frac{1}{2}, 1]$ is Malliavin differentiable. We call such processes δ-diffusions. For $\delta = \frac{1}{2}$ we have in fact a Cox-Ingersoll-Ross process and the Malliavin differentiability in this case was shown in Alos and Ewald (2008) [1]. For $\delta = 1$ we are in the situation where we have globally Lipschitz continuous coefficients with linear growth, and are able to apply Theorem 2.2.1 in Nualart (1995) [2] to conclude the Malliavin differentiability.

Feller processes with different δ coefficients may have potential applications in actuarial modeling. In particular, they may be used to model the long-term behavior of actuarial and investment variables. For example, they may be used in continuous-time versions of investment and economic variables in the Wilkie stochastic investment models, (see Wilkie (1986, 1995) [3], [4]), where the main features of these variables are mean reversion. Feller processes with different coefficients not only incorporate the mean-reverting feature of the long-term behavior of investment and economic variables, but also different degrees of conditional heteroscedasticity for these variables. Another possible application of Feller processes is modeling the force of mortality for valuing and hedging life insurance products, or longevity products.

In this article we are building the bridge from $\delta = \frac{1}{2}$ to $\delta = 1$ and show that the whole class of Feller diffusions presented above is Malliavin differentiable. The main results of this article are presented in Proposition 2.1 and Proposition 2.2 in Section 2.3. Proposition 2.1 gives a convergence result of an approximating sequence for a transformed Feller process in the $L^2(\Omega)$ space. The key to establish this convergence result is the Yamada–Watanabe comparison theorem for stochastic differential equations. Proposition 2.2 establishes the Malliavin differentiability of ν_t and gives an explicit formula for the Malliavin derivative. These results are established using Proposition 2.1, the moment stability of Feller processes, a result for the existence of a Malliavin derivative in Detemple (2005) [5] as well as the chain rule of Malliavin derivatives.

To make the article self-contained, we include a brief summary of Malliavin calculus in Section 2.2, before presenting our results in Section 2.3. The conclusions are summarized in Section 2.4.

2.2. Preliminaries on Malliavin calculus

We outline the construction of the Malliavin derivative operator. Let us consider the standard Brownian motion $W = \{W_t, t \in [0,T]\}$ defined on a complete probability space $(\Omega, \mathcal{F}, P)$. Let $\mathcal{S}$ be the set of cylindrical functionals $F : \Omega \to \mathbb{R}$, given by $F = f(W(t_1), ..., W(t_n))$ where $f \in C_b^\infty(\mathbb{R}^n)$ is a smooth function with bounded derivatives of all orders. We define the Malliavin derivative operator on $\mathcal{S}$ via

$$D_s F := \sum_{i=1}^{n} \frac{\partial f}{\partial x_i} (W_{t_1}(\omega), ..., W_{t_n}(\omega)) \cdot \mathbf{1}_{[0,t_i]}(s).$$

This operator and the iterated operators D^k are closable and unbounded from $L^p(\Omega)$ into $L^p(\Omega \times [0,T]^k)$, for all $k \geq 1$ and $p \geq 1$. Their respective domains are denoted by $\mathbb{D}^{k,p}$ and obtained as the closure of $\mathcal{S}$ with respect to the norms defined by $\|F\|_{k,p}^p = \|F\|_{L^p(\Omega)}^p + \sum_{j=1}^k \left\|D^j F\right\|_{L^2(\Omega\times[0,T]^j)}^p$. The adjoint operator of the Malliavin derivative operator $D : \mathbb{D}^{1,2} \to L^2(\Omega \times [0,T])$ is called the Skorohod integral (and is often denoted by δ). This operator has the property that its domain contains $L_a^2(\Omega \times [0,T], \mathbb{R})$, the class of square integrable adapted stochastic processes and its restriction to this class coincides with the Itô-integral. The Malliavin derivative and the divergence operator are related to each other by a duality formula, which has the interpretation of an integration-by-parts formula. We refer to Nualart (1995) [2] for further details.

2.3. Malliavin differentiability of δ-diffusions

In this section we will show that ν_t as defined in (2.1) belongs to $\mathbb{D}^{1,\infty} := \bigcap_{p\geq 1} \mathbb{D}^{1,p}$. We work on a complete probability space $(\Omega, \mathcal{F}, P)$ on which the Brownian motion W_t and the standard Brownian filtration is considered. In the case $\delta = 1$ we can apply Theorem 2.2.2 in Nualart (1995) [2] to conclude that ν_t is infinitely Malliavin differentiable. Hence, we can restrict our attention to the case $\delta \in [\frac{1}{2}, 1)$ from now on. The case $\delta = \frac{1}{2}$ has been discussed in Alos and Ewald (2008) [1] and would not demand another proof, nevertheless it would be artificial to exclude it from our consideration. We assume $\nu_0, \eta, \kappa, \theta > 0$ and for the case $\delta = \frac{1}{2}$ that the Novikov condition $2\kappa\theta > \eta^2$ holds. The SDE (2.1) admits a unique strong solution for $\delta \in [\frac{1}{2}, 1)$ by the Yamada and Watanabe Theorem (see, e.g. Proposition 2.13 in Karatzas and Shreve (1988) [6]). In order to show the Malliavin differentiability of the δ-diffusion ν_t we have to understand its boundary behavior and integrability. Using classical language we have:

Lemma 2.1. *Assume ν_t is a solution of the SDE (2.1). Then*

(1) the upper non-attachment boundary for $\delta \in [\frac{1}{2}, 1)$ is ∞,
(2) the lower non-attachment boundary for $\delta \in (\frac{1}{2}, 1)$ is 0, and
(3) the lower non-attachment boundary for $\delta = \frac{1}{2}$ provided the Novikov condition holds is 0.

Proof. The proof is a direct application of the classic Feller boundary classification criteria. See, e.g., Andersen and Piterbarg (2007) [7]. □

We shall later need the moment stability of δ-diffusions and its inverse. For this we study a slightly more general class of Feller diffusions:

$$d\nu_t = b(\nu_t)dt + \eta\nu_t^\delta dW_t$$

where $b(x)$ is a continuously differentiable function with globally bounded derivative.

Lemma 2.2.

(1) For $\delta \in [\frac{1}{2}, 1)$, $p \geq 0$, we have $\sup_{t\in[0,T]} \mathbb{E}[\nu_t^p] \leq C_1(1 + \nu_0^p)$.
(2) For $\delta \in (\frac{1}{2}, 1)$, $p \geq 0$, we have $\sup_{t\in[0,T]} \mathbb{E}[\nu_t^{-p}] \leq C_2(1 + \nu_0^{-p})$.
(3) For $\delta = \frac{1}{2}$ and $\frac{2b(0)}{\eta^2} > 2$, we have $\sup_{t\in[0,T]} \mathbb{E}[\nu_t^{-p}] \leq C_3(1 + \nu_0^{-p})$ for all $1 \leq p < \frac{2b(0)}{\eta^2} - 1$.

Here C_1, C_2, C_3 are constants only depending on p,T,η and $b(0)$.

Proof. See Bossy (2007) [8] Lemmas 2.1, 3.1 and 4.1. □

Similar to Alos and Ewald (2008) [1] we prove the Malliavin differentiability of ν_t by an approximating process, where the dynamics of the approximations have regular coefficients in the spirit of Detemple (2005) [5]. We wish to note at this point, that the particular dynamics chosen for the case $\delta = \frac{1}{2}$ by Alos and Ewald (2008) [1] is not convenient for the case $\delta \neq \frac{1}{2}$. For this reason we use a different approximation. To do so, we consider $\sigma_t := \nu_t^{1-\delta}$. While this process has a more complicated drift term than ν_t, the advantage is that it has a constant volatility term. In fact, applying the Itô formula, one can see that

$$d\sigma_t = (1-\delta)\left[\kappa\theta\sigma_t^{-\frac{\delta}{1-\delta}} - \frac{1}{2}\delta\eta^2\sigma_t^{-1} - \kappa\sigma_t\right]dt + (1-\delta)\eta dW_t. \qquad (2.2)$$

The SDE (2.2) does not a priori admit a strong solution, but we already know that the solution of (2.1) is unique. For any solution σ_t of (2.2), it follows from the Itô formula that $\sigma_t^{\frac{1}{1-\delta}}$ is a solution of (2.1) and as the latter is unique, we conclude the uniqueness of the solution for the SDE (2.2) up to a, (possibly complex), unit root. However, if σ_t solves (2.2), it is obvious that for no choice of $\lambda \in (0, 2\pi)$ does the process $e^{i\lambda}\sigma_t$ also solve the SDE (2.2).

Slightly different to Alos and Ewald (2008) [1], we approximate not only $\frac{1}{x}$ but also $x^{-\frac{\delta}{1-\delta}}$, to guarantee the appropriate growth conditions[a] for using the comparison theorem as well as Detemple (2005) [5]. Let $\varepsilon > 0$ and define the functions $\Psi_\varepsilon(x)$ and $\Phi_\varepsilon(x)$ as follows

$$\Psi_\varepsilon(x) = \begin{cases} \frac{1}{x} & \text{for} \quad x \geq \varepsilon \\ -\frac{1}{\varepsilon^2}x + \frac{2}{\varepsilon} & \text{for} \quad x < \varepsilon. \end{cases}$$

$$\Phi_\varepsilon(x) = \begin{cases} x^{-\frac{\delta}{1-\delta}} & \text{for} \quad x \geq \varepsilon \\ -\frac{\delta}{1-\delta}\varepsilon^{-\frac{\delta}{1-\delta}}x + \frac{1}{1-\delta}\varepsilon^{-\frac{\delta}{1-\delta}} & \text{for} \quad x < \varepsilon. \end{cases}$$

Both $\Psi_\varepsilon(x)$ and $\Phi_\varepsilon(x)$ are continuously differentiable with bounded derivatives. We have $\Psi_\varepsilon(x) > 0$ for all $x \in \mathbb{R}$ and $0 < x\Psi_\varepsilon(x) \leq 1$ for $x > 0$. Its first order derivative $\Psi_\varepsilon'(x)$ takes the value $-\frac{1}{x^2}$ for $x \geq \varepsilon$ and $-\frac{1}{\varepsilon^2}$ for $x < \varepsilon$. As ε goes to zero, $\Psi_\varepsilon(x)$ and $\Psi_\varepsilon'(x)$ converge uniformly on compacts to $\frac{1}{x}$ and $-\frac{1}{x^2}$ respectively. Moreover, as $|\Psi_\varepsilon'(z)| \leq \frac{1}{\varepsilon^2}$ for all $z \in \mathbb{R}$ we

[a]Note that for $\delta = \frac{1}{2}$ we have $\frac{\delta}{1-\delta} = 1$, while for general δ we have $\frac{\delta}{1-\delta} \in [1, \infty)$, which requires a more sophisticated approach to ensure various linear growth conditions.

conclude from the intermediate value theorem that for all $x, y \in \mathbb{R}$, we have $|\Psi_\varepsilon(x) - \Psi_\varepsilon(y)| \le \frac{1}{\varepsilon^2}|x-y|$.

The map $\Phi_\varepsilon(x)$ has similar properties: $\Phi_\varepsilon(x) > 0$ for all $x \in \mathbb{R}$, $0 < x^{\frac{\delta}{1-\delta}}\Phi_\varepsilon(x) \le 1$ for $x > 0$, $\Phi'_\varepsilon(x) = -\frac{\delta}{1-\delta}x^{-\frac{1}{1-\delta}}$ for $x \ge \varepsilon$, $\Phi'_\varepsilon(x) = -\frac{\delta}{1-\delta}\varepsilon^{-\frac{\delta}{1-\delta}}$ for $x < \varepsilon$, and $|\Phi_\varepsilon(x) - \Phi_\varepsilon(y)| \le \frac{\delta}{1-\delta}\varepsilon^{-\frac{\delta}{1-\delta}}|x-y|$ for all $x, y \in \mathbb{R}$.

Let us now define the approximations $\sigma_{t,\varepsilon}$ as the solutions of the SDEs

$$d\sigma_{t,\varepsilon} = (1-\delta)\left[\kappa\theta\Phi_\varepsilon(\sigma_{t,\varepsilon}) - \frac{1}{2}\delta\eta^2\Psi_\varepsilon(\sigma_{t,\varepsilon}) - \kappa\sigma_{t,\varepsilon}\right]dt + (1-\delta)\eta dW_t, \quad (2.3)$$

with $\sigma_{0,\varepsilon} = \sigma_0$ for all $\varepsilon > 0$. For $a > 0, b > 0$ and $p \in (1,\infty)$ we use the notation $B(a,b,p) := -(p-1)(\frac{b}{p})(\frac{b}{ap})^{\frac{1}{p-1}} < 0$ for the minimum of $ay^{-p} - by^{-1}$ for $y \in \mathbb{R}_+$.

Proposition 2.1. For each $t \in [0,T]$, the sequence $\sigma_{t,\varepsilon}$ converges to σ_t in $L^2(\Omega)$ as $\varepsilon \to 0$.

Proof. We first prove that $\sigma_{t,\varepsilon}$ converges to σ_t pointwise. Define a stopping time τ_ϵ via $\tau_\epsilon(\omega) := \inf\{t|\sigma_t(\omega) \le \varepsilon\}$, where $\varepsilon > 0$. With ε going to zero, the sequence (τ_ε) defines an increasing sequence of stopping times. Since ν_t is positive almost surely, so is σ_t. Lemma 2.1 gives $\lim_{\varepsilon\to 0}\tau_\varepsilon = \infty$ a.s.

Denote by $\sigma^{\tau_\varepsilon}$ the process obtained from σ by stopping at τ_ε. From the choice of the approximation function, it follows that for all $t \le \tau_\varepsilon$ we have $\sigma_t^{\tau_\varepsilon} = \sigma_{t,\varepsilon}$.

Now, for any fixed $t \in [0,T]$, letting ε go to zero, we obtain $\lim_{\varepsilon\to 0}\sigma_{t,\varepsilon} = \lim_{\varepsilon\to 0}\sigma_t^{\tau_\varepsilon} = \sigma_t$ a.s. We will now prove that $\sigma_{t,\varepsilon}$ converges to σ_t for each $t \in [0,T]$ in the space $L^2(\Omega)$. The idea is to find two square-integrable processes u_t and w_t such that $u_t \le \sigma_{t,\varepsilon} \le w_t$ satisfying $u_0 = w_0 = \sigma_0$. We look for the lower bound u_t first. It can be easily verified that

$$\kappa\theta\Phi_\varepsilon(x) - \frac{1}{2}\delta\eta^2\Psi_\varepsilon(x)$$
$$= \begin{cases} \kappa\theta x^{-\frac{\delta}{1-\delta}} - \frac{1}{2}\delta\eta^2 x^{-1}, & x \ge \varepsilon, \\ (\frac{1}{2}\delta\eta^2\varepsilon^{-2} - \frac{\kappa\theta\delta}{1-\delta}\varepsilon^{-\frac{1}{1-\delta}})x + \frac{\kappa\theta}{1-\delta}\varepsilon^{-\frac{\delta}{1-\delta}} - \delta\eta^2\varepsilon^{-1}, & x < \varepsilon. \end{cases}$$

For $\delta = \frac{1}{2}$, we have

$$\kappa\theta\Phi_\varepsilon(x) - \frac{1}{4}\eta^2\Psi_\varepsilon(x) = \begin{cases} (\kappa\theta - \frac{1}{4}\eta^2)x^{-1}, & x \geq \varepsilon, \\ -(\kappa\theta - \frac{1}{4}\eta^2)\varepsilon^{-2}x + 2(\kappa\theta - \frac{1}{4}\eta^2)\varepsilon^{-1}, & x < \varepsilon, \end{cases}$$

which is larger than 0 for all $x \in \mathbb{R}$ and $\varepsilon > 0$. For $\delta \in (\frac{1}{2}, 1)$, we notice two facts:

$$\kappa\theta x^{-\frac{\delta}{1-\delta}} - \frac{1}{2}\delta\eta^2 x^{-1} \geq B\left(\kappa\theta, \frac{1}{2}\delta\eta^2, \frac{\delta}{1-\delta}\right)$$

holds for $x \in \mathbb{R}_+$, and there always exists ε_0 such that for all $\varepsilon \leq \varepsilon_0$

$$\left(\frac{1}{2}\delta\eta^2\varepsilon^{-2} - \frac{\kappa\theta\delta}{1-\delta}\varepsilon^{-\frac{1}{1-\delta}}\right)x + \frac{\kappa\theta}{1-\delta}\varepsilon^{-\frac{\delta}{1-\delta}} - \delta\eta^2\varepsilon^{-1} > 0$$

holds for $x < \varepsilon$. Putting these two facts together, we have

$$\kappa\theta\Phi_\varepsilon(x) - \frac{1}{2}\delta\eta^2\Psi_\varepsilon(x) \geq B\left(\kappa\theta, \frac{1}{2}\delta\eta^2, \frac{\delta}{1-\delta}\right)$$

for all $x \in \mathbb{R}$ and $\varepsilon \leq \varepsilon_0$.

Let us now consider an Ornstein-Uhlenbeck process u_t satisfying $u_0 = \sigma_0$ and

$$du_t = (1-\delta)\left(B\left(\kappa\theta, \frac{1}{2}\delta\eta^2, \frac{\delta}{1-\delta}\right) - \kappa u_t\right)dt + (1-\delta)\eta dW_t.$$

By construction, its drift term is smaller than that of $\sigma_{t,\varepsilon}$. The Yamada–Watanabe comparison lemma (see Chapter 5, Proposition 2.18 in Karatzas and Shreve (1988) [6]) implies $u_t \leq \sigma_{t,\varepsilon}$ a.s. A suitable upper bound w_t can be identified as the δ-diffusion ν_t itself. This can be seen as follows. Using the Yamada–Watanabe comparison lemma once more and taking into account the positivity of ν_t, it remains to shows that

$$\kappa\theta\Phi_\varepsilon(x) - \frac{1}{2}\delta\eta^2\Psi_\varepsilon(x) \leq \kappa\theta x^{-\frac{\delta}{1-\delta}} - \frac{1}{2}\delta\eta^2 x^{-1} \tag{2.4}$$

for $x \in \mathbb{R}_+$. To do so, we define

$$G_\varepsilon(x) := \kappa\theta\Phi_\varepsilon(x) - \frac{1}{2}\delta\eta^2\Psi_\varepsilon(x) - \kappa\theta x^{-\frac{\delta}{1-\delta}} + \frac{1}{2}\delta\eta^2 x^{-1}.$$

Note that $G_\varepsilon(x) = 0$ for all $x \geq \varepsilon$ and that for all $x < \varepsilon$ we have $G_\varepsilon(x)$ equal to:

$$\left(\frac{1}{2}\delta\eta^2\varepsilon^{-2} - \frac{\kappa\theta\delta}{1-\delta}\varepsilon^{-\frac{1}{1-\delta}}\right)x + \frac{\kappa\theta}{1-\delta}\varepsilon^{-\frac{\delta}{1-\delta}} - \delta\eta^2\varepsilon^{-1} - \kappa\theta x^{-\frac{\delta}{1-\delta}} + \frac{1}{2}\delta\eta^2 x^{-1}.$$

Further note that $G_\varepsilon(\cdot)$ is continuous and $G_\varepsilon(\varepsilon) = 0$. If we are able to show that $G_\varepsilon(\cdot)$ is monotonically increasing, we obtain $G_\varepsilon(x) \leq G_\varepsilon(\varepsilon) = 0$ for all $x \in \mathbb{R}_+$, from which (2.4) follows. We calculate the first and the second derivatives of $G_\varepsilon(x)$ as follows

$$G'_\varepsilon(x) = \begin{cases} 0 & x \geq \varepsilon \\ \frac{1}{2}\delta\eta^2\varepsilon^{-2} - \frac{\kappa\theta\delta}{1-\delta}\varepsilon^{-\frac{1}{1-\delta}} + \frac{\kappa\theta\delta}{1-\delta}x^{-\frac{1}{1-\delta}} - \frac{1}{2}\delta\eta^2 x^{-2} & x < \varepsilon, \end{cases}$$

$$G''_\varepsilon(x) = \begin{cases} 0 & x \geq \varepsilon \\ -\frac{\kappa\theta\delta}{(1-\delta)^2}x^{\frac{\delta-2}{1-\delta}} + \delta\eta^2 x^{-3} & x < \varepsilon. \end{cases}$$

We observe that $G'_\varepsilon(\varepsilon) = 0$ and further, since $\frac{\delta-2}{1-\delta} \leq -3$, there exists ε_0 such that for all $\varepsilon < \varepsilon_0$ we have $G''_\varepsilon(x) \leq 0$ for $x \in \mathbb{R}$. This implies $G'_\varepsilon(x) \geq G'_\varepsilon(\varepsilon) = 0$ for $x \leq \varepsilon$. Hence, we have $G'_\varepsilon(x) \geq 0$ for $x \in \mathbb{R}$. Hence equation (2.4) holds and we obtain that $\sigma_{t,\varepsilon} \leq \nu_t =: w_t$ for all t a.s. Finally, using that $u_t \leq \sigma_{t,\varepsilon} \leq w_t$ for all t a.s. we conclude that $|\sigma_{t,\varepsilon}| \leq |u_t| + |w_t|$ for all t a.s. Since u_t and w_t belong to $L^2(\Omega)$, the dominated convergence theorem implies the desired convergence of our approximating sequence. □

We can now proceed with proving that σ_t and in effect ν_t are Malliavin differentiable. As (2.3) satisfies all the required regularity assumptions of Theorem 2.1 in Detemple (2005) [5], applying this theorem provides us with the existence and representation of the Malliavin derivative of the approximating sequence $\sigma_{t,\epsilon}$. We then obtain

$$D_r\sigma_{t,\varepsilon} = (1-\delta)\eta\exp\left\{\int_r^t (1-\delta)\left[\kappa\theta\Phi'_\varepsilon(\sigma_{s,\varepsilon}) - \tfrac{1}{2}\delta\eta^2\Psi'_\varepsilon(\sigma_{s,\varepsilon}) - \kappa\right]ds\right\} \tag{2.5}$$

for $r \leq t$ and 0 otherwise. Using the continuity of the exponential function, $\Psi_\varepsilon(\cdot)$ and $\Phi_\varepsilon(\cdot)$ we see that pointwise convergence passes onto $D_r\sigma_{t,\varepsilon}$ whose pointwise limit G is given by

$$G = (1-\delta)\eta\exp\left\{\int_r^t \left[-\delta\kappa\theta\sigma_s^{\frac{1}{1-\delta}} + \tfrac{1}{2}\delta(1-\delta)\eta^2\sigma_s^{-2} - (1-\delta)\kappa\right]ds\right\}. \tag{2.6}$$

Let us take a closer look at the integrand in the exponent of $D_r\sigma_{t,\varepsilon}$ in equation (2.5). We have

$$\kappa\theta\Phi'_\varepsilon(x) - \frac{1}{2}\delta\eta^2\Psi'_\varepsilon(x) = \begin{cases} -\frac{\kappa\theta\delta}{1-\delta}x^{-\frac{1}{1-\delta}} + \frac{1}{2}\delta\eta^2 x^{-2} & x \geq \varepsilon \\ -\frac{\kappa\theta\delta}{1-\delta}\varepsilon^{-\frac{1}{1-\delta}} + \frac{1}{2}\delta\eta^2\varepsilon^{-2} & x < \varepsilon. \end{cases}$$

Now assume $c, d \in \mathbb{R}_+$ and $q > 2$ and note that the maximum of the function $-cy^{-q} + dy^{-2}$ defined on $(0, \infty)$ is given by

$$M(c,d,q) := -c\left(\frac{2d}{cq}\right)^{\frac{q}{2-q}} + d\left(\frac{2d}{cq}\right)^{\frac{2}{q-2}}.$$

Also note that there always exists $\varepsilon_0 > 0$ such that

$$-\frac{\kappa\theta\delta}{1-\delta}\varepsilon^{-\frac{1}{1-\delta}} + \frac{1}{2}\delta\eta^2\varepsilon^{-2} < 0$$

for all $\varepsilon < \varepsilon_0$. Therefore, for $\delta \in \left(\frac{1}{2}, 1\right)$

$$\kappa\theta\Phi'_\varepsilon(x) - \frac{1}{2}\delta\eta^2\Psi'_\varepsilon(x) \leq \max\left\{M\left(\frac{\kappa\theta\delta}{1-\delta}, \frac{1}{2}\delta\eta^2, \frac{1}{1-\delta}\right), 0\right\}$$

holds for all $\varepsilon < \varepsilon_0$. For $\delta = \frac{1}{2}$ we have

$$\kappa\theta\Phi'_\varepsilon(x) - \frac{1}{2}\delta\eta^2\Psi'_\varepsilon(x) = \begin{cases} (-\kappa\theta + \frac{\eta^2}{4})x^{-2} < 0 & x \geq \varepsilon \\ (-\kappa\theta + \frac{\eta^2}{4})\varepsilon^{-2} < 0 & x < \varepsilon. \end{cases}$$

Putting things together, we see that $|D_r\sigma_{t,\varepsilon}| \leq (1-\delta)\eta e^{(t-r)C}$ for some constant C independent of ε. From the bounded convergence theorem we conclude that $D_r\sigma_{t,\varepsilon}$ converges to G in $L^2(\Omega)$. Now, $\sigma_t \in \mathbb{D}^{1,2}$ follows from Lemma 1.2.3 in Nualart (1995) [2] and we conclude $D_r\sigma_t = G$. In fact, using Lemma 2.2 and that for any $p \geq 1$

$$\begin{aligned} \|\sigma_t\|_{1,p}^p &= \mathbb{E}[\nu_t^{p(1-\delta)}] + \mathbb{E}\left[\left|\int_0^T (D_r\sigma_t)^2 dr\right|^{\frac{p}{2}}\right] \\ &\leq C_1(1 + \nu_0^{p(1-\delta)}) + (1-\delta)^p\eta^p \exp(pCT)\left(\tfrac{1}{2C}(1 - \exp(-2cT))\right)^{1/p}, \end{aligned}$$

we obtain that $\|\sigma_t\|_{1,p} < \infty$ for all $p \geq 1$, and hence that $\sigma_t \in \mathbb{D}^{1,\infty}$. Coming back to our original statement, which concerns the Malliavin differentiability of ν_t, we now show that $\nu_t \in \mathbb{D}^{1,\infty}$ for $\delta \in \left[\frac{1}{2}, 1\right)$.

Proposition 2.2. For $\delta \in \left[\frac{1}{2}, 1\right)$ the δ-diffusion ν_t is Malliavin differentiable, i.e. $\nu \in \mathbb{D}^{1,\infty}$ and

$$D_r\nu_t = \eta\nu_r^\delta \exp\left\{\int_r^t \left(-\delta\kappa\theta\nu_s^{-1} + \frac{\eta^2}{2}\delta(1-\delta)\nu_s^{2\delta-2} - (1-\delta)\kappa\right)ds\right\} \quad (2.7)$$

for $r \leq t$ and 0 otherwise.

Proof. We have $\nu_t = \sigma_t^{\frac{1}{1-\delta}}$ and know that $\sigma_t \in \mathbb{D}^{1,2}$. Application of the chain rule of Malliavin calculus (see Lemma 2.1 in Leon and Nualart (2000) [9]) yields

$$\begin{aligned}
D_r\nu_t &= D_r\sigma_t^{\frac{1}{1-\delta}} \\
&= \frac{1}{1-\delta}\sigma_t^{\frac{\delta}{1-\delta}} D_r\sigma_t \\
&= \frac{1}{1-\delta}\sigma_t^{\frac{\delta}{1-\delta}}(1-\delta)\eta \\
&\qquad \cdot \exp\left\{\int_r^t \left[-\delta\kappa\theta\sigma_s^{\frac{1}{1-\delta}} + \frac{1}{2}\delta(1-\delta)\eta^2\sigma_s^{-2} - (1-\delta)\kappa\right]ds\right\} \\
&= \eta\nu_r^{\delta}\exp\left\{\int_r^t \left(-\delta\kappa\theta\nu_s^{-1} + \frac{\eta^2}{2}\delta(1-\delta)\nu_s^{2\delta-2} - (1-\delta)\kappa\right)ds\right\}.
\end{aligned}$$

Further for all $p \geq 1$, we have

$$\begin{aligned}
\|\nu_t\|_{1,p}^p &= \mathbb{E}[\nu_t^p] + \mathbb{E}\left[\left(\int_0^T (D_r\nu_t)^2dr\right)^{\frac{p}{2}}\right] \\
&= \mathbb{E}[\nu_t^p] + \tfrac{1}{(1-\delta)^p}\mathbb{E}\left[\nu_t^{\delta p}\left(\int_0^T D_r\sigma_t)^2dr\right)^{\frac{p}{2}}\right] \\
&\leq \mathbb{E}[\nu_t^p] + \tfrac{1}{2(1-\delta)^p}\mathbb{E}\left[\nu_t^{2\delta p}\right] + \tfrac{1}{2(1-\delta)^p}\mathbb{E}\left[\left(\int_0^T (D_r\sigma_t)^2dr\right)^{p}\right].
\end{aligned}$$

Using Lemma 2.2 and $\sigma \in \mathbb{D}^{1,\infty}$, we conclude that $\|\nu_t\|_{1,p} < \infty$ for all $p \geq 1$. This completes the proof. □

2.4. Conclusions

We have extended the result by Alos and Ewald (2008) [1] on the Malliavin differentiability of the Heston volatility and shown that the diffusion $d\nu_t = \kappa(\theta - \nu_t)dt + \eta\nu_t^{\delta}dW_t$ for $\delta \in \left[\frac{1}{2}, 1\right]$ is Malliavin differentiable and in fact an element of $\mathbb{D}^{1,\infty}$, assuming the Novikov condition for the case $\delta = \frac{1}{2}$.

Acknowledgments

We would like to thank the referee for insightful comments and suggestions. The first author would like to thank Marc Yor and Elisa Alos for useful comments and acknowledges funding from the Australian Research Council, grant DP1095969.

References

[1] E. Alòs and C.-O. Ewald, Malliavin differentiability of the Heston volatility and applications to option pricing, *Adv. in Appl. Probab.* **40**(1), 144–162, (2008).

[2] D. Nualart, *The Malliavin Calculus and Related Topics* (Springer-Verlag, New York/Berlin, 1995)

[3] A. D. Wilkie, A stochastic investment model for actuarial use, *Transactions of the Faculty of Actuaries* **39**, 341–403, (1986).

[4] A. D. Wilkie, More on a stochastic asset model for actuarial use, *British Actuarial Journal* **1**, 777–964, (1995).

[5] J. Detemple, R. Garcia, M. Rindisbacher, Representation formulas for Malliavin derivatives of diffusion processes, *Finance Stoch.* **9**, 349–367, (2005).

[6] I. Karatzas and S. E. Shreve, *Brownian motion and stochastic calculus* (Springer-Verlag, Berlin-Heidelberg-New York, 1988)

[7] L. Andersen and V. Piterbarg, Moment explosions in stochastic volatility models, *Finance and Stochastics* **11**, 29–50, (2007).

[8] M. A. D. Bossy, An efficient discretization scheme for one dimensional SDEs with a diffusion coefficient function of the form $|x|^\alpha, \alpha \in [1/2, 1)$, *Res. Rep.* **5396** INRIA (2007).

[9] J. A. León and D. Nualart, Anticipating integral equations, *Potential Analysis* **413**, 249–268, (2000).

Chapter 3

A stochastic integral for adapted and instantly independent stochastic processes

Hui-Hsiung Kuo*, Anuwat Sae-Tang and Benedykt Szozda†

Department of Mathematics,
Louisiana State University,
Baton Rouge, Louisiana 70803, USA
kuo@math.lsu.edu, *anuwat@math.lsu.edu,* *benny@math.lsu.edu*

We introduce a new class of stochastic processes, called near-martingales, which arise in the study of a new stochastic integral defined by Ayed and Kuo. We obtain a special version of the Itô isometry for this new stochastic integral of certain instantly independent stochastic processes. We also prove a formula that allows us to express the new integral in terms of the classical Itô and Riemann integrals.

Contents

3.1. Introduction

Stochastic integrals of anticipating integrands with respect to a Brownian motion were introduced independently by Hitsuda [4] in 1972 and by Skorohod [19] in 1975. In 1976 Itô [7] used the ideas of enlarging the underlying filtration and quasimartingale to define such stochastic integrals. Since 1987 there has been a growing interest on anticipating integrals due

*http://www.math.lsu.edu/~kuo
†http://www.math.lsu.edu/~benny

to mathematical theory and applications to finance. Many methods have been introduced. See for instance [3, 8, 10, 11, 14, 16–18], among a rather large number of papers in the literature.

Being inspired by Itô's ideas in [7], Ayed and Kuo [1, 2] have recently introduced a new integral for integrands involving an adapted part and an instantly independent part (which serves as a counterpart to the Itô theory). In [7] Itô keeps the integrand and decomposes the integrator (with respect to an enlarged filtration), while in [1] Ayed and Kuo keep the integrator (and the filtration) and decompose the integrand (into an adapted part and an instantly independent part). Moreover, in forming a Riemann-like sum, the evaluation points for the instantly independent part are the right end-points of subintervals, while those for the adapted part are the left end-points.

Contrary to the Hitsuda–Skorohod integral, this new stochastic integral by Ayed and Kuo is exactly the Itô integral when the integrand is adapted with almost all sample paths being square integrable. Moreover, it proposes an interesting problem of developing a parallel theory for instantly independent stochastic processes. We also mention that this new stochastic integral has many advantages over the white noise methods [8, 10], e.g., it has a probabilistic interpretation (unlike a white noise integral, which in general is not a random variable).

The theory initiated in [1, 2], unlike white noise methods, provides a unified, purely probabilistic approach to stochastic integration of adapted and anticipating stochastic processes. In this paper and the forthcoming papers [12, 13], we will further develop this theory. We expect its applications to stochastic differential equations with anticipating initial conditions or coefficients, modelling of financial markets with insider trading without the use of enlarged filtrations, etc.

The purpose of this paper is to study the properties of the new stochastic integral introduced by Ayed and Kuo in [1] and [2]. We compare the properties of the new integral with the well known properties of the Itô integral, for example, the fact that the latter is a martingale (see Sections 3.2 and 3.3). We also present a version of the Itô isometry for the instantly independent stochastic processes (see Section 3.4) and investigate how the new integral relates to the classical Itô integral (see Section 3.5).

First, let us recall two definitions from [1]. The first of the definitions is that of an instantly independent stochastic process. This kind of processes serves as a counterpart to the adapted stochastic processes in the theory of the new stochastic integral. While instantly independent stochastic processes are anticipating, they have special properties that allow us to define their stochastic integrals with respect to Brownian motion.

Throughout this paper, we will denote by $\{\mathcal{F}_t\}$ the natural Brownian filtration, i.e., $\mathcal{F}_t = \sigma\{B(s) : 0 \leq s \leq t\}$.

Definition 3.1. A stochastic process $\{\varphi(t)\}$ is called *instantly independent* with respect to a filtration $\{\mathcal{F}_t\}$ if $\varphi(t)$ and $\mathcal{F}_t$ are independent for each t.

Our next example is in fact one of the examples in [1] motivating the new approach to stochastic integration. While very simple, it shows the type of stochastic processes that we are most interested in.

Example 3.1. Let $X(t) = B(1) - B(t)$. For $t \in [0, 1]$ the stochastic process $\{X_t\}$ is instantly independent with respect to $\{\mathcal{F}_t\}$. Note that $\{X_t\}$ is not a martingale for $0 \leq t \leq 1$, but it is a martingale for $t > 1$.

The second definition that we want to recall from [1] is the one of the stochastic integral of the product of adapted and instantly independent stochastic processes.

Definition 3.2. Let $\{f(t)\}$ be a stochastic process adapted with respect to $\{\mathcal{F}_t\}$, and $\{\varphi(t)\}$ be a stochastic process instantly independent with respect to $\{\mathcal{F}_t\}$. Let $\Delta = \{a = t_0 < t_1 < \cdots < t_{n-1} < t_n = b\}$ be the partition of the interval $[a, b]$. We define the *stochastic integral* of $f(t)\varphi(t)$, with respect to the Brownian motion $\{B(t)\}$, to be the limit

$$\int_a^b f(t)\varphi(t)\, dB(t) = \lim_{\|\Delta\| \to 0} \sum_{i=1}^n f(t_{i-1})\varphi(t_i)(B(t_i) - B(t_{i-1})),$$

provided that the limit exists in probability.

From now on, we will be using this definition every time we consider an integral with respect to Brownian motion. Note that if the process is purely adapted, this definition coincides with the definition of the Itô integral.

3.2. Martingales and near-martingales

In this section, we introduce a new class of stochastic processes that arise naturally in the analysis of the new stochastic integral. The motivation behind this definition is quite simple. Recall that if $f(t)$ is an adapted stochastic process such that $\mathbb{E}\big[\int_a^b f(t)^2\, dt\big] < \infty$, then the stochastic process associated with $f(t)$, namely

$$X_t = \int_a^t f(s)\, dB(s), \qquad a \le t \le b,$$

is a martingale with respect to $\{\mathcal{F}_t\}$ (see [9].). It is natural to ask whether the same holds for the new process. Of course, the immediate answer is "no," because the integral of a non-adapted process is also not adapted, hence it cannot be a martingale. However, we should investigate further if there are any properties of the new integral that are similar to the properties of the Itô integral.

Let us recall the definition of martingales.

Definition 3.3. Let $\mathbb{E}|X_t| < \infty$ for all t. We say that $\{X_t\}$ is a *martingale* with respect to filtration $\{\mathcal{F}_t\}$ if

$$\mathbb{E}\left[X_t | \mathcal{F}_s\right] = X_s, \qquad \forall s \le t. \tag{3.1}$$

Note that condition (3.1) implies that $\{X_t\}$ is adapted with respect to $\{\mathcal{F}_t\}$, so this condition can not be true for general instantly independent stochastic processes. However, we can rewrite this condition as $\mathbb{E}\left[X_t - X_s | \mathcal{F}_s\right] = 0$, which also makes sense for non-adapted processes. This motivates our definition of near-martingales (see Definition 3.5).

Following the work of Protter and Pardoux (see [17]), we will consider backward and forward filtrations.

Definition 3.4.

(1) We will say that a family $\{\mathcal{G}_t\}$ of sub-σ-fields is a *forward filtration* if $\mathcal{G}_s \subseteq \mathcal{G}_t$ for $0 \le s \le t$;
(2) We will say that a family $\left\{\mathcal{G}^{(t)}\right\}$ of sub-σ-fields is a *backward filtration* if $\mathcal{G}^{(s)} \supseteq \mathcal{G}^{(t)}$ for $0 \le s \le t$.

Definition 3.5. Let $\mathbb{E}|X_t| < \infty$ for all t. We will say that X_t is a *near-martingale* with respect to a forward filtration $\{\mathcal{G}_t\}$ if

$$\mathbb{E}\left[X_t - X_s | \mathcal{G}_s\right] = 0, \qquad \forall s \le t. \tag{3.2}$$

On the other hand, we say that X_t is a *near-martingale* with respect to a backward filtration $\{\mathcal{G}^{(t)}\}$ if

$$\mathbb{E}\left[X_t - X_s | \mathcal{G}^{(t)}\right] = 0, \qquad \forall s \le t. \tag{3.3}$$

Remark 3.1. Note that if a process is a near-martingale with respect to $\{\mathcal{G}_t\}$ and adapted to $\{\mathcal{G}_t\}$, then it is a martingale.

As we have already mentioned above, in the theory of the new stochastic integral, the instantly independent stochastic processes and the products of instantly independent processes with adapted processes play the key role. The next two theorems establish when such processes are near-martingales.

Theorem 3.1. *Suppose $\varphi(t)$ is instantly independent with respect to a forward filtration $\{\mathcal{G}_t\}$ and $\mathbb{E}|\varphi(t)| < \infty$ for all t. Then $\varphi(t)$ is a near-martingale with respect to $\{\mathcal{G}_t\}$ if and only if $\mathbb{E}[\varphi(t)] = \mathbb{E}[\varphi(s)]$ for all s, t, i.e., $\mathbb{E}[\varphi(t)]$ is a constant.*

Proof. Let $t \geq s$. First note that by the linearity and the tower property of conditional expectation,

$$\begin{aligned}\mathbb{E}[\varphi(t) - \varphi(s)|\,\mathcal{G}_s] &= \mathbb{E}[\varphi(t)|\,\mathcal{G}_s] - \mathbb{E}[\varphi(s)|\,\mathcal{G}_s] \\ &= \mathbb{E}[\,\mathbb{E}[\varphi(t)|\,\mathcal{G}_t]|\,\mathcal{G}_s] - \mathbb{E}[\varphi(s)]\,.\end{aligned}$$

Since $\varphi(t)$ is independent of $\mathcal{F}_t$, the first term in the right-hand side becomes

$$\mathbb{E}[\,\mathbb{E}[\varphi(t)]|\,\mathcal{G}_s] = \mathbb{E}[\varphi(t)]\,.$$

Therefore,

$$\mathbb{E}[\varphi(t) - \varphi(s)|\,\mathcal{G}_s] = \mathbb{E}[\varphi(t)] - \mathbb{E}[\varphi(s)]\,,$$

which shows that $\varphi(t)$ is a near-martingale with respect to $\{\mathcal{G}_t\}$ if and only if $\mathbb{E}[\varphi(t)] = \mathbb{E}[\varphi(s)]$ for all s, t. □

On the other hand, we can use similar arguments as those in the proof of Theorem 3.1 to obtain the next theorem.

Theorem 3.2. *Suppose $\varphi(t)$ is instantly independent with respect to a backward filtration $\{\mathcal{G}^{(t)}\}$ and $\mathbb{E}|\varphi(t)| < \infty$ for all t. Then $\varphi(t)$ is a near-martingale with respect to $\{\mathcal{G}^{(t)}\}$ if and only if $\mathbb{E}[\varphi(t)] = \mathbb{E}[\varphi(s)]$ for all s, t, that is, $\mathbb{E}[\varphi(t)]$ is a constant.*

Theorem 3.3. *Let $\{\mathcal{G}_t\}$ be a forward filtration. Assume that*

(1) $\{f(t)\}$ is a martingale with respect to $\{\mathcal{G}_t\}$;
(2) $\{\varphi(t)\}$ is instantly independent with respect to $\{\mathcal{G}_t\}$ and $\mathbb{E}[\varphi(t)]$ is a constant;
(3) $\mathbb{E}|f(t)\varphi(t)| < \infty$ for all t.

Then $\{\theta(t)\} = \{f(t)\varphi(t)\}$ is a near-martingale with respect to $\{\mathcal{G}_t\}$.

Remark 3.2.

(1) By Theorem 3.1, condition *(2)* implies that $\varphi(t)$ is a near-martingale.
(2) Theorem 3.3 is false if we assume that $\varphi(t)$ is a near-martingale but not instantly independent, e.g. $f(t) = \varphi(t) = B(t)$.

Proof. We have to check that equation (3.2) is satisfied. Since the conditional expectation is linear and $f(s)$ is measurable with respect to $\mathcal{G}_s$, we have

$$\begin{aligned}\mathbb{E}\left[\theta(t)-\theta(s)\middle|\,\mathcal{G}_s\right] &= \mathbb{E}\left[f(t)\varphi(t)-f(s)\varphi(s)\middle|\,\mathcal{G}_s\right]\\ &= \mathbb{E}\left[f(t)\varphi(t)\middle|\,\mathcal{G}_s\right]-f(s)\,\mathbb{E}\left[\varphi(s)\middle|\,\mathcal{G}_s\right].\end{aligned}$$

By using (i) the tower property of conditional expectation, (ii) the fact that $\{\mathcal{G}_t\}$ is a forward filtration, (iii) $\varphi(t)$ is independent of $\mathcal{G}_t$, and (iv) $\{f(t)\}$ is a martingale, we see that the above expression is equal to

$$\begin{aligned}\mathbb{E}\left[\theta(t)-\theta(s)\middle|\,\mathcal{G}_s\right] &= \mathbb{E}\Big[\mathbb{E}\left[f(t)\varphi(t)\middle|\,\mathcal{G}_t\right]\Big|\mathcal{G}_s\Big]-f(s)\,\mathbb{E}\left[\varphi(s)\middle|\,\mathcal{G}_s\right]\\ &= \mathbb{E}\Big[f(t)\,\mathbb{E}\left[\varphi(t)\middle|\,\mathcal{G}_t\right]\Big|\mathcal{G}_s\Big]-f(s)\,\mathbb{E}\left[\varphi(s)\right]\\ &= \mathbb{E}\Big[f(t)\,\mathbb{E}\left[\varphi(t)\right]\Big|\mathcal{G}_s\Big]-f(s)\,\mathbb{E}\left[\varphi(s)\right]\\ &= \mathbb{E}\left[\varphi(t)\right]\,\mathbb{E}\left[f(t)\middle|\,\mathcal{G}_s\right]-f(s)\,\mathbb{E}\left[\varphi(s)\right]\\ &= \mathbb{E}\left[\varphi(t)\right]f(s)-f(s)\,\mathbb{E}\left[\varphi(s)\right]\\ &= 0.\end{aligned}$$

Thus, by Definition 3.5, the stochastic process $\{\theta(t)\}$ is a near-martingale with respect to $\{\mathcal{G}_t\}$. □

On the other hand, we can use similar arguments as those in the proof of Theorem 3.3 to obtain the next theorem.

Theorem 3.4. *Let $\{\mathcal{G}^{(t)}\}$ be a backward filtration. Assume that*

(1) $\{f(t)\}$ *is a martingale with respect to* $\{\mathcal{G}^{(t)}\}$*;*
(2) $\{\varphi(t)\}$ *is instantly independent with respect to* $\{\mathcal{G}^{(t)}\}$ *and* $\mathbb{E}\left[\varphi(t)\right]$ *is a constant;*
(3) $\mathbb{E}|f(t)\varphi(t)| < \infty$ *for all* t*.*

Then $\theta(t) = f(t)\varphi(t)$ *is a near-martingale with respect to* $\{\mathcal{G}^{(t)}\}$*.*

3.3. Near-martingales and the new integral

In this section, we investigate the associated stochastic processes defined by the new integral. We will define two associated stochastic processes, namely, a forward stochastic process and a backward stochastic process. The forward one is well known from the theory of Itô integrals, where it is established that such processes are martingales if the integrand $f(t)$ is adapted with respect to $\{\mathcal{F}_t\}$ and $\mathbb{E}\big[\int_a^b f(t)^2\,dt\big] < \infty$ (see Chapter 4 of [9]). Throughout this section, we use the following forward and backward filtrations

$$\mathcal{F}_t = \sigma\left(B(s) : a \le s \le t\right), \qquad \mathcal{F}^{(t)} = \sigma\left(B(b) - B(s) : t \le s \le b\right).$$

The next four theorems show that the associated forward stochastic process $\{X_t\}$ defined in (3.4) is a near-martingale with respect to both forward and backward filtrations. Similarly, the associated backward stochastic process $\{Y^{(t)}\}$ defined in (3.5) is a near-martingale with respect to both forward and backward filtrations.

Theorem 3.5. *Let $f(x)$ and $\varphi(x)$ be continuous functions. Let*

$$X_t = \int_a^t f(B(s))\varphi(B(b) - B(s))\,dB(s), \qquad a \le t \le b, \tag{3.4}$$

and assume that $\mathbb{E}|X_t| < \infty$ for all $a \le t \le b$. Then $\{X_t\}$ is a near-martingale with respect to the forward filtration $\{\mathcal{F}_t\}$.

Proof. We need to show that $\mathbb{E}\left[X_t - X_s \middle| \mathcal{F}_s\right] = 0$ for $a \le s \le t$. Note that $X_t - X_s = \int_s^t f(B(u))\varphi(B(b) - B(u))\,dB(u)$. Let

$$\Delta = \{s = t_0 < t_1 < \cdots < t_{n-1} < t_n = t\}$$

be a partition of the interval $[s, t]$, and $\Delta B_i = B(t_i) - B(t_{i-1})$. By the definition of the stochastic integral we have

$$\begin{aligned}
\mathbb{E}\Big[X_t - X_s\Big|\mathcal{F}_s\Big] &= \mathbb{E}\left[\int_s^t f(B(u))\varphi(B(b) - B(u))\,dB(u)\middle|\,\mathcal{F}_s\right] \\
&= \mathbb{E}\left[\lim_{\|\Delta\|\to 0}\sum_{i=1}^n f(B(t_{i-1}))\varphi(B(b) - B(t_i))\Delta B_i\middle|\,\mathcal{F}_s\right] \\
&= \lim_{\|\Delta\|\to 0}\sum_{i=1}^n \mathbb{E}\Big[f(B(t_{i-1}))\varphi(B(b) - B(t_i))\Delta B_i\Big|\mathcal{F}_s\Big].
\end{aligned}$$

So it is enough to show that every component of the last sum is equal to zero. Using the properties of conditional expectation, the fact that $f(B(t_{i-1}))$ is $\mathcal{F}_{t_i}$-measurable and $\varphi(B(b)-B(t_i))$ is independent of $\mathcal{F}_{t_i}$, we get

$$\begin{aligned}
&\mathbb{E}\Big[f(B(t_{i-1}))\varphi(B(b)-B(t_i))\Delta B_i\Big|\mathcal{F}_s\Big]\\
&= \mathbb{E}\Big[\,\mathbb{E}\Big[f(B(t_{i-1}))\varphi(B(b)-B(t_i))\Delta B_i\Big|\mathcal{F}_{t_i}\Big]\Big|\mathcal{F}_s\Big]\\
&= \mathbb{E}\Big[f(B(t_{i-1}))\Delta B_i\,\mathbb{E}\Big[\varphi(B(b)-B(t_i))\Big|\mathcal{F}_{t_i}\Big]\Big|\mathcal{F}_s\Big]\\
&= \mathbb{E}\Big[f(B(t_{i-1}))\Delta B_i\,\mathbb{E}\Big[\varphi(B(b)-B(t_i))\Big]\Big|\mathcal{F}_s\Big].
\end{aligned}$$

By the tower property and the independence of Brownian increments, we have

$$\begin{aligned}
&\mathbb{E}\Big[f(B(t_{i-1}))\varphi(B(b)-B(t_i))\Delta B_i\Big|\mathcal{F}_s\Big]\\
&= \mathbb{E}\left[\varphi(B(b)-B(t_i))\right]\mathbb{E}\Big[\,\mathbb{E}\Big[f(B(t_{i-1}))\Delta B_i\Big|\mathcal{F}_{t_{i-1}}\Big]\Big|\mathcal{F}_s\Big]\\
&= \mathbb{E}\left[\varphi(B(b)-B(t_i))\right]\mathbb{E}\Big[f(B(t_{i-1}))\,\mathbb{E}\Big[\Delta B_i\Big|\mathcal{F}_{t_{i-1}}\Big]\Big|\mathcal{F}_s\Big]\\
&= \mathbb{E}\left[\varphi(B(b)-B(t_i))\right]\mathbb{E}\Big[f(B(t_{i-1}))\,\mathbb{E}\left[\Delta B_i\right]\Big|\mathcal{F}_s\Big]\\
&= 0.
\end{aligned}$$

Thus $\{X_t\}$ is a near-martingale with respect to $\{\mathcal{F}_t\}$. □

Theorem 3.6. *Let $f(x)$ and $\varphi(x)$ be continuous functions. Let*

$$Y^{(t)} = \int_t^b f(B(s))\varphi(B(b)-B(s))\,dB(s), \qquad a \le t \le b, \tag{3.5}$$

and assume that $\mathbb{E}|Y^{(t)}| < \infty$ for all $a \le t \le b$. Then $\{Y^{(t)}\}$ is a near-martingale with respect to the forward filtration $\{\mathcal{F}_t\}$.

Proof. Note that for $a \le s < t \le b$, we have

$$\begin{aligned}
&Y^{(t)} - Y^{(s)}\\
&= \int_t^b f(B(u))\varphi(B(b)-B(u))\,dB(u) - \int_s^b f(B(u))\varphi(B(b)-B(u))\,dB(u)\\
&= -\int_s^t f(B(u))\varphi(B(b)-B(u))\,dB(u)\\
&= -(X_t - X_s),
\end{aligned}$$

where X_t is given in Theorem 3.5. Hence $\{Y^{(t)}\}$ is a near-martingale with respect to $\{\mathcal{F}_t\}$. □

Next we consider the backward filtration $\{\mathcal{F}^{(t)}\}$ and show that both $\{X_t\}$ and $\{Y^{(t)}\}$ are near-martingales with respect to $\{\mathcal{F}^{(t)}\}$.

Theorem 3.7. *Let $f(x)$ and $\varphi(x)$ be continuous functions. Then the stochastic process $X_t = \int_a^t f(B(s))\varphi(B(b) - B(s))\, dB(s)$, $a \le t \le b$ is a near-martingale with respect to the backward filtration $\mathcal{F}^{(t)}$.*

Proof. As in the proof of Theorem 3.5, it is enough to consider

$$\mathbb{E}\Big[f(B(t_{i-1}))\varphi(B(b) - B(t_i))\Delta B_i \Big| \mathcal{F}^{(t)}\Big],$$

where $a \le s < t \le b$ and $s = t_0 < t_1 < \cdots < t_{n-1} < t_n = t$. Also note that ΔB_i is $\mathcal{F}^{(t_{i-1})}$-measurable because

$$\Delta B_i = B(t_i) - B(t_{i-1}) = (B(b) - B(t_{i-1})) - (B(b) - B(t_i)) \in \mathcal{F}^{(t_{i-1})}.$$

Thus

$$\begin{aligned}
&\mathbb{E}\Big[f(B(t_{i-1}))\varphi(B(b) - B(t_i))\Delta B_i \Big| \mathcal{F}^{(t)}\Big] \\
&= \mathbb{E}\Big[\,\mathbb{E}\Big[f(B(t_{i-1}))\varphi(B(b) - B(t_i))\Delta B_i \Big| \mathcal{F}^{(t_{i-1})}\Big]\Big| \mathcal{F}^{(t)}\Big] \\
&= \mathbb{E}\Big[\varphi(B(b) - B(t_i))\Delta B_i\, \mathbb{E}\Big[f(B(t_{i-1}))\Big| \mathcal{F}^{(t_{i-1})}\Big]\Big| \mathcal{F}^{(t)}\Big].
\end{aligned}$$

Since $f(B(t_{i-1}))$ is $\mathcal{F}_{t_{i-1}}$-measurable, and for $s > t_{i-1}$, $B(b) - B(s)$ is independent of $\mathcal{F}_{t_{i-1}}$, it follows that $f(B(t_{i-1}))$ is independent of $\mathcal{F}^{(t_{i-1})}$. Hence

$$\begin{aligned}
&\mathbb{E}\Big[f(B(t_{i-1}))\varphi(B(b) - B(t_i))\Delta B_i \Big| \mathcal{F}^{(t)}\Big] \\
&= \mathbb{E}\Big[\varphi(B(b) - B(t_i))\Delta B_i\, \mathbb{E}\left[f(B(t_{i-1}))\right] \Big| \mathcal{F}^{(t)}\Big] \\
&= \mathbb{E}\left[f(B(t_{i-1}))\right] \mathbb{E}\Big[\,\mathbb{E}\Big[\varphi(B(b) - B(t_i))\Delta B_i \Big| \mathcal{F}^{(t_i)}\Big]\Big| \mathcal{F}^{(t)}\Big] \\
&= \mathbb{E}\left[f(B(t_{i-1}))\right] \mathbb{E}\Big[\varphi(B(b) - B(t_i))\, \mathbb{E}\Big[\Delta B_i \Big| \mathcal{F}^{(t_i)}\Big]\Big| \mathcal{F}^{(t)}\Big].
\end{aligned}$$

But $\mathcal{F}^{(t_i)}$ is generated by Brownian increments following ΔB_i. It follows that ΔB_i and $\mathcal{F}^{(t_i)}$ are independent. Thus the above expression becomes

$$\mathbb{E}\left[f(B(t_{i-1}))\right] \mathbb{E}\Big[\varphi(B(b) - B(t_i))\, \mathbb{E}\left[\Delta B_i\right] \Big| \mathcal{F}^{(t)}\Big] = 0.$$

Therefore, $\{X_t\}$ is a near-martingale with respect to $\{\mathcal{F}^{(t)}\}$. □

Theorem 3.8. *Let $f(x)$ and $\varphi(x)$ be continuous functions. The stochastic process given by $Y^{(t)} = \int_t^b f(B(s))\varphi(B(b) - B(s))\, dB(s)$, for $a \le t \le b$, is a near-martingale with respect to the backward filtration $\{\mathcal{F}^{(t)}\}$.*

Proof. Recall that $Y^{(t)} - Y^{(s)} = -(X_t - X_s)$. Hence this theorem follows from Theorem 3.7. □

3.4. Instantly independent processes and Itô isometry

One of the important theorems of Itô calculus is the Itô isometry. In this section, we present a first step to find an extension of the Itô isometry for the new integral. In the proof of the first theorem in this section, we use the Itô formula for the new integral obtained by Ayed and Kuo in [1].

Theorem 3.9 (Ayed and Kuo (2008)). *Suppose $f(x)$ and $\varphi(x)$ are C^2 functions and let $\theta(x,y) = f(x)\varphi(y-x)$. Then the following equality holds for $a \le t \le b$,*

$$\theta(B(t), B(b)) = \theta(B(a), B(b)) + \int_a^t \frac{\partial \theta}{\partial x}(B(s), B(b))\, dB(s) + \int_a^t \left(\frac{1}{2} \frac{\partial^2 \theta}{\partial x^2}(B(s), B(b)) + \frac{\partial^2 \theta}{\partial x \partial y}(B(s), B(b)) \right) ds.$$

The first result presented below is a formula for the covariance between stochastic integrals of powers of $B(b) - B(s)$. We use this theorem in the proof of the Itô isometry for stochastic integrals of instantly independent stochastic processes.

Theorem 3.10. *For $a \le t \le b$, we have*

$$\mathbb{E}\left[\left(\int_a^t (B(b) - B(s))^m \, dB(s) \right) \left(\int_a^t (B(b) - B(s))^n \, dB(s) \right) \right] = \begin{cases} \dfrac{(m+n-1)!!}{\frac{m+n}{2} + 1} \left((b-a)^{\frac{m+n}{2}+1} - (b-t)^{\frac{m+n}{2}+1} \right), & \text{if } m+n \text{ is even,} \\ 0, & \text{if } m+n \text{ is odd.} \end{cases} \tag{3.6}$$

Proof. From the Taylor expansion of e^x, we get

$$e^{\tau(B(b)-B(t))} = \sum_{m=0}^{\infty} \frac{(B(b)-B(t))^m}{m!} \tau^m.$$

Thus

$$\int_a^t e^{\tau(B(b)-B(s))}\, dB(s) = \sum_{m=0}^{\infty} \frac{\tau^m}{m!} \int_a^t (B(b) - B(s))^m \, dB(s).$$

By taking the expectation, we see

$$
\begin{aligned}
&\mathbb{E}\left[\left(\int_a^t e^{\tau(B(b)-B(s))}\,dB(s)\right)\left(\int_a^t e^{\sigma(B(b)-B(s))}\,dB(s)\right)\right]\\
&= \mathbb{E}\Bigg[\sum_{n=0}^{\infty}\sum_{m=0}^{\infty}\frac{\tau^m\sigma^n}{n!m!}\left(\int_a^t (B(b)-B(s))^m\,dB(s)\right)\\
&\qquad\qquad \times\left(\int_a^t (B(b)-B(s))^n\,dB(s)\right)\Bigg]\\
&= \sum_{n=0}^{\infty}\sum_{m=0}^{\infty}\frac{\tau^m\sigma^n}{n!m!}\mathbb{E}\Bigg[\left(\int_a^t (B(b)-B(s))^m\,dB(s)\right)\\
&\qquad\qquad \times\left(\int_a^t (B(b)-B(s))^n\,dB(s)\right)\Bigg].
\end{aligned}
\tag{3.7}
$$

On the other hand, by the Itô formula in Theorem 3.9, we have

$$
\begin{aligned}
\int_a^t e^{\tau(B(b)-B(s))}\,dB(s) &= \frac{1}{\tau}e^{\tau(B(b)-B(a))} - \frac{1}{\tau}e^{\tau(B(b)-B(t))}\\
&\qquad - \frac{\tau}{2}\int_a^t e^{\tau(B(b)-B(s))}\,ds,
\end{aligned}
$$

which leads to the following expectation

$$
\begin{aligned}
&\mathbb{E}\left[\left(\int_a^t e^{\tau(B(b)-B(s))}\,dB(s)\right)\left(\int_a^t e^{\sigma(B(b)-B(s))}\,dB(s)\right)\right]\\
&= \mathbb{E}\Bigg[\left(\frac{1}{\tau}e^{\tau(B(b)-B(a))} - \frac{1}{\tau}e^{\tau(B(b)-B(t))} - \frac{\tau}{2}\int_a^t e^{\tau(B(b)-B(s))}\,ds\right)\\
&\quad \times\left(\frac{1}{\sigma}e^{\sigma(B(b)-B(a))} - \frac{1}{\sigma}e^{\sigma(B(b)-B(t))} - \frac{\sigma}{2}\int_a^t e^{\sigma(B(b)-B(s))}\,ds\right)\Bigg].
\end{aligned}
\tag{3.8}
$$

Although very tedious, it is straightforward to carry out the computations to show that the right-hand side of equation (3.8) is equal to

$$
\begin{aligned}
&\frac{2}{(\tau+\sigma)^2}\left(e^{\frac{1}{2}(b-a)(\tau+\sigma)^2} - e^{\frac{1}{2}(b-t)(\tau+\sigma)^2}\right)\\
&= \frac{2}{(\tau+\sigma)^2}\left[\sum_{n=0}^{\infty}\frac{(b-a)^n}{2^n n!}(\tau+\sigma)^{2n} - \sum_{n=0}^{\infty}\frac{(b-t)^n}{2^n n!}(\tau+\sigma)^{2n}\right]\\
&= \sum_{n=1}^{\infty}\frac{(b-a)^n-(b-t)^n}{2^{n-1}n!}(\tau+\sigma)^{2(n-1)}\\
&= \sum_{n=0}^{\infty}\frac{(b-a)^{n+1}-(b-t)^{n+1}}{2^n(n+1)!}(\tau+\sigma)^{2n}.
\end{aligned}
\tag{3.9}
$$

For simplicity let $\beta_n = \frac{(b-a)^{n+1}-(b-t)^{n+1}}{2^n(n+1)!}$. Then equation (3.9) becomes $\sum_{n=0}^{\infty}\beta_n(\tau+\sigma)^{2n}$. We use the binomial theorem and change the order of summation to show that (3.9) equals

$$\sum_{k=0}^{\infty}\sum_{n=\lceil\frac{k}{2}\rceil}^{\infty}\beta_n\binom{2n}{k}\sigma^k\tau^{2n-k}.$$

Then we split the summation into two parts: for $k=2j+1$ and $k=2j$ with $j=0,1,2,\ldots$. Hence the last summation becomes

$$\sum_{j=0}^{\infty}\left[\sum_{n=j+1}^{\infty}\beta_n\binom{2n}{2j+1}\sigma^{2j+1}\tau^{2n-2j-1}+\sum_{n=j}^{\infty}\beta_n\binom{2n}{2j}\sigma^{2j}\tau^{2n-2j}\right].$$

We substitute $m=2n-2j-1$ for the first summation and $m=2n-2j$ for the second summation. Note that m in the first summation will cover all the positive odd numbers, while m in the second summation will cover all the positive even numbers and zero. So the last summation becomes

$$\begin{aligned}&\sum_{j=0}^{\infty}\Bigg[\sum_{m:\text{ odd}}\beta_{\lceil\frac{m+2j+1}{2}\rceil}\binom{m+2j+1}{2j+1}\sigma^{2j+1}\tau^m\\&\qquad+\sum_{m:\text{ even}}\beta_{\lceil\frac{m+2j}{2}\rceil}\binom{m+2j}{2j}\sigma^{2j}\tau^m\Bigg]\\&=\sum_{n,m:\text{ odd}}\beta_{\frac{m+n}{2}}\binom{m+n}{n}\sigma^n\tau^m+\sum_{n,m:\text{ even}}\beta_{\frac{m+n}{2}}\binom{m+n}{n}\sigma^n\tau^m.\end{aligned}$$

Therefore,

$$\begin{aligned}&\mathbb{E}\left[\left(\int_a^t e^{\tau(B(b)-B(s))}\,dB(s)\right)\left(\int_a^t e^{\sigma(B(b)-B(s))}\,dB(s)\right)\right]\\&=\sum_{n,m:\text{ odd}}\beta_{\frac{m+n}{2}}\binom{m+n}{n}\tau^m\sigma^n+\sum_{n,m:\text{ even}}\beta_{\frac{m+n}{2}}\binom{m+n}{n}\tau^m\sigma^n.\qquad(3.10)\end{aligned}$$

By comparing the coefficients of $\tau^m\sigma^n$ in equations (3.10) and (3.7), we obtain the desired result. □

Theorem 3.11. *For any $a\le t\le b$ and $m,n\in\{0,1,2,\ldots\}$ we have*

$$\begin{aligned}&\mathbb{E}\left[\left(\int_a^t (B(b)-B(s))^n\,dB(s)\right)\left(\int_a^t (B(b)-B(s))^m\,dB(s)\right)\right]\\&=\int_a^t\mathbb{E}\left[(B(b)-B(s))^{n+m}\right]ds.\qquad(3.11)\end{aligned}$$

Proof. Note that $B(b)-B(s)\sim\mathcal{N}(0,b-s)$. Hence if $n+m$ is odd, then $\mathbb{E}\left[(B(b)-B(s))^{n+m}\right]$ is zero as in equation (3.6). When $n+m$ is even, we have

$$\begin{aligned}
&\int_a^t \mathbb{E}\left[(B(b)-B(s))^{n+m}\right] ds \\
&= \int_a^t (b-s)^{\frac{m+n}{2}} \frac{(m+n)!}{2^{\frac{m+n}{2}}\left(\frac{m+n}{2}\right)!}\, ds \\
&= \frac{(m+n)!}{2^{\frac{m+n}{2}}\left(\frac{m+n}{2}+1\right)!}\left((b-a)^{\frac{m+n}{2}+1}-(b-t)^{\frac{m+n}{2}+1}\right) \\
&= \frac{(m+n-1)!!}{\frac{m+n}{2}+1}\left((b-a)^{\frac{m+n}{2}+1}-(b-t)^{\frac{m+n}{2}+1}\right).
\end{aligned}$$

Hence equation (3.11) holds. □

In the next theorem we show that Itô isometry holds for stochastic integrals of certain instantly independent stochastic processes.

Theorem 3.12. *Suppose $\varphi(\cdot)$ is an analytic function on $\mathbb{R}$ such that $\int_a^b \mathbb{E}\left[\varphi(t)\right]^2 dt < \infty$ and $\int_a^b \varphi(B(b)-B(t))\, dB(t)$ exists. Then for all $a\le t\le b$,*

$$\mathbb{E}\left[\left(\int_a^t \varphi(B(b)-B(s))\, dB(s)\right)^2\right] = \int_a^t \mathbb{E}\left[\varphi(B(b)-B(s))^2\right] ds. \tag{3.12}$$

Proof. We give an informal derivation of the formula (3.12). Since $\varphi(x)$ is real analytic, we can write it as

$$\varphi(x)=\sum_{n=0}^{\infty}\frac{\varphi^{(n)}(0)}{n!}x^n, \qquad x\in\mathbb{R},$$

where $\varphi^{(n)}(0)$ is the n-th derivative of $\varphi(x)$ at zero. Writing out the left-hand side of equation (3.12), we get

$$\begin{aligned}
&\mathbb{E}\left[\int_a^t \varphi(B(b)-B(s))\, dB(s)\right]^2 \\
&\overset{?}{=} \mathbb{E}\left[\int_a^t \left(\sum_{n=0}^{\infty}\frac{\varphi^{(n)}(0)}{n!}(B(b)-B(s))^n\right) dB(s)\right]^2 \\
&\overset{?}{=} \mathbb{E}\left[\sum_{n=0}^{\infty}\frac{\varphi^{(n)}(0)}{n!}\int_a^t (B(b)-B(s))^n\, dB(s)\right]^2
\end{aligned}$$

$$= \mathbb{E}\left[\sum_{n=0}^{\infty}\sum_{m=0}^{\infty}\left\{\frac{\varphi^{(n)}(0)\varphi^{(m)}(0)}{n!m!}\right.\right.$$
$$\left.\left.\times\left(\int_a^t (B(b)-B(s))^n\,dB(s)\right)\left(\int_a^t (B(b)-B(s))^m\,dB(s)\right)\right\}\right]$$
$$= \sum_{n=0}^{\infty}\sum_{m=0}^{\infty}\left\{\frac{\varphi^{(n)}(0)\varphi^{(m)}(0)}{n!m!}\right.$$
$$\left.\times\,\mathbb{E}\left[\left(\int_a^t (B(b)-B(s))^n\,dB(s)\right)\left(\int_a^t (B(b)-B(s))^m\,dB(s)\right)\right]\right\}. \tag{3.13}$$

On the other hand, we can use similar arguments to those above to show that

$$\int_a^t \mathbb{E}\left[\varphi(B(b)-B(s))\right]^2 ds$$
$$= \int_a^t \mathbb{E}\left[\sum_{n=0}^{\infty}\frac{\varphi^{(n)}(0)}{n!}(B(b)-B(s))^n\right]^2 ds$$
$$= \int_a^t \mathbb{E}\left[\sum_{n=0}^{\infty}\sum_{m=0}^{\infty}\frac{\varphi^{(n)}(0)\varphi^{(m)}(0)}{n!m!}(B(b)-B(s))^{n+m}\right] ds$$
$$= \sum_{n=0}^{\infty}\sum_{m=0}^{\infty}\frac{\varphi^{(n)}(0)\varphi^{(m)}(0)}{n!m!}\int_a^t \mathbb{E}\left[(B(b)-B(s))^{n+m}\right] ds. \tag{3.14}$$

Therefore the assertion of the theorem follows from equations (3.13), (3.14) and Theorem 3.11. □

It is natural to ask whether Theorem 3.12 can be generalized to the stochastic integral of a stochastic process that is the product of an adapted process and an instantly independent stochastic processes. The answer is negative as shown by the next example.

Example 3.2. Direct computations show that

$$\int_0^1 \left\{\mathbb{E}\left[B(t)^2(B(1)-B(t))^2\right]^2\right\} dt = \frac{3}{10}. \tag{3.15}$$

By the Itô formula in Theorem 3.9 for the function $\theta(x,y) = \frac{1}{3}x^3(y-x)^2 + \frac{1}{6}x^4(y-x) + \frac{1}{30}x^5$, we obtain

$$\int_0^1 B(t)^2(B(1)-B(t))^2\,dB(t) = \frac{B(1)^5}{30} - \int_0^1 (B(1)B(t)(B(1)-B(t)))\,dt,$$

which yields the value

$$\mathbb{E}\left[\left(\int_0^1 B(t)^2 \big[B(1) - B(t)\big]^2\, dB(t)\right)^2\right] = \frac{11}{30}. \tag{3.16}$$

Thus the left-hand sides of equations (3.15) and (3.16) are not equal.

3.5. Itô integral and the new integral

In this section we prove two theorems to express the new integral of certain stochastic processes in terms of the Itô and Riemann integrals.

Theorem 3.13. *Let $\{f(t)\}$ be adapted with respect to $\{\mathcal{F}_t\}$ and let $\varphi(\cdot)$ be an analytic function on $\mathbb{R}$. Assume that $\int_a^b f(t)\varphi(B(b) - B(t))\, dB(t)$ exists. Then*

$$\begin{aligned}
&\int_a^b f(t)\varphi(B(b) - B(t))\, dB(t) \\
&= \sum_{n=0}^{\infty} (-1)^n \frac{\varphi^{(n)}(B(b))}{n!} \left(\int_a^b f(t)B(t)^n\, dB(t) + n\int_a^b f(t)B(t)^{n-1}\, dt\right),
\end{aligned}$$

where $\varphi^{(n)}$ is the n-th derivative of φ.

Proof. As in the proof of Theorem 3.12, we can derive the above formula informally. Using the definition of the stochastic integral and the form of $\varphi(x)$, we get

$$\begin{aligned}
&\int_a^b f(t)\varphi(B(b) - B(t))\, dB(t) \\
&= \lim_{\|\Delta\|\to 0} \sum_{i=1}^{n} f(t_{i-1})\varphi(B(b) - B(t_i))\Delta B_i \\
&= \lim_{\|\Delta\|\to 0} \sum_{i=1}^{n} \left[f(t_{i-1}) \left(\sum_{k=0}^{\infty} \frac{\varphi^{(k)}(0)}{k!}(B(b) - B(t_i))^k\right) \Delta B_i\right] \\
&= \lim_{\|\Delta\|\to 0} \sum_{i=1}^{n} \left[f(t_{i-1}) \left(\sum_{k=0}^{\infty} \frac{\varphi^{(k)}(0)}{k!} \sum_{l=0}^{k} \binom{k}{l}(-1)^l B(b)^{k-l} B(t_i)^l\right) \Delta B_i\right] \\
&= \lim_{\|\Delta\|\to 0} \sum_{i=1}^{n} \Bigg[f(t_{i-1}) \Bigg(\sum_{k=0}^{\infty} \sum_{l=0}^{k} \frac{\varphi^{(k)}(0)}{(k-l)!l!} \\
&\qquad \times (-1)^l B(b)^{k-l} (B(t_{i-1}) + \Delta B_i)^l\Bigg) \Delta B_i\Bigg].
\end{aligned} \tag{3.17}$$

Since all the terms $(\Delta B_i)^k$, $k \geq 3$ converge to 0 in probability, equation (3.17) becomes

$$\lim_{\|\Delta\|\to 0} \sum_{i=1}^{n} \Bigg[f(t_{i-1}) \Bigg(\sum_{k=0}^{\infty} \sum_{l=0}^{k} \frac{\varphi^{(k)}(0)}{(k-l)!l!} \times (-1)^l B(b)^{k-l} \left(B(t_{i-1})^l + lB(t_{i-1})^{l-1} \Delta B_i \right) \Bigg) \Delta B_i \Bigg]$$

$$= \sum_{k=0}^{\infty} \sum_{l=0}^{k} \frac{\varphi^{(k)}(0)}{(k-l)!l!} (-1)^l B(b)^{k-l} \times \lim_{\|\Delta\|\to 0} \sum_{i=1}^{n} f(t_{i-1}) \left(B(t_{i-1})^l \Delta B_i + lB(t_{i-1})^{l-1} \Delta t_i \right)$$

$$= \sum_{k=0}^{\infty} \sum_{l=0}^{k} \frac{\varphi^{(k)}(0)}{(k-l)!l!} (-1)^l B(b)^{k-l} \times \left(\int_a^t f(t) B(t)^l \, dB(s) + l \int_a^t f(t) B(t)^{l-1} \, ds \right).$$

Change the order of summation to get

$$\sum_{l=0}^{\infty} \frac{(-1)^l}{l!} \left(\int_a^t f(t) B(t)^l \, dB(s) + l \int_a^t f(t) B(t)^{l-1} \, ds \right) \sum_{k=l}^{\infty} \frac{\varphi^{(k)}(0)}{(k-l)!} B(b)^{k-l}$$

$$= \sum_{l=0}^{\infty} \frac{(-1)^l}{l!} \left(\int_a^t f(t) B(t)^l \, dB(s) + l \int_a^t f(t) B(t)^{l-1} \, ds \right) \varphi^{(l)}(B(b)).$$

Hence the proof is complete. □

Example 3.3. Suppose $\{f(t)\}$ is $\{\mathcal{F}_t\}$-adapted and $\varphi(x) = e^x$. Then $\varphi^{(n)}(x) = e^x$ and by Theorem 3.13 we have

$$\int_0^1 f(t) e^{B(1)-B(t)} \, dB(t)$$

$$= \sum_{n=0}^{\infty} (-1)^n \frac{e^{B(1)}}{n!} \left(\int_0^1 f(t) B(t)^n \, dB(t) + n \int_0^1 f(t) B(t)^{n-1} \, dt \right)$$

$$= e^{B(1)} \left(\int_0^1 \left[f(t) \sum_{n=0}^{\infty} \frac{(-B(t))^n}{n!} \right] dB(t) - \int_0^1 \left[f(t) \sum_{n=1}^{\infty} \frac{(-B(t))^{n-1}}{(n-1)!} \right] dt \right)$$

$$= e^{B(1)} \left(\int_0^1 f(t) e^{-B(t)} \, dB(t) - \int_0^1 f(t) e^{-B(t)} \, dt \right).$$

Thus we have the equality

$$\int_0^1 f(t)e^{B(1)-B(t)}\,dB(t) = e^{B(1)}\left(\int_0^1 f(t)e^{-B(t)}\,dB(t) - \int_0^1 f(t)e^{-B(t)}\,dt\right).$$

An important concept of the new integral is the fact that $B(b) - B(t)$, $a \le t \le b$ is an instantly independent stochastic process. However, this concept is motivated by the integral $\int_0^1 B(1)\,dB(t)$, which K. Itô introduced in the 1976 Kyoto Symposium on Stochastic Differential Equations [7]. Our next theorem provides the new integral for such integrands.

Theorem 3.14. *Let $\{f(t)\}$ be adapted with respect to $\{\mathcal{F}_t\}$ and let $\varphi(x)$ be an analytic function on $\mathbb{R}$. Assume that $\int_a^b f(t)\varphi(B(b))\,dB(t)$ exists. Then*

$$\begin{aligned}\int_a^b & f(t)\varphi(B(b))\,dB(t)\\ &= \sum_{k=0}^{\infty}\sum_{n=0}^{\infty}\frac{\varphi^{(n+k)}(B(b))}{n!k!}(-1)^n\\ &\quad\times\left(\int_a^b f(t)B(t)^{n+k}\,dB(t) + n\int_a^b f(t)B(t)^{n+k-1}\,dt\right)\end{aligned} \tag{3.18}$$

where $\varphi^{(n)}$ is the n-th derivative of φ.

Proof. To derive the above formula informally, we first use the analyticity of $\varphi(x)$ to write

$$\begin{aligned}\int_a^b & f(t)\varphi(B(b))\,dB(t)\\ &= \int_a^b \left(f(t)\sum_{n=0}^{\infty}\frac{\varphi^{(n)}(B(b)-B(t))}{n!}B(t)^n\right)dB(t)\\ &= \sum_{n=0}^{\infty}\frac{1}{n!}\int_a^b \left(f(t)B(t)^n\varphi^{(n)}(B(b)-B(t))\right)dB(t).\end{aligned}$$

Then we apply Theorem 3.13 for the case $f(t)B(t)^n$ and $\varphi^{(n)}(B(b)-B(t))$ to obtain the right-hand side of equation (3.18). $\square$

Example 3.4. As in Example 3.3, suppose $\{f(t)\}$ is $\{\mathcal{F}_t\}$-adapted and $\varphi(x) = e^x$. Then $\varphi^{(n+k)}(x) = e^x$ for all non-negative integers n, k, and by

Theorem 3.14 we have

$$
\begin{aligned}
&\int_0^1 f(t)e^{B(1)}\,dB(t)\\
&= \sum_{n=0}^{\infty}\sum_{k=0}^{\infty}(-1)^n\frac{e^{B(1)}}{n!k!}\left(\int_0^1 f(t)B(t)^{n+k}\,dB(t) + n\int_0^1 f(t)B(t)^{n+k-1}\,dt\right)\\
&= e^{B(1)}\sum_{n=0}^{\infty}\frac{(-1)^n}{n!}\left(\int_0^1\left[f(t)B(t)^n\sum_{k=0}^{\infty}\frac{B(t)^k}{k!}\right]dB(t)\right.\\
&\qquad\qquad\left. + n\int_0^1\left[f(t)B(t)^{n-1}\sum_{k=0}^{\infty}\frac{B(t)^k}{k!}\right]dt\right)\\
&= e^{B(1)}\sum_{n=0}^{\infty}\frac{(-1)^n}{n!}\left(\int_0^1\left[f(t)B(t)^n e^{B(t)}\right]dB(t)\right.\\
&\qquad\qquad\left. + n\int_0^1\left[f(t)B(t)^{n-1}e^{B(t)}\right]dt\right)\\
&= e^{B(1)}\left(\int_0^1\left[f(t)e^{B(t)}\sum_{n=0}^{\infty}\frac{(-B(t))^n}{n!}\right]dB(t)\right.\\
&\qquad\qquad\left. - \int_0^1\left[f(t)e^{B(t)}\sum_{n=0}^{\infty}\frac{(-B(t))^{n-1}}{(n-1)!}\right]dt\right)\\
&= e^{B(1)}\left(\int_0^1 f(t)\,dB(t) - \int_0^1 f(t)\,dt\right).
\end{aligned}
$$

Thus we have the equality

$$\int_0^1 f(t)e^{B(1)}\,dB(t) = e^{B(1)}\left(\int_0^1 f(t)\,dB(t) - \int_0^1 f(t)\,dt\right).$$

The special case with $f(t) = 1$ is given in Example 2.6 of [1].

3.6. Further results

There are certainly many open problems concerning the new stochastic integral. In particular, in the forthcoming papers [12, 13] we will prove Itô's formula, isometry properties of the new integral, and some results on anticipating stochastic differential equations.

References

[1] W. Ayed, and H.-H. Kuo, An extension of the Itô integral, *Communications on Stochastic Analysis* **2**, 323-333, (2008).

[2] W. Ayed, and H.-H. Kuo, An extension of the Itô integral: Toward a general theory of stochastic integration, *Theory of Stochastic Processes* **16** (32), N1 17-28, (2010).

[3] R. Buckdahn, Skorohod's integral and linear stochastic differential equations, *Preprint* (1987).

[4] M. Hitsuda, Formula for Brownian partial derivatives, *Second Japan-USSR Symposium Probab. Th.* **2** (1972) 111-114.

[5] M. Hitsuda, Formula for Brownian partial derivatives, *Publ. Faculty of Integrated Arts and Sciences,* Hiroshima University, Series III, Vol. 4, 1-15, (1978).

[6] K. Itô, Stochastic integral, *Proceedings of the Imperial Academy* **20** (1944) 519-524

[7] K. Itô, Extension of stochastic integrals, *Proceedings of the International Symposium on Stochastic Differential Equations,* K. Itô (ed.) Kinokuniya, 95 109, (1978).

[8] H.-H. Kuo, *White Noise Distribution Theory,* CRC Press, Boca Raton, (1996).

[9] H.-H. Kuo, *Introduction to Stochastic Integration,* Universitext (UTX), Springer, New York, (2006).

[10] H.-H. Kuo and J. Pothoff, Anticipating stochastic integrals and stochastic differential equations, *White Noise Analysis,* World Sci. Publ, River Edge, NJ, 256-273, (1990).

[11] H.-H. Kuo and A. Russek, White noise approach to stochastic integration, *Journal of Multivariate Analysis,* **24**, 218-236, (1988).

[12] H.-H. Kuo, A. Sae-Tang, and B. Szozda, Itô formula and isometry for the stochastic integral of adapted and instantly independent processes, *Preprint.*

[13] H.-H. Kuo, A. Sae-Tang, and B. Szozda, Linear SDEs with anticipating initial conditions, *Preprint.*

[14] Y. Mishura and G. Shevchenko, Approximate solutions to anticipative stochastic differential equations, *Statistics and Probability Letters* **78**, 60-66, (2008)

[15] D. Nualart, *The Malliavin Calculus and Related Topics.* Probability and its Applications (New York) (Second edition). Berlin: Springer-Verlag (2006).

[16] D. Nualart and E. Pardoux, Stochastic calculus with anticipating integrands, *Probability Theory and Related Fields* **78**, 535 581, (1988).

[17] E. Pardoux and P. Protter, A two-sided stochastic integral and its calculus, *Probability Theory and Related Fields* **76**, 15-49, (1987).

[18] F. Russo and P. Vallois, Forward, backward and symmetric integrals, *Probability Theory and Related Fields* **97**, 403-421, (1993)

[19] A. V. Skorokhod, On a generalization of a stochastic integral, *Theory of Probability and its Applications* **20**, 219-233, (1975).

Chapter 4

Independence of some multiple Poisson stochastic integrals with variable-sign kernels

Nicolas Privault

Division of Mathematical Sciences,
School of Physical and Mathematical Sciences,
Nanyang Technological University,
SPMS-MAS, 21 Nanyang Link, 637371 Singapore

It is known that the multiple Poisson stochastic integrals $I_n(f_n)$, $I_m(g_m)$ of two symmetric *constant-sign* functions are independent if and only if f_n and g_m satisfy a disjoint support condition. In this chapter we present some examples in which this property extends to some variable-sign functions.

Contents

4.1. Introduction

The independence of two Gaussian random variables can be characterized by the vanishing of their covariance, and this property has been extended to multiple Wiener integrals in [9], [10], by stating that the multiple integrals $I_n(f_n)$, $I_m(g_m)$ with respect to Brownian motion are independent if and only if the contraction

$$\int_0^\infty f_n(x_1,\ldots,x_{n-1},z)g_m(y_1,\ldots,y_{m-1},z)dz = 0,$$

vanishes, $x_1,\ldots,x_{n-1},y_1,\ldots,y_{m-1} \in \mathbb{R}$, where $L^2(\mathbb{R}_+)^{\circ n}$ denotes the subspace of $L^2(\mathbb{R}_+)^{\otimes n} = L^2(\mathbb{R}_+^n)$ made of symmetric functions.

A proof of necessity has been provided in [2] using standard probabilistic arguments, while the original proof of [9], [10] relied on the Malliavin calculus. In addition, $I_n(f_n)$ is independent of $I_m(g_m)$ if and only if

$$E[(I_n(f_n))^2(I_m(g_m))^2] = E[(I_n(f_n))^2] \times E[(I_m(g_m))^2], \qquad (4.1)$$

cf. Corollary 5.2 of [7]. Independence criteria for multiple stochastic integrals with respect to other Gaussian processes have been obtained in [1].

Coming back to the case of single integrals $I_1(f)$ with respect to Brownian motion, when $X \simeq \mathcal{N}(0, s)$ and $Y \simeq \mathcal{N}(0, t-s)$ are two independent centered Gaussian random variables it is well known that

$$X - Y \simeq I_1(f), \qquad \text{with } f = \mathbf{1}_{[0,s]} - \mathbf{1}_{[s,t]},$$

is independent of

$$X + Y \simeq I_1(g), \qquad \text{with } g = \mathbf{1}_{[0,s]} + \mathbf{1}_{[s,t]},$$

since $\langle f, g \rangle = \langle \mathbf{1}_{[0,s]} - \mathbf{1}_{[s,t]}, \mathbf{1}_{[0,s]} + \mathbf{1}_{[s,t]} \rangle = 0$.

The situation is completely different when $X \simeq \tilde{\mathcal{P}}(s)$ and $Y \simeq \tilde{\mathcal{P}}(t-s)$ are independent centered Poisson random variables with means s and $t-s$, in which case $X - Y$ is no longer independent of $X + Y$. Although we may also write

$$X - Y \simeq I_1(f) = I_1(\mathbf{1}_{[0,s]} - \mathbf{1}_{[s,t]})$$

and

$$X + Y \simeq I_1(g) = I_1(\mathbf{1}_{[0,s]} + \mathbf{1}_{[s,t]})$$

where $I_1(f)$ is the compensated Poisson integral of f, independence can occur only when $fg = 0$, as follows from Proposition 4.3 below.

When $I_n(f_n)$ and $I_m(g_m)$ are multiple Poisson stochastic integrals on $\mathbb{R}_+$, the condition

$$\begin{aligned} f_n \otimes_1^0 g_m(x_1, \ldots, x_n, y_1, \ldots, y_{n-1}) \\ := f_n(x_1, \ldots, x_n) g_m(y_1, \ldots, y_{n-1}, x_n) = 0, \end{aligned} \qquad (4.2)$$

$x_1, \ldots, x_n, y_1, \ldots, y_{n-1} \in \mathbb{R}_+$, is sufficient for the independence of $I_n(f_n)$ and $I_m(g_m)$ on the Poisson space, due to the independence of increments of the Poisson process, since by symmetry of these functions, the supports of f_n and g_m can be respectively contained in two Borel sets of the form $A^n \subset \mathbb{R}_+^n$ and $B^m \subset \mathbb{R}_+^m$ with $A \cap B = \emptyset$, cf. [5].

On the other hand, it has been claimed in [4], [5] that (4.2) is also a *necessary* condition for the independence of $I_n(f_n)$ and $I_m(g_m)$ on the Poisson space. A similar result was stated independently in [8].

However that necessity claim was proved only when the functions f_n and g_m have constant signs. More precisely, the necessity condition therein is based on the relation (4.1) which, unlike on the Wiener space, turns out to be weaker than the independence of $I_n(f_n)$ and $I_m(g_m)$ when f_n and g_m have variable signs, as was shown in [7] by a counterexample for $n = 2$ and $m = 1$, cf. also Remark 4.1 below.

In this note we present some examples in which (4.2) holds as a necessary and sufficient condition for the independence of $I_n(f_n)$, $I_m(g_m)$ when f_n and g_m are allowed to have variable signs. Namely we treat the following cases:

- tensor powers:
  f_n and g_m take the form $f_n = f^{\circ n}$ and $g_m = g^{\circ m}$,
- integrals of first and multiple orders:
  $f_n \in L^2(\mathbb{R}_+)^{\circ n}$ and g_1 has constant sign,
- integrals of first and second orders:
  $f_2 = f \circ g$, without sign restrictions on f_2 and g_1,

for which we show that $f_n \otimes_1^0 g_m = 0$ is necessary and sufficient for independence. The question whether (4.2) is a necessary condition for independence is still open in full generality.

In Section 4.2 we recall the necessary conditions for independence obtained in [4], [5]. The examples of multiple integrals with variable-sign kernels for which the disjoint support condition is necessary and sufficient are given in Section 4.3. Finally in Section 4.4 we consider the expectation of multiple stochastic integrals.

The results of this paper are stated for a standard Poisson process on the half line $\mathbb{R}_+$, however they can be extended without difficulty to Poisson measures on metric spaces.

4.2. Necessary condition for independence

We start by recalling some results of [4]. Given $f_n \in L^2(\mathbb{R})^{\circ n}$ a symmetric square-integrable function of n variables on $\mathbb{R}_+^n$, the multiple stochastic integral of f_n with respect to the standard Poisson process $(N_t)_{t\in\mathbb{R}_+}$ is defined as

$$I_n(f_n) = n! \int_0^\infty \int_0^{t_n^-} \cdots \int_0^{t_2^-} f_n(t_1,\ldots,t_n)d(N_{t_1}-t_1)\cdots d(N_{t_n}-t_n). \quad (4.3)$$

We will need the multiplication formula

$$I_n(f_n)I_m(g_m) = \sum_{s=0}^{2(n\wedge m)} I_{n+m-s}(h_{n,m,s}), \tag{4.4}$$

cf. e.g. Proposition 4.5.6 of [6], provided

$$h_{n,m,s} := \sum_{s\le 2k\le 2(s\wedge n\wedge m)} k!\binom{n}{k}\binom{m}{k}\binom{k}{s-k} f_n \circ_k^{s-k} g_m$$

is in $L^2(\mathbb{R}_+)^{\circ n+m-s}$, $0 \le s \le 2(n\wedge m)$ where $f_n \circ_k^l g_m$, $0 \le l \le k$, is the symmetrization in $n+m-k-l$ variables of the function

$$\begin{aligned} &f_n \otimes_k^l g_m(x_{l+1},\ldots,x_n,y_{k+1},\ldots,y_m) \\ &:= \int_0^\infty \cdots \int_0^\infty f_n(x_1,\ldots,x_n)g_m(x_1,\ldots,x_k,y_{k+1},\ldots,y_m)dx_1\cdots dx_l \end{aligned}$$

for $f_n \in L^2(\mathbb{R}_+)^{\circ n}$ and $g_m \in L^2(\mathbb{R}_+)^{\circ m}$ such that $f_n \circ_k^l g_m \in L^2(\mathbb{R}_+)^{\circ n+m-k-l}$, $0 \le l \le k \le n \wedge m$.

The following proposition shows that $f_n \circ_1^0 g_m = 0$ is a necessary condition for the independence of $I_n(f_n)$ and $I_m(g_m)$.

Proposition 4.1. Let $f_n \in L^2(\mathbb{R}_+)^{\circ n}$ and $g_m \in L^2(\mathbb{R}_+)^{\circ m}$ such that

$$E[(I_n(f_n))^2(I_m(g_m))^2] = E[(I_n(f_n))^2] \times E[(I_m(g_m))^2].$$

Then we have $f_n \circ_1^0 g_m = 0$.

Proof. [See [4].] If $I_n(f_n)$ and $I_m(g_m)$ are independent, then $I_n(f_n)I_m(g_m) \in L^2(\Omega)$ and using the isometry formula

$$E\left[I_n(f_n)I_m(g_m)\right] = n!\langle f_n, g_m\rangle_{L^2(\mathbb{R}_+)^{\otimes n}} \mathbf{1}_{\{n=m\}}, \tag{4.5}$$

for $f_n \in L^2(\mathbb{R}_+)^{\circ n}$, $g_m \in L^2(\mathbb{R}_+)^{\circ m}$, we get

$$\begin{aligned} &(n+m)! \mid f_n \circ g_m \mid^2_{L^2(\mathbb{R}_+)^{\otimes n+m}} \ge n!m! \mid f_n \otimes g_m \mid^2_{L^2(\mathbb{R}_+)^{\otimes n+m}} \\ &= n!m! \mid f_n \mid^2_{L^2(\mathbb{R}_+)^{\otimes n}} \mid g_m \mid^2_{L^2(\mathbb{R}_+)^{\otimes m}} \\ &= E\left[I_n(f_n)^2\right] E\left[I_m(g_m)^2\right] = E\left[(I_n(f_n)I_m(g_m))^2\right] \\ &= \sum_{r=0}^{2(n\wedge m)} (n+m-r)! \mid h_{n,m,r} \mid^2_{L^2(\mathbb{R}_+)^{\otimes n+m-r}} \\ &\ge (n+m)! \mid h_{n,m,0} \mid^2_{L^2(\mathbb{R}_+)^{\otimes n+m}} + (n+m-1)! \mid h_{n,m,1} \mid^2_{L^2(\mathbb{R}_+)^{\otimes n+m-1}} \\ &\ge (n+m)! \mid f_n \circ g_m \mid^2_{L^2(\mathbb{R}_+)^{\otimes n+m}} \\ &\qquad + nm(n+m-1)! \mid f_n \circ_1^0 g_m \mid^2_{L^2(\mathbb{R}_+)^{\otimes n+m-1}} \end{aligned}$$

from (4.4), which implies $f_n \circ_1^0 g_m = 0$. □

It follows from Proposition 4.1 that if $I_n(f_n)$ and $I_m(g_m)$ are independent then we have $f_n \circ_1^0 g_m = 0$. However, $f_n \circ_1^0 g_m = 0$ is in general only a necessary and not sufficient condition for $f_n \otimes_1^0 g_m$ to vanish, since

$$\langle f_n \circ_1^0 g_m, f_n \circ_1^0 g_m \rangle \leq \langle f_n \otimes_1^0 g_m, f_n \otimes_1^0 g_m \rangle,$$

which follows from the fact that $f_n \circ_1^0 g_m$ is the symmetrization of $f_n \otimes_1^0 g_m$ in $n + m - 1$ variables.

A counterexample for which we have $f_n \circ_1^0 g_m = f_m \circ_1^1 g_m = 0$ and $f_n \otimes_1^0 g_m \neq 0$ can be found in [7], Example 5.3, as

$$\begin{aligned} f_2(s,t) = \mathbf{1}_{\{st<0\}}(&\mathbf{1}_{[-1,-1/2]}(s) - \mathbf{1}_{[-1/2,1/2]}(s) + \mathbf{1}_{[1/2,1]}(s)) \\ &\cdot (\mathbf{1}_{[-1,-1/2]}(s) - \mathbf{1}_{[-1/2,1/2]}(s) + \mathbf{1}_{[1/2,1]}(s)) \end{aligned} \tag{4.6}$$

and $g_1(t) = \mathbf{1}_{[-1,0]}(t) + \mathbf{1}_{(0,1]}(t)$. In this example the integrals $I_2(f_2)$ and $I_1(g_1)$ are not independent, however (4.1) holds and therefore (4.1) and (4.2) cannot characterize the independence of $I_n(f_n)$ and $I_m(g_m)$.

In the case $n = 2$ and $m = 1$ we are able to show that (4.1) entails $f_2 \circ_1^1 g_1 = 0$ in addition to $f_2 \circ_1^0 g_1 = 0$.

Proposition 4.2. Let $f_2 \in L^2(\mathbb{R}_+)^{\circ 2}$ and $g_1 \in L^2(\mathbb{R}_+)$ such that

$$E[(I_2(f_2))^2(I_1(g_1))^2] = E[(I_2(f_2))^2] \times E[(I_1(g_1))^2].$$

Then we have

$$f_2 \circ_1^0 g_1 = 0, \quad \text{and} \quad f_2 \circ_1^1 g_1 = 0.$$

Proof. By the same argument as in the proof of Proposition 4.1 we find

$$h_{2,1,1} = f_2 \circ_1^0 g_1 = 0,$$

and

$$h_{2,1,2} := 2 f_2 \circ_1^1 g_1 = 0.$$

□

When $f_2 = f \circ g$ and $g_1 = h$, Proposition 4.2 yields

$$(f \circ g) \circ_1^0 h = \frac{1}{4}(f \otimes (gh) + (gh) \otimes f + g \otimes (fh) + (fh) \otimes g) = 0, \tag{4.7}$$

and

$$(f \circ g) \circ_1^1 h = f \langle g, h \rangle + g \langle f, h \rangle = 0, \tag{4.8}$$

hence

$$\langle g, h \rangle = \langle f, h \rangle = 0, \tag{4.9}$$

provided $f \circ g \neq 0$.

4.3. Variable-sign kernels

In this section we consider examples of kernels $f_n \in L^2(\mathbb{R}_+)^{\circ n}$, $g_m \in L^2(\mathbb{R}_+)^{\circ m}$ for which

$$f_n \circ_1^0 g_m = 0 \quad \Longrightarrow \quad f_n \otimes_1^0 g_m = 0,$$

in which case the condition

$$f_n \otimes_1^0 g_m = f_n(x_1, \ldots, x_{n-1}, z) g_m(y_1, \ldots, y_{m-1}, z) = 0,$$

$x_1, \ldots, x_{n-1}, y_1, \ldots, y_{m-1}, z \in \mathbb{R}$, becomes both necessary and sufficient for the independence of $I_n(f_n)$ and $I_m(g_m)$, and is equivalent to (4.1), which also becomes necessary and sufficient.

We note that

$$\begin{aligned}
&\langle f_n \circ_1^0 g_m, f_n \circ_1^0 g_m \rangle \\
&= \frac{1}{(n+m-1)!} \sum_{\sigma \in \Sigma_{n+m-1}} \int_{\mathbb{R}_+^{n+m-1}} \Big((f_n \otimes_1^0 g_m)(x_1, \ldots, x_{n+m-1}) \\
&\qquad\qquad (f_n \otimes_1^0 g_m)(x_{\sigma_1}, \ldots, x_{\sigma_{n+m-1}}) \Big) dx_1 \cdots dx_{n+m-1} \\
&= \frac{(n-1)!(m-1)!}{(n+m-1)!} \langle f_n \otimes_1^0 g_m, f_n \otimes_1^0 g_m \rangle \\
&\quad + \frac{1}{(n+m-1)!} \sum_{\sigma \in \Theta_{n,m}} \int_{\mathbb{R}_+^{n+m-1}} \Big(f_n \otimes_1^0 g_m(x_1, \ldots, x_{n+m-1}) \\
&\qquad\qquad f_n \otimes_1^0 g_m(x_{\sigma_1}, \ldots, x_{\sigma_{n+m-1}}) \Big) dx_1 \cdots dx_{n+m-1},
\end{aligned} \tag{4.10}$$

where

$$\begin{aligned}
\Theta_{n,m} = \big\{ \sigma \in \Sigma_{n+m-1} \ :\ & \sigma(\{1, \ldots, n-1\}) \neq \{1, \ldots, n-1\} \\
& \text{or } \sigma(\{n+1, \ldots, n+m-1\}) \neq \{n+1, \ldots, n+m-1\} \big\},
\end{aligned}$$

and our method of proof will rely on this decomposition to show that $f_n \otimes_1^0 g_m = 0$.

Tensor powers

In the next proposition we obtain a necessary and sufficient condition for the independence of integrals of (symmetric) tensor powers.

Proposition 4.3. Let $f, g \in L^2(\mathbb{R})$. Then $I_n(f^{\otimes n})$ is independent of $I_m(g^{\otimes m})$ if and only if $f^{\otimes n} \otimes_1^0 g^{\otimes m} = 0$, which is equivalent to $fg = 0$.

Proof. We have

$$f^{\circ n} \circ_1^0 g^{\circ n} = (fg) \circ f^{\circ(n-1)} \circ g^{\circ(m-1)},$$

hence $f^{\circ n} \circ_1^0 g^{\circ n} = 0$ is equivalent to $fg = 0$ and to

$$f^{\otimes n} \otimes_1^0 g^{\otimes n} = f^{\otimes(n-1)} \otimes (fg) \otimes g^{\otimes(m-1)} = 0.$$

□

In particular, $I_1(f)$ is independent of $I_1(g)$ if and only if $fg = 0$, for all $f, g \in L^2(\mathbb{R}_+)$.

Integrals of first and multiple orders

Here we consider the case of $I_n(f_n)$ and $I_1(g_1)$ when only g_1 has constant sign.

Proposition 4.4. Let $f_n \in L^2(\mathbb{R}_+)^{\circ n}$, and assume that $g_1 \in L^2(\mathbb{R}_+)$ only has constant sign. Then $I_n(f_n)$ is independent of $I_1(g_1)$ if and only if $f_n \otimes_1^0 g_1 = 0$, i.e.

$$f_n(x_1, \ldots, x_n) g_1(x_1) = 0, \qquad x_1, \ldots, x_n \in \mathbb{R}_+.$$

Proof. We have

$$f_n \circ_1^0 g_1(x_1, \ldots, x_n) = \frac{1}{n} f_n(x_1, \ldots, x_n) \sum_{k=1}^{n} g_1(x_k),$$

and, as in (4.10),

$$\begin{aligned}
&\langle f_n \circ_1^0 g_1, f_n \circ_1^0 g_1 \rangle \\
&= \frac{1}{n^2} \int_0^\infty \cdots \int_0^\infty f_n^2(x_1, \ldots, x_n) \sum_{i=1}^{n} \sum_{j=1}^{n} g_1(x_i) g_1(x_j) dx_1 \cdots dx_n \\
&= \frac{1}{n} \int_0^\infty \cdots \int_0^\infty f_n^2(x_1, \ldots, x_n) |g_1(x_i)|^2 dx_1 \cdots dx_n \qquad (4.11) \\
&\quad + \frac{n-1}{n} \int_0^\infty \cdots \int_0^\infty f_n^2(x_1, \ldots, x_n) g_1(x_1) g_1(x_2) dx_1 \cdots dx_n \\
&= \frac{1}{n} \langle f_n \otimes_1^0 g_1, f_n \otimes_1^0 g_1 \rangle + \frac{n-1}{n} \langle f_n \otimes_n^{n-2} f_n, g_1 \otimes g_1 \rangle,
\end{aligned}$$

hence $f_n \circ_1^0 g_1 = 0$ implies $f_n \otimes_1^0 g_1 = 0$ since $f_n \otimes_n^{n-2} f_n \geq 0$. □

Integrals of first and second orders

In the next result the necessary and sufficient condition for independence of multiple stochastic integrals is obtained without any sign assumption on the integrands in the first and second order case. In this section we work in the tensor case $f_2 = f \circ g$, which does not include Example 4.6, see (4.14) below for an example in the tensor case.

Proposition 4.5. Let $f, g, h \in L^2(\mathbb{R}_+)$. The double Poisson stochastic integral

$$I_2(f \circ g) = 2 \int_0^\infty f(t) \int_0^{t^-} g(s) d(N_s - s) d(N_t - t)$$

is independent of the single integral $I_1(h) = \displaystyle\int_0^\infty h(t) d(N_t - t)$ if and only if

$$(f \circ g) \otimes_1^0 h = 0,$$

which is equivalent to $fh = gh = 0$.

Proof. We assume for simplicity that $\langle f, f \rangle = \langle g, g \rangle = 1$. If $I_2(f \circ g)$ is independent of $I_1(h)$, Proposition 4.1 shows that $(f \circ g) \circ_1^0 h = 0$ and the conclusion follows from Lemma 4.1 in case $\langle f^2, h \rangle \langle g^2, h \rangle \geq 0$. Next, if

$$\langle f^2, h \rangle \langle g^2, h \rangle < 0,$$

Lemma 4.2 below shows that

$$h = \lambda \mathbf{1}_{\{f \neq 0\}} - \lambda \mathbf{1}_{\{g \neq 0\}}, \tag{4.12}$$

where

$$|\lambda| = \langle f^2, h^2 \rangle^{1/2} = \langle g^2, h^2 \rangle^{1/2} \neq 0. \tag{4.13}$$

In addition we have $fg = 0$ by Lemma 4.2, and by Proposition 4.2, (4.8) and (4.12), we get

$$\int_0^\infty f(x) dx = \int_0^\infty g(x) dx = 0,$$

hence, letting

$$\alpha_- = \int_0^\infty \mathbf{1}_{\{g(x) \neq 0\}} dx < \infty, \qquad \text{and} \qquad \alpha_+ = \int_0^\infty \mathbf{1}_{\{f(x) \neq 0\}} dx < \infty,$$

which are both finite since $h \in L^2(\mathbb{R}_+)$, we obtain

$$\begin{aligned}
&E[(I_2(f \circ g))^2 \mathbf{1}_{\{I_1(h)=\lambda n\}}] = E[I_1(f)^2 I_1(g)^2 \mathbf{1}_{\{I_1(h)=\lambda n\}}] \\
&= e^{-\alpha_- - \alpha_+} \sum_{k=0\vee(-n)}^{\infty} \left(\frac{1}{(2k+n)!} \int_{\mathbb{R}_+^k} \left(\sum_{i=1}^{k} f(x_i) \right)^2 dx_1 \cdots dx_k \right. \\
&\qquad \left. \cdot \int_{\mathbb{R}_+^{k+n}} \left(\sum_{i=1}^{n+k} g(y_i) \right)^2 dy_1 \cdots dy_{k+n} \right) \\
&= e^{-\alpha_- - \alpha_+} \sum_{k=0\vee(-n)}^{\infty} \left(\frac{1}{(2k+n)!} \int_{\mathbb{R}_+^k} \sum_{i=1}^{k} f^2(x_i) dx_1 \cdots dx_k \right. \\
&\qquad \left. \cdot \int_{\mathbb{R}_+^{k+n}} \sum_{i=1}^{n+k} g^2(y_i) dy_1 \cdots dy_{k+n} \right) \\
&= e^{-\alpha_- - \alpha_+} \sum_{k=0\vee(-n)}^{\infty} \frac{k(k+n)}{(2k+n)!}.
\end{aligned}$$

On the other hand, being the difference of two Poisson random variables, $\lambda^{-1} I_1(h)$ has the Skellam distribution

$$\begin{aligned}
P(I_1(h) = \lambda n) &= e^{-\alpha_- - \alpha_+} \sum_{k=0\vee(-n)}^{\infty} \frac{\alpha_+^{n+k} \alpha_-^k}{k!(n+k)!} \\
&= e^{-\alpha_- - \alpha_+} \left(\frac{\alpha_+}{\alpha_-} \right)^{n/2} \mathcal{I}_{|n|}(2\sqrt{\alpha_+ \alpha_-}),
\end{aligned}$$

$n \in \mathbb{Z}$, where

$$\mathcal{I}_n(x) = \sum_{k=0}^{\infty} \frac{(x/2)^{n+2k}}{k!(n+k)!}, \qquad x > 0,$$

is the modified Bessel function of the first kind with parameter $n \geq 0$. It follows that

$$E[(I_2(f \circ g))^2 \mid I_1(h) = \lambda n] = \left(\frac{\alpha_+}{\alpha_-} \right)^{-n/2} \frac{\sum_{k=0}^{\infty} \frac{k(k+n)}{(2k+n)!}}{\mathcal{I}_{|n|}(2\sqrt{\alpha_+ \alpha_-})},$$

which cannot be constant in n because the Bessel function $\mathcal{I}_n(x)$ is not separable in its variables x and n. Therefore, the independence of $I_2(f \circ g)$ with $I_1(h)$ imposes $\lambda = 0$ which concludes the proof by contradiction with (4.13). □

From the above proof we note again that, although independence of $I_2(f \circ g)$ and $I_1(h)$ is equivalent to $(f \circ g) \otimes_1^0 h = 0$, in general the statement $(f \circ g) \circ_1^0 h = 0$ does not imply $(f \circ g) \otimes_1^0 h = 0$, as shown by the following example:

$$f = \mathbf{1}_A, \qquad g = \mathbf{1}_B, \qquad h = \mathbf{1}_A - \mathbf{1}_B, \tag{4.14}$$

$A, B \subset \mathbb{R}_+$, with $A \cap B = \emptyset$, where we have

$$(f \circ g) \otimes_1^0 h = (\mathbf{1}_A \circ \mathbf{1}_B) \otimes_1^0 (\mathbf{1}_A - \mathbf{1}_B) = -\mathbf{1}_A \otimes \mathbf{1}_B + \mathbf{1}_B \otimes \mathbf{1}_A \neq 0,$$

and $(f \circ g) \circ_1^0 h = 0$, while $(f \circ g) \otimes_1^0 h \neq 0$, i.e. (4.2) does not hold. In this case the independence of $I_2(f \circ g)$ and $I_1(h)$ does not hold, as in Example (4.6) above.

Remark 4.1. Unlike in the Wiener case, Condition (4.1) is in general not sufficient for the independence of $I_n(f_n)$ and $I_m(g_m)$, as shown by the example (4.6). Therefore the search for a necessary and sufficient condition for independence in the general case has to involve more complex criteria than (4.1).

The next lemma has been used in the proof of Proposition 4.5.

Lemma 4.1. *Let $f, g, h \in L^2(\mathbb{R}_+)$ such that $(f \circ g) \circ_1^0 h = 0$ and*

$$\langle f^2, h\rangle \langle g^2, h\rangle \geq 0.$$

Then we have $(f \circ g) \otimes_1^0 h = 0$, and $fh = gh = 0$.

Proof. For simplicity and without loss of generality, we assume that

$$\langle f, f\rangle = \langle g, g\rangle = 1.$$

By Proposition 4.2 we have $(f \circ g) \circ_1^0 h = 0$, and from (4.11) or (4.7) we find

$$\begin{aligned} 0 &= \langle (f \circ g) \circ_1^0 h, (f \circ g) \circ_1^0 h\rangle \\ &= \frac{1}{2}\langle (f \circ g) \otimes_1^0 h, (f \circ g) \otimes_1^0 h\rangle + \frac{1}{2}\langle (f \circ g) \otimes_2^0 (f \circ g), h \otimes h\rangle \\ &= \frac{1}{2}\langle (f \circ g) \otimes_1^0 h, (f \circ g) \otimes_1^0 h\rangle + \frac{1}{4}\langle fg, h\rangle^2 + \frac{1}{4}\langle f^2, h\rangle \langle g^2, h\rangle. \end{aligned}$$

This shows that

$$(f \circ g) \otimes_1^0 h = 0,$$

provided $\langle f^2, h\rangle \langle g^2, h\rangle \geq 0$, and we conclude by Lemma 4.3 below. □

The next lemma has been used in the proof of Proposition 4.5.

Lemma 4.2. *Let $f, g, h \in L^2(\mathbb{R}_+)$ such that $(f \circ g) \circ_1^0 h = 0$ and*

$$\langle g^2, h\rangle \neq \langle f^2, h\rangle. \tag{4.15}$$

Then we have $fg = 0$ and

$$h = \lambda \mathbf{1}_{\{f \neq 0\}} - \lambda \mathbf{1}_{\{g \neq 0\}},$$

for some $\lambda \neq 0$ such that

$$\lambda^2 = \frac{\langle f^2, h^2\rangle}{\langle f, f\rangle} = \frac{\langle g^2, h^2\rangle}{\langle g, g\rangle}.$$

In particular it holds that

$$\langle g^2, h\rangle\langle f^2, h\rangle < 0.$$

Proof. For simplicity of exposition we assume that $\langle f, f\rangle = \langle g, g\rangle = 1$. The condition $(f \circ g) \circ_1^0 h = 0$ implies

$$\begin{cases} \langle (f \circ g) \circ_1^0 h, f \otimes f\rangle = \langle f, gh\rangle + \langle f^2, h\rangle\langle f, g\rangle = 0, \\ \\ \langle (f \circ g) \circ_1^0 h, g \otimes g\rangle = \langle f, gh\rangle + \langle g^2, h\rangle\langle f, g\rangle = 0, \end{cases}$$

and since $\langle g^2, h\rangle \neq \langle f^2, h\rangle$ this gives $\langle f, g\rangle = \langle f, gh\rangle = 0$, hence

$$\begin{aligned}
\langle (f \circ g) \circ_1^1 h, (f \circ g) \circ_1^1 h\rangle &= \frac{1}{2}\langle f^2, h\rangle\langle g^2, h\rangle + \langle (f \circ g) \otimes_1^0 h, (f \circ g) \otimes_1^0 h\rangle \\
&= \frac{1}{2}\langle f^2, h\rangle\langle g^2, h\rangle + \frac{1}{4}(\langle f, f\rangle\langle g^2, h^2\rangle + \langle g, g\rangle\langle f^2, h^2\rangle) \\
&\geq -\frac{1}{2}\langle f^2, h^2\rangle^{1/2}\langle g^2, h^2\rangle^{1/2} + \frac{1}{4}(\langle g^2, h^2\rangle + \langle f^2, h^2\rangle) \qquad (4.16) \\
&= \frac{1}{4}(\langle f^2, h^2\rangle^{1/2} - \langle g^2, h^2\rangle^{1/2})^2 \\
&\geq 0,
\end{aligned}$$

which shows that $\langle (f \circ g) \circ_1^1 h, (f \circ g) \circ_1^1 h\rangle = 0$ implies $\langle f^2, h^2\rangle = \langle g^2, h^2\rangle$, and, by the equality (4.16),

$$fh = \lambda f, \qquad \text{and} \qquad gh = -\lambda g,$$

for some $\lambda \in \mathbb{R}$ such that

$$|\lambda| = \langle f^2, h^2\rangle^{1/2} = \langle g^2, h^2\rangle^{1/2}.$$

Hence we have

$$h = \lambda \mathbf{1}_{\{f \neq 0\}} - \lambda \mathbf{1}_{\{g \neq 0\}}, \tag{4.17}$$

and

$$\lambda^2 fg = h^2 fg,$$

which imply that $fg = 0$ a.e. on

$$\{h^2 = 0\} = \{f = h = 0\} \cup \{fg \neq 0\},$$

hence $fg = 0$. Finally we note that $\lambda \neq 0$ by (4.15) and (4.17). □

The next lemma has been used in the proof of Lemma 4.1.

Lemma 4.3. *For any $f, g, h \in L^2(\mathbb{R}_+)$, the condition $(f \circ g) \otimes_1^0 h = 0$ is equivalent to $fh = gh = 0$.*

Proof. Assuming again that $\langle f, f\rangle = \langle g, g\rangle = 1$, we note that since

$$(f \circ g) \otimes_1^0 h = \frac{1}{2}(f \otimes (gh) + g \otimes (fh)), \tag{4.18}$$

we have

$$\begin{aligned}
\langle (f \circ g) \otimes_1^0 h, (f \circ g) \otimes_1^0 h\rangle &= \frac{1}{4}(\langle g^2, h^2\rangle + \langle f^2, h^2\rangle + 2\langle f, g\rangle\langle fg, h^2\rangle) \\
&\geq \frac{1}{4}(\langle g^2, h^2\rangle + \langle f^2, h^2\rangle - 2|\langle fg, h^2\rangle|) \qquad (4.19)\\
&\geq \frac{1}{4}(\langle g^2, h^2\rangle + \langle f^2, h^2\rangle - 2\langle g^2, h^2\rangle^{1/2}\langle f^2, h^2\rangle^{1/2}) \\
&= \frac{1}{4}(\langle f^2, h^2\rangle^{1/2} - \langle g^2, h^2\rangle^{1/2})^2 \\
&\geq 0.
\end{aligned}$$

Assuming that $(f \circ g) \otimes_1^0 h = 0$, if $\langle fg, h^2\rangle \neq 0$ then the equality (4.19) implies $f = -g$ and $fh = gh = 0$ by (4.18). On the other hand, if $\langle fg, h^2\rangle = 0$ we have

$$0 = \langle (f \circ g) \otimes_1^0 h, (f \circ g) \otimes_1^0 h\rangle = \frac{1}{4}(\langle g^2, h^2\rangle + \langle f^2, h^2\rangle),$$

hence $fh = gh = 0$. This shows that $fh = gh = 0$ in all cases, and $(f \circ g) \otimes_1^0 h = 0$ is equivalent to $fh = gh = 0$. □

4.4. Conditional expectations

In this section we present some complements on conditional expectations. Next is an adaptation of Proposition 3 in [3] to the Poisson case.

Proposition 4.6. Let $f_n \in L^2(\mathbb{R}_+)$, $n \geq 1$, and let A be a Borel set of $\mathbb{R}_+$ with finite measure. We have

$$E[I_n(f_n) \mid I_n(\mathbf{1}_A^{\otimes n})] = E[I_n(f_n) \mid I_1(\mathbf{1}_A)] = \frac{\langle f_n, \mathbf{1}_A^{\otimes n}\rangle}{\langle \mathbf{1}_A, \mathbf{1}_A\rangle^n} I_n(\mathbf{1}_A^{\otimes n}),$$

which is a polynomial in $I_1(\mathbf{1}_A)$.

Proof. For any $k \geq 1$, by (4.4) we have

$$I_1(\mathbf{1}_A)^k = \sum_{l=0}^{k} \alpha_l I_l(\mathbf{1}_A^{\otimes l}), \tag{4.20}$$

where $\alpha_0, \ldots, \alpha_k \in (0, \infty)$, hence

$$\begin{aligned} E[I_n(f_n)(I_1(\mathbf{1}_A))^k] &= \alpha_k \mathbf{1}_{\{k \geq n\}} \langle f_n, \mathbf{1}_A^{\otimes n}\rangle \\ &= \frac{\langle f_n, \mathbf{1}_A^{\otimes n}\rangle}{\langle \mathbf{1}_A, \mathbf{1}_A\rangle^n} E[(I_1(\mathbf{1}_A))^k I_n(\mathbf{1}_A^{\otimes n})], \end{aligned}$$

which shows that

$$E[I_n(f_n) \mid I_1(\mathbf{1}_A)] = \frac{\langle f_n, \mathbf{1}_A^{\otimes n}\rangle}{\langle \mathbf{1}_A, \mathbf{1}_A\rangle^n} I_n(\mathbf{1}_A^{\otimes n}).$$

By (4.4) we also have

$$I_n(\mathbf{1}_A^{\otimes n}) = \sum_{l=0}^{n} \beta_l I_l(\mathbf{1}_A^{\otimes l}),$$

where $\beta_0, \ldots, \beta_n \in \mathbb{R}$, which implies $\sigma(I_n(\mathbf{1}_A^{\otimes n})) \subset \sigma(I_1(\mathbf{1}_A))$, and

$$\begin{aligned} E[I_n(f_n) \mid I_n(\mathbf{1}_A^{\otimes n})] &= E[E[I_n(f_n) \mid I_1(\mathbf{1}_A)] \mid I_n(\mathbf{1}_A^{\otimes n})] \\ &= \frac{\langle f_n, \mathbf{1}_A^{\otimes n}\rangle}{\langle \mathbf{1}_A, \mathbf{1}_A\rangle^n} E[E[I_n(\mathbf{1}_A^{\otimes n}) \mid I_1(\mathbf{1}_A)] \mid I_n(\mathbf{1}_A^{\otimes n})] \\ &= \frac{\langle f_n, \mathbf{1}_A^{\otimes n}\rangle}{\langle \mathbf{1}_A, \mathbf{1}_A\rangle^n} I_n(\mathbf{1}_A^{\otimes n}). \end{aligned}$$

□

When $n = 1$ the above result says that

$$E[I_1(f_1) \mid I_1(\mathbf{1}_A)] = \frac{\langle f_1, \mathbf{1}_A\rangle}{\langle \mathbf{1}_A, \mathbf{1}_A\rangle} I_1(\mathbf{1}_A), \tag{4.21}$$

which follows by a direct orthogonal projection argument. In particular when $X \simeq \mathcal{P}(s)$ and $Y \simeq \mathcal{P}(t-s)$ are independent centered Poisson random variables with means s and $t-s$,

$$X \simeq I_1(\mathbf{1}_{[0,s]}) \qquad \text{and} \qquad Y \simeq I_1(\mathbf{1}_{[s,t]})$$

taking $f_1 = \mathbf{1}_{[0,s]}$ and $A = [0,t]$, Relation (4.21) follows from the fact that $X + s$ has a binomial distribution with parameter s/t given $X + Y + t$.

Finally we note that, contrary to the Wiener case and Proposition 2 of [3], the conditional expectation of an odd order integral given an even order integral is not zero in general, indeed,

$$E[I_3(f_3)(I_2(g_2))^2]$$

does not vanish in general, from the multiplication formula (4.4).

References

[1] S. Chivoret and A. Amirdjanova. Product formula and independence criterion for multiple Huang-Cambanis integrals. *Statist. Probab. Lett.*, 76(18):2037–2042, 2006.

[2] O. Kallenberg. On an independence criterion for multiple Wiener integrals. *Ann. Probab.*, 19(2):483–485, 1991.

[3] D. Nualart, A.S. Üstünel, and M. Zakai. On the moments of a multiple Wiener-Itô integral and the space induced by the polynomials of the integral. *Stochastics*, 25(4):233–240, 1988.

[4] N. Privault. On the independence of multiple stochastic integrals with respect to a class of martingales. *C. R. Acad. Sci. Paris Sér. I Math.*, 323:515–520, 1996.

[5] N. Privault. Independence of a class of multiple stochastic integrals. In R. Dalang, M. Dozzi, and F. Russo, editors, *Seminar on Stochastic Analysis, Random Fields and Applications (Ascona, 1996)*, volume 45 of *Progress in Probability*, pages 249–259. Birkhäuser, Basel, 1999.

[6] N. Privault. *Stochastic Analysis in Discrete and Continuous Settings*, volume 1982 of *Lecture Notes in Mathematics*. Springer-Verlag, Berlin, 2009.

[7] J. Rosiński and G. Samorodnitsky. Product formula, tails and independence of multiple stable integrals. In *Advances in stochastic inequalities (Atlanta, GA, 1997)*, volume 234 of *Contemp. Math.*, pages 169–194. Amer. Math. Soc., Providence, RI, 1999.

[8] C. Tudor. A product formula for multiple Poisson-Itô integrals. *Rev. Roumaine Math. Pures Appl.*, 42(3-4):339–345, 1997.

[9] A.S. Üstünel and M. Zakai. On independence and conditioning on Wiener space. *Ann. Probab.*, 17(4):1441–1453, 1989.

[10] A.S. Üstünel and M. Zakai. On the structure on independence on Wiener space. *J. Funct. Anal.*, 90(1):113–137, 1990.

PART 2
Differential and Stochastic Games

Chapter 5

Strategies for differential games

Wendell H. Fleming[1] and Daniel Hernández-Hernández[2,*]

[1] *Division of Applied Mathematics,*
Brown University,
Box F, Providence, R.I. 02912, USA
whf@dam.brown.edu

[2] *Centro de Investigación en Matemáticas,*
Apartado Postal 402, Guanajuato, Gto. 36000, México
dher@cimat.mx

This chapter considers several classes of strategies for two-player, zero-sum differential games on a finite time interval. The goal is to describe classes of strategies for the maximizer and minimizer such that a saddle point property holds for the upper differential game. Of particular interest are strategies which are called approximately Markov.

Contents

*This research was partially supported by Concayt under Grant 61423.

5.1. Introduction

The theory of two-player, zero-sum differential games began in the early 1950s with Rufus Isaacs's pioneering work on pursuit-evasion games [13]. He introduced a methodology for finding solutions, based on the method of characteristics for what is now called the Isaacs PDE. However, Isaacs did not provide a mathematically rigorous definition for the value of a differential game. In Isaacs's work, a "minimax" condition is assumed for the Hamiltonian associated with the Isaacs PDE. Without this minimax assumption, upper and lower values for the differential game must be considered.

During the 1960's several different notions of differential game values were introduced, using discrete time approximations. See, for example, [4], [5], [12]. In their landmark 1972 paper [2], Elliott and Kalton introduced a definition of upper and lower values, which avoids time discretizations. Since then the Elliott–Kalton definition has become accepted as the "standard" one.

The purpose of our paper is to revisit the notion of strategy for differential games, within the Elliott–Kalton framework. By considering suitable subclasses of Elliott–Kalton strategies, we find various features which were not recognized before.

We begin in Section 5.2 by formulating games on a finite time interval $t \leq s \leq t_1$. The game dynamics are given by the ODE (5.1) and the game payoff is (5.2). We use the same notation and assumptions as in [10, Chap 11]. The upper Isaacs PDEs is (5.3) and the lower value is defined similarly. In Section 5.3 we review the definitions of Elliott–Kalton strategies, and of the upper and lower game values $V_+(t,x)$, $V_-(t,x)$ which are functions of an initial time t and the initial game state x. These value functions turn out to be the unique bounded, Lipschitz continuous viscosity solutions to the corresponding upper and lower Isaacs PDEs, with boundary condition at the final time t_1. Throughout the paper, we will state results for the upper differential game. The lower game is similar.

In the Elliott–Kalton definition of upper value, the roles of two players appear in different ways. The weaker player (the minimizer) chooses a control function u on the interval $[t, t_1]$ and the stronger player (the maximizer) chooses a strategy β. In Section 5.4, we take a different point of view in which both players choose strategies. The minimizer must choose a strategy α from some suitable subclass $\Gamma(t)$ of the class $\Gamma_{EK}(t)$ of all Elliott–Kalton strategies. The maximizer chooses a strategy β from a class $\Delta(t)$, which might be either the class $\Delta_{EK}(t)$ of all Elliott–Kalton strategies or else a suitable subset of $\Delta_{EK}(t)$. We define a saddle point property, which is described by (5.9) – (5.10) or equivalently by (5.11) – (5.12). In Section 5.4 we describe several possible choices for $\Gamma(t)$, in case $\Delta(t) = \Delta_{EK}(t)$.

In Section 5.5, two kinds of time discretizations are described. For the first kind, the minimizing control u_s is constant on each interval $[r_j, r_{j+1})$ of a partition π of $[t, t_1]$. For the second kind of time discretization, u_s need not be constant for $r_j \leq s \leq r_{j+1}$. It can be shown using viscosity solution arguments that, for both kinds of partitions, the solutions to the corresponding discrete time dynamic programming equations converge to $V_+(t, x)$ as the mesh of the partition π tends to 0. See Theorem 5.1.

In Section 5.6, we define approximately Markov strategies, using time discretizations. Using the results of Section 5.5, we find that there is a saddle point in terms of such strategies, see Theorem 5.2. At the end of Section 5.2, we indicate that conditions for the saddle point in terms of continuous time Markov control policies could be derived formally from the upper Isaacs PDE. However, this formalism requires smoothness of the upper value function V_+ and other conditions, which can rarely be verified in practice. Theorem 5.2 gives a kind of substitute, in terms of approximately Markov strategies.

In Section 5.7 we consider another class of strategies, which we call progressively measurable. This name is suggested by the corresponding notion of progressively measurable control process, which has an important role in stochastic control theory. Every approximately Markov strategy is shown to be progressively measurable. Hence, Theorem 5.2 also provides a saddle point in the class of progressively measurable strategies.

Section 5.8 considers a special class of differential games, in which the maximizing control z_s appears linearly in the game dynamics and quadratically in the payoff. Such games arise as limits of risk sensitive stochastic control problems, as the noise intensity parameter tends to zero. The limiting differential game formally resembles a stochastic control problem, in which the usual arithmetic operations of addition and multiplication are replaced by the corresponding max-plus operations. For that reason, such differential games have been called "max-plus stochastic control problems". See [6].

Section 5.9 presents briefly some extensions of these results to stochastic differential games, with reference to [8] for details.

5.2. Differential game formulation

Fix $t_1 > t \geq 0$ and consider a controlled dynamical system for which the state process x_s at time $s \in [t, t_1]$ evolves according to the ordinary differential equation

$$\frac{d}{ds} x_s = f(s, x_s, u_s, z_s) \tag{5.1}$$

with initial condition $x_t = x \in \mathbb{R}^d$. The controls adopted by players I and II are, respectively, the control processes u_s and z_s taking values in some compact and convex subsets U and Z of $\mathbb{R}^{m_1}$ and $\mathbb{R}^{m_2}$. In this zero sum differential game Player I is the minimizing controller, while Player II is the maximizing controller.

We make the following assumptions on the coefficients of the above ODE: The function $f : [0, t_1] \times \mathbb{R}^d \times U \times Z \to \mathbb{R}^d$ is bounded, continuous, and uniformly Lipschitz continuous with respect to t, x for $(u, z) \in U \times Z$.

Let $L : [0, t_1] \times \mathbb{R}^d \times U \times Z \to \mathbb{R}$ be bounded, continuous, and uniformly Lipschitz continuous with respect to t, x for $(u, z) \in U \times Z$ and $g : \mathbb{R}^d \to \mathbb{R}$ be bounded and Lipschitz continuous. The game payoff is defined by

$$P(t, x; u, z) = \int_t^{t_1} L(s, x_s, u_s, z_s) ds + g(x_{t_1}). \tag{5.2}$$

At an intuitive level, the *upper value of the game* shall describe a certain advantage for the maximizing player in the information available to both players. It can be thought as if at each time $s \in [t, t_1]$ the maximizing player had information of the decision taken by the other player u_s before he chooses z_s. This will be made precise in the next sections introducing the concept of strategies for the players. The *lower value of the game* can be defined similarly, establishing the advantage of the information to the minimizing player.

R. Isaacs [13] found an approach to relate these values of the game with a PDE. Formally, each of the above values of the game has associated an Isaacs PDE. For the *upper value* it has the form

$$v_t + H^+(Dv(t, x), x, t) = 0, \quad v(t_1, x) = g(x), \tag{5.3}$$

with Dv the gradient of $v(t, \cdot)$,

$$H^+(p, x, t) := \min_{u \in U} \max_{z \in Z} F(p, x, t; u, z), \tag{5.4}$$

$$F(p, x, t; u, z) := [f(t, x, u, z) \cdot p + L(t, x, u, z)]. \tag{5.5}$$

The *lower Isaacs PDE* is similar, replacing the second term on the left side of (5.3) by $H^-(p, x, t) := \max_{z \in Z} \min_{u \in U} F(p, x, t; u, z)$.

To illustrate a situation in which the maximizing player has advantage in the information, consider the class of Markov policies (or control policies). For the minimizing player a Markov policy is a function $\underline{u} : [0, t_1] \times \mathbb{R}^d \to U$ while a Markov policy for the maximizing player is a function $\underline{z} : [0, t_1] \times \mathbb{R}^d \times U \to Z$. When these functions are Lipschitz continuous there exists a solution x_s to (5.1) and the controls for each player are defined, respectively, as $u_s = \underline{u}(s, x_s)$ and $z_s = \underline{z}(s, x_s, u_s)$.

In particular, when there exists a classical solution $W : [t, t_1] \times \mathbb{R}^d \to \mathbb{R}$ to the upper Isaacs equation (5.3) it is possible to define a control policy for each player $\underline{u}^*, \underline{z}^*$ such that:

$$\underline{u}^*(t, x) \in \arg\min_{u \in U} \max_{z \in Z} F(DW(t, x), x, t; u, z),$$

and

$$\underline{z}^*(t, x, u) \in \arg\max_{z \in Z} F(DW(t, x), x, t; u, z),$$

which would satisfy a *saddle point property* under suitable conditions. See [10, Thm. 11.3.1]. In Section 5.6 it will be proved that under weaker assumptions it is possible to guarantee the existence of approximately Markov strategies satisfying the *saddle point property*; see (5.11)–(5.12). These results can be extended to the stochastic differential games [8, Thm. 3.4], and are mentioned in Section 5.9.

5.3. Elliott–Kalton strategies

We define an admissible control $u : [t, t_1] \to U$ (respectively $z : [t, t_1] \to Z$) for Player I (resp. Player II) on $[t, t_1]$ as a measurable function taking values in U (resp. Z). The set of admissible controls for Player I (resp. II) is denoted by $\mathcal{U}(t)$ (resp. $\mathcal{Z}(t)$). For every $u \in \mathcal{U}(t)$ and $z \in \mathcal{Z}(t)$ there is a solution to the ODE (5.1) with given initial data. We identify the controls which are equal almost everywhere in the interval $[t, t_1]$.

Definition 5.1. An Elliott–Kalton strategy for the maximizing player II beginning at time t is a mapping β from $\mathcal{U}(t)$ into $\mathcal{Z}(t)$ provided that for each $s \in [t, t_1]$, if $u = \tilde{u}$ almost everywhere in $[t, s]$, then $\beta(u) = \beta(\tilde{u})$ a.e. in $[t, s]$. The set of these strategies is denoted as $\Delta_{EK}(t)$. The set of Elliott–Kalton strategies $\alpha : \mathcal{Z}(t) \to \mathcal{U}(t)$ for the minimizing player I can be defined in a similar way, and is denoted by $\Gamma_{EK}(t)$.

Given the initial conditions (t, x), the upper and lower value of the differential game are defined, respectively, by

$$V_+(t,x) = \sup_{\beta \in \Delta_{EK}(t)} \inf_{u \in \mathcal{U}(t)} P(t,x;u,\beta(u)) \tag{5.6}$$

and

$$V_-(t,x) = \inf_{\alpha \in \Gamma_{EK}(t)} \sup_{z \in \mathcal{Z}(t)} P(t,x;\alpha(z),z). \tag{5.7}$$

Here P is defined as in (5.2), with x_s being the solution of (5.1) with u_s and $z_s = \beta(u)_s$. Next we mention some important properties of the upper value functions; similar results can also be established for the lower value.

The upper value function $V_+(t,x)$ is bounded and Lipschitz continuous in t, x; see [3, Thm. 3.2] . Moreover, Elliott–Kalton [2] proved that $V_+(t,x)$ satisfies the (upper) dynamic programming principle: Given $x \in \mathbb{R}^d$ and $0 \le t < \tau \le t_1$,

$$V_+(t,x) = \sup_{\beta \in \Delta_{EK}(t)} \inf_{u \in \mathcal{U}(t)} \left\{ \int_t^\tau L(s, x_s, u_s, \beta(u)_s)ds + V_+(\tau, x_\tau) \right\}. \tag{5.8}$$

Using the previous results $V_+(t,x)$ can be characterized as the unique bounded, Lipschitz viscosity solution of the Isaacs PDE (5.3). See [3], [10, Thm. 11.6.1].

5.4. Saddle point property

By definition the lower value $V_-(t,x)$ is obtained when the inf in (5.7) is taken among all strategies $\alpha \in \Gamma_{EK}(t)$. In this section, we will see that the upper value $V_+(t,x)$ is obtained when α is restricted to some suitable subclass of strategies $\Gamma(t) \subset \Gamma_{EK}(t)$:

$$V_+(t,x) = \inf_{\alpha \in \Gamma(t)} \sup_{z \in \mathcal{Z}(t)} P(t,x;\alpha(z),z). \tag{5.9}$$

Moreover, we can seek suitable classes $\Delta(t) \subset \Delta_{EK}(t)$ such that

$$V_+(t,x) = \sup_{\beta \in \Delta(t)} \inf_{u \in \mathcal{U}(t)} P(t,x;u,\beta(u)). \tag{5.10}$$

Note that (5.10) is just the definition (5.6) when $\Delta(t) = \Delta_{EK}(t)$.

We call (5.9)–(5.10) a *saddle point property*. Equivalently, the saddle point property can be restated as follows: for every $\varepsilon > 0$ there exist

$\alpha_\varepsilon \in \Gamma(t)$, $\beta_\varepsilon \in \Delta(t)$ such that

$$V_+(t,x) \le \sup_{z \in \mathcal{Z}(t)} P(t,x;\alpha_\varepsilon(z),z) \le V_+(t,x) + \varepsilon \tag{5.11}$$

$$V_+(t,x) - \varepsilon \le \inf_{u \in \mathcal{U}(t)} P(t,x;u,\beta_\varepsilon(u)) \le V_+(t,x). \tag{5.12}$$

In the rest of this section, we take $\Delta(t) = \Delta_{EK}(t)$. Then (5.10), or equivalently (5.12), is just the definition of Elliott–Kalton upper value. In this case, the objective is to describe choices for $\Gamma(t)$ which satisfy (5.9) or equivalently (5.11).

As in [10, Section 11.9] let us call $\alpha \in \Gamma_{EK}(t)$ a *strictly progressive strategy* if the following holds: for every strategy $\beta \in \Delta_{EK}(t)$ the equations

$$u = \alpha(z), \quad z = \beta(u) \tag{5.13}$$

have a solution $\hat{u} \in \mathcal{U}(t)$, $\hat{z} \in \mathcal{Z}(t)$. In particular, for each partition π of $[t, t_1]$, every $\alpha \in \Gamma_\pi(t)$ is strictly progressive where $\Gamma_\pi(t)$ is to be defined later in Section 5.5.

Let $\Gamma_1(t)$ denote the class of all strictly progressive strategies for the minimizer. By [10, Theorem 11.9.1] the saddle point property holds with $\Gamma(t) = \Gamma_1(t)$ and $\Delta(t) = \Delta_{EK}(t)$.

The requirement that α is strictly progressive can be weakened as follows. Let $\Gamma_2(t)$ be the class of all $\alpha \in \Gamma_{EK}(t)$ with the following property: For every $\delta > 0$ there exist $u_\delta \in \mathcal{U}(t)$ and $z_\delta \in \mathcal{Z}(t)$ such that

$$|\alpha(z_\delta)(s) - u_\delta(s)| < \delta, \quad \beta(u_\delta) = z_\delta, \quad \text{for } t \le s \le t_1. \tag{5.14}$$

Clearly $\Gamma_1(t) \subset \Gamma_2(t)$.

Proposition 5.1. The saddle point property holds if $\Gamma(t) = \Gamma_2(t)$, $\Delta(t) = \Delta_{EK}(t)$.

Proof. Let $W(t,x)$ denote the right side of (5.9) with $\Gamma(t) = \Gamma_2(t)$. Since $\Gamma_1(t) \subset \Gamma_2(t)$ and (5.9) holds with $\Gamma(t) = \Gamma_1(t)$, we have $V_+(t,x) \ge W(t,x)$. To obtain the opposite inequality, consider any $\alpha \in \Gamma_2(t)$. Given $\theta > 0$ there exists $\beta \in \Delta_{EK}(t)$ such that

$$P(t,x;u,\beta(u)) \ge V_+(t,x) - \theta$$

for all $u \in \mathcal{U}(t)$. Given $\delta > 0$, let u_δ, z_δ satisfy (5.14). It is easily shown that

$$|P(t,x;\alpha(z_\delta),z_\delta) - P(t,x;u_\delta,z_\delta)| \le M\delta,$$

for some constant M. Then,

$$\begin{aligned}\sup_{z\in\mathcal{Z}(t)} P(t,x;\alpha(z),z) &\geq P(t,x;\alpha(z_\delta),z_\delta)\\ &\geq P(t,x;u_\delta,\beta(u_\delta)) - M\delta\\ &\geq V_+(t,x) - \theta - M\delta.\end{aligned}$$

Since α, θ and δ are arbitrary, we obtain $W(t,x) \geq V_+(t,x)$. □

Example 5.1. Let $\bar{\Gamma}_1(t)$ be the closure of $\Gamma_1(t)$ in the uniform norm. Then $\bar{\Gamma}_1(t) \subset \Gamma_2(t)$ and hence the saddle point property holds with $\Gamma(t) = \bar{\Gamma}_1(t)$. In particular, if $\underline{u}$ is a Lipschitz continuous control policy for the minimizer, let $\alpha(z)_s = u_s$, where $u_s = \underline{u}(s,x_s)$ and x_s is the solution to (5.1) with $x_t = x$. Then $\alpha \in \bar{\Gamma}_1(t)$. See [10, Remark 11.9.1].

Example 5.2. In [9, Section 3] the following class $\Gamma_3(t)$ of strategies is considered. An Elliott–Kalton strategy α is in $\Gamma_3(t)$ provided:

(a) For every $z \in \mathcal{Z}(t)$, $\alpha(z)_s$ is a right continuous function on $[t,t_1]$ with left limits at each $s \in (t,t_1]$;
(b) If $z_r = \tilde{z}_r$ almost everywhere on $[t,s]$, then $\alpha(z)_r = \alpha(\tilde{z})_r$ for all $r \in [t,s]$.

By [9, Lemma 4.8], property (5.14) holds and hence $\alpha \in \Gamma_2(t)$.

In Section 5.6 we will find that there is a saddle point property in terms of approximately Markov controls. Then in Section 5.7 we will see that a saddle point property remains valid when the strategies which are allowed are progressively measurable, as defined there.

5.5. Time discretizations

Let $\pi = \{t = t_0 < r_1 < \cdots < r_m = t_1\}$ be a partition of $[t,t_1]$ and denote its mesh by $\|\pi\| = \max_{i=1,\ldots,m}(r_i - r_{i-1})$. In this section we consider two discrete time approximations $V_+^\pi(t,x)$ and $W_+^\pi(t,x)$ to the upper value function $V_+(t,x)$. Define the class of piecewise constant minimizing controls

$$\mathcal{U}_\pi(t) = \{u \in \mathcal{U}(t) \mid u_s = u_{r_j}, \text{ for } s \in [r_j, r_{j+1}),\ j = 0,\ldots,m-1\}.$$

Define the value function

$$V_+^\pi(t,x) = \max_{\Delta_{EK}(t)} \min_{\mathcal{U}_\pi(t)} P(t,x;u,\beta(u)). \tag{5.15}$$

It satisfies a discrete time version of the dynamic programming principle [10, Lemma 11.8.1], analogous to (5.8). For $j = 1, \ldots, m-1$,

$$V_+^\pi(t,x) = \sup_{\beta \in \Delta(t,r_j)} \inf_{u \in \mathcal{U}_\pi(t,r_j)} \left\{ V_+^\pi(r_j, x_{r_j}) + \int_t^{r_j} L(s, x_s, u_s, \beta(u)_s) ds \right\}, \tag{5.16}$$

with $V_+^\pi(t_1,x) = g(x)$. In the previous equation the sets $\Delta(t,r_j)$ and $\mathcal{U}_\pi(t,r_j)$ are defined as $\Delta(t)$ and $\mathcal{U}_\pi(t)$, respectively, when we take as the final condition $t_1 = r_j$. Also, in [10, Thm. 11.8.1] it is proved that $\lim_{\|\pi\| \to 0} V_+^\pi(t,x) = V_+(t,x)$. The proof uses in an essential way viscosity solution arguments.

On the same partition π of $[t,t_1]$, define the class of strategies $\alpha \in \Gamma_\pi(t) \subset \Gamma_{EK}(t)$ such that $\alpha(z) \in \mathcal{U}_\pi$ for all $z \in \mathcal{Z}(t)$. In Section 5.6 we will consider a subclass of strategies $\alpha \in \Gamma_\pi(t)$ which will be called *approximately Markov strategies.*

In the second discrete time approximation, the minimizer chooses any $u \in \mathcal{U}(t)$. However, the maximizer can choose only strategies β which have the following form. For $s \in [r_j, r_{j+1})$,

$$\beta(u)_s = \varphi_j(u^j, u_s),$$

where $u^j = u|_{[t,r_j)}$ denotes the past of u up to time r_j and $\varphi_j(u^j, \cdot) \in H$, with $H = \{h : U \to Z \mid h \text{ is Borel measurable}\}$.

Define $W_+^\pi(t_j,x)$ by backward induction on j, as follows

$$W_+^\pi(t_1,x) = g(x) \tag{5.17}$$

$$W_+^\pi(r_j,x) = \sup_{h \in H} \inf_{u \in \mathcal{U}(r_j,r_{j+1})} \left\{ W_+^\pi(r_{j+1}, x_{r_{j+1}}) + \int_{r_j}^{r_{j+1}} L(s, x_s, u_s, h(u_s)) ds \right\}.$$

This is the discrete time dynamic programming equation, similar to (5.16), for choices of controls and strategies just indicated.

Lemma 5.1. *W_+^π is bounded and uniformly Lipschitz continuous.*

This result can be proved by verifying that W_+^π is the value function for the discrete time dynamic game described above, and then using standard arguments applied when we have a representation formula. A different way to verify this lemma is by adapting the arguments given in [8, Lemma 5.1, Corollary 5.3] to the deterministic case.

Theorem 5.1. $\lim_{\|\pi\| \to 0} W^\pi(t,x) = V_+(t,x)$ *uniformly on compact sets.*

This result is the deterministic version of [8, Thm. 5.5] and the same ideas can be adapted, since $V_+(t,x)$ is the unique bounded Lipschitz continuous viscosity solution of the upper Isaacs equation (5.3).

5.6. Approximately Markov strategies

At the end of Section 5.2 we mentioned a pair of control policies $(\underline{u}^*, \underline{z}^*)$ for both players, which would be the candidates to satisfy the saddle point property, motivated by a verification result, which would be satisfied under strong hypotheses on the existence of solutions of the upper Isaacs PDE. In this section we review the existence of approximating control policies under weak assumptions, described in the following.

Definition 5.2.

(a) $\alpha \in \Gamma_{EK}(t)$ is an *approximately Markov strategy* for the minimizing player if there exists a partition $\pi = \{t = r_0 < r_1 < \cdots < r_m = t_1\}$ of $[t, t_1]$ and Borel measurable functions $\zeta_j : \mathbb{R}^d \to U$, for $j = 0, 1, \ldots, m-1$, such that

$$\alpha(z)_s = \zeta_j(x_{r_j}), \quad r_j \leq s < r_{j+1}.$$

(b) $\beta \in \Delta_{EK}(t)$ is an *approximately Markov strategy* for the maximizing player if there exists such a partition π and Borel measurable functions $\eta_j : \mathbb{R}^d \times U \to Z$ such that

$$\beta(u)_s = \eta_j(x_{r_j}, u_s), \quad r_j \leq s < r_{j+1}.$$

The next result establishes a connection between these sets of control policies and the saddle point property.

Theorem 5.2. *The saddle point property (5.11)-(5.12) holds with* $\Gamma(t)$ *and* $\Delta(t)$ *the sets of approximately Markov control policies.*

Proof. (Sketch of proof.) The arguments are adapted to the deterministic case from the proof of [8, Theorem 3.4]. We need to show that α_ε, β_ε in (5.11)-(5.12) can be chosen as approximately Markov control strategies. The strategy β_ε is constructed successively on each subinterval $[r_j, r_{j+1})$ of the partition π using the discrete time approximation W_+^π defined in (5.17), and then we use Theorem 5.3, which states that $W_+^\pi(t,x)$ converges uniformly to $V_+(t,x)$. The form of β_ε in definition 6.1(b) is obtained using a construction in the proof of [8, Lemma 5.2].

The strategy α_ε is obtained from the discrete time approximation (5.15) and from [11, Prop. 2.5], which gives the existence of a Borel measurable function ζ_j of the form required. Then, the strategy α_ε is defined successively on each subinterval of the partition, and finally the result is obtained from the convergence of V_+^π to V_+. □

5.7. Progressively measurable strategies

In this section we introduce a different class of strategies, which are called progressively measurable. The name is suggested by the concept of progressively measurable stochastic processes, which has an important role in stochastic control theory. See [1], [10]. By Lemma 7.2, every approximately Markov strategy is progressively. Theorem 6.2 then implies that the saddle point property holds when the additional requirement that strategies are progressively measurable is added. We recall that U and Z are convex, compact subsets of $\mathbb{R}^{m_1}$ and $\mathbb{R}^{m_2}$ respectively.

Progressively measurable functions. Let Ω be a set and $\{\mathcal{F}_s\}$ an increasing family of subsets of Ω, with $t \leq s \leq t_1$. Let B_s^0 be the set of Borel subsets of $[t,s]$, and $\mathbb{B}(\mathbb{R}^l)$ be the set of Borel subset of $\mathbb{R}^l$. A function $\theta : [t,t_1] \times \Omega \to \mathbb{R}^l$ is called $\{\mathcal{F}_s\}$–progressively measurable if the map $(r,\omega) \to \theta(r,\omega)$ is $B_s^0 \times \mathcal{F}_s$, $\mathbb{B}(\mathbb{R}^l)$ measurable for $t \leq r \leq s$ and $s \in [t,t_1]$.

Remark 5.1. For fixed ω and $t \leq r \leq s$, $\theta(r,\omega)$ is B_s^0–measurable, and for fixed s, $\theta(s,\omega)$ is $\mathcal{F}_s$–measurable.

Remark 5.2. If $\theta(\cdot,\omega)$ is continuous on $[t,t_1]$ for each ω and $\theta(s,\cdot)$ is $\{\mathcal{F}_s\}$–measurable for each s, then θ is $\mathcal{F}_s$–progressively measurable. See for example [16].

Progressively measurable strategies for the maximizer. For $u \in \mathcal{U}(t)$ we define its time integral Iu by

$$(Iu)_s = \int_t^s u_r dr, \quad t \leq s \leq t_1.$$

Note that $Iu^{(1)} = Iu^{(2)}$ if and only if $u^{(1)} = u^{(2)}$ for almost all $s \in [t,t_1]$. Moreover, $(Iu)_t = 0$, Iu is Lipschitz and $\frac{d}{ds}(Iu)_s \in U$ for almost all $s \in [t,t_1]$.

Let $\Omega_1 = C_0([t,t_1];\mathbb{R}^{m_1})$, the space of continuous $\mathbb{R}^{m_1}$–valued functions η on $[t,t_1]$ such that $\eta_t = 0$. Let $\mathbb{B}_{1s}$ be the least σ–algebra containing all sets $\{\eta_{r_{j+1}} - \eta_{r_j} \in A_j,\ j = 0,1,\ldots,k-1,\ A_j \in \mathbb{B}(\mathbb{R}^{m_1})\}$, where $\eta \in \Omega_1$.

Definition 5.3. A strategy β for the maximizer is called progressively measurable if $\beta(u)_s = \theta_1(s, Iu),\ t \le s \le t_1$, where θ_1 is $\{\mathbb{B}_{1s}\}$–progressively measurable.

Note that every progressively measurable β is an Elliott–Kalton strategy. Similarly, we let $\Omega_2 = C_0([t,t_1];\mathbb{R}^{m_2})$ and define $\mathbb{B}_{2s}$ in the same way $\mathbb{B}_{1s}$ was defined above. A strategy α for the minimizer is called *progressively* measurable if $\alpha(z)_s = \theta_2(s, Iz),\ t \le s \le t_1$, where θ_2 is $\{\mathbb{B}_{2s}\}$–progressively measurable.
Progressively measurable versions of $u \in \mathcal{U}(t)$ and $z \in \mathcal{Z}(t)$. Functions $u = (u^{(1)}, \ldots, u^{(m_1)}) \in \mathcal{U}(t)$ are determined only up to subsets of $[t, t_1]$ of Lebesgue measure 0. We define a progressively measurable version $\lambda_1(u)$ as follows. Let $\eta_s^{(i)} = (Iu^{(i)})_s$, for $i = 1, \ldots, m_1$. Then $\eta_s^{(i)}$ is $\{\mathbb{B}_{1s}\}$–measurable and continuous in s, and hence $\{\mathbb{B}_{1s}\}$–progressively measurable from Remark 5.2. For $s \in [t, t_1]$ let

$$\bar{u}_s^{(i)} = \limsup_{n\to\infty} n[\eta_s^{(i)} - \eta_{s-\frac{1}{n}}^{(i)}]$$

and $\bar{u}_s = (\bar{u}_s^{(1)}, \ldots, \bar{u}_s^{(m_1)})$. Note that for almost all $s \in [t, t_1]$, $\bar{u}_s = u_s = \frac{d\eta}{ds}(s)$. Let $\pi(u) = u$, if $u \in U$, and $\pi(u)$ the point of U nearest u if $u \in \mathbb{R}^{m_1} \setminus U$. Then π is Lipschitz continuous, with Lipschitz constant 1. Let $\lambda_1(u)_s = \pi\bar{u}_s$. Then λ_1 is $\{\mathbb{B}_{1s}\}$–progressively measurable and $\lambda_1(u)_s = u_s$ for almost all $s \in [t, t_1]$, for each fixed $u \in \mathcal{U}(t)$.

The progressively measurable version $\lambda_2(z)$ of $z \in \mathcal{Z}(t)$ is defined in the same way. Let $\Gamma_{PM}(t)$, $\Delta_{PM}(t)$ denote the classes of all progressively measurable strategies α, β respectively.

Lemma 5.2. *Given $\beta \in \Delta_{PM}(t)$, $u \in \mathcal{U}(t)$ let x_s be the solution to*

$$\frac{dx_s}{ds} = F(s, x_s, \lambda_1(u)_s, \beta(u)_s), \quad t \le s \le t_1,$$

with $x_t = x$. Then x_s is $\mathbb{B}_{1s}$–measurable.

This can be proved by standard arguments, using Picard iterations. We omit the details. By Remark 7.2, x_s is $\{\mathbb{B}_{1s}\}$–progressively measurable. Similarly, given $\alpha \in \Gamma_{PM}(t)$, $z \in \mathcal{Z}(t)$, the solution to

$$\frac{dx_s}{ds} = F(s, x_s, \alpha(z)_s, \lambda_2(z)_s), \quad t \le s \le t_1,$$

with $x_t = x$ is $\{\mathbb{B}_{2s}\}$–progressively measurable.

Lemma 5.3. *Every approximately Markov strategy is progressively measurable.*

This lemma can be proved by straightforward arguments, using the definition of approximately Markov strategy and Lemma 5.2.

5.8. Max-plus stochastic control

In this section we mention an interesting class of differential games, which arise as limits of risk sensitive stochastic control problems as a noise intensity parameter tends to 0. See [10, Section 11.7], also [14], [15]. The game dynamics have the form

$$\frac{dx_s}{ds} = f(s, x_s, u_s) + \sigma(s, x_s, u_s) z_s \tag{5.18}$$

and the game payoff is

$$P(t, x; u, z) = \int_t^{t_1} \left[l(s, x_s, u_s) - \frac{1}{2}|z_s|^2 \right] ds. \tag{5.19}$$

As before, $u_s \in U$ with U compact. However, $z_s \in Z$, with $Z = \mathbb{R}^{m_2}$ which is not compact as required in previous sections of the paper. Later we will indicate how this technical difficulty can be avoided, by using a "cutoff" in which $\mathbb{R}^{m_2}$ is replaced by $Z_R = \{z : |z| \leq R\}$ for suitable R.

If $\sigma = \sigma(s, x)$, then the Isaacs minimax condition holds. However, this is not always true when $\sigma = \sigma(s, x, u)$ depends on the control variable u. The risk sensitive limit analysis leads to the upper game value $V_+(t, x)$, which may differ from the lower value $V_-(t, x)$.

This differential game can be interpreted as a "max-plus" stochastic control problem, as described in [7] [10, p.390]. Let $\xi_s = (Iz)_s$. Then (5.18) can be rewritten in the form

$$dx_s = f(s, x_s, u_s)ds + \sigma(s, x_s, u_s)d\xi_s. \tag{5.20}$$

This resembles the stochastic differential equation for a stochastic control problem in which ξ_s is replaced by a brownian motion process w_s [10, Chap. 4]. As in [7] [6] we call (5.20) a "max-plus stochastic differential equation". In the max-plus setting, ordinary expectations are replaced by the corresponding "max-plus" expectations. In particular, for the differential game described above, given a strategy α for the minimizer,

$$\sup_{z \in \mathcal{Z}(t)} P(t, x; \alpha(z), z) = \mathbb{E}^+ \left[\int_t^{t_1} l(s, x_s, u_s) ds \right], \tag{5.21}$$

where $\mathbb{E}^+$ denotes the max-plus expectation and $u_s = \alpha(z)_s$. Then (5.9) becomes

$$V_+(t,x) = \inf_{\alpha\in\Gamma(t)} \mathbb{E}^+ \left[\int_t^{t_1} l(s,x_s,u_s)ds\right] \tag{5.22}$$

provided that the class $\Gamma(t)$ of strategies is suitably chosen as discussed in Section 5.4. Thus, $V_+(t,x)$ is the max-plus counterpart of the value function for a controlled Markov diffusion process of the kind considered in [10, Chap. 4].

Cutoffs. As in [10, Sec. 11.7] assume that f, σ and l have continuous first order partial derivatives, with f, σ, l and all such derivatives bounded. Let $V_{+R}(t,x)$ denote the upper value of the differential game with $z_s \in Z_R$, where $Z_R = \{z : |z| \le R\}$. Then $V_+(t,x) = V_{+R}(t,x)$ for all sufficiently large R [10, Thm. 11.7.1]. This result reduces the game to one with compact control spaces U and Z_R, as required in this paper.

5.9. Stochastic differential games

In this section we present briefly some possible extensions to the stochastic setting of the results mentioned in Sections 5.4 and 5.6. Let $(\Omega_t, \mathcal{F}, \mathbb{P}_t)$ be the canonical probability space, where $\Omega_t = \{\omega \in C([t,t_1];\mathbb{R}^k) \mid \omega_t = 0\}$, the σ-algebra $\mathcal{F}$ is the family of Borel sets completed with respect to the Wiener measure $\mathbb{P}_t$ and the underlying filtration $\mathcal{F}_s$, $t \le s \le t_1$, is generated by the Brownian paths ω.

We consider a SDE

$$dX_s = f(s,X_s,u_s,z_s)ds + \sigma(s,X_s,u_s,z_s)dW_s, \quad 0 \le t \le s \le t_1, \tag{5.23}$$

with initial condition $X_t = x \in \mathbb{R}^d$. The function $f : [0,t_1] \times \mathbb{R}^d \times U \times Z \to \mathbb{R}^d$ and the $d \times k$ matrix $\sigma : [0,t_1] \times \mathbb{R}^d \times U \times Z \to \mathcal{M}^{d,k}$ are bounded, continuous, and Lipschitz continuous with respect to t, x uniformly for $(u,z) \in U \times Z$.

The associated payoff in this case is

$$J(t,x;u,z) = \mathbb{E}\left\{\int_t^{t_1} L(s,X_s,u_s,z_s)ds + g(X_{t_1})\right\}, \tag{5.24}$$

with L and g satisfying the same assumptions as in section 5.2. As before, Player I controlling u_s is trying to minimize J, while Player II is trying to maximize J controlling z_s. The admissible control u_s (resp. z_s) for Player I (resp. Player II) is a $\mathcal{F}_s$–progressively measurable process taking values in U (resp. Z). The sets of admissible controls are defined as $\mathcal{U}(t)$ and

$\mathcal{Z}(t)$, abusing of the previous notation, and the definition of Elliott–Kalton strategies, given in Definition 5.1, can be modified accordingly. See [8].

Denoting by $\Delta_{EK}(t)$ the set of Elliott–Kalton strategies for the minimizing player, the upper value of the SDG is defined as in (5.6) as

$$V_+(t,x) = \sup_{\beta\in\Delta_{EK}(t)} \inf_{u\in\mathcal{U}(t)} J(t,x;u,\beta(u)). \tag{5.25}$$

Let $\Gamma_1(t)$ denote the class of strictly progressive strategies for the minimizer, where strictly progressive is defined in the same way as (5.9) for deterministic games. In [8] the saddle point property with $\Delta(t) = \Delta_{EK}(t)$, $\Gamma(t) = \Gamma_1(t)$ was defined as follows.

Definition 5.4. The saddle point property for the upper game is said to hold if there exists a real valued function $V(t,x)$ such that for each $\varepsilon > 0$ and $(t,x) \in [0,t_1] \times \mathbb{R}^d$ there exist $\alpha_\varepsilon \in \Gamma_1(t)$, $\beta_\varepsilon \in \Delta_{EK}(t)$ such that

(i)

$$V(t,x) - \varepsilon \le \inf_{u\in\mathcal{U}(t)} J(t,x;u,\beta_\varepsilon(u)) \tag{5.26}$$

(ii)

$$\sup_{z\in\mathcal{Z}(t)} J(t,x;\alpha_\varepsilon(z),z) \le V(t,x) + \varepsilon.$$

The saddle point property holds with $V = V_+ = v$, where $v(t,x)$ is the unique bounded, uniformly continuous viscosity solution to the upper Isaacs PDE

$$v_t + H^+(D^2v, Dv, x, t) = 0, \quad v(t_1,x) = g(x),$$

with Dv, D^2v the gradient and matrix of second order partial derivatives of $v(t,\cdot)$,

$$H^+(A,p,x,t) := \min_{u\in U}\max_{z\in Z}\Big[\mathrm{trace}\Big(\frac{1}{2}a(t,x,u,z)A\Big) + f(t,x,u,z)\cdot p + L(t,x,u,z)\Big]$$

and $a = \sigma\cdot\sigma'$.

Moreover, α_ε, β_ε can be chosen as approximately Markov control strategies. See [8, Thm. 3.4]

Since $V = V_+$, $\Gamma(t) = \Gamma_1(t)$ and $\Delta(t) = \Delta_{EK}(t)$, the saddle point property in Definition 5.4 is equivalent to the form (5.11)–(5.12) in Section 5.4. By definition of V_+, the right side of (5.26)(i) is no more than $V_+(t,x)$.

It is easy to see from the definition of strictly progressive strategy that the left side of (5.26)(ii) is no less than $V_+(t,x)$. See [10, p. 393].

References

[1] Elliott, R.J., *Stochastic Calculus and Applications*, Springer-Verlag, New York, 1982.

[2] Elliott, R.J., Kalton, N.J., *The existence of value in differential games*, Memoirs Amer. Math. Soc. No. 126, 1972.

[3] Evans L. C., Souganidis P.E., *Differential games and representation formulas for solutions of Hamilton-Jacobi-Isaacs equations.* Indiana Univ. Math. J. 33, 773-797, 1984.

[4] Fleming W. H., *The convergence problem for differential games II*, Advances in Game Theory, Ann. of Math. Studies, 52, Princeton, New Jersey, Princeton Univ. Press, 1964.

[5] Fleming W. H., *The Cauchy problem for degenerate parabolic equations*, J. Math. Mech. 13, 987-1008, 1964.

[6] Fleming W.H., *Max-plus stochastic processes*, Applied Math. Optim. 49, 159-181, 2004.

[7] Fleming W.H., *Max-plus stochastic control*, in Stochastic Theory and Control (ed B. Pasik-Duncan) Lecture Notes in Control Info. Sci. 280, Springer-Verlag, New York, 111-119, 2002.

[8] Fleming W.H., Hernández-Hernández D., *On the value of stochastic differential games*, Comm. Stochastic Analysis. **5**, 341-351, 2011.

[9] Fleming W.H., Kaise H. Sheu S.-J., *Max-plus stochastic control and risk sensitivity.* Appl. Math. Optim. **62**, 81-144, 2010.

[10] Fleming, W. H., Soner H.M. *Controlled Markov Processes and Viscosity Solutions.* Springer-Verlag, New York, Second Edition, 2006.

[11] Fleming W. H., Souganidis P. E.: *On the existence of value functions of two player, zero sum stochastic differential games.* Indiana Univ. Math. J. **38**, 293–314, 1989.

[12] Friedman, A., *Differential Games*, Wiley, 1971.

[13] Isaacs, R., *Differential Games*, Wiley, New York, 1965 (Kruger, 1975).

[14] James, M.R., *Asymptotic analysis of nonlinear stochastic risk sensitive control and differential games*, Math. Control Signals and Syst. 5, 401-417, 1992.

[15] Kaise H., Nagai H., *Bellman-Isaacs equations of risk sensitive control with large parameters and their singular limits.* Asymptotic Analysis 16, 347-362, 1998.

[16] Karatzas I., Shreve, S., *Brownian Motion and Stochastic Calculus*, Springer-Verlag, New York, 1988.

Chapter 6

BSDE approach to non-zero-sum stochastic differential games of control and stopping

Ioannis Karatzas[1,*] and Qinghua Li[2]

[1] *INTECH Investment Management,*
One Palmer Square, Suite 441, Princeton, NJ 08542, USA
ik@enhanced.com

[2] *Statistics Department, Columbia University,*
1255 Amsterdam Avenue, 1009 SSW, New York, NY 10027, USA
qinghua@stat.columbia.edu

Dedicated to Professor Robert J. Elliott
on the occasion of his 70th Birthday

This chapter studies two non-zero-sum stochastic differential games of control and stopping. One game has interaction in the players' stopping rules, whereas the other does not. Solutions to backward stochastic differential equations (BSDEs) will be shown to provide the value processes of the first game. A multi-dimensional BSDE with reflecting barrier is studied in two cases for its solution: existence and uniqueness with Lipschitz growth, and existence in a Markovian system with linear growth rate. The extension to linear/quadratic growth rates of the equation allows the controls to observe the instantaneous volatilities of the value processes in the Markovian case.

Contents

---

*This author is on partial leave from the Department of Mathematics, Columbia University, New York, NY 10027 (e-mail *ik@math.columbia.edu*), where his research is supported by the National Science Foundation under grant NSF-DMS-09-05754.

6.1. Introduction

6.1.1. *Bibliographic notes*

We study non-zero-sum stochastic differential games, in which each of several players chooses an optimal strategy to maximize his own reward — usually the sum of a running reward and a terminal reward. The rewards of the players do not have to add up to a constant (say, zero); that is, the game is not necessarily zero-sum. The objective is to find an equilibrium point; namely, if every player's strategy is his best response to the other players' strategies, then the set of strategies is called an equilibrium point of the game, conventionally a Nash equilibrium. It is an equilibrium in the sense that no player will profit by unilaterally changing his strategy, when all the other players' strategies remain the same. The terminology "Nash equilibrium" is in deference to J. Nash's formulation of this notion of equilibrium in 1949. This notion of equilibrium for a non-zero-sum game generalizes the Von Neumann-Morgenstern notion of saddle point for a zero-sum game.

In the literature on both zero-sum and non-zero-sum stochastic differential games of control, the existence and the choice of optimal controls are shown to be equivalent to the existence of controls that satisfy the *Isaacs condition,* which is the attainability of the maximum or maxima of the associated Hamiltonians. In contrast to the counterpart optimization problem studied by Beneš in [1], [2] and by Davis (1973) in [10], the Isaacs (or Nash) condition for N-player non-zero-sum games does not typically hold. The existence, or non-existence, of an optimal control set that maximizes the Hamiltonians serves as an easy-to-verify local condition equivalent to the existence, or non-existence, of equilibrium controls.

With Markovian rewards, which are functions of the current value of an underling diffusion state process, partial differential equations are a handy tool. Over the past thirty years, Bensoussan, Frehse and Friedman built a regularity theory of PDE's to study stochastic differential games. Among their extensive works, Bensoussan and Friedman (1977) considered in [5] games of optimal stopping. The existence of optimal stopping times for such games is reduced to the study of regular solutions to quasi-variational inequalities, assuming continuous and bounded running rewards and terminal rewards. Bensoussan and Frehse (2000) in [3] solved a non-zero-sum game of optimal controls, which is terminated when the state process exits a bounded domain. Their running rewards are quadratic forms of the controls. The monograph of Fleming and Soner (1993) [17] gives a full account of controlled Markov diffusions.

As a tool for stochastic control theory, backward stochastic differential equations (BSDEs for short) were first proposed by Bismut in the 1970's. The connection between the two subjects can be viewed as a stochastic version of the verification theorems for the control of Markov diffusions. Pardoux and Peng (1990) proved in [34] existence and uniqueness of the solution to a BSDE with uniformly Lipschitz growth. Considerable attention has been devoted to studying the association between BSDEs and stochastic differential games. Cvitanić and Karatzas (1996) proved in [9] existence and uniqueness of the solution to the equation with double reflecting barriers, and associated their BSDE to a zero-sum Dynkin game. Their work generalized El Karoui, Kapoudjian, Pardoux, Peng and Quenez (1997) [13] on one-dimensional BSDE with one reflecting boundary, which captures early stopping features as that of American options. Hamadène and Lepeltier (2000) [22] and Hamadène (2006) [23] added controls to the Dynkin game studied by Cvitanić and Karatzas (1996) [9], the tool still being BSDE with double reflecting barriers. Markovian rewards of games correspond to the equations in the Markovian framework. Hamadène studied in [20] and [21] Nash equilibrium control with forward-backward SDE. In Hamadène, Lepeltier and Peng (1997) [19], the growth rates of their forward-backward SDE are linear in the value process and the volatility process, and polynomial in the state process. Their state process is a diffusion satisfying an "$\mathbb{L}^2$-dominance" condition. These three authors solve a non-zero-sum game without stopping, based on existence result of the multi-dimensional BSDE.

The martingale principle states that the value process of a control problem is a supermartingale, becoming a martingale if and only if an optimal control is applied. It facilitates the study of zero-sum and non-zero-sum games, and is particularly useful if the rewards depend on the path of the state process. This method is surveyed by Davis (1979) [11]. Elliott (1976) [15] shows that the Isaacs condition implies the existence of value and saddle strategies for a zero-sum game of control. When there are terminal rewards only, Lepeltier and Etourneau (1987) in [32] used martingale techniques to provide sufficient conditions for the existence of optimal stopping times on processes that need not be Markovian; their general theory requires some order assumption and supermartingale assumptions on the terminal reward. Karatzas and Zamfirescu (2008) in [31] took the martingale approach to characterize, then construct, saddle points for zero-sum games of control and stopping. They also characterized the value processes by the semimartingale decompositions and proved a stochastic maximum principle for a continuous, bounded running reward that can be a functional of the diffusion state process.

Zero-sum games of stopping (games of timing, Dynkin games) are connected to singular controls, in the sense that for convex cost functions, the value functions of the former games are derivatives of the value functions of the latter. This connection was first observed by Taksar (1985) [35], followed by Fukushima and Taksar (2002) [18] in a Markovian setting via solving free-boundary problems, and by Karatzas and Wang (2001) [30] in a non-Markovian setting based on weak compactness arguments.

6.1.2. *This paper*

This paper considers a non-zero-sum game with features of both stochastic control and optimal stopping, for a process of diffusion type, via the backward SDE approach. Running rewards, terminal rewards and early exercise rewards are all included. The running rewards can be functionals of the diffusion state process. Since the Nash equilibrium of an N-player non-zero-sum game is technically not more difficult than a two-player non-zero-sum game, only notationally more tedious, the number of players is assumed to be two, for concreteness.

Section 6.2 solves for the existence of Nash equilibrium for the stochastic games of control and stopping. The controls enter the drift of the underlying state process. Each player controls and stops the reward streams. In Game 6.1, a player's choice of stopping time terminates his own reward stream only. The value processes of both players are part of the solution to a multi-dimension BSDE with reflecting barrier. The instantaneous volatilities of the two players' value processes are explicitly expressed in the solution. Existence of the solution to general forms of the multi-dimensional BSDE with reflecting barrier will be proven in section 6.3 and section 6.4. Then, in the Markovian framework, the instantaneous volatilities can enter the controls as arguments, in which case the game is said to *observe volatilities* in addition to the other two arguments, namely time and the state-process.

In Game 6.2, each player can terminate the game not only for himself but also for the other player(s). We shall establish the existence of an equilibrium in a weaker sense than that for Game 6.1. It seems beyond the reach of our current ability to develop a more general theory with the present methods.

Section 6.3 proves existence and uniqueness of the solution to a multi-dimensional BSDE with reflecting barrier, a general form of the one that accompanies Game 6.1. Section 6.4 discusses extension of the existence of solutions to equations of ultra-Lipschitz growth.

The BSDE approach here proposes a multi-dimensional BSDE whose value processes in the solution provide the value processes of the non-zero-sum games. A general existence result of solutions to multi-dimensional BSDE with reflecting barrier still needs to be developed. As is proven in Hu and Peng (2006) [25], in several dimensions, the comparison theorem holds only under very restrictive conditions. Cohen, Elliott and Pearce (2010) [8] recently give a general component-wise condition for comparison of multi-dimensional BSDEs. Without invoking comparison results or penalization methods, we use Picard iterations to show the existence of solutions to equations with Lipschitz drivers. When the growth condition is ultra-Lipschitz, convergence arguments of the usual Picard type iteration cannot proceed either. In a Markovian framework, this paper proves the Markovian structure of solutions to multi-dimensional reflected BSDEs with Lipschitz growth, and uses this Markovian structure as a starting point to extend existence results to equations with growth rates which are linear in the value and volatility processes, and polynomial in the state process. Once again, the method does not rely on comparison theorems. The linear/quadratic growth rates of the equations allow the controls to observe the instantaneous volatilities of the players' value processes.

Our multi-dimensional BSDEs with reflection differ from the multi-dimensional BSDEs with oblique reflection in Hu and Tang (2010) [26], in Hamadène and Zhang (2010) [24] and in Chassagneux, Elie and Kharroubi (2010) [6], which are associated with optimal switching problems. Our equations are generalizations of the doubly reflected BSDEs in Cvitanić and Karatzas (1996) [9], which are associated with Dynkin games. The differences of these equations are determined by the essential differences between control problems, zero-sum games and non-zero-sum games.

We have tried to separate the game aspect and the BSDE aspect in the presentation of this paper, so we organize it in such a way that readers interested in stochastic differential games can read section 6.2 only, ignoring

BSDE technicalities; whereas BSDE connoisseurs can examine section 6.3 and section 6.4.

With BSDEs as one of its approaches, the theory of Stochastic Control has extensive applications in the rich fields of Mathematical Finance and Economics, like pricing and hedging of contingent claims, portfolio optimization, risk management, algorithmic trading, utility maximization, and so on. Many of these subjects are discussed in the survey paper El Karoui, Peng and Quenez (1997) [14]. The literature on Mathematical Finance tends to focus on the optimal behavior of one agent, or at most on zero-sum games between a buyer and a seller. Real-world financial phenomena, on the other hand, result inherently from the interaction of several agents, who try to optimize their own profits and whose actions form a non-zero-sum game. The present work can be seen as a contribution to the mathematical foundations of the study of the interaction among several such agents in financial markets.

6.2. The games of control and stopping

In the non-zero sum games of control and stopping to be discussed in this paper, each player receives a reward. Based on their up-to-date information, the two players I and II, respectively, first choose their controls u and v, then the times τ and ρ to stop their own reward streams. The controls u and v are two processes that enter the dynamics of the underlying state process for the rewards. The optimality criterion for our non-zero-sum games is that of a Nash equilibrium, in which each player's expected reward is maximized when the other player maximizes his. In taking conditional expectations of the rewards, the change-of-measure setup to be formulated fixes one single Brownian filtration and one single state process for all controls u and v. Hence when optimizing the expected rewards over the control sets, there is no need to keep in mind the filtration or the state process.

Let us set up the rigorous model. We start with a d-dimensional Brownian motion $B(\cdot)$ with respect to its generated filtration $\{\mathscr{F}_t\}_{0\le t\le T}$ on the canonical probability space $(\Omega, \mathscr{F}, \mathbb{P})$, in which $\Omega = C^d[0,T]$ is the set of all continuous d-dimensional functions on a finite deterministic time horizon $[0,T]$, $\mathscr{F} = \mathscr{B}\left(C^d[0,T]\right)$ is the Borel sigma algebra, and $\mathbb{P}$ is the Wiener measure.

For every $t \in [0,T]$, define a mapping $\phi_t : C[0,T] \to [0,T]$ by $\phi_t(y)(s) = y(s \wedge t)$, which truncates the function y in the set $C[0,T]$ of all one-dimensional continuous functions on $[0,T]$. For any $y^0 \in C[0,T]$, the pre-image $\phi_t^{-1}(y^0)$ collects all functions in $C[0,T]$ which are identical to y^0 up to time t. A stopping rule is a mapping $\tau : C[0,T] \to [0,T]$, such that

$$\{y \in C[0,T] : \tau(y) \le t\} \in \phi_t^{-1}\left(\mathscr{B}(C[0,T])\right).$$

The set of all stopping rules ranging between t_1 and t_2 is denoted by $\mathscr{S}(t_1, t_2)$.

In the **path-dependent** case, the state process $X(\cdot)$ solves the stochastic functional equation

$$X(t) = X(0) + \int_0^t \sigma(s, X)dB_s,\ 0 \le t \le T, \tag{6.1}$$

where the volatility matrix $\sigma : [0,T] \times \Omega \to \mathbb{R}^d \times \mathbb{R}^d$, $(t,\omega) \mapsto \sigma(t,\omega)$, is a predictable process. In particular in the **Markovian** case, the volatility matrix $\sigma : [0,T] \times \mathbb{R}^d \to \mathbb{R}^d \times \mathbb{R}^d$ is a deterministic mapping $[0,T] \times \mathbb{R}^d \ni (t,x) \mapsto \sigma(t,x) \in \mathbb{R}^d \times \mathbb{R}^d$, and then the state process equation (6.1) becomes the stochastic differential equation

$$X(t) = X(0) + \int_0^t \sigma(s, X(s))dB_s,\ 0 \le t \le T. \tag{6.2}$$

The Markovian case is indeed a special case of path-dependence. Since it will receive some extra attention later at the end of subsection 6.2.2, we describe the Markovian framework separately from the more general path-dependent case.

Assumption 6.1.

(1) The volatility matrix $\sigma(t,\omega)$ is nonsingular for every $(t,\omega) \in [0,T] \times \Omega$;
(2) there exists a positive constant A such that

$$|\sigma_{ij}(t,\omega) - \sigma_{ij}(t,\bar{\omega})| \le A \sup_{0\le s\le t} |\omega(s) - \bar{\omega}(s)|,$$

for all $1 \le i, j \le d$, for all $t \in [0,T]$, $\omega, \bar{\omega} \in \Omega$.

Under Assumption 6.1(2), for every initial value $X(0) \in \mathbb{R}^d$, there exists a pathwise unique strong solution to equation (6.1) (Theorem 14.6, Elliott (1982) [16]).

The controls u and v take values in some given separable metric spaces $\mathbb{A}_1$ and $\mathbb{A}_2$, respectively. We shall assume that $\mathbb{A}_1$ and $\mathbb{A}_2$ are countable unions of nonempty, compact subsets of these spaces, and are endowed with the σ-algebras $\mathscr{A}_1$ and $\mathscr{A}_2$ of their respective Borel subsets. The controls u and v are said

(i) to be **open loop**, if $u_t = \mu(t,\omega)$ and $v_t = \nu(t,\omega)$ are $\{\mathscr{F}_t\}_{0\le t\le T}$-adapted processes on $[0,T]$, where $\mu : [0,T]\times\Omega \to \mathbb{A}_1$ and $\nu : [0,T]\times\Omega \to \mathbb{A}_2$ are non-anticipative measurable mappings;

(ii) to be **closed loop**, if $u_t = \mu(t,X)$ and $v_t = \nu(t,X)$ are non-anticipative functionals of the state process $X(\cdot)$, for $0 \le t \le T$, where $\mu : [0,T]\times\Omega \to \mathbb{A}_1$ and $\nu : [0,T]\times\Omega \to \mathbb{A}_2$ are deterministic measurable mappings;

(iii) to be **Markovian**, if $u_t = \mu(t,X(t))$ and $v_t = \nu(t,X(t))$, for $0 \le t \le T$, where $\mu : [0,T]\times\mathbb{R}^d \to \mathbb{A}_1$ and $\nu : [0,T]\times\mathbb{R}^d \to \mathbb{A}_2$ are deterministic measurable functions.

In the path-dependent case, the set $\mathscr{U}\times\mathscr{V}$ of admissible controls are taken as all the closed loop controls. The techniques that we shall use to solve for the optimal closed loop controls also apply to the open loop controls, so the extension of the results from closed loop to open loop is only a matter of more complicated notation. The discussion will be restricted within the class of closed loop controls for clarity of the exposition. In the Markovian case, the set $\mathscr{U}\times\mathscr{V}$ of admissible controls are taken as all the Markovian controls. The collection of Markovian controls is a subset of the collection of closed loop controls.

We consider the predictable mapping

$$\begin{aligned} &f : [0,T]\times\Omega\times\mathbb{A}_1\times\mathbb{A}_2 \to \mathbb{R}^d,\\ &\quad (t,\omega,\mu(t,\omega),\nu(t,\omega)) \mapsto f(t,\omega,\mu(t,\omega),\nu(t,\omega)), \end{aligned}$$

in the path-dependent case, and the deterministic measurable mapping

$$\begin{aligned} &f : [0,T]\times\Omega\times\mathbb{A}_1\times\mathbb{A}_2 \to \mathbb{R}^d,\\ &(t,\omega,\mu(t,\omega(t)),\nu(t,\omega(t))) \mapsto f(t,\omega(t),\mu(t,\omega(t)),\nu(t,\omega(t))), \end{aligned}$$

in the Markovian case, satisfying:

Assumption 6.1. (continued)

(3) There exists a positive constant A such that

$$\left|\sigma^{-1}(t,\omega)f(t,\omega,\mu(t,\omega),\nu(t,\omega))\right| \le A,$$

and

$$|\sigma(t,\omega)|^2 \le A \sup_{0\le s\le t}\left(1+|\omega(s)|^2\right),$$

for all $0\le t\le T$, $\omega\in\Omega$, and for all the $\mathbb{A}_1\times\mathbb{A}_2$-valued representative elements $(\mu(t,\omega),\nu(t,\omega))$ of the control spaces $\mathscr{U}\times\mathscr{V}$.

For generic controls $u_t = \mu(t,\omega)$ and $v_t = \nu(t,\omega)$, we define $\mathbb{P}^{u,v}$, a probability measure equivalent to $\mathbb{P}$, via the Radon-Nikodym derivative

$$\left.\frac{d\mathbb{P}^{u,v}}{d\mathbb{P}}\right|_{\mathscr{F}_t} = \exp\left\{\int_0^t \sigma^{-1}(s,X)f(s,X,u_s,v_s)dB_s - \frac{1}{2}\int_0^t |\sigma^{-1}(s,X)f(s,X,u_s,v_s)|^2 ds\right\} \tag{6.3}$$

for $0 \le t \le T$. Then, by Girsanov's Theorem,

$$B_t^{u,v} := B_t - \int_0^t \sigma^{-1}(s,X)f(s,X,u_s,v_s)ds \tag{6.4}$$

is a $\mathbb{P}^{u,v}$-Brownian Motion on $[0,T]$ with respect to the filtration $\{\mathscr{F}_t\}_{0\le t\le T}$. In the Markovian case, the expression of (6.4) can be written as

$$B_t^{u,v} = B_t - \int_0^t \sigma^{-1}(s,X(s))f(s,X(s),\mu(s,X(s)),\nu(s,X(s)))ds. \tag{6.5}$$

On the probability space $(\Omega,\mathscr{F},\mathbb{P})$, and with respect to the filtration $\{\mathscr{F}_t\}_{0\le t\le T}$, the pair $(X, B^{u,v})$ is a weak solution to the forward stochastic functional equation

$$X(t) = X(0) + \int_0^t f(s,X,u_s,v_s)ds + \int_0^t \sigma(s,X)dB_s^{u,v} \tag{6.6}$$

in the path-dependent case, and a weak solution to the forward stochastic differential equation

$$X(t) = X(0) + \int_0^t f(s,X(s),\mu(s,X(s)),\nu(s,X(s)))ds + \int_0^t \sigma(s,X(s))dB_s^{u,v} \tag{6.7}$$

in the Markovian case, for $0 \le t \le T$.

When playing the game, the two players choose first their admissible controls u in $\mathscr{U}$ and v in $\mathscr{V}$, then for any given $t \in [0,T]$, they chose τ_t and ρ_t from $\mathscr{S}(t,T)$, times for them to quit the game. The pair of control and stopping rule (u,τ) is up to player I and the pair (v,ρ) is up to player II. For starting the game at time t, applying controls u and v, and quitting the game at τ_t and ρ_t respectively, the players receive rewards $R_t^1(\tau_t,\rho_t,u,v)$ and $R_t^2(\tau_t,\rho_t,u,v)$. To take into account the uncertainty inherent in the situation they face, their respective reward processes are measured by the conditional $\mathbb{P}^{u,v}$-expectations

$$\mathbb{E}^{u,v}[R_t^1(\tau_t,\rho_t,u,v)|\mathscr{F}_t] \quad \text{and} \quad \mathbb{E}^{u,v}[R_t^2(\tau_t,\rho_t,u,v)|\mathscr{F}_t]. \tag{6.8}$$

In the non-zero-sum games, the two players seek first admissible controls u^* in $\mathscr{U}$ and v^* in $\mathscr{V}$, and then stopping rules τ_t^* and ρ_t^* from $\mathscr{S}(t,T)$, to maximize their expected rewards, in the sense that

$$\begin{aligned} \mathbb{E}^{u^*,v^*}[R_t^1(\tau_t^*,\rho_t^*,u^*,v^*)|\mathscr{F}_t] &\geq \mathbb{E}^{u,v^*}[R_t^1(\tau_t,\rho_t^*,u,v^*)|\mathscr{F}_t], \\ &\qquad \forall \tau_t \in \mathscr{S}(t,T),\ \forall u \in \mathscr{U}; \\ \mathbb{E}^{u^*,v^*}[R_t^2(\tau_t^*,\rho_t^*,u^*,v^*)|\mathscr{F}_t] &\geq \mathbb{E}^{u^*,v}[R_t^2(\tau_t^*,\rho_t,u^*,v)|\mathscr{F}_t], \\ &\qquad \forall \rho_t \in \mathscr{S}(t,T),\ \forall v \in \mathscr{V}. \end{aligned} \tag{6.9}$$

The interpretation is as follows: when player II employs strategy (ρ_t^*, v^*), the strategy (τ_t^*, u^*) maximizes the expected reward of player I over all possible strategies on $\mathscr{S}(t,T)\times\mathscr{U}$; and vice versa, when player I employs strategy (τ_t^*, u^*), the strategy (ρ_t^*, v^*) is optimal for player II over all possible strategies on $\mathscr{S}(t,T)\times\mathscr{V}$. The set of controls and stopping rules (τ^*,ρ^*,u^*,v^*) is called "equilibrium point", or **Nash equilibrium**, for the game. We denote by

$$V^i(t) := \mathbb{E}^{u^*,v^*}[R_t^i(\tau_t^*,\rho_t^*,u^*,v^*)|\mathscr{F}_t], \tag{6.10}$$

the value process of the game for each player $i = 1, 2$.

In subsections 6.2.1–6.2.2 and subsection 6.2.3, we shall consider two games, which differ in the forms of the rewards R^1 and R^2.

Game 6.1.

$$\begin{aligned} R_t^1(\tau_t,\rho_t,u,v) &:= \int_t^{\tau_t} h_1(s,X,u_s,v_s)ds + L_1(\tau_t)\mathbb{1}_{\tau_t<T} + \xi_1\mathbb{1}_{\tau_t=T}; \\ R_t^2(\tau_t,\rho_t,u,v) &:= \int_t^{\rho_t} h_2(s,X,u_s,v_s)ds + L_2(\rho_t)\mathbb{1}_{\rho_t<T} + \xi_2\mathbb{1}_{\rho_t=T}, \end{aligned}$$

with $R_t^1(\tau_t,\rho_t,u,v) = R_t^1(\tau_t,u,v)$ not involving ρ_t, and $R_t^2(\tau_t,\rho_t,u,v) = R_t^2(\rho_t,u,v)$ not involving τ_t.

Game 6.2.

$$\begin{aligned} R_t^1(\tau_t,\rho_t,u,v) &:= \int_t^{\tau_t\wedge\rho_t} h_1(s,X,u_s,v_s)ds + L_1(\tau_t)\mathbb{1}_{\tau_t\leq\rho_t<T} \\ &\quad + U_1(\rho_t)\mathbb{1}_{\rho_t<\tau_t} + \xi_1\mathbb{1}_{\tau_t\wedge\rho_t=T}; \\ R_t^2(\tau_t,\rho_t,u,v) &:= \int_t^{\tau_t\wedge\rho_t} h_2(s,X,u_s,v_s)ds + L_2(\rho_t)\mathbb{1}_{\rho_t\leq\tau_t<T} \\ &\quad + U_2(\tau_t)\mathbb{1}_{\tau_t<\rho_t} + \xi_2\mathbb{1}_{\tau_t\wedge\rho_t=T}. \end{aligned}$$

The rewards from both games are the sums of cumulative rewards at rates $h = (h_1, h_2)'$, early exercise rewards $L = (L_1, L_2)'$ and $U = (U_1, U_2)'$, and terminal rewards $\xi = (\xi_1, \xi_2)'$. Here and throughout this paper the notation M' means transpose of some matrix M.

The cumulative reward rates h_1 and $h_2 : [0,T] \times \Omega \times \mathbb{A}_1 \times \mathbb{A}_2 \to \mathbb{R}$, $(t, \omega, \mu(t,\omega), \nu(t,\omega)) \mapsto h_i(t, \omega, \mu(t,\omega), \nu(t,\omega))$, $i = 1, 2$, are predictable processes in t, non-anticipative functionals in $X(\cdot)$, and measurable functions in $\mu(t,\omega)$ and $\nu(t,\omega)$. The early exercise rewards $L : [0,T] \times \Omega \to \mathbb{R}^2$, $(t,\omega) \mapsto L(t,\omega) =: L(t)$, and $U : [0,T] \times \Omega \to \mathbb{R}^2$, $(t,\omega) \mapsto U(t,\omega) =: U(t)$ are both $\{\mathscr{F}_t\}_{0 \le t \le T}$-adapted continuous processes. The terminal reward $\xi = (\xi_1, \xi_2)'$ is a pair of real-valued $\mathscr{F}_T$-measurable random variables.

In the Markovian case, the rewards take the form $h(t, X, u_t, v_t) = h(t, X(t), \mu(t, X(t)), \nu(t, X(t)))$, $L(t) = \bar{L}(t, X(t))$, $U(t) = \bar{U}(t, X(t))$, and $\xi = \bar{\xi}(X(T))$, for all $0 \le t \le T$ and some deterministic measurable functions $\bar{L} : [0,T] \times \mathbb{R}^d \to \mathbb{R}$, $\bar{U} : [0,T] \times \mathbb{R}^d \to \mathbb{R}$, and $\bar{\xi} : \mathbb{R}^d \to \mathbb{R}^2$.

Assumption 6.2.

(1) The early exercise reward processes L and U are continuous, progressively measurable, and $L(T) \le \xi$ holds a.e. on Ω.
(2) There exist some constants $p \ge 1$ and $C_{\text{rwd}} > 0$, such that

$$
\begin{aligned}
&|h(t,\omega,\mu(t,\omega),\nu(t,\omega))| + |L(t,\omega)| + |U(t,\omega)| + |\xi(\omega)| \\
&\quad \le C_{\text{rwd}} \left(1 + \sup_{0<s<t} |\omega(s)|^{2p} \right)
\end{aligned}
$$

holds for a.e. (t,ω) in $[0,T] \times \Omega$, and for all admissible controls $u_t = \mu(t,\omega)$ and $v_t = \nu(t,\omega)$.

From the rewards and the coefficients of the state process, we define the Hamiltonians associated with our games,

$$
\begin{aligned}
&H_1(t,\omega,z_1,u_t,v_t) = H_1(t,\omega,z_1,\mu(t,\omega),\nu(t,\omega)) \\
&\quad := z_1\sigma^{-1}(t,\omega) f(t,\omega,\mu(t,\omega),\nu(t,\omega)) + h_1(t,\omega,\mu(t,\omega),\nu(t,\omega)),
\end{aligned} \tag{6.11}
$$

and

$$
\begin{aligned}
&H_2(t,\omega,z_2,u_t,v_t) = H_2(t,\omega,z_2,\mu(t,\omega),\nu(t,\omega)) \\
&\quad := z_2\sigma^{-1}(t,\omega) f(t,\omega,\mu(t,\omega),\nu(t,\omega)) + h_2(t,\omega,\mu(t,\omega),\nu(t,\omega)),
\end{aligned} \tag{6.12}
$$

for $0 \le t \le T$, $\omega \in \Omega$, z_1 and z_2 in $\mathbb{R}^d$, and for all admissible controls $u_t = \mu(t,\omega)$ and $v_t = \nu(t,\omega)$. From Assumption 6.1(3), the Hamiltonians are Lipschitz functions in z_1 and z_2, uniformly over all $0 \le t \le T$, $\omega \in \Omega$, and all admissible controls $u_t = \mu(t, \omega)$ and $v_t = \nu(t, \omega)$.

Assumption 6.3. (Isaacs condition) There exist admissible controls $u_t^* = \mu^*(t, \omega)$ in $\mathscr{U}$ and $v_t^* = \nu^*(t, \omega)$ in $\mathscr{V}$, such that

$$H_1(t, \omega, z_1, \mu^*(t, \omega), \nu^*(t, \omega)) \ge H_1(t, \omega, z_1, \mu(t, \omega), \nu^*(t, \omega)),$$
$$H_2(t, \omega, z_2, \mu^*(t, \omega), \nu^*(t, \omega)) \ge H_2(t, \omega, z_2, \mu^*(t, \omega), \nu(t, \omega)),$$

holds for all $0 \le t \le T$, $\omega \in \Omega$, $(z_1, z_2) \in \mathbb{R}^{2\times d}$, and for all admissible controls $u_t = \mu(t, \omega)$ and $v_t = \nu(t, \omega)$.

The Isaacs conditions on the Hamiltonians are "local" optimality conditions, formulated in terms of every point (t, z_1, z_2) in Euclidean space and every path ω in the function space Ω. Theorem 6.1 takes the local conditions on the Hamiltonians and transforms them into "global" optimization statements involving each higher-dimensional object, such as a stopping time, stochastic process, etc., taking values in Euclidean space and defined over the probability space. This possibility is afforded by the continuous-time setting, in contrast to some discrete-time optimization problems where local maximization need not lead to global maximization.

When linking value processes of the games to the solutions to BSDEs, we shall discuss the solutions in the following spaces $\mathbb{M}^2(m; 0, T)$ and $\mathbb{L}^2(m \times d; 0, T)$, defined as those predictable processes satisfying

$$\mathbb{M}^k(m; t, T) := \left\{ m\text{-dimensional } \phi \text{ s.t. } \mathbb{E}\left[\sup_{[t,T]} \|\phi_s\|^2\right] \le \infty \right\}, \quad (6.13)$$

$$\mathbb{L}^k(m \times d; t, T) := \left\{ m \times d\text{-dimensional } \phi \text{ s.t. } \mathbb{E}\left[\int_t^T \|\phi_s\|^2 ds\right] \le \infty \right\}, \quad (6.14)$$

for $k = 1, 2$, and $0 \le t \le T$, where $\|\cdot\|$ is the Euclidean or Frobenius norm.

6.2.1. *The duality between Game and BSDE*

This subsection studies Game 6.1 where a player's time to quit is determined by his own decision. We shall demonstrate that the solution to a two-dimensional BSDE with reflecting barrier provides the two players' value processes. The optimal stopping rules will be derived from the reflecting conditions on the BSDE. The optimal controls will come from the Isaacs condition, namely, Assumption 6.3 on the Hamiltonians, which play here the role of the driver of the corresponding BSDE.

The solution to the following system of BSDEs

$$\begin{cases} Y_i^{u,v}(t) = \xi_i + \int_t^T H_i(s, X, Z_i^{u,v}(s), u_s, v_s)ds - \int_t^T Z_i^{u,v}(s)dB_s \\ \qquad + K_i^{u,v}(T) - K_i^{u,v}(t), \\ \text{with } Y_i^{u,v}(t) \geq L_i(t); \quad \int_0^T (Y_i^{u,v}(t) - L_i(t))dK_i^{u,v}(t) = 0; \\ \text{for } 0 \leq t \leq T, \quad i = 1, 2 \end{cases} \tag{6.15}$$

provides the players' value processes in Game 6.1, with the proper choice of controls $u = u^*$ and $v = v^*$ mandated by the Isaacs condition. From now on, a BSDE with reflecting barrier in the form of (6.15) will be denoted as $(T, \xi, H(u, v), L)$ for short. The solution to this BSDE is a triple of processes $(Y^{u,v}, Z^{u,v}, K^{u,v})$, satisfying $Y^{u,v}(\cdot) \in \mathbb{M}^2(2; 0, T)$, $Z^{u,v}(\cdot) \in \mathbb{L}^2(2\times d; 0, T)$, and $K^{u,v}(\cdot) = (K_1^{u,v}(\cdot), K_2^{u,v}(\cdot))'$ a pair of continuous increasing processes in $\mathbb{M}^2(2; 0, T)$.

We focus on the game aspect in this section, making use of such results as existence of solutions to BSDEs, one-dimensional comparison theorems, and continuous dependence properties, to be proved in section 6.3 and section 6.4. The proofs of these results will not rely on developments in this section.

Theorem 6.1. *Let $(Y^{u,v}, Z^{u,v}, K^{u,v})$ solve the BSDE (6.15) with parameters $(T, \xi, H(u, v), L)$. Define the stopping rules*

$$\tau_t^*(y; r) := \inf\{s \in [t, r] : y(s) \leq L_1(s)\} \wedge r, \tag{6.16}$$

and

$$\rho_t^*(y; r) = \inf\{s \in [t, r] : y(s) \leq L_2(s)\} \wedge r, \tag{6.17}$$

for $y \in C[0, T]$ and $r \in [t, T]$. Consider the stopping times

$$\tau_t(u, v) := \tau_t^* \left(Y_1^{u,v}(\cdot); T\right) \quad \text{and} \quad \rho_t(u, v) := \rho_t^* \left(Y_2^{u,v}(\cdot); T\right), \tag{6.18}$$

and suppose that the controls $u^* \in \mathscr{U}$ and $v^* \in \mathscr{V}$ satisfy the Isaacs condition, Assumption 6.3. Then the quadruple $(\tau(u^*, v^*), \rho(u^*, v^*), u^*, v^*)$ is a Nash equilibrium for Game 6.1, and we have $V^i(t) = Y_i^{u^*,v^*}(t)$, $i = 1, 2$.

To prove Theorem 6.1, we shall need the following result.

Lemma 6.1. *For $i = 1, 2$, the process*

$$
\begin{aligned}
M_i^{u,v}(\cdot) &:= \int_t^{\cdot} Z_i^{u,v}(s) dB_s^{u,v} \\
&= Y_i^{u,v}(\cdot) - Y_i^{u,v}(t) + \int_t^{\cdot} h_i(s, X, u_s, v_s) ds + K_i^{u,v}(\cdot) - K_i^{u,v}(t)
\end{aligned}
\tag{6.19}
$$

is a $\mathbb{P}^{u,v}$-martingale.

Proof. To show that $M_i^{u,v}(\cdot)$ is a $\mathbb{P}^{u,v}$-martingale, it suffices to show that $M_i^{u,v}(\cdot)$ is of class $\mathscr{DL}$, meaning that

$$
\lim_{c \to \infty} \sup_{\tau \in \mathscr{S} t, T} \mathbb{E}^{u,v} \left[|M_i^{u,v}(\tau)| \, \mathbb{1}_{|M_i^{u,v}(\tau)| > c} \Big| \mathscr{F}_t \right] = 0. \tag{6.20}
$$

For the fixed $t \in [0, T]$, denote

$$
\theta(s, u_s, v_s) := \sigma^{-1}(s, X) f(s, X, u_s, v_s), \ t \le s \le T. \tag{6.21}
$$

For any $\tau \in \mathscr{S}(t, T)$, from the change of measure (6.3) and Bayes' rule,

$$
\begin{aligned}
&\mathbb{E}^{u,v} \left[|M_i^{u,v}(\tau)| \, \mathbb{1}_{|M_i^{u,v}(\tau)| > c} \Big| \mathscr{F}_t \right] \\
&= \mathbb{E} \Bigg[\exp \Bigg\{ \int_t^{\tau} \theta(s, u_s, v_s) dB_s \\
&\qquad - \frac{1}{2} \int_t^{\tau} |\theta(s, u_s, v_s)|^2 ds \Bigg\} |M_i^{u,v}(\tau)| \mathbb{1}_{|M_i^{u,v}(\tau)| > c} \Big| \mathscr{F}_t \Bigg] \\
&\le \mathbb{E} \left[\sup_{t \le s \le T} \exp \left\{ \int_t^s 2\theta(r, u_r, v_r) dB_r - \frac{1}{2} \int_t^s 2 |\theta(r, u_r, v_r)|^2 dr \right\} \Big| \mathscr{F}_t \right]^{1/2} \\
&\quad \cdot \mathbb{E} \left[\sup_{t \le s \le T} |M_i^{u,v}(s)|^2 \, \mathbb{1}_{\sup_{t \le s \le T} |M_i^{u,v}(s)|^2 > c^2} \Big| \mathscr{F}_t \right]^{1/2}.
\end{aligned}
\tag{6.22}
$$

From the expression (6.19) and Assumption 6.2(2), there exists some constant C_0, such that

$$
\begin{aligned}
&\sup_{t \le s \le T} |M_i^{u,v}(s)|^2 \\
&\quad \le C_0 \left(1 + \sup_{t \le s \le T} |Y_i^{u,v}(s)|^2 + \sup_{t \le s \le T} |X(s)|^{2p} + |K_i^{u,v}(T)|^2 \right).
\end{aligned}
$$

From the definition of the solutions to reflected BSDEs, as in section 6.3, we know that

$$
\mathbb{E} \left[\sup_{t \le s \le T} (Y_i^{u,v}(s))^2 + |K_i^{u,v}(T)|^2 \right] < \infty \tag{6.23}
$$

holds. Since (X, B) is a solution to the stochastic functional equation (6.1), there exists (cf. page 306 of Karatzas and Shreve (1988) [28]) a constant C_1 such that

$$\mathbb{E}\left[\sup_{t\leq s\leq T}|X(s)|^{2p}\right]\leq C_1\left(1+|X(0)|^{2p}\right)<\infty.$$

We then apply the dominated convergence theorem to the last conditional expectations in (6.22) to get

$$\mathbb{E}\left[\sup_{t\leq s\leq T}|M_i^{u,v}(s)|^2\,\mathbb{1}_{\sup_{t\leq s\leq T}|M_i^{u,v}(s)|^2>c^2}\,\middle|\,\mathscr{F}_t\right]\to 0, \tag{6.24}$$

as $c\to 0$. It remains to show that

$$\mathbb{E}\left[\sup_{t\leq s\leq T}\exp\left\{\int_t^s 2\theta(r,u_r,v_r)dB_r-\frac{1}{2}\int_t^s 2|\theta(r,u_r,v_r)|^2dr\right\}\right]<\infty. \tag{6.25}$$

Because $|\theta(s, u_s, v_s)|$ is bounded by the constant A, from Assumption 6.1(3) and identity (6.21), we know that the process

$$\exp\left\{\int_t^{\cdot} 2\theta(s,u_s,v_s)dB_s-\frac{1}{2}\int_t^{\cdot} 2|\theta(s,u_s,v_s)|^2ds\right\}$$

is a.e. bounded by the constant e^{A^2T} times the exponential $\mathbb{P}$-martingale

$$Q(\cdot):=\exp\left\{\int_t^{\cdot} 2\theta(s,u_s,v_s)dB_s-\frac{1}{2}\int_t^{\cdot} 4|\theta(s,X,u_s,v_s)|^2ds\right\}$$

on $[0, T]$ with quadratic variation process

$$\langle Q\rangle(\cdot)=4\int_t^{\cdot} Q^2(s)|\theta(s,u_s,v_s)|^2ds. \tag{6.26}$$

But

$$\begin{aligned}&Q^2(\cdot)|\theta(\cdot,u_\cdot,v_\cdot)|^2\\&\leq A^2e^{4A^2T}\exp\left\{\int_t^{\cdot} 4\theta(s,u_s,v_s)dB_s-\frac{1}{2}\int_t^{\cdot} 16|\theta(s,u_s,v_s)|^2ds\right\}\end{aligned} \tag{6.27}$$

By the Burkholder-Davis-Gundy inequalities and inequality (6.27), there exists a constant C, such that $\mathbb{E}\left[\sup_{t\le s\le T} Q(s)\right]$ is dominated by

$$\begin{aligned}
&2CAe^{2A^2T}\mathbb{E}\Bigg[\Bigg(\int_t^T \exp\Bigg\{\int_t^s 4\theta(r,u_r,v_r)dB_r \\
&\qquad\qquad -\frac{1}{2}\int_t^s 16|\theta(r,u_r,v_r)|^2dr\Bigg\}ds\Bigg)^{1/2}\Bigg] \\
&\le 2CAe^{2A^2T}\Bigg(\int_t^T \mathbb{E}\Bigg[\exp\Bigg\{\int_t^s 4\theta(r,u_r,v_r)dB_r \\
&\qquad\qquad -\frac{1}{2}\int_t^s 16|\theta(r,u_r,v_r)|^2dr\Bigg\}\Bigg]ds\Bigg)^{1/2} \\
&= 2CAe^{2A^2T}(T-t)^{1/2}.
\end{aligned}$$

This proves (6.25), whereas the expressions (6.22), (6.24) and (6.25) together lead to (6.20). □

Proof. [Proof of Theorem 6.1] Let $(Y^{u,v}, Z^{u,v}, K^{u,v})$ solve BSDE (6.15) with parameters $(T, \xi, H(u,v), L)$. Taking a stopping rule $\tau_t \in \mathscr{S}(t,T)$, and integrating $dY_1^{u,v}$ from t to τ_t, we obtain

$$\begin{aligned}
Y_1^{u,v}(t) &= Y_1^{u,v}(\tau_t) + \int_t^{\tau_t} H_1(s,X,Z_1^{u,v}(s),u_s,v_s)ds - \int_t^{\tau_t} Z_1^{u,v}(s)dB_s \\
&\quad + K_1^{u,v}(\tau_t) - K_1^{u,v}(t) \\
&= Y_1^{u,v}(\tau_t) + \int_t^{\tau_t} h_1(s,X,u_s,v_s)ds - \int_t^{\tau_t} Z_1^{u,v}(s)dB_s^{u,v} \\
&\quad + K_1^{u,v}(\tau_t) - K_1^{u,v}(t).
\end{aligned}$$

Taking conditional expectation $\mathbb{E}^{u,v}[\cdot|\mathscr{F}_t]$, and using the comparisons $Y_1^{u,v}(\cdot) \ge L_1(\cdot)$, $Y_1^{u,v}(T) = \xi_1$, as well as the fact that $K_1^{u,v}(\cdot)$ is an increasing process, we obtain

$$\begin{aligned}
Y_1^{u,v}(t) &= \mathbb{E}^{u,v}\left[Y_1^{u,v}(\tau_t) + \int_t^{\tau_t} h_1(s,X,u_s,v_s)ds + K_1^{u,v}(\tau_t) - K_1^{u,v}(t)\,\middle|\,\mathscr{F}_t\right] \\
&\ge \mathbb{E}^{u,v}\left[L_1(\tau_t)\mathbb{1}_{\tau_t\wedge T_1^n<T} + \xi_1\mathbb{1}_{\tau_t} + \int_t^{\tau_t} h_1(s,X,u_s,v_s)ds\,\middle|\,\mathscr{F}_t\right].
\end{aligned} \tag{6.28}$$

According to the reflecting condition in BSDE (6.15), $K_1^{u,v}(\cdot)$ is flat on $\{(\omega,t) \in (\Omega\times[0,T]) : Y_1^{u,v}(t) \ne L_1(t)\}$; from the continuity of $K_1^{u,v}(\cdot)$, we see that $K_1^{u,v}(\tau_t(u,v)) = K_1^{u,v}(t)$. On $\{\tau_t(u,v) < T\}$, $Y_1^{u,v}(\tau_t(u,v)) = L_1(\tau_t(u,v))$; on $\{\tau_t(u,v) = T\}$, $Y_1^{u,v}(\tau_t(u,v)) = \xi_1$. Then,

$$\begin{aligned} Y_1^{u,v}(t) &= \mathbb{E}^{u,v}\left[Y_1^{u,v}(\tau_t(u,v)) + \int_t^{\tau_t(u,v)} h_1(s,X,u_s,v_s)ds \,\middle|\, \mathscr{F}_t \right] \\ &= \mathbb{E}^{u,v}\Bigg[L_1(\tau_t(u,v))\mathbb{1}_{\tau_t(u,v)<T} + \xi_1 \mathbb{1}_{\tau_t(u,v)=T} \\ &\qquad + \int_t^{\tau_t(u,v)} h_1(s,X,u_s,v_s)ds \,\Bigg|\, \mathscr{F}_t \Bigg]. \end{aligned} \tag{6.29}$$

The expressions (6.29) and (6.28) mean that

$$Y_1^{u,v}(t) = \mathbb{E}^{u,v}[R_t^1(\tau_t(u,v),\rho_t,u,v)|\mathscr{F}_t] \geq \mathbb{E}^{u,v}[R_t^1(\tau_t,\rho_t,u,v)|\mathscr{F}_t] \tag{6.30}$$

holds for all $\rho_t \in \mathscr{S}(t,T)$ and for all $\tau_t \in \mathscr{S}(t,T)$.

To derive the optimality of the controls (u^*,v^*) from the Isaacs condition, Assumption 6.3, an application of the comparison theorem (Theorems 6.4 and 6.8) to the first component of BSDE (6.15) gives $Y_1^{u^*,v^*}(\cdot) \geq Y_1^{u,v^*}(\cdot)$ a.e. on $[0,T]\times\Omega$. From the identity in (6.30), we have

$$\begin{aligned} &\mathbb{E}^{u,v^*}[R_t^1(\tau_t(u^*,v^*),\rho_t(u^*,v^*),u^*,v^*)|\mathscr{F}_t] = Y_1^{u^*,v^*}(t) \\ &\geq Y_1^{u,v^*}(t) = \mathbb{E}^{u,v^*}[R_t^1(\tau_t(u,v^*),\rho_t(u,v^*),u,v^*)|\mathscr{F}_t]; \end{aligned} \tag{6.31}$$

and in conjunction with (6.30), for all $\tau_t \in \mathscr{S}(t,T)$, this gives

$$\begin{aligned} &\mathbb{E}^{u^*,v^*}[R_t^1(\tau_t(u^*,v^*),\rho_t(u^*,v^*),u^*,v^*)|\mathscr{F}_t] \\ &\geq \mathbb{E}^{u,v^*}[R_t^1(\tau_t(u,v^*),\rho_t(u,v^*),u,v^*)|\mathscr{F}_t] \\ &\geq \mathbb{E}^{u,v^*}[R_t^1(\tau_t,\rho_t(u,v^*),u,v^*)|\mathscr{F}_t]. \end{aligned} \tag{6.32}$$

The above arguments proceed with arbitrary stopping times $\rho_t \in \mathscr{S}(t,T)$, because player II's stopping time ρ_t does not enter player I's reward.

By symmetry between the two players,

$$Y_2^{u^*,v^*} = \mathbb{E}^{u^*,v^*}[R_t^2(\tau_t(u^*,v^*),\rho_t(u^*,v^*),u^*,v^*)|\mathscr{F}_t], \tag{6.33}$$

and

$$\begin{aligned} &\mathbb{E}^{u^*,v^*}[R_t^2(\tau_t(u^*,v^*),\rho_t(u^*,v^*),u^*,v^*)|\mathscr{F}_t] \\ &\geq \mathbb{E}^{u^*,v}[R_t^2(\tau_t(u^*,v^*),\rho_t,u^*,v)|\mathscr{F}_t]. \end{aligned} \tag{6.34}$$

Combining (6.31), (6.32), (6.33) and (6.34), we see that the quadruple (τ^*,ρ^*,u^*,v^*) is a Nash equilibrium and their value processes $V^1(\cdot)$ and $V^2(\cdot)$ are identified with the solution to a BSDE with reflecting barrier with parameters $(T, \xi, H(u^*, v^*), L)$, as in (6.15). The optimal controls (u^*, v^*) are chosen according to the Isaacs condition, Assumption 6.3. Both players stop respectively according to the pair of rules (τ_t^*, ρ_t^*), as soon as their conditional expected rewards $V^1(\cdot)$ and $V^2(\cdot)$ hit the early stopping rewards $L_1(\cdot)$ and $L_2(\cdot)$ for the first time. □

Remark 6.1. If the deterministic time T is replaced by a bounded $\{\mathscr{F}_t\}_{0\le t\le T}$-stopping time, it technically does not make any difference to results in this subsection.

6.2.2. *Controls observing volatility*

This subsection discusses whether the inclusion of instantaneous volatilities of the value processes into the controls will expand the admissible control sets.

For the rewards considered in this paper, when using control u and v, the $\mathbb{P}^{u,v}$-conditional expected rewards are $\mathbb{P}^{u,v}$-Brownian semimartingales with respect to the filtration $\{\mathscr{F}_t\}_{0\le t\le T}$, having the decompositions

$$\begin{aligned}\mathbb{E}^{u,v}[R_t^1(\tau,\rho,u,v)|\mathscr{F}_t] &= A_1^{u,v}(t) + M_1^{u,v}(t) = A_1^{u,v}(t) + \int_0^t Z_1^{u,v}(s)dB_1^{u,v}(s);\\ \mathbb{E}^{u,v}[R_t^2(\tau,\rho,u,v)|\mathscr{F}_t] &= A_2^{u,v}(t) + M_2^{u,v}(t) = A_2^{u,v}(t) + \int_0^t Z_2^{u,v}(s)dB_2^{u,v}(s).\end{aligned} \tag{6.35}$$

The processes $A^1(\cdot)$ and $A^2(\cdot)$ are adapted and have finite variation. The processes $M^1(\cdot)$ and $M^2(\cdot)$ are $\mathbb{P}^{u,v}$-local martingales with respect to $\{\mathscr{F}_t\}_{0\le t\le T}$. The predictable, square-integrable processes $Z_1^{u,v}(\cdot)$ and $Z_2^{u,v}(\cdot)$ from the martingale representation are called the instantaneous volatility processes, the integrand processes of the stochastic integrals in the BSDE (6.15). Because they naturally show up in the BSDEs solved by value process of the game, we may include the instantaneous volatilities $Z_1^{u,v}(\cdot)$ and $Z_2^{u,v}(\cdot)$ as arguments of the controls u and v, in the hope of making more informed decisions. Going one step further, in the case of risk-sensitive controls initiated by Whittle, Bensoussan and coworkers, among others, for example Bensoussan, Frehse and Nagai (1998) [4], the players are sensitive not only to the expectations, but also to the variances of their rewards; we emphasize sensitivity to volatilities by including them as arguments of the controls. El Karoui and Hamadène (2003) identified in [12] risk-sensitive controls to BSDEs with quadratic growth in $Z_1^{u,v}(\cdot)$ and $Z_2^{u,v}(\cdot)$, which made the problem very tractable. Their value processes are different from the risk-indifferent case only up to an exponential transformation.

Apply the controls

$$u_t = \mu(t, X, Z_1(t), Z_2(t)) \quad \text{and} \quad v_t = \nu(t, X, Z_1(t), Z_2(t))$$

for some deterministic measurable functionals $\mu : [0,T] \times \Omega \times \mathbb{R}^d \times \mathbb{R}^d \to \mathbb{A}_1$ and $\nu : [0,T] \times \Omega \times \mathbb{R}^d \times \mathbb{R}^d \to \mathbb{A}_2$, and for some $\{\mathscr{F}_t\}_{0\le t\le T}$-predictable processes $Z_1(\cdot)$ and $Z_2(\cdot)$. If the resulting instantaneous volatilities $Z_1^{u,v}(\cdot)$ and $Z_2^{u,v}(\cdot)$ in the semimartingale decomposition (6.35) coincide with the arguments $Z_1(\cdot)$ and $Z_2(\cdot)$ of the functionals μ and ν, then u and v are said to be *a pair of closed loop controls that observe the instantaneous volatilities.*

Apply the controls $u_t = \mu(t, X(t), Z_1(t), Z_2(t))$ and $v_t = \nu(t, X(t), Z_1(t), Z_2(t))$, for some deterministic measurable functions $\mu : [0,T] \times \mathbb{R}^d \times \mathbb{R}^d \times \mathbb{R}^d \to \mathbb{A}_1$ and $\nu : [0,T] \times \mathbb{R}^d \times \mathbb{R}^d \times \mathbb{R}^d \to \mathbb{A}_2$, and for some $\{\mathscr{F}_t\}_{0\le t\le T}$-predictable processes $Z_1(\cdot)$ and $Z_2(\cdot)$. If the resulting instantaneous volatilities $Z_1^{u,v}(\cdot)$ and $Z_2^{u,v}(\cdot)$ in the semimartingale decomposition (6.35) coincide with the arguments $Z_1(\cdot)$ and $Z_2(\cdot)$ of the functions μ and ν, then u and v are said to be *a pair of Markovian controls that observe the instantaneous volatilities.* This is the case about which we are going to have more to say.

The Hamiltonians in this case become

$$\begin{aligned} &H_1(t, \omega(t), z_1, (\mu,\nu)(t, \omega(t), z_1, z_2)) \\ &= z_1\sigma^{-1}(t, \omega(t)) f(t, \omega(t), (\mu,\nu)(t, \omega(t), z_1, z_2)) \\ &\quad + h_1(t, \omega(t), (\mu,\nu)(t, \omega(t), z_1, z_2)) \end{aligned} \tag{6.36}$$

and

$$\begin{aligned} &H_2(t, \omega(t), z_2, (\mu,\nu)(t, \omega(t), z_1, z_2)) \\ &= z_2\sigma^{-1}(t, \omega(t)) f(t, \omega(t), (\mu,\nu)(t, \omega(t), z_1, z_2)) \\ &\quad + h_2(t, \omega(t), (\mu,\nu)(t, \omega(t), z_1, z_2)), \end{aligned} \tag{6.37}$$

for $0 \le t \le T$, $\omega \in \Omega$, z_1 and z_2 in $\mathbb{R}^d$, and $\mathbb{A}_1 \times \mathbb{A}_2$-valued measurable functions (μ, ν). From Assumption 6.1(3) and Assumption 6.2(2), the Hamiltonians are linear in z_1 and z_2, and polynomial in $\sup_{0\le s\le t} |\omega(s)|$. To be more specific, we have

$$|H_i(t, \omega(t), z_1, z_2, (\mu,\nu)(t, \omega(t), z_1, z_2))| \le A|z_i| + C_{\mathrm{rwd}} \left(1 + \sup_{0\le s\le t} |\omega(s)|^{2p}\right), \tag{6.38}$$

for $i = 1, 2$, all $0 \le t \le T$, $\omega \in \Omega$, z_1 and z_2 in $\mathbb{R}^d$, and $\mathbb{A}_1 \times \mathbb{A}_2$-valued measurable functions (μ, ν). The growth rates of the Hamiltonians (6.36) and (6.37) satisfy Assumption 6.7(2) for the driver of the BSDE (6.64). With all other assumptions on the coefficients also satisfied, by Theorem 6.7, there exists a solution $(Y^{\mu,\nu}, Z^{\mu,\nu}, K^{\mu,\nu})$ to the following equation

$$\begin{cases} Y_i^{\mu,\nu}(t) = \xi_i + \int_t^T H_i(s, X(s), Z_i^{\mu,\nu}(s), (\mu,\nu)(s, X(s), Z_1^{\mu,\nu}(s), Z_2^{\mu,\nu}(s)))ds \\ \qquad - \int_t^T Z_i^{\mu,\nu}(s)dB_s + K_i^{\mu,\nu}(T) - K_i^{\mu,\nu}(t), \\ \quad \text{with } Y_i^{\mu,\nu}(t) \geq L_i(t); \quad \int_0^T (Y_i^{\mu,\nu}(t) - L_i(t))dK_i^{\mu,\nu}(t) = 0; \\ \quad \text{for } 0 \leq t \leq T; i = 1, 2. \end{cases} \tag{6.39}$$

Assumption 6.4. (Isaacs condition) There exist deterministic functions $\mu^* : [0,T] \times \mathbb{R}^d \times \mathbb{R}^d \times \mathbb{R}^d \to \mathbb{A}_1$ and $\nu^* : [0,T] \times \mathbb{R}^d \times \mathbb{R}^d \times \mathbb{R}^d \to \mathbb{A}_2$, such that

$$H_1(t, x, z_1, (\mu^*, \nu^*)(t, x, z_1, z_2)) = \sup_{\bar{z}_1, \bar{z}_2 \in \mathbb{R}^d} H_1(t, x, z_1, (\mu, \nu^*)(t, x, \bar{z}_1, \bar{z}_2));$$
$$H_2(t, x, z_2, (\mu^*, \nu^*)(t, x, z_1, z_2)) = \sup_{\bar{z}_1, \bar{z}_2 \in \mathbb{R}^d} H_2(t, x, z_2, (\mu^*, \nu)(t, x, \bar{z}_1, \bar{z}_2)),$$

for all $0 \leq t \leq T$, x, z_1 and z_2 in $\mathbb{R}^d$, and all $\mu : [0,T] \times \mathbb{R}^d \times \mathbb{R}^d \times \mathbb{R}^d \to \mathbb{A}_1$ and $\nu : [0,T] \times \mathbb{R}^d \times \mathbb{R}^d \times \mathbb{R}^d \to \mathbb{A}_2$.

Associated with the coefficients f and σ of the state process $X(\cdot)$ and with the rewards h, $L(\cdot)$ and ξ, the admissible set $\mathscr{U} \times \mathscr{V} = \{(u, v)\}$ of Markovian controls that observe volatilities are defined as the collection of all

$$(u_t, v_t) = (\mu, \nu)(t, X(t), Z_1^{\mu,\nu}(t), Z_2^{\mu,\nu}(t)), \tag{6.40}$$

for measurable functions $\mu : [0,T] \times \mathbb{R}^d \times \mathbb{R}^d \times \mathbb{R}^d \to \mathbb{A}_1$ and $\nu : [0,T] \times \mathbb{R}^d \times \mathbb{R}^d \times \mathbb{R}^d \to \mathbb{A}_2$. In particular,

$$(u_t^*, v_t^*) = (\mu^*, \nu^*)(t, X(t), Z_1^{\mu^*,\nu^*}(t), Z_2^{\mu^*,\nu^*}(t)),$$
$$(u_t, v_t^*) = (\mu, \nu^*)(t, X(t), Z_1^{\mu,\nu^*}(t), Z_2^{\mu,\nu^*}(t)),$$
$$(u_t^*, v_t) = (\mu^*, \nu)(t, X(t), Z_1^{\mu^*,\nu}(t), Z_2^{\mu^*,\nu}(t)).$$

Assumption 6.4 implies the Isaacs condition of Assumption 6.3. Then we reach the same statements as in Theorem 6.1, the only difference being that $(Y^{u,v}, Z^{u,v}, K^{u,v})$ is now replaced by $(Y^{\mu,\nu}, Z^{\mu,\nu}, K^{\mu,\nu})$, and the BSDE (6.15) is replaced by the BSDE (6.39).

In fact, by Theorem 6.6, there exist deterministic measurable mappings $\beta_1^{\mu,\nu}$ and $\beta_2^{\mu,\nu} : [0,T] \times \mathbb{R}^d \to \mathbb{R}^d$, such that $Z_1^{\mu,\nu}(t) = \beta_1^{\mu,\nu}(t, X(t))$, and $Z_2^{\mu,\nu}(t) = \beta_2^{\mu,\nu}(t, X(t))$ hold for all $0 \le t \le T$. Hence (6.40) becomes

$$(u_t, v_t) = (\mu, \nu)(t, X(t), \beta_1^{\mu,\nu}(t, X(t)), \beta_2^{\mu,\nu}(t, X(t))), \tag{6.41}$$

a pair of Markovian controls.

6.2.3. *Rewards terminated by both players*

In this subsection, we shall discuss the existence of equilibrium for Game 6.2 when rewards can be terminated by both players, in a sense weaker than (6.9). The existence of an equilibrium as in (6.9) is still not quite clear in this case. What also remains open is to write down a dual BSDE associated with this game, the solution to which can then be shown to exist.

By entering Game 6.2 at time t, player I receives reward

$$R_t^1(\tau, \rho, u, v) = \int_t^{\tau\wedge\rho} h_1(s, X, u_s, v_s)ds + \begin{cases} L_1(\tau) & \text{if player I stops first;} \\ U_1(\rho) & \text{if player II stops first;} \\ \xi_1 & \text{if neither stops before } T. \end{cases}$$

at the time when the game is terminated, whereas player II receives reward

$$R_t^2(\tau, \rho, u, v) = \int_t^{\tau\wedge\rho} h_2(s, X, u_s, v_s)ds + \begin{cases} U_2(\tau) & \text{if player I stops first;} \\ L_2(\rho) & \text{if player II stops first;} \\ \xi_2 & \text{if neither stops before } T. \end{cases}$$

when the game is terminated.

To prove the existence of equilibrium for Game 6.2, we shall first study the following game.

Game 6.2'.

$$\begin{aligned} \bar{R}_t^1(\tau_t, \rho_t, u, v) &:= \int_t^{\tau_t\wedge\rho_t} h_1(s, X, u_s, v_s)ds + (L_1 \vee U_1)(\rho_t \wedge \tau_t)\mathbb{1}_{\tau_t\wedge\rho_t<T} \\ &\quad + \xi_1 \mathbb{1}_{\tau_t\wedge\rho_t=T}; \\ \bar{R}_t^2(\tau_t, \rho_t, u, v) &:= \int_t^{\tau_t\wedge\rho_t} h_2(s, X, u_s, v_s)ds + (L_2 \vee U_2)(\rho_t \wedge \tau_t)\mathbb{1}_{\tau_t\wedge\rho_t<T} \\ &\quad + \xi_2 \mathbb{1}_{\tau_t\wedge\rho_t=T}, \end{aligned}$$

where $(L_i \vee U_i)(t) := \max\{L_i(t), U_i(t)\}$, for $0 \le t \le T$, and $i = 1, 2$.

Lemma 6.2. *Under Assumptions 6.1–6.3, there exists an equilibrium* $(\bar{\tau}^*, \bar{\rho}^*, u^*, v^*)$ *for Game 6.2', in the sense of (6.9) with* R^1 *and* R^2 *replaced by* $\bar{R}^1$ *and* $\bar{R}^2$.

Proof. The proof follows the inductive scheme proposed in Karatzas and Sudderth (2006) [29]. Suppose that the controls $u^* \in \mathscr{U}$ and $v^* \in \mathscr{V}$ satisfy the Isaacs condition, Assumption 6.3. Let $\tau_t^0 = \rho_t^0 = T$. Define the stopping rules

$$\bar{\tau}_t(y; r) = \inf\{s \in [t, r]\,|y(s) \le (L_1 \vee U_1)(s)\} \wedge r, \tag{6.42}$$

and

$$\bar{\rho}_t(y; r) = \inf\{s \in [t, r]\,|y(s) \le (L_2 \vee U_2)(s)\} \wedge r, \tag{6.43}$$

for $y \in C[0,T]$ and $r \in [t, T]$. For $n = 0, 1, 2, \cdots$, define the value functions as

$$\bar{V}^1_{n+1}(t) := \sup_{\tau_t \in \mathscr{S}(t,T)} \mathbb{E}^{u^*,v^*}[\bar{R}^1_t(\tau_t, \rho_t^n, u^*, v^*)|\mathscr{F}_t], \tag{6.44}$$

and

$$\bar{V}^2_{n+1}(t) := \sup_{\rho_t \in \mathscr{S}(t,T)} \mathbb{E}^{u^*,v^*}[\bar{R}^2_t(\tau_t^n, \rho_t, u^*, v^*)|\mathscr{F}_t]. \tag{6.45}$$

The stopping times

$$\tau_t^{n+1} := \bar{\tau}_t(\bar{V}^1_{n+1}; \rho_t^n), \tag{6.46}$$

and

$$\rho_t^{n+1} := \bar{\rho}_t(\bar{V}^2_{n+1}; \tau_t^n) \tag{6.47}$$

achieve the suprema in (6.44) and in (6.45), respectively. By applying Theorem 6.1 in dimension one to each individual player, we know that the inequalities

$$\mathbb{E}^{u^*,v^*}[\bar{R}^1_t(\tau_t^{n+1}, \rho_t^n, u^*, v^*)|\mathscr{F}_t] \ge \mathbb{E}^{u,v^*}[\bar{R}^1_t(\tau_t, \rho_t^n, u, v^*)|\mathscr{F}_t], \tag{6.48}$$

for all $\tau_t \in \mathscr{S}(t,T)$, all $u \in \mathscr{U}$, and

$$\mathbb{E}^{u^*,v^*}[\bar{R}^2_t(\tau_t^n, \rho_t^{n+1}, u^*, v^*)|\mathscr{F}_t] \ge \mathbb{E}^{u^*,v}[\bar{R}^2_t(\tau_t^n, \rho_t, u^*, v)|\mathscr{F}_t], \tag{6.49}$$

for all $\rho_t \in \mathscr{S}(t,T)$, all $v \in \mathscr{V}$, hold for a.e. (t, ω) in $[0,T] \times \Omega$.

The comparisons $\tau_t^1 \le \tau_t^0$ and $\rho_t^1 \le \rho_t^0$ imply that $\bar{V}^1_2(t) \le \bar{V}^1_1(t)$ and $\bar{V}^2_2(t) \le \bar{V}^2_1(t)$. Inductively, we know that $\tau_t^{n+1} \le \tau_t^n$, $\rho_t^{n+1} \le \rho_t^n$, $\bar{V}^1_{n+1}(t) \le \bar{V}^1_n(t)$ and $\bar{V}^2_{n+1}(t) \le \bar{V}^2_n(t)$, for all $0 \le t \le T$, and all $n = 0, 1, 2, \cdots$. The decreasing sequences of the value functions $\{\bar{V}^1_n(t)\}_{n\in\mathbb{N}}$, $\{\bar{V}^2_n(t)\}_{n\in\mathbb{N}}$, and the stopping times $\{\tau_t^n\}_{n\in\mathbb{N}}$, $\{\rho_t^n\}_{n\in\mathbb{N}}$, have limits $\bar{V}_*^1(t)$, $\bar{V}_*^2(t)$, $\bar{\tau}_t^*$ and $\bar{\rho}_t^*$. Furthermore, the stopping times $\bar{\tau}_t^*$ and $\bar{\rho}_t^*$ satisfy

$$\bar{\tau}_t^* = \bar{\tau}_t(\bar{V}_*^1; \bar{\rho}_t^*) \leq \bar{\rho}_t^*,$$

and

$$\bar{\rho}_t^* = \bar{\rho}_t(\bar{V}_*^2; \bar{\tau}_t^*) \leq \bar{\tau}_t^*,$$

hence

$$\bar{\tau}_t^* = \bar{\rho}_t^*.$$

By the continuity of the early exercise rewards $L(\cdot)$ and $U(\cdot)$, sending $n \to \infty$ in (6.48) and (6.49), we conclude that

$$\mathbb{E}^{u^*,v^*}[\bar{R}_t^1(\bar{\tau}_t, \bar{\rho}_t, u^*, v^*)|\mathscr{F}_t] \geq \mathbb{E}^{u,v^*}[\bar{R}_t^1(\tau_t, \bar{\rho}_t, u, v^*)|\mathscr{F}_t], \tag{6.50}$$

for all $\tau_t \in \mathscr{S}(t,T)$, all $u \in \mathscr{U}$, and

$$\mathbb{E}^{u^*,v^*}[\bar{R}_t^2(\bar{\tau}_t, \bar{\rho}_t, u^*, v^*)|\mathscr{F}_t] \geq \mathbb{E}^{u^*,v}[\bar{R}_t^2(\bar{\tau}_t, \rho_t, u^*, v)|\mathscr{F}_t], \tag{6.51}$$

for all $\rho_t \in \mathscr{S}(t,T)$, all $v \in \mathscr{V}$. □

Assumption 6.5. There exist adapted processes $\gamma_i : [0,T] \times \Omega \to \{0,1\}$, $(t,\omega) \mapsto \gamma_i(t,\omega) =: \gamma_i(t)$, such that

$$\gamma_i(t)L_i(t) + (1-\gamma_i(t))U_i(t) = (L_i \vee U_i)(t),$$

for $i = 1, 2$, and

$$\gamma_1(t) + \gamma_2(t) > 1,$$

for all $0 \leq t \leq T$ and $\omega \in \Omega$.

Theorem 6.2. *In addition to Assumptions 6.1–6.3, if Assumption 6.5 also holds, then there exists an equilibrium $(\tau^*, \rho^*, u^*, v^*)$ for Game 6.2, in the sense that*

$$\begin{aligned} \mathbb{E}^{u^*,v^*}[R_t^1(\tau_t^*, \rho_t^*, u^*, v^*)|\mathscr{F}_t] &\geq \mathbb{E}^{u,v^*}[R_t^1(\tau_t, \rho_t^*, u, v^*)|\mathscr{F}_t], \\ \mathbb{E}^{u^*,v^*}[R_t^2(\tau_t^*, \rho_t^*, u^*, v^*)|\mathscr{F}_t] &\geq \mathbb{E}^{u^*,v}[R_t^2(\tau_t^*, \rho_t, u^*, v)|\mathscr{F}_t], \end{aligned} \tag{6.52}$$

for all $\tau_t, \rho_t \in \mathscr{S}(t, \tau_t^* \wedge \rho_t^*)$, all $u \in \mathscr{U}$, all $v \in \mathscr{V}$.

Proof. Let $(\bar{\tau}^*, \bar{\rho}^*, u^*, v^*)$ be the equilibrium for Game 6.2', the quadruple specified in Lemma 6.2. Denote by $T_t^* := \bar{\tau}_t^* \wedge \bar{\rho}_t^*$. Define the stopping rules

$$\tau_t^* := \begin{cases} \bar{\tau}_t^* = \bar{\tau}_t(\bar{V}_*^1; T_t^*) \text{ if } \gamma_1(T_t^*) = 1 \text{ or } T_t^* = T; \\ \text{any stopping rule with values in } (T_t^*, T] \text{ if } \gamma_1(T_t^*) = 0, \end{cases} \tag{6.53}$$

and

$$\rho_t^* := \begin{cases} \bar{\rho}_t^* = \bar{\rho}_t(\bar{V}_*^2; T_t^*) & \text{if } \gamma_2(T_t^*) = 1 \text{ or } T_t^* = T; \\ \text{any stopping rule with values in } (T_t^*, T] & \text{if } \gamma_2(T_t^*) = 0. \end{cases} \tag{6.54}$$

By Lemma 6.2, one can verify that the quadruple $(\tau^*, \rho^*, u^*, v^*)$ is an equilibrium for Game 6.2, in the sense of (6.52). Furthermore, $\bar{V}_*^i(t) = V^i(t)$ holds for all $0 \le t \le T$, $\omega \in \Omega$, and $i = 1, 2$. □

6.3. A multi-dimensional reflected BSDE with Lipschitz growth

Starting with this section, we solve multi-dimensional BSDEs with reflecting barriers, the type of BSDEs associated with Game 6.1, and provide two useful properties of the equations: the comparison theorem in dimension one, and the theorem about continuous dependence of the solution on the terminal values. We have postponed the study of the BSDEs up until this point, in order to discuss the Stochastic Game aspects first. The proofs of the results to be stated from now onwards in this paper, do not depend on any earlier arguments or developments.

This section assumes the following Lipschitz and integrability conditions on the parameters of the equations.

Assumption 6.6.

(1) The driver g is a mapping $g : [0, T] \times \mathbb{R}^m \times \mathbb{R}^{m \times d} \to \mathbb{R}^m$, $(t, y, z) \mapsto g(t, y, z)$. For every fixed $y \in \mathbb{R}^m$ and $z \in \mathbb{R}^{m \times d}$, the process $\{g(t, y, z)\}_{0 \le t \le T}$ is $\{\mathscr{F}_t\}_{0 \le t \le T}$-predictable. For all $t \in [0, T]$, $g(t, y, z)$ is uniformly Lipschitz in y and z, i.e. there exists a constant $b > 0$, such that

$$|g(t, y, z) - g(t, \bar{y}, \bar{z})| \le b(|y - \bar{y}| + |z - \bar{z}|),$$

for all $t \in [0, T]$, $y \in \mathbb{R}^m$ and $z \in \mathbb{R}^{m \times d}$. Furthermore, $\int_0^T g(t, 0, 0)^2 dt < \infty$.

(2) The random variable ξ is $\mathscr{F}_T$-measurable and square-integrable. The lower reflecting boundary L is continuous, progressively measurable, and satisfies

$$\mathbb{E}\left[\sup_{[0,T]} L^+(t)^2\right] < \infty.$$

Also, $L(T) \le \xi$, a.e. on Ω.

Under Assumption 6.6, this section proves the existence and uniqueness of the solution (Y, Z, K) to the following BSDE

$$\begin{cases} Y(t) = \xi + \int_t^T g(s, Y(s), Z(s))ds - \int_t^T Z(s)dB_s + K(T) - K(t); \\ \quad Y(t) \geq L(t); \quad \int_0^T (Y(t) - L(t))dK(t) = 0, \end{cases} \tag{6.55}$$

for $0 \leq t \leq T$, in the spaces (cf. (6.13), (6.14))

$Y(\cdot) = (Y_1(\cdot), \cdots, Y_m(\cdot))' \in \mathbb{M}^2(m; 0, T)$

$Z(\cdot) = (Z_1(\cdot), \cdots, Z_m(\cdot))' \in \mathbb{L}^2(m \times d; 0, T)$

$K(\cdot) = (K_1(\cdot), \cdots, K_m(\cdot))'$: continuous, increasing process in $\mathbb{M}^2(m; 0, T)$,

where the positive integer m is the dimension of the equation. The backward equation and the reflecting condition in (6.55) should be interpreted component-wise; for every $i = 1, \cdots, m$, we have

$$\begin{cases} Y_i(t) = \xi_i + \int_t^T g_i(s, Y(s), Z(s))ds - \int_t^T Z_i(s)dB_s + K_i(T) - K_i(t); \\ Y_i(t) \geq L_i(t), \qquad \int_0^T (Y_i(t) - L_i(t))dK_i(t) = 0 \end{cases}$$

for $0 \leq t \leq T$. The value process $Y_i(\cdot)$ is driven by the Brownian noise $B(\cdot)$, whose intensity is modulated by a "control" $Z_i(\cdot)$. The driver g_i leads the value $Y_i(\cdot)$ towards the "final destination" ξ_i. Whenever the ith component $Y_i(\cdot)$ drops to the lower reflecting boundary $L_i(\cdot)$, it receives a force $K_i(\cdot)$ that kicks it upwards. When $Y_i(\cdot)$ stays above level $L_i(\cdot)$, the force $K_i(\cdot)$ does not apply. The process $K_i(\cdot)$ stands for the minimum cumulative exogenous energy required to keep $Y_i(\cdot)$ above level $L_i(\cdot)$. The m equations compose a system of m "vehicles" whose "drivers" track each other. For notational simplicity, the vector form (6.55) is used as a shorthand.

Lemma 6.3. *For any processes $(Y^0(\cdot), Z^0(\cdot)) \in \mathbb{L}^2(m; 0, T) \times \mathbb{L}^2(m \times d; 0, T)$, there exist unique $(Y^1(\cdot), Z^1(\cdot)) \in \mathbb{M}^2(m; 0, T) \times \mathbb{L}^2(m \times d; 0, T)$, and $K^1(\cdot) \in \mathbb{M}^2(m; 0, T)$, such that*

$$\begin{cases} dY^1(t) = -g(t, Y^0(t), Z^0(t))dt + Z^1(t)dB_t - dK^1(t), \\ Y^1(T) = \xi; \\ Y^1(t) \geq L(t), \qquad \int_0^T (Y^1(t) - L(t))dK^1(t) = 0, \end{cases} \tag{6.56}$$

for $0 \le t \le T$.

Proof. For any $i = 1, \cdots, m$, in the ith dimension, by Corollary 3.7 of El Karoui, Kapoudjian, Pardoux, Peng and Quenez (1997) [13], there exists a unique solution $(Y_i^1(\cdot), Z_i^1(\cdot)) \in \mathbb{M}^2(1; 0, T) \times \mathbb{L}^2(d; 0, T)$, and a continuous, increasing process $K_i^1(\cdot) \in \mathbb{M}^2(1; 0, T)$, to the one-dimensional reflected BSDE

$$\begin{cases} dY_i^1(t) = -g_i(t, Y^0(t), Z^0(t))dt + Z_i^1(t)dB_t - dK_i^1(t), \\ Y_i^1(T) = \xi_i; \\ Y_i^1(t) \ge L_i(t), \qquad \displaystyle\int_0^T (Y_i^1(t) - L_i(t))dK_i^1(t) = 0, \end{cases} \tag{6.57}$$

for $0 \le t \le T$. The processes $Y^1(\cdot) := (Y_1^1(\cdot), \cdots, Y_m^1(\cdot))'$, $Z^1(\cdot) := (Z_1^1(\cdot), \cdots, Z_m^1(\cdot))'$, and $K^1(\cdot) := (K_1^1(\cdot), \cdots, K_m^1(\cdot))'$ form the desired triple. □

To prove existence and uniqueness of the solution to the multi-dimensional BSDE (6.55) with reflecting barrier, it suffices to show that the mapping

$$\begin{aligned} \Lambda : \mathbb{L}^2(m; 0, T) \times \mathbb{L}^2(m \times d; 0, T) \to \mathbb{L}^2(m; 0, T) \times \mathbb{L}^2(m \times d; 0, T); \\ (Y^0, Z^0) \mapsto (Y^1, Z^1) \end{aligned} \tag{6.58}$$

is a contraction.

Theorem 6.3. *The mapping* Λ *is a contraction from* $\mathbb{L}^2(m; 0, T) \times \mathbb{L}^2(m \times d; 0, T)$ *to* $\mathbb{L}^2(m; 0, T) \times \mathbb{L}^2(m \times d; 0, T)$.

Proof. For a progressively measurable process $\phi(\cdot)$, the norm $||\phi||_2 := \sqrt{\mathbb{E}\Big[\int_0^T \phi_t^2 dt\Big]}$ is equivalent to the norm $||\phi||_{2,\beta} := \sqrt{\mathbb{E}\Big[\int_0^T e^{\beta t}\phi_t^2 dt\Big]}$. We prove the contraction statement under the norm $||\cdot||_{2,\beta}$. Suppose $(Y^0(\cdot), Z^0(\cdot))$ and $(\bar{Y}^0(\cdot), \bar{Z}^0(\cdot))$ are both in $\mathbb{M}^2(m; 0, T) \times \mathbb{L}^2(m \times d; 0, T)$. Denote $(Y^1(\cdot), Z^1(\cdot)) = \Lambda(Y^0(\cdot), Z^0(\cdot))$ and $(\bar{Y}^1(\cdot), \bar{Z}^1(\cdot)) = \Lambda(\bar{Y}^0(\cdot), \bar{Z}^0(\cdot))$. Applying Itô's rule to $e^{\beta t}(Y^1(t) - \bar{Y}^1(t))^2$, integrating the derivative from t to T, using the uniform Lipschitz condition, Assumption 6.6(1) of g, and applying some elementary inequalities, we find that after taking expectation,

$$||Y^1 - \bar{Y}^1||_{2,\beta}^2 + ||Z^1 - \bar{Z}^1||_{2,\beta}^2 \le \frac{1}{2}||Y^0 - \bar{Y}^0||_{2,\beta}^2 + \frac{1}{2}||Z^0 - \bar{Z}^0||_{2,\beta}^2.$$

□

Proposition 6.1. The BSDE (6.55) with reflecting barrier has a unique solution in $\mathbb{M}^2(m;0,T) \times \mathbb{L}^2(m \times d;0,T)$.

Proof. The solution is the unique fixed-point, say $(Y(\cdot), Z(\cdot))$, of the contraction Λ. Since $(Y(\cdot), Z(\cdot)) \in \mathbb{L}^2(m;0,T) \times \mathbb{L}^2(m \times d;0,T)$, $(Y(\cdot), Z(\cdot)) = \Lambda(Y(\cdot), Z(\cdot))$ is also in $\mathbb{M}^2(m;0,T) \times \mathbb{L}^2(m \times d;0,T)$ by Lemma 6.3. □

Theorem 6.4 (Comparison Theorem, El Karoui et al. [13]). *Suppose (Y, Z, K) solves (6.55) with parameter set (ξ, g, L), and $(\bar{Y}, \bar{Z}, \bar{K})$ solves (6.55) with parameter set $(\bar{\xi}, \bar{g}, \bar{L})$. Let the dimension of the equations be $m = 1$. Under Assumption 6.6, except that the uniform Lipschitz condition is only needed for either g or $\bar{g}$, if*

(1) $\xi \leq \bar{\xi}$, a.e.;
(2) $g(t, y, z) \leq \bar{g}(t, y, z)$, a.e. $(t, \omega) \in [0,T] \times \Omega$, $\forall (y, z) \in \mathbb{R} \times \mathbb{R}^d$; and
(3) $L(t) \leq \bar{L}(t)$, a.e. $(t, \omega) \in [0,T] \times \Omega$,

then

$$Y(t) \leq \bar{Y}(t), \text{ a.e. } (t, \omega) \in [0,T] \times \Omega.$$

Theorem 6.5 (Continuous Dependence Property). *Under Assumption 6.6, suppose that (Y, Z, K) solves the reflected BSDE (6.55), and that $(\bar{Y}, \bar{Z}, \bar{K})$ solves*

$$\begin{cases} \bar{Y}(t) = \bar{\xi} + \displaystyle\int_t^T g(s, \bar{Y}(s), \bar{Z}(s))ds - \int_t^T \bar{Z}(s)dB_s + \bar{K}(T) - \bar{K}(t); \\ \bar{Y}(t) \geq L(t); \qquad \displaystyle\int_0^T (\bar{Y}(t) - L(t))d\bar{K}(t) = 0 \end{cases} \tag{6.59}$$

for $0 \leq t \leq T$. Then there exists a constant number C, such that for all $0 \leq t \leq T$,

$$\begin{aligned} &\mathbb{E}[(Y(t) - \bar{Y}(t))^2] + \mathbb{E}\left[\int_0^T (Y(s) - \bar{Y}(s))^2 ds\right] \\ &\quad + \mathbb{E}\left[\int_0^T (Z(s) - \bar{Z}(s))^2 ds\right] + \mathbb{E}[(K(t) - \bar{K}(t))^2] \\ &\leq C\mathbb{E}[(\xi - \bar{\xi})^2]. \end{aligned}$$

Proof. Repeating the methods in the proof of Theorem 6.3, we can show that both

$$\mathbb{E}[(Y(t) - \bar{Y}(t))^2] \leq e^{\beta T}\mathbb{E}[(\xi - \bar{\xi})^2], \text{ for all } 0 \leq t \leq T, \tag{6.60}$$

and

$$\mathbb{E}\left[\int_0^T (Y(s)-\bar{Y}(s))^2 ds\right] + \mathbb{E}\left[\int_0^T (Z(s)-\bar{Z}(s))^2 ds\right] \le 2e^{\beta T}\mathbb{E}\left[(\xi-\bar{\xi})^2\right]. \tag{6.61}$$

From the expressions

$$K(t) = Y(0) - Y(t) - \int_0^t g(s, Y(s), Z(s))ds + \int_0^t Z(s)dB_s,$$

and

$$\bar{K}(t) = \bar{Y}(0) - \bar{Y}(t) - \int_0^t g(s, \bar{Y}(s), \bar{Z}(s))ds + \int_0^t \bar{Z}(s)dB_s,$$

the Lipschitz condition, Assumption 6.6(1), and Itô's isometry, for all $0 \le t \le T$, we derive the following estimate for the $\mathbb{L}^2$-norm of $(K(t) - \bar{K}(t))$,

$$\begin{aligned}
&\mathbb{E}\left[(K(t)-\bar{K}(t))^2\right] \\
&\le C_1\Bigg(\mathbb{E}\left[(Y(0)-\bar{Y}(0))^2\right] + \mathbb{E}\left[(Y(t)-\bar{Y}(t))^2\right] \\
&\quad + 2Tb^2\mathbb{E}\left[\int_0^T (Y(t)-\bar{Y}(t))^2 dt\right] + (2Tb^2+1)\mathbb{E}\left[\int_0^T (Z(t)-\bar{Z}(t))^2 dt\right]\Bigg) \\
&\le 4C_1(Tb^2+1)e^{\beta T}\mathbb{E}\left[(\xi-\bar{\xi})^2\right],
\end{aligned} \tag{6.62}$$

where the last inequality follows from (6.60) and (6.61). □

6.4. Markovian system with linear growth rate

This section shows the existence of solution to the multi-dimensional BSDE with reflecting barrier within a Markovian framework. The growth rate of the forward equation is assumed polynomial in the state process X, and linear in both the value process Y and the volatility process Z. The comparison theorem in dimension one, and the continuous dependence property of the value process and the volatility process on the terminal condition, are also provided.

The Markovian system of forward-backward SDEs in question is the following pair of equations.

$$\begin{cases} X^{t,x}(s) = x, \ 0 \le s \le t; \\ dX^{t,x}(s) = f(s, X^{t,x}(s))ds + \sigma(s, X^{t,x}(s))dB_s, \ t < s \le T. \end{cases} \tag{6.63}$$

$$
\begin{cases}
Y^{t,x}(s) = \xi(X^{t,x}(T)) + \displaystyle\int_s^T g(r, X^{t,x}(r), Y^{t,x}(r), Z^{t,x}(r))dr \\
\qquad\qquad - \displaystyle\int_s^T Z^{t,x}(r)dB_r + K^{t,x}(T) - K^{t,x}(s); \\
Y^{t,x}(s) \geq L(s, X^{t,x}(s)), \qquad \displaystyle\int_t^T (Y^{t,x}(s) - L(s, X^{t,x}(s)))dK^{t,x}(s) = 0
\end{cases}
\tag{6.64}
$$

for $t \leq s \leq T$. For any $x \in \mathbb{R}^l$, the SDE (6.63) has a unique strong solution, under Assumption 6.7(1) below (cf. Karatzas and Shreve (1988) [28, p. 287]). A solution to the forward-backward system (6.63) and (6.64) is a triple of processes $(Y^{t,x}, Z^{t,x}, K^{t,x})$ satisfying (6.64), where $Y^{t,x} \in \mathbb{M}^2(m; 0, T)$, $Z^{t,x} \in \mathbb{L}^2(m \times d; 0, T)$, and $K^{t,x}$ is a continuous, increasing process in $\mathbb{M}^2(m; 0, T)$. The superscript (t, x) on X, Y, Z, and K indicates the state x of the underlying process X at time t. It will be omitted for notational simplicity.

Assumption 6.7.

(1) In (6.63), the drift $f : [0,T] \times \mathbb{R}^l \to \mathbb{R}^l$, and volatility $\sigma : [0,T] \times \mathbb{R}^l \to \mathbb{R}^{l \times d}$, are deterministic, measurable mappings, locally Lipschitz in x uniformly over all $t \in [0,T]$. And for all $(t, x) \in [0,T] \times \mathbb{R}^l$, $|f(t,x)|^2 + |\sigma(t,x)|^2 \leq C(1 + |x|^2)$, for some constant C.
(2) In (6.64), the driver g is a deterministic measurable mapping $g : [0,T] \times \mathbb{R}^l \times \mathbb{R}^m \times \mathbb{R}^{m \times d} \to \mathbb{R}^m$, $(t, x, y, z) \mapsto g(t, x, y, z)$. And for all $(t, x, y, z) \in [0,T] \times \mathbb{R}^l \times \mathbb{R}^m \times \mathbb{R}^{m \times d}$, $|g(t, x, y, z)| \leq b(1 + |x|^p + |y| + |z|)$, for some positive constant b.
(3) For every fixed $(t, x) \in [0,T] \times \mathbb{R}^l$, the mapping $g(t, x, \cdot, \cdot)$ is continuous.
(4) The terminal value $\xi : \mathbb{R}^l \to \mathbb{R}^m$, $x \mapsto \xi(x)$, is a deterministic measurable mapping. The lower reflecting boundary $L : [0,T] \times \mathbb{R}^l \to \mathbb{R}^m$, $(s, x) \mapsto L(s, x)$ is deterministic measurable mapping continuous in (s, x). They satisfy $\mathbb{E}[\xi(X(T))^2] < \infty$, $\mathbb{E}\big[\sup_{[0,T]} L^+(s, X(s))^2\big] < \infty$, and $L(T, X(T)) \leq \xi(X(T))$, a.e. on Ω.

Theorem 6.8. *Suppose that Assumption 6.7 holds, except for the growth rate condition on g. If the driver $g(s, x, y, z)$ in the reflected BSDE (6.64) is Lipschitz in y and z, uniform for all $s \in [0,T]$ and all $x \in \mathbb{R}^l$, then there exist measurable deterministic functions $\alpha : [0,T] \times \mathbb{R}^l \to \mathbb{R}^m$, and $\beta : [0,T] \times \mathbb{R}^l \to \mathbb{R}^{m \times d}$, such that for any $0 \leq t \leq s \leq T$, $Y^{t,x}(s) = \alpha(s, X^{t,x}(s))$, and $Z^{t,x}(s) = \beta(s, X^{t,x}(s))$. The solutions to the BSDE are functions of the state process X.*

Proof. First, the one-dimensional case $m = 1$. There exist measurable, deterministic functions $a^n : [0,T] \times \mathbb{R}^l \to \mathbb{R}$, $b^n : [0,T] \times \mathbb{R}^l \to \mathbb{R}^d$, such that for any $0 \le t \le s \le T$, the solution $(Y^{(t,x),n}, Z^{(t,x),n})$ to the penalized equation

$$\begin{aligned} Y^{(t,x),n}(s) &= \xi(X^{t,x}(T)) + \int_s^T g(r, X^{t,x}(r), Y^{(t,x),n}(r), Z^{(t,x),n}(r))dr \\ &\quad - \int_s^T Z^{(t,x),n}(r)dB_r + n\int_s^T (Y^{(t,x),n}(r) - L(r, X^{t,x}(r)))^- dr \end{aligned}$$

can be expressed as $Y^{(t,x),n}(s) = a^n(s, X^{t,x}(s))$, and $Z^{(t,x),n}(s) = b^n(s, X^{t,x}(s))$; in particular, $Y^{(t,x),n}(t) = a^n(t,x)$. This is the Markovian property of solutions to one one-dimensional forward-backward SDEs with Lipschitz driver, stated as Theorem 4.1 in El Karoui, Peng and Quenez (1997) [14]. Their proof uses the Picard iteration and the Markov property of the iterated sequence of solutions, the latter being an interpretation of Theorem 6.27 on page 206 of Çinlar, Jacod, Protter and Sharpe (1980) [7].

As analyzed in section 6, El Karoui, Kapoudjian, Pardoux, Peng and Quenez (1997) [13], the solution $(Y^{(t,x),n}, Z^{(t,x),n})$ converges to some limit $(Y^{t,x}, Z^{t,x})$ in $\mathbb{M}^2(m;t,T) \times \mathbb{L}^2(m \times d; t, T)$. The penalization term $n\int_0^s (Y^{(t,x),n}(r) - L(r, X^{t,x}(r)))^- dr$ also has an $\mathbb{M}^2(m;0,T)$-limit $K^{t,x}(s)$. The triple $(Y^{t,x}, Z^{t,x}, K^{t,x})$ solves the system (6.63) and (6.64). But the convergences are also almost everywhere on $\Omega \times [t,T]$, so

$$\begin{aligned} Y^{t,x}(s) &= \lim_{n\to\infty} Y^{(t,x),n}(s) = \limsup_{n\to\infty}(a^n(s, X^{t,x}(s))) \\ &= \limsup_{n\to\infty}(a^n)(s, X^{t,x}(s)) =: a(s, X^{t,x}(s)), \\ Z^{t,x}(s) &= \lim_{n\to\infty} Z^{(t,x),n}(s) = \limsup_{n\to\infty}(b^n(s, X^{t,x}(s))) \\ &= \limsup_{n\to\infty}(b^n)(s, X^{t,x}(s)) =: b(s, X^{t,x}(s)). \end{aligned}$$

Back to a general dimension m. By Theorem 6.3 and Proposition 6.1, the sequence $(Y^{n+1}, Z^{n+1}) = \Lambda(Y^n, Z^n)$, $n = 0, 1, 2, \cdots$, iterated via the mapping Λ as in (6.3), converges to (Y, Z) a.e. on $\Omega \times [t,T]$ and in $\mathbb{M}^2(m;0,T) \times \mathbb{L}^2(m \times d; 0, T)$. If one can prove $Y^1(s)$ and $Z^1(s)$ are functions of $(s, X(s))$, so is every $(Y^n(s), Z^n(s))$ by induction. Then the theorem holds, because (Y, Z) is the pointwise limit of $\{(Y^n(s), Z^n(s))\}_{n\in\mathbb{N}}$.

Starting with $Y^{(t,x),0}(s) = \alpha^0(s, X(s))$, and $Z^{(t,x),0}(s) = \beta^0(s, X(s))$, for any measurable, deterministic functions $\alpha^0 : [0,T] \times \mathbb{R}^l \to \mathbb{R}^m$, and $\beta^0 : [0,T] \times \mathbb{R}^l \to \mathbb{R}^{m\times d}$ satisfying $\alpha^0(\cdot, X^{t,x}(\cdot)) \in \mathbb{M}^2(m;0,T)$, and

$\beta^0(\cdot, X^{t,x}(\cdot)) \in \mathbb{L}^2(m \times d; 0, T)$. In an arbitrary ith dimension, $1 \le i \le m$,

$$\begin{cases} Y_i^1(s) = \xi_i(X^{t,x}(T)) + \int_s^T g_i(r, X^{t,x}(r), \alpha^0(r, X(r)), \beta^0(r, X(r)))dr \\ \qquad - \int_s^T Z_i^1(r)dB_r + K_i^1(T) - K_i^1(s); \\ Y_i^1(s) \ge L_i(s, X^{t,x}(s)), \qquad \int_t^T (Y_i^1(s) - L_i(s, X^{t,x}(s)))dK_i^1(s) = 0 \end{cases} \tag{6.65}$$

for $t \le s \le T$. From the one-dimensional result, there exist measurable, deterministic functions $\alpha_i^1 : [0,T] \times \mathbb{R}^l \to \mathbb{R}$, and $\beta_i^1 : [0,T] \times \mathbb{R}^l \to \mathbb{R}^d$, such that $Y_i^{(t,x),1}(s) = \alpha_i^1(s, X^{t,x}(s))$, and $Z_i^{(t,x),1}(s) = \beta_i^1(s, X^{t,x}(s))$, for all $0 \le t \le s \le T$. Let $\alpha^1 = (\alpha_1^1, \cdots, \alpha_m^1)'$, and $\beta^1 = (\beta_1^1, \cdots, \beta_m^1)'$, then $Y^{(t,x),1}(s) = \alpha^1(s, X^{t,x}(s))$, and $Z^{(t,x),1}(s) = \beta^1(s, X^{t,x}(s))$, for all $0 \le t \le s \le T$. □

Remark 6.2. To prove the above theorem, besides using the notion of additive martingales as in Çinlar et al (1980) [7], the two deterministic functions can also be obtained by solving a multi-dimensional variational inequality following the four-step-scheme proposed by Ma, Protter and Yong (1994) [33].

The rest of this section will be devoted to proving the existence of solutions to the reflected forward-backward system (6.63) and (6.64) under the Assumption 6.7. We shall construct a specific sequence of Lipschitz drivers $\{g^n\}_{n\in\mathbb{N}}$ to approximate the linear-growth driver g. The corresponding sequence of solutions will turn out to converge to the system (6.63) and (6.64). We then approximate the continuous linear growth driver g by a sequence of Lipschitz functions g^n.

Let $\bar{\psi}$ be an infinitely differentiable mapping from $\mathbb{R}^m \times \mathbb{R}^{m\times d}$ to $\mathbb{R}$, such that

$$\bar{\psi}(y, z) = \begin{cases} 1, & |y|^2 + |z|^2 \le 1; \\ 0, & |y|^2 + |z|^2 \ge 4, \end{cases} \tag{6.66}$$

and ψ a rescaling of $\bar{\psi}$ by a multiplicative constant such that

$$\int_{\mathbb{R}^m \times \mathbb{R}^{m\times d}} \psi(y, z)dydz = 1. \tag{6.67}$$

The function ψ is a kernel conventionally used to smooth out non-differentiability, for example, by Karatzas and Ocone (1992) [27], or to approximate functions of higher growth rate, for example, by Hamadène, Lepeltier and Peng (1997) [19].

The approximating sequence g^n is defined as

$$\begin{aligned} &g^n(t,x,y,z) \\ &= n^2\psi\left(\frac{y}{n},\frac{z}{n}\right)\int_{\mathbb{R}^m\times\mathbb{R}^{m\times d}} g(t,x,y_1,z_1)\bar{\psi}(n(y-y_1),n(z-z_1))dy_1dz_1. \end{aligned} \tag{6.68}$$

According to Hamadène, Lepeltier and Peng (1997) [19], the sequence of functions g^n has the properties:

(a) g^n is Lipschitz with respect to (y,z), uniformly over all $(t,x) \in [0,T]\times\mathbb{R}^l$;

(b) $|g^n(t,x,y,z)| \le b(1+|x|^p+|y|+|z|)$, for all $(t,x,y,z) \in [0,T]\times\mathbb{R}^l\times\mathbb{R}^m\times\mathbb{R}^{m\times d}$, for some positive constant b;

(c) $|g^n(t,x,y,z)| \le b_n(1+|x|^p)$, for all $(t,x,y,z)\in[0,T]\times\mathbb{R}^l\times\mathbb{R}^m\times\mathbb{R}^{m\times d}$, for some positive constant b_n;

(d) for any $(t,x)\in[0,T]\times\mathbb{R}$, and for any compact set $S\subset\mathbb{R}^m\times\mathbb{R}^{m\times d}$,

$$\sup_{(y,z)\in S}|g^n(t,x,y,z)-g(t,x,y,z)|\to 0, \text{ as } n\to 0.$$

Proposition 6.2. The BSDE with reflecting barrier

$$\begin{cases} Y^n(s) = \xi(X(T)) + \int_s^T g^n(r,X(r),Y^n(r),Z^n(r))dr - \int_s^T Z^n(r)dB_r \\ \qquad + K^n(T) - K^n(s); \\ Y^n(s)\ge L(s,X(s)), \quad \int_t^T (Y^n(s)-L(s,X(s)))dK^n(s) = 0 \end{cases} \tag{6.69}$$

for $t\le s\le T$, has a unique solution (Y^n,Z^n,K^n). Furthermore, there exist measurable, deterministic functions α^n and β^n, such that $Y^n(s) = \alpha^n(s,X(s))$, and $Z^n(s)=\beta^n(s,X(s))$, for all $0\le s\le T$.

Proof. This is a direct consequence of the uniform Lipschitz property of g^n, Proposition 6.1 and Theorem 6.6. □

Lemma 6.4. *Suppose (Y,Z,K) solves the BSDE (6.64) with reflecting barrier. Under the conditions (2) and (4) of Assumption 6.7, there exists a* *positive constant C such that*

$$\mathbb{E}\left[\sup_{0\le s\le T} Y(s)^2 + \int_t^T Z(r)^2 ds + K(T)^2\right] \le C(1+|x|^{2(p\vee 1)}).$$

The constant C does not depend on t, but depends on m, b, T, $\mathbb{E}[\xi(X(T))^2]$ and $\mathbb{E}\big[\sup_{[0,T]} L^+(t,X(t))^2\big]$.

Proof. First prove that, for some constant C', we have

$$\mathbb{E}\left[Y(s)^2 + \int_t^T Z(r)^2 ds + K(T)^2\right] \le C'(1+|x|^{2(p\vee 1)}), \text{ for all } 0\le s\le T. \tag{6.70}$$

Applying Itô's rule to $Y(\cdot)^2$, and integrating from s to T, we get

$$\begin{aligned}
&Y(s)^2 + \int_s^T Z(r)^2 dr \\
&= \xi(X(T))^2 + 2\int_s^T Y(r) g(r,X(r),Y(r),Z(r))dr - 2\int_s^T Y(r)Z(r)dB(r) \\
&\quad + 2\int_s^T L(r,X(r))dK(r).
\end{aligned} \tag{6.71}$$

Taking expectations of (6.71), and using Assumption 6.7(2), we obtain

$$\begin{aligned}
&\mathbb{E}\left[Y(s)^2 + \int_s^T Z(r)^2 dr\right] \\
&\le \mathbb{E}[\xi(X(T))^2] + 2b\mathbb{E}\left[\int_s^T |Y(r)|(1+|X(r)|^p + |Y(r)| + |Z(r)|)dr\right] \\
&\quad + 2\mathbb{E}\left[\int_s^T L(r,X(r))dK(r)\right] \\
&\le \mathbb{E}[\xi(X(T))^2] + 2\mathbb{E}\left[\int_s^T (1+|X(r)|^{2p})dr\right] + C_1(b)\mathbb{E}\left[\int_s^T |Y(r)|^2 dr\right] \\
&\quad + \frac{1}{4}\mathbb{E}\left[\int_s^T |Z(r)|^2 dr\right] + 2\mathbb{E}\left[\int_s^T L(r,X(r))dK(r)\right].
\end{aligned} \tag{6.72}$$

For any $\alpha > 0$,

$$2\int_t^T L(s,X(s))dK(s) \le 2\left(\sup_{[0,T]} L(s,X(s))\right)K(T) \le \frac{1}{\alpha}K(T)^2 + \alpha \sup_{[0,T]} L^+(s,X(s))^2. \tag{6.73}$$

Combine (6.72) and (6.73), and apply Gronwall's Lemma to $Y(\cdot)$,

$$\begin{aligned}&\mathbb{E}\left[Y(s)^2 + \frac{3}{4}\int_s^T Z(r)^2 dr\right]\\ &\le C_2(b,T)\Big(1 + \mathbb{E}[\xi(X(T))^2] + \mathbb{E}\Big[\int_s^T |X(r)|^{2p}dr\Big]\\ &\quad + \frac{1}{\alpha}K(T)^2 + \alpha \sup_{[0,T]} L^+(s,X(s))^2\Big).\end{aligned} \tag{6.74}$$

Rewriting (6.64) from t to T, $K(\cdot)$ can be expressed in terms of $Y(\cdot)$ and $Z(\cdot)$ by

$$K(T) = Y(t) - \xi(X(T)) - \int_t^T g(s,X(s),Y(s),Z(s))ds + \int_t^\cdot Z(s)dB_s, \tag{6.75}$$

and hence because of the linear growth Assumption 6.7 (2), we have

$$\begin{aligned}&\mathbb{E}[K(T)^2]\\ &= C_3\mathbb{E}\left[Y(t)^2 + \xi(X(T))^2 + \int_t^T g(s,X(s),Y(s),Z(s))^2 ds + \int_t^T Z(s)^2 ds\right]\\ &\le C_4(b)\Big(\mathbb{E}\Big[Y(t)^2 + \xi(X(T))^2 + 1 + \int_t^T |X(s)|^{2p}ds\Big] + \mathbb{E}\Big[\int_t^T |Y(s)|^2 ds\Big]\\ &\quad + \mathbb{E}\Big[\int_t^T |Z(s)|^2 ds\Big]\Big).\end{aligned} \tag{6.76}$$

Using the strength of the bounds on $\mathbb{E}[|Y(s)|^2]$ and $\mathbb{E}\left[\int_t^T |Z(s)|^2 ds\right]$ obtained in (6.74), we deduce from (6.76):

$$\begin{aligned}\mathbb{E}[K(T)^2] &\le C_5(b,t,T)\Big(\mathbb{E}\Big[\xi(X(T))^2 + 1 + \int_t^T |X(s)|^{2p}ds\Big]\\ &\quad + \frac{1}{\alpha}\mathbb{E}[K(T)^2] + \alpha\mathbb{E}\Big[\sup_{[0,T]} L^+(s,X(s))^2\Big]\Big).\end{aligned} \tag{6.77}$$

Let $\alpha = 4C_5(b,t,T)$, and collect $\mathbb{E}[K(T)^2]$ terms on both sides of (6.77),

$$\begin{aligned}\mathbb{E}[K(T)^2] \le C_6(b,t,T)\mathbb{E}\Big[\xi(X(T))^2 + 1 + \int_t^T |X(s)|^{2p} ds \\ + \sup_{[0,T]} L^+(s,X(s))^2\Big].\end{aligned} \tag{6.78}$$

Finally, (6.74) and (6.78) altogether gives

$$\begin{aligned}&\mathbb{E}\left[Y(s)^2 + \int_s^T Z(r)^2 ds + K(T)^2\right] \\ &\le C_7(b,t,T)\bigg(1 + \mathbb{E}[\xi(X(T))^2] + \mathbb{E}\left[\int_t^T |X(r)|^{2p} dr\right] \\ &\quad + \mathbb{E}\left[\sup_{[0,T]} L^+(s,X(s))^2\right]\bigg).\end{aligned} \tag{6.79}$$

From Karatzas and Shreve (1988) [28, p. 306], for $p \ge 1$,

$$\mathbb{E}\left[\sup_{[0,T]} |X^{t,x}(s)|^{2p}\right] \le C_8(1 + |x|^{2p}). \tag{6.80}$$

Then the constant C' in (6.70) can be chosen as

$$\begin{aligned}C' = \Big(\sup_{0\le t\le T} C_7(b,t,T)\Big) \max\Big\{1 + \mathbb{E}[\xi(X(T))^2] \\ + \mathbb{E}\left[\sup_{[0,T]} L^+(s,X(s))^2\right], C_8 T\Big\} < \infty.\end{aligned} \tag{6.81}$$

To bound the $\mathbb{L}^2$ supremum norm of $Y(\cdot)$, first take the supremum over $s \in [0,T]$, then the expectation, on both sides of (6.71). Using the Burkholder-

Davis-Gundy inequality, and combining with (6.73),

$$
\begin{aligned}
&\mathbb{E}\left[\sup_{[0,T]} Y(s)^2 + \int_t^T Z(r)^2 dr\right] \\
&\le \mathbb{E}[\xi(X(T))^2] + 2b\mathbb{E}\left[\sup_{[0,T]} \int_s^T |Y(r)|(1 + |X(r)|^p + |Y(r)| + |Z(r)|)dr\right] \\
&\quad + C_9(m)\mathbb{E}\left[\sqrt{\int_t^T |Y(r)|^2 \cdot |Z(r)|^2 dr}\right] + 2\mathbb{E}\left[\int_s^T L(r, X(r))dK(r)\right] \\
&\le \mathbb{E}[\xi(X(T))^2] + b\mathbb{E}\left[\int_t^T (1 + |X(r)|^p + |Y(r)| + |Z(r)|)^2 dr\right] \\
&\quad + b\mathbb{E}\left[\int_t^T |Y(r)|^2 dr\right] + C_9(m)\mathbb{E}\left[\sup_{[0,T]} |Y(s)| \sqrt{\int_t^T |Z(r)|^2 dr}\right] \\
&\quad + \mathbb{E}[K(T)^2] + \mathbb{E}\left[\sup_{[0,T]} L^+(s, X(s))^2\right] \\
&\le \mathbb{E}[\xi(X(T))^2] + C_{10}(b)\mathbb{E}\left[\int_t^T (1 + |X(r)|^{2p} + |Y(r)|^2 + |Z(r)|^2)dr\right] \\
&\quad + \frac{1}{2}\mathbb{E}\left[\sup_{[0,T]} |Y(s)|^2\right] + 2C_9(m)^2\mathbb{E}\left[\int_t^T |Z(r)|^2 dr\right] + \mathbb{E}[K(T)^2] \\
&\quad + \mathbb{E}\left[\sup_{[0,T]} L^+(s, X(s))^2\right].
\end{aligned}
\tag{6.82}
$$

Equation (6.82) implies that

$$
\begin{aligned}
&\frac{1}{2}\mathbb{E}\left[\sup_{[0,T]} Y(s)^2\right] \\
&\le \mathbb{E}[\xi(X(T))^2] + C_{10}(b)\mathbb{E}\left[\int_t^T (1 + |X(r)|^{2p} + |Y(r)|^2 + |Z(r)|^2)dr\right] \\
&\quad + 2C_9(m)^2\mathbb{E}\left[\int_t^T |Z(r)|^2 dr\right] + \mathbb{E}[K(T)^2] + \mathbb{E}\left[\sup_{[0,T]} L^+(s, X(s))^2\right].
\end{aligned}
\tag{6.83}
$$

Inequalities (6.70), (6.80) and (6.83) conclude the lemma. □

Proposition 6.3. There exists a positive constant C, such that for $0 \leq t \leq T$, $n = 1, 2, \cdots$,

$$\alpha^n(t,x) = Y^{(t,x),n}(t) = \mathbb{E}[Y^{(t,x),n}(t)|\mathscr{F}_t] \leq C(1 + |x|^{p\vee 1}). \tag{6.84}$$

Proposition 6.4. The sequence $\{g^n(\cdot, X(\cdot), Y^n(\cdot), Z^n(\cdot))\}_{n\in\mathbb{N}}$ is uniformly bounded in the $\mathbb{L}^2(m;t,T)$-norm, and the sequence $\{K^n(\cdot)\}_{n\in\mathbb{N}}$ is uniformly bounded in the $\mathbb{M}^2(m;t,T)$-norm, both uniformly over all n. As $n \to \infty$, $g^n(\cdot, X(\cdot), Y^n(\cdot), Z^n(\cdot))$ weakly converges to some limit $G(\cdot)$ in $\mathbb{L}^2(m;t,T)$ along a subsequence, and $K^n(\cdot)$ weakly converges to some limit $K(\cdot)$ in $\mathbb{M}^2(m;t,T)$ along a subsequence, for every $s \in [t,T]$.

Proof. It suffices to show the uniform boundedness of

$$\{g^n(\cdot, X(\cdot), Y^n(\cdot), Z^n(\cdot))\}_{n\in\mathbb{N}}$$

in $\mathbb{L}^2(m;t,T)$ and of $\{K^n(T)\}_{n\in\mathbb{N}}$ in $\mathbb{L}^2(m)$, which is a result of the linear growth property (b) and Lemma 6.4. The $\mathbb{L}^2(m)$ uniform boundedness of $\{K^n(T)\}_{n\in\mathbb{N}}$ means that there exists a constant $C < \infty$, such that $\mathbb{E}[|K^n(T)|^2] < C$. Since $K^n(\cdot)$ is required to be an increasing process starting from $K^n(t) = 0$, then for all $t \leq s \leq T$,

$$\mathbb{E}[|K^n(s)|^2] \leq \mathbb{E}[|K^n(T)|^2] < C.$$

□

With the help of weak convergence along a subsequence, we proceed to argue that the weak limits are also strong, thus deriving a solution to BSDE (6.64). For notational simplicity, the weakly convergent subsequences are still indexed by n. The passing from weak to strong convergence makes use of the Markovian structure of the system described by Theorem 6.6, which states that the valued process $Y^n(s)$ is a deterministic function of time s and state process $X(s)$ only.

Lemma 6.5. *The approximating sequence of solutions*

$$\{(Y^{(t,x),n}, Z^{(t,x),n})\}_{n\in\mathbb{N}}$$

is Cauchy in $\mathbb{L}^2(m;t,T) \times \mathbb{L}^2(m \times d;t,T)$, *thus has a limit* $(Y^{t,x}, Z^{t,x})$ *in* $\mathbb{L}^2(m;t,T) \times \mathbb{L}^2(m \times d;t,T)$ *and a.e. on* $[t,T] \times \Omega$.

Proof. For any $t \in [0,T]$, any $x \in \mathbb{R}^l$, and any $n = 1, 2, \cdots$, $Y^{(t,x),n}(t) = \alpha^n(t,x)$ is deterministic. First prove the convergence of $\{\alpha^n(t,x)\}_{n\in\mathbb{N}}$ by showing it is Cauchy. For $n_1, n_2 = 1, 2, \cdots$, from equation (6.69) comes the following inequality,

$$\begin{aligned} &|\alpha^{n_1}(t,x) - \alpha^{n_2}(t,x)| = |Y^{n_1}(t) - Y^{n_2}(t)| \\ \leq &\left|\mathbb{E}\left[\int_t^T (g^{n_1}(s, X(s), Y^{n_1}(s), Z^{n_1}(s)) - g^{n_2}(s, X(s), Y^{n_2}(s), Z^{n_2}(s)))ds\right]\right| \\ &+ |\mathbb{E}[K^{n_1}(T) - K^{n_2}(T)]| + |\mathbb{E}[K^{n_1}(t) - K^{n_2}(t)]|. \end{aligned} \tag{6.85}$$

By the weak convergence from Proposition 6.4, all the three summands on the right hand side of the above inequality converge to zero, as n_1 and n_2 both go to infinity. Denote the limit of $\alpha^n(t,x)$ as $\alpha(t,x)$, which is consequently deterministic and measurable, because $\alpha^n(\cdot,\cdot)$ is measurable. Theorem 6.6 states that for any $t \leq s \leq T$, $Y^{(t,x),n}(s) = \alpha^n(s, X^{t,x}(s))$. Because of the pointwise convergence of $\alpha^n(\cdot,\cdot)$, $Y^{(t,x),n}(s)$ converges to some $Y^{(t,x)}(s)$, a.e. $(s,\omega) \in [t,T] \times \Omega$, as $n \to \infty$. Proposition 6.3 states that there exists a positive constant C, such that for $0 \leq t \leq T$, $n = 1, 2, \cdots$,

$$|Y^{(t,x),n}(s)| = |\alpha^n(s, X_s^{t,x})| \leq C(1 + |X_s^{t,x}|^{p\vee 1}), \tag{6.86}$$

the last term of which is square-integrable by (6.80). Then it follows from the dominated convergence theorem that the convergence of $Y^{(t,x),n}(s)$ is also in $\mathbb{L}^2(m; t, T)$.

Apply Itô's rule to $(Y^{(t,x),n_1}(s) - Y^{(t,x),n_2}(s))^2$, and integrate from s to T. The reflection conditions in (6.64) gives

$$\begin{aligned} &(Y^{n_1}(s) - Y^{n_2}(s))^2 + \int_s^T (Z^{n_1}(r) - Z^{n_2}(r))^2 dr \\ &\leq \int_s^T (Y^{n_1}(r) - Y^{n_2}(r))\Big(g^{n_1}(r, X(r), Y^{n_1}(r), Z^{n_1}(r)) \\ &\quad - g^{n_2}(r, X(r), Y^{n_2}(r), Z^{n_2}(r))\Big) dr \\ &\quad + \int_s^T (Y^{n_1}(r) - Y^{n_2}(r))(Z^{n_1}(r) - Z^{n_2}(r)) dB_r. \end{aligned} \tag{6.87}$$

Taking the expectation of (6.87),

$$\begin{aligned}
&\mathbb{E}[(Y^{n_1}(s) - Y^{n_2}(s))^2] + \mathbb{E}\left[\int_s^T (Z^{n_1}(r) - Z^{n_2}(r))^2 dr\right] \\
&\le \mathbb{E}\bigg[\int_s^T (Y^{n_1}(r) - Y^{n_2}(r))\big(g^{n_1}(r, X(r), Y^{n_1}(r), Z^{n_1}(r)) \\
&\qquad - g^{n_2}(r, X(r), Y^{n_2}(r), Z^{n_2}(r))\big)\, dr\bigg] \\
&\le \mathbb{E}\bigg[\int_s^T (Y^{n_1}(r) - Y^{n_2}(r))^2 dr\bigg]^{\frac{1}{2}} \cdot \mathbb{E}\bigg[\int_s^T \big(g^{n_1}(r, X(r), Y^{n_1}(r), Z^{n_1}(r)) \\
&\qquad\qquad - g^{n_2}(r, X(r), Y^{n_2}(r), Z^{n_2}(r))\big)^2 dr\bigg]^{\frac{1}{2}}.
\end{aligned} \tag{6.88}$$

In order to prove convergence of $\{Z^n(\cdot)\}_{n\in\mathbb{N}}$, it suffices to prove uniform boundedness of

$$\mathbb{E}\bigg[\int_t^T g^n(s, X(s), Y^n(s), Z^n(s))^2 ds\bigg],$$

for all n, which is part of Proposition 6.4. The $\mathbb{L}^2(m\times d; t, T)$-convergence of $\{Z^n\}_{n\in\mathbb{N}}$ implies almost sure convergence along a subsequence, also denoted as $\{Z^n\}_{n\in\mathbb{N}}$ to simplify the notation. □

We have identified a strongly convergent subsequence of $\{(Y^n, Z^n)\}_{n\in\mathbb{N}}$, also denoted as $\{(Y^n, Z^n)\}_{n\in\mathbb{N}}$. Let us remind ourselves that (Y^n, Z^n) solves the system (6.63) and (6.69), so if the weak limit $G(\cdot)$ of $g^n(\cdot, X(\cdot), Y^n(\cdot), Z^n(\cdot))$ is also the strong limit, and if $G(\cdot)$ has the form $g(\cdot, X(\cdot), Y(\cdot), Z(\cdot))$, then the limit (Y, Z, K) indeed solves the forward-backward system (6.63) and (6.64).

Lemma 6.6. *As $n \to \infty$,*

$$g^n(s, X(s), Y^n(s), Z^n(s)) \to g(s, X(s), Y(s), Z(s)),$$

in $\mathbb{L}^2(m; t, T)$ and a.e. on $[t, T] \times \Omega$.

Proof. The method is the same as that on page 122 of Hamadène, Lepeltier and Peng (1997) [19]. The proof is briefly repeated here for completeness.

$$
\begin{aligned}
&\mathbb{E}\left[\int_t^T |g^n(s,X(s),Y^n(s),Z^n(s)) - g(s,X(s),Y(s),Z(s))|ds\right] \\
&\quad \le \mathbb{E}\Bigg[\int_t^T \Big|g^n(s,X(s),Y^n(s),Z^n(s)) \\
&\qquad - g(s,X(s),Y^n(s),Z^n(s))\Big|\mathbb{1}_{|Y^n(s)+Z^n(s)|\ge A}ds\Bigg] \\
&\quad + \mathbb{E}\Bigg[\int_t^T \Big|g^n(s,X(s),Y^n(s),Z^n(s)) \\
&\qquad - g(s,X(s),Y^n(s),Z^n(s))\Big|\mathbb{1}_{|Y^n(s)+Z^n(s)|\le A}ds\Bigg] \\
&\quad + \mathbb{E}\left[\int_t^T |g(s,X(s),Y^n(s),Z^n(s)) - g(s,X(s),Y(s),Z(s))|ds\right].
\end{aligned} \tag{6.89}
$$

By the linear growth Assumption 6.7(2) for g and property (b) for g^n, and Lemma 6.4, both

$$|g^n(s,X(s),Y^n(s),Z^n(s)) - g(s,X(s),Y^n(s),Z^n(s))|$$

and

$$|g(s,X(s),Y^n(s),Z^n(s)) - g(s,X(s),Y(s),Z(s))|$$

are uniformly bounded in $\mathbb{L}^2(m;0,T)$ for all n. The first term on the right hand side of (6.89) is at most of the order $\frac{1}{A}$, thus vanishing as A goes to infinity. Recalling property (d), for fixed A, the second term vanishes as $n \to \infty$. Because of its uniform boundedness in $\mathbb{L}^2(m;t,T)$, the integrand in the third term is uniformly integrable for all n, so expectation of the integral again goes to 0 as $n \to \infty$.

The a.e. convergent subsequence of $g^n(s,X(s),Y^n(s),Z^n(s))$ is also indexed by n to simplify the notation. □

Proposition 6.5. The $\mathbb{L}^2(m;t,T)$ convergence and the a.e. convergence of $\{Y^{(t,x),n}(s)\}_{n\in\mathbb{N}}$ to $Y^{(t,x)}(s)$ are uniform over all $s \in [t,T]$.

Proof. To see uniform convergence of $\{Y^n\}$, applying Itô's rule to $(Y^n(s) - Y(s))^2$, integrating from s to T, taking supremum over $0 \le s \le T$ and then expectation, by the Burkholder-Davis-Gundy inequality,

$$
\begin{aligned}
&\mathbb{E}\left[\sup_{[0,T]}(Y^n(s)-Y(s))^2\right]+\mathbb{E}\left[\int_t^T(Z^n(r)-Z(r))^2dr\right]\\
&\leq \mathbb{E}\Bigg[\sup_{[0,T]}\int_s^T(Y^n(r)-Y(r))(g^n(r,X(r),Y^n(r),Z^n(r))\\
&\qquad -g(r,X(r),Y(r),Z(r)))dr\Bigg]\\
&\quad +\mathbb{E}\left[\left(\int_t^T(Y^n(r)-Y(r))^2(Z^n(r)-Z(r))^2dr\right)^{\frac{1}{2}}\right]\\
&\leq \mathbb{E}\Bigg[\left(\int_s^T(g^n(s,X(r),Y^n(r),Z^n(r))-g(r,X(r),Y(r),Z(r)))^2dr\right)^{\frac{1}{2}}\\
&\qquad \cdot\sup_{s\in[0,T]}\left(\int_s^T(Y^n(r)-Y(r))^2dr\right)^{\frac{1}{2}}\Bigg]\\
&\quad +\mathbb{E}\left[\sup_{s\in[0,T]}\{|Y^n(s)-Y(s)|\}\left(\int_t^T(Z^n(r)-Z(r))^2dr\right)^{\frac{1}{2}}\right]\\
&\leq\left(\mathbb{E}\left[\int_t^T(g^n(r,X(r),Y^n(r),Z^n(r))-g(r,X(r),Y(r),Z(r)))^2dr\right]\right)^{\frac{1}{2}}\\
&\quad\cdot\left(\mathbb{E}\left[\int_t^T(Y^n(r)-Y(r))^2dr\right]\right)^{\frac{1}{2}}+\frac{1}{4}\mathbb{E}\left[\sup_{s\in[0,T]}|Y^n(s)-Y(s)|^2\right]\\
&\quad+\mathbb{E}\left[\int_t^T(Z^n(r)-Z(r))^2dr\right].
\end{aligned}\tag{6.90}
$$

Equation (6.90) implies

$$
\begin{aligned}
&\frac{3}{4}\mathbb{E}\left[\sup_{s\in[0,T]}(Y^n(s)-Y(s))^2\right]\\
&<\left(\mathbb{E}\left[\int_t^T(Y^n(r)-Y(r))^2dr\right]\right)^{\frac{1}{2}}\\
&\quad\cdot\left(\mathbb{E}\left[\int_t^T(g^n(r,X(r),Y^n(r),Z^n(r))-g(r,X(r),Y(r),Z(r)))^2dr\right]\right)^{\frac{1}{2}}.
\end{aligned}\tag{6.91}
$$

By Proposition 6.4, by linear growth properties (b) of g^n and Assumption 6.7(2) on g, and by Lemma 6.4, the second multiplier on the right hand side of (6.91) is bounded by a constant, uniformly over all n. By Lemma 6.5, the first multiplier on the right hand side of (6.91) converges to zero as $n \to \infty$. Hence

$$\lim_{n\to\infty} \mathbb{E}\left[\sup_{[0,T]}(Y^n(s) - Y(s))^2\right] = 0.$$

□

Proposition 6.6. The process $K^n(\cdot)$ converges to some limit $K(\cdot)$ in $\mathbb{M}^1(m; t, T)$, uniformly over all $s \in [t, T]$, and a.e. on $[t, T] \times \Omega$.

Proof. Define

$$\bar{K}(s) := Y(t) - Y(s) - \int_t^s g(r, X(r), Y(r), Z(r))dr + \int_t^s Z(r)dB_r,$$

for $t \le s \le T$, where $Y(\cdot)$, $Z(\cdot)$ and g are the limits of $Y^n(\cdot)$, $Z^n(\cdot)$ and g^n. From (6.69),

$$K^n(s) = Y^n(t) - Y^n(s) - \int_t^s g^n(r, X(r), Y^n(r), Z^n(r))dr + \int_t^s Z^n(r)dB_r.$$

We need to show that

$$\mathbb{E}\left[\sup_{s\in[0,T]} \left|K^n(s) - \bar{K}(s)\right|\right] \to 0,$$

as $n \to \infty$.

For all $n = 1, 2, \cdots$,

$$\begin{aligned}
&\mathbb{E}\left[\sup_{s\in[0,T]} \left|K^n(s) - \bar{K}(s)\right|\right] \\
&\le \mathbb{E}\left[|Y^n(t) - Y(t)|\right] + \mathbb{E}\left[\sup_{s\in[0,T]} |Y^n(s) - Y(s)|\right] \\
&\quad + \mathbb{E}\left[\sup_{s\in[0,T]} \left|\int_t^s (Z^n(r) - Z(r))dB_r\right|\right] \\
&\quad + \mathbb{E}\left[\int_t^T |g^n(r, X(r), Y^n(r), Z^n(r)) - g(r, X(r), Y(r), Z(r))|\, dr\right].
\end{aligned} \tag{6.92}$$

As $n \to \infty$, the first three summands in (6.92) go to zero, by Lemma 6.5, Proposition 6.5 and Lemma 6.6. From the Burkholder-Davis-Gundy inequality, there exists a constant C universal for all n, such that

$$\mathbb{E}\left[\sup_{s\in[0,T]}\left|\int_t^s (Z^n(r)-Z(r))dB_r\right|\right] \le C\mathbb{E}\left[\left(\int_t^T |Z^n(r)-Z(r)|^2 dr\right)^{\frac{1}{2}}\right], \tag{6.93}$$

the right hand side of which converges to zero as $n \to \infty$, by Lemma 6.5.

The a.e. convergent subsequence is still denoted as $\{K^n(\cdot)\}_{n\in\mathbb{N}}$ to simplify notation. The strong limit $\bar{K}(\cdot)$ coincides with the weak limit $K(\cdot)$ in Proposition 6.4. □

Proposition 6.7. The processes $Y(\cdot)$ and $K(\cdot)$ satisfy the reflection conditions $Y(\cdot) \ge L(\cdot, X(\cdot))$ and $\int_t^T (Y(s) - L(s, X(s)))dK(s) = 0$.

Proof. Since (Y^n, Z^n, K^n) solves (6.69), $Y^n(\cdot)$ and $K^n(\cdot)$ satisfy the reflecting conditions $Y^n(s) \ge L(s, X(s))$, $t \le s \le T$, and $\int_t^T (Y^n(s) - L(s, X(s)))dK^n(s) = 0$. Since $Y^n(\cdot)$ converges to $Y(\cdot)$ pointwise on $[0,T] \times \Omega$, that $Y(\cdot) \ge L(\cdot, X(\cdot))$ holds true. It remains to prove

$$\int_t^T (Y(s) - L(s, X(s)))dK(s) = \int_t^T (Y^n(s) - L(s, X(s)))dK^n(s).$$

To wit,

$$\begin{aligned}
&\left|\int_t^T (Y^n(s) - L(s, X(s)))dK^n(s) - \int_t^T (Y(s) - L(s, X(s)))dK(s)\right| \\
&\le \left|\int_t^T (Y^n(s) - Y(s))dK^n(s)\right| \\
&\quad + \left|\int_t^T (Y(s) - L(s, X(s)))d(K(s) - K^n(s))\right| \qquad (6.94)\\
&\le \left|\sup_{s\in[0,T]}\{Y^n(s) - Y(s)\}K^n(T)\right| \\
&\quad + \left|\int_t^T (Y(s) - L(s, X(s)))d(K(s) - K^n(s))\right|,
\end{aligned}$$

Let n tend to zero. By Proposition 6.5, the first summand in the last line of (6.94) converges to $|0 \cdot K(T)| = 0$, a.e. on Ω. Proposition 6.6 implies that $K^n(s)$ converges to $K(s)$ in probability, uniformly over all $s \in [t, T]$, so the measure $dK^n(s)$ weakly converges to $dK(s)$ in probability, uniformly over all $s \in [t, T]$. It follows that the second summand in the last line of (6.94) converges to zero, a.e. on Ω. □

We may now conclude the following existence result.

Theorem 6.7. *Under Assumption 6.7, there exists a solution (Y, Z, K) to the BSDE (6.64) with reflecting barrier in the Markovian framework.*

Proof. The solutions $\{(Y^n, Z^n, K^n)\}_{n\in\mathbb{N}}$ to the approximating equations (6.69) have limits (Y, Z, K). The triple (Y, Z, K) is a solution to the Markovian system (6.63) and (6.64). □

Theorem 6.8 (Comparison Theorem). *Suppose $(Y^{t,x}, Z^{t,x}, K^{t,x})$ solves the forward-backward system (6.63) and (6.64) with parameter set (ξ, g, L), and $(\bar{Y}^{t,x}, \bar{Z}^{t,x}, \bar{K}^{t,x})$ solves the forward-backward system (6.63) and (6.64) with parameter set $(\bar{\xi}, \bar{g}, \bar{L})$. Let dimension of the equations be $m = 1$. Under Assumption 6.7 for both sets of parameters, if*

(1) $\xi(x) \le \bar{\xi}(x)$, a.e., for all $x \in \mathbb{R}^l$;
(2) $g(s, x, y, z) \le \bar{g}(s, x, y, z)$, for all $t \le s \le T$, and all $(x, y, z) \in \mathbb{R}^l \times \mathbb{R} \times \mathbb{R}^d$; and
(3) $L(s, x) \le \bar{L}(s, x)$, for all $t \le s \le T$, and all $x \in \mathbb{R}^l$,

then

$$Y^{t,x}(s) \le \bar{Y}^{t,x}(s), \text{ for all } t \le s \le T.$$

Proof. Let $\{g^n\}_{n\in\mathbb{N}}$ and $\{\bar{g}^n\}_{n\in\mathbb{N}}$ be, respectively, the uniform Lipschitz sequences approximating g and $\bar{g}$ as in (6.68). According to Property (a), both g^n and $\bar{g}^n$ are Lipschitz in (y, z), for all t and x. We notice that (2) in the conditions of this theorem implies that

$$g^n(s, x, y, z) \le \bar{g}^n(s, x, y, z),$$

for all $t \leq s \leq T$, and all $(x, y, z) \in \mathbb{R}^l \times \mathbb{R}^m \times \mathbb{R}^{m\times d}$, via construction (6.68). Let $(Y^{(t,x),n}, Z^{(t,x),n}, K^{(t,x),n})$ be solution to system (6.63) and (6.64) with parameter set (ξ, g^n, L), and $(\bar{Y}^{(t,x),n}, \bar{Z}^{(t,x),n}, \bar{K}^{(t,x),n})$ be solution to system (6.63) and (6.64) with parameter set $(\bar{\xi}, \bar{g}^n, \bar{L})$. By Theorem 6.4,

$$Y^{(t,x),n}(s) \leq \bar{Y}^{(t,x),n}(s),\ t \leq s \leq T.$$

But as $n \to \infty$, proven earlier in this section,

$$Y^{(t,x),n}(\cdot) \to Y^{t,x}(\cdot),\ \bar{Y}^{(t,x),n}(\cdot) \to \bar{Y}^{t,x}(\cdot),$$

a.e. on $[t, T] \times \Omega$ and in $\mathbb{L}^2(m; t, T)$, so

$$Y^{t,x}(s) \leq \bar{Y}^{t,x}(s),\ t \leq s \leq T.$$

□

Theorem 6.9 (Continuous Dependence Property). *Under Assumption 6.7, if $(Y^{t,x}, Z^{t,x}, K^{t,x})$ solves the system (6.63) and (6.64), and $(\bar{Y}^{t,x}, \bar{Z}^{t,x}, \bar{K}^{t,x})$ solves the system (6.63) and*

$$\begin{cases} \bar{Y}^{t,x}(s) = \bar{\xi}(X^{t,x}(T)) + \int_s^T g(r, X^{t,x}(r), \bar{Y}^{t,x}(r), \bar{Z}^{t,x}(r))dr \\ \qquad - \int_s^T \bar{Z}^{t,x}(r)dB_r + \bar{K}^{t,x}(T) - \bar{K}^{t,x}(s); \\ \bar{Y}^{t,x}(s) \geq L(s, X^{t,x}(s)), \quad \int_t^T (\bar{Y}^{t,x}(s) - L(s, X^{t,x}(s)))d\bar{K}^{t,x}(s) = 0, \end{cases} \tag{6.95}$$

for $t \leq s \leq T$, then

$$\mathbb{E}[(Y^{t,x}(s) - \bar{Y}^{t,x}(s))^2] + \mathbb{E}\left[\int_s^T (Z^{t,x}(r) - \bar{Z}^{t,x}(r))^2 dr\right]$$
$$< \mathbb{E}[|\zeta - \bar{\zeta}|^2] + C\mathbb{E}\left[\int_s^T (Y^{t,x}(r) - \bar{Y}^{t,x}(r))^2 dr\right]^{\frac{1}{2}}$$

for $0 \leq t \leq s \leq T$.

Proof. Apply Itô's rule to $(Y^{t,x} - \bar{Y}^{t,x})^2$, and integrate from s to T. Use Lemma 6.4 and Assumption 6.7(2). □

Remark 6.3. When the driver g is concerned about in Assumption 6.7, 6.7(2) (linear growth rates in y and z, and polynomial growth rate in x) is crucial in bounding the $\mathbb{L}^2$-norms thus proving convergence of a Lipschitz approximating sequence. It is likely that the continuity Assumption 6.7(3) can be relaxed, because a measurable function can always be approximated by continuous functions of the same growth rate.

Remark 6.4. The results in section 6.3 and section 6.4 are valid for any arbitrary filtered probability space that can support a d-dimensional Brownian motion. In particular, in the canonical space set up at the beginning of section 6.2, we may replace Assumption 6.7(1) and (2) with the more general Assumption 6.1A(1') and (2'), while still getting exactly the same statements in section 6.4 with tiny modifications of the proofs. Assumption 6.1A corresponds to Assumption 6.1 on the state process $X(\cdot)$ in (6.2). The growth rate (6.38) of the Hamiltonians (6.36) and (6.37) satisfies Assumption 6.1A(2').

Assumption 6.1A.

(1') In (6.63), the drift $f : [0,T] \times C^l[0,\infty) \to \mathbb{R}^l$, $(t,\omega) \mapsto f(t,\omega(t))$, and volatility $\sigma : [0,T] \times C^l[0,\infty) \to \mathbb{R}^{l\times d}$, $(t,\omega) \mapsto \sigma(t,\omega(t))$, are deterministic, measurable mappings such that

$$|f(t,\omega(t)) - f(t,\bar{\omega}(t))| + |\sigma(t,\omega(t)) - \sigma(t,\bar{\omega}(t))| \leq C \sup_{0\leq s\leq t} |\omega(s) - \bar{\omega}(s)|,$$

and

$$|f(t,\omega(t))|^2 + |\sigma(t,\omega(t))|^2 \leq C\left(1 + \sup_{0\leq s\leq t} |\omega(s)|^2\right),$$

with some constant C for all $0 \leq t \leq T$, ω and $\bar{\omega}$ in $C^l[0,\infty)$.

(2') In (6.64), the driver g is a deterministic measurable mapping $g : [0,T]\times C^l[0,\infty) \times \mathbb{R}^m \times^{m\times d} \to \mathbb{R}^m$, $(t,\omega,y,z) \mapsto g(t,\omega(t),y,z)$. And

$$|g(t,\omega(t),y,z)| \leq b\left(1 + \sup_{0\leq s\leq t} |\omega(s)|^p + |y| + |z|\right),$$

with some positive constant b for all $(t,\omega,y,z) \in [0,T] \times C^l[0,\infty) \times \mathbb{R}^m \times \mathbb{R}^{m\times d}$.

Acknowledgements

We would like to thank Professor Said Hamadène for sending to us the paper Hamadène, Lepeltier and Peng (1997) [19]. Thanks go to Professor Shige

Peng for helpful conversations with the second author during his visit to Columbia University in Fall 2008; in particular, the proof of Theorem 6.6 in this paper is due to Professor Peng. We would also like to thank Professor Daniel Ocone for his many very helpful comments and suggestions.

References

[1] Beneš, V. E. (1970). Existence of Optimal Strategies Based on Specified Information for a Class of Stochastic Decision Problems. *SIAM Journal on Control and Optimization*. Vol. 8, 179-188.

[2] Beneš, V. E. (1971). Existence of Optimal Stochastic Control Laws. *SIAM Journal on Control and Optimization*. 9, 446-472.

[3] Bensoussan, A. and J. Frehse (2000). Stochastic Games for N Players. *Journal of Optimization Theory and Applications*, Vol. 105, No. 3, 543-565.

[4] Bensoussan, A., J. Frehse and H. Nagai (1998). Some Results on Risk-Sensitive Control with Full Observation. *Applied Mathematics and Optimization*. 37 (1), 1-41.

[5] Bensoussan, A. and A. Friedman (1977). Nonzero-Sum Stochastic Differential Games With Stopping Times and Free Boundary Problems. *Transactions of the American Mathematical Society*, Vol. 231, No. 2, 275-327.

[6] Chassagneux J.-F., R. Elie and I. Kharroubi (2010). A Note on Existence and Uniqueness for Solutions of Multidimensional Reflected BSDEs. *Elect. Comm. in Probab.* 16, 120-128.

[7] Çinlar, E., J. Jacod, P. Protter and M. J. Sharpe (1980). Semimartingales and Markov Processes. *Z. Warscheinlichkeitstheorie verw. Gebiete*. 54, 161-219.

[8] Cohen, S. N., R. J. Elliott and C. E. M. Pearce (2010). A General Comparison Theorem for Backward Stochastic Differential Equations. *Adv. Appl. Prob.* 42, 878-898.

[9] Cvitanić, J. and I. Karatzas (1996). Backward Stochastic Differential Equations with Reflection and Dynkin Games. *The Annals of Probability*. Vol. 24, No. 4, 2024-2056.

[10] Davis, M. H. A. (1973). On the Existence of Optimal Policies in Stochastic Control. *SIAM Journal on Control and Optimization*. 11, 587-594.

[11] Davis, M. H. A. (1979). Martingale Methods in Stochastic Control. *Lecture Notes in Control and Information Sciences*. 16. Springer-Verlag, Berlin.

[12] El Karoui, N. and S. Hamadène (2003). BSDEs and Risk-Sensitive Control, Zero-Sum and Nonzero-Sum Game Problems of Stochastic Functional Differential Equations. *Stochastic Processes and their Applications* 107, 145-169.

[13] El Karoui, N., C. Kapoudjian, E. Pardoux, S. Peng and M. C. Quenez (1997). Reflected Solutions of Backward SDE's, and Related Obstacle Problems for PDE's. *The Annals of Probability*. Vol. 25, No. 2, 702-737.

[14] El Karoui, N., S. Peng, M. C. Quenez (1997). Backward Stochastic Differential Equations in Finance. *Mathematical Finance*. Vol. 7, No. 1, 1-71.

[15] Elliott, R. J. (1976). The Existence of Value in Stochastic Differential Games. *SIAM Journal on Control and Optimization.* Vol. 14, No. 1.

[16] Elliott, R. J. (1982). *Stochastic Calculus and Applications.* Springer, New York.

[17] Fleming, W. H., and H. M. Soner (1993). *Controlled Markov Processes and Viscosity Solutions.* Springer, New York.

[18] Fukushima, M. and M. I. Taksar (2002). Dynkin Games via Dirichlet Forms and Singular control of One-Dimensional Diffusions. *SIAM Journal on Control Optimization.* Vol 41, No. 3, 682-699.

[19] Hamadène, S., J-P. Lepeltier and S. Peng (1997). BSDEs with Continuous Coefficients and Stochastic Differential Games. *Pitman Research Notes in Mathematics Series* 364.

[20] Hamadène, S. (1998). Backward-forward SDE's and Stochastic Differential Games. *Stochastic Processes and their Applications* 77, 1-15.

[21] Hamadène, S. (1999). Nonzero Sum Linear-Quadratic Stochastic Differential Games and Backward-forward Equations. *Stochastic Analysis and Applications.* Vol. 17, No. 1, 117-130.

[22] Hamadène, S. and J-P. Lepeltier (2000). Reflected BSDEs and Mixed Game Problem. *Stochastic Processes and their Applications* 85, 177-188.

[23] Hamadène, S. (2006). Mixed Zero-sum Stochastic Differential Game and American Game Options. *SIAM Journal on Control and Optimization.* Vol. 45, No. 2, 496-518.

[24] Hamadène, S. and J. Zhang (2010). Switching Problem and Related System of Reflected Backward SDEs. *Stochastic Processes and their Applications* 120, 403-426.

[25] Hu, Y. and S. Peng. On the Comparison Theorem for Multidimensional BSDEs. *Comptes Rendus de l'Acadmie des Sciences.* Série I. Mathématique 343 (2006) 135-140.

[26] Hu, Y. and S. Tang (2010). Multi-dimensional BSDE with Oblique Reflection and Optimal Switching. *Probability Theory and Related Fields.* Vol. 147, No. 1-2, 89-121.

[27] Karatzas, I. and D. L. Ocone (1992). The resolvent of a Degenerate Diffusion on the Plane, with Applications to Partially Observed Control. *The Annals of Applied Probability.* Vol. 2, No. 3, 629-668.

[28] Karatzas, I. and S. E. Shreve. *Brownian Motion and Stochastic Calculus.* Springer-Verlag, 1988.

[29] Karatzas, I. and W. Sudderth (2006). Stochastic Games of Control and Stopping for a Linear Diffusion. *Random Walk, Sequential Analysis and Related Topics: A Festschrift in Honor of Y.S. Chow* (A. Hsiung, Z.L. Ying and C.H. Zhang, eds.), 100-117. World Scientific Publishers.

[30] Karatzas, I. and H. Wang (2001). Connections Between Bounded-Variation Control and Dynkin Games. *IOS Press, Amsterdam.* 353-362.

[31] Karatzas, I. and I-M. Zamfirescu (2008). Martingale Approach to Stochastic Differential Games of Control and Stopping. *The Annals of Probability.* Vol. 36, No. 4, 1495-1527.

[32] Lepeltier, J.P. and E. Etourneau (1987). A Problem of Non-zero Sum Stopping Game. *Lecture notes in control and information sciences; 96*. Springer-Verlag, 1987.

[33] Ma, J., P. Protter and J. M. Yong (1994). Solving Forward-Backward Stochastic Differential Equations Explicitly - a Four Step Scheme. *Probab. Theory Related Fields* 98. 339-359.

[34] Pardoux, E. and S. Peng (1990). Adapted Solution of a Backward Stochastic Differential Equation. *Systems & Control Letters* 14 55-61.

[35] Taksar, M. I. (1985) Average Optimal Singular Control and a Related Stopping Problem. *Mathematics of Operations Research*. Vol. 10, No. 1, 63-81.

PART 3
Mathematical Finance

Chapter 7

On optimal dividend strategies in insurance with a random time horizon

Hansjörg Albrecher[1,2] and Stefan Thonhauser[2]

[1] *Department of Actuarial Science, Faculty of Business and Economics, University of Lausanne, Quartier UNIL-Dorigny, Bâtiment Extranef, 1015 Lausanne, Switzerland*

[2] *Swiss Finance Institute, University of Lausanne, Switzerland*

For the classical compound Poisson surplus process of an insurance portfolio we investigate the problem of how to optimally pay out dividends to shareholders if the criterion is to maximize the expected discounted dividend payments until the time of ruin or a random time horizon, whichever is smaller. We explicitly solve this problem for an exponential time horizon and exponential claim sizes. Furthermore, we study the case of an Erlang(2) time horizon by introducing an external state process and derive the solution under the assumption that the external state process is observable. The results are illustrated by numerical examples.

Contents

7.1. Introduction and model

Let $(\Omega, \mathcal{F}, \mathbb{P})$ be a probability space, sufficiently large to carry all the stochastic quantities that will be introduced in the sequel. Consider the classical collective risk model (also called Cramér-Lundberg model) for the surplus process $R = (R_t)_{t\geq 0}$ of an insurance portfolio, i.e.

$$R_t = x + c\,t - \sum_{i=1}^{N_t} Y_i,$$

with initial surplus $R_t = x \geq 0$, a homogeneous Poisson process $N = (N_t)_{t\geq 0}$ with parameter $\lambda > 0$ and a sequence of independent identically distributed claim amounts $\{Y_i\}_{i\in\mathbb{N}}$ with $F_Y(x) = \mathbb{P}(Y_1 \leq x)$, where $\{Y_i\}_{i\in\mathbb{N}}$ and $N = (N_t)_{t\geq 0}$ are independent. Here $c > \lambda\mathbb{E}(Y)$ is a constant premium intensity.

Let $\tau = \inf\{t > 0 \,|\, R_t < 0\}$ denote the time of ruin. One of the classical concepts in risk theory is the probability of ruin $\psi(x) = \mathbb{P}_x(\tau < \infty)$, which is a measure for the safety of the insurance portfolio, see Rolski et al. [16] and Asmussen & Albrecher [4] for an overview of ruin probabilities and related quantities.

In [10], de Finetti proposed the use of expected discounted dividend payments until ruin as a different criterion for measuring the performance of an insurance portfolio. This leads to a stochastic control problem of how to optimally pay out dividends in order to maximize this quantity. In the present model, an admissible dividend strategy is a càglàd non-decrasing process $L = (L_t)_{t\geq 0}$, and the controlled reserve becomes

$$R_t^L = x + c\,t - \sum_{i=1}^{N_t} Y_i - L_t.$$

The process L describes the cumulated dividend payments over time. One has the restrictions $\Delta L_t \leq R_{t-}^L$ (i.e. dividend payments cannot lead to ruin) and $L_t = L_{\tau^L}$ for $t \geq \tau^L$ (i.e. no payments after the event of ruin). Here τ^L denotes the time of ruin of R^L. The value of an admissible dividend strategy L is given by

$$\widetilde{V}(x, L) = \mathbb{E}_x\left(\int_0^{\tau^L} e^{-\delta t}\, dL_t\right),$$

where $\delta > 0$ is a constant discount rate (or simply a rate that reflects the preference to receive payments earlier rather than later).

The stochastic control problem to determine the optimal dividend strategy has been studied under various model assumptions and constraints, see Albrecher & Thonhauser [3] and Avanzi [6] for recent surveys. One should note that that there are similarities, but also subtle differences to other stochastic control problems in mathematical finance.

In this contribution, we would like to address a variation of the classical dividend problem. Usually the time horizon of the dividend maximization problems is the time of ruin (unless one includes the possibility of reinvestments whenever the surplus turns negative, in which case the time horizon is infinite, see Kulenko & Schmidli [14]). An exception is the paper of Grandits [13], which deals with the problem on a finite time horizon for a diffusion approximation of the risk reserve process R_t (i.e. dividends can be collected up to the minimum of the time of ruin and a prespecified fixed time horizon). Such a finite time horizon is a reasonable assumption in view of finite planning horizons for portfolio management. However, the introduction of the finite time horizon makes the problem technically highly involved, and one can expect this to be even worse when generalizing the underlying portfolio process to a compound Poisson process with drift (which is the more natural model for the insurance context). Here we propose a different route towards the solution of the dividend problem in the classical collective risk model with finite horizon, namely to randomize a deterministic time horizon. Such a randomization was successfully used in other finite horizon problems in insurance and mathematical finance (see for instance Asmussen et al. [5] and Carr [8]) and the results served as a reasonable approximation to the original problem. In Section 2 we will focus on the case of an exponential time horizon. In Section 3 we will then study the extension to an Erlang(2) time horizon, which is a crucial step towards an approximation for the solution of the optimal dividend problem in the Cramér-Lundberg model with deterministic finite time horizon. In Section 4 we finally indicate some further research directions.

7.2. Exponential time horizon

Let us assume in this section that the time horizon is an exponentially distributed random variable $\zeta \sim Exp(\gamma)$, independent of $\{Y_i\}$ and N. In other words, we apply exponential killing at a constant rate γ to the risk reserve and want to study its effect on the dividend maximization problem. Under this assumption the value of a strategy L is given by

$$\tilde{V}(x, L) = \mathbb{E}_x \left(\int_0^{\tau^L \wedge \zeta} e^{-\delta t} \, dL_t \right).$$

In view of the setting of the original de Finetti problem, where the emphasis is on maximizing profit, it is reasonable to assume that if the process is stopped at a time $\zeta < \tau^L$, then the remaining positive reserve can be paid out as dividends as well (otherwise the problem is simpler, see Remark 7.1). This leads to the modification

$$V(x,L) = \mathbb{E}_x\left(\int_0^{\tau^L \wedge \zeta} e^{-\delta t}\, dL_t + e^{-\delta\zeta} R_\zeta^L \, I_{\{\zeta<\tau^L\}}\right).$$

Let $\mathcal{L}$ be the set of admissible dividend strategies with the constraint that there are no payments after time $\zeta \wedge \tau^L$ and that the reserve process is stopped from $\zeta \wedge \tau^L$ onwards. The value function of the dividend maximization problem under the random time horizon is then given by

$$V(x) = \sup_{L\in\mathcal{L}} V(x,L).$$

We first observe that, due to $P(\zeta > t) = e^{-\gamma t}$, for an admissible strategy L in the infinite horizon problem the strategy $\widetilde{L} = (e^{-\gamma t} L_t)_{t\geq 0}$ corresponds to an admissible strategy for the killed reserve. Secondly we have that $P(\zeta \in dt)$ is $\gamma e^{-\gamma t} dt$, so that the expectation of $e^{-\delta\zeta} R_\zeta^L \, I_{\{\zeta<\tau^L\}}$ with respect to ζ translates to $\int_0^{\tau^L} \gamma\, e^{-(\delta+\gamma)t} R_t^L\, dt$. As a consequence we can rewrite $V(x,L)$ in the form

$$V(x,L) = \mathbb{E}_x\left(\int_0^{\tau^L} e^{-(\delta+\gamma)t} dL_t\right) + \mathbb{E}_x\left(\int_0^{\tau^L} e^{-(\delta+\gamma)t}\gamma\, R_t^L\, dt\right).$$

Similar arguments for interchanging killing and discounting apply for state-dependent killing rates, see for instance Davis [9, Theorem 31.9]. We can now write down (for the moment informally) the Hamilton-Jacobi-Bellman (HJB) equation associated to V in terms of a suitable function $f : [0,\infty) \to [0,\infty)$:

$$\begin{aligned} \max\Big\{cf'(x) + \lambda \int_0^x f(x-y)dF_Y(y) - (\delta+\gamma+\lambda)f(x) + \gamma x,& \\ 1 - f'(x)\Big\} = 0.& \end{aligned} \tag{7.1}$$

Note that $V(x) = 0$ for $x < 0$. The derivation of (7.1) follows standard arguments, see Fleming & Soner [11] or Schmidli [17].

If we define Δ as the death state, the generator of the killed risk reserve (as a part of (7.1)) is given by

$$cf'(x) + \lambda \int_0^x f(x-y)dF_Y(y) - (\gamma+\lambda)f(x) + \gamma f(\Delta).$$

In our situation the value function evaluated at the death state is just equal to the *terminal* payment of the reserve.

Remark 7.1. Considering the killing procedure for the dividend problem without possible terminal payment (i.e. using $\widetilde{V}(x,L)$ instead of $V(x,L)$) leads to the original problem without killing, but with the modified discounting factor $\delta+\gamma$ instead of δ. So in this case the effect of a finite random time horizon is just a simple shift of the discounting parameter (the solution to this problem can correspondingly be found in Azcue & Muler [7] and Schmidli [17]).

7.2.1. *Explicit solution in case of exponential claims*

In order to quantitatively study the consequences of the finite (random) time horizon on the dividend problem, we will restrict ourselves in the following to the case of exponentially distributed claim amounts ($Y_1 \sim Exp(\alpha)$). Under this assumption one can expect an explicit expression for the value function and the optimal strategy. Concretely, we can then construct a concave solution to (7.1) with the consequence that a so-called barrier strategy is optimal (at the end of the section we will give a verification result for a formal justification). A barrier strategy with barrier $b \geq 0$ pays every excess of the reserve over b immediately as dividend.

Barrier at zero.

Suppose a barrier strategy with level $b=0$ is applied and denote the associated dividend process by L^0, then we have

$$V(x,L^0) = x + \int_0^\infty \int_0^T e^{-(\delta+\gamma)t} c\,dt\,\lambda e^{-\lambda T}\,dT = x + \frac{c}{\delta+\gamma+\lambda}.$$

Define the constant $A = (\gamma+\delta+\lambda) - \alpha c$.

Lemma 7.1. *For $Y_1 \sim Exp(\alpha)$, $V(x,L^0)$ is a solution to (7.1) either if $A \geq 0$ or $A < 0$ and $\alpha\lambda c - (\delta+\lambda)(\gamma+\delta+\lambda) \leq 0$*

Proof. We only need to check if

$$cV'(x,L^0) + \lambda \int_0^x V(x-y,L^0)dF_Y(y) - (\delta+\gamma+\lambda)V(x,L^0) + \gamma x \leq 0 \tag{7.2}$$

for $x \geq 0$. Define $G(x) := \lambda(e^{-\alpha x} - 1)A - x\alpha\delta(\gamma + \delta + \lambda)$, then the left hand side of (7.2) can be written as

$$\frac{G(x)}{\alpha(\delta + \gamma + \lambda)},$$

so that we only need to check the sign of $G(x)$.
If $A > 0$, then $G(x)$ is decreasing, with $G(0) = 0$ as its maximum.
If $A < 0$, then $G(x)$ is strictly concave. Furthermore $G'(0) \leq 0$ which is equivalent to $\alpha\lambda c - (\delta + \lambda)(\gamma + \delta + \lambda) \leq 0$. Hence $G(0) = 0$ is a maximum of G for $x \geq 0$. □

Barrier larger than zero.

From now on assume $A < 0$ and $\alpha\lambda c - (\delta + \lambda)(\gamma + \delta + \lambda) > 0$. Trying the construction of a concave solution $f : [0, \infty) \to [0, \infty)$ to (7.1) implies the existence of a point x^* such that $f'(x) \geq 1$ for $0 \leq x \leq x^*$ and $f'(x) < 1$ for $x^* < x$.

At first we are going to construct a differentiable solution f_b with $b > 0$ to the problem

$$\begin{cases} cf_b'(x) + \lambda \int_0^x f_b(x-y)dF_Y(y) - (\delta + \gamma + \lambda)f_b(x) + \gamma x = 0, & 0 \leq x \leq b, \\ f_b(x) = x - b + f_b(b), & x > b. \end{cases} \tag{7.3}$$

Correspondingly, $f_b(x) = V(x, L^b)$ refers to the expected discounted dividends until ruin under a barrier strategy with barrier b. Some calculations show that for $0 \leq x \leq b$ the function

$$f_b(x) = A_1(b)\, e^{R_1 x} + A_2(b)\, e^{R_2 x} + \frac{\gamma x}{\gamma + \delta} + \frac{\gamma(\alpha c - \lambda)}{\alpha(\gamma + \delta)^2}, \tag{7.4}$$

together with the linear continuation in b is a differentiable solution to (7.3). The corresponding coefficients are given by

$$A_1(b) = \frac{(R_1 + \alpha)(\alpha^2\delta(\gamma + \delta) - R_2(R_2 + \alpha)\gamma A e^{R_2 b})}{\alpha^2(\gamma + \delta)^2(R_1(R_1 + \alpha)e^{R_1 b} - R_2(R_2 + \alpha)e^{R_2 b})},$$

$$A_2(b) = -\frac{(R_2 + \alpha)(\alpha^2\delta(\gamma + \delta) - R_1(R_1 + \alpha)\gamma A e^{R_1 b})}{\alpha^2(\gamma + \delta)^2(R_1(R_1 + \alpha)e^{R_1 b} - R_2(R_2 + \alpha)e^{R_2 b})}$$

and the exponents are

$$R_1 = \frac{A - \sqrt{A^2 + 4\alpha c(\gamma + \delta)}}{2c} < 0, \quad R_2 = \frac{A + \sqrt{A^2 + 4\alpha c(\gamma + \delta)}}{2c} > 0.$$

The principle of smooth fit suggests to derive the maximizing level b^* by the condition $f''_{b^*}(b^*) = 0$. Intuitively this is in accordance with pasting the linear function with the highest curve from below. Define $H(b) := f''_b(b)$.

Lemma 7.2. *If $A < 0$ and $\alpha\lambda c - (\delta+\lambda)(\gamma+\delta+\lambda) > 0$, then there exists a unique positive solution b^* to $H(b^*) = 0$.*

Proof. By our assumption, $H(0) = -\frac{\alpha\lambda c-(\delta+\lambda)(\gamma+\delta+\lambda)}{c(\delta+\gamma+\lambda)} < 0$. Furthermore, $\lim_{b\to\infty} H(b) = \frac{\delta R_2}{\delta+\gamma} > 0$. Since $R_1 + \alpha > 0$, there is no positive root of the denominator and therefore $H(b)$ is continuous for $b \geq 0$. One can further show by straight-forward but cumbersome calculations that $H'(b) > 0$, which proves that there exists a unique $b^* > 0$ such that $H(b^*) = 0$. □

Remark 7.2. One observes that at $b = b^*$ the coefficients simplify as follows

$$A_1(b^*) = \frac{\delta R_2 e^{-b^* R_1}}{R_1(R_2 - R_1)(\delta+\gamma)}, \quad A_2(b^*) = \frac{\delta R_1 e^{-b^* R_2}}{R_2(R_1 - R_2)(\delta+\gamma)}.$$

In particular, we have

$$f_{b^*}(b^*) = \frac{\alpha(c+\gamma b) - (\delta+\lambda)}{\alpha(\delta+\gamma)}. \tag{7.5}$$

Lemma 7.3. *For $Y_1 \sim Exp(\alpha)$ and under the conditions $A < 0$ and $\alpha\lambda c - (\delta+\lambda)(\gamma+\delta+\lambda) > 0$, the function $f_{b^*}(x)$ defined by (7.4) and (7.3) with b^* from Lemma 7.2 is an increasing concave solution to (7.2).*

Proof. Since $A < 0$ we have that $A_1(b^*) < 0$ and from the definition of b^*,

$$R_1^2 A_1(b^*)\, e^{R_1 b^*} = -R_2^2 A_2(b^*)\, e^{R_2 b^*},$$

which implies $A_2(b^*) > 0$, such that in total for $0 \leq x \leq b^*$ we have $f'_{b^*}(x) > 0$. Furthermore this implies that $f'''_{b^*}(x) > 0$ for $0 \leq x \leq b^*$, since $f''_{b^*}(b^*) = 0$ the function is concave, and that $f'_{b^*}(b^*) = 1$ is a strict minimum of the first derivative. Because of the linear continuation above b^* these properties remain valid for $x > b^*$ as well.

By construction, the function f_{b^*} solves (7.2) on $(0, b^*]$. Since on (b^*, ∞) we have $f'_{b^*}(x) = 1$ we finally need to check if

$$c + \lambda \int_0^{x-b^*} (x - b^* + f_{b^*}(b^*) - y)\alpha e^{-\alpha y}\, dy$$
$$- (\delta+\gamma+\lambda)(x - b^* + f_{b^*}(b^*)) + \gamma x + \lambda \int_{x-b^*}^{x} f_{b^*}(x-y)\alpha e^{-\alpha y}\, dy \leq 0.$$

From $\lambda\int_{x-b^*}^{x} f_{b^*}(x-y)\alpha e^{-\alpha y}\,dy \le \lambda f_{b^*}(b^*)(e^{-\alpha(x-b^*)}-e^{-\alpha x})$, (7.5) and using $\alpha\lambda c-(\delta+\lambda)(\gamma+\delta+\lambda)>0$, we obtain the result. □

The following standard verification theorem gives the formal basis for the above calculations. Since in both parameter constellations above, we have concave and increasing solutions to (7.2) which correspond to admissible dividend strategies, we are going to prove:

Theorem 7.1. *Let $h:\mathbb{R}\to\mathbb{R}^+$ with $h(x)=0$ for $x<0$ be a differentiable, concave and increasing solution to (7.2) which corresponds to an admissible dividend strategy $L^*=(L_t^*)_{t\ge 0}$. Then $h=V$, and the strategy L^* is optimal.*

Proof. Let $L=(L_t)_{t\ge0}$ be an arbitrary admissible dividend strategy (L^c denoting its absolutely continuous component). Then for $t>0$, writing $\mathcal{S}=\{s: 0\le s\le t\wedge\tau^L, R_{s+}^L\ne R_s^L\}$, we have

$$e^{-(\delta+\gamma)t\wedge\tau^L}h(R_{t\wedge\tau^L}^L)=h(x)+\int_0^{t\wedge\tau^L}e^{-(\delta+\gamma)s}\Bigg[ch'(R_{s-}^L)$$

$$+\lambda\int_0^{R_{s-}^L}h(R_{s-}^L-y)\alpha e^{-\alpha y}dy-(\delta+\gamma+\lambda)h(R_{s-}^L)\Bigg]ds$$

$$-\int_0^{t\wedge\tau^L}e^{-(\delta+\gamma)s}h'(R_{s-}^L)dL_s^c-\sum_{s\in\mathcal{S}}e^{-(\delta+\gamma)s}[h(R_{s+}^L)-h(R_s^L)].$$

Since $R_{s+}^L-R_s^L=L_s-L_{s+}$ and $1-h'(x)\ge0$ we have

$$-\int_0^{t\wedge\tau^L}e^{-(\delta+\gamma)s}h'(R_{s-}^L)dL_s^c-\sum_{s\in\mathcal{S}}e^{-(\delta+\gamma)s}[h(R_{s+}^L)-h(R_s^L)]$$

$$\le-\int_0^{t\wedge\tau^L}e^{-(\delta+\gamma)s}dL_s^c-\sum_{s\in\mathcal{S}}e^{-(\delta+\gamma)s}[L_{s+}-L_s]$$

$$=-\int_0^{t\wedge\tau^L}e^{-(\delta+\gamma)s}dL_s.$$

Together with

$$ch'(R_{s-}^L)+\lambda\int_0^{R_{s-}^L}h(R_{s-}^L-y)\alpha e^{-\alpha y}dy-(\delta+\gamma+\lambda)h(R_{s-}^L)\le-\gamma R_{s-}^L,$$

we arrive at

$$\mathbb{E}_x\left(e^{-(\delta+\gamma)t\wedge\tau^L}h(R^L_{t\wedge\tau^L})\right)$$
$$+\mathbb{E}_x\left(\int_0^{t\wedge\tau^L}e^{-(\delta+\gamma)s}dL_s+\int_0^{t\wedge\tau^L}e^{-(\delta+\gamma)s}\gamma R^L_s\ ds\right)\leq h(x), \quad (7.6)$$

where we can take R^L_s instead of R^L_{s-} in the last integral, since with probability zero there occurs a claim exactly at time ζ.
Now we use dominated convergence for the limit $t\to\infty$ to deduce

$$V(x,L)\leq h(x).$$

Using strategy L^*, we derive an equality in (7.6) which proves its optimality. □

We summarize our results in the following theorem.

Theorem 7.2. *If $A\geq 0$ or $A<0$ and $\alpha\lambda c-(\delta+\lambda)(\gamma+\delta+\lambda)\leq 0$, then the value function is given by*

$$V(x)=x+\frac{c}{\delta+\gamma+\lambda}.$$

The optimal dividend strategy is then a barrier strategy with barrier $b^*=0$. If on the other hand $A<0$ and $\alpha\lambda c-(\delta+\lambda)(\gamma+\delta+\lambda)>0$, then the value function is $V(x)=f_{b^*}(x)$ where f_{b^*} is determined by (7.3) and (7.4). In this case the optimal dividend strategy is a barrier strategy with barrier $b^*>0$, which is determined in Lemma 7.2.

Example 7.1. Figure 7.1 shows the optimal barrier as a function of γ for the parameter set $c=6.5$, $\alpha=1$, $\lambda=5$ and $\delta=0.04$. Due to the conditions of Theorem 7.2 it tends to zero if γ increases. Figure 7.2 displays the value function for different values of γ. The dotted line corresponds to $\gamma=0$ with $b^*(0)=15.7448$, the dotted-dashed line to $\gamma=0.1$ with $b^*(0.1)=11.6348$, the solid line to $\gamma=0.5$ with $b^*(0.5)=7.1230$ and finally the dashed line to $\gamma=1.5$ with $b^*(1.5)=0$. As expected the value function decreases with increasing γ (due to expected earlier killing of the surplus process).

Remark 7.3. If the underlying process R_t is a diffusion, the above result will again hold, with analogous barrier strategies to be optimal. This can either be shown directly by similar calculations, or also by a suitable limiting procedure of the compound Poisson process (see e.g. Albrecher, Bäuerle and Thonhauser [1] for a related procedure in another control problem).

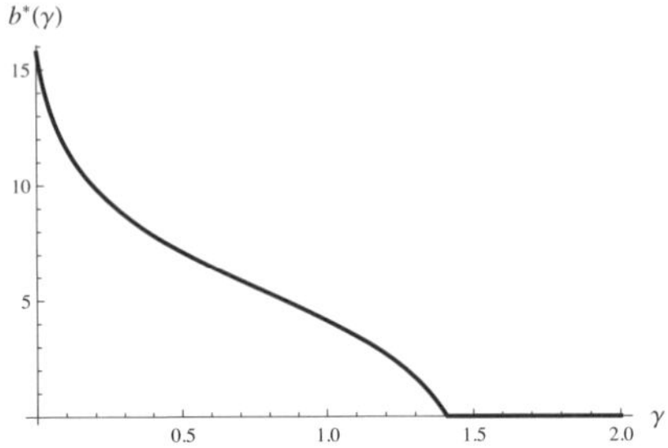

Fig. 7.1. b^* as function of γ

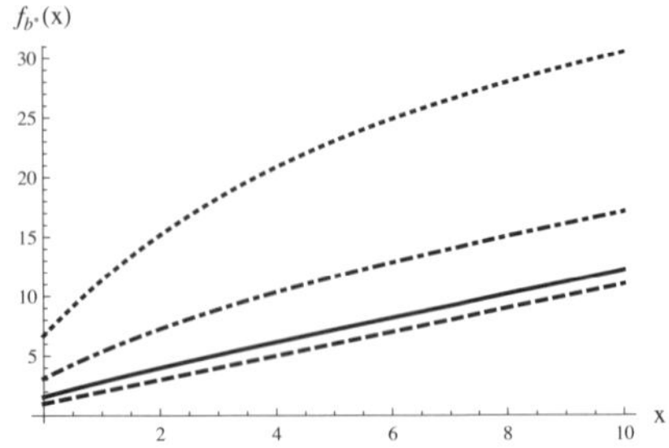

Fig. 7.2. f_{b^*} for $\gamma = 0, 0.1, 0.5, 1.5$

Remark 7.4. The HJB equation corresponding to $\widetilde{V}(x) = \sup_{L \in \mathcal{L}} \widetilde{V}(x, L)$ is given by

$$\max\left\{cf'(x) + \lambda \int_0^x f(x-y)dF_Y(y) - (\delta + \gamma + \lambda)f(x), 1 - f'(x)\right\} = 0.$$

Although in the infinite horizon case $\gamma = 0$, barrier strategies are not always optimal beyond exponential claims (see Loeffen & Renaud [15] for more detailed criteria) it is surprisingly hard to identify explicit optimal non-barrier strategies. There is one concrete example in Azcue & Muler [7] for Erlang(2) claims, in which a barrier strategy is not optimal: For the parameters $F_Y(y) = 1 - (1+y)e^{-y}$, $\lambda = 10$, $\delta = 0.1$, $c = 21.4$ the value function turns out to be

$$\widetilde{V}(x) = \begin{cases} x + 2.11881, & 0 \le x \le 1.80303, \\ g(x), & 1.80303 < x \le 10.2162, \\ x + 2.45582, & 10.2162 < x, \end{cases}$$

with

$$g(x) = 11.2571e^{0.039567x} - 9.43151e^{-0.079355x} + 0.094314e^{-1.48825x}.$$

The optimal strategy is of band type, i.e. for $0 \le x \le 1.80303$ a barrier strategy in zero is used, for $1.80303 < x \le 10.2162$ no dividends are paid and for $10.2162 < x$ a barrier strategy with height 10.2162 is applied.
Since band strategies are sensitive to slight modifications of the model (see e.g. Albrecher et al. [1]), let us check the effect of $\gamma > 0$ on the optimal strategy with respect to the infinite horizon case $\gamma = 0$. Figures 7.3 and 7.4 illustrate that for the same parameter set, but $\gamma = 0.1$, a barrier strategy in zero is already optimal (i.e., the bands collapse to a barrier in zero). Figure 7.3 shows the left part of the maximum in the HJB equation ("IDE part")

evaluated at the function $x + \frac{c}{\delta+\lambda}$ (which corresponds to a barrier in zero) for $\gamma = 0$, which clearly shows that it cannot be the value function. On the other hand, Figure 7.4 shows the IDE part evaluated at the value of a barrier strategy equal to zero for $\gamma = 0.1$, which shows that now $x + \frac{c}{\delta+\lambda+\gamma}$ solves the HJB equation. Figure 7.5 shows $\widetilde{V}(x)$ in the case $\gamma = 0$ (solid line) and $f_0 = x + \frac{c}{\delta+\gamma+\lambda}$ which is the value function for $\gamma = 0.1$ (dashed line).

On the other hand, for $\gamma = 0.01$ a band strategy is still optimal (namely to use a barrier in zero for $x \in [0, 2.18793]$, pay no dividends for $x \in (2.18793, 9.07912]$ and use another barrier in 9.07912 for $x > 9.07912$). Figure 7.6 depicts the associated value function (solid line) together with the performance under the barrier strategy f_0 (dashed line).

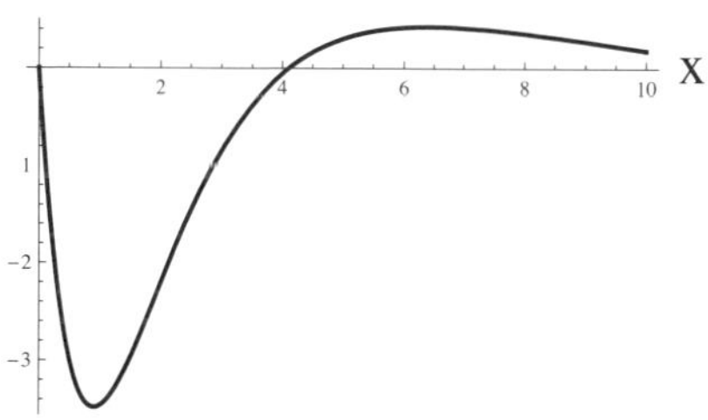

Fig. 7.3. IDE part of HJB eq. for f_0 and $\gamma = 0$

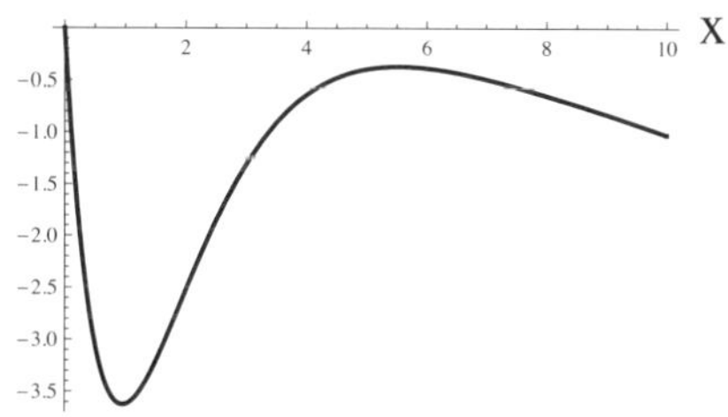

Fig. 7.4. IDE part of HJB eq. for f_0 and $\gamma = 0.1$

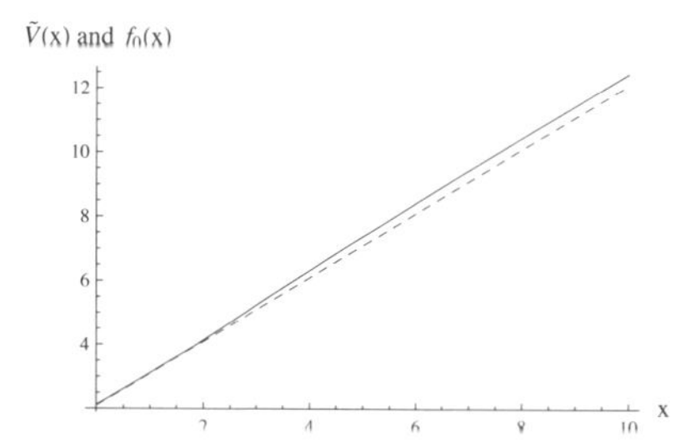

Fig. 7.5. $\widetilde{V}(x)$ for $\gamma = 0$, $f_0(x)$ for $\gamma = 0.1$

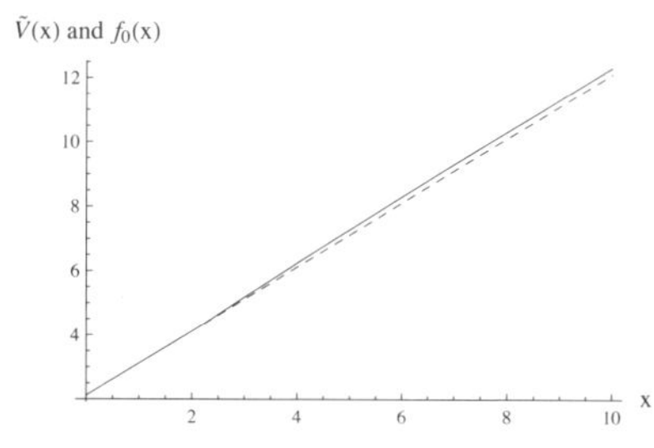

Fig. 7.6. $\widetilde{V}(x)$ and $f_0(x)$ for $\gamma = 0.01$

7.3. Erlang time horizon

A natural extension of an exponential random variable for the random time horizon is an $Erlang(n, \gamma)$ distributed random variable. Although one then loses the attractive lack-of-memory property of ζ, choosing $n > 1$

is a substantial step forward towards the case of deterministic finite time horizon, since for fixed expected value the $Erlang(n)$ distribution converges weakly to a point mass at $\mathbb{E}(\zeta)$. This so-called Erlangization method was for instance exploited to approximate finite-horizon ruin probabilities in Asmussen et al. [5], where it turned out that small values of n often already give very good approximations for the deterministic fixed time horizon counterpart. As mentioned in the introduction, the dividend maximization problem with a finite time horizon was up to now only studied for a diffusion risk reserve process in Grandits [13]. It turns out that even for this relatively simple case (with constant coefficients), the problem becomes technically very complicated and it is only possible to construct ε-optimal strategies, given by time-dependent barrier strategies.

In this section we will study the case $\zeta \sim Erlang(2, \gamma)$ to see the effect of the Erlangization on the calculations and to provide an alternative route towards finite horizon dividend problems in the framework of the classical risk model. In principle, the methodology will be the same when using larger values for n, but for reasons of compactness we restrict ourselves to the case $n = 2$. As in the previous section, we consider exponentially distributed claim amounts. In the sequel, the flow of information over time is supposed to be given by the filtration generated by the reserve process and an external state process.

7.3.1. *State-dependent barrier strategies*

Let us recall the simple property that a random variable $\zeta \sim Erlang(2, \gamma)$ can be written as $\zeta \sim \zeta_1 + \zeta_2$ where $\zeta_1, \zeta_2 \overset{iid}{\sim} Exp(\gamma)$. Suppose the process $\varepsilon = (\varepsilon_t)_{t \geq 0}$ is an external state process governed by the rule

$$\varepsilon_t = \begin{cases} 1, & 0 \leq t < \zeta_1, \\ 2, & \zeta_1 \leq t < \zeta_1 + \zeta_2, \\ \Delta, & t \geq \zeta_1 + \zeta_2. \end{cases}$$

I.e., at time zero it starts in state 1, after a time interval of length ζ_1 it moves to state 2 and after a further time interval of length ζ_2, it moves to the death state Δ. Note that this is a particular form of a Markov switching problem, where the underlying process has the same dynamics across states, but the control at time t depends on the value of ε_t. In particular, we assume here that we can observe the process (ε_t), whereas in practice one will not be able to determine the current state of the process (this is somewhat akin to the approach of Albrecher & Hartinger [2] in a different context). Recently, Sotomayor & Cadenillas [18] considered a Markov switching environment for a diffusion process with state-dependent coefficients.

Note that the two-dimensional process (R_t, ε_t) has the strong Markov property. Let $L = (L_t)_{t\geq 0}$ be an admissible strategy, the value of which, when starting in state 1, is defined as

$$V_1(x, L) = \mathbb{E}_{(x,1)}\left(\int_0^{\tau^L\wedge\zeta} e^{-\delta t}dL_t + e^{-\delta\zeta}R^L_\zeta I_{\{\zeta<\tau^L\}}\right).$$

If starting in state 2 we write,

$$V_2(x, L) = \mathbb{E}_{(x,2)}\left(\int_0^{\tau^L\wedge\zeta_2} e^{-\delta t}dL_t + e^{-\delta\zeta_2}R^L_{\zeta_2}I_{\{\zeta<\tau^L\}}\right).$$

When starting in state 2, there is only one exponentially distributed time span left until moving to the death state, and so $V_2(x, L)$ coincides with the value of L under exponential killing from Section 7.2.
The next lemma connects $V_1(x, L)$ with $V_2(x, L)$.

Lemma 7.4. *Let $L = (L_t)_{t\geq 0}$ be an admissible dividend strategy such that (R^L, ε) is a strong Markov process. Then*

$$V_1(x, L) = \mathbb{E}_{(x,1)}\left(\int_0^{\tau^L} e^{-(\delta+\gamma)t}dL_t + \int_0^{\tau^L} e^{-(\delta+\gamma)t}\gamma V_2(R^L_t, L)\,dt\right).$$

Proof. At first we write $\zeta \overset{d}{=} \zeta_1 + \zeta_2$ so that

$$\begin{aligned} V_1(x, L) = \mathbb{E}_{(x,1)}\Bigg(&\int_0^{\tau^L\wedge\zeta_1} e^{-\delta t}dL_t \\ &+ \int_{\tau^L\wedge\zeta_1}^{\tau^L\wedge(\zeta_1+\zeta_2)} e^{-\delta t}dL_t + e^{-\delta(\zeta_1+\zeta_2)}R^L_{\zeta_1+\zeta_2}I_{\{\zeta_1+\zeta_2<\tau^L\}}\Bigg). \end{aligned}$$

Since the process (R^L_t, ε_t) is a strong Markov process, we can write the second integral together with the terminal payment as $e^{-\delta\zeta_1}V_2(R_{\zeta_1}, \widehat{L})I_{\{\zeta_1<\tau^L\}}$. The strategy $\widehat{L}$ is given by shifting the strategy L by the starting time ζ_1, i.e. $\hat{L}_t = L_{\zeta_1+t}$. Taking the expectation with respect to ζ_1 then gives the result. □

The value functions of this state-dependent dividend maximization problem are defined as

$$V_1(x) = \sup_{L\in\mathcal{L}} V_1(x, L),$$
$$V_2(x) = \sup_{L\in\mathcal{L}} V_2(x, L).$$

As mentioned above, $V_2(x)$ is determined by Theorem 7.2. Combining Lemma 7.4 with the proof of a related dynamic programming principle (see Azcue & Muler [7]) and replacing the arbitrary stopping time by ζ_1, one gets that

$$V_1(x) = \sup_{L\in\mathcal{L}} \mathbb{E}_{(x,1)} \left(\int_0^{\tau^L} e^{-(\delta+\gamma)t} dL_t + \int_0^{\tau^L} e^{-(\delta+\gamma)t} \gamma V_2(R_t^L)\, dt \right),$$

where $L \in \mathcal{L}$ can be taken as an admissible strategy (which in state 1 does not need to be a Markov control). The associated HJB equation for state 1 is then given by

$$\max\left\{ cf'(x) + \int_0^x f(x-y)\alpha e^{-\alpha y} dy - (\delta+\gamma+\lambda)f(x) + \gamma V_2(x), \right.$$
$$\left. 1 - f'(x) \right\} = 0. \quad (7.7)$$

The remainder of this section is devoted to the construction of an increasing and concave solution to (7.7).

It will turn out that it is without loss of generality to fix the strategies under consideration to be of state-dependent barrier type. I.e., if $\varepsilon_t = 1$ we apply a barrier strategy at height b_1, and if $\varepsilon_t = 2$ we apply a barrier strategy at height b_2. From the formulation of the switching problem, if $\varepsilon_t = 2$ it will be preferable to distribute dividends in a more risky manner than in state 1 since killing is closer and because of discounting (i.e. in general $b_1 > b_2$). This also matches the observation in Grandits [13] that ε-optimal barriers are decreasing in time.
We have to distinguish the following cases.

Case $b_1 = b_2 = 0$.

In this case the initial capital is immediately paid out as a dividend and all the incoming premium is also immediately paid out as dividends up to time $\min\{T_1, \zeta\}$, where $T_1 \sim Exp(\lambda)$ denotes the time of the first claim occurrence. Let $F_\zeta(s)$ denote the distribution function of ζ, then the associated value is given by:

$$\begin{aligned} f_{(0,0)}(x) &= x + c\int_0^\infty \int_0^t e^{-\delta s}(1 - F_\zeta(s))\, ds\, \lambda e^{-\lambda t}\, dt \\ &= x + \frac{c(2\gamma + \delta + \lambda)}{(\gamma + \delta + \lambda)^2}. \end{aligned}$$

At first we observe that this function can only be a candidate for the value function if $b_2 = 0$ is optimal for the second state.

Lemma 7.5. *$f_{(0,0)}$ is a solution to (7.7) if $V_2(x) = x + \frac{c}{\delta+\gamma+\lambda}$ and if $(\delta+\gamma+\lambda)^2 - \alpha c(2\gamma+\delta+\lambda) \geq 0$ or if $(\delta+\gamma+\lambda)^2 - \alpha c(2\gamma+\delta+\lambda) < 0$ and $\alpha\lambda c(2\gamma+\delta+\lambda) - (\delta+\lambda)(\delta+\gamma+\lambda)^2 \leq 0$.*

Proof. Since $f'_{(0,0)}(x) = 1$, we need to verify that the integro-differential part is non-positive. Plugging in $f_{(0,0)}(x)$, this part can be written as

$$\frac{\lambda((\delta+\gamma+\lambda)^2 - \alpha c(2\gamma+\delta+\lambda))(e^{-\alpha x} - 1) - \alpha\delta x(\delta+\gamma+\lambda)^2}{\alpha(\delta+\gamma+\lambda)^2} =: J(x)$$

for $x \geq 0$. Obviously $J(x) \leq 0$ if $(\delta+\gamma+\lambda)^2 - \alpha c(2\gamma+\delta+\lambda) \geq 0$. Now suppose $(\delta+\gamma+\lambda)^2 - \alpha c(2\gamma+\delta+\lambda) < 0$, then $J(x)$ is concave. Therefore $J(0) = 0$ is an upper bound for $J(x)$ if $J'(0) \leq 0$. This is true if and only if

$$\alpha\lambda c(2\gamma+\delta+\lambda) - (\delta+\lambda)(\delta+\gamma+\lambda)^2 \leq 0.$$

□

Case $b_1 > 0$ and $b_2 = 0$.

Since the second barrier $b_2 = 0$, the value of starting in state 2 at level x is $x + \frac{c}{\delta+\gamma+\lambda}$. Therefore the value of this particular strategy $f_{(b_1,0)}$, if starting in state 1, fulfills

$$\begin{cases} cf'_{(b_1,0)}(x) + \lambda\int_0^x f_{(b_1,0)}(x-y)\alpha e^{-\alpha y}dy & \\ \qquad -(\delta+\gamma+\lambda)f_{(b_1,0)}(x) + \gamma(x + \frac{c}{\delta+\gamma+\lambda}) = 0, & 0 \leq x \leq b_1, \\ f_{(b_1,0)}(x) = x - b_1 + f_{(b_1,0)}(b_1), & x > b_1. \end{cases}$$

Similar to the calculations in Section 7.2, we get that for $0 \leq x \leq b_1$:

$$f_{(b_1,0)}(x) = D_1(b_1)e^{R_1 x} + D_2(b_1)e^{R_2 x} + \frac{\gamma x}{\delta+\gamma} + D_3,$$

where $R_1 < 0$, $R_2 > 0$ are the exponents determined in Section 7.2. Using $B := \alpha c(2(\delta+\gamma)+\lambda) - (\delta+\gamma+\lambda)^2$, the coefficients are

$$D_1(b_1) = \frac{(R_1+\alpha)(\alpha^2\delta(\delta+\gamma)(\delta+\gamma+\lambda) + R_2(R_2+\alpha)\gamma B e^{R_2 b_1})}{\alpha^2(\delta+\gamma)^2)(\delta+\gamma+\lambda)(R_1(R_1+\alpha)e^{R_1 b_1} - R_2(R_2+\alpha)e^{R_2 b_1}},$$

$$D_2(b_1) = -\frac{(R_2+\alpha)(\alpha^2\delta(\delta+\gamma)(\delta+\gamma+\lambda) + R_1(R_1+\alpha)\gamma B e^{R_1 b_1})}{\alpha^2(\delta+\gamma)^2)(\delta+\gamma+\lambda)(R_1(R_1+\alpha)e^{R_1 b_1} - R_2(R_2+\alpha)e^{R_2 b_1}},$$

$$D_3 = \frac{\gamma(-\lambda(\delta+\gamma+\lambda) + \alpha c(2(\delta+\gamma)+\lambda)}{\alpha(\delta+\gamma)^2)(\delta+\gamma+\lambda)}.$$

This particular choice of coefficients gives that $f_{(b_1,0)}$ is differentiable on $\mathbb{R}^+$.

Lemma 7.6. *Assume that* $V_2(x) = x + \frac{c}{\delta+\gamma+\lambda}$ *and that* $f_{(0,0)}$ *does not solve (7.7). Let* $H(b) := f''_{(b,0)}(b)$, *then there exists a unique* $b_1^* > 0$ *such that* $H(b_1^*) = 0$.

Proof. Similarly to the proof of Lemma 7.2, we get that

$$H(0) = \frac{(\delta+\lambda)(\delta+\gamma+\lambda)^2 - \alpha\lambda c(2\gamma+\delta+\lambda)}{c(\delta+\gamma+\lambda)^2}$$

and $\lim_{b\to\infty} H(b) = \frac{\delta R_2}{\delta+\gamma} > 0$. We observe that $H(0) < 0$ in this case and that $H(b)$ is continuous for $b \geq 0$. Further, one can see that using $\delta+\gamma+\lambda-\alpha c < 0$ (which holds since a barrier in zero is optimal for state 2) one gets $H'(b) > 0$ for $b \geq 0$, which completes the proof. □

In the same way as in Lemma 7.3 one can derive that $f_{(b_1^*,0)}$ is an increasing and concave function for $x \geq 0$. But nevertheless it is hard to analytically establish that

$$c + \int_0^x f_{(b_1^*,0)}(x-y)\alpha e^{-\alpha y}dy - (\delta+\gamma+\lambda)f_{(b_1^*,0)}(x) + \gamma\left(x + \frac{c}{\delta+\gamma+\lambda}\right) \leq 0, \quad x > b_1^*,$$

which would prove that $f_{(b_1^*,0)}$ fulfills (7.7). However there is numerical evidence that this is fulfilled in general.

Case $b_1 > b_2 > 0$.

In this situation the area below b_1 splits into two regions (due to the form of V_2). The value $f_{(b_1,b_2)}(x)$ of this strategy is a solution to

$$\begin{cases} cf'_{(b_1,b_2)}(x) + \lambda \int_0^x f_{(b_1,b_2)}(x-y)\alpha e^{-\alpha y}dy \\ \qquad -(\delta+\gamma+\lambda)f_{(b_1,b_2)}(x) + \gamma f_{b_2}(x) = 0, & 0 \le x \le b_2, \\ cf'_{(b_1,b_2)}(x) + \lambda \int_0^x f_{(b_1,b_2)}(x-y)\alpha e^{-\alpha y}dy \\ \qquad -(\delta+\gamma+\lambda)f_{(b_1,b_2)}(x) + \gamma(x - b_2 + f_{b_2}(b_2)) = 0, & b_2 < x \le b_1, \\ f_{(b_1,b_2)}(x) = x - b_1 + f_{(b_1,b_2)}(b_1), & x > b_1. \end{cases}$$

One needs to be careful when dealing with the second integro-differential equation because the integration area includes the region below b_2. Nevertheless there are enough conditions (from smooth-pasting and the integrals) which can be translated into a linear system of equations, such that one can determine a differentiable solution of the following form:

$$\begin{aligned} f_{(b_1,b_2)}(x) &= (B_1(b_1,b_2) + xB_2(b_2))e^{R_1x} + (B_3(b_1,b_2) + xB_4(b_2))e^{R_2x} \\ &\qquad + B_5x + B_6, \quad 0 \le x \le b_2, \\ f_{(b_1,b_2)}(x) &= C_1(b_1,b_2)e^{R_1x} + C_2(b_1,b_2)e^{R_2x} + C_3x + C_4, \quad b_2 < x \le b_1. \end{aligned}$$

Although available in closed form, we cannot display the voluminous explicit expressions for the involved coefficients here (they also include the coefficients of $f_{b_2}(x)$).
One can see that the right- and left-hand second derivatives of $f_{(b_1,b_2)}(x)$ at $x = b_2$ coincide since $f'_{b_2}(b_2) = 1$. Therefore the function $f_{(b_1,b_2)}(x)$ is twice differentiable on the whole interval $(0, b_1)$.

Case $b_1 = b_2 > 0$.

For completeness we look at this case explicitly, which can be solved like the previous one, but the integro-differential part for $b_2 < x \le b_1$ disappears now and correspondingly calculations simplify. Setting $b = b_1 = b_2 > 0$, the value of the associated strategy for $0 < x < b$ is given by

$$f_{(b,b)}(x) = (E_1(b) + xE_2(b))e^{R_1x} + (E_3(b) + xE_4(b))e^{R_2x} + xE_5 + E_6.$$

Again the coefficients are determined by solving a linear system of equations such that $f'_{(b,b)}(b) = 1$.

Remark 7.5. Obviously a simple (b_1, b_2)-type strategy can only be a candidate for an optimal dividend strategy, if b_2 is optimal for the second state.

Now suppose that $b_2^* > 0$ (characterized via Theorem 7.2) is optimal, then a good candidate for b_1 would be the solution to $f''_{(b_1,b_2^*)}(b_1) = 0$. Like in the proof of Theorem 7.2, this condition together with $f'''_{(b_1,b_2^*)}(b_1) > 0$ would imply that at b_1 the derivative attains a minimum with value 1. This relates to the construction procedure for optimal band strategies in the infinite horizon case as described in Schmidli [17] , where one searches for minima equal to 1 of derivatives of solutions to the IDE part of the associated HJB equation.

Because of the elaborate structure of the coefficients of $f_{(b_1,b_2)}$ it seems out of reach to prove that a solution to $f''_{(b_1,b_2)}(b_1) = 0$ exists in general and that the associated function solves (7.7) (in particular to verify that for $x > b_1^*$ the IDE part of the HJB equation evaluates to a non-positive value is intricate). We therefore propose the following procedure for deriving a candidate for V_1. For state 2 we already know the solution which is given in Theorem 7.2.

- If for state 2 the optimal barrier level $b_2^* = 0$, from the conditions above either check whether $b_1^* = 0$, or calculate $b_1^* > 0$ using Lemma 7.6 and check whether

$$c + \int_0^x f_{(b_1^*,0)}(x-y)\alpha e^{-\alpha y}dy - (\delta+\gamma+\lambda)f_{(b_1^*,0)}(x) + \gamma\left(x + \frac{c}{\delta+\gamma+\lambda}\right) \leq 0, \quad x > b_1^*.$$

  If this is true, $f_{(b_1^*,0)}$ is an increasing and concave solution to (7.7).
- If $b_2^* > 0$, use the case $b_1 \geq b_2 > 0$. Determine $b_1^* > b_2^*$ as a solution to $f''_{(b_1,b_2)}(b_1) = 0$ and check whether it is a concave solution to (7.7).

If a simple state-dependent barrier strategy solves the dividend problem, the above procedure will deliver the solution. Under this type of strategy the controlled reserve process is a strong Markov process with respect to the filtration generated by the risk reserve process and the external state process ε_t.

We conclude this section with the following verification theorem.

Theorem 7.3. *Let $h : \mathbb{R} \to \mathbb{R}^+$ with $h(x) = 0$ for $x < 0$, be an increasing, differentiable and concave solution to (7.7) which corresponds to an admissible strategy L^* for state 1 and the optimal strategy for state 2. Then $h(x) = V_1(x)$ and L^* is optimal if starting in state 1.*

Proof. Let L be an admissible strategy. We can follow the proof of Theorem 7.1 to arrive at

$$\mathbb{E}_{(x,1)}\left(\int_0^{\tau^L} e^{-(\delta+\gamma)s}dL_s + \int_0^{\tau^L} e^{-(\delta+\gamma)s}\gamma V_2(R_{s-}^L)ds\right) \le h(x). \qquad (7.8)$$

Since $V_2(x) \ge V_2(x, L)$ we get, by incorporating the time split $\zeta \sim \zeta_1 + \zeta_2$,

$$V_1(x,L) = \mathbb{E}_{(x,1)}\left(\int_0^{\tau^L\wedge\zeta} e^{-\delta s}dL_s + e^{-\delta\zeta}R_\zeta^L I_{\{\zeta<\tau^L\}}\right) \le h(x).$$

Now using the strategy L^* while being in state 1 and the optimal strategy for state 2 we have an equality in (7.8) which completes the proof. □

7.3.2. *Illustrations*

Example 7.2. For the first example we choose a parameter set which delivers positive barriers for both states: $c = 6.5$, $\lambda = 5$, $\alpha = 1$, $\gamma = 1$, $\delta = 0.04$. For state 2 we numerically calculate $b_2^* = 4.13685$ and have that $V_2(x) = f_{b_2^*}(x)$. One then determines $b_1^* = 7.17286$ as a solution to $f''_{(b_1,b_2^*)}(b_1) = 0$.

Figure 7.7 shows $f_{(b_1^*,b_2^*)}(x)$. Figure 7.10 displays the IDE part of the HJB equation evaluated at $f_{(b_1^*,b_2^*)}$ for $x > b_1^*$. Since it is negative the HJB equation is fulfilled and one can deduce from Theorem 7.3 that it is the value function of the dividend maximization problem, if starting in state 1. Furthermore Figure 7.8 illustrates that the minimum of the derivative is exactly equal to 1 at b_1^*. Figure 7.9 contains the graph of $f''_{(b_1,b_2^*)}(b_1)$ as a function of the barrier for state 1. The optimal state-dependent strategy, namely to use the barrier strategy b_1^* up to the random time ζ_1, and from then on b_2^*, is depicted in Figure 7.11.

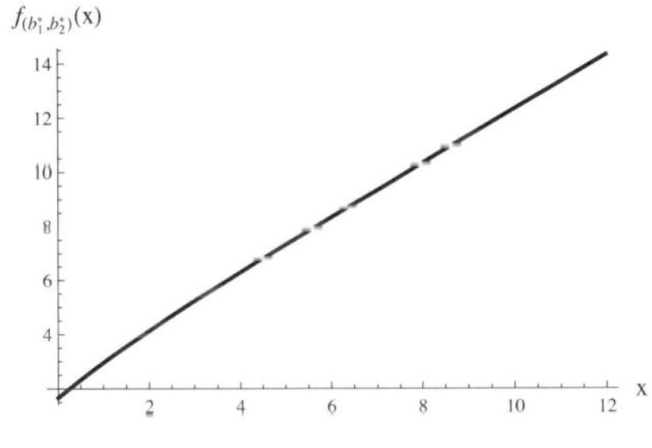

Fig. 7.7. $f_{(b_1^*,b_2^*)}$ for $x \in [0, 10)$

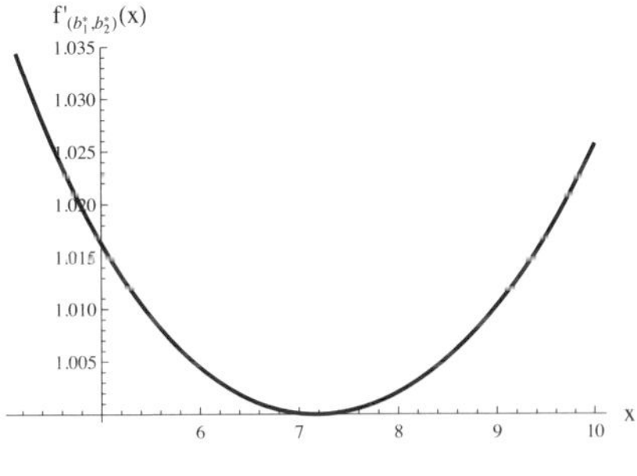

Fig. 7.8. $f'_{(b_1^*,b_2^*)}(x)$, for $x \in (b_2^*, 10)$

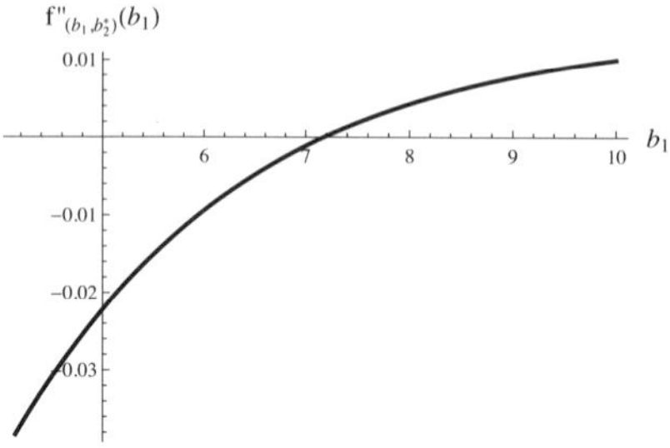

Fig. 7.9. $f''_{(b_1,b_2^*)}(b_1)$ for $b_1 \in (b_2^*, 10)$

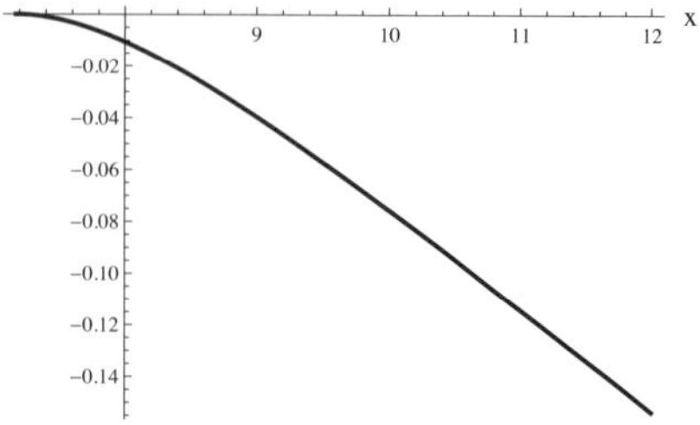

Fig. 7.10. IDE part of HJB eq., $x \in (b_1^*, 12)$

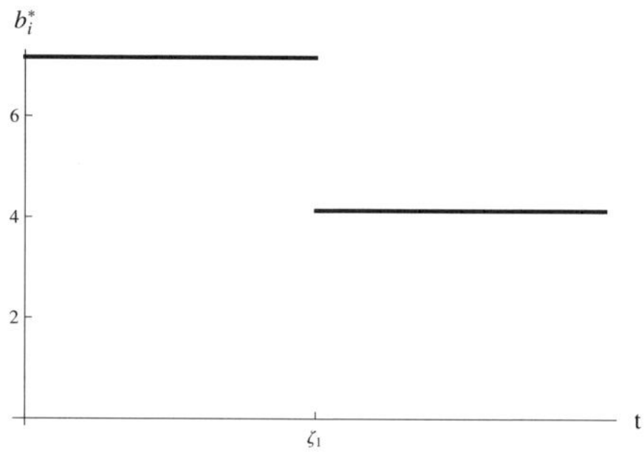

Fig. 7.11. b_i^* for $i = 1, 2$

Example 7.3. As a second example we reduce the premium intensity to $c = 5.5$ and leave all other parameters unchanged. For state 2 the optimal barrier is now zero, i.e. $V_2(x) = x + \frac{c}{\delta+\gamma+\lambda}$ (one can observe the same effect by changing $\gamma = 1$ to $\gamma = 2$ (earlier killing) while keeping $c = 6.5$). One then calculates $b_1^* = 3.70785$ as a solution to $f''_{(b_1,0)}(b_1) = 0$.
Here the crucial plot is Figure 7.15. As the IDE part is negative to the right of the barrier, we have that $f_{(b_1^*,0)}$ is the value function, a plot of which is given in Figure 7.12. Figures 7.13 and 7.14 show $f'_{(b_1^*,0)}(x)$ and $f''_{(b_1,0)}(b_1)$, illustrating that the minimum of the derivative is equal to 1 and the unique root of this specific second derivative.

Example 7.4. If we keep $c = 5.5$ and $\lambda = 5$, but choose $\gamma = 3$, we get $V_2 = x + \frac{c}{\delta+\gamma+\lambda}$ and a barrier in zero is then also optimal for state 1, i.e. $V_1(x) = x + \frac{c(2\gamma+\delta+\lambda)}{(\gamma+\delta+\lambda)^2}$.

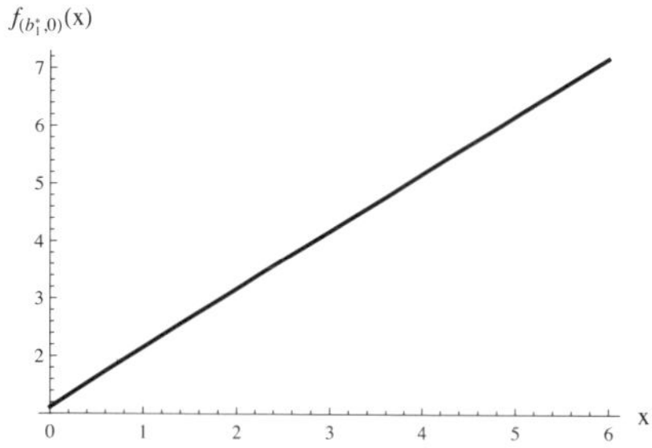

Fig. 7.12. $f_{(b_1^*,0)}$ for $x \in [0, 6)$

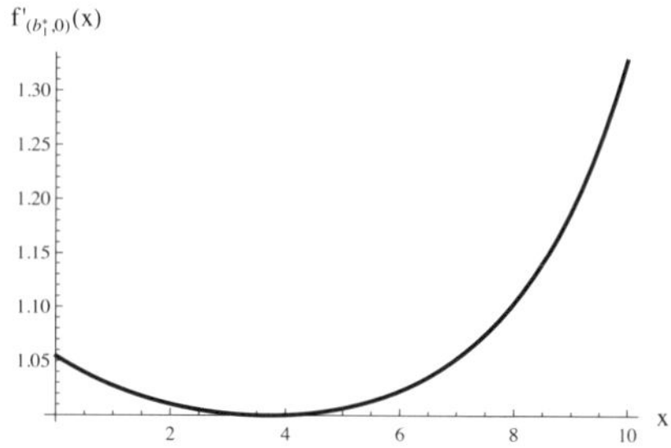

Fig. 7.13. $f'_{(b_1^*,0)}(x)$, for $x \in (0, 10)$

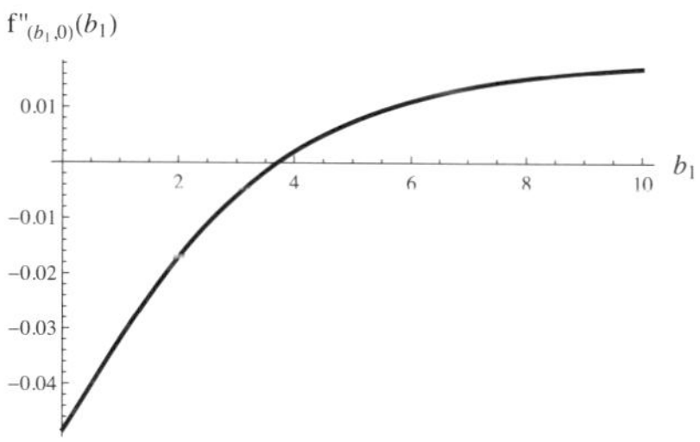

Fig. 7.14. $f''_{(b_1,0)}(b_1)$ for $b_1 \in (0, 10)$

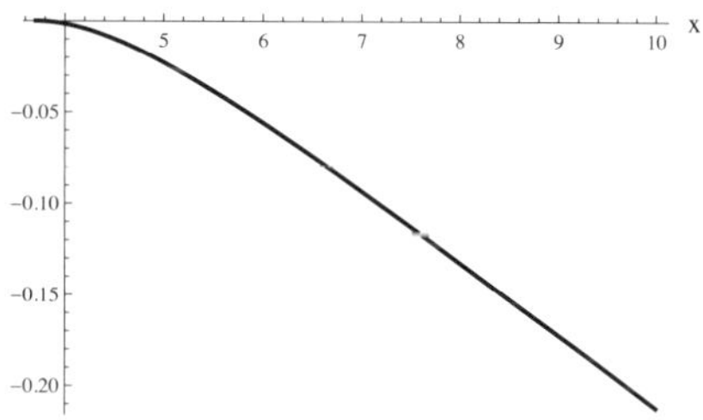

Fig. 7.15. IDE-HJB for $x \in (b_1^*, 10)$

7.4. Conclusion and outlook

In this contribution we identified and discussed a control problem in insurance, namely we extended the classical optimal dividend problem in insurance by the non-standard feature of having a random time horizon. We explicitly solved this problem for the case when both the time horizon and the individual claim sizes are exponentially distributed. We then extended the solution procedure to the case of an Erlang(2) time horizon by introducing an external state process and under the assumption that the state of this external process is observable, in which case the problem gets the flavor of a Markov regime switching problem. A natural next step for future research will be to apply some filtering technique to remove the (somewhat artificial) assumption of observability of the state process to explicitly solve the Erlang(2) time horizon problem (in which case there is an interpretation in terms of a Hidden Markov Model). The optimal strategy may then be expected to be of a time-dependent barrier form, where the value of the barrier at each time point arises from mixing over the optimal

barrier values of the two states and the mixing probabilities are given by the probability to be in the respective state at that time point. This procedure should then naturally extend to the Erlang(n) case, and the latter can then also be used to approximate the deterministic finite time horizon dividend problem. Since the diffusion approximation for the risk reserve process is a limit of the classical insurance risk process, this program may also give additional tools for analyzing the behavior of the optimal time-dependent barrier for the diffusion model studied by Grandits [13].
However, in view of the complexity of the involved expressions, already the Erlang(2) case is intricate and correspondingly the explicit identification of the optimal solution candidate for general Erlang(n) horizons and the formal verification of its optimality are expected to be a challenging problem.

Acknowledgement

The authors were supported by the Swiss National Science Foundation Project 200021-124635/1.

References

[1] H. Albrecher, N. Bäuerle and S. Thonhauser. Optimal dividend pay-out in random discrete time. *Statistics and Risk Modeling.* **28** (3), 251–276, (2011).

[2] H. Albrecher and J. Hartinger. On the non-optimality of horizontal barrier strategies in the Sparre Andersen model. *Hermis Journal on Computer Mathematics and Applications.* **7**, 109–122, (2006).

[3] H. Albrecher and S. Thonhauser. Optimality results for dividend problems in insurance. *Rev. R. Acad. Cienc. Exactas Fís. Nat. Ser. A Mat. RACSAM.* **103** (2), 295–320, (2009).

[4] S. Asmussen and H. Albrecher. *Ruin probabilities.* (World Scientific, New Jersey, 2010).

[5] S. Asmussen, F. Avram, and M. Usabel. Erlangian approximations for finite-horizon ruin probabilities. *ASTIN Bulletin.* **32** (2), 267–281, (2002).

[6] B. Avanzi. Strategies for dividend distribution: A review. *N. Am. Actuar. J..* **13** (2), 217–251, (2009).

[7] P. Azcue and N. Muler. Optimal reinsurance and dividend distribution policies in the Cramér-Lundberg model. *Math. Finance.* **15** (2), 261–308, (2005).

[8] P. Carr. Randomization and the American put. *Rev. Financ. Stud..* **11** (3), 597–626, (1998).

[9] M. H. A. Davis. *Markov models and optimization.* (Chapman & Hall, London, 1993).

[10] B. De Finetti. Su un' impostazione alternativa dell teoria collettiva del rischio. *Transactions of the XVth congress of actuaries.* **(II)**, 433–443, (1957).
[11] W. H. Fleming and H. M. Soner. *Controlled Markov processes and viscosity solutions.* (Springer, New York, 1993).
[12] J. Grandell. *Aspects of risk theory.* (Springer, New York, 1991).
[13] P. Grandits. Optimal consumption in a Brownian model with absorption and finite time horizon. *Preprint,* 31 pages, (2009).
[14] N. Kulenko and H. Schmidli. Optimal dividend strategies in a Cramér-Lundberg model with capital injections. *Insurance Math. Econom..* **43** (2), 270–278, (2008).
[15] R. Loeffen and J-F. Renaud. De Finetti's optimal dividends problem with an affine penalty function at ruin. *Insurance Math. Econom..* **46** (1), 98–108, (2010).
[16] T. Rolski, H. Schmidli, V. Schmidt, and J. Teugels. *Stochastic processes for insurance and finance.* (Wiley, 1999).
[17] H. Schmidli. *Stochastic Control in Insurance.* (Springer, New York, 2008).
[18] L. R. Sotomayor and A. Cadenillas. Classical and singular stochastic control for the optimal dividend policy when there is regime switching. *Insurance Math. Econom..* **48** (3), 344–354, (2011).

Chapter 8

Counterparty risk and the impact of collateralization in CDS contracts

Tomasz R. Bielecki*, Igor Cialenco* and Ismail Iyigunler

Department of Applied Mathematics,
Illinois Institute of Technology,
Chicago, 60616 IL, USA
bielecki@iit.edu, *igor@math.iit.edu,* *iiyigunl@iit.edu*

We analyze the counterparty risk embedded in CDS contracts, in presence of a bilateral margin agreement. First, we investigate the pricing of collateralized counterparty risk and we derive the bilateral Credit Valuation Adjustment (CVA), Unilateral Credit Valuation Adjustment (UCVA) and Debt Valuation Adjustment (DVA). We propose a model for the collateral by incorporating all related factors such as the thresholds, haircuts and margin period of risk. We derive the dynamics of the bilateral CVA in a general form with related jump martingales. We also introduce the Spread Value Adjustment (SVA) indicating the counterparty risk adjusted spread. Counterparty risky and the counterparty risk-free spread dynamics are derived and the dynamics of the SVA is found as a consequence. We finally employ a Markovian copula model for default intensities and illustrate our findings with numerical results.

Contents

*TRB and IC acknowledge support from the NSF grant DMS-0908099. The authors would like to thank the anonymous referee and the editors for their helpful comments and suggestions which improved greatly the final manuscript.

8.1. Introduction

Not very long after the collapse of prestigious institutions like Long-Term Capital Management, Enron and Global Crossing, the financial industry has again witnessed dramatic downfalls of financial institutions such as Lehman Brothers, Bear Stearns and Wachovia. These recent collapses have stressed out the importance of measuring, managing and mitigating counterparty risk appropriately.

Counterparty risk is defined as the risk that a party in an over-the-counter (OTC) contract will default and will not be able to honor its contractual obligations. Since the exchange-traded derivative contracts are subject to clearing by the exchange, counterparty risk arises from OTC derivatives only. The main challenge in the counterparty risk assessment and hedging is that the exposures of OTC derivatives are stochastic and involve dependencies and systemic risk factors such as wrong way risks; the additional level of complexity is introduced by risk mitigation techniques such as collateralization and netting. Therefore, one needs to model potential future exposures and to price the counterparty risk appropriately according to margin agreements that underlie the collateralization procedures.

Brigo and Capponi [1] focuses on a Gaussian copula model and study bilateral counterparty risk using a CIR++ intensity model. Recently, Brigo et al. [2] extended this framework to the collateralized contracts with an application to interest rate swaps under bilateral margin agreements. Hull and White [3] propose a static copula model and study unilateral counterparty risk on credit default swaps. Bielecki et al. [4] study unilateral counterparty risk with the absence of any margin agreements. Assefa et al. [5] consider the portfolio of credit default swaps under a Markovian copula model and consider only fully collateralized contracts. Jarrow and Yu [6] deal with the counterparty risk by using a dependence structure based on the default intensities of the counterparties. This approach, that also addresses the contagion risk issue, is considered in Leung and Kwok [7]. All these works mentioned above employ the reduced form modeling technology. However, structural models have also been used to model counterparty risk. Good examples of this approach are papers by Lipton and Sepp [8] and Blanchet-Scalliet and Patras [9]. Moreover, Stein and Lee [10] study and illustrate credit valuation adjustment computations in the fixed income markets.

Various issues regarding the simulation of credit valuation adjustments under margin agreements are studied by Pykhtin in [11]. Furthermore, Cesari et al. [12] and Gregory [13] provide thorough treatments of the methods and the applications used in practice regarding the counterparty risk.

In this paper, we analyze the counterparty risk in a Credit Default Swap (CDS) contract in presence of a bilateral margin agreement. There are three risky names associated with the contract: the reference entity, protection seller (the counterparty) and the protection buyer (the investor). Contrary to the common approach which starts with defining the Potential Future Exposure (PFE) and derives the Credit Valuation Adjustment (CVA) as the price of the counterparty risk, we find the CVA as the difference between the market values of a counterparty risk-free and a counterparty risky CDS contracts and deduct the relevant credit exposures accordingly. We consider the problem of bilateral counterparty risk assessment; that is, we consider the situation where the two counterparties of the CDS contract, i.e. the investor and the counterparty, are subject to default risk in a counterparty risky CDS contract.

We focus on collateralized contracts, where a bilateral margin agreement is in force as a vital risk mitigation tool, and it requires the counterparty and the investor to post collateral in case their exposure exceeds specific threshold values. We propose a model for the collateral by incorporating all related factors, such as thresholds, margin period of risk and minimum transfer amount. We derive the dynamics of the bilateral CVA, which are essential for dynamic hedging of counterparty risk. We also compute the decomposition of the fair spread for the CDS, and we analyze the so called Spread Value Adjustment (SVA). Essentially, SVA represents the adjustment to be made to the fair spread to incorporate counterparty risk into the CDS contract.

Using the bilateral CVA formula, we derive relevant formulas for assessment of credit exposures, such as PFE, Expected Positive Exposure (EPE) and Expected Negative Exposure (ENE).

In our model, the dependence between defaults and the wrong way risk is represented by a Markovian copula framework that accounts for simultaneous defaults among the three names represented in a CDS contract. In this way, our model takes broader systemic risk factors into account and quantifies the wrong way risk and the double defaults in a tangible manner.

This paper is organized as follows. In Section 8.2, we first define the dividend processes regarding the counterparty risky and the counterparty risk-free CDS contract in case of a bilateral margin agreement. We also define the CVA, UCVA and the DVA terms as well as the credit exposures such as PFE, EPE and ENE. We then prove the dynamics of the CVA in Section 8.2.3. Moreover, we find the fair spread adjustment term and its dynamics in Section 8.2.4. In Section 8.3, we simulate the collateralized exposures and the CVA using our Markovian copula model of default dependence.

8.2. Pricing counterparty risk: CVA, UCVA and DVA

We consider a standard CDS contract, and we label by 1 the reference name, by 2 the counterparty, and by 3 the investor. Each of the three names may default before the maturity of the CDS contract, and we denote by τ_1, τ_2 and τ_3 their respective default times. These times are modeled as non-negative random variables given on a underlying probability space $(\Omega, \mathcal{G}, \mathbb{Q})$. We let T and κ denote the maturity and the spread of our CDS contract, respectively. We assume the recovery at default covenant; that is, we assume that recoveries are paid at times of default.

We introduce right-continuous processes H_t^i by setting $H_t^i = \mathbb{I}_{\{\tau_i \le t\}}$ and we denote by $\mathbb{H}^i$ the associated filtrations so that $\mathcal{H}_t^i = \sigma\left(H_u^i : u \le t\right)$ for $i = 1, 2, 3$.

We assume that we are given a market filtration $\mathbb{F}$, and we define the enlarged filtration $\mathbb{G} = \mathbb{F} \vee \mathbb{H}^1 \vee \mathbb{H}^2 \vee \mathbb{H}^3$, that is $\mathcal{G}_t = \sigma\left(\mathcal{F}_t \cup \mathcal{H}_t^1 \cup \mathcal{H}_t^2 \cup \mathcal{H}_t^3\right)$ for any $t \in \mathbb{R}_+$. For each $t \in \mathbb{R}_+$ total information available at time t is captured by the σ-field $\mathcal{G}_t$. In particular, processes H^i are $\mathbb{G}$-adapted and the random times τ_i are $\mathbb{G}$-stopping times for $i = 1, 2, 3$.

Next, we define the first default time as the minimum of τ_1, τ_2 and τ_3: $\tau = \tau_1 \wedge \tau_2 \wedge \tau_3$; the corresponding indicator process is $H_t = \mathbb{I}_{\{\tau \le t\}}$. In addition, we define the first default time of the two counterparties: $\hat{\tau} = \tau_2 \wedge \tau_3$, and the corresponding indicator process $\hat{H}_t = \mathbb{I}_{\{\hat{\tau} \le t\}}$.

We also denote by B_t the savings account process, that is

$$B_t = e^{\int_0^t r_s ds},$$

where the $\mathbb{F}$-progressively measurable process r models the short-term interest rate. We also postulate that $\mathbb{Q}$ represents a martingale measure associated with the choice of the savings account B as a discount factor (or numeraire).

8.2.1. *Dividend processes and marking-to-market*

In this paper, all cash flows and the prices are considered from the perspective of the investor.

We start by introducing the *counterparty-risk-free* dividend process D, which describes all cash flows associated with a counterparty-risk-free CDS contract;[a] that is, D does not account for the counterparty risk.

Definition 8.1. The cumulative dividend process D of a counterparty risk-free CDS contract maturing at time T is given as,

$$D_t = \int_{]0,t]} \delta^1_u dH^1_u - \kappa \int_{]0,t]} \left(1 - H^1_u\right) du, \tag{8.1}$$

for every $t \in [0,T]$, where $\delta^1 : [0,T] \to \mathbb{R}$ is an $\mathbb{F}$-predictable processes.

Process δ^1 represents the loss given default (LGD); that is $\delta^1 = 1 - R^1_t$, where R^1 is the fraction of the nominal that is recovered in case of the default of the reference name. We assume unit nominal, for simplicity.

The ex-dividend price processes of the counterparty risk-free CDS contract, say S, describes the current market value, or the Mark-to-Market (MtM) value of this contract.

Definition 8.2. The ex-dividend price process S of a counterparty risk-free CDS contract maturing at time T is given by,

$$S_t = B_t \mathbb{E}_{\mathbb{Q}} \left(\int_{]t,T]} B_u^{-1} dD_u \middle| \mathcal{G}_t \right), \quad t \in [0,T]. \tag{8.2}$$

Remark 8.1. Accordingly, we define the cumulative (dividend) price process, say $\widehat{S}$, of a counterparty risk-free CDS contract as

$$\widehat{S}_t = S_t + B_t \int_{]0,t]} B_u^{-1} dD_u, \quad t \in [0,T].$$

Now, we are in position to define the dividend process D^C of a counterparty risky CDS contract, that is the CDS contract that accounts for the counterparty risk associated with the two counterparties of the contract.

Definition 8.3. The dividend process D^C of a T-maturity counterparty

[a] We shall sometimes refer to such contract as to the clean contract.

risky CDS contract is given as

$$\begin{aligned} D_t^C = &\int_{]0,t]} C_u dH_u + \int_{]0,t]} \widetilde{\delta}_u^1 (1 - H_{u-}) \, dH_u^1 + \int_{]0,t]} \widetilde{\delta}_u^2 (1 - H_{u-}) \, dH_u^2 \\ &+ \int_{]0,t]} \widetilde{\delta}_u^3 (1 - H_{u-}) \, dH_u^3 + \int_{]0,t]} \widetilde{\delta}_u^4 (1 - H_{u-}) \, d[H^2, H^3]_u \qquad (8.3) \\ &+ \int_{]0,t]} \widetilde{\delta}_u^5 (1 - H_{u-}) \, d[\hat{H}, H^1]_u - \kappa \int_{]0,t]} (1 - H_u) \, du, \quad t \in [0,T], \end{aligned}$$

where $\widetilde{\delta}^i : [0,T] \to \mathbb{R}$ is an $\mathbb{F}$-predictable processes for $i = 1, \dots, 5$ and $C : [0,T] \to \mathbb{R}$ is an $\mathbb{F}$-predictable process representing the collateral amount kept in the margin account.

A margin account is a contractual tool that supplements the CDS contract so as to reduce potential losses that may be incurred by one of the counterparties in case of the default of the other counterparty, while the CDS contract is still alive. For a detailed description of the mechanics of the collateral formation in the margin account we refer to Section 8.2.1.1 (see also [14]).

In case of any credit event associated with the collateralized CDS contract, the first cash flow that takes place is the "transfer" of the collateral amount; for example, in case when the underlying entity defaults at time $t = \tau = \tau_1$, (before any of the counterparties defaults) the collateral in the margin account is acquired by one of the counterparties (depending on the sign of C_τ). Thus, consistently with the convention of so called close-out cash flows (cf. [14]) we define $\widetilde{\delta}^i$s as follows:

- We set $\widetilde{\delta}_t^1 = \delta_t^1 - C_t$. This is because after the collateral transfer the counterparty pays the remaining recovery amount $\delta_t^1 - C_t$.
- At time $t = \tau = \tau_2$, when the counterparty defaults, then, after the collateral transfer takes place, if the uncollateralized mark-to-market (MtM) of the CDS contract, that is $S_t + \mathbb{I}_{\{t=\tau_1\}} \delta_t^1 - C_t$ [b], is negative, the investor closes out the position by paying the defaulting counterparty the uncollateralized MtM. If the uncollateralized MtM is positive, the investor closes out the position and receives a fraction R_2 of the uncollateralized MtM from the counterparty. Therefore, in this case, the close-out payment is defined as,

$$\widetilde{\delta}_t^2 = R_2 \left(S_t + \mathbb{I}_{\{t=\tau_1\}} \delta_t^1 - C_t \right)^+ - \left(S_t + \mathbb{I}_{\{t=\tau_1\}} \delta_t^1 - C_t \right)^- .$$

[b] The term $\mathbb{I}_{\{t=\tau_1\}} \delta_t^1$ represents the exposure in case when the counterparty and the underlying entity default simultaneously.

- In the case of investor default, that is, at time $t = \tau = \tau_3$, if the uncollateralized MtM is positive, that is, if $S_t + \mathbb{I}_{\{t=\tau_1\}}\delta_t^1 - C_t > 0$, the counterparty closes out the position by paying the uncollateralized MtM. If the uncollateralized MtM is negative, the counterparty receives a fraction R_3 of the uncollateralized MtM. Hence, the close-out payment is defined as,

$$\widetilde{\delta}_t^3 = \left(S_t + \mathbb{I}_{\{t=\tau_1\}}\delta_t^1 - C_t\right)^+ - R_3\left(S_t + \mathbb{I}_{\{t=\tau_1\}}\delta_t^1 - C_t\right)^- .$$

- If the investor and the counterparty default simultaneously at time $t = \tau = \tau_2 = \tau_3$, and if the uncollateralized MtM is negative, the counterparty receives a fraction R_3 of the uncollateralized MtM; however, if the uncollateralized MtM is positive, the investor receives a fraction R_2 of the uncollateralized MtM. Therefore, we set,

$$\widetilde{\delta}_t^4 = -\left(S_t + \mathbb{I}_{\{t=\tau_1\}}\delta_t^1 - C_t\right) .$$

- If $t = \tau = \hat{\tau} = \tau_1$, that is, when the investor or the counterparty default simultaneously with the reference entity, the investor receives a fraction R_2 of the remaining recovery amount, $\left(\delta_t^1 - C_t\right)^+$, when the counterparty defaults. Likewise, if the investor defaults, the counterparty receives a portion R_3 of the remaining recovery amount, $\left(\delta_t^1 - C_t\right)^-$. The close-out payment in joint defaults including the underlying entity has the form,

$$\widetilde{\delta}_t^5 = -\left(\delta_t^1 - C_t\right) .$$

We are now ready to define the price processes associated with a counterparty risky CDS contract.

Definition 8.4. The ex-dividend price process S^C of a counterparty risky CDS contract maturing at time T is given as,

$$S_t^C = B_t \mathbb{E}_{\mathbb{Q}}\left(\int_{]t,T]} B_u^{-1} dD_u^C \,\middle|\, \mathcal{G}_t\right), \quad t \in [0,T]. \tag{8.4}$$

The cumulative price process $\widehat{S}^C$ of a counterparty risky CDS contract is given by,

$$\widehat{S}_t^C = S_t^C + B_t \int_{]0,t]} B_u^{-1} dD_u^C, \quad t \in [0,T].$$

8.2.1.1. *Bilateral margin agreement and collateral modeling*

Collateralization is one of the most important techniques of mitigation of counterparty risk, and modeling the collateral process (also termed the margin call process) is of great practical importance (cf. [15]). In this section, we propose a model to describe the formation of the required collateral amount at every time $t \in [0, T]$ with regard to bilateral margin agreements.[c] The following contractual parameters are essential in bilateral margin agreements and they are precisely defined in CSA documents.

Margin Period of Risk: The margin period of risk consists of several components. Firms usually monitor their exposure on a periodic basis and receive or make appropriate margin calls considering other collateral parameters. The frequency of this process is called the margin call period and it is typically one day. This period includes a number of phases such as computation, negotiation, verification and settlement of the margin call also with possible disputes during the process. According to the ISDA Master Agreement, in case of a potential default, the defaulting counterparty enters into a short forbearance period to recover from a potential default event where the collateral is pledged by the other firm. This time interval is called the cure period. If the default is uncured, liquidation process of the collateral assets starts (cf. [17], page 26). This period mainly depends on the collateral portfolio selection, precise assessment of asset correlation and concentration risks as well as their liquidity, volatility and credit quality parameters. Therefore, the time interval from the last margin call plus the cure period until all collateral assets are liquidated and the resulting market risk is re-hedged is called the margin period of risk (cf. [11]); we shall denote it as Δ.

Threshold: The threshold is the unsecured credit exposure that both counterparties are willing to tolerate without holding any collateral. Bilateral margin agreements specify thresholds for both counterparties and require them to post collateral whenever the current credit exposure exceeds their thresholds (cf. [17], page 11). These threshold amounts are defined in the related CSA documents and often set to react to the changes in the credit rating of the counterparties (cf. [18], page 13). We will denote the counterparty and the investor's thresholds by Γ_{cpty} and Γ_{inv}, respectively. Since we perform our analysis from the point of view of the investor, we set

[c] A bilateral margin agreement is a contractual agreement governed by a Credit Support Annex (CSA), which is a regulatory part of the ISDA Master Agreement (cf. [16], page 34) describing the use of collateral which is either directly transferred between counterparties or held by a third party such as a clearing house (cf. [16], page 68).

the counterparty's threshold Γ_{cpty} to be a non-negative constant, and the investor's threshold Γ_{inv} to be a non-positive constant.

Minimum Transfer Amount: Margin calls for amounts smaller than the MTA are not allowed. The purpose of the MTA is to prevent calling small amounts; this is done so to avoid the operational costs involved in small transactions (cf. [17], page 13). The minimum transfer amount is usually the same for the investor and the counterparty.

Re-hypothecation Risk and Segregation: Collateral assets can be reused as a funding source on other derivatives transactions. This is known as re-hypothecation. An investor (counterparty) can rehypothecate the collateral received from the counterparty (investor) by selling or lending out the assets to a third party, which dramatically increases the credit risk associated with the collateral. Elimination of this rehypothecation risk is essentially done by segregating the collateral to a third party, such as a clearing house. This procedure carries certain funding risks, since the counterparties will not be recieving a funding benefit from the collateral posted, so they need to raise funding in connection with their transactions using their own funding rates.

The construction of the collateral process presented below builds upon the construction given in [14]. Let us denote the margin call dates by $0 < t_1 < \cdots < t_n < T$. On each margin call date, if the exposure is above the counterparty's threshold, Γ_{cpty}, and the difference between the current exposure and the collateral amount is greater than the MTA the counterparty posts collateral and updates the margin account; otherwise, no collateral exchange takes place since the transfer amount is less than the MTA. Likewise, the investor delivers collateral on each margin call date, if the exposure is below investor's threshold, Γ_{inv}, and the difference between the current exposure and the collateral amount is greater than MTA (cf. [16], pages 52-56). Note that in this model a collateral transfer are allowed only if it is greater than the MTA amount.

In accordance with the above discussion the collateral process is modeled as follows:

We set $C_0 = 0$. Then, for $i = 1, 2, \ldots, n$ we postulate that

$$\begin{aligned} C_{t_i+} &= \mathbb{I}_{\{S_{t_i}-\Gamma_{cpty}-C_{t_i}>MTA\}}(S_{t_i} - \Gamma_{cpty} - C_{t_i}) \\ &\quad + \mathbb{I}_{\{S_{t_i}-\Gamma_{inv}-C_{t_i}<-MTA\}}(S_{t_i} - \Gamma_{inv} - C_{t_i}) + C_{t_i}, \end{aligned}$$

on the set $\{t_i < \hat{\tau}\}$, and it is constant on interval $(t_i, t_{i+1}]$. Moreover, $C_t = C_{\hat{\tau}}$ on the set $\{\hat{\tau} < t < \hat{\tau} + \Delta\}$.

Observe that the collateral increments at each margin call date $t_i < \hat{\tau}$ can now be represented as,

$$\begin{aligned}\Delta C_{t_i} :&= C_{t_i+} - C_{t_i}\\ &= \mathbb{I}_{\{S_{t_i}-\Gamma_{cpty}-C_{t_i}>MTA\}}(S_{t_i} - \Gamma_{cpty} - C_{t_i})\\ &\quad + \mathbb{I}_{\{S_{t_i}-\Gamma_{inv}-C_{t_i}<-MTA\}}(S_{t_i} - \Gamma_{inv} - C_{t_i}).\end{aligned}$$

One should also note that the collateral construction given in [11], which reads

$$C_t = \mathbb{I}_{\{S_t>\Gamma_{cpty}+MTA\}}\left(S_t - \Gamma_{cpty}\right) + \mathbb{I}_{\{S_t<\Gamma_{buy}-MTA\}}\left(S_t - \Gamma_{inv}\right),$$

allows intermediate collateral updates that are smaller than MTA. In our case, we avoid this intricacy by defining the collateral process as a left-continuous, piecewise constant process.

Remark 8.2. The collateral construction described above is cash based. The net cash value of the collateral portfolio is determined using haircuts.

The haircut (or, valuation percentage) describes the amount that will be charged from a particular collateral asset. Effective value of the collateral asset is determined by subtracting the mark-to-market value of the asset multiplied by an appropriate haircut (cf. [16], page 67). Therefore, the haircuts applied to collateral assets should reflect the market risk on those assets. The haircut is defined as a percentage, where 0% haircut implies complete mark-to-market value of the asset to be used as collateral without any discounting. Government securities having high credit rating such as Treasury bonds and Treasury bills are usually subjected to 1% to 10% haircut, while for more risky, volatile or illiquid securities, such as a stock option, the haircut might be as high as 30%. The only asset that is not subjected to any haircut as collateral is cash, where usually both parties mutually agree on the use of an overnight index rate (cf. [17], page 27). The term valuation percentage is also used in Credit Support Annex (CSA) documents. The valuation percentage defines the amount that the market value of the asset is multiplied by to yield the effective collateral value of the asset. Hence, we have $\mathrm{VP}_t = 1 - h_t$, where VP_t is the valuation percentage and h_t is the total haircut applied to the collateral assets at time t. We will not go into the details of the formation of the haircut since it is either pre-determined in the CSA documents or related to market risk measures such as VaR of the collateral assets. (cf. [16], page 68). The main purpose of the haircut is to mitigate amortization or depreciation in the collateral asset value at the time of a default and in the margin period of risk. Moreover, the haircut should be updated as frequently as possible to reflect the changes in the volatility or liquidity of the collateral assets (cf. [16], page 63).

Therefore, the total value of the collateral portfolio at time t is equal to $(1+h_t)C_t$, where h_t is the appropriate haircut applied to the collateral portfolio.

8.2.2. *Bilateral Credit Valuation Adjustment*

In this section, we shall compute the CVA on a CDS contract, subject to a bilateral margin agreement.

Definition 8.5. The bilateral Credit Valuation Adjustment process on a CDS contract maturing at time T is defined as

$$\mathrm{CVA}_t = S_t - S_t^C, \tag{8.5}$$

for every $t \in [0,T]$.

We now present an alternative representation for the bilateral CVA, which is convenient for computational purposes.

Proposition 8.1. The bilateral CVA process on a CDS contract maturing at time T satisfies

$$\begin{aligned}\mathrm{CVA}_t = {}& B_t \mathbb{E}_{\mathbb{Q}}\left(\mathbb{I}_{\{t<\tau=\tau_2\le T\}} B_\tau^{-1} (1-R_2) \left(S_\tau + \mathbb{I}_{\{\tau=\tau_1\}}\delta_\tau^1 - C_\tau\right)^+ \Big| \mathcal{G}_t \right) \\ & - B_t \mathbb{E}_{\mathbb{Q}}\left(\mathbb{I}_{\{t<\tau=\tau_3\le T\}} B_\tau^{-1} (1-R_3) \left(S_\tau + \mathbb{I}_{\{\tau=\tau_1\}}\delta_\tau^1 - C_\tau\right)^- \Big| \mathcal{G}_t \right),\end{aligned} \tag{8.6}$$

for every $t \in [0,T]$.

Proof. We begin by observing that

$$\int_{]t,T]} B_u^{-1} \widetilde{\delta}_u^i (1-H_{u-})\, dH_u^i = B_\tau^{-1} \widetilde{\delta}_\tau^i \mathbb{I}_{\{t<\tau=\tau_i\le T\}},$$

for $i = 1, 2, 3$. Consequently,

$$\begin{aligned}\int_{]t,T]} B_u^{-1} dD_u^C = {} & B_\tau^{-1}\widetilde{\delta}_\tau^1 \mathbb{I}_{\{t<\tau=\tau_1\le T\}} + B_\tau^{-1}\widetilde{\delta}_\tau^2 \mathbb{I}_{\{t<\tau=\tau_2\le T\}} \\ & + B_\tau^{-1}\widetilde{\delta}_\tau^3 \mathbb{I}_{\{t<\tau=\tau_3\le T\}} + B_\tau^{-1}\widetilde{\delta}_\tau^4 \mathbb{I}_{\{t<\tau=\tau_2=\tau_3\le T\}} \\ & + B_\tau^{-1}\widetilde{\delta}_\tau^5 \mathbb{I}_{\{t<\tau=\tau^*=\tau_1\le T\}} + B_\tau^{-1} C_\tau \mathbb{I}_{\{t<\tau\le T\}} \\ & - \kappa \int_{]t,T]} B_u^{-1} \mathbb{I}_{\{\tau>u\}} du. \end{aligned} \tag{8.7}$$

Using the definitions of the close-out cash-flows $\widetilde{\delta}_\tau^i$, $i = 1, \ldots, 5$, we get from (8.7)

$$\begin{aligned}\int_{]t,T]} B_u^{-1} dD_u^C = {} & B_\tau^{-1}\left(\delta_\tau^1 - C_\tau\right) \mathbb{I}_{\{t<\tau=\tau_1\le T\}} \\ & - \kappa \int_{]t,T]} B_u^{-1} \mathbb{I}_{\{\tau>u\}} du + B_\tau^{-1} C_\tau \mathbb{I}_{\{t<\tau\le T\}} \\ & + B_\tau^{-1}\Big(R_2\left(S_\tau + \mathbb{I}_{\{\tau=\tau_1\}}\delta_\tau^1 - C_\tau\right)^+ \\ & \qquad - \left(S_\tau + \mathbb{I}_{\{\tau=\tau_1\}}\delta_\tau^1 - C_\tau\right)^-\Big) \mathbb{I}_{\{t<\tau=\tau_2\le T\}} \\ & + B_\tau^{-1}\Big(\left(S_\tau + \mathbb{I}_{\{\tau=\tau_1\}}\delta_\tau^1 - C_\tau\right)^+ \\ & \qquad - R_3\left(S_\tau + \mathbb{I}_{\{\tau=\tau_1\}}\delta_\tau^1 - C_\tau\right)^-\Big) \mathbb{I}_{\{t<\tau=\tau_3\le T\}} \\ & - B_\tau^{-1}\left(S_\tau + \mathbb{I}_{\{\tau=\tau_1\}}\delta_\tau^1 - C_\tau\right) \mathbb{I}_{\{t<\tau=\tau_2=\tau_3\le T\}} \\ & - B_\tau^{-1}\left(\delta_\tau^1 - C_\tau\right) \mathbb{I}_{\{t<\tau=\tau^*=\tau_1\le T\}}. \end{aligned} \tag{8.8}$$

Since

$$\begin{aligned}\mathbb{I}_{\{t<\tau\le T\}} = {} & \mathbb{I}_{\{t<\tau=\tau_1\le T\}} + \mathbb{I}_{\{t<\tau=\tau_2\le T\}} + \mathbb{I}_{\{t<\tau=\tau_3\le T\}} \\ & - \mathbb{I}_{\{t<\tau=\tau_2=\tau_3\le T\}} - \mathbb{I}_{\{t<\tau^*=\tau_1\le T\}}, \end{aligned}$$

using the equality

$$R_i\left(S_\tau - C_\tau\right)^+ - \left(S_\tau - C_\tau\right)^- + C_\tau = S_\tau - \left(1 - R_i\right)\left(S_\tau - C_\tau\right)^+$$

and observing that $\mathbb{I}_{\{\tau=\tau_1\}} S_\tau = 0$, we can rearrange the terms in (8.8) as follows,

$$\int_{]t,T]} B_u^{-1} dD_u^C = B_\tau^{-1}\delta_\tau^1 \mathbb{I}_{\{t<\tau=\tau_1\le T\}} - \kappa \int_{]t,T]} B_u^{-1}\mathbb{I}_{\{\tau>u\}} du \tag{8.9}$$
$$+ B_\tau^{-1} S_\tau \left(\mathbb{I}_{\{t<\tau=\tau_2\le T\}} + \mathbb{I}_{\{t<\tau=\tau_3\le T\}} - \mathbb{I}_{\{t<\tau=\tau_2=\tau_3\le T\}}\right)\mathbb{I}_{\{\tau\neq\tau_1\}}$$
$$- B_\tau^{-1}(1-R_2)\left(S_\tau + \mathbb{I}_{\{\tau=\tau_1\}}\delta_\tau^1 - C_\tau\right)^+ \mathbb{I}_{\{t<\tau=\tau_2\le T\}}$$
$$+ B_\tau^{-1}(1-R_3)\left(S_\tau + \mathbb{I}_{\{\tau=\tau_1\}}\delta_\tau^1 - C_\tau\right)^- \mathbb{I}_{\{t<\tau=\tau_3\le T\}}.$$

Now, combining (8.9) with (8.1) we see that

$$S_t^C = B_t \mathbb{E}_{\mathbb{Q}}\left(\left(\mathbb{I}_{\{t<\tau=\tau_1\le T\}} + \mathbb{I}_{\{\tau>T\}}\right) \int_{]t,T]} B_u^{-1} dD_u \,\middle|\, \mathcal{G}_t\right)$$
$$+ B_t \mathbb{E}_{\mathbb{Q}}\Big(\left(\mathbb{I}_{\{t<\tau=\tau_2\le T\}} + \mathbb{I}_{\{t<\tau=\tau_3\le T\}} - \mathbb{I}_{\{t<\tau=\tau_2=\tau_3\le T\}}\right)\mathbb{I}_{\{\tau\neq\tau_1\}}\Big) \mathbb{E}_{\mathbb{Q}}\left(\left(\int_{]\tau,T]} B_u^{-1} dD_u \,\middle|\, \mathcal{G}_\tau\right) \,\middle|\, \mathcal{G}_t\right)$$
$$- B_t \mathbb{E}_{\mathbb{Q}}\left(\mathbb{I}_{\{t<\tau=\tau_2\le T\}} B_\tau^{-1}(1-R_2)\left(S_\tau + \mathbb{I}_{\{\tau=\tau_1\}}\delta_\tau^1 - C_\tau\right)^+ \,\middle|\, \mathcal{G}_t\right)$$
$$+ B_t \mathbb{E}_{\mathbb{Q}}\left(\mathbb{I}_{\{t<\tau=\tau_3\le T\}} B_\tau^{-1}(1-R_3)\left(S_\tau + \mathbb{I}_{\{\tau=\tau_1\}}\delta_\tau^1 - C_\tau\right)^- \,\middle|\, \mathcal{G}_t\right).$$

From here, observing that

$$\mathbb{I}_{\{\tau\le t\}} + \mathbb{I}_{\{\tau>T\}} + \mathbb{I}_{\{t<\tau=\tau_1\le T\}} + \Big(\mathbb{I}_{\{t<\tau=\tau_2\le T\}} + \mathbb{I}_{\{t<\tau=\tau_3\le T\}} - \mathbb{I}_{\{t<\tau=\tau_2=\tau_3\le T\}}\Big)\mathbb{I}_{\{\tau\neq\tau_1\}} = 1,$$

we get

$$S_t^C = B_t \mathbb{E}_{\mathbb{Q}}\left(\int_{]t,T]} B_u^{-1} dD_u \,\middle|\, \mathcal{G}_t\right)$$
$$- B_t \mathbb{E}_{\mathbb{Q}}\left(\mathbb{I}_{\{t<\tau=\tau_2\le T\}} B_\tau^{-1}(1-R_2)(S_\tau - C_\tau)^+ \,\middle|\, \mathcal{G}_t\right)$$
$$+ B_t \mathbb{E}_{\mathbb{Q}}\left(\mathbb{I}_{\{t<\tau=\tau_3\le T\}} B_\tau^{-1}(1-R_3)(S_\tau - C_\tau)^- \,\middle|\, \mathcal{G}_t\right), \tag{8.10}$$

which is

$$S_t^C = S_t - B_t \mathbb{E}_{\mathbb{Q}}\left(\mathbb{I}_{\{t<\tau=\tau_2\le T\}} B_\tau^{-1}(1-R_2)(S_\tau - C_\tau)^+ \,\middle|\, \mathcal{G}_t\right)$$
$$+ B_t \mathbb{E}_{\mathbb{Q}}\left(\mathbb{I}_{\{t<\tau=\tau_3\le T\}} B_\tau^{-1}(1-R_3)(S_\tau - C_\tau)^- \,\middle|\, \mathcal{G}_t\right).$$

This proves the result. □

Remark 8.3. The above results shows that the value of the bilateral CVA is the same as the sum of the value of a long position in a zero-strike call option on the uncollateralized amount and the value of a short position in a zero-strike put option on the uncollateralized amount.

8.2.2.1. *Unilateral CVA and debt value adjustment*

The bilateral nature of the counterparty risk is a consequence of possible default of the counterparty and the possible default of the investor. The values of potential losses associated with these two components are called unilateral CVA (UCVA) and debt value adjustment (DVA), respectively, and defined below.

Definition 8.6. The Unilateral Credit Value Adjustment is defined as,

$$\mathrm{UCVA}_t = B_t \mathbb{E}_{\mathbb{Q}} \left(\mathbb{I}_{\{t<\tau=\tau_2\le T\}} B_\tau^{-1} (1-R_2) \left(S_\tau + \mathbb{I}_{\{\tau=\tau_1\}} \delta_\tau^1 - C_\tau\right)^+ \Big| \mathcal{G}_t \right),$$

for $t \in [0,T]$, and symmetrically the Debt Value Adjustment is defined as,

$$\mathrm{DVA}_t = B_t \mathbb{E}_{\mathbb{Q}} \left(\mathbb{I}_{\{t<\tau=\tau_3\le T\}} B_\tau^{-1} (1-R_3) \left(S_\tau + \mathbb{I}_{\{\tau=\tau_1\}} \delta_\tau^1 - C_\tau\right)^- \Big| \mathcal{G}_t \right),$$

for $t \in [0,T]$.

Remark 8.4. DVA accounts for the risk of investor's own default, and it represents the value of any potential outstanding liabilities of the investors that will not be honored at the time of the investor's default:

In fact, at time of his/her default, the investor only pays to the counterparty the recovery amount, that is $R_3 \left(S_\tau + \mathbb{I}_{\{\tau=\tau_1\}} \delta_\tau^1 - C_\tau\right)^-$. Therefore, the investor gains the remaining amount, which is equal to $(1-R_3) \left(S_\tau + \mathbb{I}_{\{\tau=\tau_1\}} \delta_\tau^1 - C_\tau\right)^-$, on his/her outstanding liabilities by defaulting. Risk management of this component is of great importance for financial institutions.

When considering the unilateral counterparty risk DVA is set to zero.

In view of Proposition 8.1 and of the above definition we have that

$$\mathrm{CVA}_t = \mathrm{UCVA}_t - \mathrm{DVA}_t, \quad t \in [0,T].$$

Note that the bilateral CVA amount may be negative for the investor due to "own default risk". This also indicates that the price S^C of counterparty risky CDS contract may be greater than the price S of counterparty risk-free contract.

Remark 8.5. (Upfront CDS Conversion) After the "CDS Big Bang" (cf. [19]) a process originated to replace standard CDS contracts with so called *upfront CDS contracts.* An upfront CDS contract is composed of an upfront payment, which is an amount to be exchanged upon the inception of the contract, and a fixed spread. The fixed spread, say $\widehat{\kappa}$, will be 100bps for investment grade CDS contracts, and 500bps for high yield CDS contracts. The recovery rate is also standardized to two possible values: 20% or 40%, depending on the credit worthiness of the reference name. The corresponding cumulative dividend process of a counterparty-risk-free CDS contract is described in the following definition.

Definition 8.7. The cumulative dividend process $\widehat{D}$ of a counterparty-risk-free upfront CDS contract, maturing at time T, is given as

$$\widehat{D}_t = \int_{]0,t]} \delta_u^1 dH_u^1 - \mathrm{UP} - \widehat{\kappa} \int_{]0,t]} \left(1 - H_u^1\right) du\,, \quad t \in [0,T]\,,$$

where UP is the upfront payment, and $\widehat{\kappa}$ is the fixed spread.

Recall that the spread κ_0 of a standard CDS contract is set such that the protection leg PL_0 and fixed leg $\kappa_0 DV01_0$ are equal at initiation (making the price of the contract equal to zero). Similarly, in the case of an upfront CDS contract, with $\widehat{\kappa}$ being fixed, the upfront payment UP is chosen such that the contract has zero value at initiation. It is easy to convert the conventional spread κ_0 into an upfront payment PU and vise versa. Indeed, directly from the Definition 8.7, and definitions of PL_0 and $DV01_0$, we have

$$PL_0 - UP - \widehat{\kappa} DV01_0 = PL_0 - \kappa_0 DV01_0 = 0\,,$$

which implies the following representations

$$UP = (\kappa_0 - \widehat{\kappa})\, DV01_0, \qquad \kappa_0 = \frac{UP}{DV01_0} + \widehat{\kappa}\,.$$

In view of the conversion formulae presented above the discussion of CVA, DVA and UCVA done for standard CDS contracts can be adopted to the case of the upfront CDS contracts in a straightforward manner.

8.2.2.2. *CVA via credit exposures*

Credit exposure is defined as the potential loss that may be suffered by either one of the counterparties due to the other party's default. Here, we discuss some measures commonly used to quantify credit exposure, such as

Potential Future Exposure (PFE), *Expected Positive Exposure* (EPE) and *Expected Negative Exposure* (ENE), and their relation to CVA.

Potential Future Exposure is the basic measure of credit exposure:

Definition 8.8. The Potential Future Exposure of a CDS contract with a bilateral margin agreement is defined as follows,

$$\begin{aligned} PFE = &\mathbb{I}_{\{\tau=\tau_2\}}\left(1-R_2\right)\left(S_\tau+\mathbb{I}_{\{\tau=\tau_1\}}\delta_\tau^1-C_\tau\right)^+ \\ &-\mathbb{I}_{\{\tau=\tau_3\}}\left(1-R_3\right)\left(S_\tau+\mathbb{I}_{\{\tau=\tau_1\}}\delta_\tau^1-C_\tau\right)^- . \end{aligned}$$

Note that there exist several forms by which the potential future exposure is defined by financial institutions. The PFE definition given above, as a random variable, is in line with the PFE definitions in (cf. [20]), as opposed to the rather classical definition of the PFE as the quantile of the exposure distribution (cf. [12]).

Remark 8.6. Observe that the CVA is related to PFE as follows,

$$CVA_t = B_t\mathbb{E}_{\mathbb{Q}}\left(\mathbb{I}_{\{t<\tau\le T\}}B_\tau^{-1}PFE\big|\,\mathcal{G}_t\right),\ t\in[0,T].$$

Expected Positive Exposure is defined as the expected amount the investor will lose if the counterparty default happens at time t, and Expected Negative Exposure is defined as the expected amount the investor will lose if his own default happens at time t. Note that the losses are conditional on default at time t. EPE and ENE are necessary quantities to price and hedge counterparty risk.

Definition 8.9. The Expected Positive Exposure of a CDS contract with a bilateral margin agreement is defined as,

$$EPE_t = \mathbb{E}_{\mathbb{Q}}\left(\left(1-R_2\right)\left(S_\tau+\mathbb{I}_{\{\tau=\tau_1\}}\delta_\tau^1-C_\tau\right)^+\Big|\,\tau=\tau_2=t\right),$$

and the Expected Negative Exposure is defined as,

$$ENE_t = \mathbb{E}_{\mathbb{Q}}\left(\left(1-R_3\right)\left(S_\tau+\mathbb{I}_{\{\tau=\tau_1\}}\delta_\tau^1-C_\tau\right)^-\ \Big|\,\tau=\tau_3=t\right)$$

for every $t\in[0,T]$.

8.2.3. *Dynamics of CVA*

In this section we derive the dynamics for CVA. This is important for deriving formulae for dynamic hedging of counterparty risk, however this issue will be discussed in a different paper.

We begin with defining some auxiliary stopping times, that will be useful later on:

$$\tau^{\{1\}} := \begin{cases} \tau_1 & \text{if } \tau_1 \neq \tau_2,\ \tau_1 \neq \tau_3 \\ \infty & \text{otherwise} \end{cases}, \qquad \tau^{\{2\}} := \begin{cases} \tau_2 & \text{if } \tau_2 \neq \tau_1, \tau_2 \neq \tau_3 \\ \infty & \text{otherwise} \end{cases},$$

$$\tau^{\{3\}} := \begin{cases} \tau_3 & \text{if } \tau_3 \neq \tau_1, \tau_3 \neq \tau_2 \\ \infty & \text{otherwise} \end{cases}, \qquad \tau^{\{4\}} := \begin{cases} \tau_2 & \text{if } \tau_2 = \tau_3,\ \tau_2 \neq \tau_1 \\ \infty & \text{otherwise} \end{cases},$$

$$\tau^{\{5\}} := \begin{cases} \tau_1 & \text{if } \tau_1 = \tau_2,\ \tau_1 \neq \tau_3 \\ \infty & \text{otherwise} \end{cases}, \qquad \tau^{\{6\}} := \begin{cases} \tau_1 & \text{if } \tau_1 = \tau_3,\ \tau_1 \neq \tau_2 \\ \infty & \text{otherwise} \end{cases},$$

$$\tau^{\{7\}} := \begin{cases} \tau_1 & \text{if } \tau_1 = \tau_2 = \tau_3 \\ \infty & \text{otherwise} \end{cases}.$$

Accordingly, we define the default indicator processes

$$\begin{aligned}
H_t^{\{1\}} &:= \mathbb{I}_{\{\tau_1 \le t, \tau_1 \neq \tau_2, \tau_1 \neq \tau_3\}} = \mathbb{I}_{\{\tau^{\{1\}} \le t\}}, \\
H_t^{\{2\}} &:= \mathbb{I}_{\{\tau_2 \le t, \tau_2 \neq \tau_1, \tau_2 \neq \tau_3\}} = \mathbb{I}_{\{\tau^{\{2\}} \le t\}}, \\
H_t^{\{3\}} &:= \mathbb{I}_{\{\tau_3 \le t, \tau_3 \neq \tau_1, \tau_3 \neq \tau_2\}} = \mathbb{I}_{\{\tau^{\{3\}} \le t\}}, \\
H_t^{\{4\}} &:= \mathbb{I}_{\{\tau_2 = \tau_3 \le t, \tau_1 \neq \tau_2\}} = \mathbb{I}_{\{\tau^{\{4\}} \le t\}}, \\
H_t^{\{5\}} &:= \mathbb{I}_{\{\tau_1 = \tau_2 \le t, \tau_1 \neq \tau_3\}} = \mathbb{I}_{\{\tau^{\{5\}} \le t\}}, \\
H_t^{\{6\}} &:= \mathbb{I}_{\{\tau_1 = \tau_3 \le t, \tau_1 \neq \tau_2\}} = \mathbb{I}_{\{\tau^{\{6\}} \le t\}}, \\
H_t^{\{7\}} &:= \mathbb{I}_{\{\tau_1 = \tau_2 = \tau_3 \le t\}} = \mathbb{I}_{\{\tau^{\{7\}} \le t\}}.
\end{aligned}$$

Remark 8.7. Note that one can represent processes $H_t^{\{i\}}, i = 1, \ldots, 7$, as follows

$$\begin{aligned}
H_t^{\{7\}} &= \left[\left[H^1, H^2\right], H^3\right]_t, \ H_t^{\{6\}} = \left[H^1, H^3\right]_t - H_t^{\{7\}}, \\
H_t^{\{5\}} &= \left[H^1, H^2\right]_t - H_t^{\{7\}}, \ H_t^{\{4\}} = \left[H^2, H^3\right]_t - H_t^{\{7\}}, \\
H_t^{\{3\}} &= H_t^3 - H_t^{\{4\}} - H_t^{\{6\}} - H_t^{\{7\}}, \\
H_t^{\{2\}} &= H_t^2 - H_t^{\{4\}} - H_t^{\{5\}} - H_t^{\{7\}}, \\
H_t^{\{1\}} &= H_t^1 - H_t^{\{5\}} - H_t^{\{6\}} - H_t^{\{7\}}.
\end{aligned}$$

In particular, these processes are $\mathbb{G}$-adapted processes.

Let $G(t) = \mathbb{Q}(\tau > t | \mathcal{F}_t)$ be the survival probability process of τ with respect to filtration $\mathbb{F}$. It is a $\mathbb{F}$ supermartingale and it admits unique Doob-Meyer decomposition $G = \mu - \nu$ where μ is the martingale part and ν is a predictable increasing process. We assume that G is a continuous process and v is absolutely continuous with respect to the Lebesgue measure, so that $d\nu_t = v_t dt$ for some $\mathbb{F}$-progressively measurable, non-negative process v. We denote by l the $\mathbb{F}$-progressively measurable process defined as $l_t = G(t)^{-1} v_t$. Finally, we assume that all $\mathbb{F}$ martingales are continuous.

We assume that hazard process of each stopping time $\tau^{\{i\}}$ admits an $(\mathbb{F}, \mathbb{G})$-intensity process q^i for every $i = 1, \ldots, 7$, so that the process $M^{\{i\}}$, given by the formula,

$$M_t^{\{i\}} = H_t^{\{i\}} - \int_0^t \left(1 - H_u^{\{i\}}\right) q_u^i du$$

is a $\mathbb{G}$-martingale for every $t \in [0, T]$ and $i = 1, \ldots, 7$.

We now have the following technical result,

Lemma 8.1. *The processes*

$$M_t^i := M_{t\wedge\tau}^{\{i\}} = H_{t\wedge\tau}^{\{i\}} - \int_0^{t\wedge\tau} l_u^i du, \quad t \geq 0, i = 1, 2, \ldots, 7,$$

and

$$M_t := H_{t\wedge\tau} - \int_0^{t\wedge\tau} l_u du, \quad t \geq 0,$$

where

$$l_t^i = \mathbb{I}_{\{\tau \geq t\}} q_t^i \text{ and } l_t = \sum_{i=1}^{7} l_t^i \quad t \geq 0, i = 1, 2, \ldots, 7,$$

are $\mathbb{G}$-martingales

Proof. Fix $i = 1, \ldots, 7$. Process M^i follows a $\mathbb{G}$-martingale, since it is $\mathbb{G}$-martingale $M^{\{i\}}$ stopped at the $\mathbb{G}$ stopping time τ. Moreover, we have that $M_t = \sum_{i=1}^{7} M_t^i$, so that process M is also a $\mathbb{G}$-martingale. □

We shall now proceed with deriving some useful representations for the processes S^C and S.

Lemma 8.2. *The ex-dividend price process,* S^C, *of a counterparty risky CDS contract, given in* (8.4), *can be represented as follows,*

$$S_t^C = B_t \mathbb{E}_{\mathbb{Q}} \left(B_\tau^{-1} \sum_{i=1}^{7} \mathbb{I}_{\{t<\tau=\tau^{\{i\}}\leq T\}} \bar{\delta}_\tau^i - \kappa \int_{]t,T]} B_u^{-1} \mathbb{I}_{\{\tau>u\}} du \,\middle|\, \mathcal{G}_t \right) \tag{8.11}$$

where

$$\begin{aligned}
\bar{\delta}_t^1 &= \delta_t^1, \ \bar{\delta}_t^2 = S_t - (1-R_2)(S_t - C_t)^+ \\
\bar{\delta}_t^3 &= S_t + (1-R_3)(S_t - C_t)^-, \\
\bar{\delta}_t^4 &= S_t - (1-R_2)(S_t - C_t)^+ + (1-R_3)(S_t - C_t)^- \\
\bar{\delta}_t^5 &= \delta_t^1 - (1-R_2)\left(\delta_t^1 - C_t\right)^+, \ \bar{\delta}_t^6 = \delta_t^1 + (1-R_3)\left(\delta_t^1 - C_t\right)^- \\
\bar{\delta}_t^7 &= \delta_t^1 - (1-R_2)\left(\delta_t^1 - C_t\right)^+ + (1-R_3)\left(\delta_t^1 - C_t\right)^-.
\end{aligned}$$

Proof. Let us rewrite (8.9) in the following form,

$$\begin{aligned}
S_t^C = B_t \mathbb{E}_{\mathbb{Q}} \Bigg(& B_\tau^{-1} \delta_\tau^1 \sum_{i=1,5,6,7} \mathbb{I}_{\{t<\tau=\tau^{\{i\}}\leq T\}} + B_\tau^{-1} S_\tau \sum_{i=2,3,4} \mathbb{I}_{\{t<\tau=\tau^{\{i\}}\leq T\}} \\
& - B_\tau^{-1}(1-R_2)\left(S_\tau + \mathbb{I}_{\{\tau=\tau_1\}}\delta_\tau^1 - C_\tau\right)^+ \sum_{i=2,4,5,7} \mathbb{I}_{\{t<\tau=\tau^{\{i\}}\leq T\}} \\
& + B_\tau^{-1}(1-R_3)\left(S_\tau + \mathbb{I}_{\{\tau=\tau_1\}}\delta_\tau^1 - C_\tau\right)^- \\
& \times \sum_{i=3,4,6,7} \mathbb{I}_{\{t<\tau=\tau^{\{i\}}\leq T\}} - \kappa \int_{]t,T]} B_u^{-1} \mathbb{I}_{\{\tau>u\}} du \,\Bigg|\, \mathcal{G}_t \Bigg),
\end{aligned}$$

which, after rearranging terms, leads to

$$
\begin{aligned}
S_t^C = B_t \mathbb{E}_{\mathbb{Q}} \Bigg(& B_\tau^{-1} \delta_\tau^1 \mathbb{I}_{\{t<\tau=\tau^{\{1\}}\le T\}} \\
& + B_\tau^{-1} \left(S_t - (1-R_2)\left(S_\tau - C_\tau\right)^+ \right) \mathbb{I}_{\{t<\tau=\tau^{\{2\}}\le T\}} \\
& + B_\tau^{-1} \left(S_\tau + (1-R_3)\left(S_\tau - C_\tau\right)^- \right) \mathbb{I}_{\{t<\tau=\tau^{\{3\}}\le T\}} \\
& + B_\tau^{-1} \Big(S_\tau - (1-R_2)\left(S_\tau - C_\tau\right)^+ \\
& \qquad + (1-R_3)\left(S_\tau - C_\tau\right)^- \Big) \mathbb{I}_{\{t<\tau=\tau^{\{4\}}\le T\}} \\
& + B_\tau^{-1} \left(\delta_\tau^1 - (1-R_2)\left(\delta_\tau^1 - C_\tau\right)^+ \right) \mathbb{I}_{\{t<\tau=\tau^{\{5\}}\le T\}} \\
& + B_\tau^{-1} \left(\delta_\tau^1 + (1-R_3)\left(\delta_\tau^1 - C_\tau\right)^- \right) \mathbb{I}_{\{t<\tau=\tau^{\{6\}}\le T\}} \\
& + B_\tau^{-1} \Big(\delta_\tau^1 - (1-R_2)\left(\delta_\tau^1 - C_\tau\right)^+ \\
& \qquad + (1-R_3)\left(\delta_\tau^1 - C_\tau\right)^- \Big) \mathbb{I}_{\{t<\tau=\tau^{\{7\}}\le T\}} \\
& - \kappa \int_{]t,T]} B_u^{-1} \mathbb{I}_{\{\tau>u\}} du \,\Bigg|\, \mathcal{G}_t \Bigg).
\end{aligned}
$$

This proves the result. □

In case when $R_2 = R_3 = 1$ process S is the same as process S^C. Thus, we obtain from the above

Corollary 8.1. *The ex-dividend price process S of a counterparty risk-free CDS contract, can be represented as follows*[d],

$$
S_t = B_t \mathbb{E}_{\mathbb{Q}} \left(B_\tau^{-1} \sum_{i=1}^{7} \mathbb{I}_{\{t<\tau=\tau^{\{i\}}\le T\}} \hat{\delta}_\tau^i - \kappa \int_{]t,T]} B_u^{-1} \mathbb{I}_{\{\tau>u\}} du \,\middle|\, \mathcal{G}_t \right), \tag{8.12}
$$

[d]We note that formula (8.13) provides a representation of S_t, which is convenient for our purposes. The traditional representation of S_t, typically used in the context of counterparty risk free CDS contracts is

$$
S_t = B_t \mathbb{E}_{\mathbb{Q}} \left(B_{\tau_1}^{-1} \mathbb{I}_{\{t<\tau_1\le T\}} \delta_{\tau_1}^1 - \kappa \int_{]t,T]} B_u^{-1} \mathbb{I}_{\{\tau_1>u\}} du \,\middle|\, \mathcal{G}_t \right).
$$

where $\hat{\delta}_t^1 = \hat{\delta}_t^5 = \hat{\delta}_t^6 = \hat{\delta}_t^7 = \delta_t^1$, *and* $\hat{\delta}_t^2 = \hat{\delta}_t^3 = \hat{\delta}_t^4 = S_t$. *Thus,*

$$\begin{aligned} S_t = B_t \mathbb{E}_{\mathbb{Q}} \Bigg(& B_\tau^{-1} \mathbb{I}_{\{t<\tau=\tau_1\le T\}} \delta_\tau^1 + B_\tau^{-1} \sum_{i=2}^{4} \mathbb{I}_{\{t<\tau=\tau^{\{i\}}\le T\}} S_\tau \\ & -\kappa \int_{]t,T]} B_u^{-1} \mathbb{I}_{\{\tau>u\}} du \Bigg| \mathcal{G}_t \Bigg). \end{aligned} \tag{8.13}$$

The following result is borrowed from [21] (see Lemma 3.1 therein)

Lemma 8.3. *The following equality holds ($\mathbb{Q}$ a.s.)*

$$\begin{aligned} & B_t \mathbb{E}_{\mathbb{Q}} \left(\mathbb{I}_{\{t<\tau=\tau^{\{i\}}\le T\}} B_\tau^{-1} \overline{\delta}_\tau^i \middle| \mathcal{G}_t \right) \\ & = \mathbb{I}_{\{t<\tau\}} \frac{B_t}{G(t)} \mathbb{E}_{\mathbb{Q}} \left(\int_t^T B_u^{-1} l_u^i \overline{\delta}_u^i G(u)\, du \middle| \mathcal{F}_t \right), \end{aligned} \tag{8.14}$$

for every $t \in [0, T]$.

The pre-default ex-dividend price processes, say $\widetilde{S}$ and $\widetilde{S}^C$, are defined as the (unique) $\mathbb{F}$-adapted processes (cf. [21]) such that

$$S_t^C = \mathbb{I}_{\{t<\tau\}} \widetilde{S}_t^C, \quad S_t = \mathbb{I}_{\{t<\tau\}} \widetilde{S}_t.$$

In view of the above we thus obtain the following result

Lemma 8.4. *We have that, for every* $t \in [0, T]$,

$$\widetilde{S}_t^C = \frac{B_t}{G(t)} \mathbb{E}_{\mathbb{Q}} \left(\int_t^T B_u^{-1} G(u) \left(\sum_{i=1}^{7} l_u^i \overline{\delta}_u^i - \kappa \right) du \middle| \mathcal{F}_t \right), \tag{8.15}$$

and

$$\widetilde{S}_t = \frac{B_t}{G(t)} \mathbb{E}_{\mathbb{Q}} \left(\int_t^T B_u^{-1} G(u) \left(\sum_{i=1}^{7} l_u^i \hat{\delta}_u^i - \kappa \right) du \middle| \mathcal{F}_t \right). \tag{8.16}$$

Proof. From Lemma 8.2 we have that

$$\begin{aligned} S_t^C = B_t \mathbb{E}_{\mathbb{Q}} & \left(B_\tau^{-1} \sum_{i=1}^{7} \mathbb{I}_{\{t<\tau=\tau^{\{i\}}\le T\}} \overline{\delta}_t^i \middle| \mathcal{G}_t \right) \\ & - \kappa B_t \mathbb{E}_{\mathbb{Q}} \left(\int_{]t,T]} B_u^{-1} \mathbb{I}_{\{\tau>u\}} du \middle| \mathcal{G}_t \right). \end{aligned}$$

Now, in view of (8.14) we see that

$$B_t \mathbb{E}_{\mathbb{Q}} \left(B_{\tau}^{-1} \sum_{i=1}^{7} \mathbb{I}_{\left\{t<\tau=\tau^{\{i\}} \leq T\right\}} \overline{\delta}_t^i \middle| \mathcal{G}_t \right)$$

$$= \mathbb{I}_{\{t<\tau\}} \frac{B_t}{G(t)} \mathbb{E}_{\mathbb{Q}} \left(\sum_{i=1}^{7} \int_t^T B_u^{-1} l_u^i \overline{\delta}_u^i G(u)\, du \middle| \mathcal{F}_t \right).$$

Let us now fix $t \geq 0$, and define $Y_s := -\kappa \int_{]t,s]} B_u^{-1} du$ for $s \geq t$. Thus, we get

$$-\kappa B_t \mathbb{E}_{\mathbb{Q}} \left(\int_{]t,T]} B_u^{-1} \mathbb{I}_{\{\tau>u\}} du \middle| \mathcal{G}_t \right) = B_t \mathbb{E}_{\mathbb{Q}} \left(\mathbb{I}_{\{\tau>T\}} Y_T \middle| \mathcal{G}_t \right) + B_t \mathbb{E}_{\mathbb{Q}} \left(\mathbb{I}_{\{t<\tau\leq T\}} Y_\tau \middle| \mathcal{G}_t \right).$$

It is known from [21], that

$$B_t \mathbb{E}_{\mathbb{Q}} \left(\mathbb{I}_{\{t<\tau\leq T\}} Y_\tau \middle| \mathcal{G}_t \right) = -\mathbb{I}_{\{t<\tau\}} \frac{B_t}{G(t)} \mathbb{E}_{\mathbb{Q}} \left(\int_t^T Y_u dG(u) \middle| \mathcal{F}_t \right)$$

and

$$B_t \mathbb{E}_{\mathbb{Q}} \left(\mathbb{I}_{\{\tau>T\}} Y_T \middle| \mathcal{G}_t \right) = \mathbb{I}_{\{t<\tau\}} \frac{B_t}{G(t)} \mathbb{E}_{\mathbb{Q}} \left(G(T) Y_T \middle| \mathcal{F}_t \right).$$

Finally, since Y is of finite variation, (8.15) follows by applying the integration by parts formula

$$G(t)Y_T - \int_t^T Y_s dG(s) = \int_t^T G(s)\, dY_s = -\kappa \int_t^T G(s)\, B_u^{-1} du.$$

Equality (8.16) is obtained as a special case of (8.15), by setting the recovery rates $R_2 = R_3 = 1$. □

We are ready now to derive dynamics of the pre-default price processes, that we shall use in order to derive the dynamics of the CVA process.

Lemma 8.5.

(i) *The pre-default ex-dividend price of a counterparty risky CDS contract follows the dynamics given as*

$$d\widetilde{S}_t^C = \left((r_t + l_t)\, \widetilde{S}_t^C - \left(\sum_{i=1}^{7} l_t^i \overline{\delta}_t^i - \kappa \right) \right) dt + G^{-1}(t) \left(B_t dm_t^C - \widetilde{S}_t^C d\mu \right) + G^{-2}(t) \left(\widetilde{S}_t^C d\left\langle \mu \right\rangle_t - B_t d\left\langle \mu, m^C \right\rangle_t \right),$$

for $t \in [0,T]$ *where*

$$m_t^C = \mathbb{E}_{\mathbb{Q}}\left(\int_0^T B_u^{-1} G(u)\left(\sum_{i=1}^{7} l_u^i \overline{\delta}_u^i - \kappa\right) du \,\middle|\, \mathcal{F}_t\right).$$

(ii) *The pre-default ex-dividend price of a counterparty risk-free CDS contract follows the dynamics given as*

$$\begin{aligned} d\widetilde{S}_t =& \left(\left(r_t + l_t\right)\widetilde{S}_t - \left(\sum_{i=1}^{7} l_t^i \hat{\delta}_t^i - \kappa\right)\right) dt \\ &+ G^{-1}(t)\left(B_t dm_t - \widetilde{S}_t d\mu\right) + G^{-2}(t)\left(\widetilde{S}_t d\left\langle\mu\right\rangle_t - B_t d\left\langle\mu, m\right\rangle_t\right), \end{aligned}$$

for $t \in [0,T]$, *where*

$$m_t = \mathbb{E}_{\mathbb{Q}}\left(\int_0^T B_u^{-1} G(u)\left(\sum_{i=1}^{7} l_u^i \hat{\delta}_u^i - \kappa\right) du \,\middle|\, \mathcal{F}_t\right).$$

Proof. The argument below follows the one in the proof of Proposition 1.2 in [21].

In view of (8.15) we may write $\widetilde{S}_t^C$ as

$$\widetilde{S}_t^C = B_t G^{-1}(t) U_t,$$

where

$$U_t = m_t^C - \int_0^t B_u^{-1} G(u)\left(\sum_{i=1}^{7} l_u^i \overline{\delta}_u^i - \kappa\right) du.$$

Since $G = \mu - v$, then applying Itô's formula one obtains

$$\begin{aligned} d\left(G^{-1}(t)U_t\right) =& G^{-1}(t) dm_t^C - B_t^{-1}\left(\sum_{i=1}^{7} l_t^i \overline{\delta}_t^i - \kappa\right) dt \\ &+ U_t\left(G^{-3}(t) d\left\langle\mu\right\rangle_t - G^{-2}(t)\left(d\mu_t - dv_t\right)\right) \\ &- G^{-2}(t) d\left\langle\mu, m^C\right\rangle_t. \end{aligned}$$

Consequently,

$$\begin{aligned}
d\widetilde{S}_t^C &= B_t G^{-1}(t) dm_t^C - \left(\sum_{i=1}^{7} l_t^i \overline{\delta}_t^i - \kappa \right) dt \\
&\quad + B_t U_t \left(G^{-3}(t) d\langle \mu \rangle_t - G^{-2}(t) \left(d\mu_t - l_t G(t) dt \right) \right) \\
&\quad - B_t G^{-2}(t) d\left\langle \mu, m^C \right\rangle + r_t B_t G^{-1}(t) U_t dt \\
&= \left((r_t + l_t) \widetilde{S}_t^C - \left(\sum_{i=1}^{7} l_t^i \overline{\delta}_t^i - \kappa \right) \right) dt + G^{-1}(t) \left(B_t dm_t^C - \widetilde{S}_t d\mu \right) \\
&\quad + G^{-2}(t) \left(\widetilde{S}_t d\langle \mu \rangle_t - B_t d\left\langle \mu, m^C \right\rangle_t \right),
\end{aligned}$$

which verifies the result stated in (i).

Starting from (8.16), and using computations analogous to the ones done in (i), one can derive the result stated in (ii). □

The dynamics of the CVA process are easily derived with help of the above lemma,

Proposition 8.2. The bilateral CVA process satisfies,

$$\begin{aligned}
d\text{CVA}_t &= r_t \text{CVA}_t dt - \text{CVA}_{t-} dM_t - (1 - H_t) \left(\sum_{i=1}^{7} l_t^i \xi_t^i \right) dt \\
&\quad + (1 - H_t) B_t G^{-1}(t) dn_t - G^{-1}(t) \text{CVA}_t d\mu_t + G^{-2}(t) \text{CVA}_t d\langle \mu \rangle_t \\
&\quad - (1 - H_t) G^{-2}(t) B_t \left(d\langle \mu, m \rangle_t - d\left\langle \mu, m^C \right\rangle_t \right),
\end{aligned}$$

where

$$n_t = \mathbb{E}_{\mathbb{Q}} \left(\int_0^T B_u^{-1} G(u) \left(\sum_{i=1}^{7} l_u^i \xi_u^i \right) du \,\middle|\, \mathcal{F}_t \right), \quad t \in [0, T],$$

with

$$\begin{aligned}
&\xi_t^1 = 0, \ \xi_t^2 = (1 - R_2)(S_t - C_t)^+, \ \xi_t^3 = -(1 - R_3)(S_t - C_t)^-, \\
&\xi_t^4 = (1 - R_2)(S_t - C_t)^+ \ - (1 - R_3)(S_t - C_t)^-, \\
&\xi_t^5 = (1 - R_2)\left(\delta_t^1 - C_t\right)^+, \ \xi_t^6 = -(1 - R_3)\left(\delta_t^1 - C_t\right)^-, \\
&\xi_t^7 = (1 - R_2)\left(\delta_t^1 - C_t\right)^+ - (1 - R_3)\left(\delta_t^1 - C_t\right)^-.
\end{aligned}$$

Proof. Applying the integration by parts formula we get that

$$d\text{CVA}_t = (1 - H_t) \left(d\widetilde{S}_t - d\widetilde{S}_t^C \right) - \left(\widetilde{S}_t - \widetilde{S}_t^C \right) dH_t.$$

This together with Lemma 8.5 implies

$$\begin{aligned} d\mathrm{CVA}_t = & -\left(S_{t-} - S^C_{t-}\right) dM_t + (1-H_t)\left(r_t\left(S_t - S^C_t\right) - \sum_{i=1}^{7} l^i_t\left(\hat{\delta}^i_t - \overline{\delta}^i_t\right)\right) dt \\ & + (1-H_t)\, B_t G^{-1}(t)\left(dm_t - dm^C_t\right) - (1-H_t)\, G^{-1}(t)\left(S_t - S^C_t\right) d\mu_t \\ & + (1-H_t)\, G^{-2}(t)\left(S_t - S^C_t\right) d\langle\mu\rangle_t \\ & - (1-H_t)\, G^{-2}(t) B_t \left(d\langle\mu, m\rangle_t - d\left\langle\mu, m^C\right\rangle_t\right), \end{aligned}$$

which proves the result. □

8.2.3.1. *Dynamics of CVA when the immersion property holds*

Here we adapt the results derived above to the case when the immersion property holds between filtrations $\mathbb{F}$ and $\mathbb{G}$, that is the case when every $\mathbb{F}$-martingale is a $\mathbb{G}$-martingale under $\mathbb{Q}$. In this case, the continuous martingale μ in the Doob-Meyer decomposition of G vanishes, so that the survival process G is a non-increasing process represented as $G = -v$. Frequently, the immersion property is referred to as Hypothesis $(\mathcal{H})$. For an excellent discussion of the immersion property we refer to [22].

Assumption 8.1. Hypothesis $(\mathcal{H})$ holds between the filtrations $\mathbb{F}$ and $\mathbb{G}$ under $\mathbb{Q}$.

In view of the results (and the notation) from Proposition 8.2 we obtain

Corollary 8.2. *Assume that Assumption 8.1 is satisfied. Then,*

$$\begin{aligned} dCVA_t = & \; r_t CVA_t dt - CVA_{t-} dM_t - (1-H_t)\left(\sum_{i=1}^{7} l^i_t \xi^i_t\right) dt \\ & + (1-H_t)\, B_t G^{-1}(t) dn_t, \;\; t \in [0,T]. \end{aligned}$$

Remark 8.8. If we assume that the filtration $\mathbb{F}$ is generated by a Brownian motion, then, in view of the Brownian martingale representation theorem, there exists an $\mathbb{F}$-predictable process ζ such that $dn_t = \zeta_t dW_t$ and

$$\begin{aligned} d\mathrm{CVA}_t = & \; r_t \mathrm{CVA}_t dt - \mathrm{CVA}_{t-} dM_t - (1-H_t)\left(\sum_{i=1}^{7} l^i_t \xi^i_t\right) dt \\ & + (1-H_t)\, B_t G^{-1}(t) \zeta_t dW_t. \end{aligned}$$

8.2.4. *Fair Spread Value Adjustment*

Let us fix $t \in [0, T]$, and let us denote by κ_t the market spread of the counterparty risk-free CDS contract at time t; that is, κ_t is this level of spread that makes the pre-default values of the two legs of a counterparty risk-free CDS contract equal to each other at time t,

$$\widetilde{S}_t(\kappa_t) = 0. \tag{8.17}$$

It is convenient to write the above equation in the form that is common in practice:

$$PL_t - \kappa_t RDV01_t = 0, \tag{8.18}$$

where PL and $RDV01$ are processes representing (pre-default) values of the protection leg[e] and the risky annuity, respectively, so that

$$PL_t = \frac{B_t}{G^1(t)} \mathbb{E}_{\mathbb{Q}} \left(\int_{]t,T]} B_u^{-1} G^1(u) \delta_u^1 \left(\sum_{i=1,5,6,7} l_u^i \right) du \middle| \mathcal{F}_t \right), \tag{8.19}$$

and

$$RDV01_t = \frac{B_t}{G^1(t)} \mathbb{E}_{\mathbb{Q}} \left(\int_{]t,T]} B_u^{-1} G^1(u)\, du \middle| \mathcal{F}_t \right), \tag{8.20}$$

where

$$G^1(t) = \mathbb{Q}(\tau_1 > t | \mathcal{F}_t).$$

Therefore, we get,

$$\kappa_t = \frac{\mathbb{E}_{\mathbb{Q}} \left(\int_{]t,T]} B_u^{-1} G^1(u) \delta_u^1 \left(\sum_{i=1,5,6,7} l_u^i \right) du \middle| \mathcal{F}_t \right)}{\mathbb{E}_{\mathbb{Q}} \left(\int_{]t,T\wedge\tau_1]} B_u^{-1} G^1(u)\, du \middle| \mathcal{F}_t \right)}. \tag{8.21}$$

We denote by κ_t^C the spread which makes the values of the two pre-first-default legs of a counterparty risky CDS contract equal to each other at every $t \in [0, T]$ as

$$\widetilde{S}_t^C(\kappa_t^C) = PL_t^C - \kappa_t^C RDV01_t^C = 0. \tag{8.22}$$

[e]We note that formula (8.19) provides a representation of PL_t which is convenient for our purposes. The traditional representation of PL_t, typically used in the context of counterparty risk free CDS contracts, is

$$PL_t = \frac{B_t}{G^1(t)} \mathbb{E}_{\mathbb{Q}} \left(\int_{]t,T]} B_u^{-1} G^1(u) \delta_u^1 \lambda_u^1 du \middle| \mathcal{F}_t \right),$$

where λ^1 is the $\mathbb{F}$ intensity of τ_1.

Similarly, we use the spread κ_0^C initiated at time $t = 0$ in order to compute the fair price of a counterparty risky CDS contract at any time $t \in [0,T]$. Using Lemma 3.1, κ_t^C admits the following representation for every $t \in [0,T]$,

$$\kappa_t^C = \frac{PL_t^C}{RDV01_t^C},$$

where

$$PL_t^C = \frac{B_t}{G(t)} \mathbb{E}_{\mathbb{Q}} \left(\int_t^T B_u^{-1} G(u) \left(\sum_{i=1}^{7} l_u^i \overline{\delta}_u^i \right) du \middle| \mathcal{F}_t \right) \tag{8.23}$$

and

$$RDV01_t^C = \frac{B_t}{G(t)} \mathbb{E}_{\mathbb{Q}} \left(\int_{]t,T]} B_u^{-1} G(u)\, du \middle| \mathcal{F}_t \right). \tag{8.24}$$

We may now introduce the following definition,

Definition 8.10. The Spread Value Adjustment process of a counterparty risky CDS contract maturing at time T is defined as,

$$\mathrm{SVA}_t = \kappa_t - \kappa_t^C$$

for every $t \in [0,T]$.

Monitoring SVA is of great importance since it provides a more practical way to quantify the counterparty risk. Moreover, the spread difference is a very useful indicator for the trading decisions in practice (cf. [13]).

Proposition 8.3. The SVA of a counterparty risky CDS contract maturing at time T equals,

$$\mathrm{SVA}_t = \frac{\widetilde{\mathrm{CVA}}_t}{B_t G^{-1}(t) \mathbb{E}_{\mathbb{Q}} \left(\int_{]t,T]} B_u^{-1} G(u)\, du \middle| \mathcal{F}_t \right)}$$

for every $t \in [0,T]$, where the pre-first-default bilateral Credit Valuation Adjustment process $\widetilde{\mathrm{CVA}}$ is given as

$$\widetilde{\mathrm{CVA}}_t = \widetilde{S}_t - \widetilde{S}_t^C, \tag{8.25}$$

for every $t \in [0,T]$.

Proof. Let us rewrite PL^C as

$$PL_t^C = PL_t^C - \kappa_t RDV01_t^C + \kappa_t RDV01_t^C$$

by a simple modification. Now, using (8.5) and (8.22), we conclude that

$$\begin{aligned} PL_t^C &= \widetilde{S}_t^C(\kappa_t) + \kappa_t RDV01_t^C \\ &= \widetilde{S}_t(\kappa_t) - \widetilde{\mathrm{CVA}}_t + \kappa_t RDV01_t^C. \end{aligned}$$

Since $\widetilde{S}_t(\kappa_t) = 0$, then κ_t^C has the following form,

$$\kappa_t^C = \frac{-\widetilde{\mathrm{CVA}}_t + \kappa_t RDV01_t^C}{RDV01_t^C},$$

which is

$$\kappa_t^C = -\frac{\widetilde{\mathrm{CVA}}_t}{RDV01_t^C} + \kappa_t.$$

□

8.2.4.1. *SVA dynamics*

Applying Itô formula one obtains the dynamics of the fair spread process and of the counterparty risk adjusted spread process as

$$\begin{aligned} d\kappa_t = &\frac{1}{\widetilde{RDV01}_t}\left(B_t^{-1}G^1(t)\left(\kappa_t - \delta_t^1 l_t^1\right)dt + \frac{\kappa_t}{\widetilde{RDV01}_t}d\left\langle \eta^2\right\rangle_t \right. \\ &\left. - \frac{1}{\widetilde{RDV01}_t}d\left\langle \eta^1, \eta^2\right\rangle_t\right) + \frac{1}{\widetilde{RDV01}_t}\left(d\eta_t^1 - \kappa_t d\eta_t^2\right), \qquad t \in [0,T], \end{aligned} \tag{8.26}$$

where

$$\widetilde{RDV01}_t := \mathbb{E}_{\mathbb{Q}}\left(\int_{]t,T]} B_u^{-1}G^1(u)\,du \,\middle|\, \mathcal{F}_t\right),$$

$$\eta_t^1 := \mathbb{E}_{\mathbb{Q}}\left(\int_{]0,T]} B_u^{-1}G^1(u)\,\delta_u^1 l_u^1 du \,\middle|\, \mathcal{F}_t\right),$$

$$\eta_t^2 = \mathbb{E}_{\mathbb{Q}}\left(\int_{]0,T]} B_u^{-1}G^1(u)\,du \,\middle|\, \mathcal{F}_t\right) = \widetilde{RDV01}_t + \int_{]0,t]} B_u^{-1}G^1(u)\,du,$$

and

$$d\kappa_t^C = \frac{1}{\widetilde{RDV01}_t^C}\left(B_t^{-1}G(t)\left(\kappa_t^C - \sum_{i=1}^{7}\widetilde{\delta}_t^i l_t^i\right)dt + \frac{\kappa_t^C}{\widetilde{RDV01}_t^C}d\left\langle \zeta^2\right\rangle_t \right. \tag{8.27}$$

$$\left. -\frac{1}{\widetilde{RDV01}_t^C}d\left\langle \zeta^1, \zeta^2\right\rangle_t\right) + \frac{1}{\widetilde{RDV01}_t^C}\left(d\zeta_t^1 - \kappa_t^C d\zeta_t^2\right),$$

where

$$\widetilde{RDV01}_t = \mathbb{E}_{\mathbb{Q}}\left(\int_{]t,T]} B_u^{-1}G^1(u)\,du \,\middle|\, \mathcal{F}_t\right),$$

with

$$\zeta_t^1 = \mathbb{E}_{\mathbb{Q}}\left(\int_{]0,T]} B_u^{-1}G(u)\left(\sum_{i=1}^{7} l_u^i\overline{\delta}_u^i\right)du \,\middle|\, \mathcal{F}_t\right),$$

and

$$\zeta_t^2 = \mathbb{E}_{\mathbb{Q}}\left(\int_{]0,T]} B_u^{-1}G(u)\,du \,\middle|\, \mathcal{F}_t\right) = \widetilde{RDV01}_t^C + \int_{]0,t]} B_u^{-1}G(u)\,du.$$

Combining the above results, we find the dynamics of the SVA process:

$$d\,\mathrm{SVA}_t = d\kappa_t - d\kappa_t^C, \quad t \in [0,T].$$

Dynamics of the SVA is of great importance for observing the behavior of the difference between the fair spread and the counterparty risk adjusted spread. Counterparty risk dynamics can be assessed in a more intuitive manner by computing the SVA dynamics.

8.3. Multivariate Markovian default model

In this section, we propose an underlying stochastic model following the lines of [4]. Towards this end we define a Markovian model of multivariate default times with factor processes $X = \left(X^1, X^2, X^3\right)$ which will have the following key features,

- The pair (X, H) is Markov in its natural filtration,
- Each pair $\left(X^i, H^i\right)$ is a Markov process,
- At every instant, either each counterparty defaults individually or simultaneously with other counterparties.

Note that the second property grants quick valuation of the CDS and independent calibration of each model marginal $\left(X^i, H^i\right)$, whereas the third property will allow us to account for dependence between defaults. We present here some numerical results as an application of above theory. The default intensities are assumed to be of the affine form

$$l_i\left(t, X_t^i\right) = a_i + X_t^i,$$

where a_i is a constant and X^i is a homogenous CIR process generated by,

$$dX_t^i = \zeta_i\left(\mu_i - X_t^i\right)dt - \sigma_i\sqrt{X_t^i dW_t^i},$$

for $i = 1, 2, 3$. Each collection of the parameters $(\zeta_i, \mu_i, \sigma_i)$ may take values corresponding to a low, a medium or a high regime which are given as follows.

Table 8.1. Various risk profiles

Credit Risk Level	ζ	μ	σ	X_0
Low	0.9	0.001	0.01	0.001
Medium	0.8	0.02	0.1	0.02
High	0.5	0.05	0.2	0.05

Moreover, following the methodology in [4], we specify the marginal default intensity processes as follows

$$q_t^1 = l_t^1 + l_t^5 + l_t^6 + l_t^7, \ q_t^2 = l_t^2 + l_t^4 + l_t^5 + l_t^7, \ q_t^3 = l_t^3 + l_t^4 + l_t^6 + l_t^7$$

where the related survival probabilities are found as

$$\mathbb{Q}\left(\tau_i > t\right) = \mathbb{E}_{\mathbb{Q}}\left(e^{-\int_0^t q_u^i du}\right) \text{ and } \mathbb{Q}\left(\tau > t\right) = \mathbb{E}_{\mathbb{Q}}\left(e^{-\int_0^t l_u du}\right).$$

For a detailed discussion including implementation and the calibration of the model, we refer to [4] and [5].

8.3.1. *Results*

Our aim here is to assess by means of numerical experiments the impact of collateralization on the counterparty risk exposure. We present numerical results for different collateralization regimes distinguished by different threshold values. The numerical experiments below have been done using the three factor (2F) parametrization given in [4], the recovery rates are fixed to 40%, the risk-free rate r is taken as 0 and the maturity is set to $T = 5$ years.

Table 8.2 shows the values of CVA_0 and SVA_0 for different threshold regimes. Threshold values are chosen as a fraction of the notional (cf. [11]). Computations are done assuming that (refer to Table 8.1) the underlying entity, the counterparty, and the investor has high risk levels. Simulated fair spread without counterparty risk is found as 153bps. Case A represents the uncollateralized regime where there is no collateral exchanged (this is done by setting the thresholds infinity), whereas other Case F corresponds to the full collateralization where the thresholds are set to 0. In each case, computations are done by setting MTA to zero and assuming there is no margin period. One can observe that decreasing threshold value dramatically decreases the initial CVA and therefore the SVA values.

Table 8.2. Initial CVA and SVA values for different thresholds

	Γ_{cpty}	Γ_{inv}	CVA_0	SVA_0
Case A	∞	$-\infty$	1.01×10^{-4}	0.2153
Case B	1.5×10^{-3}	0.4×10^{-3}	6.13×10^{-5}	0.1305
Case C	1×10^{-3}	0.2×10^{-3}	4.36×10^{-5}	0.0931
Case D	0.5×10^{-3}	0.1×10^{-3}	2.18×10^{-5}	0.0464
Case E	0.25×10^{-3}	0.05×10^{-3}	1.14×10^{-5}	0.0243
Case F	0	0	0	0

In Figure 8.1, we present the EPE and ENE curves for each case A to F, and we also plot the mean collateral values. Computations are carried out by running 10^4 Monte Carlo simulations. It is apparent that the behavior of the EPE and ENE values decreases as a result of increased collateralization. Note that there are peaks in the collateral value in the very beginning and through the maturity. This effect can be explained as follows: Observe from Table 1 that the investor has lower threshold than the counterparty in each cases from A to F. As a result, having a lower threshold value, investor will be posting collateral before the counterparty. Therefore, until the counterparty's exposure reaches the threshold, the collateral value remains negative; meaning that there will be margin calls for the investor before the counterparty.

Figure 8.2 plots the mean of sample CVA paths. Starting from CVA_0 we compute the mean sample paths in each case. The behavior of CVA as a credit hybrid option, as indicated in Remark 8.3, can be clearly observed in the graphs. CVA values decrease over time as a result of time decay since the expected loss decreases close to the expiration. The effect of collateralization on the CVA values is apparent in the graphs. Observe that increased initial threshold values are of great importance since one can significantly reduce the future CVA values by appropriately setting the collateral thresholds. Moreover, one can also use dynamic thresholds by linking the threshold values to the counterparties' default intensities or credit ratings. In this way, counterparties will have more control on the future values of the CVA of the CDS contract and dynamically manage the CVA, since the collateral thresholds will be reacting to the changes in the default intensities or credit ratings. This approach will be further investigated in future research.

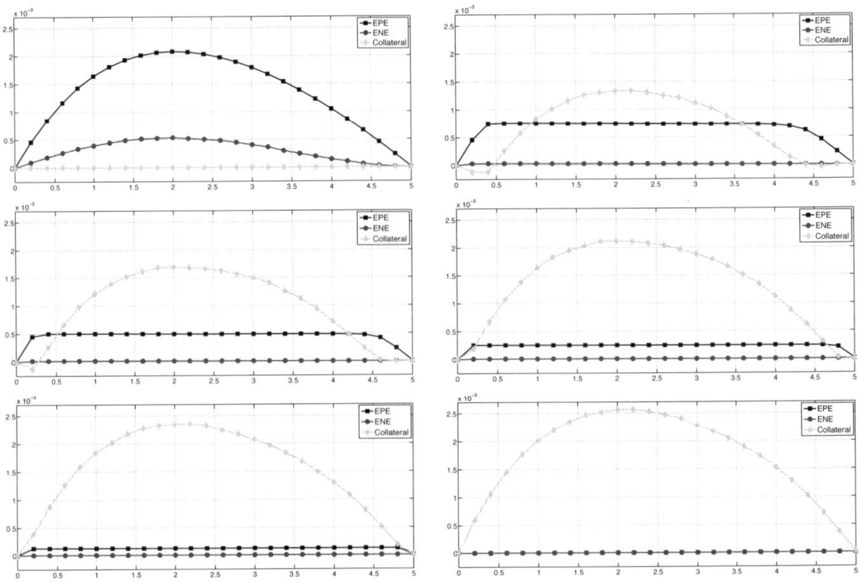

Fig. 8.1. EPE, ENE and the mean Collateral curves for Cases A, B, C, D, E, and F

8.4. Conclusion

In this paper, we discussed the modeling of counterparty risk in the presence of bilateral margin agreements. We defined an appropriate collateral process which takes various margin agreement parameters into account. The dynamics of the counterparty risk adjustment, CVA, has been found for the bilateral case. This achievement helps us to better understand and monitor the behavior of the bilateral CVA as well as the unilateral CVA and the DVA.

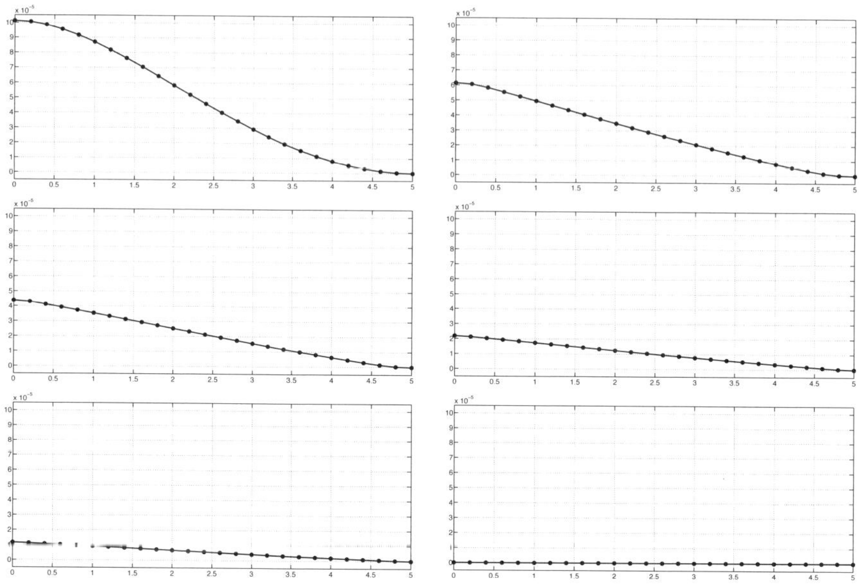

Fig. 8.2. Forward CVA curves for Cases A, B, C, D, E, and F

We observed the impact of collateral agreements on counterparty risk adjustments as well as the credit exposures such as the EPE and the ENE. The presence of simultaneous defaults in our model represents the wrong way risk involved in the CDS contracts. We formulate the fair spread value adjustment, which is named as SVA, that indicates the additional spread value to incorporate the counterparty risk into the fair spread value. Moreover, we derive the dynamics of the fair spread and the counterparty risky spread and therefore the spread value adjustment, SVA. Finally, as in [4] and [5], we present our numerical results using a Markovian model of counterparty credit risk.

References

[1] D. Brigo and A. Capponi, Bilateral counterparty risk valuation with stochastic dynamical models and application to credit default swaps, *Working Paper.* (2009).

[2] D. Brigo, A. Capponi, A. Pallavicini, and V. Papatheodorou, Collateral margining in arbitrage-free counterparty valuation adjustment including rehypotecation and netting, *Working Paper.* (2011).

[3] J. Hull and A. White, Valuing credit default swaps ii: Modeling default correlation, *The Journal of derivatives.* **8**(3), 12–22, (2001).

[4] T. R. Bielecki, S. Crepey, M. Jeanblanc, and B. Zargari, Valuation and Hedging of CDS Counterparty Exposure in a Markov Copula Model, *Forthcoming in International Journal of Theoretical and Applied Finance.* (2011).

[5] S. Assefa, T. R. Bielecki, S. Crepey, and M. Jeanblanc. CVA computation for counterparty risk assessment in credit portfolios. In *Credit Risk Frontiers: Subprime Crisis, Pricing and Hedging, CVA, MBS, Ratings, and Liquidity.* Bloomberg, (2011).

[6] R. Jarrow and F. Yu, Counterparty risk and the pricing of defaultable securities, *the Journal of Finance.* **LVI**(5), 1765–1799, (Oct. 2001).

[7] S. Leung and Y. Kwok, Cds valuation with counterparty risk, *The Kyoto Economic Review.* **74**(1), 24–45, (2005).

[8] A. Lipton and A. Sepp, Credit value adjustment for credit default swaps via the structural default model, *The Journal of Credit Risk.* **5**(2), (Summer 2009).

[9] C. Blanchet-Scalliet and F. Patras. Structural counterparty risk valuation for cds. In eds. B. D. Bielecki, T.R. and F. Patras, *Credit Risk Frontiers: Subprime crisis, Pricing and Hedging, CVA, MBS, Ratings and Liquidity.* Wiley, (2011).

[10] H. Stein and K. Lee. Counterparty valuation adjustments. In eds. B. D. Bielecki, T.R. and F. Patras, *Credit Risk Frontiers: Subprime crisis, Pricing and Hedging, CVA, MBS, Ratings and Liquidity,*. Wiley, (2011).

[11] M. Pykhtin, Modeling Credit Exposure for Collateralized Counterparties, *Journal of Credit Risk.* **5**(4), 3–27, (2009).

[12] G. Cesari, J. Aquilina, and N. Charpillon, *Modelling, Pricing, and Hedging Counterparty Credit Exposure.* (Springer Finance, 2010).

[13] J. Gregory, *Counterparty Credit Risk: The New Challenge for Global Financial Markets.* (Wiley, 2009).

[14] T. R. Bielecki and S. Crepey. Dynamic hedging of counterparty exposure. In eds. T. Zariphopoulou, M. Rutkowski, and Y. Kabanov, *The Musiela Festschrift.* Spring, (2011).

[15] Algorithmics, Credit Value Adjustment and the changing environment for pricing and managing counterparty risk. (2009). URL `http://www.algorithmics.com/EN/media/pdfs/Algo-WP1209-CVASurvey.pdf`.

[16] ISDA, ISDA Collateral Guidelines. (2005).

[17] ISDA, ISDA Market Review of OTC Derivative Bilateral Collateralization Practices. (2010).

[18] ISDA, ISDA Margin Survey. (2010).

[19] Markit, The CDS Big Bang: Understanding the Changes to the Global CDS Contract and North American Conventions. (2009).

[20] B. De Prisco and D. Rosen, Modelling stochastic counterparty credit exposures for derivatives portfolios, *Edited By Michael Pykhtin, RISK Books, London, 2005.* **Chapter 1 of** *Counterparty Credit Risk Modelling: Risk Management, Pricing and Regulation.*

[21] T. R. Bielecki, M. Jeanblanc, and M. Rutkowski, Pricing and trading credit default swaps in a hazard process model, *Annals of Applied Probability.* **18** (6), 2495–2529, (2008).

[22] M. Jeanblanc and Y. Le Cam. Immersion property and credit risk modelling. In *Optimality and Risk - Modern Trends in Mathematical Finance*, pp. 99–132. Springer, (2009).

Chapter 9

A modern view on Merton's jump-diffusion model

Gerald H.L. Cheang[1] and Carl Chiarella[2]

[1] *Centre for Industrial and Applied Mathematics,*
School of Mathematics and Statistics, University of South Australia,
GPO Box 2471, City West Campus, Adelaide, SA 5001, Australia
gerald.cheang@unisa.edu.au

[2] *The Finance Discipline Group,*
UTS Business School, University of Technology, Sydney,
PO Box 123, Broadway, NSW 2007, Australia
carl.chiarella@uts.edu.au

Merton has provided a formula for the price of a European call option on a single stock where the stock price process contains a continuous Poisson jump component, in addition to a continuous log-normally distributed component. In Merton's analysis, the jump-risk is not priced. Thus the distribution of the jump-arrivals and the jump-sizes do not change under the change of measure. We introduce a Radon-Nikodým derivative process that induces the change of measure from the market measure to an equivalent martingale measure. The choice of parameters in the Radon-Nikodým derivative allows us to price the option under different financial-economic scenarios. We give a hedging argument that eliminates the jump-risk in some sort of averaged sense, and derive an integro-partial differential equation of the option price that is related to the one obtained by Merton.

Contents

9.1. Introduction

[Merton (1976)] has provided a formula for the price of a European call option on a single stock where the stock price process contains a continuous Poisson jump component, in addition to a continuous log-normally distributed component. Merton's analysis in essence does not price the jump-risk. In this paper, we extend the results of Merton to the case where the market price of jump-risk is priced in the hedging portfolio. In Merton's case, financial economic arguments relating to systematic and unsystematic risk allows one to argue that the distribution of the Poisson jump components does not change under the change of measure. He considered a constant arrival intensity, log normally distributed jump sizes, set the market price of jump risk to zero and obtained a Poisson weighted sum of a Black-Scholes type formulae. He also considered the same hedge portfolio used by [Black and Scholes (1973)], namely one consisting of a position in the stock, the option and the risk-free asset only. In this case a perfect hedge does not exist and hedging was achieved by Merton by averaging out idiosyncratic risk. However, this leaves the market price of jump-risk unpriced, and also the distribution of the jump components remain unchanged. Further extensions to the [Merton (1976)] model include those by [Anderson (1984)] and [Aase (1988)]. However these authors also make assumptions that amount to leaving the jump risk unpriced. Furthermore these later derivations do not appeal to the traditional hedging argument but rather appeal directly to the risk-neutral valuation principle and change of measure arguments.

However, the market which contains stocks with jump components is inherently incomplete in the [Harrison and Pliska (1981)] sense. When the market price of the jump-risk is accounted for, there are many equivalent martingale measures, and hence different prices for the option. For a single stock market, one could for example, apply a local risk-minimizing trading strategy in the manner of [Schweizer (1991)] or [Colwell and Elliott (1993)], or a minimum entropy martingale measure approach in the manner of [Miyahara (2001)]. [Jeanblanc-Picqué and Pontier (1990)] applied a general equilibrium model to the problem and used two assets driven by the same Wiener and Poisson noise factors. [Jarrow and Madan (1995)] included additional traded assets in order to hedge away the jump-risk in interest rate term-structure-related securities. [Mercurio and Runggaldier

(1993)], and [Runggaldier (2003)], suggested that other assets driven by the same Wiener and Poisson noise factors as the stock be included in the hedge portfolio. [Elliott and Kopp (1990)] considered a portfolio of call options on stocks driven by fixed jump-size Poisson processes but in their case, whenever a jump occurs, the jump-size for each stock can only jump by a certain discrete range of relative jump-size factors. In [Colwell *et al.* (1991)], a hedging portfolio of call options on stocks driven by geometric Brownian motion is considered. However, in both cases, the market is complete.

In our case, we set up a hedging portfolio in which two options of different maturities are required, so that the jump-risk is properly priced in the portfolio. We use a hedging argument that eliminates the jump-risk by averaging over the jump-sizes, where the range of the jump-sizes can be infinite. We then obtain the standard integro-partial differential equation for the option price and interpret economically the various parameters.

The paper develops as follows. In Section 9.2 we specify the asset pricing model. In Section 9.3 we introduce a Radon-Nikodým derivative that induces the change of measure from the market measure to an equivalent martingale measure for an option on the underlying asset. Section 9.4 then applies the results of Section 9.3 in the derivation of an integro-partial differential equation for the option price via the martingale approach. In Section 9.5 we develop a hedging portfolio that is used in the derivation of the same integro-partial differential equation as in Section 9.4. The hedging portfolio also takes into account the market price of jump-risk. In Section 9.6, a general pricing formula is obtained for a European style call option and we recover Merton's (1976) call option formula as a special case. Section 9.7 concludes.

9.2. Merton's jump-diffusion model

Throughout this paper, as in [Merton (1976)], we assume that S_t is the price of a financial asset whose return dynamics are given by

$$\frac{dS_t}{S_{t-}} = (\mu - \lambda\kappa)dt + \sigma dB_t + [e^J - 1]dN_t, \tag{9.1}$$

where μ is the instantaneous expected return per unit time, and σ is the instantaneous volatility per unit time. The stochastic process B_t is a standard Wiener process under the market measure $\mathbb{P}$. The process N_t is a Poisson process, independent of the jump-sizes J and the Wiener process

B_t, with arrival intensity λ per unit time under the measure $\mathbb{P}$, so that its increments satisfy

$$dN_t = \begin{cases} 1 & \text{with probability } \lambda dt, \\ 0 & \text{with probability } 1 - \lambda dt. \end{cases} \tag{9.2}$$

The expected proportional jump size is

$$\kappa \equiv \mathbb{E}_{\mathbb{P}}\left[e^J - 1\right]. \tag{9.3}$$

The stochastic differential equation (9.1) yields a solution for the stock price of the form

$$S_t = S_0 \exp\left[\left(\mu - \lambda\kappa - \frac{\sigma^2}{2}\right)t + \sigma B_t + \sum_{n=1}^{N_t} J_n\right]. \tag{9.4}$$

Jumps arriving at different times are assumed to be independent of each other. A filtered probability measure space $(\Omega, \mathcal{F}, \{\mathcal{F}_t\}, \mathbb{P})$ is assumed where the filtration $\{\mathcal{F}_t\}$ is the natural filtration generated by the Wiener process B_t and the compound Poisson process $\sum_{n=1}^{N_t} J_n$. The moment generating function of all the jump-sizes is given by

$$M_{\mathbb{P},J}(u) = \mathbb{E}_{\mathbb{P}}\left[e^{uJ}\right]. \tag{9.5}$$

We also assume that the stock pays a continuous dividend yield at rate q. For simplicity, we assume that all the parameters in our model are constants, although the model can easily be extended to one with time-varying but non-stochastic parameters.

9.3. A Radon-Nikodým derivative

The stock price model (9.1) is driven by both a Wiener component B_t and a compound Poisson process $\sum_{n=1}^{N_t} J_n$. We consider what happens to both the Wiener component and the jump component when we do a change of measure from $\mathbb{P}$ to another measure $\mathbb{Q}$. In the context of our model, the Radon-Nikodým derivative restricted to any time $t \leq T$ takes the form

$$\left.\frac{d\mathbb{Q}}{d\mathbb{P}}\right|_t = \exp\left[-\theta B_t - \frac{\theta^2}{2}t + \sum_{n=1}^{N_t}(\gamma J_n + \nu) - \lambda\kappa' t\right], \tag{9.6}$$

where

$$\kappa' \equiv e^{\nu} M_{\mathbb{P},J}(\gamma) - 1 = e^{\nu}\mathbb{E}_{\mathbb{P}}(e^{\gamma J}) - 1. \tag{9.7}$$

As summarized in Lemma 9.1 below, the values of the parameters γ and ν give rise to different equivalent martingale measures that change the distribution of the jump-sizes and the jump-arrival intensity accordingly. Certain special cases also arise with certain specific choices of γ and ν as outlined below.

It can easily be verified that (9.6) satisfies the properties of a Radon-Nikodým derivative and it is a martingale with respect to the measure $\mathbb{P}$. The form of the Radon-Nikodým derivative in (9.6) is a more compact representation and illustrates the Esscher transform nature of the representation. In our application, the form (9.6) allows us to manipulate the parameters to produce families of equivalent martingale measures in the stock price model. More general representations expressed in terms of Lévy measures, jump measures or compensated jump measures can be found in [Colwell and Elliott (1993)], [Cont and Tankov (2004)] and in [Runggaldier (2003)].

Lemma 9.1. *Let $\mathbb{P}$ and $\mathbb{Q}$ be equivalent measures. Consider a probability measure space $(\Omega, \mathcal{F}, \{\mathcal{F}_t\}, \mathbb{P})$ such that $\{\mathcal{F}_t\}$ is the natural filtration generated by a Wiener process B_t and a compound Poisson process $\sum_{n=0}^{N_t} J_n$ and a Radon-Nikodým derivative given by (9.6) where $\gamma \in \mathbb{R}$, $\nu \in \mathbb{R}$ and κ' is given by (9.7).*

Then the Wiener process B_t has drift $-\theta$ under the measure $\mathbb{Q}$ and the compound Poisson process $\sum_{n=0}^{N_t} J_n$ under the measure $\mathbb{Q}$ has a new intensity rate $\tilde{\lambda} = \lambda(1+\kappa')$ and a new distribution for the jump-sizes. The moment generating function of the new jump-size distribution is given by

$$M_{\mathbb{Q},J}(u) = \frac{M_{\mathbb{P},J}(\gamma + u)}{M_{\mathbb{P},J}(\gamma)}. \tag{9.8}$$

Proof. See e.g., [Cont and Tankov (2004)] or [Runggaldier (2003)]. This is a special case of Theorem 2.5 in [Runggaldier (2003)]. □

Note that if the distribution of the jumps J in the measure $\mathbb{P}$ comes from an exponential family, then the distribution of J under the measure $\mathbb{Q}$ as given by the moment generating function in (9.8), also comes from the same exponential family but with different parameters, e.g. see [Gerber and Shiu (1994)]. It can be observed that even if $\nu = 0$, as long as γ is non-zero, there will be a change in the jump-size distribution under the change of measure, and this in turn leads to a change in the intensity of the jump arrivals. On the other hand, if γ is zero, but ν is non-zero, then there is a change in the intensity of the jump-arrivals only. Conversely, if $\nu = -\ln M_{\mathbb{P},J}(\gamma)$, then the Radon-Nikodým derivative (9.6) takes the form

$$\left.\frac{d\mathbb{Q}}{d\mathbb{P}}\right|_t = e^{-\theta B_t - \frac{\theta^2}{2}t} \times \prod_{n=1}^{N_t} \frac{e^{\gamma J_n}}{M_{\mathbb{P},J}(\gamma)}. \tag{9.9}$$

In this case, the intensity of the jump-arrivals does not change under the change of measure.

In our context, the asset price dynamics contain both Wiener components and jump components given by (9.1). In the Radon-Nikodým derivative (9.6), the choice of the market prices of risk for the Wiener process, θ, will be determined by the choice of the other parameters γ and ν, which determine the market prices of risk for the jump component. Lemma 9.1 indicates the new intensity rate $\tilde{\lambda}$ and the new distributions of the jump-sizes under the martingale measure $\mathbb{Q}$. The Wiener process B_t now has drift $-\theta$ under the measure $\mathbb{Q}$.

9.4. An integro-partial differential equation for the option price

Consider a European style option, e.g. a call option with payoff $X_T(S_T) = (S_T - K)^+$ at maturity time T. As stated in Section 2, we shall assume that the underlying asset pays a continuous dividend at the rate q per unit time, so that the yield process of the stock is $S_t e^{qt}$. We assume the presence of a money market account e^{rt} where r is the risk-free rate. We require that both the discounted stock yield process $\left\{\frac{S_t e^{qt}}{e^{rt}}\right\}$ and the discounted option price process $\left\{\frac{X_t(S_t)}{e^{rt}}\right\}$ be martingales under an equivalent martingale measure $\mathbb{Q}$. In the presence of jumps, it is well-known that the martingale measure $\mathbb{Q}$ is not unique. Instead, the martingale measure $\mathbb{Q}$ can be determined by the choice of the parameters γ and ν in the Radon-Nikodým derivative (9.6), with the market price of Wiener risk θ determined by the martingale condition after the choices of γ and ν have been made.

The dynamics of the discounted yield process is given by

$$d\left(\frac{S_t e^{qt}}{e^{rt}}\right) = \left(\frac{S_{t-}\, e^{qt}}{e^{rt}}\right)\Big[(\mu + q - r - \lambda\kappa)dt + \sigma dB_t + [e^J - 1]dN_t\Big]. \tag{9.10}$$

After choosing the parameters γ and ν in the Radon-Nikodyḿ derivative (9.6), the jump-arrival process N_t is Poisson with arrival intensity $\tilde{\lambda} = \lambda(1 + \kappa')$ and the distribution of the jump-sizes is given by the moment generating function

$$M_{\mathbb{Q},J}(u) = \frac{M_{\mathbb{P},J}(u+\gamma)}{M_{\mathbb{P},J}(\gamma)}$$

from Lemma 9.1. As a consequence of the martingale condition on the discounted stock yield process, the dynamics for the discounted stock yield process can be expressed as

$$d\left(\frac{S_t e^{qt}}{e^{rt}}\right) = \left(\frac{S_{t-}\, e^{qt}}{e^{rt}}\right)\left[\sigma d\widetilde{B}_t - \tilde{\lambda}\tilde{\kappa}dt + [e^J - 1]dN_t\right], \tag{9.11}$$

where $\widetilde{B}_t$ is a standard Brownian motion under the martingale measure $\mathbb{Q}$ and the expected jump-increment is $\tilde{\kappa} = \mathbb{E}_{\mathbb{Q}}[e^J] - 1$. From Girsanov's Theorem,

$$d\widetilde{B}_t = \theta dt + dB_t, \tag{9.12}$$

and the market price of Wiener risk θ satisfies the risk-premium equation

$$\mu + q - r = \sigma\theta + \lambda\kappa - \tilde{\lambda}\tilde{\kappa}, \tag{9.13}$$

which can be obtained by the substitution of the expression for $d\widetilde{B}_t$ in (9.12) into (9.11), and equating the resulting dt coefficient with the dt coefficient of (9.10).

In (9.13), the premium for Wiener risk is $\sigma\theta$, and the premium for jump-risk is $\lambda\kappa - \tilde{\lambda}\tilde{\kappa}$ which can be manipulated into the form

$$\lambda\kappa\left[1 - e^{\nu}\frac{M_{\mathbb{P},J}(\gamma+1) - M_{\mathbb{P},J}(\gamma)}{M_{\mathbb{P},J}(1) - 1}\right], \tag{9.14}$$

in which the term

$$1 - e^{\nu}\frac{M_{\mathbb{P},J}(\gamma+1) - M_{\mathbb{P},J}(\gamma)}{M_{\mathbb{P},J}(1) - 1} \tag{9.15}$$

can be interpreted as the market price of jump-risk. If the market price of jump-risk ψ is specified, then the parameters γ and ν in the Radon-Nikodým derivative (9.6) can be chosen such that

$$1 - e^{\nu}\frac{M_{\mathbb{P},J}(\gamma+1) - M_{\mathbb{P},J}(\gamma)}{M_{\mathbb{P},J}(1) - 1} = \psi. \tag{9.16}$$

Through the removal of the discount factor in (9.11), the stock price dynamics (9.1) can also be expressed as

$$\begin{aligned} dS_t = S_{t-}(r - q - \tilde{\lambda}\tilde{\kappa})dt + \sigma S_{t-}\, d\widetilde{B}_t \\ + S_{t-}[e^J - 1]dN_t. \end{aligned} \tag{9.17}$$

The following theorem gives the integro-partial differential equation for the option price.

Theorem 9.1. *The European call option price satisfies the integro-partial differential equation*

$$\frac{\partial X_t(S_{t-})}{\partial t} + S_{t-}(r - q - \tilde{\lambda}\tilde{\kappa})\frac{\partial X_t(S_{t-})}{\partial S} + \frac{1}{2}\sigma^2 S_{t-}^2 \frac{\partial^2 X_t(S_{t-})}{\partial S^2} + \tilde{\lambda}\mathbb{E}_{\mathbb{Q}}^J\Big[X_t(S_{t-}\, e^J) - X_t(S_{t-})\Big] = rX_t(S_{t-}), \tag{9.18}$$

with terminal condition $X_T(S_T) = (S_T - K)^+$.

Proof. We require the discounted option price process $\left\{\frac{X_t(S_t)}{e^{rt}}\right\}$ to be a martingale under the martingale measure $\mathbb{Q}$. Since the option price $X_t(S_t)$ is a function of S_t, through the application of Itô's Lemma for jump-diffusion processes, the stochastic differential equation satisfied by this quantity is

$$\begin{aligned} d\left(\frac{X_t(S_t)}{e^{rt}}\right) = \Bigg[&e^{-rt}\frac{\partial X_t(S_{t-})}{\partial t} - r\frac{X_t(S_{t-})}{e^{rt}} + \frac{S_{t-}}{e^{rt}}(r - q - \tilde{\lambda}\tilde{\kappa})\frac{\partial X_t(S_{t-})}{\partial S} \\ &+ \frac{1}{2}\sigma^2\frac{S_{t-}^2}{e^{rt}}\frac{\partial^2 X_t(S_{t-})}{\partial S^2} + \frac{\tilde{\lambda}}{e^{rt}}\mathbb{E}_{\mathbb{Q}}^J\Big[X_t(S_{t-}e^J) - X_t(S_{t-})\Big]\Bigg]dt \\ &+ \frac{\sigma S_{t-}}{e^{rt}}\frac{\partial X_t(S_{t-})}{\partial S}d\widetilde{B}_t - \frac{\tilde{\lambda}}{e^{rt}}\mathbb{E}_{\mathbb{Q}}^J\Big[X_t(S_{t-}e^J) - X_t(S_{t-})\Big]dt \\ &+ \frac{1}{e^{rt}}\Big[X_t(S_{t-}e^J) - X_t(S_{t-})\Big]dN_t. \end{aligned} \tag{9.19}$$

In (9.19), the coefficient of the first dt term must be zero in order that $\left\{\frac{X_t(S_t)}{e^{rt}}\right\}$ be a martingale under the martingale measure $\mathbb{Q}$, hence we obtain the integro-partial differential equation (9.18). □

Using the fact that $\left\{\frac{X_t(S_t)}{e^{rt}}\right\}$ is a martingale under $\mathbb{Q}$, the solution for the option price in the form of a conditional expectation of the discounted final payoff can be written as

$$X_t(S_t) = e^{rt}\mathbb{E}_{\mathbb{Q}}\left[\frac{X_T(S_T)}{e^{rT}}\middle|\mathcal{F}_t\right]. \tag{9.20}$$

9.5. The hedging portfolio

Now in order to obtain some economic intuition, we derive the integro-partial differential equation (9.18) by use of a hedging argument in some appropriate sense to be made clear below. Following Runggaldier (2003), where two options expiring at different maturities are needed to hedge an option where the underlying stock price return follows jump-diffusion dynamics, here our portfolio also consists of two similar options with differing maturities, that is, $X_{1,t}(S_t)$ and $X_{2,t}(S_t)$, expiring at arbitrary maturity dates T_1 and T_2 respectively. The value of the portfolio is

$$\Pi_t = Q_1 X_{1,t}(S_t) + Q_2 X_{2,t}(S_t) + Q_S S_t, \tag{9.21}$$

where Q_1 and Q_2 are time-varying positions in the respective options, and Q_S is a time-varying position in the stock. Adopting the short-form notation $X_{i,t} = X_{i,t}(S_t)$ $(i = 1, 2)$ for the option price at time t when the underlying stock price is S_t, and $X_{i,t-} = X_{i,t}(S_{t-})$, the application of Itô's Lemma for jump-diffusion processes yields the dynamics for the option prices

$$\frac{dX_{i,t}}{X_{i,t-}} = [\mu_{X_i} - \lambda\kappa_{X_i}]dt + \sigma_{X_i} dB_t + [e^{J_{X_i}} - 1]dN_t, \tag{9.22}$$

where the drift of the option X_i is denoted by

$$\begin{aligned}\mu_{X_i} X_{i,t} = {} & \frac{\partial X_{i,t-}}{\partial t} + (\mu_i - \lambda\kappa_{X_i}) S_{t-} \frac{\partial X_{i,t}}{\partial S} \\ & + \sigma^2 \frac{S_{t-}^2}{2} \frac{\partial^2 X_{i,t-}}{\partial S^2} + \lambda\, \kappa_{X_i} X_{i,t-},\end{aligned}$$

the option price volatility is denoted by

$$\sigma_{X_i} X_{i,t-} = \sigma S_{t-} \frac{\partial X_{i,t-}}{\partial S},$$

the expected jump-increment in the option price is

$$X_{i,t-}\, \kappa_{X_i} = \mathbb{E}_{\mathbb{P}}^{J}\left[X_{i,t}(S_{t-}e^{J}) - X_{i,t-}\right],$$

with the option price increment being given by

$$[e^{J_{X_i}} - 1] X_{i,t-} = X_{i,t}(S_{t-}e^{J}) - X_{i,t-}.$$

The infinitesimal change in the portfolio value $d\Pi_t$ over a time interval $[t, t+dt)$ evolves according to

$$d\Pi_t = Q_1 dX_{1,t} + Q_2 dX_{2,t} + Q_S dS_t + Q_S q S_{t-} dt. \tag{9.23}$$

The change in the value of the portfolio Π_t is thus given by

$$\begin{aligned} d\Pi_t &= \sum_{i=1}^{2}\Big[(\mu_{X_i} - \lambda\kappa_{X_i})dt + \sigma_{X_i}dB_t + [e^{J_{X_i}} - 1]dN_t\Big]Q_i X_{i,t-} \\ &\quad + \Big[(\mu + q - \lambda\kappa)dt + \sigma dB_t + [e^{J} - 1]dN_t\Big]Q_S S_{t-} \\ &= \Big[Q_S S_{t-}(\mu + q - \lambda\kappa) + Q_1 X_{1,t-}(\mu_{X_1} - \lambda\,\kappa_{X_1}) \\ &\qquad + Q_2 X_{2,t-}(\mu_{X_2} - \lambda\,\kappa_{X_2})\Big]dt \\ &\quad + \Big[Q_S S_{t-}\sigma + Q_1 X_{1,t-}\sigma_{X_1} + Q_2 X_{2,t-}\sigma_{X_2}\Big]dB_t \\ &\quad + \Big[Q_S S_{t-}[e^{J} - 1] + Q_1 X_{1,t-}[e^{J_{X_1}} - 1] + Q_2 X_{2,t-}[e^{J_{X_2}} - 1]\Big]dN_t. \end{aligned} \tag{9.24}$$

As first explained by [Merton (1976)], in a jump-diffusion model it is not possible to hedge away the idiosyncratic risks due to the jumps. Merton chose to leave the jump-risks unpriced, resulting in no change in the distribution of the jump components of his model under the change of measure. We choose an approach that considers a martingale measure under which the distribution of the jump components may change. Thus the market price of jump-risk remains present in our model. The choice of our equivalent martingale measure corresponds to a perfect hedge only when the jump-size J takes on some particular value, say $J = J^*$, or if the jump-sizes have been averaged out with respect to some equivalent martingale measure so that some pre-determined value for the average relative jump-size increment is obtained. Note that if the stock pricing model is such that only one jump-size is possible, say, J^* whenever a jump occurs, then the jump-risk is perfectly hedged.

We opt for the procedure of averaging out the jump-sizes according to some equivalent martingale measure $\hat{\mathbb{Q}}$ so that the portfolio (9.21) with the jumps averaged out under $\hat{\mathbb{Q}}$ becomes

$$\mathbb{E}^J_{\hat{\mathbb{Q}}}\Pi_t = \hat{Q}_1\mathbb{E}^J_{\hat{\mathbb{Q}}}X_{1,t}(S_t) + \hat{Q}_2\mathbb{E}^J_{\hat{\mathbb{Q}}}X_{2,t}(S_t) + \hat{Q}_S\mathbb{E}^J_{\hat{\mathbb{Q}}}S_t, \tag{9.25}$$

where $\mathbb{E}^J_{\hat{\mathbb{Q}}}$ denotes averaging the jumps under $\hat{\mathbb{Q}}$. The time-varying weights $\hat{Q}_1$, $\hat{Q}_2$ and $\hat{Q}_S$ will be the optimal weights over the interval $[t, t+dt)$ after the jumps have been averaged out. They correspond to a particular choice for the market price of jump risk to be discussed later. It will also be shown that these weights do not depend on the jump-sizes over the interval $[t, t+dt)$. Then by replacing the related quantities in (9.21) and (9.24) with their averaged out equivalent, we obtain

$$\begin{aligned} d(\mathbb{E}^J_{\hat{\mathbb{Q}}}\Pi_t) &= \Big[\hat{Q}_S S_{t-}(\mu + q - \lambda\kappa) + \sum_{i=1}^{2} \hat{Q}_i X_{i,t-}(\mu_{X_i} - \lambda\,\kappa_{X_i})\Big]dt \\ &\quad + \Big[\hat{Q}_S S_{t-}\sigma + \hat{Q}_1 X_{1,t-}\sigma_{X_1} + \hat{Q}_2 X_{2,t-}\sigma_{X_2}\Big]dB_t \\ &\quad + \Big[\hat{Q}_S S_{t-}\hat{\kappa} + \hat{Q}_1 X_{1,t-}\hat{\kappa}_{X_1} + \hat{Q}_2 X_{2,t-}\hat{\kappa}_{X_2}\Big]dN_t, \end{aligned} \tag{9.26}$$

where

$$\begin{aligned} \hat{\kappa} &= \mathbb{E}^J_{\hat{\mathbb{Q}}}[e^{J^*} - 1], \\ X_{i,t-}\hat{\kappa}_{X_i} &= \mathbb{E}^J_{\hat{\mathbb{Q}}}[X_{i,t}(S_{t-}e^J) - X_{i,t-}], \\ X_{t-}\hat{\kappa}_X &= \mathbb{E}^J_{\hat{\mathbb{Q}}}[X_t(S_{t-}e^J) - X_{t-}]. \end{aligned}$$

We need to choose the weights in (9.26) to remove the Wiener risk and the jump-risk. Thus the condition for removing Wiener risk is

$$\hat{Q}_1 X_{1,t-}\sigma_{X_1} + \hat{Q}_2 X_{2,t-}\sigma_{X_2} = -\hat{Q}_S S_{t-}\sigma, \tag{9.27}$$

and the condition for removing jump risk is

$$\hat{Q}_1 X_{1,t-}\hat{\kappa}_{X_1} + \hat{Q}_2 X_{2,t-}\hat{\kappa}_{X_2} = -\hat{Q}_S S_{t-}\hat{\kappa}. \tag{9.28}$$

Solving (9.27) and Eq. (9.28) for Q_1, Q_2 we obtain

$$\hat{Q}_1 = -\frac{\hat{Q}_S S_{t-}}{X_{1,t-}}\left(\frac{\sigma\hat{\kappa}_{X_2} - \sigma_{X_2}\hat{\kappa}}{\sigma_{X_1}\hat{\kappa}_{X_2} - \sigma_{X_2}\hat{\kappa}_{X_1}}\right), \tag{9.29}$$

and

$$\hat{Q}_2 = \frac{\hat{Q}_S S_{t-}}{X_{2,t-}}\left(\frac{\sigma\hat{\kappa}_{X_1} - \sigma_{X_1}\hat{\kappa}}{\sigma_{X_1}\hat{\kappa}_{X_2} - \sigma_{X_2}\hat{\kappa}_{X_1}}\right), \tag{9.30}$$

which are independent of the jump-size over the interval $[t, t + dt)$. Hence the portfolio with the jumps averaged out evolves according to

$$d(\mathbb{E}^J_{\hat{\mathbb{Q}}}\Pi_t) = \Big[\hat{Q}_S S_{t-}(\mu + q - \lambda\kappa) + \sum_{i=1}^{2} \hat{Q}_i X_{i,t-}(\mu_{X_i} - \lambda\kappa_{X_i})\Big]dt. \tag{9.31}$$

We note that the weights were chosen to make $d(\mathbb{E}^J_{\hat{\mathbb{Q}}}\Pi_t)$ risk-less and hence the averaged portfolio (obtained by setting $Q_1 = \hat{Q}_1$, $Q_2 = \hat{Q}_2$ and $Q_S = \hat{Q}_S$ in Eq. (9.21)) has to grow at the risk-free rate to avoid arbitrage opportunities on average, so that

$$d(\mathbb{E}^J_{\hat{\mathbb{Q}}}\Pi_t) = r[\hat{Q}_S S_{t-} + \hat{Q}_1 X_{1,t-} + \hat{Q}_2 X_{2,t-}]dt. \tag{9.32}$$

Hence, equating the right hand sides of Eq. (9.31) and Eq. (9.32) we obtain

$$\hat{Q}_S S_{t-}(\mu + q - \lambda\kappa - r) + \sum_{i=1}^{2} \hat{Q}_i X_{i,t-}(\mu_{X_i} - \lambda\kappa_{X_i} - r) = 0. \tag{9.33}$$

Substituting for $\hat{Q}_1$ and $\hat{Q}_2$ in (9.29) and (9.30) into (9.33) we obtain

$$\frac{(\mu_{X_1} - \lambda\,\kappa_{X_1} - r) - \frac{\sigma_{X_1}}{\sigma}(\mu + q - \lambda\kappa - r)}{\sigma\hat{\kappa}_{X_1} - \sigma_{X_1}\hat{\kappa}} = \frac{(\mu_{X_2} - \lambda\,\kappa_{X_2} - r) - \frac{\sigma_{X_2}}{\sigma}(\mu + q - \lambda\kappa - r)}{\sigma\hat{\kappa}_{X_2} - \sigma_{X_2}\hat{\kappa}}. \tag{9.34}$$

The left hand side and right hand side of (9.34) must be independent of the respective option's maturity time, so we can conclude that for any option X_t based on the stock S_t, we must have

$$\frac{(\mu_X - \lambda\kappa_X - r) - \frac{\sigma_X}{\sigma}(\mu + q - \lambda\kappa - r)}{\sigma\hat{\kappa}_X - \sigma_X\hat{\kappa}} = -\xi \tag{9.35}$$

for some ξ independent of maturity. Thus, by rearranging, we obtain

$$\frac{\left(\mu_X - r - \lambda\kappa_X + \hat{\kappa}_X\sigma\xi\right)}{\sigma_X} = \frac{\left(\mu + q - r - \lambda\kappa + \hat{\kappa}\sigma\xi\right)}{\sigma}. \tag{9.36}$$

We have let $\xi > 0$ in Eq. (9.35) in order to ensure the correct sign for the market premium for the jump-risk in Eq. (9.36). Of course mathematically there is no reason to guarantee that the RHS of Eq. (9.35) should be negative. It is in fact an empirical issue as to whether the risk premium for bearing jump-risk is positive. Whilst there are a lot of empirical results in this area supporting the positivity of the ex-ante risk premium (e.g. [Fama and French (2002)]), we should also point to the studies of [Boudoukh *et al.* (1973)] and [Walsh (2006)] who report that the ex-ante market risk premium can on occasions be negative.

Now recall the form of the Radon-Nikodým derivative (9.6). For (9.36) to make economic sense, the LHS has to be the risk premium less the jump-risk per unit volatility for the option, and the RHS, the risk premium less the jump-risk per unit volatility for the stock. After the jump-risk is removed from both sides of (9.36), then either expression is the market price of the Wiener risk θ in the Radon-Nikodým derivative (9.6). Hence it follows that the new jump intensity under $\hat{\mathbb{Q}}$ has to be $\hat{\lambda} = \sigma\xi$.

In order to solve for the parameters ν and γ in the Radon-Nikodým derivative (9.6), we need to solve simultaneously

$$1 + \hat{\kappa} = \frac{\mathbb{E}_{\mathbb{P}}[e^{(\gamma+1)J}]}{\mathbb{E}_{\mathbb{P}} e^{\gamma J}} \tag{9.37}$$

and

$$1 + \hat{\kappa} = \frac{e^{\nu}\mathbb{E}_{\mathbb{P}}[e^{(\gamma+1)J}]}{\hat{\lambda}}. \tag{9.38}$$

The values of the parameters $\hat{\nu}$ and $\hat{\gamma}$ from Eq. (9.37) and Eq. (9.38) yield the Radon-Nikodým derivative $\frac{d\hat{\mathbb{Q}}}{d\mathbb{P}}\Big|_T$. Thus the risk premium of the option in (9.36) with the jumps averaged out in the market measure $\mathbb{P}$ is

$$\frac{\left(\mu_X - r - \lambda\kappa_X + \hat{\lambda}\mathbb{E}^J_{\mathbb{P}}\left[\frac{d\hat{\mathbb{Q}}}{d\mathbb{P}}\Big|_T [e^{J_X} - 1]\right]\right)}{\sigma_X} = \frac{\left(\mu + q - r - \lambda\kappa + \hat{\lambda}\mathbb{E}^J_{\mathbb{P}}\left[\frac{d\hat{\mathbb{Q}}}{d\mathbb{P}}\Big|_T (e^J - 1)\right]\right)}{\sigma}, \tag{9.39}$$

which is equivalent to

$$\frac{\left(\mu_X - r - \lambda\kappa_X + \hat{\lambda}\mathbb{E}^J_{\hat{\mathbb{Q}}}[e^{J_X} - 1]\right)}{\sigma_X} = \frac{\left(\mu + q - r - \lambda\kappa + \hat{\lambda}\mathbb{E}^J_{\hat{\mathbb{Q}}}[e^J - 1]\right)}{\sigma}, \tag{9.40}$$

after expressing the expectations under $\mathbb{P}$ in Eq. (9.39) as expectations under the martingale measure $\hat{\mathbb{Q}}$ in Eq. (9.40).

Recall that $\hat{\kappa}$ is the expected jump-increment of the stock, and $\hat{\kappa}_X$ is the expected jump-increment in the option price, both in the equivalent martingale measure $\hat{\mathbb{Q}}$. Then the risk premium of the option from (9.40) satisfies

$$\mu_X - r - \lambda\kappa_X + \hat{\lambda}\hat{\kappa}_X = \frac{\sigma_X}{\sigma}(\mu + q - r - \lambda\kappa + \hat{\lambda}\hat{\kappa}). \tag{9.41}$$

From the expression θ for the market price of Wiener risk (9.13), we see that (9.41) expresses the risk premium of the option less the jump risk as the market price of Wiener risk θ scaled up by the option price volatility σ_X, so that (9.41) can be rewritten as

$$\mu_X - r - (\lambda\kappa_X - \hat{\lambda}\hat{\kappa}_X) = \sigma_X\theta. \tag{9.42}$$

Finally we multiply both sides of (9.41) by X_{t-} and substitute into the left side of (9.41) for

$$\begin{aligned} X_{t-}\mu_X &= \frac{\partial X_{t-}}{\partial t} + (\mu - \lambda\kappa)S_{t-}\frac{\partial X_{t-}}{\partial S} \\ &\quad + \sigma^2\frac{S_{t-}^2}{2}\frac{\partial^2 X_{t-}}{\partial S^2} + \lambda\kappa_X X_{t-}, \\ \sigma_X X_{t-} &= \sigma S_{t-}\frac{\partial X_{t-}}{\partial S}, \\ X_{t-}\kappa_X &= \mathbb{E}_{\mathbb{P}}^J\Big[X_t(S_{t-}e^J) - X_{t-}\Big], \end{aligned}$$

and

$$X_{t-}\hat{\kappa}_X = \mathbb{E}_{\hat{\mathbb{Q}}}^J\Big[X_t(S_{t-}e^J) - X_{t-}\Big].$$

Thus (9.41) becomes

$$\begin{aligned} &\frac{\partial X_{t-}}{\partial t} + (r - q - \hat{\lambda}\hat{\kappa})S_{t-}\frac{\partial X_{t-}}{\partial S} + \sigma^2\frac{S_{t-}^2}{2}\frac{\partial^2 X_{t-}}{\partial S^2} \\ &\qquad + \hat{\lambda}\mathbb{E}_{\hat{\mathbb{Q}}}^J\Big[X_t(S_{t-}e^J) - X_{t-}\Big] = rX_{t-}, \end{aligned}$$

which is again the integro-partial differential equation (9.18), where the expectations are taken under the equivalent martingale measure $\hat{\mathbb{Q}}$. Note that if the values $\hat{\nu}$ and $\hat{\gamma}$ from Eq. (9.37) and Eq. (9.38) were used as the values of ν and γ in the Radon-Nikodým derivative in Lemma 9.1, and applied to the martingale approach for the derivation of the integro-partial differential equation (9.18) in Section 4, then the martingale measure $\hat{\mathbb{Q}}$ is the same as $\mathbb{Q}$.

9.6. Pricing the call option

Let us now price the call option at time $t = 0$. Following [Merton (1976)], we now assume that the jump-sizes are normally distributed with mean α and variance δ^2 under the market measure $\mathbb{P}$. For a chosen γ and ν in the Radon-Nikodým derivative (9.6), we see from the application of Lemma 9.1 that the jump-sizes will be normally distributed with mean $\tilde{\alpha} = \alpha + \gamma\delta^2$ with the same variance δ^2, and the new intensity of the jump-arrivals is $\tilde{\lambda} = \lambda(1 + \kappa')$, under the equivalent martingale measure $\mathbb{Q}$.

The next theorem gives the European call option price.

Theorem 9.2. *Suppose the stock price follows the dynamics (9.1), and the stock pays a continuous dividend at the rate q. The European call option* *price $X_0(S_0)$ at time $t = 0$ is given by*

$$X_0(S_0) = \sum_{n=0}^{\infty} e^{-\hat{\lambda}T} \frac{(\hat{\lambda}T)^n}{n!} \times \left[S_0 e^{-qT} \Phi(d_{1,n}) - K e^{-r_n T} \Phi(d_{2,n}) \right], \quad (9.43)$$

where

$$d_{1,n} = \frac{\ln \frac{S_0}{K} + \left(r_n - q + \frac{\sigma_n^2}{2} \right) T}{\sigma_n \sqrt{T}}$$

and

$$d_{2,n} = d_{1,n} - \sigma_n \sqrt{T};$$

with

$$\begin{aligned}
\tilde{\alpha} &\equiv \alpha + \gamma\delta^2, \\
\tilde{\lambda} &\equiv \lambda \exp\left[\nu + \gamma\alpha + \frac{\gamma^2\delta^2}{2} \right], \\
\tilde{\kappa} &\equiv \exp\left[\tilde{\alpha} + \frac{\delta^2}{2} \right] - 1, \\
\hat{\lambda} &\equiv \lambda e^{\nu + \gamma^2\alpha + \frac{\gamma^2\delta^2}{2}} \times e^{\tilde{\alpha} + \frac{\delta^2}{2}}, \\
r_n &\equiv r - \tilde{\lambda}\tilde{k} + \frac{n\tilde{\alpha}}{T} + \frac{n\delta^2}{2T},
\end{aligned}$$

and

$$\sigma_n^2 \equiv \sigma^2 + \frac{n\delta^2}{T}.$$

Proof. Note that the option price discounted by the money market account e^{rt} is a martingale in the martingale measure $\mathbb{Q}$. Let $A = \{S_T > K\}$ be the event that the option is in the money at maturity. The event A is equivalent to the event that

$$\left\{ \sigma \widetilde{B}_T + \sum_{n=1}^{N_T} J_n > \ln \frac{S_0}{K} - \left(r - q - \tilde{\lambda}\tilde{\kappa} - \frac{\sigma^2}{2} \right) T \right\}.$$

Hence from (9.20), the call option price is

$$\begin{aligned}
X_0(S_0) &= e^{-rT} \mathbb{E}_{\mathbb{Q}} \left[(S_T - K) \mathbb{1}_A \right] \\
&= S_0 e^{-qT} \mathbb{E}_{\mathbb{Q}} \left[e^{-\frac{1}{2}\sigma^2 T + \sigma \widetilde{B}_T - \tilde{\lambda}\tilde{\kappa}T + \sum_{n=1}^{N_T} J_n} \mathbb{1}_A \right] \\
&\quad - K e^{-rT} \mathbb{E}_{\mathbb{Q}} \mathbb{1}_A \\
&= S_0 e^{-qT} \widetilde{\mathbb{Q}}(A) - K e^{-rT} \mathbb{Q}(A)
\end{aligned} \quad (9.44)$$

where in (9.44), the Radon-Nikodým derivative is

$$\left.\frac{d\widetilde{\mathbb{Q}}}{d\mathbb{Q}}\right|_T = e^{-\frac{1}{2}\sigma^2 T + \sigma\widetilde{B}_T - \tilde{\lambda}\tilde{\kappa}T + \sum_{n=1}^{N_T} J_n}. \tag{9.45}$$

We note that

$$\mathbb{Q}(A) = \sum_{n=0}^{\infty} e^{-\tilde{\lambda}T}\frac{(\tilde{\lambda}T)^n}{n!} \times \Phi(d_{2,n}) \tag{9.46}$$

where

$$d_{2,n} = \frac{\ln\frac{S_0}{K} + \left(r - q - \tilde{\lambda}\tilde{\kappa} - \frac{\sigma^2}{2}\right)T + n(\alpha + \gamma\delta^2)}{\sqrt{\left(\sigma^2 + \frac{n\delta^2}{T}\right)T}}.$$

Under the measure $\widetilde{\mathbb{Q}}$ and the application of Lemma 9.1, the Wiener component $\sigma\widetilde{B}_T$ is normally distributed as $N(\sigma^2 T, \sigma^2 T)$ and J is normally distributed as $N(\alpha + \gamma\delta^2 + \delta^2, \delta^2)$ and the Poisson process N_t has intensity $\hat{\lambda} = \tilde{\lambda}(1 + \tilde{\kappa})$. Hence

$$\widetilde{\mathbb{Q}}(A) = \sum_{n=0}^{\infty} e^{-\hat{\lambda}T}\frac{(\hat{\lambda}T)^n}{n!} \times \Phi(d_{1,n}) \tag{9.47}$$

where

$$d_{1,n} = \frac{\ln\frac{S_0}{K} + \left(r - q - \tilde{\lambda}\tilde{\kappa} + \frac{\sigma^2}{2}\right)T + n(\alpha + \gamma\delta^2 + \delta^2)}{\sqrt{\left(\sigma^2 + \frac{n\delta^2}{T}\right)T}}.$$

□

Remark 9.1. In the proof of Theorem 9.1, the decomposition of the option price in (9.44) is analogous to that obtained by [Geman *et al.* (1995)] for the pure-diffusion case. The measure $\widetilde{\mathbb{Q}}$ corresponds to the equivalent martingale measure with the stock price as the numéraire.

We conclude this section by noting that when the parameters γ and ν are set to zero in the Radon-Nikodým derivative (9.6), we recover Merton's (1976) model as a special case. In this case since the parameters γ and ν are set to zero, the market price of jump-risk in (9.13) to (9.15) is not priced.

9.7. Conclusion

This paper has extended the analysis of [Merton (1976)] to the case where the distribution of the jump-arrivals and the jump-sizes change under the change of measure. A Radon-Nikodým derivative process that induces the change of measure through the choice of suitable parameters has been introduced. We have shown how the non-uniqueness of the option price manifests itself through variations in the parameters of the Radon-Nikodým derivative that induces the change of measure. Through a hedging portfolio that averages over the jumps, we relate the change of the distributions of the jump components to the market price of jump-risk. We also derive the standard pricing integro-partial differential equation.

Acknowledgement

The authors acknowledge the helpful discussions with Wolfgang Runggaldier over an earlier draft of a multi-asset version of this paper. The usual caveat applies.

References

[Aase (1988)] Aase, K. K. (1988). Contingent claim valuation when the security price is a combination of an itô process and a random point process, *Stochatic Processes and Applications* **28**, pp. 185–220.

[Anderson (1984)] Anderson, A. J. (1984). Hedge portfolios and the black-scholes equations, *Stochastic Analysis and Applications* **2**, 1, pp. 1–11.

[Black and Scholes (1973)] Black, F. and Scholes, M. (1973). The pricing of corporate liabilities, *Journal of Political Economy* **381**, pp. 637–659.

[Boudoukh *et al.* (1973)] Boudoukh, J., Richardson, M. and Smith, T. (1973). Is the ex-ante risk premium always positive? a new approach to testing conditional asset pricing models, *Journal of Financial Economics* **34**, pp. 387–408.

[Colwell and Elliott (1993)] Colwell, D. B. and Elliott, R. J. (1993). Discontinuous asset prices and non-attainable contingent claims, *Mathematical Finance* **3**, 3, pp. 295–308.

[Colwell *et al.* (1991)] Colwell, D. B., Elliott, R. J. and Kopp, P. E. (1991). Martingale representation and hedging policies, *Stochastic Processes and their Applications* **38**, pp. 335–345.

[Cont and Tankov (2004)] Cont, R. and Tankov, P. (2004). *Financial modelling with jump processes* (Chapman and Hall/CRC, London).

[Elliott and Kopp (1990)] Elliott, R. J. and Kopp, P. E. (1990). Option pricing and hedge portfolios for Poisson processes, *Stochastic Analysis and Applications* **8**, 2, pp. 157–167.

[Fama and French (2002)] Fama, E. F. and French, K. R. (2002). The equity premium, *Journal of Finance* **57**, 2, pp. 637–655.

[Geman *et al.* (1995)] Geman, H., El-Karoui, N. and Rochet, J. C. (1995). Change of numéraire, changes of probability measure and option pricing, *Journal of Applied Probability* **32**, 2, pp. 443–459.

[Gerber and Shiu (1994)] Gerber, H. U. and Shiu, E. S. W. (1994). Option pricing by Esscher transforms, *Transactions, Society of Actuaries* **46**, pp. 99–191.

[Harrison and Pliska (1981)] Harrison, J. M. and Pliska, S. R. (1981). Martingales and stochastic integrals in the theory of continuous trading, *Stochastic Processes and Applications* **11**, 3, pp. 215–260.

[Jarrow and Madan (1995)] Jarrow, R. and Madan, D. B. (1995). Option pricing using the term structure of interest rates to hedge systematic discontinuities in asset returns, *Mathematical Finance* **5**, 4, pp. 311–356.

[Jeanblanc-Picqué and Pontier (1990)] Jeanblanc-Picqué, M. and Pontier, M. (1990). Optimal portfolio for a small investor in a market model with discontinuous prices, *Applied Mathematical Optimization* **22**, pp. 287–310.

[Mercurio and Runggaldier (1993)] Mercurio, F. and Runggaldier, W. J. (1993). Optimal portfolio for a small investor in a market model with discontinuous prices, *Mathematical Finance* **3**, 2, pp. 191–200.

[Merton (1976)] Merton, R. C. (1976). Option pricing when underlying stock returns are discontinuous, *Journal of Financial Economics* **3**, 1, pp. 125–144.

[Miyahara (2001)] Miyahara, Y. (2001). Geometric Lévy process & MEMM pricing model and related estimation problems, *Asia-Pacific Financial Markets* **8**, pp. 45–60.

[Runggaldier (2003)] Runggaldier, W. J. (2003). *Jump diffusion models,* In ed. S. T. Rachev, *Handbook of heavy tailed distributions in finance*, pp. 169–209, Elsevier, North-Holland.

[Schweizer (1991)] Schweizer, M. (1991). Option hedging for semimartingales, *Stochastic Processes and Applications* **37**, pp. 339–363.

[Walsh (2006)] Walsh, K. D. (2006). Is the ex-ante risk premium always positive? further evidence, *Australian Journal of Management* **31**, 1, pp. 93–113.

Chapter 10

Hedging portfolio loss derivatives with CDS's

Areski Cousin[1,*] and Monique Jeanblanc[2]

[1] *Université de Lyon, Université Lyon 1, ISFA, France*

[2] *Laboratoire analyse et probabilité,*
Université d'Evry-Val-D'Essonne, France;
Institut Europlace de Finance, France

In this chapter, we consider the hedging of portfolio loss derivatives using single-name credit default swaps as hedging instruments. The hedging issue is investigated in a general pure jump dynamic setting where default times are assumed to admit a joint density. In a first step, we compute default intensities adapted to the global filtration of defaults. In particular, we stress the impact of a default event on the price dynamics of non-defaulted names. In a two defaults setting, we also fully describe the hedging of a loss derivative with single name instruments. The methodology can be applied recursively to the case of a multidefault setting. We completely characterize the hedging strategies for general n-dimensional credit portfolios when default times are assumed to be ordered. The computation of the hedging strategies does not require any Markovian assumption.

Contents

*This research is a part of CRIS program and Chaire Risque de Crédit of French Banking Federation. The first author also benefited from the support of the ANR project Ast&Risk. We thank the anonymous referee for helpful comments.

10.1. Introduction

The hedging of loss derivatives such as CDO tranches or basket default swaps is a prominent risk-management issue especially given the recent *revisions to the Basel II market risk framework, Dec 2010.* Indeed, according to [1], "correlation trading portfolios" including tranches on standard indices and their associated liquid hedging positions will continue to be charged as hedge-sets under an internal VaR-based method. The practice of hedging is still recognized as a risk mitigation technics for these "correlation products" as far as computation of trading book capital requirement is concerned. As a result, the performance and efficiency of underlying hedging methods will have a direct impact on the amount of capital required for loss derivatives. Cousin and Laurent (2010) discuss various issues related to the use of models in designing hedging strategies for CDO tranches and back-testing or assessing hedging performance.

In this paper, we consider the hedging of loss derivatives using single-name credit default swaps as hedging instruments. The hedging issue is investigated in a general pure jump setting where default times are assumed to admit a joint density which is the only input of the model – so that our results can be considered as model independent – and we compute default intensities adapted to the global filtration of defaults. We check that, if CDSs on each default are traded, the market is complete. The hedging strategies can be found by identifying the terms associated with the fundamental default martingales.

We extend some recent results by Laurent, Cousin and Fermanian (2010) and Cousin, Jeanblanc and Laurent (2010). In particular, we stress the impact of a default event on the price dynamics of non-defaulted names. Moreover, in a two defaults setting, we fully describe the hedging of a loss derivative with single name instruments. The generalization to a multidefault setting can be done following the same methodology. Furthermore, we are able to completely characterize the hedging strategies in single-name CDS for general n-dimensional credit portfolios when default times are assumed to be ordered. The computation of the hedging strategies does not require any Markovian assumption.

The paper is organized as follows. The first section aims at presenting the general setting of the model and it recalls the predictable representation theorem. In the second section, we investigate the case where only one name is considered. In particular, we exhibit the intensity of the default time and the dynamics of CDS prices. Section three is devoted to the case where the credit portfolio is composed of two names. The extension to a multivariate setting can be done using a recursive procedure. In particular, we highlight the contagion effect occurring at default time of one of the two names on the CDS price dynamics of the other name. We also compute the dynamics of the hedging strategies at any time in all the possible default situations. In Section four, we consider the hedging of a loss derivatives written on a general n-dimensional portfolio. We stress that when default times are assumed to be ordered, i.e., CDS are kth-to-default swaps, the hedging strategies can be simply characterized as the solutions of a linear system.

10.2. Mathematical tools: The general case

In what follows, we consider n default times τ_i, $i = 1, \dots, n$, that is, nonnegative and finite random variables defined on the same probability space $(\Omega, \mathcal{G}, \mathbb{P})$. For any $i = 1, \dots, n$, we denote by $(H_t^i = \mathbb{1}_{\{\tau_i \le t\}}, t \ge 0)$ the i-th default process, and by $\mathcal{H}_t^i = \sigma(H_s^i, s \le t)$ the natural filtration of H^i (after completion and regularization on right). We introduce $\mathbb{H}$, the filtration generated by the processes $H^i, i = 1, \dots, n$, defined as $\mathbb{H} = \mathbb{H}^1 \vee \cdots \vee \mathbb{H}^n$, i.e., $\mathcal{H}_t = \vee_{i=1}^n \mathcal{H}_t^i$ (after regularization on right).
We assume that $G(t_1, \dots, t_n) := \mathbb{P}(\tau_1 > t_1, \dots, \tau_n > t_n)$ is twice differentiable with respect to $(t_1, \dots, t_n)$ and that G and its derivatives do not vanish. Then, as we shall prove in the next section, for any $i = 1, \dots, n$, there exists a non-negative $\mathbb{H}$-adapted process $(\lambda_t^i, t \ge 0)$ such that the process

$$M_t^i := H_t^i - \int_0^t \lambda_s^i ds$$

is an $\mathbb{H}$-martingale. The process λ^i is called the $\mathbb{H}$-intensity of τ_i. This process vanishes after τ_i (otherwise, after τ_i, the martingale M^i would be continuous and decreasing) and can be written $\lambda^i_t = (1 - H^i_t)\widetilde{\lambda}^i_t$ for some $\mathbb{H}^1 \vee \cdots \vee \mathbb{H}^{i-1} \vee \mathbb{H}^{i+1} \vee \cdots \vee \mathbb{H}^n$-adapted process $\widetilde{\lambda}^i$. In terms of the process $\widetilde{\lambda}^i$, one has

$$M^i_t = H^i_t - \int_0^{t\wedge\tau_i} \widetilde{\lambda}^i_s ds = H^i_t - \int_0^t (1 - H^i_s)\widetilde{\lambda}^i_s ds\,.$$

In particular, denoting by $\tau_{(i)}$, $i = 1, \ldots, n$, the ranked sequence of default times, the process $\widetilde{\lambda}^1$ is a deterministic function on the time interval $[0, \tau_{(1)}[$ (i.e., $\widetilde{\lambda}^1_t = \bar{\lambda}^1(t)$ where $\bar{\lambda}^1$ is deterministic), a deterministic function evaluated at time $\tau_{(1)}$ on the time interval $[\tau_{(1)}, \tau_{(2)}[$ (i.e., $\widetilde{\lambda}^1_t = \bar{\lambda}^{1,2}(t, \tau_{(1)})$ where $\bar{\lambda}^{1,2}$ is deterministic), and a deterministic function evaluated at times $\tau_{(j)}$, $j \leq i$ on the time interval $[\tau_{(i)}, \tau_{(i+1)}[$. In particular, the value of the intensity depends not only on the number of defaults which have occurred in the past, but also on the times where the defaults have taken place, which is more realistic.

The following predictable representation theorem holds true (see Brémaud [5]).

Theorem 10.1. *Let $B \in \mathcal{H}_T$ be an integrable random variable. Then, there exist $\mathbb{H}$-predictable processes $\vartheta^i, i = 1, \ldots, n$ such that*

$$B = \mathbb{E}(B) + \sum_{i=1}^n \int_0^T \vartheta^i_s dM^i_s\,,$$

and $\mathbb{E}(\int_0^T |\vartheta^i_s|\lambda^i_s ds) < \infty$. Moreover, if B is square integrable, these processes are unique in the class of processes which satisfy $\mathbb{E}(\int_0^T |\vartheta^i_s|^2\lambda^i_s ds) < \infty$.

Due to the integrability assumption and the predictable property of the ϑ's, the processes $\int_0^t \vartheta^i_s dM^i_s$ are $\mathbb{H}$-martingales.
In what follows, for a bounded variation left-continuous process A, we denote $\int_t^T \ldots dA_s$ for $\int_{(t,T]} \ldots dA_s$.

We shall in the first part present computations for the intensity in terms of the density of τ in the case $n = 1$. Then, we shall study the case $n = 2$ and we determine the hedging strategy of any payoff, when the hedging instruments are CDSs. The methodology can be easily extended to other hedging instruments, such as defaultable zero-coupons and digital CDSs.

The multidefault case can be studied along the same lines. For simplicity, we restrict our attention to the particular situation of ranked times.

10.3. The single default case

In this section, we present some well known results concerning the price dynamics of a CDS written on a single default, working in the filtration $\mathbb{H}^1$. As we shall see in the next section, the price dynamics of a CDS with the same recovery will be different in the filtration taking into account knowledge of other defaults.

10.3.1. *Some important martingales*

We recall some well known results (see Elliott [10], Dellacherie [9] and Bielecki and Rutkowski [2]). Here, τ is a non-negative random variable on the probability space $(\Omega, \mathcal{G}, \mathbb{P})$ with survival function $G(t) := \mathbb{P}(\tau > t) = 1 - \mathbb{P}(\tau \leq t) = 1 - F(t)$ where F is the cumulative distribution function of τ. We assume that $G(t) > 0$, $\forall t$, and that G is differentiable, i.e., that τ admits a density f, so that $G'(t) = -f(t)$. The filtration is $\mathbb{H} = \mathbb{H}^1$.

Proposition 10.1. For any (integrable) random variable X

$$\mathbb{1}_{\{t<\tau\}}\mathbb{E}_{\mathbb{P}}(X|\mathcal{H}_t) = \mathbb{1}_{\{t<\tau\}}\frac{1}{G(t)}\mathbb{E}_{\mathbb{P}}(X\mathbb{1}_{\{t<\tau\}}) \tag{10.1}$$

and for any bounded, Borel-measurable function h

$$\mathbb{E}_{\mathbb{P}}(h(\tau)|\mathcal{H}_t) = \mathbb{1}_{\{\tau\leq t\}}h(\tau) - \mathbb{1}_{\{t<\tau\}}\frac{1}{G(t)}\int_t^\infty h(u)dG(u)\,.$$

The process $(M_t, t \geq 0)$ defined as

$$M_t = H_t - \int_0^{\tau\wedge t}\frac{f(s)}{G(s)}ds = H_t - \int_0^t (1-H_s)\frac{f(s)}{G(s)}ds$$

is a $(\mathbb{P}, \mathbb{H})$-martingale. In other terms, the $\mathbb{H}$-intensity of τ is $(1 - H_t)\lambda(t)$ where λ is the deterministic function $\lambda(t) = \frac{f(t)}{G(t)}$.

Note that the survival probability G can be expressed in terms of the deterministic function λ: indeed we have proved that

$$\lambda(t) = f(t)/G(t) = -G'(t)/G(t)\,.$$

Solving this ODE with initial condition $G(0) = 1$ leads to

$$G(t) = \mathbb{P}(\tau > t) = \exp\left(-\int_0^t \lambda(u)\,du\right).$$

Note that $\lambda(t)dt = \mathbb{P}(\tau \in dt | \tau > t)$. The default intensity can be interpreted as the instantaneous conditional default probability given that default has not yet occurred.

10.3.2. *Price of a traditional single-name CDS*

We assume that $\mathbb{P}$ is the pricing measure.

We denote by B the savings account, henceforth the price process of any tradeable security, paying no coupons or dividends, is a $(\mathbb{P}, \mathbb{H})$-martingale, when discounted by B. The ex-dividend price of an asset paying dividends is

$$B_t \mathbb{E}\left(\int_t^T B_s^{-1} dD_s | \mathcal{H}_t \right)$$

where D represents the cumulative dividend. In that case, the discounted cumulative dividend price V_t^{cum} is such that

$$V_t^{cum} B_t^{-1} = \mathbb{E}\left(\int_t^T B_s^{-1} dD_s | \mathcal{H}_t \right) + \int_0^t B_s^{-1} dD_s$$

is a $(\mathbb{P}, \mathbb{H})$-martingale. As usual, B is given by

$$B_t = \exp\left(\int_0^t r_u \, du \right), \quad \forall t \in \mathbb{R}_+,$$

where the short-term interest rate r is here a deterministic process.

Let us recall that a credit default swap is a bilateral contract involving a protection seller and a protection buyer. We consider a CDS maturing at time T. If a default event occurs at time $\tau < T$, then the protection seller delivers to the protection buyer the unrecovered portion of the loss $\delta(\tau)$ where δ is a deterministic function[a]. As for the premium leg, we assume for simplicity that the fee is paid to the protection seller in continuous time, i.e., the amount κdt is paid by the protection buyer during the time interval dt, till time $\tau \wedge T$. The time-t market value of a CDS with payment at default δ and with a contractual spread κ is equal to

$$V_t(\kappa) = \tilde{D}_t - \kappa P_t,$$

[a]More precisely, the quantity $\delta(\tau)$ is equal to the loss given default associated with the reference entity times the CDS notional amount.

where $\tilde{D}_t$ and P_t, the default leg and the premium leg, are given by

$$\tilde{D}_t = B_t\,\mathbb{E}\Big(B_\tau^{-1}\delta(\tau)\mathbb{1}_{\{t<\tau\le T\}}\,\Big|\,\mathcal{H}_t\Big)$$

$$P_t = B_t\mathbb{E}\Big(\int_{t\wedge\tau}^{T\wedge\tau} B_u^{-1}\,du\Big|\,\mathcal{H}_t\Big)$$

and the cumulative dividend price is

$$V_t^{\mathrm{cum}}(\kappa) = B_t\mathbb{E}\left(B_\tau^{-1}\delta(\tau)\mathbb{1}_{\{\tau\le T\}} - \kappa\int_0^{T\wedge\tau} B_u^{-1}\,du|\mathcal{H}_t\right).$$

In the case of a zero interest rate,

$$V_t(\kappa) = \mathbb{1}_{\{t<\tau\}}\mathbb{E}(\delta(\tau)\mathbb{1}_{\{\tau\le T\}} - \kappa((T\wedge\tau)-t)|\mathcal{H}_t),$$
$$V_t^{\mathrm{cum}}(\kappa) = \mathbb{E}(\delta(\tau)\mathbb{1}_{\{\tau\le T\}} - \kappa(T\wedge\tau)|\mathcal{H}_t)\,.$$

It is worthwhile noting that the ex-dividend price is not a martingale under the pricing measure, despite the fact that the interest rate is null. However, the cumulative dividend price is a martingale, this will be useful latter on. In what follows, we restrict our attention to the case of zero interest rate.

We recall a well-known result (see, e.g., [4]).

Proposition 10.2. The price at time $t \in [0,T]$ of a credit default swap with spread κ is

$$V_t(\kappa) = \mathbb{1}_{\{t<\tau\}}\widetilde{V}_t(\kappa), \quad \forall\, t\in[0,T],$$

where $\widetilde{V}_t(\kappa)$ is a deterministic function associated with the pre-default value of the CDS and equals

$$\widetilde{V}_t(\kappa) = \frac{1}{G(t)}\left(-\int_t^T \delta(u)\,dG(u) - \kappa\int_t^T G(u)\,du\right).$$

Proof. From Proposition 10.1, we have, on the set $\{t<\tau\}$,

$$V_t(\kappa) = -\frac{\int_t^T \delta(u)\,dG(u)}{G(t)} - \kappa\left(\frac{-\int_t^T u\,dG(u) + TG(T)}{G(t)} - t\right)$$
$$= \frac{1}{G(t)}\left(-\int_t^T \delta(u)\,dG(u) - \kappa\Big(TG(T) - tG(t) - \int_t^T u\,dG(u)\Big)\right).$$

where, in the last equality, we have used an integration by parts to obtain

$$\int_t^T G(u)\,du = TG(T) - tG(t) - \int_t^T u\,dG(u).$$

□

10.3.3. *Dynamics of CDS prices in a single default setting*

Here, we compute the dynamics of the CDS's price. It is useful (see [3]) to obtain the hedging strategy of a defaultable claim based on CDS and savings account.

Proposition 10.3. The dynamics of the (ex-dividend) price $V_t(\kappa)$ on $[0,T]$ are

$$dV_t(\kappa) = -V_{t-}(\kappa)\, dM_t + (1-H_t)(\kappa - \delta(t)\widetilde{\lambda}(t))\, dt,$$

where the $(\mathbb{P},\mathbb{H})$-martingale M is given in Proposition 10.1.

Proof. It suffices to note that

$$V_t(\kappa) = (1-H_t)\widetilde{V}_t(\kappa)$$

with $\widetilde{V}(\kappa)$ given in Proposition 10.2, so that, using integration by parts formula,

$$dV_t(\kappa) = (1-H_t)\, d\widetilde{V}_t(\kappa) - \widetilde{V}_{t-}(\kappa)\, dH_t.$$

Using the explicit expression of $\widetilde{V}_t(\kappa)$, we find easily that

$$d\widetilde{V}_t(\kappa) = \widetilde{\lambda}(t)\widetilde{V}_t(\kappa)\, dt + (\kappa - \delta(t)\widetilde{\lambda}(t))\, dt.$$

The SDE for V follows. □

Comment. It is well known that the risk neutral dynamics of a dividend paying asset is $dS_t = dm_t - \delta_t dt$, where m is a martingale and δ is the dividend rate. Here, the premium κ is similar to a dividend to be paid up to time t, hence the quantity $\kappa(1-H_t)dt$ appears. The $\delta(t)$ can be interpreted as a dividend to be received, at time t, with probability $\widetilde{\lambda}(t)dt$. At default time, the price jumps from $V_{\tau-}(\kappa)$ to 0, as can be seen in the right-hand side of the dynamics.

Corollary 10.1. *The dynamics of the cum-dividend price V^{cum} on $[0,T]$ are*

$$dV_t^{\text{cum}}(\kappa) = (\delta(t) - V_{t-}(\kappa))\, dM_t. \tag{10.2}$$

Proof. The cumdividend price is

$$\begin{aligned} V^{\text{cum}}{}_t(\kappa) &= V_t(\kappa) + \mathbb{1}_{\{t<\tau\}}\delta(\tau) - \kappa(t\wedge\tau) \\ &= V_t(\kappa) + \int_0^t \delta(s)dH_s - \kappa t(1-H_t) - \int_0^t \kappa s dH_s\,. \end{aligned}$$

The result follows. □

10.4. Two default times

Let us first study the case with two random times τ_1, τ_2. For $i = 1, 2$, we denote by $(H_t^i, t \geq 0)$ the default process associated with τ_i. The filtration generated by the process H^i is denoted $\mathbb{H}^i$ and the filtration generated by the two processes H^1, H^2 is $\mathbb{H} = \mathbb{H}^1 \vee \mathbb{H}^2$.

Note that, since $\mathcal{H}_t^i = \sigma(\tau_i \vee t)$, an $\mathcal{H}_t^1 \vee \mathcal{H}_t^2$-measurable random variable is

- a constant on the set $\{t < \tau_1 \wedge \tau_2\}$,
- a $\sigma(\tau_1 \wedge \tau_2)$-measurable random variable on the set $\{\tau_1 \wedge \tau_2 \leq t < \tau_1 \vee \tau_2\}$, i.e., a $\sigma(\tau_1)$-measurable random variable on the set $\{\tau_1 \leq t < \tau_2\}$, and a $\sigma(\tau_2)$-measurable random variable on the set $\{\tau_2 \leq t < \tau_1\}$. We recall that a $\sigma(\tau_1)$-measurable random variable is a Borel function of τ_1.
- a $\sigma(\tau_1, \tau_2)$-measurable random variable (i.e., a Borel function $h(\tau_1, \tau_2)$) on the set $\{\tau_1 \vee \tau_2 \leq t\}$.

To summarize, for a fixed t, any $\mathcal{H}_t^1 \vee \mathcal{H}_t^2$-measurable random variable Z admits a representation as

$$Z = h_0 \mathbb{1}_{\{t<\tau_1\wedge\tau_2\}} + h_1(\tau_1)\mathbb{1}_{\{\tau_1\leq t<\tau_2\}} + h_2(\tau_2)\mathbb{1}_{\{\tau_2\leq t<\tau_1\}} + h(\tau_1, \tau_2)\mathbb{1}_{\{\tau_1\vee\tau_2\leq t\}}.$$

We denote by $G(t, s) = \mathbb{P}(\tau_1 > t, \tau_2 > s)$ the survival probability of the pair (τ_1, τ_2) and we assume that this function is twice differentiable. We denote by $\partial_i G$, the partial derivative of G with respect to the i-th variable, $i = 1, 2$ and by $\partial_{1,2} G$, the second order partial derivative of G. The joint density of the pair (τ_1, τ_2) is denoted by f. Simultaneous defaults are precluded in this framework, i.e., $\mathbb{P}(\tau_1 = \tau_2) = 0$.

Even if the case of two default times is more involved, closed form expressions for the intensities are available. It is important to take into account that the choice of the filtration is very important. Indeed, in general, an $\mathbb{H}^1$-martingale is not an $\mathbb{H}^1 \vee \mathbb{H}^2$-martingale. We shall illustrate this important fact below.

10.4.1. *Intensities*

We present the computation of martingales associated with default times τ_i, $i = 1, 2$, in different filtrations. In particular, we shall obtain the computation of the intensities in various filtrations.

10.4.1.1. *Filtration* $\mathbb{H}^i$

We study, for any fixed i, the Doob–Meyer decomposition of the submartingale H^i in the filtration $\mathbb{H}^i$. In other terms, we compute the $\mathbb{H}^i$-compensator of H^i. From Proposition 10.1, the process

$$M_t^{(i)} := H_t^i - \int_0^{t\wedge\tau_i} \frac{f_i(s)}{G_i(s)} ds \tag{10.3}$$

is an $\mathbb{H}^i$-martingale. Here, $1 - G_i(s) = F_i(s) = \mathbb{P}(\tau_i \le s) = \int_0^s f_i(u)du$. Hence, the process $(1 - H_t^i)\frac{f_i(t)}{G_i(t)}$ is the $\mathbb{H}^i$-intensity of τ^i.
Note that, thanks to Theorem 10.1, any $\mathbb{H}^i$-martingale can be written as a stochastic integral with respect to $M^{(i)}$.

10.4.1.2. *Filtration* $\mathbb{H}$

We recall a result proved in Bielecki et al. [4].

Proposition 10.4. The process M^1 defined as

$$M_t^1 := H_t^1 - \int_0^{t\wedge\tau_1\wedge\tau_2} \frac{\partial_1 G(s,s)}{G(s,s)} ds - \int_{t\wedge\tau_1\wedge\tau_2}^{t\wedge\tau_1} \frac{\partial_{1,2} G(s,\tau_2)}{\partial_2 G(s,\tau_2)} ds$$

is an $\mathbb{H}$-martingale.

The process M^2 defined as

$$M_t^2 := H_t^2 - \int_0^{t\wedge\tau_1\wedge\tau_2} \frac{\partial_2 G(s,s)}{G(s,s)} ds - \int_{t\wedge\tau_1\wedge\tau_2}^{t\wedge\tau_2} \frac{\partial_{1,2} G(\tau_1,s)}{\partial_1 G(\tau_1,s)} ds$$

is an $\mathbb{H}$-martingale.

Proof. The proof relies on some Itô's calculus to obtain the Doob–Meyer decomposition of $\mathbb{Q}(\tau_1 > t|\mathcal{H}_t^2)$. We refer the reader to [4] for details. □

This means that the $\mathbb{H}$-intensity of τ_1 takes into account knowledge of τ_2 and is equal to the deterministic function $-\frac{\partial_1 G(t,t)}{G(t,t)}$ on the set $t < \tau_1 \wedge \tau_2$ and to the random quantity $\varphi(t,\tau_2)$ where $\varphi(t,s) = -\frac{\partial_{1,2}G(t,s)}{\partial_2 G(t,s)}$ on the set $\tau_2 \le t < \tau_1$.

In a closed form, the processes $H_t^i - \int_0^t \lambda_s^i ds$, $i = 1, 2$, are $\mathbb{H}$-martingales, where

$$\begin{aligned}
\lambda_t^1 &= (1 - H_t^1)\left((1 - H_t^2)\frac{-\partial_1 G(t,t)}{G(t,t)} - H_t^2 \frac{\partial_{1,2}G(t,\tau_2)}{\partial_2 G(t,\tau_2)}\right) \\
&= (1 - H_t^1)(1 - H_t^2)\widetilde{\lambda}_t^1 + (1 - H_t^1)H_t^2\widetilde{\lambda}_t^{1|2}(\tau_2), \\
\lambda_t^2 &= (1 - H_t^2)\left((1 - H_t^1)\frac{-\partial_2 G(t,t)}{G(t,t)} - H_t^1 \frac{\partial_{1,2}G(\tau_1,t)}{\partial_1 G(\tau_1,t)}\right) \\
&= (1 - H_t^1)(1 - H_t^2)\widetilde{\lambda}_t^2 + H_t^1(1 - H_t^2)\widetilde{\lambda}_t^{2|1}(\tau_1).
\end{aligned}$$

Here

$$\widetilde{\lambda}_t^i = -\frac{\partial_i G(t,t)}{G(t,t)}, \tag{10.4}$$

$$\widetilde{\lambda}_t^{1|2}(s) = -\frac{\partial_{1,2}G(t,s)}{\partial_2 G(t,s)} = -\frac{f(t,s)}{\partial_2 G(t,s)}, \tag{10.5}$$

$$\widetilde{\lambda}_t^{2|1}(s) = -\frac{\partial_{1,2}G(s,t)}{\partial_1 G(s,t)} = -\frac{f(s,t)}{\partial_1 G(s,t)}.$$

Note that the minus signs in the value of the intensity are due to the fact that G is decreasing with respect to its components, hence the first derivatives are non-positive and the second order derivative $\partial_{1,2}G$ – equal to the density of the pair (τ_1, τ_2) – is non-negative. The quantity $\widetilde{\lambda}_t^1 dt$ is equal to $\mathbb{P}(\tau_1 \in dt | \tau_1 \wedge \tau_2 > t)$, that is the probability that τ_1 occurs in the time interval $[t, t+dt]$, knowing that neither τ_1 nor τ_2 have occurred before time t.

The quantity $\widetilde{\lambda}_t^{1|2}(s) = -\frac{f(t,s)}{\partial_2 G(t,s)}$ evaluated at $s = \tau_2$, represents the value of the default intensity process of τ_1 with respect to the filtration $\mathbb{H}$ on the event $\{\tau_2 < t\}$. This quantity $\widetilde{\lambda}_t^{1|2}(s)dt$ is also the probability that τ_1 occurs in the time interval $[t, t+dt]$, knowing that τ_1 has not occurred before t and that $s = \tau_2$.

Since we are working in the filtration[b] $\mathbb{H}^1 \vee \mathbb{H}^2$ for both martingales M^i, the compensated martingale of the counting process $H_t = H_t^1 + H_t^2 =$

[b] The sum of two martingales in the same filtration is a martingale.

$\sum_{i=1}^2 \mathbb{1}_{\{\tau_i \le t\}}$ is $M_t := H_t - \int_0^t \lambda_s ds$ where

$$\begin{aligned}\lambda_t &= \lambda_t^1 + \lambda_t^2 \\ &= (1-H_t^1)(1-H_t^2)\left(-\frac{\partial_1 G(t,t) + \partial_2 G(t,t)}{G(t,t)}\right) \\ &\quad - (1-H_t^1)H_t^2 \frac{\partial_{1,2}G(t,\tau_2)}{\partial_2 G(t,\tau_2)} - (1-H_t^2)H_t^1 \frac{\partial_{1,2}G(\tau_1,t)}{\partial_2 G(\tau_1,t)}.\end{aligned}$$

10.4.2. *Dynamics of prices of default contingent claims*

In this section, our aim is to find the dynamics of the price of a contingent claim with payoff $h(\tau_1,\tau_2)$. This contains in particular the case of first or second to default claim, with payoff associated with $h(u,v) = \mathbb{1}_{\{u<v\}}\varphi(u)$ or $h(u,v) = \mathbb{1}_{\{u<v<T\}}\psi(v)$. The goal is to find the dynamics of $Z_t := \mathbb{E}(h(\tau_1,\tau_2)|\mathcal{H}_t)$.

The first step is to prove that

$$\begin{aligned}Z_t &= h(\tau_1,\tau_2)H_t^1 H_t^2 + \psi_{1,0}(\tau_1,t)H_t^1(1-H_t^2) \\ &\quad + \psi_{0,1}(t,\tau_2)H_t^2(1-H_t^1) + (1-H_t^1)(1-H_t^2)\psi_{0,0}(t)\end{aligned}$$

with

$$\begin{aligned}\psi_{1,0}(u,t) &:= \frac{-1}{\partial_1 G(u,t)}\int_t^\infty h(u,v)f(u,v)dv, \\ \psi_{0,1}(t,v) &:= \frac{-1}{\partial_2 G(t,v)}\int_t^\infty h(u,v)f(u,v)du, \\ \psi_{0,0}(t) &:= \frac{1}{G(t,t)}\int_t^\infty du \int_t^\infty dv h(u,v)f(u,v).\end{aligned}$$

The proof follows from iterative conditioning and use of Proposition 10.1. We leave the details to the reader.

One notes that, on the one hand, for any function ϕ,

$$\begin{aligned}H_t^1 H_t^2 \phi(\tau_1,\tau_2) &= \int_0^t dH_u^1 \int_0^t dH_v^2 \phi(u,v), \\ H_t^1 \phi(\tau_1,t) &= \int_0^t \phi(u,t)dH_u^1,\end{aligned}$$

so that, using integration by parts formula and re-arranging the terms

$$
\begin{aligned}
dZ_t = & \Big(\big(h(t,\tau_2) - \psi_{0,1}(t,\tau_2)\big) H_t^2 + \big(\psi_{1,0}(t,t) - \psi_{0,0}(t)\big)(1 - H_t^2) \Big) dH_t^1 \\
& + \Big(\big(h(\tau_1,t) - \psi_{1,0}(\tau_1,t)\big) H_t^1 + \big(\psi_{0,1}(t,t) - \psi_{0,0}(t)\big)(1 - H_t^1) \Big) dH_t^2 \\
& + \Big((1 - H_t^2) \int_0^t \partial_2 \psi_{1,0}(u,t) dH_u^1 + (1 - H_t^1) \int_0^t \partial_1 \psi_{0,1}(t,v) dH_v^2 \\
& \quad + (1 - H_t^1)(1 - H_t^2) \frac{d}{dt} \psi_{0,0}(t) \Big) dt.
\end{aligned}
$$

On the other hand, one checks with easy computation that

$$
\begin{aligned}
\partial_1 \psi_{0,1}(t,v) &= \lambda^{1|2}(t,v)\left(\psi_{0,1}(t,v) - h(t,v)\right), \\
\partial_2 \psi_{1,0}(u,t)) &= \lambda^{2|1}(u,t)\left(\psi_{1,0}(u,t) - h(u,t)\right), \\
\frac{d}{dt} \psi_{0,0}(t) &= (\widetilde{\lambda}_1(t) + \widetilde{\lambda}_2(t)) \psi_{0,0}(t) \\
& \qquad + \frac{1}{G(t,t)} (\partial_1 G(t,t) \psi_{1,0}(t,t) + \partial_2 G(t,t) \psi_{0,1}(t,t)).
\end{aligned}
$$

It follows that

$$
\begin{aligned}
dZ_t = & \Big(\big(h(t,\tau_2) - \psi_{0,1}(t,\tau_2)\big) H_t^2 + \big(\psi_{1,0}(t,t) - \psi_{0,0}(t)\big)(1 - H_t^2) \Big) dM_t^1 \\
& + \Big(\big(h(\tau_1,t) - \psi_{1,0}(\tau_1,t)\big) H_t^1 + \big(\psi_{0,1}(t,t) - \psi_{0,0}(t)\big)(1 - H_t^1) \Big) dM_t^2.
\end{aligned}
$$

10.4.3. *Dynamics of CDS prices*

Let us now examine the valuation of a single-name CDS written on name 1, in the case of null interest rate. Our aim is to show that the dynamics of this CDS will be affected by the information on τ_2: when τ_2 occurs, the intensity of τ_1 changes, and this will change the parameters of the price dynamics. We reproduce some results appearing in Bielecki et al. [4].

We consider a CDS

- with a constant spread κ_1
- which delivers $\delta(\tau_1)$ at time τ_1 if $\tau_1 < T$, where δ is a deterministic function.

The value of the CDS takes the form

$$
V_t(\kappa_1) = \widetilde{V}_t(\kappa_1) \mathbb{1}_{\{t < \tau_1 \wedge \tau_2\}} + \widehat{V}_t(\kappa_1) \mathbb{1}_{\{\tau_1 \wedge \tau_2 \le t < \tau_1\}} .
$$

First, we restrict our attention to the case $t < \tau_1 \wedge \tau_2$.

Proposition 10.5. On the set $\{t < \tau_1 \wedge \tau_2\}$, the value of the CDS is

$$\widetilde{V}_t(\kappa_1) = \frac{1}{G(t,t)} \left(- \int_t^T \delta(u) \partial_1 G(u,t)\, du - \kappa_1 \int_t^T G(u,t)\, du \right).$$

Proof. The value $V(\kappa_1)$ of this CDS, computed in the filtration $\mathbb{H}$, i.e., taking care on the information on the second default contained in that filtration, is

$$V_t(\kappa_1) = \mathbb{1}_{\{t<\tau_1\}} \mathbb{E}\left(\delta(\tau_1) \mathbb{1}_{\{\tau_1 \le T\}} - \kappa_1((T \wedge \tau_1) - t) | \mathcal{H}_t \right).$$

Let us denote by $\tau = \tau_1 \wedge \tau_2$ the first default time. Then, $\mathbb{1}_{\{t<\tau\}} V_t(\kappa_1) = \mathbb{1}_{\{t<\tau\}} \widetilde{V}_t(\kappa_1)$, where

$$\begin{aligned}
\widetilde{V}_t(\kappa_1) &= \frac{1}{\mathbb{P}(\tau > t)} \mathbb{E}\left(\delta(\tau_1) \mathbb{1}_{\{\tau_1 \le T\}} \mathbb{1}_{\{t<\tau\}} - \kappa_1((T \wedge \tau_1) - t) \mathbb{1}_{\{t<\tau\}} \right) \\
&= \frac{1}{G(t,t)} \mathbb{E}\left(\delta(\tau_1) \mathbb{1}_{\{\tau_1 \le T\}} \mathbb{1}_{\{t<\tau\}} - \kappa_1((T \wedge \tau_1) - t) \mathbb{1}_{\{t<\tau\}} \right) \\
&= \frac{1}{G(t,t)} \left(\int_t^T \delta(u) \mathbb{P}(\tau_1 \in du, \tau_2 > t) \right. \\
&\quad - \kappa_1 \int_t^T (u-t) \mathbb{P}(\tau_1 \in du, \tau_2 > t) \\
&\quad \left. -(T-t)\kappa_1 \int_T^\infty \mathbb{P}(\tau_1 \in du, \tau_2 > t) \right).
\end{aligned}$$

For other terms, using the integration by parts formula, we end up with

$$\widetilde{V}_t(\kappa_1) = \frac{1}{G(t,t)} \left(- \int_t^T \delta(u) \partial_1 G(u,t)\, du - \kappa_1 \int_t^T G(u,t)\, du \right).$$

□

Proposition 10.6. On the event $\{\tau_2 \le t < \tau_1\}$, the CDS price is given by $V_t(\kappa_1) = V_t^{1|2}(\tau_2)$ where

$$V_t^{1|2}(s) = \frac{1}{\partial_2 G(t,s)} \left(- \int_t^T \delta(u) f(u,s)\, du - \kappa_1 \int_t^T \partial_2 G(u,s)\, du \right).$$

Proof. One has

$$\begin{aligned}
V_t(\kappa_1) = \widehat{V}_t(\kappa_1) &= \mathbb{E}\left(\delta(\tau_1) \mathbb{1}_{\{\tau_1 \le T\}} - \kappa_1((T \wedge \tau_1) - t) | \sigma(\tau_2) \right) \\
&= \frac{1}{\partial_2 G(t,\tau_2)} \left(- \int_t^T \delta(u) f(u,\tau_2)\, du - \kappa_1 \int_t^T \partial_2 G(u,\tau_2)\, du \right).
\end{aligned}$$

□

In the financial interpretation, $V_t^{1|2}(s)$ is the market price at time t of a CDS on the first credit name, under the assumption that the default τ_2 occurs at time s and the first name has not yet defaulted (recall that simultaneous defaults are excluded, since we have assumed that G is differentiable).

The price of a CDS is $V_t = \widehat{V}_t(\kappa_1)\mathbb{1}_{\{t<\tau_2\wedge\tau_1\}} + \widehat{V}_t(\kappa_1)\mathbb{1}_{\{\tau_2\wedge\tau_1\le t<\tau_1\}}$. Differentiating the deterministic function which gives the value of the CDS, we obtain

$$d\widetilde{V}_t(\kappa_1) = \left(\left(\widetilde{\lambda}_1(t) + \widetilde{\lambda}_2(t)\right)\widetilde{V}_t(\kappa_1) + \kappa_1 - \widetilde{\lambda}_1(t)\delta(t) - \widetilde{\lambda}_2(t)V_t^{1|2}(t)\right)\,dt,$$

where for $i = 1, 2$ the function $\widetilde{\lambda}_i(t)$ is the (deterministic) *pre-default intensity* of τ_i given in (10.4) and

$$d\widehat{V}_t(\kappa_1) = \left(\widetilde{\lambda}_t^{1|2}(\tau_2)\left(\widehat{V}_t(\kappa_1) - \delta(t)\right) + \kappa_1\right)\,dt$$

where $\widetilde{\lambda}_t^{1|2}(u)$ is given in (10.5).

Proposition 10.7. The price of a CDS follows

$$\begin{aligned} dV_t(\kappa_1) = &(1 - H_t^1)(1 - H_t^2)(\kappa_1 - \delta(t)\widetilde{\lambda}^1(t))dt \\ &+(1 - H_t^1)H_t^2(\kappa_1 - \delta(t)\widetilde{\lambda}_t^{1|2}(\tau_2))dt \\ &-V_{t-}(\kappa_1)dM_t^1 + (1 - H_t^1)(V_t^{1|2}(t) - V_{t-}(\kappa_1))dM_t^2. \end{aligned} \tag{10.6}$$

Proof. Using integration by parts formula for $V_t(\kappa_1) = \widetilde{V}_t(\kappa_1)(1 - H_t^1)(1 - H_t^2) + \widehat{V}_t(\kappa_1)(1 - H_t^1)H_t^2$, one obtains

$$\begin{aligned} dV_t(\kappa_1) = &(1 - H_t^1)(1 - H_t^2)d\widetilde{V}_t(\kappa_1) + (1 - H_t^1)H_t^2 d\widehat{V}_t(\kappa_1) - V_{t-}(\kappa_1)dH_t^1 \\ &+(1 - H_t^1)(V_t^{1|2}(t) - \widetilde{V}_t(\kappa_1))dH_t^2, \end{aligned}$$

which leads to the result after light computations. □

Comment. As for a single name CDS, the quantity $-\delta(t)\widetilde{\lambda}^1(t)$ corresponds to the dividend δ to be paid at time t with probability $\widetilde{\lambda}^1(t)dt$ on the set $t < \tau_1 \wedge \tau_2$ and $-\delta(t)\widetilde{\lambda}_t^{1|2}$ corresponds to the dividend δ to be paid at time t with probability $\widetilde{\lambda}_t^{1|2}dt$ on the set $\tau_2 < t < \tau_1$. The quantity $V_t^{1|2}(t) - \widetilde{V}_t(\kappa_1)$ represents the jump in the value of the CDS, when default τ_2 occurs at time t.

The cumulative dividend price of the CDS is

$$V_t^{\text{cum}}(\kappa_1) = \mathbb{E}(\delta(\tau_1)\mathbb{1}_{\{\tau_1\le T\}} - \kappa_1(T\wedge\tau_1)|\mathcal{H}_t).$$

It follows that

$$dV_t^{\text{cum}}(\kappa_1) = dV_t(\kappa_1) + \delta(t)dH_t^1 - \kappa_1(1 - H_t^1)dt,$$

hence, since the cumulative price is a martingale

$$dV_t(\kappa_1) = dm_t - \delta(t)\lambda_t^1 dt + \kappa_1(1 - H_t^1)dt,$$

where $dm_t = dV_t^{\mathrm{cum}}(\kappa_1) - \delta(t)dM_t^1$. This is an easy way to obtain the drift term in Equation (10.6). It turns out that the cum-dividend CDS price process has the following dynamics

$$\begin{aligned} dV_t^{\mathrm{cum}}(\kappa_1) &= (1 - H_t^1)(1 - H_t^2)(\kappa_1 - \delta(t)\widetilde{\lambda}^1(t))dt \\ &\quad +(1 - H_t^1)H_t^2(\kappa_1 - \delta(t)\widetilde{\lambda}_t^{1|2})dt - V_{t-}(\kappa_1)dM_t^1 \\ &\quad +(1 - H_t^1)(V_t^{1|2}(t) - V_{t-}(\kappa_1))dM_t^2 + \delta(t)dH_t^1 - \kappa_1(1 - H_t^1)dt \\ &= (\delta(t) - V_{t-}(\kappa_1))dM_t^1 + (1 - H_t^1)(V_t^{1|2}(t) - V_{t-}(\kappa_1))dM_t^2. \end{aligned}$$

10.4.4. *CDSs as hedging assets*

Assume now that a CDS written on τ_2 is also traded in the market. We denote by $\delta^i, i = 1, 2$ the recovery (assumes to be deterministic) and $V^i, i = 1, 2$ the prices of the two CDSs with spreads κ_i. We assume that these CDSs are traded in the market. Since the CDS are paying dividends, a self financing strategy consisting in ϑ^i shares of CDS's has value $X_t = \vartheta_t^0 + \vartheta_t^1 V_t^1 + \vartheta_t^2 V_t^2$ and dynamics

$$\begin{aligned} dX_t &= \vartheta_t^1 dV_t^{1,cum} + \vartheta_t^2 dV_t^{2,cum} \\ &= \vartheta_t^1 \left((\delta^1(t) - V_{t-}^1)dM_t^1 + (1 - H_t^1)(V_t^{1|2}(t) - \widetilde{V}_t^1)dM_t^2 \right) \\ &\qquad + \vartheta_t^2 \left((\delta^2(t) - V_{t-}^2)dM_t^2 + (1 - H_t^2)(V_t^{2|1}(t) - \widetilde{V}_t^2)dM_t^1 \right) \\ &= \left(\vartheta_t^1(\delta^1(t) - V_{t-}^1) + \vartheta_t^2(1 - H_t^2)(V_t^{2|1}(t) - \widetilde{V}_t^2) \right) dM_t^1 \\ &\qquad + \left(\vartheta_t^1(1 - H_t^1)(V_t^{1|2}(t) - \widetilde{V}_t^1) + \vartheta_t^2(\delta^2(t) - V_{t-}^2) \right) dM_t^2. \end{aligned}$$

The position ϑ_t^0 in the savings account (which is worth a constant in this zero interest-rate set-up) is necessary to make the strategy self-financing. Note that, due to liquidity issue, one needs to use rolling CDS-s in practice so as to construct market feasible hedging portfolios. We refer the reader to [3] for more details on rolling-CDS. Mathematically, there is little difference between portfolios consisting of CDS-s[c], and portfolios consisting of rolling CDS-s, so a portfolio consisting of CDS-s is chosen for illustrative purposes.

[c] A rolling-CDS has a fixed maturity T and a time-dependent contractual spread equal to the current CDS market spread.

Let $A \in \mathcal{H}_T$ be a terminal payoff with price $A_t = \mathbb{E}(A \mid \mathcal{H}_t)$, then from Theorem 10.1 there exist predictable processes π^1 and π^2 such that

$$A_t = \mathbb{E}(A) + \int_0^t \pi_s^1 dM_s^1 + \int_0^t \pi_s^2 dM_s^2 .$$

In order to hedge that claim, it remains to solve the linear system

$$\begin{aligned} \vartheta_t^1(\delta^1(t) - V_{t-}^1) + \vartheta_t^2(1 - H_t^2)(V_t^{2|1}(t) - \widetilde{V}_t^2) &= \pi_t^1, \\ \vartheta_t^1(1 - H_t^1)(V_t^{1|2}(t) - \widetilde{V}_t^1) + \vartheta_t^2(\delta^2(t) - V_{t-}^2) &= \pi_t^2. \end{aligned}$$

Hence, on the set $t < \tau_1 \wedge \tau_2$, noting that $V_t^i = \widetilde{V}_t^i$ on that set,

$$\begin{aligned} \vartheta_t^1 &= \frac{\pi_t^1(\delta^2(t) - \widetilde{V}_t^2) - \pi_t^2(V_t^{2|1}(t) - \widetilde{V}_t^2)}{(\delta^1(t) - \widetilde{V}_t^1)(\delta^2(t) - \widetilde{V}_t^2) - (V_t^{1|2} - \widetilde{V}_t^1)(V_t^{2|1} - \widetilde{V}_t^2)}, \\ \vartheta_t^2 &= \frac{\pi_t^2(\delta^1(t) - \widetilde{V}_t^1) - \pi_t^1(V_t^{1|2} - \widetilde{V}_t^1)}{(\delta^1(t) - \widetilde{V}_t^1)(\delta^2(t) - \widetilde{V}_t^2) - (V_t^{1|2} - \widetilde{V}_t^1)(V_t^{2|1}(t) - \widetilde{V}_t^2)}. \end{aligned}$$

On the set $\tau_1 < t < \tau_2$

$$\vartheta_t^1 = \frac{\pi_t^1(\delta^2(t) - V_t^2) - \pi_t^2(V_t^{2|1} - V_t^2)}{(\delta^1(t) - V_t^1)(\delta^2(t) - V_t^2)}, \quad \vartheta_t^2 = \frac{\pi_t^2}{\delta^2(t) - V_t^2}.$$

On the set $\tau_2 < t < \tau_1$

$$\vartheta_t^1 = -\frac{\pi_t^1}{\delta^1(t) - V_t^1}, \quad \vartheta_t^2 = \frac{\pi_t^2(\delta^1(t) - V_t^1) - \pi_t^1(V_t^{1|2} - V_t^2)}{(\delta^1(t) - V_t^1)(\delta^2(t) - V_t^2)}.$$

On the set $\tau_1 \vee \tau_2 < t$

$$\vartheta_t^1 = \frac{\pi_t^1}{\delta^1(t) - V_t^1}, \quad \vartheta_t^2 = \frac{\pi_t^2}{\delta^2(t) - V_t^2}.$$

As we saw above, for the case $A = h(\tau_1, \tau_2)$, one has a closed form for the coefficients π:

$$\begin{aligned} \pi_t^1 &= \big(h(t, \tau_2) - \psi_{0,1}(t, \tau_2)\big)H_t^2 + \big(\psi_{1,0}(t, t) - \psi_{0,0}(t)\big)(1 - H_t^2), \\ \pi_t^2 &= \big(h(\tau_1, t) - \psi_{1,0}(\tau_1, t)\big)H_t^1 + \big(\psi_{0,1}(t, t) - \psi_{0,0}(t)\big)(1 - H_t^1) \end{aligned}$$

10.4.5. *Multidefault setting*

The same methodology can be applied in a multidefault setting. The only difficulty is that one has to rank the defaults and to distinguish the various possibilities. For example, for three defaults, the dynamics of the cumdividend price of the CDS written on the default τ_1 is

$$\begin{aligned} dV_t^{\text{cum}} &= (\delta^1(t) - V_{t-})dM_t^1 \\ &\quad + (1 - H_t^1)\left((1 - H_t^3)V_t^{1|2}(t) + H_t^3 V_t^{1|32}(t, \tau_3) - V_{t-}\right) dM_t^2 \\ &\quad + (1 - H_t^1)\left((1 - H_t^2)V_t^{1|3}(t) + H_t^2 V_t^{1|23}(\tau_2, t) - V_{t-}\right) dM_t^3, \end{aligned}$$

where V_t is the (ex-dividend) price of the CDS at time t, $V_t^{1|2}(u)$ is the price of the CDS on the set $\{\tau_2 = u\}$ for $u < t < \tau_3$, $V_t^{1|3}(v)$ is the price of the CDS on the set $\{\tau_3 = v\}$ for $v < t < \tau_2$, $V_t^{1|23}(u, v)$ is the price of the CDS, on the set $\{\tau_2 = u, \tau_3 = v\}$ for $u < v < t$ and $V_t^{1|32}(u, v)$ is the price of the CDS, on the set $\{\tau_2 = u, \tau_3 = v\}$ for $v < u < t$.

10.5. Particular case: Ordered defaults

In this section, we consider the particular case where default times are ordered, i.e., $\tau_1 < \tau_2 < \cdots < \tau_n$. Recall that simultaneous defaults are precluded in our setting. Then, single-name CDS k offers credit protection against the kth default occurring in the portfolio so that it can be viewed as a kth-to-default swap. We first consider a setting with two names only, then we investigate the hedging of loss derivatives written on a multivariate n-dimensional credit portfolio.

10.5.1. *Dynamics of CDS prices in a two defaults setting*

Let us now assume that $\tau_1 < \tau_2$, a.s. In that case, $G(t, s) = G(t, t)$ for $s \le t$, hence the martingale M^1 defined in Proposition 10.4 simplifies:

$$M_t^1 = H_t^1 - \int_0^{t\wedge\tau_1} \frac{\partial_1 G(s, s)}{G(s, s)} ds = H_t^1 - \int_0^{t\wedge\tau_1} \frac{f_1(s)}{G_1(s)} ds$$

where

$$G_1(s) = \mathbb{P}(\tau_1 > s) = G(s, s) = \int_s^\infty f_1(u)du = \int_s^\infty \partial_1 G(u, u)du\,.$$

The $\mathbb{H}$-martingale M^1 is $\mathbb{H}^1$-adapted, hence is an $\mathbb{H}^1$-martingale. From Theorem 10.1 applied to the case $n = 1$, any $\mathbb{H}^1$-martingale is a stochastic integral w.r.t. M^1, hence is a $\mathbb{H}$-martingale. Furthermore, the intensity of τ_2 vanishes on the set $t < \tau_1$ and

$$M_t^2 = H_t^2 - \int_{t\vee\tau_1}^{t\wedge\tau_2} \frac{f(\tau_1, s)}{\partial_1 G(\tau_1, s)} ds = H_t^2 - \int_0^t \mathbb{1}_{\{\tau_1 < s < \tau_2\}} \frac{f(\tau_1, s)}{\partial_1 G(\tau_1, s)} ds\,.$$

Proposition 10.8. Let $V^i, i = 1,2$ be the price of a CDS on name i, with contractual spread κ_i and payment at default given by a deterministic function δ_i. The $\mathbb{H}$-dynamics of V^1 are

$$dV_t^1 = -V_{t-}^1 dM_t^1 + (1 - H_t^1)(\kappa_1 - \delta_1(t)\widetilde{\lambda}_1(t))dt \tag{10.7}$$

with $\widetilde{\lambda}_1(t) = \frac{f_1(t)}{G_1(t)}$. The $\mathbb{H}$-dynamics of V^2 are

$$\begin{aligned} dV_t^2 &= -V_{t-}^2 dM_t^2 \\ &\quad +(1 - H_t^2)\kappa_2 dt - (1 - H_t^2)H_t^1 \delta_2(t)\widetilde{\lambda}_t^{2|1}(\tau_1)dt + (V_t^{2|1}(t) - V_{t-}^2)dM_t^1 \,. \end{aligned} \tag{10.8}$$

Proof. Apply Proposition 10.3 or (10.6) to obtain (10.7), and (10.6) to obtain (10.8). □

10.5.2. *Multidefault setting*

Let G be the survival function of the joint defaults, assumed to be differentiable

$$G(t_1, \ldots, t_n) = \mathbb{P}(\tau_1 > t_1, \ldots, \tau_n > t_n)$$

and G_j be the survival function of the j-first defaults

$$G_j(t_1, \ldots, t_j) = \mathbb{P}(\tau_1 > t_1, \ldots, \tau_j > t_j) \,.$$

We shall denote by f the density of the n-uple $(\tau_i, i \leq n)$ and by f_j the density of the j-uple $(\tau_i, i \leq j)$. Since the defaults are ordered, setting $t_j^1 = t_1, \ldots, t_j$ one has

$$\begin{aligned} G_j(t_j^1) &:= \mathbb{P}(\tau_1 > t_1, \ldots, \tau_j > t_j) \\ &= \mathbb{P}(\tau_1 > t_1, \ldots, \tau_j > t_j, \tau_{j+1} > t_j, \ldots, \tau_n > t_j) = G(t_j^1, t_j, \ldots, t_j). \end{aligned}$$

From an immediate extension of Proposition 10.4, noting that the density of the defaults is null outside the set $\{t_1 \leq t_2 \leq \cdots \leq t_n\}$, the fundamental martingales are

$$M_t^j = H_t^j - \int_{t \vee \tau_{j-1}}^{t \wedge \tau_j} \widetilde{\lambda}_s^{j|j-1}(\tau_1, \ldots, \tau_{j-1}) ds$$

where

$$\widetilde{\lambda}_t^{j|j-1}(t_{j-1}^1) = -\frac{\partial_{1,j} G_j(t_{j-1}^1, t)}{\partial_{1,j-1} G_j(t_{j-1}^1, t)}$$

and $\partial_{1,j} = \partial_1 \ldots \partial_j$.

Proposition 10.9. If V^i is the price process of a CDS with maturity T, written on the i-th default, with spread κ_i and payment at default given by a deterministic function δ_i, then

$$\begin{aligned} dV_t^i &= -V_{t-}^i dM_t^i - (1 - H_t^i) H_t^{i-1} \delta_i(t) \widetilde{\lambda}_t^{i|i-1}(\tau_1, \ldots, \tau_{i-1}) dt \\ &\quad + \sum_{j=1}^{i-1} (V_t^{i|j}(\tau_1, \ldots, \tau_j) - V_{t-}^i) dM_t^j + (1 - H_t^i) \kappa_i dt, \end{aligned} \tag{10.9}$$

where

$$V_t^{i|j}(t_j^1) = \frac{-\int_t^T \delta_i(u) f_{j+1}(t_j^1, u)\, du - \kappa_i \int_t^T \partial_{1,j} G_{j+1}(t_j^1, u)\, du}{\partial_{1,j} G_j(t_j^1)}.$$

10.5.3. *Hedging of a loss*

In order to hedge the payoff B, one proceeds in two steps. The first step is to compute the martingale representation of $\mathbb{E}(B|\mathcal{H}_t)$, i.e., identify the predictable processes π such that

$$\mathbb{E}(B|\mathcal{H}_t) = \mathbb{E}(B) + \sum_{j=1}^{n} \int_0^t \pi_s^j dM_s^j .$$

We denote by D^i the dividend part associated with the CDS written on τ_i. A self-financing strategy with value

$$V_t = \vartheta_t^0 + \sum_{i=1}^{n} \vartheta_t^i V_t^i$$

satisfies

$$\begin{aligned} dV_t &= \sum_{i=1}^{n} \vartheta_t^i (dV_t^i + dD_t^i) \\ &= \sum_{i=1}^{n} \vartheta_t^i \left((\delta^i(t) - V_{t-}^i) dM_t^i + \sum_{j=1}^{i-1} (V_t^{i|j} - V_{t-}^i) dM_t^j \right) \\ &= \sum_{j=1}^{n} dM_t^j \sum_{i=j}^{n} \vartheta_t^i (V_t^{i|j} - V_{t-}^i) \end{aligned} \tag{10.10}$$

where we set $V_t^{i|i} = \delta^i(t)$. It remains to solve the linear system (with unknown ϑ)

$$\sum_{i=j}^{n} \vartheta_t^i (V_t^{i|j} - V_{t-}^i) = \pi_t^j,\ j = 1, \ldots, n .$$

As an example, we now compute the conditional law of the loss, i.e., $\mathbb{E}(f(L_T)|\mathcal{H}_t)$ where $L_T = \sum_{k=1}^n \mathbb{1}_{\{\tau_k \le T\}}$. Let $B_t^k(T) = \mathbb{E}(\mathbb{1}_{\{T<\tau_k\}}|\mathcal{H}_t)$ (or simply B_t^k) be the price of a defaultable zero-coupon written on the k-th default, with maturity T, then

$$\mathbb{E}(f(L_T)|\mathcal{H}_t) = \sum_{k=1}^{n} f(k)\left(B_t^{k+1} - B_t^k\right) = f(n) + \sum_{k=1}^{n} B_t^k(f(k-1) - f(k)).$$

Obviously, setting $\tau_0 = 0$

$$\begin{aligned} B_t^k &= \sum_{j=1}^{k} \mathbb{1}_{\{\tau_{j-1}\le t<\tau_j\}} \mathbb{E}(\mathbb{1}_{\{T<\tau_k\}}|\mathcal{H}_t) \\ &= \sum_{j=1}^{k} \mathbb{1}_{\{\tau_{j-1}\le t<\tau_j\}} \frac{\mathbb{P}(T<\tau_k, t<\tau_j|\mathcal{H}_t^{j-1})}{\mathbb{P}(t<\tau_j|\mathcal{H}_t^{j-1})}. \end{aligned}$$

Now, on the set $\tau_{j-1} \le t$

$$\mathbb{P}(T<\tau_k, t<\tau_j|\mathcal{H}_t^{j-1}) = \Phi_{k,j-1}(\tau_1, \ldots, \tau_{j-1}, t, T)$$

where, for $j \le k-1$

$$\begin{aligned} \Phi_{k,j-1}(t_1,\ldots,t_{j-1},t,T) &= \frac{\mathbb{P}(t<\tau_j, T<\tau_k, \tau_1 \in dt_1, \ldots, \tau_{j-1} \in dt_{j-1})}{\mathbb{P}(\tau_1 \in dt_1, \ldots, \tau_{j-1} \in dt_{j-1})} \\ &= \frac{\partial_{1,j-1} G_k(t_{j-1}^1, t, \ldots, t, T)}{\partial_{1,j-1} G_k(t_{j-1}^1, \ldots, t_{j-1})}. \end{aligned}$$

On the set $\tau_{k-1} \le t$ (for $j = k$)

$$\frac{\mathbb{P}(T<\tau_k, t<\tau_k|\mathcal{H}_t^{k-1})}{\mathbb{P}(t<\tau_k|\mathcal{H}_t^{k-1})} = \frac{\mathbb{P}(T<\tau_k|\mathcal{H}_t^{k-1})}{\mathbb{P}(t<\tau_k|\mathcal{H}_t^{k-1})} = \frac{\partial_{1,k-1} G_k(t_{k-1}^1, T)}{\partial_{1,k-1} G_k(t_{k-1}^1, t)}.$$

Since

$$dB_t^k = -B_{t-}^k dM_t^k + \sum_{j=1}^{k-1} \nu_t^{j,k} dM_t^j = \sum_{j=1}^{k} \nu_t^{j,k} dM_t^j \tag{10.11}$$

where, for $j < k$, we have set $\nu_t^{j,k} = \frac{\Phi_{k,j}(\tau_1,\ldots,\tau_{j-1},t,t,T)}{\Phi_{k,j}(\tau_1,\ldots,\tau_{j-1},t,t,t)} - \frac{\Phi_{k,j-1}(\tau_1,\ldots,\tau_{j-1},t,T)}{\Phi_{k,j-1}(\tau_1,\ldots,\tau_{j-1},t,t)}$ and $\nu_t^{k,k} = -B_{t-}^k$. It follows that, setting $X_t = \mathbb{E}(f(L_T)|\mathcal{H}_t)$, one has

$$dX_t = \sum_{j=1}^{n} dM_t^j \sum_{k=j}^{n} (f(k-1) - f(k))\nu_t^{i,k}$$

and the hedging strategy for the contingent claim $f(L_T)$ is the solution ϑ of the triangular system

$$\sum_{i=j}^{n} \vartheta_t^i (V_t^{i|j} - V_{t-}^i) = \sum_{k=j}^{n} (f(k-1) - f(k))\nu_t^{j,k}, j = 1, \ldots, n.$$

Conclusion

We investigate a quite general pure jump setting where the density of joint default is known. We compute the default intensities in the filtration of all the default times. In particular, at each instant when a default event occurs, default intensities of non-defaulted names are dynamically updated. This leads to a dependence structure among default times which is regularly updated as defaults arrive. We have seen that the hedging of loss derivatives such as CDO tranches or basket default swaps can be fully described in this framework with no Markovian assumption. The hedging strategies with respect to single-name CDS can be derived analytically in a two-defaults setting. Even if similar ideas can be exploited in higher dimension, the construction of dynamic hedging strategies would involve very cumbersome computations if one wishes to consider all possible default scenarios. Interestingly, in the particular case of ranked default times, the hedging issue can be solved explicitly. All these results admit an extension to the case where there exists a reference filtration, leading to intensity processes that depend on a factor process driven by a Brownian motion.

References

[1] Basel Committee on Banking Supervision (Dec 2010) *Revisions to the Basel II market risk framework.*

[2] Bielecki, T.R. and Rutkowski, M. (2001), *Credit risk: Modelling Valuation and Hedging*, Springer Verlag, Berlin.

[3] Bielecki, T.R., Jeanblanc, M., and Rutkowski, M. (2008) *Pricing and trading credit default swaps in a hazard process model.* Annals of Applied Probability 18, 2495–2529.

[4] Bielecki, T.R., Jeanblanc, M., and Rutkowski, M. (2009) *Credit Risk Modelling*, Lectures Notes, Osaka University.

[5] Brémaud, P. (1981) *Point Processes and Queues: Martingale Dynamics*, Springer Verlag.

[6] Cousin, A. and Laurent, J.-P. (2010) *Dynamic hedging of synthetic CDO tranches: bridging the gap between theory and practice*, in Credit Risk Frontiers: Subprime crisis, Pricing and Hedging, CVA, MBS, Ratings and Liquidity, Wiley.

[7] Cousin, A., Jeanblanc, M. and Laurent, J.-P. (2010) *Hedging CDO tranches in a Markovian environment*, in Paris-Princeton Lectures in Mathematical Finance, Lecture Notes in Mathematics, Springer.
[8] Laurent, J.P., Cousin, A. and Fermanian, J.D (2011) *Hedging default risk for CDOs in Markovian contagion models*, Quantitative Finance 11(12), 1773-1791.
[9] Dellacherie, C. (1972) *Capacités et processus stochastiques*, Springer.
[10] Elliott, R.J. (1982) *Stochastic Calculus and Applications*, Springer, Berlin
[11] Kusuoka, S. (1999), *A remark on default risk models*, Adv. Math. Econ. 1, 69-82.

Chapter 11

New analytic approximations for pricing spread options

John van der Hoek and Małgorzata W. Korolkiewicz*

Centre for Industrial and Applied Mathematics,
School of Mathematics and Statistics,
University of South Australia,
GPO Box 2471, Adelaide, SA 5001, Australia

We propose new analytic approximations for the pricing of spread options. A *spread option* is an option whose payoff depends on the price spread between two underlying assets. Let $S_1(t)$ and $S_2(t)$ be the prices of the two assets. The payoff of a European spread call option with a strike price K and maturity T is $(S_1(T) - S_2(T) - K)^+$. Valuation of such options, widely used in a number of financial markets, is fundamentally different from the valuation of other types of options. Since the behaviour of the spread is affected by the behaviour of two traded assets, it cannot be modelled as if it were simply a single asset. Moreover, the value of an option on a spread depends critically on the correlation between the price changes of the two underlying assets. Working under a risk-neutral probability measure, we assume that the underlying asset price processes are geometric Brownian motions with constant non-zero correlation. Using a perturbation expansion of the solution to the spread option price partial differential equation, we derive analytic formulae for approximations up to second order. We show that Kirk's (1995) approximation for spread options on energy futures, commonly used by traders, is an exact solution to a related problem, and corresponds to the $n = 0$ term in our perturbation expansion. Results from a numerical implementation of our approximations to spread option prices are then compared against those obtained by other approximation methods available in the literature.

---

*The first author is a long time research collaborator and the second author a former PhD student. Both authors wish Robert Elliott many happy returns.

Contents

11.1. Introduction

Spread options are a fundamental class of derivative contracts written on multiple assets, and are widely used in a range of financial markets. A European *spread option* is an option with a payoff that depends on the price spread between two underlying assets. Let $S_1(t)$ and $S_2(t)$ be the prices of the two assets. The payoff to a spread call option with a strike price K and maturity T is $(S_1(T) - S_2(T) - K)^+$. The valuation of options on spreads is fundamentally different from the valuation of other types of options. First, the behaviour of the spread is affected by the behaviour of two traded contracts; a spread cannot be simply modelled as if it were a single asset. Second, the spread value may become negative and third, the value of an option on a spread depends critically on the correlation between the price changes of the two underlying assets.

The literature on spread options is quite extensive, and is reviewed by Carmona and Durrleman [3] who explore applications of spread options beyond the case of equities. Closed formulas for pricing of spread options are known only for a limited set of cases. In the Bachelier stock model, explored for example in Shimko [15], Poitras [13] and Schaefer [14], the dynamics of the underlying assets are modelled as arithmetic Brownian motions, and in this case a Black-Scholes type formula is obtained for a spread option. In the special case when $K = 0$ and the underlying assets are geometric Brownian motions, the price of a spread option is given by the Margrabe [12] formula. In the important case where $K > 0$, no explicit pricing formula is known and approximation methods are used instead. One approach is to employ numerical methods such as Monte Carlo simulation, numerical integration and fast Fourier transforms. See for example Dempster and Hong [5], as well as Hurd and Zhou [10], where numerical integration methods based on double fast Fourier transforms (FFT) are employed and prove to be efficient for geometric Brownian motion and certain other more general asset price processes. Another stream of research develops analytical methods applicable to log-normal models. Kirk [11] presented an analytical approximation that performs well in practice and is commonly utilized by practitioners. Carmona and Durrleman [4] obtain an analytical approximation formula using a family of lower bounds for the spread option price and extend these results to derive approximate values for the Greeks. Deng, Li and Zhou [6] develop another systematic analytical approximation scheme and compare its performance to that of other available methods. Bjerksund and Stensland [2] in turn derive their spread option formula using a specific exercise strategy as the starting point and compare accuracy of their method to the Carmona and Durrleman [4] lower bound procedure. In Alexander and Venkatramanan [1], analytic approximations are obtained by representing spread options as compound exchange options. As yet, there is no preferred approach that is accurate, efficient and flexible enough to apply in general models. Numerical methods tend to be more accurate, but their computing times can be much longer compared to analytical approximation methods.

In this paper we propose a new approach to deriving analytical approximations for a spread option price in the geometric Brownian motion case. Working under a risk-neutral probability measure, we assume that the underlying asset price processes are geometric Brownian motions with constant non-zero correlation. Using the perturbation option pricing techniques of [7], we derive analytic formulae for approximations to the spread option price up to second order. We show that Kirk's [11] approximation for spread options on energy futures, commonly used by traders, is an exact solution to a related problem, and corresponds to the $n = 0$ term in our perturbation expansion. Rather than rely on Monte Carlo simulations to obtain exact spread option prices numerically, we use the recombining binomial tree model introduced in van der Hoek [8], where approximations are derived for weak solutions of stochastic differential equations. Resulting spread option prices are the benchmark against which our closed-form analytic approximations are compared. Perturbation methods have been introduced in other contexts, but their application to pricing spread options appears to be new.

The paper is organized as follows. In Section 11.2 we describe our framework and derive our analytic approximation formulae. Section 11.3 is devoted to numerical results and Section 11.4 concludes the paper.

11.2. Spread option pricing

Assume that under a risk-neutral probability measure P, the dynamics of the underlying assets are

$$\begin{cases} dS_i(t) = (r - q_i)S_i(t)dt + \sigma_i S_i(t)dB_i(t), \quad i = 1, 2 \\ dB_1(t)\, dB_2(t) = \rho\, dt, \end{cases}$$

where r is the risk-free rate, q_i, $i = 1, 2$, are dividend yields, σ_i, $i = 1, 2$, are volatilities and ρ is the correlation between the two Brownian motions. We wish to calculate the value of the spread option

$$\begin{aligned} C(t) &= e^{-r(T-t)} E\left[(S_1(T) - S_2(T) - K)^+ \Big| \mathcal{F}_t \right] \\ &= e^{-r(T-t)} E\left[(S_2(T) + K) \left(\frac{S_1(T)}{S_2(T) + K} - 1 \right)^+ \Bigg| \mathcal{F}_t \right], \end{aligned}$$

where $\mathcal{F}_t = \sigma\{B_1(s), B_2(s) : 0 \le s \le t\}$. Define a new probability measure $\hat{P}$ with density

$$\Lambda_T := \frac{d\hat{P}}{dP}\bigg|_{\mathcal{F}_T} = \frac{S_2(T) + K}{E[S_2(T) + K]}$$

and introduce

$$\begin{aligned} \hat{G}(t) = \hat{E}\left[\frac{S_1(T)}{S_2(T) + K} \bigg| \mathcal{F}_t \right] &= \frac{E\left[\Lambda_T \frac{S_1(T)}{S_2(T)+K} \Big| \mathcal{F}_t \right]}{E\left[\Lambda_T | \mathcal{F}_t \right]} \\ &= \frac{E\left[S_1(T) | \mathcal{F}_t \right]}{E\left[S_2(T) | \mathcal{F}_t \right] + K} := \frac{G_1(t,T)}{G_2(t,T) + K}. \end{aligned}$$

Similarly, define

$$\hat{I}(t) = \hat{E}\left[\frac{S_2(T)}{S_2(T) + K} \bigg| \mathcal{F}_t \right] = \frac{E\left[S_2(T) | \mathcal{F}_t \right]}{E\left[S_2(T) | \mathcal{F}_t \right] + K} := \frac{G_2(t,T)}{G_2(t,T) + K}.$$

Clearly, both $\hat{G}(t)$ and $\hat{I}(t)$ are martingales under $\hat{P}$ and we have

$$G_i(t,T) = S_i(t)\, e^{(r-q_i)(T-t)}$$

and hence

$$dG_i(t,T) = \sigma_i G_i(t,T)\, dB_i(t), \quad i = 1, 2.$$

The application of Itô's lemma gives the dynamics of $\hat{I}(t)$ as follows:

$$\begin{aligned} d\hat{I}(t) &= (1-\hat{I}(t))\hat{I}(t)\sigma_2 dB_2(t) - (1-\hat{I}(t))\hat{I}(t)^2\sigma_2^2 dt \\ &= \left(1-\hat{I}(t)\right)\hat{I}(t)\sigma_2\left[dB_2(t) - \sigma_2\hat{I}(t)dt\right]. \end{aligned}$$

With the Brownian motion $\hat{B}_2(t)$ defined such that $d\hat{B}_2(t) = dB_2(t) - \sigma_2\hat{I}(t)dt$, the dynamics of $\hat{I}(t)$ become simply

$$d\hat{I}(t) = \sigma_2\left(1-\hat{I}(t)\right)\hat{I}(t)d\hat{B}_2(t).$$

Applying Itô's lemma to $\hat{G}(t)$ as a function of $G_1(t,T)$ and $G_2(t,T)$ gives

$$\begin{aligned} d\hat{G}(t) &= \sigma_1\hat{G}(t)\,dB_1(t) - \sigma_2\hat{G}(t)\hat{I}(t)dB_2(t) + \sigma_2^2\hat{G}(t)\hat{I}(t)^2 dt \\ &\quad - \rho\,\sigma_1\sigma_2\hat{G}(t)\hat{I}(t)dt \\ &= \sigma_1\hat{G}(t)\left[dB_1(t) - \rho\,\sigma_2\hat{I}(t)dt\right] - \sigma_2\hat{G}(t)\hat{I}(t)\left[dB_2(t) - \sigma_2\hat{I}(t)dt\right] \\ &= \sigma_1\hat{G}(t)d\hat{B}_1(t) - \sigma_2\hat{G}(t)\hat{I}(t)d\hat{B}_2(t) \\ &= \hat{G}(t)\left[\sigma_1 d\hat{B}_1(t) - \sigma_2\hat{I}(t)d\hat{B}_2(t)\right], \end{aligned}$$

where $\hat{B}_1(t)$ is defined such that $d\hat{B}_1(t) = dB_1(t) - \sigma_2\rho\,\hat{I}(t)dt$. Let $\hat{\sigma}(t)\,d\hat{B}(t) := \sigma_1 d\hat{B}_1(t) - \sigma_2\hat{I}(t)d\hat{B}_2(t)$, where

$$\hat{\sigma}(t) = \sqrt{\sigma_1^2 + \sigma_2^2\hat{I}(t)^2 - 2\rho\,\sigma_1\sigma_2\hat{I}(t)}$$

The dynamics of $\hat{G}(t)$ then become simply

$$d\hat{G}(t) = \hat{\sigma}(t)\,\hat{G}(t)\,d\hat{B}(t).$$

Now, the value of the spread option is

$$\begin{aligned} C(t) &= e^{-r(T-t)}E\left[(S_2(T)+K)\left(\frac{S_1(T)}{S_2(T)+K} - 1\right)^+ \middle| \mathcal{F}_t\right] \\ &= e^{-r(T-t)}E\left[S_2(T)+K|\mathcal{F}_t\right]\hat{E}\left[\left(\frac{S_1(T)}{S_2(T)+K} - 1\right)^+ \middle| \mathcal{F}_t\right] \\ &= e^{-r(T-t)}E\left[S_2(T)+K|\mathcal{F}_t\right]\hat{E}\left[\left(\hat{G}(T) - 1\right)^+ \middle| \mathcal{F}_t\right] \\ &= e^{-r(T-t)}\left[G_2(t,T)+K\right]\hat{E}\left[\left(\hat{G}(T) - 1\right)^+ \middle| \mathcal{F}_t\right]. \end{aligned}$$

Clearly, $\mathcal{F}_t^{S_1,S_2} = \mathcal{F}_t^{B_1,B_2} = \mathcal{F}_t^{\hat{B}_1,\hat{B}_2} = \mathcal{F}_t^{\hat{G},\hat{I}}$. We shall therefore need to calculate the martingale

$$\begin{aligned} V(t) &:= \hat{E}\left[\left(\hat{G}(T)-1\right)^+ \Big| \mathcal{F}_t\right] = \hat{E}\left[\left(\hat{G}(T)-1\right)^+ \Big| \hat{G}(t), \hat{I}(t)\right] \\ &= u(t,\hat{G}(t),\hat{I}(t)) \end{aligned}$$

for some $u = u(t,x,y)$ sufficiently smooth. Using Itô's lemma we have

$$\begin{aligned} dV(t) &= u_t(t,\hat{G}(t),\hat{I}(t))\,dt + u_x(t,\hat{G}(t),\hat{I}(t))\,d\hat{G}(t) \\ &\quad + u_y(t,\hat{G}(t),\hat{I}(t))\,d\hat{I}(t) + u_{xy}(t,\hat{G}(t),\hat{I}(t))\,d\hat{G}(t)\,d\hat{I}(t) \\ &\quad + \frac{1}{2}u_{xx}(t,\hat{G}(t),\hat{I}(t))\,d\hat{G}(t)^2 + \frac{1}{2}u_{yy}(t,\hat{G}(t),\hat{I}(t))\,d\hat{I}(t)^2. \end{aligned}$$

Substituting

$$\begin{aligned} d\hat{I}(t) &= \sigma_2(1-\hat{I}(t))\hat{I}(t)d\hat{B}_2(t) \\ d\hat{G}(t)\,d\hat{I}(t) &= \left[\rho\,\sigma_1\sigma_2(1-\hat{I}(t))\hat{I}(t) - \sigma_2^2\hat{I}(t)^2(1-\hat{I}(t))\right]\hat{G}(t)dt \\ d\hat{I}(t)^2 &= \sigma_2^2(1-\hat{I}(t))^2\hat{I}(t)^2dt \\ d\hat{G}(t)^2 &= \hat{\sigma}(t)^2\hat{G}(t)^2dt \end{aligned}$$

we have

$$\begin{aligned} dV(t) &= u_t(t,\hat{G}(t),\hat{I}(t))\,dt + u_x(t,\hat{G}(t),\hat{I}(t))\,d\hat{G}(t) \\ &\quad + u_y(t,\hat{G}(t),\hat{I}(t))\,d\hat{I}(t) \\ &\quad + u_{xy}(t,\hat{G}(t),\hat{I}(t))\,(\rho\sigma_1\sigma_2 - \sigma_2^2\hat{I}(t))(1-\hat{I}(t))\hat{I}(t)\hat{G}(t) \\ &\quad + \frac{1}{2}u_{xx}(t,\hat{G}(t),\hat{I}(t))\,\hat{\sigma}(t)^2\hat{G}(t)^2 \\ &\quad + \frac{1}{2}u_{yy}(t,\hat{G}(t),\hat{I}(t))\,(1-\hat{I}(t))^2\hat{I}(t)^2\sigma_2^2dt, \end{aligned}$$

so $V(t) = u(t,\hat{G}(t),\hat{I}(t))$ where u satisfies the boundary value problem

$$\begin{cases} u_t(t,x,y) + \frac{1}{2}\left[x^2\sigma(y)^2u_{xx}(t,x,y) + 2(1-y)xy(\rho\,\sigma_1\sigma_2 - \sigma_2^2y)u_{xy}(t,x,y)\right. \\ \quad \left. + \sigma_2^2y^2(1-y)^2u_{yy}(t,x,y)\right] = 0, \\ u(T,x,y) = (x-1)^+, \end{cases}$$

and

$$\sigma(y) = \sqrt{\sigma_1^2 + \sigma_2^2y^2 - 2\rho\sigma_1\sigma_2y}.$$

This equation does not have a closed-form solution and so a number of analytic and numerical approximations have been proposed.

11.2.1. *Kirk's approximation*

One approximation popular with practitioners is due to Kirk [11], who considered spread options on futures traded in energy markets. He proposed an approximation to volatility of the spread that allows for the use of a standard Black-Scholes model to price a spread option of interest. We show how this approximation is obtained.

Consider the partial differential equation derived in the previous section. If, as assumed by Kirk [11], we take the strike price K to be small relative to $G_2(t,T)$, then $y \approx 1$ and the boundary value problem is approximated by the simpler boundary value problem

$$\begin{cases} u_t(t,x,y) + \frac{1}{2}\sigma(y)^2 x^2 u_{xx}(t,x,y) = 0 \\ u(T,x,y) = (x-1)^+. \end{cases}$$

We note that this is a Black-Scholes type boundary value problem for which the solution is

$$u(t,x,y) = x\,\mathcal{N}(d_1(t,x,y)) - \mathcal{N}(d_2(t,x,y))\,,$$

where

$$d_1(t,x,y) = \frac{\log x + \frac{1}{2}\sigma(y)^2(T-t)}{\sigma(y)\sqrt{T-t}},$$
$$d_2(t,x,y) = \frac{\log x - \frac{1}{2}\sigma(y)^2(T-t)}{\sigma(y)\sqrt{T-t}},$$

and

$$\mathcal{N}(d) = \frac{1}{\sqrt{2\pi}}\int_{-\infty}^{d} e^{-\frac{z^2}{2}}\,dz.$$

The value of the call spread option is then

$$\begin{aligned} C(t) &= e^{-r(T-t)}\,[G_2(t,T)+K]\,\hat{E}\left[\left(\hat{C}(T)-1\right)^+ \middle| \mathcal{F}_t\right] \\ &= e^{-r(T-t)}\,[G_2(t,T)+K]\,V(t) \\ &= e^{-r(T-t)}\,[G_2(t,T)+K]\,u(t,\hat{G}(t),\hat{I}(t)) \\ &= e^{-r(T-t)}\,[G_2(t,T)+K]\left[\frac{G_1(t,T)}{G_2(t,T)+K}\mathcal{N}(d_1) - \mathcal{N}(d_2)\right] \\ &= e^{-r(T-t)}\,[G_1(t,T)\,\mathcal{N}(d_1) - (G_2(t,T)+K)\,\mathcal{N}(d_2)]\,, \end{aligned}$$

where

$$d_1 = \frac{\log \frac{G_1(t,T)}{G_2(t,T)+K} + \frac{1}{2}\sigma_{\text{Kirk}}^2(T-t)}{\sigma_{\text{Kirk}}\sqrt{T-t}},$$

$$d_2 = d_1 - \sigma_{\text{Kirk}}\sqrt{T-t},$$

$$\sigma_{\text{Kirk}} = \sqrt{\sigma_1^2 + \sigma_2^2\left(\frac{G_2(t,T)}{G_2(t,T)+K}\right)^2 - 2\rho\,\sigma_1\sigma_2\left(\frac{G_2(t,T)}{G_2(t,T)+K}\right)},$$

which agrees with the Kirk approximation.

11.2.2. *First-order analytic approximation*

Recall that the price at time t of a spread option is

$$C(t) = e^{-r(T-t)}\,(G_2(t,T)+K)\,u(t,\hat{G}(t),\hat{I}(t)),$$

where u satisfies the boundary value problem

$$\begin{cases} u_t(t,x,y) + \frac{1}{2}\big[x^2\sigma(y)^2 u_{xx}(t,x,y) + 2(1-y)xy(\rho\,\sigma_1\sigma_2 - \sigma_2^2 y)u_{xy}(t,x,y) \\ \qquad + \sigma_2^2 y^2(1-y)^2 u_{yy}(t,x,y)\big] = 0 \\ u(T,x,y) = (x-1)^+, \end{cases} \tag{11.1}$$

with

$$\sigma(y)^2 = \sigma_1^2 + \sigma_2^2 y^2 - 2\rho\,\sigma_1\sigma_2 y$$
$$\alpha(y) = y\,(\rho\,\sigma_1\sigma_2 - \sigma_2^2 y)$$
$$\beta(y) = \sigma_2^2 y^2$$

on $\mathcal{D} = \{(t,x,y) : t \geq 0,\ x > 0,\ 0 < y \leq 1\}$.

Write u as a perturbation expansion

$$u(t,x,y) = \sum_{n=0}^{\infty}(1-y)^n u^{(n)}(t,x,y), \tag{11.2}$$

where $u^{(n)}(t,x,y)$ are found recursively as we now illustrate for $n = 1$ and later in generality. The first-order approximation is then

$$u(t,x,y) \cong u^{(0)}(t,x,y) + (1-y)u^{(1)}(t,x,y). \tag{11.3}$$

For $n = 0$ we solve

$$\begin{cases} u_t^{(0)}(t,x,y) + \frac{1}{2}\sigma(y)^2 x^2 u_{xx}^{(0)}(t,x,y) = 0, \\ u^{(0)}(T,x,y) = (x-1)^+, \end{cases} \tag{11.4}$$

which is the Kirk approximation.

To define $u^{(1)} = u^{(1)}(t, x, y)$, plug the approximation (11.3) into (11.1) and set the coefficient of $(1 - y)^2$ to zero. Using (11.4):

$$\begin{aligned}
&(1-y)u_t^{(1)}(t,x,y) + \frac{1}{2}\big[\sigma(y)^2x^2(1-y)u_{xx}^{(1)}(t,x,y) \\
&\quad + 2\alpha(y)x(1-y)u_{xy}^{(0)}(t,x,y) - 2\alpha(y)x(1-y)u_x^{(1)}(t,x,y) \\
&\quad + 2\alpha(y)x(1-y)^2u_{xy}^{(1)}(t,x,y) + \beta(y)(1-y)^2u_{yy}^{(0)}(t,x,y) \\
&\quad + \beta(y)(1-y)^3u_{yy}^{(1)}(t,x,y) - 2\beta(y)(1-y)^2u_y^{(1)}(t,x,y)\big] = 0.
\end{aligned}$$

Divide this equation by $(1 - y)$ and delete terms with remaining $(1 - y)$ and $(1 - y)^2$ to obtain on $\mathcal{D}$ the boundary value problem for $u^{(1)}$:

$$\begin{cases}
u_t^{(1)}(t,x,y) + \frac{1}{2}\sigma(y)^2x^2u_{xx}^{(1)}(t,x,y) - \alpha(y)xu_x^{(1)}(t,x,y) \\
\quad = -\alpha(y)xu_{xy}^{(0)}(t,x,y), \\
u^{(1)}(T,x,y) = 0.
\end{cases}$$

We now obtain an analytic expression for $u^{(1)}(t, x, y)$. Write $u^{(1)}(t, x, y) = \alpha(y)v(t, x, y)$; then v satisfies the boundary value problem

$$\begin{cases}
v_t(t,x,y) + \frac{1}{2}\sigma(y)^2x^2v_{xx}(t,x,y) - \alpha(y)xv_x(t,x,y) = -xu_{xy}^{(0)}(t,x,y), \\
v(T,x,y) = 0.
\end{cases} \tag{11.5}$$

Lemma 11.1. *The solution of (11.5) is given by*

$$v(t,x,y) = E\left[\int_t^T X(s)u_{xy}^{(0)}(s,X(s),y)ds \middle| X(t) = x\right],$$

where X is an Itô process with dynamics

$$dX(t) = -\alpha(y)X(t)dt + \sigma(y)X(t)dW(t).$$

Proof. Applying Itô's lemma to $v(s, X(s), y)$ we obtain

$$\begin{aligned}
&dv(s, X(s), y)\\
&= v_s(s, X(s), y)ds + v_x(s, X(s), y)dX(s) + \frac{1}{2}v_{xx}(s, X(s), y)dX(s)^2\\
&= v_s(s, X(s), y)ds + v_x(s, X(s), y)\left[-\alpha(y)X(s)ds + \sigma(y)X(s)dW(s)\right]\\
&\qquad + \frac{1}{2}\sigma(y)^2X(s)^2v_{xx}(s, X(s), y)ds\\
&= \big[v_s(s, X(s), y) - \alpha(y)X(s)v_x(s, X(s), y)\\
&\qquad + \frac{1}{2}\sigma(y)^2X(s)^2v_{xx}(s, X(s), y)\big]ds + v_x(s, X(s), y)\sigma(y)X(s)dW(s)\\
&= -X(s)u^{(0)}_{xy}(s, X(s), y)ds + v_x(s, X(s), y)\sigma(y)X(s)dW(s).
\end{aligned}$$

This leads to

$$v(t, x, y) = \int_t^T X(s)u^{(0)}_{xy}(s, X(s), y)ds - \int_t^T v_x(s, X(s), y)\sigma(y)X(s)dW(s).$$

Taking expectations gives

$$\begin{aligned}
v(t, x, y) &= E\left[\int_t^T X(s)u^{(0)}_{xy}(s, X(s), y)ds\big|X(t) = x\right]\\
&\qquad - E\left[\int_t^T \sigma(y)X(s)v_x(s, X(s), y)dW(s)\big|X(t) = x\right],
\end{aligned}$$

and the result follows. □

Remark 11.1. We note that

$$\begin{aligned}
v(t, x, y) &= E\left[\int_t^T X(s)u^{(0)}_{xy}(s, X(s), y)ds\big|X(t) = x\right]\\
&= \frac{\partial}{\partial y}E\left[\int_t^T X(s)u^{(0)}_{x}(s, X(s), y)ds\big|X(t) = x\right].
\end{aligned}$$

We shall now calculate

$$E\left[\int_t^T X(s)u^{(0)}_{x}(s, X(s), y)ds\big|X(t) = x\right].$$

Lemma 11.2. *Using the same process X as in Lemma 11.1 we obtain*

$$\begin{aligned}
&\alpha(y)\, E\Big[\int_t^T X(s)u^{(0)}_{x}(s, X(s), y)ds\Big|X(t) = x\Big]\\
&\qquad = u^{(0)}(t, x, y) - E[(X(T) - 1)^+|X(t) = x].
\end{aligned}$$

Proof. Applying Itô's lemma to $u^{(0)}(s, X(s), y)$ we obtain

$$u^{(0)}(T, X(T), y) - u^{(0)}(t, x, y) = -\int_t^T \alpha(y)X(s)u_x^{(0)}(s, X(s), y)ds + \int_t^T \sigma(y)X(s)u_x^{(0)}(s, X(s), y)dW(s)$$

and taking expectations we have

$$E[u^{(0)}(T, X(T), y)|X(t) = x] - u^{(0)}(t, x, y) = -E\left[\int_t^T \alpha(y)X(s)u_x^{(0)}(s, X(s), y)ds\big|X(t) = x\right].$$

But as $u^{(0)}(T, X(T), y) = (X(T) - 1)^+$, the result follows. □

Remark 11.2. Lemma 11.2 demonstrates that $u^{(1)}(t, x, y)$ can be calculated from knowing the difference between two Black-Scholes option prices, one for the case of zero interest rates and zero dividends, and the other for the case of zero interest rates but a dividend yield $\alpha(y)$.

Let $\hat{u}^{(0)}(t, x, y) = E[(X(T)-1)^+|\mathcal{F}_t] = E[(X(T)-1)^+|X(t) = x]$. Then

$$\hat{u}^{(0)}(t, x, y) = e^{-\alpha(y)(T-t)}x\mathcal{N}(\hat{d}_1(t, x, y)) - \mathcal{N}(\hat{d}_2(t, x, y))$$

where

$$\hat{d}_1(t, x, y) = \frac{\log x + \left(-\alpha(y) + \frac{1}{2}\sigma(y)^2\right)(T-t)}{\sigma(y)\sqrt{T-t}},$$

$$\hat{d}_2(t, x, y) = \frac{\log + \left(-\alpha(y) - \frac{1}{2}\sigma(y)^2\right)(T-t)}{\sigma(y)\sqrt{T-t}},$$

by Lemma 11.3 in Appendix with

$$X(T) = x\exp\left[-(\alpha(y) + \frac{1}{2}\sigma(y)^2)(T-t) + \sigma(y)[W(T) - W(t)]\right].$$

When $\alpha(y) = 0$, we have $\hat{u}^{(0)}(t, x, y) = u^{(0)}(t, x, y)$. Therefore,

$$\begin{aligned} u^{(1)}(t, x, y) = \alpha(y)\,v(t, x, y) &= \alpha(y)\frac{\partial}{\partial y}\left[\frac{1}{\alpha(y)}\left(u^{(0)}(t, x, y) - \hat{u}^{(0)}(t, x, y)\right)\right] \\ &= \alpha(y)\left[\frac{\partial}{\partial y}\left(\frac{1}{\alpha(y)}\right)\right]\left(u^{(0)}(t, x, y) - \hat{u}^{(0)}(t, x, y)\right) \\ &\quad + \left[\frac{\partial}{\partial y}\left(u^{(0)}(t, x, y) - \hat{u}^{(0)}(t, x, y)\right)\right], \end{aligned} \tag{11.6}$$

where we assume $y \neq \rho\frac{\sigma_1}{\sigma_2}$, else $\alpha(y) \equiv 0$, and an explicit formula is thus obtained. The first-order approximation for the spread option price is then given by

$$\begin{aligned} C(t) &= e^{-r(T-t)}\,[G_2+K]\,u(t,\hat{G}(t),\hat{I}(t)) \\ &= e^{-r(T-t)}\,[G_2+K]\left[u^{(0)}(t,\hat{G}(t),\hat{I}(t)) + (1-\hat{I}(t))u^{(1)}(t,\hat{G}(t),\hat{I}(t))\right], \end{aligned}$$

for which values can be computed. Numerical results based on this first-order approximation are provided in Section 11.3.

11.2.3. *Higher order analytic approximations*

Recall the boundary value problem for $V(t) = \hat{E}[(\hat{G}(t)-1)^+] = u(t,\hat{G}(t),\hat{I}(t))$, namely

$$\begin{cases} u_t(t,x,y) + \frac{1}{2}\big[x^2\sigma(y)^2u_{xx}(t,x,y) + 2(1-y)xy(\rho\,\sigma_1\sigma_2 - \sigma_2^2 y)u_{xy}(t,x,y) \\ \quad + \sigma_2^2y^2(1-y)^2u_{yy}(t,x,y)\big] = 0, \\ u(T,x,y) = (x-1)^+, \end{cases} \tag{11.7}$$

where

$$\begin{aligned} \sigma(y)^2 &= \sigma_1^2 + \sigma_2^2y^2 - 2\rho\,\sigma_1\sigma_2 y, \\ \alpha(y) &= y\left(\rho\,\sigma_1\sigma_2 - \sigma_2^2 y\right), \\ \beta(y) &= \sigma_2^2y^2. \end{aligned}$$

We now derive a general-form boundary value problem for $u^{(n)}(t,x,y)$, a perturbation coefficient of order n. The partial derivatives of $u(t,x,y)$ are

$$\begin{aligned} u_t(t,x,y) &= \sum_{n=0}^{\infty}(1-y)^n u_t^{(n)}(t,x,y) \\ u_x(t,x,y) &= \sum_{n=0}^{\infty}(1-y)^n u_x^{(n)}(t,x,y) \\ u_y(t,x,y) &= -\sum_{n=1}^{\infty} n(1-y)^{n-1}u^{(n)}(t,x,y) + \sum_{n=0}^{\infty}(1-y)^n u_y^{(n)}(t,x,y) \end{aligned}$$

$$u_{xx}(t,x,y) = \sum_{n=0}^{\infty}(1-y)^n u_{xx}^{(n)}(t,x,y)$$

$$u_{xy}(t,x,y) = -\sum_{n=1}^{\infty} n(1-y)^{n-1}u_x^x(t,x,y) + \sum_{n=0}^{\infty}(1-y)^n u_{xy}^{(n)}(t,x,y)$$

$$\begin{aligned} u_{yy}(t,x,y) &= \sum_{n=2}^{\infty} n(n-1)(1-y)^{n-2}u^{(n)}(t,x,y) \\ &\quad - 2\sum_{n=1}^{\infty} n(1-y)^{n-1}u_y^{(n)}(t,x,y) + \sum_{n=0}^{\infty}(1-y)^n u_{yy}^{(n)}(t,x,y) \end{aligned}$$

Substituting these into (11.7) and suppressing the arguments gives

$$\begin{aligned} &u_t^{(0)} + \frac{1}{2}x^2\sigma(y)^2u_{xx}^{(0)} \\ &+ (1-y)\left[u_t^{(1)} + \frac{1}{2}x^2\sigma(y)^2u_{xx}^{(1)} + x\alpha(y)u_{xy}^{(0)} - x\alpha(y)u_x^{(1)}\right] \\ &+ \sum_{n=2}^{\infty}(1-y)^n\Big[u_t^{(n)} + \frac{1}{2}x^2\sigma(y)^2u_{xx}^{(n)} + x\alpha(y)u_{xy}^{n-1} - nx\alpha(y)u_x^{(n)} \\ &\qquad + \frac{1}{2}n(n-1)\beta(y)u^{(n)} - n\beta(y)u_y^{n-1} + \frac{1}{2}\beta(y)u_{yy}^{n-2}\Big] = 0. \end{aligned}$$

Therefore, the partial differential equations for successive perturbations' coefficients are:

$$\begin{aligned} 0 &= u_t^{(0)} + \frac{1}{2}\sigma(y)^2x^2u_{xx}^{(0)} \\ 0 &= u_t^{(1)} + \frac{1}{2}\sigma(y)^2x^2u_{xx}^{(1)} + \alpha(y)xu_{xy}^{(0)} - \alpha(y)xu_x^{(1)} \\ 0 &= u_t^{(n)} + \frac{1}{2}\sigma(y)^2x^2u_{xx}^{(n)} + \alpha(y)xu_{xy}^{(n-1)} - n\alpha(y)xu_x^{(n)} \\ &\quad + \frac{1}{2}n(n-1)\beta(y)u^{(n)} - n\beta(y)u_y^{(n-1)} + \frac{1}{2}\beta(y)u_{yy}^{(n-2)}, \qquad n > 2. \end{aligned}$$

For $n \geq 2$ we write this as

$$\begin{aligned} &u_t^{(n)}(t,x,y) + \frac{1}{2}\sigma(y)^2x^2u_{xx}^{(n)}(t,x,y) - n\alpha(y)xu_x^{(n)}(t,x,y) \\ &\quad + \frac{1}{2}n(n-1)\beta(y)u^{(n)}(t,x,y) \\ &= -\alpha(y)xu_{xy}^{(n-1)}(t,x,y) + n\beta(y)u_y^{(n-1)}(t,x,y) - \frac{1}{2}\beta(y)u_{yy}^{(n-2)}(t,x,y). \end{aligned}$$

Set

$$\begin{aligned} u^{(n)}(t,x,y) &= e^{-\frac{1}{2}n(n-1)\beta(y)t}w(t,x,y) \\ f(t,x,y) &= -\alpha(y)xu_{xy}^{(n-1)}(t,x,y) + n\beta(y)u_y^{(n-1)}(t,x,y) \\ &\quad - \frac{1}{2}\beta(y)u_{yy}^{(n-2)}(t,x,y). \end{aligned}$$

The partial differential equation for $w(t,x,y)$ is then

$$\begin{cases} w_t(t,x,y) + \frac{1}{2}\sigma(y)^2x^2w_{xx}(t,x,y) - n\alpha(y)xw_x(t,x,y) \\ \quad = e^{\frac{1}{2}n(n-1)\beta(y)t}f(t,x,y), \\ w(T,x,y) = 0. \end{cases}$$

Let $dX_n(t) = -n\alpha(y)X_n(t)dt + \sigma(y)X_n(t)dW_n(t)$ be an Itô process where $\{W_n\}$ is an independent collection of standard Brownian motions. Applying Itô's lemma to $w(t,X_n(t),y)$ gives

$$dw(t,X_n(t),y) = e^{\frac{1}{2}n(n-1)\beta(y)t}f(t,x,y)dt + w_x(t,x,y)\sigma(y)X_n(t)dW_n(t),$$

which leads to

$$w(t,x,y) = -E\left[\int_t^T e^{\frac{1}{2}n(n-1)\beta(y)s}f(s,X_n(s),y)ds\middle|X_n(t)=x\right],$$

and

$$\begin{aligned} w(t,x,y) = \alpha(y)E&\left[\int_t^T e^{\frac{1}{2}n(n-1)\beta(y)s}X_n(s)u_{xy}^{n-1}(s,X_n(s),y)ds\middle|X_n(t)=x\right] \\ - \beta(y)E&\left[\int_t^T e^{\frac{1}{2}n(n-1)\beta(y)s}nu_y^{n-1}(s,X_n(s),y)ds\middle|X_n(t)=x\right] \\ + \frac{1}{2}\beta(y)E&\left[\int_t^T e^{\frac{1}{2}n(n-1)\beta(y)s}u_{yy}^{n-2}(s,X_n(s),y)ds\middle|X_n(t)=x\right] \end{aligned}$$

for which we need to calculate

$$\begin{aligned} &E\left[X_n(s)u_x^{(n-1)}(s,X_n(s),y)\middle|X_n(t)=x\right], \\ &E\left[u^{(n-1)}(s,X_n(s),y)\middle|X_n(t)=x\right], \\ &E\left[u^{(n-2)}(s,X_n(s),y)\middle|X_n(t)=x\right]. \end{aligned}$$

11.2.4. *Second-order analytic approximation*

We now perform the necessary calculations for $n = 2$. In order to calculate $u^{(2)}(t, x, y)$ we require the valuation of four expressions:

$$E\left[X_2(s)u_x^{(0)}(s, X_2(s), y)\Big|\, X_2(t) = x\right],$$
$$E\left[X_2(s)\hat{u}_x^{(0)}(s, X_2(s), y)\Big| X_2(t) = x\right],$$
$$E\left[u^{(0)}(s, X_2(s), y)\Big|\, X_2(t) = x\right],$$
$$E\left[\hat{u}^{(0)}(s, X_2(s), y)\Big|\, X_2(t) = x\right],$$

where X_2 is an Itô process with dynamics

$$dX_2(t) = -2\alpha(y)X_2(t)dt + \sigma(y)X_2(t)\, dW_2(t).$$

We note that $u^{(0)}(s, x, y) = xu_x^{(0)}(s, x, y) - E[I(X_0(T) > 1)\big|X_0(s)]$. Then for $t \leq s$,

$$\begin{aligned}
&E\left[X_2(s)u_x^0(s, X_2(s), y)\Big|X_2(t) = x\right]\\
&= E\left[u^{(0)}(s, X_2(s), y) + E[I(X_0(T) > 1)|X_0(s) = X_2(s)]\Big|X_2(t) = x\right]\\
&= E\left[u^{(0)}(s, X_2(s), y)|X_2(t) = x\right]\\
&\quad + E\left[E\left[I\left(X_0(T) > 1\right)|X_0(s) = X_2(s)\right]|X_2(t) = x\right].
\end{aligned}$$

Now,

$$\begin{aligned}
&E\left[u^{(0)}(s, X_2(s), y)|X_2(t) = x\right]\\
&= E\left[E\left[(X_0(T) - 1)^+ \,|X_0(s) = X_2(s)\right]|X_2(t) = x\right]\\
&= E\left[(Y - 1)^+\right].
\end{aligned}$$

But

$$X_2(s) = x \exp\left[\left(-2\alpha(y) - \frac{1}{2}\sigma(y)^2\right)(s - t) + \sigma(y)\Big(W_2(s) - W_2(t)\Big)\right]$$

so

$$
\begin{aligned}
Y &= X_2(s)\exp\left[-\frac{1}{2}\sigma(y)^2(T-s)+\sigma(y)\Big(W_0(T)-W_0(s)\Big)\right] \\
&= x\exp\Big[-2\alpha(y)(s-t)-\frac{1}{2}\sigma(y)^2(T-t)+\sigma(y)\big(W_0(T)-W_0(s)\big) \\
&\qquad +\sigma(y)\big(W_2(s)-W_2(t)\big)\Big] \\
&= x\exp\left[\hat{\mu}+\hat{\sigma}Z\right],
\end{aligned}
$$

where we recall that W_0 and W_2 are independent Brownian motions to get $Z \sim \mathcal{N}(0,1)$ and

$$
\begin{aligned}
\hat{\mu} &= -2\alpha(y)(s-t)-\frac{1}{2}\sigma(y)^2(T-t), \\
\hat{\sigma}^2 &= \sigma(y)^2(T-s)+\sigma(y)^2(s-t)=\sigma(y)^2(T-t).
\end{aligned}
$$

Using Lemma 11.3 of the Appendix,

$$
\begin{aligned}
E\left[(Y-1)^+\right] &= xe^{-2\alpha(y)(s-t)}\mathcal{N}\left(\frac{-2\alpha(y)(s-t)+\frac{1}{2}\sigma(y)^2(T-t)+\log x}{\sigma(y)\sqrt{T-t}}\right) \\
&\qquad -\mathcal{N}\left(\frac{-2\alpha(y)(s-t)-\frac{1}{2}\sigma(y)^2(T-t)+\log x}{\sigma(y)\sqrt{T-t}}\right).
\end{aligned}
$$

We now calculate

$$
\begin{aligned}
P(Y>1) &= P\left(\hat{\mu}+\hat{\sigma}Z > -\log x\right) \\
&= P\left(Z > \frac{-\hat{\mu}-\log x}{\hat{\sigma}}\right) \\
&= P\left(Z < \frac{\hat{\mu}+\log x}{\hat{\sigma}}\right) \\
&= \mathcal{N}\left(\frac{-2\alpha(y)(s-t)-\frac{1}{2}\sigma(y)^2(T-t)+\log x}{\sigma(y)\sqrt{T-t}}\right).
\end{aligned}
$$

We therefore have

$$
\begin{aligned}
&E\left[X_2(s)u_x^0(s,X_2(s),y)\Big|X_2(t)=x\right] \\
&= xe^{-2\alpha(y)(s-t)}\mathcal{N}\left(\frac{-2\alpha(y)(s-t)+\frac{1}{2}\sigma(y)^2(T-t)+\log x}{\sigma(y)\sqrt{T-t}}\right).
\end{aligned}
$$

Recall that

$$
\hat{u}^{(0)}(t,x,y) = E\left[(X_1(T)-1)^+\right]
$$

and

$$dX_1(t) = -\alpha(y)X_1(t)dt + \sigma(y)X_1(t)dW_1(t).$$

Then

$$\begin{aligned}
&E\left[X_2(s)\hat{u}_x^0(s, X_2(s), y)ds \middle| X_2(t) = x\right] \\
&= E\left[e^{\alpha(y)(T-s)}\hat{u}^{(0)}(s, X_2(s), y)|X_2(t) = x\right] \\
&\quad + E\left[e^{\alpha(y)(T-s)}E\left[I(X_1(T) > 1)|X_1(s) = X_2(s)\right]|X_2(t) = x\right] \\
&= e^{\alpha(y)(T-s)}E\left[(\hat{Y} - 1)^+\right] + e^{\alpha(y)(T-s)}P(\hat{Y} > 1).
\end{aligned}$$

Recalling that W_1 and W_2 are assumed to be independent,

$$\begin{aligned}
\hat{Y} &= X_2(s)\exp\left[-\alpha(y)(T-s) - \frac{1}{2}\sigma(y)^2(T-s) + \sigma(y)\big(W_1(T) - W_1(s)\big)\right] \\
&= x\exp\Big[-2\alpha(y)(s-t) - \frac{1}{2}\sigma(y)^2(s-t) + \sigma(y)\big(W_2(s) - W_2(t)\big) \\
&\qquad - \alpha(y)(T-s) - \frac{1}{2}\sigma(y)^2(T-s) + \sigma(y)\big(W_1(T) - W_1(s)\big)\Big] \\
&= x\exp\left[\hat{\mu} + \hat{\sigma}Z\right],
\end{aligned}$$

where now

$$\begin{aligned}
\hat{\mu} &= -2\alpha(y)(s-t) - \alpha(y)(T-s) - \frac{1}{2}\sigma(y)^2(T-t) \\
&= -\alpha(y)\left(T - s + 2s - 2t\right) - \frac{1}{2}\sigma(y)^2(T-t) \\
&= -\alpha(y)\left(T + s - 2t\right) - \frac{1}{2}\sigma(y)^2(T-t)
\end{aligned}$$

and

$$\hat{\sigma}^2 = \sigma(y)^2(T-t).$$

As before, $Z \sim \mathcal{N}(0,1)$. Using Lemma 11.3 again, we have

$$\begin{aligned}
&E\left[\left(\hat{Y} - 1\right)^+\right] \\
&= xe^{-\alpha(y)(T+s-2t)}\mathcal{N}\left(\frac{-\alpha(y)(T+s-2t) + \frac{1}{2}\sigma(y)^2(T-t) + \log x}{\sigma(y)\sqrt{T-t}}\right) \\
&\qquad - \mathcal{N}\left(\frac{-\alpha(y)(T+s-2t) - \frac{1}{2}\sigma(y)^2(T-t) + \log x}{\sigma(y)\sqrt{T-t}}\right).
\end{aligned}$$

Also,

$$P(\hat{Y} > 1) = \mathcal{N}\left(\frac{-\alpha(y)(T+s-2t) - \frac{1}{2}\sigma(y)^2(T-t) + \log x}{\sigma(y)\sqrt{T-t}}\right).$$

Therefore,

$$\begin{aligned}
&E\left[X_2(s)\hat{u}_x^0(s, X_2(s), y)\Big|X_2(t) = x\right] \\
&= e^{\alpha(y)(T-s)}E\left[(\hat{Y}-1)^+\right] + e^{\alpha(y)(T-s)}P(\hat{Y} > 1) \\
&= xe^{-2\alpha(y)(s-t)}\mathcal{N}\left(\frac{-\alpha(y)(T+s-2t) + \frac{1}{2}\sigma(y)^2(T-t) + \log x}{\sigma(y)\sqrt{T-t}}\right).
\end{aligned}$$

Indirectly we have also found expressions for

$$\begin{aligned}
&E\left[u^0(s, X_2(s), y)\Big|X_2(t) = x\right] \\
&= xe^{-2\alpha(y)(s-t)}\mathcal{N}\left(\frac{-2\alpha(y)(s-t) + \frac{1}{2}\sigma(y)^2(T-t) + \log x}{\sigma(y)\sqrt{T-t}}\right) \\
&\qquad -\mathcal{N}\left(\frac{-2\alpha(y)(s-t) - \frac{1}{2}\sigma(y)^2(T-t) + \log x}{\sigma(y)\sqrt{T-t}}\right).
\end{aligned}$$

and

$$\begin{aligned}
&E\left[\hat{u}^{(0)}\left(s, X_2(s), y)\Big|X_2(t) = x\right)\right] \\
&= xe^{-\alpha(y)(T+s-2t)}\mathcal{N}\left(\frac{-\alpha(y)(T+s-2t) + \frac{1}{2}\sigma(y)^2(T-t) + \log x}{\sigma(y)\sqrt{T-t}}\right) \\
&\qquad -\mathcal{N}\left(\frac{-\alpha(y)(T+s-2t) - \frac{1}{2}\sigma(y)^2(T-t) + \log x}{\sigma(y)\sqrt{T-t}}\right).
\end{aligned}$$

To compute

$$E\left[X_2(s)u_x^{(1)}(s, X_2(s), y)\Big|X_2(t) = x\right]$$

and

$$E\left[u^{(1)}(s, X_2(s), y)\Big|X_2(t) = x\right]$$

we recall equation (11.6).

The equation for $w(t,x,y)$ and hence $u^{(2)}(t,x,y)$ is then expressed as a deterministic integral of explicitly computed functions, which can be evaluated numerically. That is,

$$u^{(2)}(t,x,y) = \int_t^T g(t,x,y,s)\,ds.$$

11.3. Numerical results

In this section we discuss the performance of our approximation method.

11.3.1. *Binomial approximation*

Recall from Section 11.2 the dynamics of the underlying assets under a risk-neutral probability measure P,

$$\begin{cases} dS_i(t) = (r - q_i)S_i(t)dt + \sigma_i S_i(t)dB_i(t), \quad i = 1, 2 \\ dB_1(t)\, dB_2(t) = \rho\, dt, \end{cases}$$

where r is the risk-free rate, q_i, $i = 1, 2$, are dividend yields, σ_i, $i = 1, 2$, are volatilities and ρ is the correlation between the two Brownian motions. We shall use the methodology described in van der Hoek [8] to construct a recombining binomial tree approximation for the solution to this two-dimensional stochastic differential equation.

Let

$$S_i(t) = S_i(0) \exp\left(\gamma_i t + \sigma_i B_i(t)\right), \quad i = 1, 2.$$

Then, using Itô's Lemma,

$$dS_i(t) = \left(\gamma_i + \frac{1}{2}\sigma_i^2\right) S_i(t)dt + \sigma_i S_i(t)dB_i(t), \quad i = 1, 2.$$

Now set

$$\gamma_i = r - q_i - \frac{1}{2}\sigma_i^2, \quad i = 1, 2$$

to adjust the drifts to match those in the assumed underlying asset dynamics given above.

Write

$$B_2(t) = \rho B_1(t) + \sqrt{1 - \rho^2} B_1^*(t)$$

where B_1 and B_1^* are independent Brownian motions.

For the approximations, let N be a positive integer and let $\Delta t = \frac{T}{N}$. Define

$$S_1(n, j, k) = S_1(0) \exp\left(\gamma_1 n\Delta t + \sigma_1(2j - n)\sqrt{\Delta t}\right)$$

and

$$S_2(n, j, k) = S_2(0) \exp\Big(\gamma_2 n\Delta t + \sigma_2 \rho(2n - j)\sqrt{\Delta t} \\ + \sigma_2\sqrt{1 - \rho^2}(2k - n)\sqrt{\Delta t}\Big)$$

for states $j = 0, 1, \ldots, n$, and $k = 0, 1, \ldots, n$, and times $n = 0, 1, \ldots, N$.

From each node (n, j, k), we can move to one of the four nodes

$$\begin{aligned}&(n+1, j+1, k+1)\\&(n+1, j+1, k)\\&(n+1, j, k+1)\\&(n+1, j, k)\end{aligned}$$

with equal probabilities, where we employ the approximation

$$B(t+\Delta t) - B(t) = \pm\sqrt{\Delta t}$$

with equal probabilities under P for each of the independent Brownian motions B_1 and B_1^*. For spread option valuation, set

$$V(N, j, k) = (S_1(N, j, k) - S_2(N, j, k) - K)^+$$

for the option value at a terminal node (N, j, k). Then at a node (n, j, k),

$$\begin{aligned}V(n, j, k) = \exp(-r\Delta t)\frac{1}{4}\big[&V(n+1, j+1, k+1) + V(n+1, j+1, k)\\&+ V(n+1, j, k+1) + V(n+1, j, k)\big]\end{aligned}$$

and we compute $V(0, 0, 0)$.

We now apply this binomial approximation method to compute spread option prices for the special case when the strike price $K = 0$, when exact values are given by the Margrabe [12] formula. Table 11.1 shows Margrabe option values for $S_1(0) = 100$, $q_1 = 0.05$, $\sigma_1 = 0.20$, $S_2(0) = 96$, $q_2 = 0.05$, $\sigma_2 = 0.1$, $r = 0.10$, $T = 1$ and correlation ρ ranging from -1 to $+1$.

Table 11.2 compares Monte Carlo results[a] reported in Dempster and Hong [5] with binomial tree approximations for spread options with the same underlying assets for strike prices K ranging from 0 to 4 and correlation $\rho = 0.5$. It is evident that the van der Hoek [8] binomial approximation method produces equally accurate results with much less effort.

Remark 11.3. While we do not provide error bounds for binomial approximations in this paper, it is possible to do so based on results of [16] and related papers. One of our authors will present these detailed results in a future paper. Error bounds are not generally available for Monte Carlo

[a] As noted in [5], to compute spread option prices via Monte Carlo it is not necessary to discretise the time horizon $[0, T]$, as terminal values of the two assets can be simulated directly. In [5] this discretisation was performed to develop an understanding of how sensitive the accuracy and computation time are to changes in the underlying assumptions.

Table 11.1. Margrabe option values

ρ	**Exact**	**Binomial**
−1.0	13.1196	13.1219
−0.9	12.8740	12.8748
0.8	12.6226	12.6233
−0.7	12.3650	12.3657
−0.6	12.1006	12.1013
−0.5	11.8291	11.8297
−0.4	11.5498	11.5504
−0.3	11.2619	11.2625
−0.2	10.9647	10.9652
−0.1	10.6572	10.6577
0.0	10.3383	10.3385
0.1	10.0067	10.0071
0.2	9.6606	9.6610
0.3	9.2982	9.2985
0.4	8.9168	8.9171
0.5	8.5132	8.5135
0.6	8.0831	8.0834
0.7	7.6205	7.6208
0.8	7.1169	7.1171
0.9	6.5590	6.5593
1.0	5.9251	5.9247

$N = 1000$

simulations. For our purposes it was necessary to arrive at reasonably accurate values of the true option price against which to compare the numerical methods. The two methods provided corroborate the approximate true values.

11.3.2. *First- and second-approximation results*

In this section we compare our first and second-order approximation prices to Kirk's approximations and 'exact' values given by the binomial model with $N = 1000$ steps.[b] Table 11.3 shows spread option values for $S_1(0) = 110$, $q_1 = 0.03$, $\sigma_1 = 0.10$, $S_2(0) = 100$, $q_2 = 0.02$, $\sigma_2 = 0.15$, $r = 0.05$, $T = 1$, strike prices K ranging from -20 to $+25$, and correlation ρ ranging from -1 to $+1$. Errors, relative to the 'exact' spread option values obtained using the van der Hoek [8] binomial approximation method, are shown in Figure 11.1 for strike prices ranging from 0 to 10.

[b]Spread option values were calculated using the van der Hoek [8] binomial approximation method with $N = 1000$. The control variate technique of [9] was then applied, together with Margrabe option values in Table 11.1, to obtain more 'exact' spread option values.

Table 11.2. Comparison of Monte Carlo and binomial spread option values

Strike K	Binomial	Monte Carlo
0.0	8.5135	8.5166
0.4	8.3128	8.3158
0.8	8.1153	8.1183
1.2	7.9211	7.9241
1.6	7.7302	7.7332
2.0	7.5426	7.5455
2.4	7.3583	7.3611
2.8	7.1772	7.1801
3.2	6.9994	7.0022
3.6	6.8248	6.8277
4.0	6.6534	6.6564

Binomial tree with $N = 1000$. Monte Carlo with 2000 time steps, 80000 simulations.

As observed by other authors, Kirk's approximation tends to underestimate spread option prices for strikes close to zero, and overestimate spread option prices for strikes further away from zero. As can be seen in Figure 11.1, our first- and second-order approximations do not follow the same pattern as the Kirk approximation, and produce much smaller errors, in absolute value, particularly for strike prices close to zero. It is also worth noting that improvement due to the second-order approximation is generally very small, suggesting that the first-order approximation may be sufficient, particularly for spread options with strike prices that are close to zero.

11.4. Conclusion

In this paper we have developed a new approach to the valuation of spread options. The method is based on a perturbation expansion and leads to explicit formulas for successive approximations to spread option prices. The method is easily implemented and provides a clear improvement over Kirk's approximation popular with practitioners, particularly for small strike prices.

Table 11.3. Comparison of Kirk, first-order, second-order and binomial approximation spread option prices

	K	Kirk	First	Second	Binomial
$\rho = -1$	-20	29.7203	29.6670	29.6721	29.6565
	-10	21.8930	21.8735	21.8740	21.8701
	0	15.1332	15.1332	15.1332	15.1339
	5	12.2425	12.2357	12.2358	12.2443
	15	7.5376	7.4834	7.4837	7.5217
	25	4.2475	4.1284	4.1286	4.2018
$\rho = -0.5$	-20	29.0501	28.9946	29.0000	28.9947
	-10	20.9279	20.9040	20.9045	20.9050
	0	13.9180	13.9180	13.9180	13.9183
	5	10.9543	10.9505	10.9506	10.9566
	15	6.2559	6.2116	6.2119	6.2426
	25	3.1686	3.0721	3.0728	3.1302
$\rho = 0$	-20	28.4242	28.3747	28.3810	28.3809
	-10	19.9097	19.8833	19.8840	19.8888
	0	12.5237	12.5237	12.5237	12.5237
	5	9.4431	9.4420	9.4421	9.4456
	15	4.7562	4.7207	4.7211	4.7448
	25	1.9923	1.9192	1.9206	1.9622
$\rho = 0.3$	-20	28.1030	28.0634	28.0705	28.0698
	-10	19.2890	19.2626	19.2634	19.2699
	0	11.5618	11.5618	11.5618	11.5619
	5	8.3649	8.3655	8.3655	8.3677
	15	3.6907	3.6594	3.6600	3.6801
	25	1.2441	1.1865	1.1886	1.2200
$\rho = 0.8$	-20	27.7783	27.7726	27.7814	27.7700
	-10	18.3924	18.3746	18.3759	18.3809
	0	9.6325	9.6325	9.6325	9.6325
	5	5.9628	5.9682	5.9683	5.9672
	15	1.3545	1.3243	1.3254	1.3427
	25	0.1124	0.0921	0.0955	0.1040
$\rho = 1$	-20	27.7542	27.7013	27.7084	27.7538
	-10	18.2458	18.2439	18.2457	18.2438
	0	8.8212	8.8212	8.8212	8.8210
	5	4.4420	4.4638	4.4636	4.4540
	15	0.0724	0.0193	0.0208	0.0488
	25	0.0000	0.0000	0.0000	0.0000

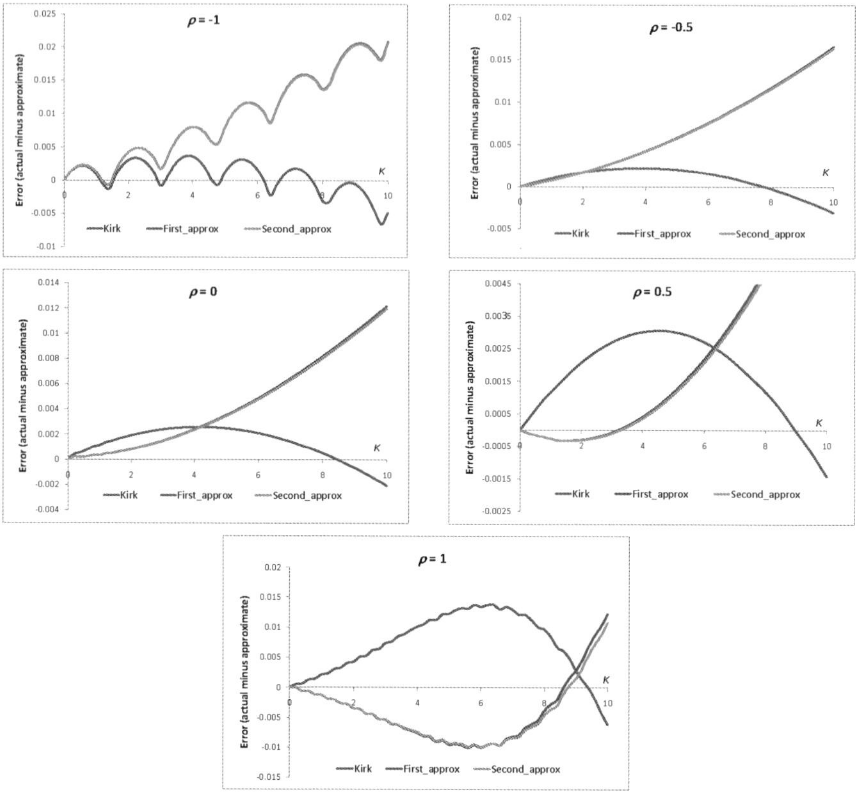

Fig. 11.1. Approximation errors. 'Exact' value obtained from the van der Hoek [8] binomial method minus value obtained from an analytic approximation method

References

[1] Alexander C and Venkatramanan A, Analytic Approximations for Spread Options (August 1, 2007, revised June 2009). Available at SSRN: http://ssrn.com/abstract=1012521

[2] Bjerksund P and Stensland G, Closed Form Spread Option Valuation (December 01, 2006). NHH Dept. of Finance & Management Science Discussion Paper No. 2006/20. Available at SSRN: http://ssrn.com/abstract=1145206

[3] Carmona R and Durrleman V (2003a). Pricing and hedging spread options, SIAM Review, 45(4), 627-685.

[4] Carmona R and Durrleman V (2003b). Pricing and hedging spread options in a lognormal model, working paper. Available at http://orfe.princeton.edu/~rcarmona/download/fe/spread.pdf.

[5] Dempster MAH and Hong SSG (2000). Spread option valuation and the fast

Fourier transform. manuscript, tech. report, Cambridge University. Available at http://www.jbs.cam.ac.uk/research/working_papers/2000/wp0026.pdf.
[6] Deng S, Li M and Zhou J (2008). Closed-form approximations for spread options and greeks. Journal of Derivatives, 16(4), 58-80.
[7] Hagan PS, Kumar D, Lesniewski AS and Woodward DE (September 2002). Managing Smile Risk. Wilmott, 84-108.
[8] van der Hoek J (2009). Recombining Binomial Tree Approximations for Diffusions, in A. Bensoussan and Q. Zhang eds, Mathematical Modelling and Numerical Methods in Finance. Elsevier, 361-368.
[9] Hull J and White A (1988). The Use of the Control Variate Technique to in Option Pricing. Journal of Financial and Quantitative Analysis, 23(3), 237-251.
[10] Hurd TR and Zhou Z (2009). A Fourier transform method for spread option pricing. Arxiv preprint arXiv:0902.3643 available at http://arxiv.org/PS_cache/arxiv/pdf/0902/0902.3643v1.pdf.
[11] Kirk E (1995). Correlation in the Energy Markets, in Managing Energy Price Risk. London: Risk Publications and Enron.
[12] Margrabe W (1978). The value of an option to exchange one asset for another. The Journal of Finance, 33, 177-186.
[13] Poitras G (1998). Spread Options, Exchange Options, and Arithmetic Brownian Motion. The Journal of Futures Markets, 18, 487-517.
[14] Schaefer MP (2002). Pricing and Hedging European Options on Futures Spreads Using the Bachelier Spread Option Model. Available at http://econpapers.repec.org/paper/agsncrtwo/19055.htm
[15] Shimko D (1994). Options on Futures Spreads: Hedging, Speculation, and Valuation, The Journal of Futures Markets, 14(2), 183-213.
[16] Szabados T and Szekely B (2010). An elementary approach to Brownian local time based on simple, symmetric random walks. Arxiv preprint arXiv:1008.1701 available at http://arxiv.org/PS_cache/arxiv/pdf/1008/1008.1701v1.pdf

Appendix

Lemma 11.3. *Let* $X = a\exp(\mu + \sigma Z)$, *where* $Z \sim \mathcal{N}(0,1)$. *Then*

$$E\left[(X-K)^+\right] = a\exp\left(\mu + \frac{1}{2}\sigma^2\right)\mathcal{N}(\sigma + d) - K\mathcal{N}(d)$$

where

$$d = \frac{\mu + \log\left(\frac{a}{K}\right)}{\sigma}.$$

Proof.

$$\begin{aligned}
E\left[(X-K)^+\right] &= E\left[I\left(X > K\right)\left(X-K\right)\right] \\
&= E\left[XI\left(X > K\right)\right] - KE\left[I\left(X > K\right)\right] \\
&= ae^{\mu}E\left[e^{\sigma Z}I\left(Z > -\frac{\mu + \log\left(\frac{a}{K}\right)}{\sigma}\right)\right] \\
&\quad - KE\left[I\left(Z > -\frac{\mu + \log\left(\frac{a}{K}\right)}{\sigma}\right)\right].
\end{aligned}$$

Let

$$d = \frac{\mu + \log\left(\frac{a}{K}\right)}{\sigma}.$$

Then,

$$\begin{aligned}
E\left[(X-K)^+\right] &= ae^{\mu}E\left[e^{\sigma Z}I\left(Z > -d\right)\right] - KE\left[I\left(Z > -d\right)\right] \\
&= ae^{\mu}e^{\frac{1}{2}\sigma^2}\mathcal{N}(\sigma + d) - K\mathcal{N}(d).
\end{aligned}$$

□

Chapter 12

On the Polynomial–Normal model and option pricing

Hao Li and Alexander Melnikov*

Department of Mathematical and Statistical Sciences,
University of Alberta, CAB 632, University of Alberta,
Edmonton AB T6G 2G1 Canada
melnikov@ualberta.ca

The Polynomial–Normal model is an extension of the well-known Black–Scholes model, which is widely used to simulate stock price behaviors. One unrealistic assumption of the Black–Scholes model is that the stock return is normally distributed. We shall use the Polynomial–Normal distribution to alleviate this assumption. In the framework of this model the probability density function (p.d.f.) of logarithmic stock returns becomes the product of the standard normal distribution p.d.f. and a polynomial. We approach the model by using the Hermite series expansion to derive the closed-form solutions for pricing European call and put options along with the power and polynomial options.

Contents

12.1. Introduction

One common assumption regarding the stock price behavior is that it the stock price returns are normally distributed. The Black–Scholes model, which has become one of the most popular models for analyzing stock behavior and pricing derivatives due to its computational simplicity, is based

*Corresponding author.

upon this assumption.

The deficiencies of the Black–Scholes model has become apparent in recent decades, as the stock price behavior has been shown to be too complex for the returns to be described by a simple normal model. After Black's paper (Ref. [1]) provided practical and theoretical evidence that the assumption of log-normal distributed stock price is incorrect, a number of researchers started looking for a better model to describe stock price behavior.

One significant drawback of the Black–Scholes model is that it does not take into account the fact that the skewness and kurtosis of stock returns may vary considerably. To allow for non-constant skewness and kurtosis, and hence make the model more precise, the famous Gram–Charlier expansion was introduced. Under this model, the p.d.f. of logarithmic returns of the underlying stock price may be expressed as the product $h(x)p(x)$, where $h(x)$ is the standard normal distribution p.d.f. and $p(x)$ is a 4th degree polynomial. (See Refs. [2, 3].)

Even though the Gram–Charlier model is an improvement over the Black–Scholes model, in the way that the skewness and kurtosis of returns are no longer arbitrary, its higher order moments can still be made to fit better. A natural way to further improve the model is to allow $p(x)$ to be a higher order polynomial. We shall name such an extension of the Black–Scholes and the Gram–Charlier models as the "Polynomial–Normal" model and pose it as the main subject of the paper. The paper is organized as follows. Section 12.2 is devoted to the Polynomial–Normal model, with the Black–Scholes and Gram–Charlier models being its particular cases. To analyze the Polynomial–Normal model, the Hermite series technique is used (see Ref. [4] for instance). Section 12.3 deals with option pricing and generalization of the Black–Scholes formula for call/put options. It also covers power and polynomial options (see Ref. [5]). Section 12.4 is devoted to the numerical estimation of the models in question. The method of maximum likelihood is used to provide numerical specifications of the models; a numerical illustration is also provided. It is shown that the Polynomial–Normal model provides a better fit than the Black–Scholes model and the Gram–Charlier model.

12.2. Polynomial–Normal model and related considerations

Let $\{\Omega, F, P\}$ be a probability space which will be used in models below. Among the models in our discussion, the following standard assumptions are supposed to hold true (see Ref. [6]):

(1) the market consists of two components: a non-risky asset B (bank account) and a risky asset S (stock); both assets are perfectly divisible;
(2) the interest rate is a non-negative constant r;
(3) the stock pays no dividends;
(4) no short selling;
(5) no transaction costs;
(6) the market is complete; Q is the unique risk-neutral probability measure.

One well-known market model that satisfies the conditions above is the Black–Scholes model. Under the risk-neutral probability measure Q (martingale measure), the variation of stock price on the time interval $[0, T]$ is restricted by the following equality,

$$E_Q(S_T) = S_0 e^{rT}. \tag{12.1}$$

Define the (logarithmic) stock return as $R_T = \ln(S_T/S_0)$ and the moment generating function of R_T as $M(\theta) = E_Q(e^{\theta R_T})$. Thus

$$M(1) = E_Q(e^{R_T}) = E_Q\left(\frac{S_T}{S_0}\right) = e^{rT}.$$

In the Black–Scholes model R_T is normally distributed (for given S_0 and T): $R_T \sim N(m, \tilde{\sigma}^2)$, where m and $\tilde{\sigma}$ are the mean and the standard deviation of R_T respectively. Then $M(\theta) = \exp(m\theta + \theta^2\tilde{\sigma}^2/2)$ and $M(1) = \exp(m + \tilde{\sigma}^2/2) = \exp(rT)$. Therefore

$$m = rT - \frac{\tilde{\sigma}^2}{2}. \tag{12.2}$$

Due to the Gaussian nature of the Black–Scholes model, both the skewness and kurtosis of the return are fixed constants, no matter what values m and $\tilde{\sigma}$ take. However, many real-world examples confirm that skewness and kurtosis are different from those provided by the Black–Scholes model (see for instance, Ref. [7]).

To provide better a model of the stock price, the Black–Scholes model may be generalized with the help of other distributions. It turns out that the Gram–Charlier series expansion presents a particularly good example (see Refs. [2, 8]). Applying the Gram–Charlier series to model R_T under Q, the p.d.f. of R_T under Q can be expressed in the following way,

$$f_{R_T}(x) = h\Big(\frac{x-m}{\tilde{\sigma}}\Big)p\Big(\frac{x-m}{\tilde{\sigma}}\Big)/\tilde{\sigma}, \tag{12.3}$$

where $h(x) = \exp(-x^2/2)/\sqrt{2\pi}$ is the p.d.f. of the standard normal distribution $N(0,1)$. The function $p(x)$ is a linear combination of Hermite polynomials

$$p(x) = 1 + \frac{\xi}{6}H_3(x) + \frac{\kappa-3}{24}H_4(x),$$
$$H_k(x) = (-1)^k \frac{d^k h(x)}{dx^k}/h(x).$$

Denote P_R the set of all polynomials on real line R. Let $L : P_R \to P_R$ be an operator such that $L(l(x)) = -(l(x)h(x))'/h(x)$ for any polynomial $l(x)$. Using this, we can rewrite Hermite polynomials as follows

$$H_k(x) = L^k(1_R(x)),$$

where $1_R(x) \equiv 1$ for any $x \in R$. Let us define the function

$$\tilde{p}(x) = 1 + \xi x^3/6 + (\kappa-3)x^4/24,$$

then $p(x) = (\tilde{p}(L)(1_R))(x)$.

The probabilistic nature of parameters m and $\tilde{\sigma}$ remains the same as $E_Q(R_T) = m$, and $Var_Q(R_T) = \tilde{\sigma}^2$. It can also be determined from (12.3) that $E_Q((x-m)^3) = \tilde{\sigma}^3\xi$ and $E_Q((x-m)^4) = \tilde{\sigma}^4\kappa$. So ξ and κ are the skewness and kurtosis of R_T respectively.

Generalizing the equality (12.2), we can find that in the model (12.3)

$$m = rT - \frac{\tilde{\sigma}^2}{2} - \ln(\tilde{p}(\tilde{\sigma})). \tag{12.4}$$

The Hermite polynomial series is a very common type of orthogonal polynomials in the theory of probability and functional analysis, featuring a number of useful properties (see Ref. [4]).

Being an extension of the Black–Scholes model, the Gram–Charlier model allows for arbitrary skewness and kurtosis while the higher moments of the stock return are constants determined by input parameters. The Gram–Charlier model is further expanded using the Polynomial–Normal distribution.

Assume the p.d.f. of R_T is defined by (12.3) where

$$p(x) = \sum_{k=0}^{N} b_k H_k(x) = (\tilde{p}(L)(1_R))(x),$$

and $\tilde{p}(x) = \sum_{k=0}^{N} b_k x^k$. Parameter $b_0 = 1$ because $f_{R_T}(x)$ is a p.d.f. Assume for simplicity that $b_1 = b_2 = 0$ since there exists a unique solution in the parameter set $(m, \tilde{\sigma}, b_3, \cdots, b_N)$. Similar to the Gram–Charlier model, $E_Q(R_T) = m$ and $Var_Q(R_T) = \tilde{\sigma}^2$.

We named this model the Polynomial–Normal model. When the polynomial degree $N = 0$, the model becomes the Black–Scholes model. When $N = 4$, the model becomes the Gram–Charlier model. The Polynomial–Normal model provides a further extension of the above models and also presents a step towards fat tail models.

Considering the Polynomial–Normal model, one must be careful regarding the sign of f_{R_T}. Generally this function is not globally positive. In other words, the function $f_{R_T}(x)$ defined by (12.3) is positive for every x, and therefore is a valid p.d.f., only for parameters b_i in a special region of $\mathbb{R}^N$.

In the Gram–Charlier Model ($N = 4$), Barton and Dennis (see Ref. [10]) obtained the parameter conditions for positivity with the help of a numerical method. Later Jondeau and Rockinger (See Ref. [11]) obtained the border of this region using an analytical method. In the following section we provide the region of (ξ, κ) that ensures positivity for the Gram–Charlier distribution, and this method is also extended to the case of a general Polynomial–Normal distribution.

In the Gram–Charlier distribution, the density function of $\ln(S_T/S_0)$ is $f_{R_T}(x) = h(y)p(y)/\tilde{\sigma}$ with $y = (x - m)/\tilde{\sigma}$. We have to find a region of (ξ, κ) where the following condition is satisfied

$$p(y) = 1 + \frac{\xi}{6} H_3(y) + \frac{\kappa - 3}{24} H_4(y) \geq 0 \text{ for every } y\,.$$

Let us specify the border of this region. If a point (ξ_0, κ_0) is on the border, the function $z = p(y)$ should be tangent to the line $z = 0$ on the $y-z$ plane. It is because any small change of (ξ, κ) would make the function above or below the y-axis at a neighborhood of tangent point, and affect the existence of a negative $p(y)$ value. Therefore, the equation $p(y) = 0$ should have a root of multiplicity 2 or higher. Denote this root by y_0, then

(ξ_0, κ_0) should be the solution of following system of linear equations

$$\begin{cases} \left.\left(1+\frac{\xi_0}{6}H_3(y)+\frac{\kappa_0-3}{24}H_4(y)\right)\right|_{y=y_0} = 0, \\ \left.\frac{d}{dy}\left(1+\frac{\xi_0}{6}H_3(y)+\frac{\kappa_0-3}{24}H_4(y)\right)\right|_{y=y_0} = 0. \end{cases}$$

Solving these equations, we have

$$(\xi(y), \kappa(y)) = \left(\frac{-24H_3(y)}{4H_3^2(y)-3H_4(y)H_2(y)}, \quad 3+\frac{72H_2(y)}{4H_3^2(y)-3H_4(y)H_2(y)}\right).$$

Figure 12.1 below shows the parametric curve $(\xi(y), \kappa(y))$ and the shaded region represents the acceptable area that generates a positive density function.

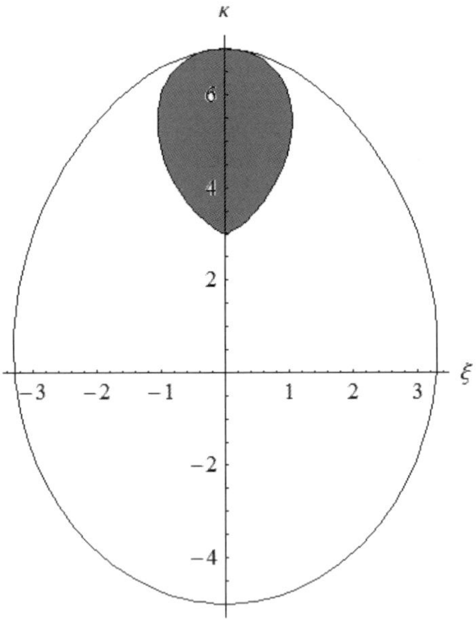

Fig. 12.1. Acceptable skewness-kurtosis region

The similar method can be used for a general Polynomial–Normal distribution where $p(x) = \sum_{k=0}^{N} b_k H_k(x)$. Any parameter set $(b_3, b_4, \cdots, b_N)$ that makes $p(x) = 0$ and $p\prime(x) = 0$ for some x would correspond to a point on the border, if 0 is the global minimum of the polynomial. Denote

$A = \{(b_3, b_4, \cdots, b_N) : p(x) = p\prime(x) = 0 \text{ for some } x\}$. The set A generates a complicated manifold in $\mathbb{R}^N$ dividing the space into many parts; the part containing $(0, 0, \cdots, 0)$ is the region we are looking for.

A numerical example of constructing such an area is presented in Section 12.4.

12.3. Option pricing in the framework of the Polynomial–Normal model

This section demonstrates how to calculate call, put, power and polynomial option prices in the Polynomial–Normal model. We also provide the analysis of sensitivities to model parameters (the Greeks).

In a complete market any option can be hedged by a specific strategy and the initial price of the option must be equal to the initial investment to avoid arbitrage. The price of the option turns out to be the discounted expected payoff under the risk-neutral measure.

Theorem 12.1. *In the Polynomial–Normal model the price of a call option is given by*

$$\begin{aligned} C = &- Ke^{-rT}\left(\Phi(-D_2) + \sum_{k=1}^{n} b_k H_{k-1}(D_2)h(D_2)\right) \\ &+ S_0\Phi(-D_2) + \frac{S_0 h(D_1)}{\tilde{p}(\tilde{\sigma})}\sum_{k=1}^{n} b_k \sum_{j=1}^{k}\binom{k}{j} H_{j-1}(D_1)\tilde{\sigma}^{k-j}, \quad (12.5) \end{aligned}$$

where

$$\begin{aligned} D_1 &= (\ln(\frac{K}{S_0}) - rT - \frac{\tilde{\sigma}^2}{2} + \ln(\tilde{p}(\tilde{\sigma})))/\tilde{\sigma}, \\ D_2 &= D_1 + \tilde{\sigma}. \end{aligned}$$

Proof. To prove the formula we recall some properties of the Hermite polynomials:

$$\begin{aligned} &\int_{x=-\infty}^{+\infty} h(x)H_i(x)H_j(x)dx = \begin{cases} 0 & \text{if } i \neq j, \\ j! & \text{if } i = j, \end{cases} \\ &H_n{}'(x) = nH_{n-1}{}'(x), \\ &H_n(x+t) = \sum_{i=0}^{n}\binom{n}{i} t^i H_i(x). \end{aligned}$$

In the Polynomial–Normal model the risk-neutral option price is determined, as the discounted expected payoff of the option with respect to the risk-neutral measure:

$$\begin{aligned}
C &= e^{-rT}E_Q(S_T - K)^+ \\
&= e^{-rT}\int_{-\infty}^{\infty}(S_0e^x - K)^+ h\left(\frac{x-m}{\tilde{\sigma}}\right)p\left(\frac{x-m}{\tilde{\sigma}}\right)\frac{1}{\tilde{\sigma}}dx \\
&= S_0e^{-rT}\int_{D_2}^{\infty}\left(e^{\tilde{\sigma}y+m} - \frac{K}{S_0}\right)h(y)p(y)dy \\
&= S_0e^{-rT}\int_{D_2}^{\infty}e^{\tilde{\sigma}y+m}h(y)p(y)dy - Ke^{-rT}\int_{D_2}^{\infty}h(y)p(y)dy \\
&= -Ke^{-rT}(\Phi(y) - \sum_{k=1}^{n}b_kH_{k-1}(y)h(y))\Big|_{D_2}^{+\infty} \\
&\quad +S_0e^{-rT+\frac{\tilde{\sigma}^2}{2}+m}\int_{D_2}^{\infty}h(y-\tilde{\sigma})p(y)dy \\
&= -Ke^{-rT}\left(\Phi(-D_2) + \sum_{k=1}^{n}b_kH_{k-1}(D_2)h(D_2)\right) \\
&\quad +S_0e^{-\ln(\tilde{p}(\tilde{\sigma}))}\int_{D_1}^{\infty}h(z)p(z+\tilde{\sigma})dz \\
&= -Ke^{-rT}\left(\Phi(-D_2) + \sum_{k=1}^{n}b_kH_{k-1}(D_2)h(D_2)\right) \\
&\quad +\frac{S_0}{\tilde{p}(\tilde{\sigma})}\int_{D_1}^{\infty}h(z)\left(1 + \sum_{k=1}^{n}b_k\sum_{j=0}^{k}\binom{k}{j}H_j(z)\tilde{\sigma}^{k-j}\right)dz \\
&= -Ke^{-rT}\left(\Phi(-D_2) + \sum_{k=1}^{n}b_kH_{k-1}(D_2)h(D_2)\right) + \frac{S_0}{\tilde{p}(\tilde{\sigma})}\Bigg\{\Phi(-D_1) \\
&\qquad +\sum_{k=1}^{n}b_k\left(\tilde{\sigma}^k\Phi(-D_1) + h(D_1)\sum_{j=1}^{k}\binom{k}{j}H_{j-1}(D_1)\tilde{\sigma}^{k-j}\right)\Bigg\} \\
&= -Ke^{-rT}\left(\Phi(-D_2) + \sum_{k=1}^{n}b_kH_{k-1}(D_2)h(D_2)\right) \\
&\quad +S_0\Phi(-D_1) + \frac{S_0h(D_1)}{\tilde{p}(\tilde{\sigma})}\sum_{k=1}^{n}b_k\sum_{j=1}^{k}\binom{k}{j}H_{j-1}(D_1)\tilde{\sigma}^{k-j}.
\end{aligned}$$

□

In addition, put-call parity can be used in order to derive the explicit formula for the put price:

$$P = -Ke^{-rT}\left(-\Phi(D_2) + \sum_{k=1}^{n} b_k H_{k-1}(D_2)h(D_2)\right) - S_0\Phi(D_2) + \frac{S_0 h(D_1)}{p(\tilde{\sigma})}\sum_{k=1}^{n} b_k \sum_{j=1}^{k}\binom{k}{j} H_{j-1}(D_1)\tilde{\sigma}^{k-j}. \quad (12.6)$$

Sensitivity of option price is usually measured with the Greeks (See Ref. [12]) which are defined as derivatives with respect to the model parameters.

In the Black–Scholes formula, call and put option prices depend on 5 parameters: initial price, time to maturity, interest rate, volatility and strike price. In the Gram–Charlier and Polynomial–Normal models they also depend on the polynomial parameters. With option formulae at hand, we can calculate the Greeks simply by differentiating the price. Using Theorem (12.1) we arrive at the following expressions for delta ($\Delta = \partial C/\partial S_0$):

$$\Delta_{\text{call}} = \Phi(-D_1) + \frac{S_0 h(D_1)}{\tilde{p}(\tilde{\sigma})}\sum_{k=1}^{n} b_k \sum_{j=1}^{k}\binom{k}{j} H_{j-1}(D_1)\tilde{\sigma}^{k-j}, \quad (12.7)$$

$$\Delta_{\text{put}} = -\Phi(D_1) + \frac{S_0 h(D_1)}{\tilde{p}(\tilde{\sigma})}\sum_{k=1}^{n} b_k \sum_{j=1}^{k}\binom{k}{j} H_{j-1}(D_1)\tilde{\sigma}^{k-j}. \quad (12.8)$$

Note that $\Delta_{\text{call}} - \Delta_{\text{put}} = 1$ because of put-call parity. Other Greeks can be calculated in a similar way.

Let us demonstrate our approach for other type of options. An α-power call is a call option which pays the asset price raised to the power α ($\alpha > 0$) less the strike price. An α-power put is a put option which pays the strike price less the asset price raised to the power α ($\alpha > 0$). The payoff function of an α-power call with maturity T and strike price K is $g = (S_T^\alpha - K)^+$, and the payoff function for the corresponding put option is $g = (K - S_T^\alpha)^+$. A polynomial option is an option whose payoff at maturity is a specific polynomial of the asset price, provided the payoff is positive. The payoff function is $g = (l(S_T))^+$, where $l(x)$ is a polynomial specified by the option. These kinds of exotic options present a powerful tools for investors, since the payoff has a non-linear relationship with the stock price.

Power and polynomial options are used in practice: for instance, squared LIBOR notes in the interest rate market, compensation options for top managers, etc (see, for example, Ref. [13]). Another example involving polynomial options may be found in portfolio choice theory (see Ref. [14]).

The fair price of a power option equals the expected value of the discounted payoff with respect to the risk-neutral measure which leads to the following theorem.

Theorem 12.2. *In the Polynomial–Normal model, the fair prices of call/put power options are given by*

$$\begin{aligned} C = &- Ke^{-rT}\left(\Phi(-D_2) + \sum_{k=1}^{N} b_k H_{k-1}(D_2)h(D_2)\right) \\ &+ S_0^\alpha e^{-rT+(\alpha\tilde{\sigma})^2/2+\alpha m}\left(\Phi(-D_1)\tilde{p}(\alpha\tilde{\sigma}) + \sum_{k=1}^{N} b_k \sum_{j=1}^{k}\binom{k}{j} H_{j-1}(D_1)(\alpha\tilde{\sigma})^{k-j}\right), \end{aligned} \tag{12.9}$$

and

$$\begin{aligned} P = &- Ke^{-rT}\left(-\Phi(D_2) + \sum_{k=1}^{N} b_k H_{k-1}(D_2)h(D_2)\right) \\ &+ S_0^\alpha e^{-rT+(\alpha\tilde{\sigma})^2/2+\alpha m}\left(-\Phi(D_1)\tilde{p}(\alpha\tilde{\sigma}) + \sum_{k=1}^{N} b_k \sum_{j=1}^{k}\binom{k}{j} H_{j-1}(D_1)(\alpha\tilde{\sigma})^{k-j}\right). \end{aligned} \tag{12.10}$$

Here $D_2 = (\ln K/\alpha - \ln S_0 - m)/\tilde{\sigma}$ *and* $D_1 = D_2 - \alpha\tilde{\sigma}$.

Proof.

$$\begin{aligned} C &= e^{-rT} E_Q({S_T}^\alpha - K)^+ \\ &= e^{-rT}\int_{-\infty}^{\infty} ({S_0}^\alpha e^{\alpha x} - K)^+ h\left(\frac{x-m}{\tilde{\sigma}}\right) p\left(\frac{x-m}{\tilde{\sigma}}\right)\frac{1}{\tilde{\sigma}}dx \\ &= {S_0}^\alpha e^{-rT}\int_{D_2}^{\infty}\left(e^{\alpha\tilde{\sigma}y+\alpha m} - \frac{K}{{S_0}^\alpha}\right) h(y)p(y)dy \\ &= {S_0}^\alpha e^{-rT}\int_{D_2}^{\infty} e^{\alpha\tilde{\sigma}y+\alpha m} h(y)p(y)dy - Ke^{-rT}\int_{D_2}^{\infty} h(y)p(y)dy \\ &= -Ke^{-rT}\left(\Phi(y) - \sum_{k=1}^{N} b_k H_{k-1}(y)h(y)\right)\Bigg|_{D_2}^{+\infty} \\ &\quad + {S_0}^\alpha e^{-rT+\frac{(\alpha\tilde{\sigma})^2}{2}+\alpha m}\int_{D_2}^{\infty} h(y-\alpha\tilde{\sigma})p(y)dy \end{aligned}$$

$$
\begin{aligned}
&= -Ke^{-rT}\left(\Phi(-D_2) + \sum_{k=1}^{N} b_k H_{k-1}(D_2)h(D_2)\right)\\
&\quad + {S_0}^{\alpha} e^{-rT+\frac{(\alpha\tilde{\sigma})^2}{2}+\alpha m} \int_{D_1}^{\infty} h(z)p(z+\alpha\tilde{\sigma})dz\\
&= -Ke^{-rT}\left(\Phi(-D_2) + \sum_{k=1}^{N} b_k H_{k-1}(D_2)h(D_2)\right)\\
&\quad + {S_0}^{\alpha} e^{-rT+\frac{(\alpha\tilde{\sigma})^2}{2}+\alpha m} \int_{D_1}^{\infty} h(z)\left(1 + \sum_{k=1}^{N} b_k \sum_{j=0}^{k} \binom{k}{j} H_j(z)(\alpha\tilde{\sigma})^{k-j}\right)dz\\
&= -Ke^{-rT}\left(\Phi(-D_2) + \sum_{k=1}^{N} b_k H_{k-1}(D_2)h(D_2)\right) + {S_0}^{\alpha} e^{-rT+\frac{(\alpha\tilde{\sigma})^2}{2}+\alpha m}\\
&\quad \cdot\left\{\Phi(-D_1) + \sum_{k=1}^{N} b_k \left((\alpha\tilde{\sigma})^k \Phi(-D_1) + \sum_{j=1}^{k} \binom{k}{j} H_{j-1}(D_1)(\alpha\tilde{\sigma})^{k-j}\right)\right\}\\
&= -Ke^{-rT}\left(\Phi(-D_2) + \sum_{k=1}^{N} b_k H_{k-1}(D_2)h(D_2)\right) + {S_0}^{\alpha} e^{-rT+\frac{(\alpha\tilde{\sigma})^2}{2}+\alpha m}\\
&\quad \cdot\left\{\Phi(-D_1)\tilde{p}(\alpha\tilde{\sigma}) + \sum_{k=1}^{N} b_k \sum_{j=1}^{k} \binom{k}{j} H_{j-1}(D_1)(\alpha\tilde{\sigma})^{k-j}\right\}.
\end{aligned}
$$

□

If $p(x) = 1$, the model is essentially reduced to the Black–Scholes case, and the prices (12.9), (12.10) have the following form (see Ref. [13]):

$$
\begin{aligned}
C &= -Ke^{-rT}\Phi(-D_2) + S_0^{\alpha} e^{-rT+(\alpha\tilde{\sigma})^2/2+\alpha m}\Phi(-D_1),\\
P &= Ke^{-rT}\Phi(D_2) - S_0^{\alpha} e^{-rT+(\alpha\tilde{\sigma})^2/2+\alpha m}\Phi(D_1).
\end{aligned}
$$

To simplify the calculation of polynomial option prices, we introduce the notion of a gap power option. The payoff function of this option is given by

$$
g = \begin{cases} S_T^{\alpha} - K & \text{if } S_T > \lambda,\\ 0 & \text{if } S_T \le \lambda. \end{cases}
$$

We can price this option using the same technique as before. Denote the risk neutral price of this option by $C(S_0, \alpha, K, \lambda)$.

Theorem 12.3. *The gap power option price can be calculated as*

$$C(S_0, \alpha, K, \lambda) = \beta(S_0, \alpha, \lambda) - \gamma(K, \alpha, \lambda), \tag{12.11}$$

where

$$\beta(S_0, \alpha, \lambda) = S_0^\alpha e^{-rT+(\alpha\tilde{\sigma})^2/2+\alpha m}\Bigg(\Phi(-D_1) + \sum_{k=1}^{K} b_k \left((\alpha\tilde{\sigma})^k \Phi(-D_1) + \sum_{j=1}^{k}\binom{k}{j} H_{j-1}(D_1)(\alpha\tilde{\sigma})^{k-j}\right)\Bigg)$$

represents the present value of S_T^α, *and*

$$\gamma(K, \lambda) = Ke{-}rT\left(\Phi(-D_2) - \sum_{k=1}^{N} b_k H_{k-1}(D_2)h(D_2)\right)$$

represents the present value of the strike price, with

$$D_2 = (\ln(\lambda/S_0) - m)/\tilde{\sigma},$$
$$D_1 = D_2 - \alpha\tilde{\sigma}.$$

Proof. We know

$$\begin{aligned} C(S_0, \alpha, K, \lambda) &= e^{-rT} E_Q(({S_T}^\alpha - K) I(S_T > \lambda)) \\ &= e^{-rT} E_Q({S_T}^\alpha I(S_T > \lambda)) - Ke^{-rT} Q(S_T > \lambda). \end{aligned}$$

Let $\beta(S_0, \alpha, \lambda) = e^{-rT} E_Q({S_T}^\alpha I(S_T > \lambda))$ and $\gamma(K, \alpha, \lambda) = Ke^{-rT} Q(S_T > \lambda)$, then

$$\begin{aligned} \beta(S_0, \alpha, \lambda) &= e^{-rT} E_Q({S_T}^\alpha I(S_T > \lambda)) \\ &= e^{-rT} \int_{\ln(\lambda/S_0)}^{\infty} {S_0}^\alpha e^{\alpha x} h\left(\frac{x-m}{\tilde{\sigma}}\right) p\left(\frac{x-m}{\tilde{\sigma}}\right) \frac{1}{\tilde{\sigma}} dx \\ &= {S_0}^\alpha e^{-rT} \int_{D_2}^{\infty} e^{\alpha\tilde{\sigma}y + \alpha m} h(y) p(y) dy \\ &= {S_0}^\alpha e^{-rT + \frac{(\alpha\tilde{\sigma})^2}{2} + \alpha m} \int_{D_2}^{\infty} h(y - \alpha\tilde{\sigma}) p(y) dy \\ &= {S_0}^\alpha e^{-rT + \frac{(\alpha\tilde{\sigma})^2}{2} + \alpha m} \int_{D_1}^{\infty} h(z) p(z + \alpha\tilde{\sigma}) dz \\ &= {S_0}^\alpha e^{-rT + \frac{(\alpha\tilde{\sigma})^2}{2} + \alpha m} \int_{D_1}^{\infty} h(z)(1 + \sum_{k=1}^{N} b_k \sum_{j=0}^{k} \binom{k}{j} H_j(z)(\alpha\tilde{\sigma})^{k-j}) dz \end{aligned}$$

$$
\begin{aligned}
&= {S_0}^{\alpha} e^{-rT+\frac{(\alpha\tilde{\sigma})^2}{2}+\alpha m}\Bigg\{\Phi(-D_1) \\
&\qquad + \sum_{k=1}^{N} b_k\left((\alpha\tilde{\sigma})^k \Phi(-D_1) + \sum_{j=1}^{k}\binom{k}{j} H_{j-1}(D_1)(\alpha\tilde{\sigma})^{k-j}\right)\Bigg\} \\
&= {S_0}^{\alpha} e^{-rT+\frac{(\alpha\tilde{\sigma})^2}{2}+\alpha m}\Bigg\{\Phi(-D_1)\tilde{p}(\alpha\tilde{\sigma}) \\
&\qquad + \sum_{k=1}^{N} b_k \sum_{j=1}^{k}\binom{k}{j} H_{j-1}(D_1)(\alpha\tilde{\sigma})^{k-j}\Bigg\}.
\end{aligned}
$$

and

$$
\begin{aligned}
\gamma(K,\alpha,\lambda) &= Ke^{-rT}Q(S_T > \lambda) \\
&= Ke^{-rT}\int_{\ln(\lambda/S_0)}^{\infty} h(\frac{x-m}{\tilde{\sigma}})p(\frac{x-m}{\tilde{\sigma}})\frac{1}{\tilde{\sigma}}dx \\
&= Ke^{-rT}\int_{D_2}^{\infty} h(y)p(y)dy \\
&= Ke^{-rT}(\Phi(y) - \sum_{k=1}^{N} b_k H_{k-1}(y)h(y))\Bigg|_{D_2}^{+\infty} \\
&= Ke^{-rT}(\Phi(-D_2) + \sum_{k=1}^{N} b_k H_{k-1}(D_2)h(D_2)).
\end{aligned}
$$

□

Let the payoff function of the option be $g = (l(S_T))^+$, where $l(x) = \sum_{i=1}^{n} c_i x^i$ is a polynomial specified by the option. Since a polynomial is a linear sum of power functions, this option can be seen as a linear sum of gap power options. Assume $l(x) > 0$ in the region $(x_1, x_2) \cup (x_3, x_4) \cup \cdots \cup (x_{n-1}, x_n)$, where x_1 and x_n can be $\pm\infty$, and other x_i are roots of $l(x)$. Thus the price of such a polynomial option is given as follows:

$$
\begin{aligned}
C &= e^{-rT}E_Q(g(S_T))^+ \\
&= e^{-rT}E_Q(\sum_{i=1}^{n} c_i S_T^i)I((x_1,x_2)\cup\cdots\cup(x_{n-1},x_n)) \\
&= \sum_{j=1}^{n/2}\sum_{i=1}^{n} c_i(\beta(S_0,i,x_{2j-1}) - \beta(S_0,i,x_{2j})). \qquad (12.12)
\end{aligned}
$$

The proof of (12.12) is a corollary of the formula for the gap power option.

12.4. Numerical illustrations for the Polynomial–Normal model

In this section we will use the real world data to compare three different models: the Black–Scholes model, the Gram–Charlier model and the Polynomial–Normal model. The comparison analysis is done based on the historical logarithmic weekly returns of S&P 500 Index from January 1996 to February 2011 (see http://finance.yahoo.com). We consider the model under both physical and risk-neutral measures. The variable m is chosen according to equation (12.4) to identify the risk-neutral measure, while m is free under the physical measure. We consider Polynomial–Normal models up to degree 10, and use the method of maximum likelihood to find the optimal polynomial. Furthermore, we use Bayesian Information Criteria (BIC) to balance complexity and accuracy of the model (see Ref. [15]). To determine statistical properties of parameter b_i we use non-parametric and parametric bootstrap methods (see Ref. [7]). We find the confidence interval for b_i and test the hypothesis $b_i = 0$ using the following formula:

$$b_i = \frac{1}{i!} E_Q\left(H_i\left(\frac{x-m}{\tilde{\sigma}}\right)\right).$$

If hypothesis $b_i = 0$ cannot be rejected for every b_i, we accept the Black–Scholes model there is no strong statistical evidence for us to use a more complex model like the Polynomial–Normal. If only $b_3 = 0$ and/or $b_4 = 0$ can be rejected, the Cram-Charlier model appears to be the best one. Otherwise, the Polynomial–Normal model is the best. We shall calculate option prices in different models and observe the difference between them.

Let us consider the model under the physical measure. According to Table 12.1, the optimal model with the best BIC value is the Polynomial–Normal model with the following polynomial:

$$p(x) = 1 + b_4 H_4(x) + b_6 H_6(x) + b_8 H_8(x). \tag{12.13}$$

The optimal parameters in this case are $b_4 = 0.0975$, $b_6 = 0.0143$, $b_8 = 0.00172$. It may be easily verified that these parameters satisfy the positivity condition mentioned in Section 12.2. The border of the positive region with the polynomial (12.13) is

Table 12.1. Max likelihood and BIC value under physical measure

	Black–Scholes	Gram–Charlier	Polynomial–Normal
$p(x)$	1	$1 + b_3H_3 + b_4H_4$	$1 + b_4H_4 + b_6H_6 + b_8H_8$
Max Likelihood	1764.78	1798.91	1804.43
BIC Value	1764.78	1792.23	1794.42

$$\begin{cases} b_6 = \dfrac{b_4H_3(x)H_8(x) - b_4H_4(x)H_7(x) - H_7(x)}{H_6(x)H_7(x) - H_5(x)H_8(x)}, \\ b_8 = \dfrac{b_4H_3(x)H_6(x) - b_4H_4(x)H_5(x) - H_5(x)}{H_5(x)H_8(x) - H_6(x)H_7(x)}, \end{cases} \quad \text{for any } x \text{ and } b_4.$$

According to Table 12.2, in both non-parametric and parametric bootstrap methods, we can reject $b_4 = 0$ at a 95% confidence level, and we can reject $b_4 = 0$ and $b_6 = 0$ at a 90% confidence level. Therefore, we have enough evidence that b_4 is non-zero and b_6 is also not likely to be zero. So, the Polynomial–Normal model is more appropriate than both the Black–Scholes and the Gram–Charlier models.

Table 12.2. 95% and 90% confidence interval when using non-parametric and parametric bootstrap methods

	Non-parametric 95% CI	90% CI	Parametric 95% CI	90% CI
b_3	(-0.3534,0,1893)	(-0.3011,0.0058)		
b_4	(0.0430,0.6199)	(0.0503,0.5799)	(0.0457,0.1657)	(0.0520,0.1496)
b_5	(-0.7688,0.0230)	(-0.7366,0,0181)		
b_6	(-0.0013,0.9152)	(0.0004,0.8743)	(-0.0014,0.0409)	(0.0002,0.0371)
b_7	(-0.9018,0.0024)	(-0.8701,0.0016)		
b_8	(-0.0002,0.7697)	(-0.0001,0.7186)	(-0.0004,0.0095)	(-0.0003,0.0070)
b_9	(-0.5735,0.0002)	(-0.5400,0.0001)		
b_{10}	(-0.0002,0.3674)	(-0.0002,0.3420)		

Now we consider the model under the risk-neutral measure. We use the average U.S. 3 month treasury bill rate as the risk free return rate, which should be the same as the expected return under risk-neutral measure. According to Table 12.3, the optimal model with the best BIC value is the Polynomial Normal model with the polynomial

$$p(x) = 1 + b_4H_4(x) + b_6H_6(x) + b_{10}H_{10}(x). \tag{12.14}$$

Table 12.3. Max likelihood and BIC value for the three models under risk-neutral measure

	Black–Scholes Model	Gram–Charlier Model	Polynomial–Normal
$p(x)$	1	$1 + b_3H_3 + b_4H_4$	$1 + b_4H_4 + b_6H_6 + b_{10}H_{10}$
Max Likelihood	1764.48	1765.73	1789.25
BIC Value	1764.48	1759.05	1779.24

The optimal parameters in this case are $b_4 = 0.0422$, $b_6 = 0.00493$, $b_{10} = 0.0000565$. The positivity condition (see Section 2) is also satisfied and we can find the border of the positive region as in the previous case.

According to Table 12.4, using a non-parametric bootstrap, we can reject $b_4 = 0$ at a 95% confidence level, and we can reject $b_6 = 0$ at a 90% confidence level. Using a parametric bootstrap, we can reject $b_4 = 0$ at a 95% confidence level. So it is evident that b_4 is non-zero and b_6 is also not likely to be zero. In this case, the Polynomial–Normal model is more preferable than both the Black–Scholes and Gram–Charlier models.

Table 12.4. 95% and 90% Confidence Interval of parameters using non-parametric and parametric bootstrap method

	Non-parametric 95% CI	90% CI	Parametric 95% CI	90% CI
b_3	(-0.3564,0.0190)	(-0.3134,0.0056)		
b_4	(0.0466,0.6283)	(0.0540,0.5950)	(0.0154,0.0753)	(0.0190,0.0684)
b_5	(-0.7634,0.0266)	(-0.7475,0,0210)		
b_6	(-0.0007,0.9082)	(0.0013,0.8998)	(-0.0010,0.0195)	(-0.0004,0.0164)
b_7	(-1.1731,0.0021)	(-0.8846,0.0016)		
b_8	(-0.0002,0.8788)	(-0.0001,0.7484)		
b_9	(-0.5485,0.0002)	(-0.5484,0.0001)		
b_{10}	(-0.0002,0.3517)	(-0.0002,0.3516)	(-0.00004,0.00011)	(-0.00003,0.00007)

Table 12.5 shows the call/put option prices in these models. When the initial index price and the strike price are 1000, and the time to maturity is one week, and the risk-free rate is 0.03, we can see that the option prices are lowest under the Polynomial–Normal model. Note, the option prices for the Gram–Charlier and Black–Scholes models are almost equal. A power option price is also calculated. Taking the initial index price as 1000, the strike price as 1000000 and the power degree $\alpha = 2$, we can observe that the prices are also the lowest under the Polynomial–Normal model, and are slightly lower in the Gram–Charlier model than in the Black–Scholes model.

Table 12.5. Option price under different models

	Black–Scholes	Gram–Charlier	Polynomial–Normal
classical call	10.63	10.63	10.41
classical put	10.06	10.06	9.84
power call	21623	21620	21181
power put	19796	19794	19354

Acknowledgements

The research was supported by NSERC grant 261855, and authors are grateful for this support. The authors also thank to Ivan Smirnov and anonymous reviewer for their reasonable comments.

References

[1] Black, F. How to Use the Holes in Black–Scholes, *Journal of Applied Corporate Finance.* **1** (4), 67–73, (1989).

[2] Jarrow, R. and Rudd, A. Approximate Option Valuation for Arbitrary Stochastic Processes, *Journal of Financial Economics.* **10** (3), 347-369, (1982).

[3] Madan, D. and Milne, F. Contingent Claims Valued and Hedged by Pricing and Investing in a Basis, *Mathematical Finance.* **4** (3), 223-245, (1994).

[4] Fedoryuk, M. V. Hermite Polynomial, (Editor: Hazewinkel), *M. Encyclopedia of Mathematics, Springer Verlag* (2001).

[5] de Jager, G., Winsen, J. Curved Option Payoffs, Series working paper, **19**, (1992), (provided by School of Finance and Economics, University of Technology, Sydney).

[6] Black, F., Scholes, M. The Pricing of Options and Corporate Liabilities, *The Journal of Political Economy.* **81** (3), 637-654, (1973).

[7] Kling, J. L., Alles, L. A. Regularities in the Variation of Skewness in Asset Returns, *Journal of Financial Research.* **17** (3), 427-438, (1994).

[8] Abken, P. A., Madan, D. B., Ramamurtie, S. Estimation of Risk Neutral and Statistical Densities by Hermite Polynomial Approximation: With Application to Eurodollar Futures Options, Working Paper 96-5 (1996), (Federal Reserve Bank of Atlanta).

[9] Abken, P. A., Madan, D. B.; Ramamurtie, S. Pricing S&P 500 Index options using a Hilbert space basis, Working Paper 96-21 (1996), (Federal Reserve Bank of Atlanta).

[10] Barton, D. E. and Dennis, K. E., The Conditions under Which Gram Charlier and Edgeworth Curves are Positive Definite and Unimodal, *Biometrika.* **39** (3-4), 425-427, (1952).

[11] Jondeau, E., Rockinger, M. Estimating Gram–Charlier Expansions with Positive Constraints, *Documents de Travail*, **56** (1998), (provided by Banque de France).
[12] Hull, J. Options, Futures and Other Derivative Securities, 8th edition, Prentice-Hall, (2011).
[13] Macovschi, S, Quittard-Pinon, F. On the Pricing of Power and Other Polynomial Options, *The Journal of Derivatives.* **13** (4), 61-71, (2006).
[14] Bertrand, P., Prigent, J.-L. Portfolio Insurance Strategies: OBPI versus CPPI, *Finance.* **26** (1), 5-32, (2005).
[15] Liddle, A. R. Information Criteria for Astrophysical Model Selection. *Monthly Notices of the Royal Astronomical Society: Letters.* **377** (1), 74-78, (2007).
[16] Backus, D, Foresi, S., Wu, L. Accounting for Biases in Black–Scholes, (Finance provided by EconWPA), No. 0207008 (2002).
[17] Jondeau, E., Rockinger, M. Gram–Charlier Densities, *Journal of Economic Dynamics and Controls.* **25** (10), 1457-1483, (2001).
[18] Ki, H., Choi, B., Chang, K., Lee, M. Option Pricing Under Extended Normal Distribution, *Journal of Future Markets.* **25** (9), 845-871, (2005).
[19] Knight, J., Satchell, S. Pricing Derivatives Written on Assets with Arbitrary Skewness and Kurtosis. *Return Distributions in Finance*, Butterworth Heinemann, 2000.
[20] Lien, D. A Note on the Relationships between Some Risk-Adjusted Performance Measures, *The Journal of Future Markets.* **22** (5), 483-495, (2002).
[21] Lo, A. W. The Statistics of Sharpe Ratio, *Financial Analysts Journal.* **58** (4), 36-52, (2002).
[22] Rachev, S. T., Handbook of Heavy-Tailed Distribution in Finance, North Holland, (2003).
[23] Saphores, J. M. The Density of Bounded Diffusions, *Journal of Economics Letters.* **86** (1), 87-93, (2005).

Chapter 13

A functional transformation approach to interest rate modelling

Shunlong Luo[1], Jia-an Yan[1] and Qiang Zhang[2]
[1] *Academy of Mathematics and Systems Science,*
Chinese Academy of Sciences, Beijing, 100190, P. R. China
luosl@amt.ac.cn, jayan@amt.ac.cn

[2] *Department of Mathematics,*
City University of Hong Kong, Kowloon, Hong Kong
mazq@cityu.edu.hk

In this chapter, we model interest rates as functional transformations of a basic Markov state variable, which couples the Ornstein–Uhlenbeck process and the Bessel process. This simple approach unifies many previously proposed diffusion models of interest rates such as the Vasicek, Black–Derman–Toy, Black–Karasinski, Cox–Ingersoll-Ross, Hull–White, Marsh–Rosenfeld, Constantinides–Ingersoll, Pearson–Sun, Duffie–Kan models, and provides an easy scheme for establishing other interest rate models.

Contents

13.1. Introduction

Interest rate structure is a fundamental factor of financial markets, and its modelling is of both theoretical and practical relevance. From the probabilistic perspective, the dynamics of short term (instantaneous, spot) interest rates r_t for risk free borrowing or lending prevailing at time t over the infinitesimal time interval $[t, t + dt]$ is usually modelled by the Itô diffusion

$$dr_t = \mu(t, r_t)dt + \sigma(t, r_t)dB_t, \tag{13.1}$$

where $\{B_t\}_{t \geq 0}$ is a standard one-dimensional Brownian motion on some probability space $\{\Omega, \mathcal{F}, \mathrm{P}\}$. This model has the Markov property in the sense that the current yield curve depends only on the short rate r_t.

In the last thirty years, researchers have proposed a variety of concrete models of (13.1) and studied the implications of these models. We now have standard one-factor models such as the Merton (1973), Vasicek (1977), Marsh–Rosenfeld (1983), Constantinides–Ingersoll (1984), Cox–Ingersoll–Ross (1985), Ho–Lee (1986), Black-Derman–Toy (1990), Hull–White (1990), Black–Karasinski (1991), Pearson–Sun (1994), and affine term structure models, among others. Aït-Sahalia (1996) has tested many of these models empirically.

A functional transformation approach to interest rate modelling has been used by some researchers. See Constantinides (1992), Jamshidian (1996) for very specific examples, and Rogers (1995) for a general Markov viewpoint. This approach consists in specifying a state variable and modelling the interest rate as certain functional transformation of the underlying state variable. In this paper we follow this approach and choose a particular diffusion process as state variable. This process is a combination of the Ornstein–Uhlenbeck process and the Bessel process. This functional transformation approach not only provides a unified way to represent many of the existing single factor interest rate models in the literature, but more importantly, provides a simple and analytic scheme for constructing other new models with great flexibility. In addition, it has advantages in numerical computations.

The paper is structured as follows. In Section 13.2, a stochastic process which serves as the state variable and a functional transformation approach for interest rate modelling are proposed. This approach unifies all the interest rate models listed above. In Section 13.3, we discuss briefly the computational advantages of this functional transformation approach in pricing interest rate derivatives. The basic properties of our underlying state variable are reviewed in the appendix.

13.2. A functional transformation approach to interest rate modelling

Let $\{B_t\}_{t\geq 0}$ be a one-dimensional Brownian motion, and $\epsilon(t)$, $\eta(t)$ and $\sigma(t)$ be deterministic functions of t. We consider a stochastic process $\{X_t\}_{t\geq 0}$ governed by the following stochastic differential equation

$$dX_t = \Big(- \eta(t)X_t + \frac{\epsilon(t)}{X_t}\Big)dt + \sigma(t)dB_t, \qquad \epsilon(t) > -\frac{1}{2}\sigma(t)^2, \quad t \geq 0. \tag{13.2}$$

When $\epsilon(t) = 0$ the process becomes a (generalized) Ornstein–Uhlenbeck process. When $\eta(t) = 0$ the process reduces to a (generalized) Bessel process, see, e.g., Revuz and Yor (1991). The condition $\epsilon(t) > -\frac{1}{2}\sigma(t)^2$ is inherited from the Bessel process. In the appendix we will show that, by a change of time, the process $\{X_t\}$ can be reduced to the $\eta = 0$ case. Moreover, if $\epsilon(t) \geq \epsilon > 0$, σ is a constant and $X_0 \geq 0$, then almost all paths of the process X are non-negative and take value zero only on a time set with Lebesgue measure zero.

In our approach, we model the interest rate r_t as a function of the time t and the Markov state variable X_t, namely, we take

$$r_t = f(t, X_t), \qquad t \geq 0, \tag{13.3}$$

where f is an appropriate function to be specified according to concrete contexts. When $f(t,x)$ is twice continuously differentiable in x and continuously differentiable in t, applying the Itô formula, we obtain the stochastic differential equation for r_t:

$$\begin{aligned} dr_t &= df(t, X_t) \\ &= f_t(t, X_t)dt + f_x(t, X_t)dX_t + \frac{1}{2}f_{xx}(t, X_t)d[X]_t \\ &= \Big\{ f_t(t, X_t) + \frac{\sigma(t)^2}{2} f_{xx}(t, X_t) + \Big(\frac{\epsilon(t)}{X_t} - \eta(t)X_t\Big) f_x(t, X_t)\Big\}dt \\ &\quad + \sigma(t) f_x(t, X_t)dB_t. \end{aligned} \tag{13.4}$$

If we model the interest rate r_t as a function of the single state variable X_t, then we write simply $r_t = f(X_t)$.

Eqs. (13.2), (13.3) and (13.4) constitute our interest rate model. In this setting, a good model depends on the judicious choice of f, η, ϵ and σ. If the range of f is confined to positive real values, the usual drawback

Table 13.1. Common interest rate models

Model	Reference
$dr_t = \alpha dt + \sigma dB_t$	Merton, 1973
$dr_t = \alpha(t)dt + \sigma dB_t$	Ho–Lee, 1986
$dr_t = ar_t dt + \sigma r_t dB_t$	Dothan, 1978
$dr_t = \beta(\alpha - r_t)dt + \sigma dB_t$	Vasicek, 1977
$dr_t = (a(t) - b(t)r_t)dt + \sigma(t)dB_t$	Hull–White, 1990
$d\ln r_t = (a(t) - b(t)\ln r_t)dt + \sigma(t)dB_t$	Black–Karasinski, 1991
(with $b(t) = -\frac{\sigma'(t)}{\sigma(t)}$)	Black–Derman–Toy, 1990
$dr_t = (\alpha r_t^{\delta-1} + \beta r_t)dt + \sigma r_t^{\delta/2}dB_t$	Marsh–Rosenfeld, 1983
(with $\delta = 3, \beta = 0$)	Constantinides–Ingersoll, 1984
$dr_t = \beta(\alpha - r_t)dt + \sigma\sqrt{r_t}dB_t$	Cox–Ingersoll–Ross (CIR), 1985
$dr_t = (\alpha(t) - \beta(t)r_t)dt + \sigma(t)r_t^{1/2}dB_t$	Hull–White (extended CIR), 1990
$dr_t = \beta(\alpha - r_t)dt + \sigma(\gamma + r_t)^{1/2}dB_t$	Pearson–Sun, 1994

of obtaining a negative interest rate, such as occurs in the Vasicek model, is avoided.

Below are listed some widely recognized interest rate models.

The calibration of model parameters is typically done using numerical methods such as binomial trees, Monte-Carlo simulation, and finite-difference approximation of PDEs.

The Merton, Ho–Lee, and Dothan models are trivially represented as functional transformations of the Brownian motion. For example, when we take $X_t = \ln r_0 + \sigma B_t, f(t, x) = \exp\{x + (a - \frac{\sigma^2}{2})t\}$ and $r_t = f(t, X_t)$, we get the Dothan model. On the other hand, it suffices to take

$$X_t = r_0 + \sigma B_t, \quad f(t, x) = x + \int_0^t \alpha(s)ds, \quad r_t = f(t, X_t),$$

to get the Ho–Lee model.

The following two models can be represented as functional transformations of the Ornstein–Uhlenbeck process.

13.2.1. *Vasicek (1977), Hull–White (1990)*

The Hull–White (generalized Vasicek) model is

$$dr_t = (a(t) - b(t)r_t)dt + \sigma(t)dB_t.$$

Let $l(t) = \int_0^t b(u)du$. Then we have

$$d(e^{l(t)}r_t) = e^{l(t)}\Big(a(t)dt + \sigma(t)dB_t\Big),$$

so that

$$r_t = e^{-l(t)}\Big(r_0 + \int_0^t e^{l(u)}a(u)du + \int_0^t e^{l(u)}\sigma(u)dB_u\Big).$$

Thus, we can write r_t in the form $r_t = X_t + g(t)$ with $g(t) = e^{-l(t)}\int_0^t e^{l(u)}a(u)du$, and

$$dX_t = -b(t)X_t dt + \sigma(t)dB_t, \quad X_0 = r_0.$$

The Vasicek model is

$$dr_t = \beta(\alpha - r_t)dt + \sigma dB_t.$$

According to the above result, we have $f(t,x) = x+\alpha(1-e^{\beta t})$ and $r(t,X_t) = f(t,X_t) = X_t + \alpha(1-e^{-\beta t})$, where

$$dX_t = -\beta X_t dt + \sigma dB_t, \quad X_0 = r_0.$$

13.2.2. *Black–Derman–Toy (1990), Black–Karasinski (1991)*

The Black–Karasinski model is

$$d\ln r_t = (a(t) - b(t)\ln r_t)dt + \sigma(t)dB_t.$$

From the above investigation for the Hull–White (generalized Vasicek) model, we get immediately $\ln r_t = X_t + g(t)$, where $g(t) = e^{-l(t)}\int_0^t e^{l(u)}a(u)du$, and

$$dX_t = -b(t)X_t dt + \sigma(t)dB_t, \quad X_0 = \ln r_0.$$

If we set $b(t) = -\frac{\sigma'(t)}{\sigma(t)}$, then we obtain the Black–Derman–Toy model.

The following models can be represented as functional transformations of the Markov process X_t in Eq. (13.2).

13.2.3. *Cox–Ingersoll–Ross (1985), Hull–White (1990)*

The Hull–White generalized CIR model is

$$dr_t = (a(t) - b(t)r_t)dt + \sigma(t)r_t^{1/2}dB_t.$$

If we let $r_t = g(X_t)$, then by Eq. (13.4) we have

$$dr_t = \Big\{\frac{\sigma^2(t)}{2}g''(X_t) + \Big(\frac{\epsilon(t)}{X_t} - \eta(t)X_t\Big)g'(X_t)\Big\}dt + \sigma(t)g'(X_t)dB_t.$$

Thus if we let $g(x) = \frac{1}{4}x^2, \eta(t) = \frac{b(t)}{2}, \epsilon(t) = 2a(t) - \frac{1}{2}\sigma^2(t)$, and $X_0 = 2\sqrt{r_0}$, we recover the Hull–White model. If a, b and σ are constants, we obtain the well-known CIR model.

When $a(t) = \sigma^2/4$ (i.e., $\epsilon = 0$ in our model), the CIR model can be mapped from the Ornstein–Uhlenbeck process. This observation is known in the literature, see e.g., Rogers (1995) and Maghsoodi (1996). In general cases, the CIR model has to be mapped from the process given by Eq. (13.2).

13.2.4. *Marsh–Rosenfeld (1983), Constantinides–Ingersoll (1984)*

The Marsh–Rosenfeld model is

$$dr_t = (\alpha r_t^{\delta-1} + \beta r_t)dt + \sigma r_t^{\delta/2} dB_t.$$

Let $\{X_t\}$ be defined by Eq. (13.2) with constant parameters η, ϵ and σ. The condition $\epsilon > -\frac{1}{2}\sigma^2$ guarantees that the process $\{X_t\}$ is non-negative (see Appendix). We consider separately the cases $\delta < 2, \delta > 2$ and $\delta = 2$. If $\delta \neq 2$, we let $f(x) = ax^{\frac{2}{2-\delta}}$ and put $r_t = aX_t^{\frac{2}{2-\delta}}$. Then Eq. (13.4) becomes

$$dr_t = \left\{(\delta\sigma^2 + 2(2-\delta)\epsilon)\frac{a^{2-\delta}}{(2-\delta)^2} r_t^{\delta-1} - \frac{2\eta}{2-\delta} r_t\right\}dt + \frac{2\sigma a^{1-\delta/2}}{2-\delta} r_t^{\delta/2} dB_t.$$

First, assume $\delta < 2$. In this case we should take a such that $\frac{2a^{1-\delta/2}}{2-\delta} = 1$, i.e., $a = (\frac{2-\delta}{2})^{\frac{2}{2-\delta}}$. By identifying η, ϵ through the equation $(\delta\sigma^2 + (2-\delta)\epsilon)\frac{a^{2-\delta}}{(2-\delta)^2} = \alpha$, $-\frac{2\eta}{2-\delta} = \beta$, that is, putting $\eta = -\frac{\beta(2-\delta)}{2}$ and $\epsilon = \frac{4\alpha - \delta\sigma^2}{2(2-\delta)}$, we recover the Marsh–Rosenfeld model. In order that the condition $\epsilon > -\frac{1}{2}\sigma^2$ to be satisfied one should assume $\alpha > \frac{1}{2}(\delta - 1)\sigma^2$.

Next, assume $\delta > 2$. In this case, if we solve the equation $\frac{2a^{1-\delta/2}}{2-\delta} = 1$ we will obtain a negative a. This is not reasonable for an interest rate model. In order to overcome this difficulty we should consider the following state process:

$$dX_t = \left(-\eta X_t + \frac{\epsilon}{X_t}\right)dt - \sigma dB_t, \qquad \epsilon > -\frac{1}{2}\sigma^2.$$

After that we solve the equation $\frac{2a^{1-\delta/2}}{2-\delta} = -1$ and obtain $a = (\frac{\delta-2}{2})^{\frac{2}{2-\delta}}$. The parameters η and ϵ are determined as above. In order that the condition $\epsilon > -\frac{1}{2}\sigma^2$ to be satisfied one should assume $\alpha < \frac{1}{2}(\delta - 1)\sigma^2$. In particular, when we set $\delta = 3$ and $\beta = 0$, we recover the Constantinides–Ingersoll model. In this model, one should assume that $\alpha < \sigma^2$.

Finally, we consider the case $\delta = 2$. In this case the Marsh–Rosenfeld model becomes

$$dr_t = (\alpha + \beta)r_t dt + \sigma r_t dB_t.$$

If we choose $X_t = \ln r_0 + \sigma B_t$ and $f(t,x) = \exp\{x + (\alpha + \beta - \frac{\sigma^2}{2})t\}$, then we get $r_t = f(t, X_t)$.

13.2.5. *Pearson–Sun (1994)*

The Pearson–Sun model is

$$dr_t = \beta(\alpha - r_t)dt + \sigma(\gamma + r_t)^{1/2} dB_t.$$

Let $f(x) = \frac{1}{4}x^2 - \gamma$ and put $r_t = \frac{1}{4}X_t^2 - \gamma$ with $X_0 = 2\sqrt{\gamma + r_0}$. Then Eq. (13.4) becomes

$$dr_t = \left(\frac{\sigma^2}{4} + \frac{\epsilon}{2} - 2\eta\gamma - 2\eta r_t\right)dt + \sigma(r_t + \gamma)^{1/2} dB_t.$$

By identifying $\eta = \frac{1}{2}\beta$, $\epsilon = 2\beta(\alpha + \gamma) - \frac{\sigma^2}{2}$, we recover the Pearson–Sun model.

13.2.6. *Duffie and Kan (1996)*

If the bond price $\{P(t,T); 0 \le t \le T, T > 0\}$ has the form $P(t,T) = \exp\{A(t,T) - B(t,T)r_t\}$, where A and B are deterministic functions, then the model is said to possess an affine term structure. If

$$dr_t = (\alpha(t)r_t + \beta(t))dt + \sqrt{\gamma(t)r_t + \delta(t)}\, dB_t,$$

then the model possesses an affine term structure (see Duffie and Kan (1996)).

We are going to show that if $\gamma(t)$ and $\delta(t)$ are differentiable and $\gamma(t)$ never vanishes, then we can express r_t by $f(t, X_t)$, where X_t is given by Eq. (13.2). In fact, if $r_t = f(t, X_t)$, then we must have

$$\sigma(t) f_x(t, X_t) = \sqrt{\gamma(t) f(t, X_t) + \delta(t)}.$$

Thus, we can take

$$f(t,x) = \frac{\gamma(t)}{4\sigma^2(t)}x^2 - \frac{\delta(t)}{\gamma(t)}.$$

Using this f we get

$$\begin{aligned} f_t(t, X_t) &= \Big[\frac{\gamma'(t)}{4\sigma^2(t)} - \frac{\gamma(t)\sigma'(t)}{2\sigma^3(t)}\Big]X_t^2 - \frac{\delta'(t)}{\gamma(t)} + \frac{\delta(t)\gamma'(t)}{\gamma(t)^2} \\ &= \Big(\frac{\gamma'(t)}{\gamma(t)} - \frac{2\sigma'(t)}{\sigma(t)}\Big)r_t - \frac{2\sigma'(t)\delta(t)}{\sigma(t)\gamma(t)} - \frac{\delta'(t)}{\gamma(t)} + \frac{2\delta(t)\gamma'(t)}{\gamma(t)^2}, \end{aligned} \tag{13.5}$$

$$\frac{1}{X_t} f_x(t, X_t) = \frac{\gamma(t)}{2\sigma^2(t)}; \quad f_{xx}(t, X_t) = \frac{\gamma(t)}{2\sigma^2(t)}, \tag{13.6}$$

$$X_t f_x(t, X_t) = \frac{\gamma(t)}{2\sigma^2(t)} X_t^2 = 2r_t + \frac{2\delta(t)}{\gamma(t)}. \tag{13.7}$$

Identifying the coefficients in Eq. (13.4) with those in Eqs. (13.5), (13.6) and (13.7) we get

$$\eta(t) = -\frac{\alpha(t)}{2} + \frac{\gamma'(t)}{\gamma(t)} - \frac{\sigma'(t)}{\sigma(t)},$$

$$\epsilon(t) = \frac{2\sigma^2(t)}{\gamma(t)}\Big[\beta(t) - \frac{\delta(t)}{\gamma(t)}\Big(\alpha(t) + \frac{\gamma'(t)}{\gamma(t)}\Big) - \frac{\gamma(t)}{4} + \frac{\delta'(t)}{\gamma(t)}\Big].$$

Now the condition $\epsilon(t) > -\frac{1}{2}\sigma^2(t)$ becomes the following condition:

$$\beta(t)\gamma(t) - \delta(t)\Big(\alpha(t) + \frac{\gamma'(t)}{\gamma(t)}\Big) + \delta'(t) > 0.$$

This shows that in general, we cannot solve the problem for arbitrarily given $\alpha(t), \beta(t), \gamma(t)$ and $\delta(t)$.

13.3. Advantages in numerical computations

To price interest rate derivatives, we need to know not only the dynamics of the interest rates, but also the associated market price of risk (the excess return required for an investor to be able to bear an extra unit of risk) or an equivalent martingale measure (risk-neutral measure). Since the existence of a risk-neutral measure is, under purely technical conditions, equivalent to the absence of arbitrage, we can assume without loss of generality that from the beginning, we are modelling the basic stochastic process given by Eq. (13.2) in a risk-neutral world.

Consider an interest rate derivative which pays $h(r_T)$ at maturity T. Options on swaps (swaptions), on caps (captions), and on floors (floortions) are examples of these types of interest rate derivative securities. From our interest rate model, the payoff can be expressed as $h(f(T, X_T))$. Thus the price of the interest rate derivative is given by $v(t, X_t)$, where

$$v(t,x) = \mathrm{E}\left(e^{-\int_t^T f(s,X_s)ds} h(f(T,X_T)) \Big| X_t = x\right), \qquad t < T.$$

From the Feynman–Kac formula, v satisfies the following partial differential equation

$$\begin{aligned}(D - f)v &= 0,\\ v(T,x) &= h(f(T,x)),\end{aligned}$$

where

$$D = \frac{\partial}{\partial t} + \frac{1}{2}\sigma^2 \frac{\partial^2}{\partial x^2} + \left(-\eta x + \frac{\epsilon}{x}\right)\frac{\partial}{\partial x}$$

is a rather simple differential operator. For swaps, caps and floors, the corresponding equation is of the form

$$(D - f)v + F(f) = 0.$$

Where $F(f)$ does not involve a differential operator. In most cases, one does not have analytical expressions for interest rate derivatives. Numerical computation is the main tool used in the determination of the value of interest rate derivatives. An important issue in numerical computations is the stability of a numerical scheme used in computations. When a numerical scheme is unstable, the subsequent numerical solutions will never converge to an accurate result. In the traditional approach of interest rate modelling, for each model, one must design a specific numerical method, perform a stability analysis and implement that method. Our functional transformation approach to interest rate modelling offers a great advantage in numerical computations.

It is well known, from the mathematical theory of numerical analysis, that to check the stability condition of a numerical scheme for the above equation, one only needs to check the stability of the numerical scheme for the case $F(f) = 0$. Furthermore, for many numerical schemes, such as the forward-time central-space scheme, the backward-time central-space scheme, and the Crank–Nicolson scheme, the lower order term $-fv$ does not affect the stability condition. Therefore, we only need to check whether such scheme is stable for the equation

$$Dv = 0.$$

Note that the above equation does not depend on f. Therefore such a class of numerical schemes will be stable for all functional forms of f (and consequently, for all interest rate models). For example, the interest rate derivatives based on the various interest rate models given in the introduction can be computed using the same numerical method. For each different model, i.e., different function f, one only needs to make a one-line change in the program, namely the place where the definition of f is made. This feature is very important. It significantly simplifies our task in numerical computations and adds considerable power to choose a desirable functional form for an interest rate model.

Appendix

We review in this appendix some fundamental properties of the stochastic differential equation (13.2). According to the general theory of stochastic calculus, Eq. (13.2) has a unique strong solution under rather general conditions.

We first assume that all the coefficients in Eq. (13.2) are constants and consider two particular cases.

(1) When $\epsilon = 0$, the equation becomes the famous Langevin equation frequently appeared in engineering and physics. The corresponding process is the Ornstein–Uhlenbeck process. The explicit solution is

$$X_t = e^{-\eta t}\Big(X_0 + \int_0^t e^{-\eta(t-s)}\sigma dB_s\Big).$$

The transition function is

$$p_{\text{O.U.}}(t;x,y) = \frac{1}{\sqrt{2\pi}\sigma\sqrt{\frac{1-e^{-2\eta t}}{2\eta}}} e^{-\frac{\eta(y-e^{-\eta t}x)^2}{\sigma^2(1-e^{-2\eta t})}}.$$

(2) When $\eta = 0$, and the state space is restricted to $[0,\infty)$, the corresponding process is the Bessel process with transition function

$$p_{\text{B}}(t;x,y) = \frac{e^{-(x^2+y^2)/2\sigma^2 t} y^{\frac{2\epsilon}{\sigma^2}}}{\sigma^2 t (xy)^{\frac{\epsilon}{\sigma^2}-\frac{1}{2}}} I_{\frac{\epsilon}{\sigma^2}-\frac{1}{2}}\left(\frac{xy}{\sigma^2 t}\right), \qquad x,\ y > 0.$$

Here $I_\nu(x) = \left(\frac{x}{2}\right)^\nu \sum_{n=0}^\infty \frac{1}{n!\Gamma(\nu+n+1)}\left(\frac{x}{2}\right)^{2n}$ is the modified Bessel function of the first kind of order ν. When $-\frac{1}{2} < \frac{\epsilon}{\sigma^2} < \frac{1}{2}$, the point 0 is instantaneously reflect; when $\frac{\epsilon}{\sigma^2} > \frac{1}{2}$, it is polar.

Now we turn to time-varying case. First of all we show that by a change of time our basic process $\{X_t\}$, defined in Eq. (13.2), can be reduced to the case where $\eta = 0$. In particular, if η and ϵ are constants, and $\epsilon > -\frac{1}{2}\sigma^2$, then the reduced process is a Bessel process. More precisely, put

$$\beta(t) = \int_0^t \exp\{2\int_0^s \eta(u)du\}ds,$$
$$\alpha(t) = \inf\{s > t : \beta(s) > t\},$$
$$Z_t = \frac{1}{\sqrt{\alpha'(t)}} X_{\alpha(t)},$$
$$W_t = \int_0^{\alpha(t)} \sqrt{\beta'(s)} dB_s.$$

Then $\{W_t\}_{t\geq 0}$ is a standard Brownian motion, and we have

$$dZ_t = \frac{\epsilon(\alpha(t))}{Z_t} dt + \sigma(\alpha(t))dW_t, \qquad \epsilon \neq 0$$

with state space $[0, \infty)$.

In fact, assume that $\mathcal{F}_t = \sigma(B_s, s \leq t)$. Put $\mathcal{G}_t = \mathcal{F}_{\alpha(t)}$. Clearly, $\{W_t\}_{t\geq 0}$ is a continuous martingale w.r.t. $\{\mathcal{G}_t\}$. Its quadratic variation is

$$[W]_t = \int_0^{\alpha(t)} \beta'(s)ds = \beta(\alpha(t)) = t.$$

Consequently, by Lévy's characterization theorem, $\{W_t\}_{t\geq 0}$ is a standard Brownian motion. By the time change formula for the Itô integral (cf. Øksendal, 1998, Theorem 8.5.7), we have

$$\int_0^{\alpha(t)} \sigma(s)dB_s = \int_0^t \sigma(\alpha(s))\sqrt{\alpha'(s)}dW_s.$$

Consequently, $\{X_{\alpha(t)}\}$ is a $\mathcal{G}_t$-semimartingale. Since

$$dX_{\alpha(t)} = -\eta(\alpha(t))X_{\alpha(t)}\alpha'(t)dt + \frac{\epsilon(\alpha(t))}{X_{\alpha(t)}}\alpha'(t)dt + \sigma(\alpha(t))\sqrt{\alpha'(t)}dW_t,$$

and

$$d\frac{1}{\sqrt{\alpha'(t)}} = d\sqrt{\beta'(\alpha(t))} = \sqrt{\beta'(\alpha(t))}\eta(\alpha(t))\alpha'(t)dt = \eta(\alpha(t))\sqrt{\alpha'(t)}dt,$$

we have

$$\begin{aligned} dZ_t &= \frac{1}{\sqrt{\alpha'(t)}}dX_{\alpha(t)} + X_{\alpha(t)}d\sqrt{\beta'(\alpha(t))} \\ &= \frac{\epsilon(\alpha(t))}{X_{\alpha(t)}}\sqrt{\alpha'(t)}dt + \sigma(\alpha(t))dW_t \\ &= \frac{\epsilon(\alpha(t))}{Z_t}dt + \sigma(\alpha(t))dW_t. \end{aligned}$$

Next we show that if $\epsilon(t) \geq \epsilon > 0$, σ is constant and $X_0 \geq 0$, then $\{X_t\}$ is non-negative and the point 0 is instantaneously reflecting. In fact, by the well-known comparison theorem in the theory of stochastic differential equations, the above reduced process $\{Z_t\}$ dominates the following Bessel process $\{Y_t\}$:

$$dY_t = \frac{\epsilon}{Y_t}dt + \sigma dW_t.$$

Acknowledgments

Luo and Yan were supported by the Science Fund for Creative Research Groups (No.10721101) and the Key Lab of Random Complex Structures and Data Science, CAS (No. 2008DP173182). Yan was also supported by the National Basic Research Program of China (973 Program, No.2007CB814902). Zhang was supported by Research Grant Council of Hong Kong Special Administrative Region, China (CityU 103807).

References

[1] Y. Aït-Sahalia, Testing continuous-time models of the spot interest rates, *Review of Financial Studies*, **9**, 385-426, (1996).

[2] F. Black, E. Derman, and W. Toy, A one factor model of interest rates and its application to treasury bond options, *Financial Analysts Journal*, **46**, 33-39, (1990).

[3] F. Black and P. Karasinski, Bond and option pricing when short rates are lognormal, *Financial Analysts Journal*, **47**, 52-59, (1991).

[4] G. M. Constantinides and J. E. Ingersoll, Optimal bond trading with personal taxes, *Journal of Financial Economics*, **13**, 299-335, (1984).

[5] G. M. Constantinides, A theory of the nominal term structure of interest rates, *Review of Financial Studies*, **5**, 531-552, (1992).

[6] J. C. Cox, J. E. Ingersoll Jr, and S. A. Ross, A theory of term structure of interest rates, *Econometrica*, **53**, 385-407, (1985).

[7] J. C. Cox, S. A. Ross, and M. Rubinstein, Option pricing: a simplified approach, *Journal of Financial Economics*, **7**, 229-263, (1979).

[8] D. Duffie and R. Kan, A yield-factor model of interest rates, *Math. Finance*, **6**, 379-406, (1996).

[9] U. Dothan, On the term structure of interest rates, *Journal of Financial Economics*, **6**, 59-69, (1978).

[10] T. S. Ho and S.-B Lee, Term structure movements and pricing interest rate contingent claims, *Journal of Finance*, **41**, 1011-1029, (1986).

[11] J. Hull and A. White, Pricing interest-rate-derivative securities, *Review of Financial Studies*, **3**, 573-592, (1990).

[12] N. Ikeda and S. Watanabe, *Stochastic Differential Equations and Diffusion Process*, (North-Holland, 1989), 2nd Edition.
[13] F. Jamshidian, An exact bond option pricing formula, *Journal of Finance*, **44**, 205-209, (1984).
[14] F. Jamshidian, Bond, futures and option evaluation in the quadratic interest rate models, *Applied Mathematical Finance*, **3**, 93-115, (1996).
[15] Y. Maghsoodi, Solution of the extended CIR term structure and bond option valuation, *Mathematical Finance*, **6**, 89-109, (1996).
[16] T. Marsh and E. R. Rosenfeld, Stochastic processes for interest rates and equilibrium bond prices, *Journal of Finance*, **38**, 635-646, (1983).
[17] R. C. Merton, The theory of rational option pricing, *Bell Lab. Journal of Economic and Management Sciences*, **4**, 141-183, (1973).
[18] B. Øksendal, *Stochastic Differential Equations.* (Springer, 1998), 5th Edition.
[19] N. D. Pearson and T.-S, Sun, Exploiting the conditional density in estimating the term structure: an application to the Cox, Ingersoll, and Ross model, *Journal of Finance*, **49**, 1279-1304, (1994).
[20] D. Revuz and M. Yor, *Continuous Martingales and Brownian Motion.* (Springer, 1991)
[21] L. C. G. Rogers, Which model for term structure of interest rates should one use. In eds. M. H. A. Davis *et al.* *Mathematical Finance*, pp. 93-115, Springer, 1995.
[22] O. Vasicek, An equilibrium characterization of the term structure, *Journal of Financial Economics*, **5**, 177-188, (1977).

Chapter 14

S&P 500 index option surface drivers and their risk neutral and real world quadratic covariations

Dilip B. Madan

Robert H. Smith School of Business,
University of Maryland,
College Park, MD 20742, USA
dbm@rhsmith.umd.edu

Summarizing daily option surfaces by four parameters of a Sato process a time series is constructed for a five dimensional set of factors driving the surface. We then observe that movements in parameters from day to day can be hedged by a position in a portfolio of options. Given that delta exposures to parameter movements are hedgeable, the theta of the option is seen as compensation for quadratic exposures to parameter movements and the underlying spot price. On regressing the theta at a point of time on all the second order derivatives, we note that the coefficients of this regression form estimates for the risk neutral quadratic covariations of the factors driving movements in the option surface. A constrained optimization forcing positive semidefinite coefficients is then employed to estimate risk neutral covariations. These risk neutral covariations are then contrasted with their statistical counterparts. It is learned that squared returns not associated with movements in the option surface are positively priced with most of the generally negative pricing of squared returns being associated with comovements in the option surface.

Contents

14.1. Introduction

The surface of option prices on a stock in general, and the S&P 500 index in particular, is quite a rich data source with over a hundred option prices continuously quoted across strikes and maturities. It is well known that these hundred prices cannot be reduced to a single number like the volatility of the underlying stock, as would be implied by the Black and Scholes (1973) and Merton (1973) models. Yet given all the arbitrage restrictions between these options it is unlikely that they can all move independently. Carr and Madan (2005) observe that in the absence of static arbitrage option prices at a single point of time must be consistent with a one dimensional Markov martingale model for the stock price. This is not to say that option prices over time are consistent with the dynamic law of this one dimensional Markov martingale. All we know from the conditions associated with excluding static arbitrage, is that at each point of time the option prices are in line with the law of a one dimensional Markov martingale model. This one dimensional Markov martingale may well change from day to day or moment to moment, it is just that at each moment there is one. Hopefully then the number of degrees of freedom in describing the movement of the surface of option prices over time, though greater than unity may possibly be captured by a few parameters needed to describe this one dimensional Markov martingale law.

Such a parsimonious one dimensional Markov martingale model was formulated by Carr, Geman, Madan and Yor (2007) with just four parameters. The underlying process for the logarithm of the stock price is based on scaling a self decomposable law at unit time. The existence of a process of independent but inhomogeneous increments supporting the scaled unit time distributions as the marginal distributions of the process is demonstrated in Sato (1991) and the resulting process is termed the Sato process in Carr, Geman, Madan and Yor (2007). It is shown in Carr, Geman, Madan and Yor (2007) that an adequate characterization of the surface of option prices may be obtained by scaling the particular self decomposable law of the variance gamma process (Madan and Seneta (1990), Madan, Carr and Chang (1998)). The variance gamma process is obtained by time changing Brownian motion with drift θ, and volatility σ by a gamma process with a mean rate of unity and a variance rate of ν. The addition of the scaling parameter γ then yields the four parameters for the Sato process.

The gamma time change makes the actual time in a unit of calendar time a random variable and hence it introduces a volatility of volatility leading to the presence of excess kurtosis represented by a positive value for ν the volatility of the gamma time change. We therefore interpret this parameter as representing the volatility of volatility. The drift θ of the Brownian motion after the gamma time change, and an associated renormalization, introduces asymmetry in log returns and accounts for skewness in the return distributions. Risk neutrally these skews are very important and subject to movement. Finally the scaling parameter γ calibrates the term structure of volatility or the relationship between at the money volatilities at two different maturities. Coupled with basic volatility σ, these four parameters adequately synthesize most option surfaces on most days (Carr, Geman, Madan and Yor (2007)). In addition to these surface parameters, option prices clearly also reflect movements in the underlying spot price S.

The five entities $(S_t, \sigma_t, \nu_t, \theta_t, \gamma_t)$ of the spot price and the estimated Sato parameters then give us all the option prices on a single underlier at the time point t. One may then represent the sequence of option prices through calendar time by this five dimensional vector $(S_t, \sigma_t, \nu_t, \theta_t, \gamma_t)$. For a latent factor model approach to the evolution of option surfaces we refer to Cont and Fonseca (2002). For a nonparametric approach to the joint law of the stock and its volatility we refer to Werker, Boes and Drost (2007).

In this study we report on the estimates for these four parameters for data on SPX options from January 4, 1996 to August 11, 2010. In addition to movements in the stock price, option prices are affected by movements in these underlying attributes of the option surface. The Sato process extracted from option prices each day just provides us with a dimensional reduction that enables us to represent the continuum of option prices across strikes and maturities by the four surface parameters. Though the process has the stated dynamics, these are not the dynamics for options in real time. The latter experience variations in the underlying parameters that are constant in the dynamics of the Sato process. Hence the Sato process does not deliver any dynamics for the surface of option prices, and is merely a static object summarizing the static option surface each day. We now address the nature of the real world and risk neutral dynamics for this five dimensional vector that synthesizes option prices each day.

We do not, in this study, build a five dimensional real world or risk neutral model for the time variation of option prices. Instead we have a more limited objective of just comparing the real world and risk neutral covariance matrices. We enquire into what are the main directions of variation in the real world and risk neutral dynamics and we focus attention on just estimating second-order moments and cross-moments under the two distributions. Stylized observations on stock options and the time series of stock price data show that risk neutral variances are higher than their real world counterparts. We ask if this is also true with respect to the other three dimensions of skewness, volatility of volatility, and the term structure of volatility. In this regard we supplement the recent studies of Bakshi, Madan and Panayotov ((2010a), (2010b)) that address similar questions for the stock price and its volatility, but with a different methodology.

It is worthwhile reminding ourselves at this point that the world is a much more complex environment than what one encounters in the relatively more elementary models of financial economics. Markets will trade movements in skewness, leverage and the volatility of volatility, or its term structure, with relative ease. Markets then potentially deliver differences between the real world expected covariations between the stock and the option skew, seen as entities embedded daily in static risk neutral distributions, and their risk neutral counterparts. This can happen long before financial economics comes to grips with what theoretically these differences may represent in terms of theoretical economic primitives.

With these considerations in mind we proceed to study the principal components for changes in these five variables synthesizing the movements of the option surface. This is done by decomposing estimates for the covariation in these entities. Apart from the real world variation in these entities there is the risk neutral covariation of these entities. We estimate this risk neutral covariation by a novel method developed here. We postulate that the theta of the options or its time decay measured by the derivative of price with respect to the passage of calendar time is compensation for exposure to the second order terms, priced at the risk neutrally expected covariation. Such a formulation endogenizes the fact that option trading permits one to hedge out all the delta exposures to all parameters. The resulting expression for the theta generalizes the theta gamma relationship developed by Black and Scholes (1973) and Merton (1973). The generalization incorporates movements in other factors driving the option surface. We then project the option thetas onto the second order terms to recover the risk neutral matrix for the covariation of our five surface drivers. Similar studies, albeit in a lower dimension, have been conducted in a stochastic volatility context, see for example, Lipton and McGhee (2002), Wystup (2003), and Arslan, Eid, El Khoury and Roth (2009).

Having extracted both the real world and risk neutral covariations for the drivers of the option surface we comment on the structure of the two covariations. We find that risk neutrally the highest contribution to surface variation comes from movements in the volatility of volatility, followed by the term structure of volatility, the level of skewness, movements in the spot price and finally changes in base volatility. The corresponding real world order is first the movements in the spot price, followed by changes in the volatility of volatility, the term structure of volatility, the skewness and the finally the base volatility. Hence risk neutrally movements in the spot are less important than surface movements, while under the real world measure they are dominant.

With regard to a comparison of risk neutral and real world volatilities, we find that risk neutrally they are larger for every dimension excepting the spot price. This leads us to conclude that though risk neutrally volatilities are higher than real world ones, the opposite is the case when we control for the other dimensions of movements in the surface. Put another way squared moves in the spot unaccompanied by movements in the option surface have a lower risk neutral expectation than the corresponding real world expectation. The differences between these volatilities are informative about the covariation between the market pricing kernel and the factors driving the option surface.

Finally we conclude by illustrating the construction of an option portfolio that sells the largest eigenvalue of the difference between risk neutral and real world covariations as they have a high price and buys the lowest eigenvalue that is cheap and has a good squared real world outcome.

We outline the following sections. Section 2 summarizes the parameter estimation and the associated results. Section 3 introduces the methodology behind the estimation of risk neutral covariations. Statistical and risk neutral covariation results are described in Section 4. Section 5 reports on covariations between quadratic risk exposures and the market pricing kernel. Section 6 concludes.

14.2. The S&P 500 index option surface synthesized

We present in this section our results in synthesizing the option surface every day using the Sato process. A brief history leading up to the development of the Sato process is first presented as a helpful guide. Issues related to the identification of parameters are also addressed.

The modeling of risk-neutral stock price dynamics recognizes that, once one focuses attention on the dynamics of the forward or futures price, one is essentially modeling a positive martingale, denoted by $M = (M(t), t \geq 0)$. Every such positive martingale is known to be the stochastic exponential of another martingale $m = (m(t), t \geq 0)$, that is, the solution to the equation

$$dM(t) = M(t_{-})dm(t), \tag{14.1}$$

where we signify the left limit by $M(t_{-})$ when there are jumps in the process for M. When we take $m = \sigma W$, where $W = (W(t), t \geq 0)$ is a standard Brownian motion and σ is a volatility parameter, we obtain the classical Black and Scholes (1973) and Merton (1973) model of geometric Brownian motion. This model has a single parameter the volatility and the process m is an infinitely divisible process of independent, identically distributed increments over intervals of regular length. The increments have a normal distribution with zero mean and variance $\sigma^2 \Delta t$ over any interval of length Δt.

A generalization is obtained with richer unit time distributions capable of displaying skewness and excess kurtosis by taking m to be a Lévy process. A number of Lévy processes have been introduced in the option pricing literature as time changes of Brownian motion with drift and we mention the variance gamma process (Madan and Seneta (1990), Madan, Carr and Chang (1998)), the $CGMY$ model of (Carr, Geman, Madan and Yor (2002)), the normal inverse Gaussian model (Barndorff-Nielsen(1998)) and generalized hyperbolic model (Eberlein (2001)). It has been observed in these papers among others that a Lévy process can adequately synthesize option prices at a single maturity. However, it was observed in Konikov and Madan (2002) that no Lévy process would fit the option surface across strike and maturity as all such processes have skewness decreasing like the reciprocal of the square root of maturity while excess kurtosis (kurtosis-3) decreases like the reciprocal of maturity. For observed option surfaces for maturities below 2 years these moments may be estimated in a model independent way and are either constant or slightly increasing. Hence a synthesis of the surface required one to move away from a Lévy process as a model for m. An interesting property of a Lévy process is the specification of the entire probability law of the process from the infinitely divisible marginal law at unit time. We mention this property as we shall encounter it again shortly in the formulation of the Sato process.

Meanwhile, it was observed (Carr and Madan (2005)) that a surface of option prices was free of static arbitrage just if there existed a one dimensional Markov martingale that matches all the option prices. Hence we expect a model along the lines of equation (14.1) to fit the option surface. A solution was presented in Carr, Geman, Madan and Yor (2007) where one takes for m a Sato process. A subclass of the infinitely divisible laws known as self decomposable laws were identified as limit laws of sums of independent effects, much like the Gaussian law is the limit law via the central limit theorem, by Lévy (1937) and Khintchine (1938). Sato (1991) showed that one may construct a self similar additive process associated with any self decomposable random variable X at unit time. The self similarity leads in particular to marginal distributions at time t equal to the distribution of $t^{\gamma}X$. Sato (1991) then showed that one may form a process with independent but time inhomogeneous increments with these marginal laws. Hence much like the Lévy process, what Carr, Geman, Madan and Yor (2007) termed the Sato process also has a full specification of the process originating from the choice of a self decomposable law at unit time. An example of a self decomposable law is the unit time distribution of the variance gamma process and though the Lévy process associated with this unit time law fails to synthesize the option surface, the associated Sato process works well. This was observed in Carr, Geman, Madan and Yor (2007) and here we use this Sato process that is also referred to as the $VCSSD$ process.

One may alternatively synthesize the surface with a Heston (1993) type stochastic volatility model incorporating jumps for a better fit to shorter maturity options. The underlying dynamics are then no longer in the one dimensional Markov class represented by equation (14.1). The number of parameters goes up to eight or more depending on how the jump components are modeled. There is also an issue with the identification of parameters as the matrix of second derivatives of the estimation criterion with respect to the parameters often has at least two near zero eigenvalues generally associated with the eigenvectors representing mean reversion and the volatility of volatility.

We use, for dimensional reduction of the information content of the option surface, the Sato process associated with the variance gamma model at unit time. More formally the variance gamma process (VG), (Madan and Seneta (1990), Madan, Carr and Chang (1998)) $(X(t;\sigma,\nu,\theta), t \geq 0)$ is obtained on time changing a Brownian motion $(W(t), t \geq 0)$ with drift θ and volatility σ, where the time change is given by the gamma process $(g(t;\nu), t \geq 0)$ with mean rate unity and variance rate ν. Specifically, one may write

$$X(t;\sigma,\nu,\theta) = \theta g(t;\nu) + \sigma W(g(t;\nu)). \tag{14.2}$$

The variance gamma process is a process of independent identically distributed increments and is thus a Lévy process.

Let X be a random variable with the probability law of the variance gamma process at unit time. Define by $Y(t)$ a family of random variables indexed by t with the property that

$$Y(t;\sigma,\nu,\theta,\gamma) \stackrel{(d)}{=} t^{\gamma}X \tag{14.3}$$

whereby the probability law of $Y(t)$ is that of X scaled by t^{γ} for some scaling parameter γ.

The model for the stock price $S(t)$ at time t is obtained from the $VGSSD$ laws for an interest rate of r and a dividend yield of q as

$$\begin{aligned} S(t) &= S(0)\exp((r-q)t + \omega(t) + Y(t;\sigma,\nu,\theta,\gamma)) \\ \omega(t) &= -\log\left(E\left[\exp(Y(t;\sigma,\nu,\theta,\gamma))\right]\right). \end{aligned} \tag{14.4}$$

The characteristic function for the logarithm of the stock price at any maturity is obtainable in closed form and the methods of Carr and Madan (1999) then permit the derivation of $VGSSD$ option prices using the fast Fourier transform. One may then estimate the parameters each day from market quotes and we did this for 3664 days over the period *January 4, 1996* to *August 11, 2010*. The calibration criterion was the root mean square error between model and market prices.

We present first the result of a sample calibration on the single day, July 15, 2010. We also present information on parameter identification for this day. The number of options used here was 354 with maturities ranging from one month to 2.5 years. The estimated parameters were

$$\begin{aligned} \sigma &= 0.2215 \\ \nu &= 1.3026 \\ \theta &= -0.2549 \\ \gamma &= 0.6526. \end{aligned}$$

The fit statistics given by the root mean squared error ($rmse$), the average absolute error (aae) and the average percentage error (ape) defined as the average absolute error relative to the average option price were

$$\begin{aligned} rmse &= 1.6102 \\ aae &= 1.2851 \\ ape &= 0.0308. \end{aligned}$$

Figure 14.2 presents a graph of the market prices and the model fit.

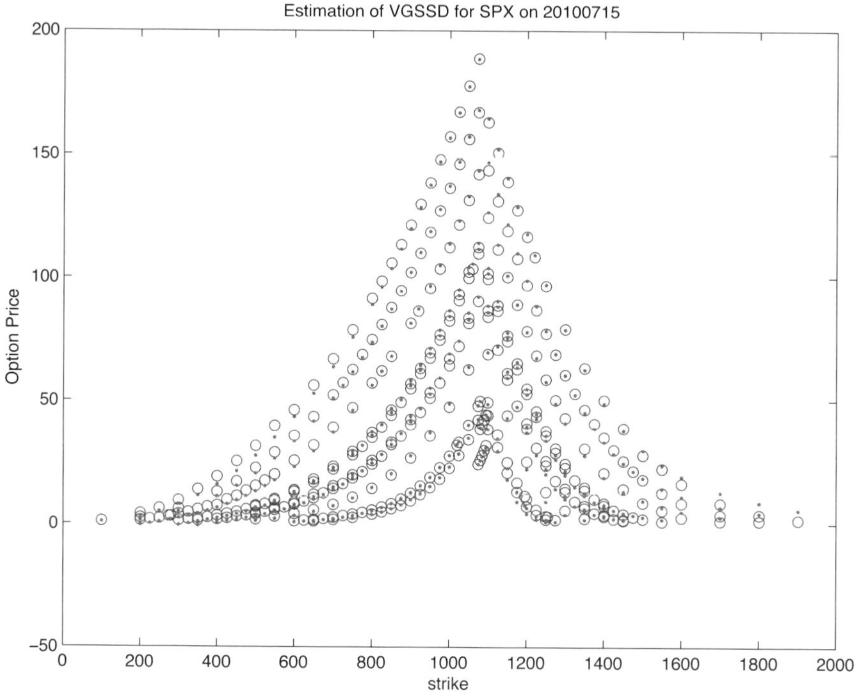

Fig. 14.1. Market prices are shown by circles while model prices are represented by dots. The model is VGSSD and the data is for 354 out of the money option prices on the S&P 500 index at market close on July 15 2010.

We now comment on a typical identification of the parameters. We do this in two ways. We first evaluate the vector of absolute values of the derivative of the pricing errors with respect to each parameter. We could compute the average value across all options of the absolute error derivatives and then sort this vector of averages. The highest value goes with the most identified parameter. If this value is small then the parameter is not being identified.

Performing this analysis we obtain the following vector of means for the absolute value of the derivative of pricing errors with respect to each of the four parameters. This is the computation of

$$\left| \frac{\partial \frac{1}{2} \left(w_i - \widehat{w}_i \right)^2}{\partial \Theta_k} \right| .$$

We then average across all options to obtain the following four values for our four parameters.

σ	ν	θ	γ
166.4686	8.8886	166.5251	42.5897

We may also see which options are doing the work in identifying what parameter. For this we graph, in Figure 14.2, for four maturities .2110, .6740, .9589, and 2.4384, the absolute value of the sensitivity of the squared error to each of the four parameters. We see that θ is being identified by the long dated puts, σ and ν are being identified by out of the money options on both sides and γ is identified by at the money short maturity options and out of the money long maturity options.

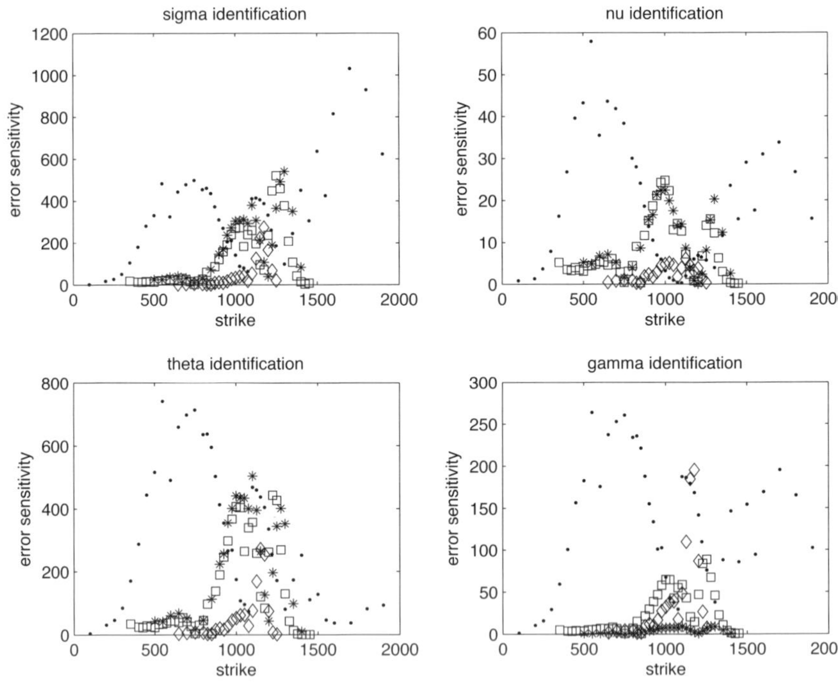

Fig. 14.2. Shown are the absolute size of the derivative of the squared error with respect to each of the four option surface parameters. Option strikes are presented on the x axis while the different maturities are represented by different shapes. The maturities are 0.2110 in diamond, 0.6740 in squares, 0.9589 in stars and 2.4384 in dots.

The second approach is to compute the matrix of second order derivatives of the criterion with respect to the parameters. This is the matrix

$$\frac{\partial^2}{\partial \Theta_k \partial \Theta_l} \sum_i \frac{1}{2} \left(w_i - \widehat{w}_i\right)^2 .$$

For our example we obtain the following four by four matrix.

$$\begin{bmatrix} 32641.2 & 390.29 & -28948.4 & 5486.3 \\ 390.29 & 149.6 & -1024.2 & -83.58 \\ -28948.4 & -1024.2 & 33482.9 & -4548.9 \\ 5486.3 & -83.58 & -4548.9 & 3471.0 \end{bmatrix} .$$

We perform an eigenvalue analysis of this matrix to get four eigenvalues in decreasing order

$$\begin{bmatrix} 62875.8 \\ 4449.3 \\ 2340.3 \\ 79.2 \end{bmatrix} .$$

We may list the four eigenvectors as follows.

0.6978	0.6262	−0.3431	0.0560
0.0157	−0.1119	−0.0099	0.9936
−0.7062	0.6781	−0.1848	0.0857
0.1185	0.3682	0.9209	0.0487

We see that in the first eigenvector, a combination of σ and θ, is most identified. Next we have a combination of σ, θ, γ, followed by γ and then ν. The identification order is therefore $\sigma, \theta, \gamma, \nu$.

We present, in Table 14.1, the summary statistics for the parameter estimated over the entire 15 year period. We present in Table 14.2 the fit statistics along with the number of options used in the estimation. We observe from these tables that the parameters are well estimated with reasonably small standard deviations or inter quartile ranges. One is essentially capturing around 120 option prices to within 3% with four parameters regularly daily for some 15 years. Hence we take as the drivers for the option surface the level of the spot coupled with volatility σ, volatility of volatility ν, skewness θ and the term structure of volatility γ. There are then five underliers for the option surface.

Table 14.1. Summary statistics for VGSSD parameter estimates

	Parameter σ	ν	θ	γ
mean	0.1917	0.8893	-0.1618	0.5565
std	0.0515	0.2640	0.0723	0.0672
skew	1.1014	2.4275	-1.3007	-0.4503
kurtosis	5.3244	48.8275	5.9915	3.6107
quantiles				
1%	0.1157	0.3943	-0.3932	0.3852
5%	0.1234	0.4956	-0.2932	0.4389
10%	0.1296	0.5657	-0.2498	0.4710
25%	0.1538	0.6883	-0.2043	0.5138
50%	0.1864	0.9024	-0.1536	0.5591
75%	0.2185	1.0487	-0.1062	0.6034
90%	0.2512	1.2088	-0.0863	0.6421
95%	0.2791	1.2934	-0.0791	0.6604
99%	0.3629	1.5148	-0.0639	0.6891

3664 days from Jan. 4 1996 to Aug. 11 2010

14.3. Methodology for estimating risk neutral quadratic covariations

This section generalizes the Black and Scholes (1973) and Merton (1973) partial differential equation with a view towards permitting estimation of risk neutral quadratic covariations. We first show that in the presence of options trading, parameter variations may be hedged out and therefore only second or higher order risks are held and they are then compensated by the option theta.

For any strike K, maturity T and type u, equal to one for calls and zero otherwise, on any date $t < T$ we synthesize the price of the option of this strike and maturity on this date as

$$w_t(K,T,u) = \Phi_t\left(S_t, \sigma_t, \nu_t, \theta_t, \gamma_t; K, T, u\right), \tag{14.5}$$

where the function Φ_t is given by the option pricing formula of the $VGSSD$ model. We therefore see the option surface as driven by these five underlying variables. If we view the underlying stock as an underlying futures level then we know that the first order movement in this variable may be delta hedged at zero cost. Option traders have known for some time now how to hedge movements in the parameters of any model. We briefly identify two ways of describing the cash flow to be accessed in hedging a specific

Table 14.2. Fit statistics for VGSSD estimation

	rmse	aae	ape	nop
mean	0.5881	0.4571	0.0326	120.47
std	0.3433	0.2272	0.0165	28.52
quantiles				
1%	0.1789	0.1402	0.0101	79
5%	0.2383	0.1889	0.0133	87
10%	0.2906	0.2326	0.0159	91
25%	0.3836	0.3106	0.0218	101
50%	0.5030	0.3994	0.0300	118
75%	0.7168	0.5628	0.0396	135
90%	0.9469	0.7481	0.0501	154
95%	1.0944	0.8501	0.0611	169
99%	1.9725	1.1884	0.0971	206

rmse is the root mean square error;
aae is the average absolute error
ape is the average absolute error over the average option price;
nop is the number of options used in the fit.

parameter. Such a construction is of interest in its own right and has links to the role played by the score function in econometrics. Furthermore we are able to see that this hedge can be accomplished at zero cost.

Say we wish to construct a cash flow $c(S)$ that rises in value when a parameter η is moved. The value of $c(S)$, ignoring discounting issues, for a risk neutral stock price density $f(S)$ appropriate for the cash flow maturity being considered is

$$w_C = \int c(S)f(S)dS. \tag{14.6}$$

The partial derivative of this value with respect to the parameter η is then

$$\begin{aligned}\frac{\partial w_C}{\partial \eta} &= \int c(S)f_\eta(S)dS \\ &= \int c(S)\frac{f_\eta(S)}{f(S)}f(S)dS. \end{aligned} \tag{14.7}$$

To ensure a positive sensitivity of the value to the cash flow held it is suggested that the we hold the cash flow

$$c(S) = \frac{f_\eta(S)}{f(S)}. \tag{14.8}$$

One may scale the position by the constant a, where

$$a = \left(\int \left(\frac{f_\eta(S)}{f(S)} \right)^2 f(S)dS \right)^{-1}$$

in order to exactly earn the parameter innovation $d\eta$. The initial cost of this cash flow is

$$\int \frac{f_\eta(S)}{f(S)} f(S)dS = 0. \tag{14.9}$$

Hence one may in principle hedge out all the delta exposures at zero cost.

Alternatively one may just compute the derivative of option prices with respect to a parameter η say the vector w_η for a set of strikes and maturities. Let the vector of option prices be w for the same set of strikes and maturities. The vector α of positions in these options, with

$$\alpha = w_\eta - \frac{w_\eta' w}{w' w} w$$

has zero cost and a positive exposure to η. More generally one may simultaneously organize in addition to being self financed, that the position is neutral to movements in other parameters and the spot price. Further details for such positions are described in Section 14.5.

We show in Figure 14.3 the positions in 21 strikes at two maturities .25 and .5 that trade the four parameters $\sigma, \nu, \theta, \gamma$ of the $VGSSD$ model in a self financed way that is also neutral to moves in the other drivers. The associated cash flows are presented in Figure 14.3.

Suppose now for example that all the underliers are in fact diffusing in some manner. An application of Ito's lemma would imply on generalizing the Black–Merton–Scholes arbitrage argument that the option theta must compensate for all the second order risks being held. Specifically that a background partial differential equation of the form

$$\begin{aligned} \Phi_T = -\Phi_t = \mathrm{Trace}\Big(&\nabla^2 \Phi_{(S,\sigma,\nu,\theta,\gamma)} \\ &\cdot E_t^Q \left[(\Delta S, \Delta\sigma, \Delta\nu, \Delta\theta, \Delta\gamma) (\Delta S, \Delta\sigma, \Delta\nu, \Delta\theta, \Delta\gamma)^T \right] \Big), \end{aligned} \tag{14.10}$$

holds through time, for some underlying risk neutral measure.

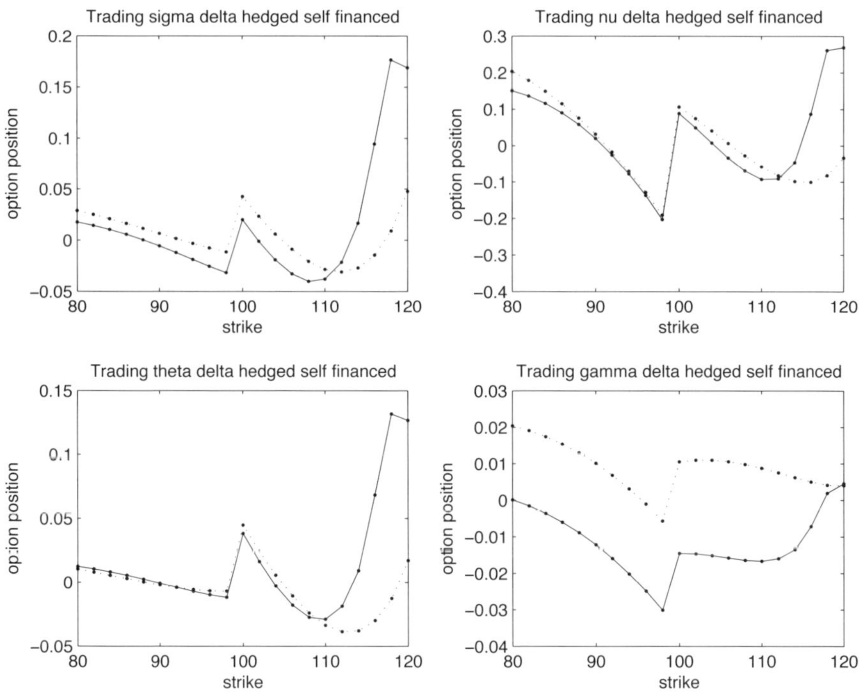

Fig. 14.3. Option positions in three month options in solid lines with six month positions in dashed lines. The graphs display positions in options for taking positions in each of four parameters delta hedged with respect to the other four surface drivers.

Under a true underlying multidimensional diffusion we have as a consequence of Girsanov's theorem that

$$\begin{aligned} &E_t^Q\left[(\Delta S, \Delta\sigma, \Delta\nu, \Delta\theta, \Delta\gamma)(\Delta S, \Delta\sigma, \Delta\nu, \Delta\theta, \Delta\gamma)^T\right] \\ &= E_t^P\left[(\Delta S, \Delta\sigma, \Delta\nu, \Delta\theta, \Delta\gamma)(\Delta S, \Delta\sigma, \Delta\nu, \Delta\theta, \Delta\gamma)^T\right] \end{aligned} \tag{14.11}$$

where the measure P is the underlying physical measure.

More generally for special semimartingale underliers driven by pure jump processes the background equation is a partial integro-differential equation of the form

$$\Phi_T = \int_{\mathbb{R}^n} (\Phi(X(t_- + x)) - \Phi(X(t_-)) - \Phi_X(X(t_-))x) k_t^Q(x) dx, \tag{14.12}$$

where $k_t^Q(x)dx$ is the risk neutral jump compensating measure. Approximating the integrand on the right to second order we may approximate

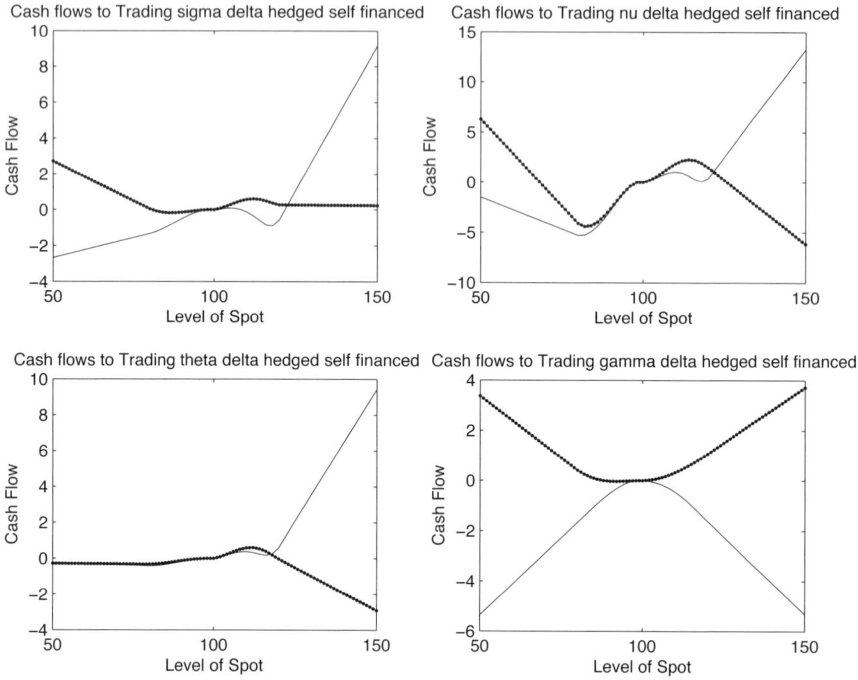

Fig. 14.4. Cash flows to three month positions in solid line and six month cash flows in dotted line. The graphs display cash flows accessed for taking positions in each of four parameters delta hedged with respect to the other four surface drivers.

as

$$\int_{\mathbb{R}^n} (\Phi(X(t_- + x)) - \Phi(X(t_-) - \Phi_X(X(t_-))x)k_t^Q(x)dx \tag{14.13}$$

$$\approx \frac{1}{2}\sum_{ij} \left(\nabla^2\Phi\right)_{ij} \int_{-\infty}^{\infty}\int_{-\infty}^{\infty} x_i x_j k_t^Q(x_i, x_j)dx_i dx_j \tag{14.14}$$

$$= \text{Trace}\Big(\nabla^2\Phi_{(S,\sigma,\nu,\theta,\gamma)} \cdot E_t^Q\left[(\Delta S, \Delta\sigma, \Delta\nu, \Delta\theta, \Delta\gamma)(\Delta S, \Delta\sigma, \Delta\nu, \Delta\theta, \Delta\gamma)^T\right]\Big).$$

But now in such jump contexts the risk neutral quadratic covariation compensations may differ from their physical counterparts. We shall specifically investigate and report on the differences.

To recover the risk neutral quadratic compensating covariations in equation (14.10) one may consider regressing at a point of time the theta of a number of options across the strike maturity spectrum on the vectorized

Table 14.3. Unconstrained regression results for VGSSD option θs projected[a] on second order VGSSD Greeks

	Coeff.	t-stat
constant	0.2730	2.3013
SS	552.0142	32.4805
$\sigma\sigma$	-9.3172	-3.5476
$\nu\nu$	-4.7520	-2.4295
$\theta\theta$	-10.0053	-4.5849
$\gamma\gamma$	5.1561	24.8002
$S\sigma$	-0.0775	-8.7767
$S\nu$	0.0066	0.0968
$S\theta$	-0.0638	-5.3177
$S\gamma$	0.0273	4.5583
$\sigma\nu$	-0.6372	-7.1996
$\sigma\theta$	-0.2671	-11.0458
$\sigma\gamma$	-0.3005	-4.4313
$\nu\theta$	-0.2883	-13.6287
$\nu\gamma$	-0.1194	-3.6721
$\theta\gamma$	-0.3137	-4.8464

[a]Projection at a sample parameter set of $\sigma = 0.2$, $\nu = 0.6$, $\theta = -0.4$ and $\gamma = 0.5$. The R^2 was 99.97%.

matrix of second order derivatives. The coefficients of these second order terms would then provide us with estimates of our risk neutral quadratic covariations for the particular day. We ran such a regression of thetas on the fifteen second order derivatives and cross derivatives for our five underliers.

We did this for 126 options for 21 strikes from 80 to 120 in steps of 2 dollars and 6 maturities from *one* to *six* months. We took the parameter values at $(.2, .6, -.4, .5)$. We then regressed the theta on all fifteen second order terms with the result presented in Table 14.3. The R^2 was 99.97% and the $t-stats$ were as follows. All but one of the second order terms are significant.

We know from our equation that the theta is linear in the second order derivatives but that the coefficients refer to the risk neutral quadratic covariations,

$$E_t^Q\left[(\Delta S, \Delta\sigma, \Delta\nu, \Delta\theta, \Delta\gamma)(\Delta S, \Delta\sigma, \Delta\nu, \Delta\theta, \Delta\gamma)^T\right]$$

and this matrix is by construction a symmetric positive definite matrix. When we run a regression we may increase the R^2 on occasion by placing a negative value on a diagonal term and this is potentially the effect of noise. So we choose to minimize the deviation between the true theta and the theta predicted by our model, which is linear in the second order Greeks, but we force the coefficients to form a symmetric positive definite matrix. This may be done via a Cholesky decomposition whereby we model the matrix L by

$$L = \begin{bmatrix} x_2 & x_7 & x_8 & x_9 & x_{10} \\ 0 & x_3 & x_{11} & x_{12} & x_{13} \\ 0 & 0 & x_4 & x_{14} & x_{15} \\ 0 & 0 & 0 & x_5 & x_{16} \\ 0 & 0 & 0 & 0 & x_6 \end{bmatrix}. \tag{14.15}$$

We then form

$$C = L'L \tag{14.16}$$

and generate the predicted theta by

$$\widehat{w}_T = x_1 + Tr\left([\nabla^2 w]C\right). \tag{14.17}$$

We then find the vector x that minimizes

$$\theta_{\text{crit}} = (w_T - \widehat{w}_T)'(w_T - \widehat{w}_T) \tag{14.18}$$

and we store the upperdiagonal of the matrix C as our estimated coefficients. We define the R^2 by 1− residual variance divided by the total variance.

For the 3664 days from January 4, 1996 to August 11, 2010 we present in Figure 14.3 a graph of the resulting $R^{2\prime}s$. We see that the $R^{2\prime}s$ have come down relative to running a regression but they are still at very good levels and we have in addition a positive definite matrix of coefficients that is the market's risk neutral expectation of quadratic moves. This is essentially the market price of the quadratic moves that is built into the option theta.

We present in Figure 14.3 a time series graph of the prices of quadratic risks for the fifteen quadratic risk components.

We also present a table of the statistics on the prices of quadratic risk prices and the cross prices. We present the quantiles at 1, 5, 10, 25, 50, 75, 90, 95 and 99 percentage points along with the mean, the standard deviation, skewness and kurtosis. We observe that the price of squared price moves is quite skewed with a high level of kurtosis indicating substantial variability. A similar pattern is observed for the prices of squared changes in skewness or the parameter θ. The prices of squared changes in ν and γ are more stable.

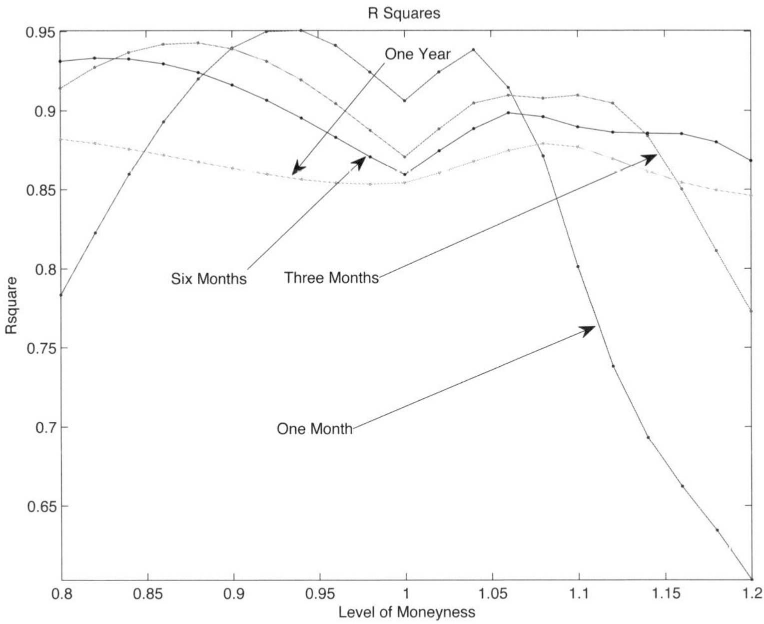

Fig. 14.5. R^2 from 3664 constrained least squares estimations enforcing positive semidefiniteness conditions on coefficients.

14.4. Principal components analysis of real-world and risk-neutral quadratic covariations

In addition to the five dimensional risk neutral quadratic covariations estimated by our constrained procedure, reported in Section 14.3, we may evaluate real world covariations from the data on the time series of the level of the spot and the four parameters calibrated each day. We report in this section the results of such an exercise. For the real world estimation we start in day 127 and use the data on six months of changes in the spot price taken as returns with an initial value at 100, and the changes in the four parameters. We compute each day a five dimensional real world covariance matrix for these five underliers. We then have for days 127 to 3664 two five dimensional covariance matrices, one risk neutral and one real world.

We first report on the sum of the eigenvalues for these two matrices. This gives us the total risk neutral five dimensional variance and its real world counterpart. We present in Figure 14.4 a graph of this total variance.

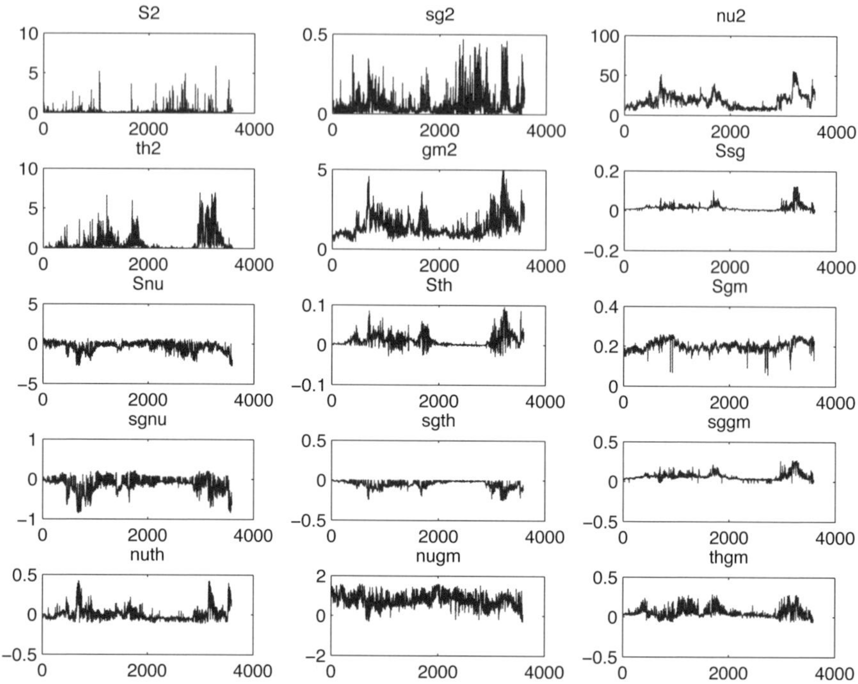

Fig. 14.6. Time series of quadratic risk prices implied by option thetas on SPX options. The prices are from constrained projections of thetas on second order Greeks. The projections are run daily for 3664 days covering the period January 4 1996 to August 11 2010.

We observe that the total risk neutral variance is substantially higher than its real world counterpart and is on average 6.16 times its real world counterpart. We next present the proportion of the total variance captured by each of the five eigenvalues both risk neutrally and under the real world measure. We consider eight quantile ranges for the largest eigenvalue with end points given by 1, 5, 10, 25, 50, 75, 90, 95 and 99 percent levels. In each of these eight ranges we report the proportion of the variance covered by the five eigenvalues. There are then two 8 by 5 matrices of variance proportions, one risk neutral and the other real world. These are presented in Tables 14.3 and 14.4 respectively.

We see that the first eigenvalue explains the bulk of the total variance in each case with the real world proportion attributed to the first eigenvalue being higher than in the risk neutral case. To understand the structure of the eigenvectors associated with these eigenvalues we first constructed the

Table 14.4. Quantiles of risk neutral variations and covariations estimated daily from θ regressions constrained for positive definiteness.

Second Order Risk Price Statistics for SPX options 19960104 to 20100811

	S^2	σ^2	ν^2	θ^2	γ^2
0.01	0.0202	0.0009	5.5461	0.0027	0.4560
0.05	0.0333	0.0035	7.4951	0.0057	0.7230
0.1	0.0373	0.0064	8.6690	0.0091	0.8156
0.25	0.0473	0.0121	10.7653	0.0342	0.9620
0.5	0.0741	0.0296	17.4635	0.1804	1.1746
0.75	0.1505	0.0704	24.1860	0.7303	1.7318
0.9	0.3101	0.1385	29.2994	2.0655	2.4092
0.95	0.4924	0.1997	35.8822	3.5062	2.7784
0.99	6.6539	0.4742	55.6990	6.9711	4.9741
mean	0.1568	0.0540	18.5786	0.6463	1.4185
std	0.3360	0.0654	8.9145	1.1036	0.6758
skw	8.4087	2.5096	1.1154	2.7239	1.7834
krt	95.0214	10.8882	4.7635	11.0247	7.0239

Spot Surface Cross Statistics

	$S\sigma$	$S\nu$	$S\theta$	$S\gamma$
0.01	-0.0129	-2.7633	-0.0281	0.0545
0.05	0.0036	-1.6456	-0.0025	0.1674
0.1	0.0045	-1.2348	-0.0004	0.1743
0.25	0.0065	-0.6689	0.0012	0.1872
0.5	0.0121	-0.2657	0.0079	0.2011
0.75	0.0174	-0.0300	0.0192	0.2183
0.9	0.0289	0.2394	0.0319	0.2366
0.95	0.0450	0.3543	0.0432	0.2439
0.99	0.1223	0.6252	0.0959	0.2577
mean	0.0154	-0.3996	0.0120	0.2027
std	0.0157	0.5758	0.0155	0.0232
skw	3.5429	-1.2574	1.7615	-0.2493
krt	19.1847	5.0166	7.7392	4.5588

Surface Cross Statistics

	$\sigma\nu$	$\sigma\theta$	$\sigma\gamma$	$\nu\theta$	$\nu\gamma$	$\theta\gamma$
0.01	-0.8577	-0.2518	-0.0191	-0.1145	-0.2993	-0.0653
0.05	-0.5533	-0.1362	0.0217	-0.0740	0.2072	-0.0248
0.1	-0.4072	-0.1086	0.0282	-0.0611	0.3680	-0.0064
0.25	-0.2143	-0.0720	0.0363	-0.0449	0.5483	0.0179
0.5	-0.0989	-0.0394	0.0552	-0.0165	0.7173	0.0340
0.75	-0.0517	-0.0136	0.0762	0.0269	0.9677	0.0700
0.9	0.0575	-0.0057	0.1167	0.0946	1.1845	0.1302
0.95	0.0957	-0.0008	0.1631	0.1883	1.3518	0.1830
0.99	0.2078	0.0066	0.2721	0.4323	1.5927	0.2870
mean	0.1446	-0.0490	0.0645	0.0039	0.7500	0.0497
std	0.1776	0.0438	0.0429	0.0773	0.3186	0.0562
skw	-1.3123	-1.4224	2.0480	2.2158	0.0447	1.4216
krt	5.1931	5.5428	8.5213	9.3067	3.2483	5.5403

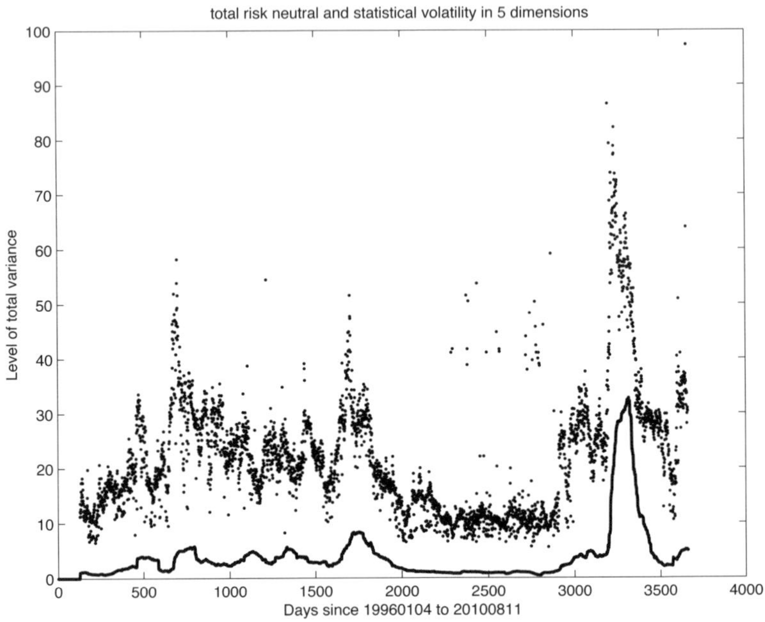

Fig. 14.7. Total risk neutral variance in dotted lines and total risk real world variance in a solid line. These are measured by the sum of the respective covariation matrix eigenvalues.

average of the squared eigenvector and determined the dominant squared component. We then signed this component positively adjusting the signs of the other components to do this. We then averaged the eigenvectors to determine what variables are at play in each eigenvector on average. Tables 14.5 and 14.6 present the averaged eigenvectors for the risk neutral and real world covariances.

We observe from Table 14.5 that the first eigenvector risk neutrally prices movements in the volatility of volatility, the second is related to increases in the term structure of volatility coupled with increases in the spot and the level of skewness. The third eigenvector represents an increase in skewness with decreasing volatility and volatility spreads. The fourth is associated with a rise in the spot price and volatility and a decline in the volatility spreads. The last one is associated with increased volatility and skewness.

Under the real world measure the first eigenvector is for a movement in the spot price. Then we have the volatility of volatility followed by the term structure of volatility, the skewness coupled with volatility and finally

Table 14.5. Risk neutral proportion of variance due to each eigenvalue sorted by quantiles for the largest eigenvalue

quantile	Order of eigenvalue 1	2	3	4	5
1 – 5	0.7916	0.1489	0.0528	0.0066	0.0001
5 – 10	0.8383	0.1073	0.0492	0.0051	0.0001
10 – 25	0.8734	0.0873	0.0357	0.0035	0.0002
25 – 50	0.8967	0.0804	0.0202	0.0025	0.0001
50 – 75	0.9131	0.0716	0.0132	0.0020	0.0001
75 – 90	0.9238	0.0641	0.0107	0.0014	0.0001
90 – 95	0.9310	0.0577	0.0100	0.0012	0.0001
95 – 99	0.9403	0.0499	0.0085	0.0013	0.0000

There were 3538 risk neutral 5 by 5 covariation matrices in the sample extracted by daily constrained projections of option thetas on the second order Greeks.

Table 14.6. Real World Proportion of variance due to each eigenvalue

quantile	Order of eigenvalue 1	2	3	4	5
1-5	0.8572	0.1420	0.0007	0.0001	0.0000
5-10	0.9689	0.0307	0.0004	0.0000	0.0000
10-25	0.9853	0.0141	0.0005	0.0001	0.0000
25-50	0.9921	0.0076	0.0003	0.0000	0.0000
50-75	0.9954	0.0043	0.0002	0.0000	0.0000
75-90	0.9982	0.0017	0.0001	0.0000	0.0000
90-95	0.9988	0.0011	0.0001	0.0000	0.0000
95-99	0.9990	0.0009	0.0001	0.0000	0.0000

There were 3538, 5 by 5 real world covariation matrices in the sample estimated from 126 days of data on moves in the five drivers. The eigenvalues are sorted by quantiles for the largest eigenvalue.

volatility coupled with a decrease in skewness and the term structure of volatility.

14.5. Quadratic risk premia or the difference in risk-neutral and real-world covariation matrices

Risk premia has been discussed on asset returns in financial economics for quite some time. Our interest here is in the difference between the real world and risk neutral quadratic and cross covariations. We have already

Table 14.7. Risk Neutral averaged eigenvectors

eigenvector	S	σ	ν	θ	γ
1	−0.0268	−0.0067	0.9874	−0.0011	0.0581
2	0.1733	0.0368	−0.0440	0.1926	0.8323
3	−0.0647	−0.2749	−0.0023	0.6147	−0.1415
4	0.6818	0.1664	0.0217	−0.0348	−0.1378
5	−0.0539	0.6422	0.0101	0.3201	−0.0503

noted that in a pure diffusion context these are the same but more generally they may differ. We observed in Section 14.4 that the structure of risk neutral covariations is very different from that of its real world counterpart.

We now wish to analyze the difference between the risk neutral covariation and its real world counterpart. For a vector of underlying risks X we wish to consider

$$E_t^Q[(\Delta X)(\Delta X)^T] - E_t^P\left[(\Delta X)(\Delta X)^T\right].$$

Now for a pricing kernel Λ we know that

$$\begin{aligned} E_t^Q\left[(\Delta X)(\Delta X)^T\right] &= E_t^P\left[\Lambda(\Delta X)(\Delta X)^T\right] \\ &= E_t^P\left[(\Delta X)(\Delta X)^T\right] + Cov^P\left((\Lambda - 1)(\Delta X)(\Delta X)^T\right). \end{aligned}$$

So we may infer covariations of quadratic risks with the pricing kernel by evaluating the difference

$$Cov^P\left((\Lambda-1)(\Delta X)(\Delta X)^T\right) = E_t^Q\left[(\Delta X)(\Delta X)^T\right] - E_t^P\left[(\Delta X)(\Delta X)^T\right].$$

Though both the risk neutral and real world quadratic covariation matrices are symmetric and positive definite the difference is just symmetric and may possess both positive and negative eigenvalues. We might enquire whether risk neutral covariations exceed real world covariations in all directions or are there directions in which risk neutral quadratic variations are lower. To evaluate this proposition we computed the eigenvalues and eigenvectors for the difference between our risk neutral and real world covariations. We then determined the dominant term in each eigenvector on the basis of the dominant squared term and signed the vector for this to be positive. We then average the eigenvalues and signed eigenvectors to present in Table 14.7 these averages.

We see from Table 14.7 and an examination of the eigenvectors and eigenvalues that only one eigenvalue is negative and all the others are pos-

Table 14.8. Real world averaged eigenvectors

eigenvector	S	σ	ν	θ	γ
1	0.9993	−0.0009	0.0212	0.0055	0.0152
2	−0.0221	0.0428	0.9844	0.0964	0.0372
3	−0.0128	0.0848	−0.0442	−0.0840	0.8879
4	−0.0038	0.1108	−0.0770	0.8360	0.0373
5	0.0035	0.9131	−0.0153	−0.1270	−0.1007

Table 14.9. Average eigenvalues and signed eigenvectors for the difference between risk neutral and real world covariation matrices.

eigenvector	S	σ	ν	θ	γ
18.8468	−0.0274	−0.0067	0.9876	−0.0012	0.0582
1.6865	0.0613	0.0383	−0.0484	0.1951	0.8513
0.5065	−0.0103	−0.3685	−0.0024	0.6679	−0.1617
0.0237	−0.0023	0.8142	0.0090	0.3551	−0.0542
−3.5792	0.9772	0.0009	0.0234	0.0210	0.0578

itive. Hence risk neutral quadratic variations exceed real world quadratic variations in all directions in a four dimensional subspace. The only negative value is associated with the quadratic variation in the stock when there are no other movements in the surface. The highest premium is for variations in the volatility of volatility followed by the term structure of volatility then skewness and finally the volatility itself.

Finally we illustrate positioning to sell the largest eigenvalue and buy the lowest eigenvalue of the difference between the risk neutral and real world covariation matrices, keeping the position hedged with respect to all five dimensions of motion. Since in general we sell exposures with high price and low probability and buy cash flows with high probability and low price, Table 14.7 suggests that one takes a position in $(\Delta S)^2$ at the same time as one sells $(\Delta \nu)^2$. We may wish to do this at zero cost and simultaneously have all deltas zeros. Hence we seek a position α in say 21 strikes at maturities .25 and .5 that satisfies $a'w_\eta = 0$ for $\eta = (S, \sigma, \nu, \theta, \gamma)$ while $\alpha'(w_{SS} - w_{\nu\nu}) > 0$. This is simply constructed by defining the matrix

$$M = [w, w_\sigma, w_\nu, w_\theta, w_\gamma, w_S, w_{SS} - w_{\nu\nu}]$$

with

$$\alpha = M(M'M)^{-1}(0,0,0,0,0,0,1)'.$$

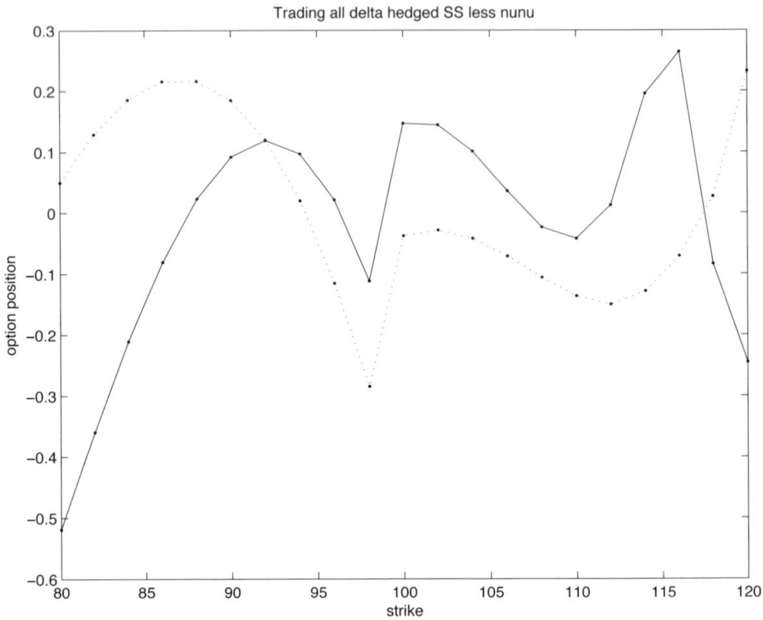

Fig. 14.8. Option Positions at 3 months in solid line and six months in dashed line. The positions are for buying exposure to squared moves in the stock less squared moves in the volatility of volatility delta hedged with respect to the five drivers.

We present in Figure 14.5 the positions taken in the options traded and in Figure 14.5 the cash flow accessed by such a trade for the parameter setting $\sigma = .25, \nu = 1.2$, $\theta = -.4$, $\gamma = .6$.

14.6. Conclusion

We first observe that the option surface may be effectively synthesized daily using the level of the spot price and the four parameters of the Sato process based on the variance gamma process introduced by Carr, Geman, Madan and Yor (2007). A time series is constructed for this five dimensional set of factors driving the surface of S&P 500 index option prices.

We then observe that parameter movements can be hedged at zero cost and a delta hedged position then earns in its theta, compensation for the exposure to the fifteen second order squared and cross product terms. The coefficients of compensation form an estimate for the risk neutral quadratic covariation between the five surface drivers. A constrained optimization forcing positive semidefinite coefficients is then employed to estimate risk

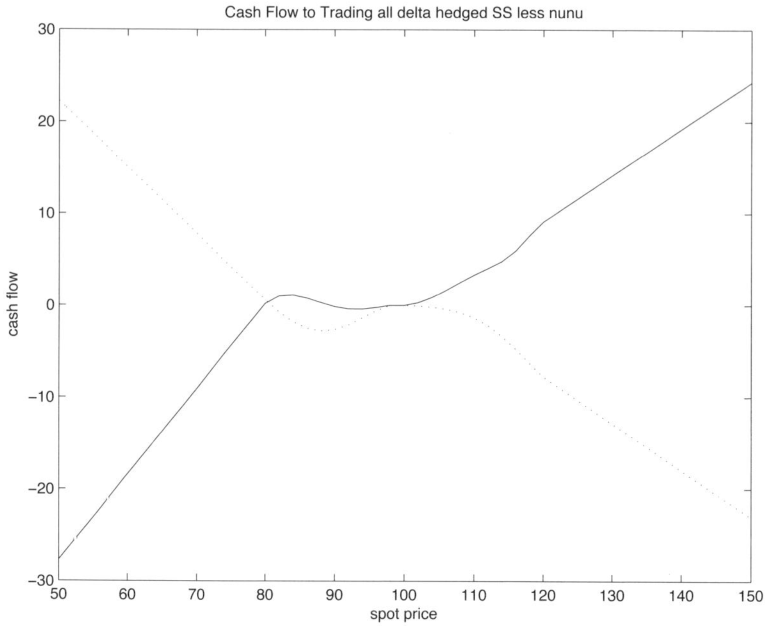

Fig. 14.9. Cash flows at three months in solid line and six months in dashed line. The cash flows are for exposure to squared moves in the spot price less squared moves in the volatility of volatility delta hedged for moves in all the five surface drivers.

neutral covariations. Simultaneously real world covariations are estimated and it is shown that the two sets of covariations are quite different.

We then ask if quadratic variations in all directions risk neutrally exceed their real world counterparts. This is not the case and quadratic variations of the stock price unaccompanied by other movements of the surface have a lower risk neutral expectation. All other directions have a higher risk neutral quadratic variation from their real world counterparts. We then illustrate an option portfolio that takes a zero cost position hedged with respect to all drivers that goes long the lowest eigenvalue and shorts the largest one.

This is a first paper exploring the structure of risk neutral and real world quadratic covariations. We anticipate much theoretical and empirical work remains to be done and will prove to be useful in exploring these issues.

References

[1] Arslan, M., G. Eid, Jean El Khoury and J. Roth (2009), "The Gamma-Vanna-Volga Cost Framework for constructing Implied Volatility Curves," Working Paper, Deutsche Bank.

[2] Bakshi, G., D. Madan and G. Panayotov (2010a),"Returns of claims on the upside and the viability of U-shaped pricing kernels," *Journal of Financial Economics*, 97, 130-154.

[3] Bakshi, G., D. Madan and G. Panayotov (2010b),"What do Returns on VIX Puts reveal about the Volatility Market," Working Paper, Robert H. Smith School of Business, University of Maryland.

[4] Barndorff-Nielsen, O.E. (1998), "Processes of normal inverse Gaussian type,"*Finance and Stochastics, 2, 41-68.*

[5] Black, F. and M. Scholes (1973), "The pricing of corporate liabilities," *Journal of Political Economy*, 81, 637-654.

[6] Carr, P., H. Geman, D. Madan and M. Yor (2002), "The fine structure of asset returns: An empirical investigation," *Journal of Business*, 75, 2, 305-332.

[7] Carr, P., H. Geman, D. Madan and M. Yor (2007), "Self Decomposability and Option Pricing," *Mathematical Finance*, 17, 31-57.

[8] Carr, P. and D. Madan (1999), "Option valuation using the fast Fourier transform," *Journal of Computational Finance*, 2, 61-73.

[9] Carr, P. and D. Madan (2005), "A Note on Sufficient Conditions for No Arbitrage," *Finance Research Letters,* 2, 125-130.

[10] Cont, R. and J. da Fonseca (2002), "Dynamics of implied volatility surfaces," *Quantitative Finance*, 2, 45-60.

[11] Eberlein, E. (2001), "Application of generalized hyperbolic Lévy motions to finance," In *Lévy Processes: Theory and Applications*, (Eds), O.E. Barndorff-Nielsen, T. Mikosch, and S. Resnick, Birkhäuser Verlag, 319-326.

[12] Galloway, M. L., and C. A. Nolder (2008), "Subordination, selfsimilarity and option pricing," *Journal of Applied Mathematics and Decision Sciences*, Article Number 397028.

[13] Heston, S. (1993), "A closed-form solution for options with stochastic volatility with applications to bond and currency options." *Review of Financial Studies*, 6, 327-343.

[14] Khintchine, A. Ya. (1938), "Limit laws of sums of independent random variables," ONTI, Moscow, (Russian).

[15] Konikov, M. and D. B. Madan (2002), "Option Pricing Using Variance Gamma Markov Chains," *Review of Derivatives Research*, 5, 81-115.

[16] Lévy, P. (1937), *Théorie de l'Addition des Variables Aléatoires*, Gauthier-Villars, Paris.

[17] Lipton, A. and W. McGhee (2002) "Universal Barriers," *Risk*, May, 81-85.

[18] Madan, D., P. Carr and E. Chang (1998), "The variance gamma process and option pricing," *European Finance Review*, 2, 79-105.

[19] Madan, D. and E. Seneta (1990), "The variance gamma (V.G.) model for share market returns," *Journal of Business*, 63, 511-524.

[20] Merton, R.C. (1973), "Theory of rational option pricing," *Bell Journal of Economics and Management*, 4, 141-183.

[21] Nolder, C. A. (2003), "Selfsimilar additive processes and financial modelling," *Proceedings of the Financial Engineering and Applications Conference,* IASTED, Banff, Alberta.

[22] Sato, K. (1991), "Self similar processes with independent increments," *Probability Theory and Related Fields*, 89, 285-300.

[23] Werker, B., M. J. Boes and F. Drost (2007), "Nonparameteric risk-neutral Joint return and volatility distributions," Working Paper, VU University Amsterdam.

[24] Wystup, U. (2003), "The Market Price of One-Touch Options in Foreign Exchange Markets," *Derivatives Week*, 12, 1-4.

Chapter 15

A dynamic portfolio approach to asset markets and monetary policy

Eckhard Platen[1] and Willi Semmler[2]

[1]*School of Finance & Economics and Department of Mathematical Sciences, University of Technology, Sydney, PO Box 123, Broadway, NSW, 2007, Australia*
Eckhard.Platen@uts.edu.au

[2]*Department of Economics, New School, New York, 79 Fifth Ave, New York, NY 10003, USA, and Center for Empirical Macroeconomics, Bielefeld University, Germany*
semmlerw@newschool.edu

Monetary policy in many countries has pursued the concept of inflation targeting. Here interest rates are supposed to respond to an inflation gap and output gap. Despite long term continuing growth of the world financial assets, recently, monetary policy, in particular in the U.S. after the subprime credit crisis, was challenged by severe disruptions and a meltdown of the financial market. Subsequently, one has been in search of a type of monetary policy that does allow to influence in an appropriate manner the investor's behavior and, thus, the dynamics of the economy and its financial market. Based on a dynamic portfolio approach this chapter allows one to study the interaction between investors' strategic behavior and monetary policy. The article derives relationships that explain how monetary authorities could support a targeted total wealth evolution by setting the short term interest rate appropriately in relation to inflation rate, expected return from equity, risk aversion, equity volatility, leverage of the market and consumption rate. Important is the finding that the inflation rate needs to have a certain minimal level to allow the interest rate to be a viable control instrument. A particular interest rate policy has been identified for an optimal regime. If the proposed interest rate rule is applied, then the consumption rate will remain stable and the inflation rate can be kept close to a minimal possible level. Additionally, in the case of a financial market meltdown and crisis the revealed relationships indicate in which direction to act to bring the economy back on track.

Contents

15.1. Introduction

It seems appropriate to acknowledge the fact that the financial market and economic activity strongly interact, representing some challenge for monetary policy. The question is what the quantitative relationships are and what appropriate monetary policy rules should be. In times of normal economic activity and asset price movements it has turned out that monetary policy, which sets the interest rate appropriately, is sufficient to keep the economy on a prosperous path.

One of the most influential monetary policy rules has been the Taylor rule. Taylor [22] has put forward the concept of inflation targeting, suggesting the short term interest rate to respond to the inflation and output gap. This has been made popular by economic research, for example by the works of [20, 21] and [25]. The latter, among others, extended the concept of inflation targeting to an optimizing framework, introducing some optimal Taylor rule.

Yet, from early on, the question came up of whether the monetary authorities should also give some weight to asset price movements. In the context of the Bernanke, Gertler and Gilchrist [2] framework of an imperfect capital market, net worth moves procyclically and the financial accelerator can magnify disturbances. On the other hand, it has been proposed that asset price movements should only be taken into account as much as they affect the inflation rate. It was assumed that a central bank that responds strongly to inflation pressures eliminates much of the distortionary effects of asset price movements on economic activity. Some monetary authorities refer to the Tinbergen rule which says that one needs as many control instruments in monetary policy as there are targets.

Cecchetti et al [5] argued that monetary authorities would obtain beneficial effects when responding to movements in asset prices, that is return on capital. This, in particular, should hold if there are exogenous bubbles or non-fundamental asset price movements. In this context [11] emphasizes that capital valuations are impacted by asset price bubbles which cause distortions in aggregate demand. Thus, the monetary authority, by reducing distortions owing to variations in the return on capital, may generate beneficial effects on real economic activity by also giving weight to movements of asset prices.

Others have argued that these asset price bubbles could be seen as resulting from asset price booms that occur in conjunction with changes in the underlying economic fundamentals, see [1]. Along those lines, some literature has attempted to construct models where bubbles arise due to misperceived technology shocks, see [6], suggesting that asset prices should not be targeted by central banks.

Asset prices have become also important in studies on the Japanese deflation of the 1990s. In the last ten years, the literature has focused on the Japanese deflationary experience and argued that when asset prices and product prices decline and inflation rates approach zero, or become negative, there will be a zero bound of the nominal interest rate that may create a, so called, liquidity trap for monetary authorities. Central banks may no longer be able to effectively steer the economy with monetary policy to some targets, see [8] and [12]. When interest rates hit the zero level bound, Bernanke, Reinhard and Sack [3] have brought forward the monetary policy of "quantitative easing"[a], which is to provide the private economy—suffering from a burst of an asset price bubble—with liquidity by providing funds for credit. Fiscal policy has been added as policy tool in a Keynesian sense with the aim to revive the economy through government spending, see also [14].

The experience of the recent banking crisis and its serious effects on real economic activity has shifted the emphasis to the study of monetary policy in the context of asset markets. Asset price movements since the beginning of 2008 led to extensive rebalancing of portfolios, namely movements into treasury bonds and cash. The monetary authority in the U.S., but also

[a]Quantitative easing improves banks' and firms' balance sheets and is supposed to unlock the credit channel.

other central banks and governments worldwide, reacted with extreme interest rate reductions, support for banks and substantial government spending plans. In the U.S. interest rates were reduced in 2008/2009 to virtually zero. Moreover, the asset price deflation, the large losses in portfolios, and the danger of a catastrophic economic meltdown, has led monetary policy to react strongly, not only in terms of interest rate reduction by the monetary authority, but also in terms of "quantitative easing".

Under these extreme conditions the study of the effects of monetary policy in the context of asset accumulation and dynamic portfolio models have become a challenging issue. The approach of this paper uses optimal asset allocation in a dynamic setting. The resulting optimal total wealth evolution implies certain relationships between key financial and economic quantities. This dynamic portfolio approach makes visible not only the transmission mechanism through which the financial meltdown is translated into a real economic meltdown, but will allow us to study the effectiveness and limitations of potential monetary policies. Such policies should be conducted recognisant of the optimal relationships we are revealing. Some of these relationships coincide with common thinking. However, since the interest rate enters into these relationships in a nonlinear manner new insights will emerge.

The application of a portfolio model to macroeconomics has already been favored by [23, 24] and [13], to name a few. But this was mainly a portfolio approach of static type. What is needed is a dynamic portfolio optimization that permits the investors to strategically respond to expected returns from capital, inflation rate, interest rate, and other factors. Campbell and Viceira [4] have contributed in a seminal way in the economic literature to enhance our appreciation of the potential of dynamic portfolio optimization for the understanding of economic activity. They give an important role to nominal and real bonds in financial investments and show how different types of investors, that is investors with different risk aversion and different time horizon, would react to a change of an investment environment created by monetary policy. They also show that inflation rates, which the central banks try to target, affect strongly the composition of asset holdings. They, for example, demonstrate how the anti-inflation interest rate policy of the Volcker–Greenspan era had substantial effects on the composition of asset holdings and, thus, on the economy.

Whereas [4], when performing dynamic portfolio optimization, use assumptions such as constant variances, constant expected risk premia,[b] a constant consumption-wealth ratio and the existence of an equivalent risk neutral probability measure, these assumptions do not seem to be realistic enough if one studies the interaction of monetary policy and the asset market in more general terms over long periods of time.

Our approach suggested here is related to the work of [4]. The key economic factors like inflation rate, consumption rate and expected return and volatility of equity can be general stochastic processes. Moreover, we pursue a nonlinear approach when we spell out an interest rate rule and inflation rate rule as a consequence of attempting to achieve an optimal wealth evolution with stable consumption rate. As we will show, there can arise important knife edge problems in the interrelation of expected return on equity, inflation rate, interest rate, risk aversion, volatility of the equity index and consumption rate. These relations are nonlinear ones and will be studied as such. It will roughly turn out that in an optimal regime the interest rate should increase when the expected return on equity increases or the squared volatility decreases. Furthermore, the inflation rate will increase for increasing interest rate, decreasing consumption rate and increasing volatility.

Overall, what we aim for is a better understanding about the impact of monetary policy under extraordinary and, of course, normal circumstances. It will turn out that if monetary policy is not cautiously implemented with a long term view for a balanced economic and financial development, then situations may arise where the economy cannot be controlled via monetary policy and can suffer for long periods of time even under strong government intervention.

The paper is organized as follows. Section 15.2 considers the asset market in the context of a dynamic portfolio model with the usual budget equation. Section 15.3 studies the optimal market dynamics and Section 15.4 derives formulas for the optimal interest rate and Section 15.5 for the optimal inflation rate. In Section 15.6 the optimal change in the consumption rate is discussed. Section 15.7 studies various economic regimes. Section 15.8 provides comments on monetary policy, pursued during the current and previous financial market meltdowns. Section 15.9 then discusses some empirical facts illustrating our proposed interest rate rule. Section 15.10 concludes the paper.

[b] An example is given in Chapter 3 of the book by [4].

15.2. Asset market and budget equation

To simplify our modeling we aggregate all stocks of the economy in an equity index $P = \{P_t,\, t \geq 0\}$ with value P_t at time $t \geq 0$. A common hypothesis about the behavior of equity indices is that they follow stochastic differential equations (SDEs), see [15, 16] and [7]. The total return equity index value P_t shall satisfy the SDE

$$dP_t = P_t\,(a_t\,dt + \sigma_t\,dz_t) \tag{15.1}$$

for $t \geq 0$ with $P_0 > 0$. Here $z = \{z_t,\, t \geq 0\}$ denotes a standard Wiener process. We assume that the equity index volatility process $\sigma = \{\sigma_t,\, t \geq 0\}$ is strictly positive. The process $a = \{a_t,\, t \geq 0\}$ models the expected rate of return on capital invested in equity and, thus, in some sense the productivity of the underlying economy. This provides a crucial link between macro economy and the financial market. In the long term the average rate of return from producing goods and services in the economy should match the expected rate of return from capital invested in the equity index.

For the efficient functioning of an economy the credit market is essential. To simplify the modeling of the credit market we employ a savings account process $\beta = \{\beta_t,\, t \geq 0\}$, where

$$\beta_t = \exp\left\{\int_0^t i_s\,ds\right\}, \tag{15.2}$$

with nominal interest rate i_t at time $t \geq 0$. We assume the same interest rate to be applied for borrowing and lending, which is set by the monetary authority. The investors of the economy invest in aggregation a fraction α_t of their total wealth at time $t \geq 0$ in the equity index and the remainder in the savings account. The leverage α_t equals one if there is neither outside borrowing nor lending. The leverage can reflect with $\alpha_t > 1$ some borrowing of the economy from foreign economies or with $\alpha_t < 1$ some lending of capital to foreign economies.

Furthermore, the investors take funds out continuously from their total wealth W_t for consumption at a rate $c_t > 0$ at time $t \geq 0$. The consumption rate process $c = \{c_t,\, t \geq 0\}$ is a process that can here be modeled freely. The total wealth process $W = \{W_t,\, t \geq 0\}$ of the investors satisfies then the budget equation

$$dW_t = W_t\big((1 - \alpha_t)\,i_t\,dt - c_t\,dt + \alpha_t(a_t\,dt + \sigma_t\,dz_t)\big) \tag{15.3}$$

for $t \geq 0$ and $W_0 > 0$.

If we denote by

$$\theta_t = \frac{a_t - i_t}{\sigma_t} \tag{15.4}$$

the market price of risk at time $t \geq 0$, then we can formulate the stochastic discount factor M_t at time $t \geq 0$, see [7] and [19], in the form

$$M_t = \exp\left\{-\int_0^t \left(i_s + \frac{\theta_s^2}{2}\right)ds - \int_0^t \theta_s\, dz_s\right\}. \tag{15.5}$$

By excluding arbitrage, see [7], it follows that we reinvest the consumed wealth in the total wealth process and multiply the resulting security with the stochastic discount factor, then the process $\tilde{W} = \{\tilde{W}_t,\, t \geq 0\}$ given by

$$\tilde{W}_t = W_t \exp\left\{\int_0^t c_s\, ds\right\} M_t, \tag{15.6}$$

has to be a martingale[c], that is

$$\tilde{W}_t = E_t(\tilde{W}_s) \tag{15.7}$$

for all $0 \leq t \leq s < \infty$. Here $E_t(\cdot)$ denotes the conditional expectation under the information available at time t. As shown in [18], the martingale property (15.7) ensures in a very general modeling framework that the wealth process $\tilde{W}$ reaches its future values in the least expensive manner. Moreover, it follows from (15.6), (15.5), (15.4) and (15.3) by the Itô formula that

$$d\tilde{W}_t = \tilde{W}_t\,(\alpha_t\, \sigma_t - \theta_t)\, dz_t \tag{15.8}$$

for all $t \geq 0$.

15.3. Optimal wealth evolution

We are now going to formulate a family of dynamic portfolio problems. These will exhibit some similarity with the Merton problem, see [15, 16], and various problems treated in [4].

To model the average cost of living we introduce at time $t \geq 0$ the consumer price index

$$I_t = \exp\left\{\int_0^t \pi_s\, ds\right\} \tag{15.9}$$

determined by the inflation rate process $\pi = \{\pi_t,\, t \geq 0\}$. For modeling the investor's preferences we employ the utility function $U(\cdot)$, which we assume

[c] For a martingale the current value is the best forecast for any of its future values.

to be sufficiently smooth, such that $U'(\cdot) > 0$ and $U''(\cdot) < 0$, with inverse function $U'^{-1}(\cdot)$ of its first derivative $U'(\cdot)$. Our approach would still work if the utility function were time dependent. We assume at a given time $t \geq 0$ that investors aim to maximize consistently for all future times $s > t$ the expected utility from consumed real wealth per unit of time, which means

$$v_{t,s} = \max E_t\Big(U\big(\frac{c_s W_s}{I_s}\big)\Big). \tag{15.10}$$

The above maximization will reveal an optimal evolution of the value of the total wealth process. For given t and different times $s > t$ the optimal expected utility $v_{t,s}$ may be rather different. This will allow us to draw conclusions for the design of monetary policies that balance the expected utility for consumption at different future times.

Similar as in [4] (Ch.5), the objective (15.10) translates with equation (15.7) into the constraint optimization problem

$$v_t = \max\left[E_t\left(U\left(\frac{c_s W_s}{I_s}\right)\right) - l_t E_t(\tilde{W}_s - \tilde{W}_t)\right] \tag{15.11}$$

$$= \max E_t\left(U\left(\frac{c_s W_s}{I_s}\right) - l_t(W_s M_s G_s - W_t M_t G_t)\right) \tag{15.12}$$

with

$$G_t = \exp\left\{\int_0^t c_s\, ds\right\} \tag{15.13}$$

for all $0 \leq t < s < \infty$. Here ℓ_t is the Lagrange multiplier which is independent of s. It captures the budget constraint and ensures that $\tilde{W}_t$ will form a martingale. We emphasize that the conditional expectation $E_t(\cdot)$ is taken under the real world probability measure.

To identify the candidate for the optimal wealth process we interchange in (15.12) maximization and expectation and consider for each $t \geq 0$ and $s > t$ the static optimization problem

$$\max_{W_s}\left[U\left(\frac{c_s W_s}{I_s}\right) - l_t(W_s M_s G_s - M_t\, W_t\, G_t)\right],$$

and maximize with respect to the value W_s of the total wealth process at time s. This leads us to the first order condition

$$U'\left(\frac{c_s W_s}{I_s}\right)\frac{c_s}{I_s} - l_t M_s G_s = 0.$$

By using the inverse function $U'^{-1}(\cdot)$ of the first derivative $U'(\cdot)$ of the utility function one obtains

$$\frac{c_s W_s}{I_s} = U'^{-1}\left(U'\left(\frac{c_s W_s}{I_s}\right)\right) = U'^{-1}\left(l_t M_s G_s \frac{I_s}{c_s}\right).$$

Consequently, the candidate for the optimal total wealth process is of the form

$$W_s = \frac{I_s}{c_s}\, U'^{-1}(l_t \psi_s) \tag{15.14}$$

with

$$\psi_s = \frac{M_s G_s I_s}{c_s}. \tag{15.15}$$

By using the initial total wealth W_0 and the initial consumption rate c_0 we identify the Lagrange multiplier as the constant $l_t = l_0 = c_0 U'(c_0 W_0)$ for all $0 \leq t < \infty$. Note that the initial total wealth $W_0 > 0$ acts as budget constraint.

Now, we assume that the relation (15.14) describes the evolution of the optimal total wealth process that achieves for any given t and all $s > t$ the maximum expected utility $v_{t,s}$ in (15.12). For a wide range of particular models and utility functions, this can be verified by using the methods described in [15, 16], [9] and [4]. When we study the total wealth process given in (15.14) we will find two crucial relationships between key financial and economic quantities that have to hold under an optimal wealth evolution. These two relationships will allow us to formulate an interest rate rule and also an inflation rate rule. Moreover, they will predict where the consumption rate will drift also in extreme market situations.

To reveal the first relation note that for ψ_t given in (15.15) the corresponding SDE follows by the Itô formula as

$$d\psi_t = \psi_t([r_t + \pi_t - e_t - i_t]\, dt - \theta_t\, dz_t).$$

By (15.6), (15.13) and the explicit relation (15.14) we can express the martingale

$$\tilde{W}_t = F(\psi_t) \tag{15.16}$$

via the function

$$F(\psi_t) = \psi_t\, U'^{-1}(l_0\, \psi_t). \tag{15.17}$$

Therefore, it follows, by the Itô formula, that we have the SDE

$$\begin{aligned} d\tilde{W}_t &= \frac{\partial}{\partial \psi} F(\psi_t)\psi_t([c_t + \pi_t - e_t - i_t]\, dt - \theta_t\, dz_t) \\ &+ \frac{1}{2}\frac{\partial^2}{\partial \psi^2} F(\phi_t)\psi_t^2\, \theta_t^2\, dt. \end{aligned} \tag{15.18}$$

A comparison of the drift coefficients of the SDEs (15.8) and (15.18) yields for the optimal wealth evolution the *first key relation* in the form

$$0 = c_t + \pi_t - e_t - i_t - \frac{\theta_t^2}{2\,\gamma_t}, \tag{15.19}$$

where

$$\gamma_t = -\frac{\frac{\partial}{\partial \psi} F(\phi_t)}{\psi_t \frac{\partial^2}{\partial \psi^2} F(\phi_t)}. \tag{15.20}$$

For the case of a power utility function

$$U(x) = \frac{1}{1-\gamma}\, x^{(1-\gamma)} \tag{15.21}$$

the quantity γ_t in (15.20) coincides with the risk aversion parameter γ, assuming $\gamma > 0$ and $\gamma \neq 1$, since

$$F(\psi_s) = l_0^{-\frac{1}{\gamma}}\, \psi_s^{1-\frac{1}{\gamma}}. \tag{15.22}$$

The function $F(\phi_s)$ covers also the case of logarithmic utility $U(x) = \ln(x)$, which emerges for the choice $\gamma = 1$ in (15.22).

The second crucial relation results from a comparison of the diffusion coefficients of the SDEs (15.8) and (15.18), which yields the equation

$$F(\psi_t)(\alpha_t\, \sigma_t - \theta_t) = -\frac{\partial}{\partial \psi_t} F(\psi_t)\psi_t\, \theta_t.$$

Thus, the optimal fraction α_t of total wealth at time t invested in the risky asset P_t is given in the *second key relation*

$$\alpha_t = \frac{\theta_t}{\bar{\gamma}_t\, \sigma_t}, \tag{15.23}$$

where

$$\frac{1}{\bar{\gamma}_t} = 1 - \psi_t \frac{\frac{\partial}{\partial \psi} F(\phi_t)}{F(\phi_t)}. \tag{15.24}$$

Note that in the special case of power utility, see (15.21), one obtains $\bar{\gamma}_t = \gamma$, which generalizes for $\gamma = 1$ to the case of logarithmic utility.

The optimal fraction in (15.23) is consistent with the familiar optimal fraction that arises when maximizing expected power utility from terminal wealth, see for instance [15] and [4].

If the two key relationships are satisfied, then the wealth evolution is optimal given the objective to maximize expected utility from consumed real wealth at future times.

15.4. Optimal interest rate

Since in the second key relation (15.23) the market price of risk θ_t, see (15.4), involves the interest rate, we can identify the corresponding optimal interest rate. Indeed, equations (15.4) and (15.23) lead us to the *optimal interest rate* $\tilde{i}_t$, which is obtained in the form

$$\tilde{i}_t = a_t - \bar{\gamma}_t \, \alpha_t \, \sigma_t^2. \tag{15.25}$$

We call this the *interest rate rule* and note that this choice is necessary but not sufficient to obtain an optimal wealth evolution. Similar expressions for some optimal interest rate choices can be found, for instance, in [17] and [4].

The monetary authority has the power to set the interest rate which could be far from being optimal. Knowing which choice supports an optimal wealth evolution is valuable for the understanding of the interaction between the financial market and the economy. The optimal interest rate $\tilde{i}_t$ in (15.25) increases when the expected return a_t of the equity index increases, which is what one would reasonably expect. If the other quantities remain unchanged, then $\tilde{i}_t$ increases when the fraction α_t of wealth invested in the equity market decreases. The optimal interest rate $\tilde{i}_t$ increases in the case when the squared equity index volatility σ_t^2 decreases provided that the other quantities remain unchanged. In this case there is less risk in the equity market, which makes borrowing to invest in equity more attractive. A higher optimal interest rate offsets this effect. The relation (15.25) gives theoretical guidance for setting the interest rate when the economy aims to allocate its resources towards an optimal wealth evolution.

In reality the nominal interest rate i_t cannot become negative. This imposes a constraint on monetary policy. Therefore, the theoretical optimal interest rate $\tilde{i}_t$, as suggested in (15.25), has to be modified towards an *adjusted interest rate*

$$i_t = (\tilde{i}_t)^+, \tag{15.26}$$

which represents the positive part of the optimal interest rate. Consequently, in the case of a negative optimal interest rate $\tilde{i}_t < 0$ the wealth evolution cannot be an optimal one since the interest rate can then only rest at zero. For this reason the monetary authority should avoid market scenarios that lead to negative optimal interest rates, see also [14]. The expected utility from consumption of real wealth will be less than optimal as a consequence of such a monetary policy failure. Unfortunately, such failures are typical in periods after extreme asset bubbles.

An example where in history zero interest rates have occurred is the Japanese stagnation in the 1990s until recently. This market situation emerged after the burst of the Japanese equity bubble in the 1980s, see [21]. In the U.S. the period during the great depression in the 1930s after the 1929 market crash provides another example. Finally, the recent financial crisis resulting in virtually zero interest rates in the U.S. gives a most recent example. The mentioned crises led to severe downturns in real economic activity and, thus, in the expected rate of return a_t of the equity index. This is why the optimal interest rate is by (15.25) most likely negative in the aftermath of the burst of a major asset bubble.

In market situations when negative optimal interest rates arise, the expected return of the equity index is by (15.25) too low compared to the product of its squared volatility σ_t^2, the risk aversion $\bar{\gamma}_t$ and the fraction α_t invested in the equity index. Note that an increase in debt, that is an increase in the fraction α_t of wealth invested in the equity index, does not easily resolve the issue of negative optimal interest rates. The expected rate of return from equity needs to be increased sufficiently and sustainably to bring the optimal interest rate back into positive territory.

In the period after a financial market crash the market may be forced into non-optimal wealth dynamics due to a non-optimal zero interest rate policy. The consequences for the real economy, in particular employment, can be dramatic. Potential credit shortage because of looming defaults in highly leveraged activities could delay a necessary increase in the expected return on equity that could make the optimal interest rate positive again. Only over time volatility may decrease and government stimulus may help to sufficiently increase the expected rate of return on equity a_t, so that the optimal interest rate $\tilde{i}_t$ can finally turn back to positivity.[d] We emphasize that positive optimal interest rates are necessary but not sufficient for an optimal wealth evolution and, thus, a successful monetary policy.

[d]Therefore, a responsible monetary policy should avoid equity bubbles in the first place.

By exploiting the first key relation (15.19) together with (15.4) and assuming the growth rate e_t of the consumption rate to be zero, which means a stable consumption rate, the optimal consumption rate satisfies the equation

$$c_t = i_t - \pi_t + \frac{(a_t - i_t)^2}{2\,\gamma_t\,\sigma_t^2}.$$

It is important to note that under optimal wealth evolution the consumption rate is maximized with respect to the following choice of the interest rate

$$i_t = a_t - \gamma_t\,\sigma_t^2, \tag{15.27}$$

yielding the consumption rate

$$c_t = a_t - \pi_t - \frac{\gamma_t\,\sigma_t^2}{2}. \tag{15.28}$$

This means, by (15.25) and (15.27), the monetary authority has the additional information that the optimal interest rate $\tilde{i}_t$ maximizes the consumption rate if the ratio of wealth invested in the equity market equals

$$\alpha_t = \frac{\gamma_t}{\bar{\gamma}_t}. \tag{15.29}$$

We call this choice the case of credit clearing since for power utility, see (15.21), one has $\gamma_t = \bar{\gamma}_t = \gamma$ and, thus, $\alpha_t = 1$. We conclude that for an optimal total wealth evolution the case of credit clearing, where there is only inside debt, borrowing and lending in the economy offset each other, yields the maximum sustainable consumption rate. This coincides also with common sense, since excessive borrowing forces borrowers to pay substantial interest to lenders. On the other hand, excessive lending brings only interest in return and not the extra premium that equity provides in the long term.

15.5. Optimal inflation rate

Let us now deduce the optimal inflation rate from the first key relationship (15.19). Assuming an optimal interest rate $\tilde{i}_t$ to be set by the monetary authority it follows from (15.19) and (15.25) the following formula for the optimal inflation rate

$$\tilde{\pi}_t = a_t - c_t + e_t - \bar{\gamma}_t\,\alpha_t\,\sigma_t^2\left(1 - \frac{\alpha_t\,\bar{\gamma}_t}{2\,\gamma_t}\right) \tag{15.30}$$

$$= \tilde{i}_t - c_t + e_t + \frac{(\bar{\gamma}_t\,\alpha_t\,\sigma_t)^2}{2\,\gamma_t}. \tag{15.31}$$

It may be not as obvious to see how the monetary authority can influence the inflation rate. However, there exist measures that can move the inflation rate in an economy. The most obvious is an increase in money supply, or the "printing" of money by the government. This measure increases inflation because more money is chasing the same amount of goods. The opposite is far more difficult to achieve by strict monetary policy. One notes in (15.30) that if the optimal interest rate increases, then the optimal inflation rate will increase as well, assuming no change in the other quantities. Interestingly, if an economy is increasing its lending to other economies, which means α_t is set lower, then its optimal inflation rate is decreasing. This can become a problem when the optimal inflation rate $\tilde{\pi}_t$ turns negative and the economy will face deflation. In this case it is rational for investors to delay consumption. This, however, slows down economic activity and the expected return a_t on equity is likely to decrease, causing by (15.30) even further decline in the optimal inflation rate. The economy finds itself in a trap. Such a situation seems to have arisen in the Japanese stagnation that has been lasting already for almost two decades. By allowing the inflation rate to increase so that the deflationary period could be overcome, the interest rate can turn finally positive.

To avoid significant delays in consumption one should not only avoid deflation but ensure also time value of money. This means, to target an optimal interest rate slightly above the inflation rate such that the optimal real interest rate

$$\tilde{i}_t - \tilde{\pi}_t = c_t - e_t - \frac{(\bar{\gamma}_t\, \alpha_t\, \sigma_t)^2}{2\, \gamma_t} \tag{15.32}$$

will be positive. In reality this situation may not always be achieved, however, it represents a desirable additional target for monetary policy.

For those who have to rely in an economy on savings in fixed income, they would appreciate the lowest possible inflation rate and, of course, a positive optimal real interest rate (15.32). From (15.30), by using the first order condition, we obtain the minimal optimal inflation rate with respect to the fraction α_t for the choice of the fraction

$$\alpha_t = \frac{\gamma_t}{\bar{\gamma}_t}. \tag{15.33}$$

Note that this choice is the case of credit clearing, which already maximizes the consumption rate, as shown in (15.28). In this case the optimal interest rate equals

$$\tilde{i}_t = a_t - \gamma_t\, \sigma_t^2, \tag{15.34}$$

see also (15.27). The minimal value of the optimal inflation rate is then

$$\tilde{\pi}_t = a_t - c_t + e_t - \frac{\gamma_t \sigma_t^2}{2}, \tag{15.35}$$

and the optimal real interest rate (15.32) takes the form

$$\tilde{i}_t - \tilde{\pi}_t = c_t - e_t - \frac{\gamma_t \sigma_t^2}{2}. \tag{15.36}$$

To make the latter positive for the case $e_t = 0$ one needs a stable consumption rate c_t greater than or equal to $\frac{\gamma_t}{2} \sigma_t^2$. This provides the monetary authority with a critical consumption rate level that should be surpassed.

Now, let us see what happens to the expected utility when applying an optimal interest and inflation rate policy in the special case of power and logarithmic utility, see (15.21).

With the choice of an optimal interest rate $\tilde{i}_t$, as given in (15.25), and that of an optimal inflation rate $\tilde{\pi}_t$, as described in (15.30), one obtains for power utility the SDE

$$dU\left(\frac{c_t W_t}{I_t}\right) = U\left(\frac{c_t W_t}{I_t}\right)(1-\gamma)\, \alpha_t\, \sigma_t\, dz_t \tag{15.37}$$

for $t \geq 0$. In the usual case when $U(\frac{c_t W_t}{I_t})$ is not only having a driftless SDE but is also a true martingale, we can draw from the optimal wealth evolution under power utility the conclusion that the expected utilities are equal for different time horizons, that is, for given $t \geq 0$ we have

$$v_{t,s} = v_{t,t} \tag{15.38}$$

for all $s \geq t$. Note that (15.38) holds also for the case of logarithmic utility. This means, the various groups of investors that for generational reasons consume now and others that consume later have the same expected utility from their consumed real wealth and no group is disadvantaged. This would be a desirable outcome of a wise monetary policy.

We obtain from (15.29) and (15.33) in the case of credit clearing for power utility the fraction $\alpha_t = 1$ that provides the maximum consumption rate and minimal optimal inflation rate. This is an important and also rather intuitive finding, which says that the optimal wealth evolution with maximum consumption and minimal inflation is reached under power and logarithmic utility when the credit market of the economy fully clears. This means, for the economy there is no outside borrowing or lending. It uses its wealth to advance its own activities. From (15.30) it follows under power utility that outside borrowing will increase the optimal inflation.

15.6. Optimal change in consumption rate

In general, the interest rate and the inflation rate are not optimal. Assuming a constant consumption rate the monetary authority can only aim to influence the market such that these come close to their optimal values. We now discuss the impact of interest rate and inflation rate on the growth rate e_t of the consumption rate, assuming an optimal wealth evolution. This growth rate comes into consideration when we exploit the first key relation (15.19) together with (15.4). Under an optimal wealth evolution we obtain for the growth rate of the consumption rate the equation

$$\frac{d\ln(c_t)}{dt} = e_t = c_t + \pi_t - i_t - \frac{(a_t - i_t)^2}{2\gamma_t\,\sigma_t^2},$$

which we can rewrite as a quadratic form

$$e_t = \pi_t - \hat{\pi}_t - \frac{(i_t - \phi_t)^2}{2\gamma_t\,\sigma_t^2}, \tag{15.39}$$

using the *critical inflation rate*

$$\hat{\pi}_t = a_t - c_t - \frac{\gamma_t\,\sigma_t^2}{2} \tag{15.40}$$

and the *critical interest rate*

$$\phi_t = a_t - \gamma_t\,\sigma_t^2. \tag{15.41}$$

Note for a constant consumption rate the critical inflation rate $\hat{\pi}_t$ coincides with the minimal possible value for the optimal inflation rate $\tilde{\pi}_t$ and the critical interest rate with the corresponding optimal interest rate, see (15.35), (15.40), (15.34) and (15.41).

The sign of the expression on the right hand side of (15.39) determines the direction in which the consumption rate is moving, when assuming an optimal wealth evolution.

From (15.39) it follows for $e_t = 0$ that

$$\pi_t = \hat{\pi}_t + \frac{(i_t - \phi_t)^2}{2\,\gamma_t\,\sigma_t^2}.$$

In Fig. 15.1 we show areas where we have an increase or decrease of the consumption rate, when we plot on the horizontal axis the interest rate and on the vertical axis the inflation rate. Assuming the growth rate e_t to be zero such that the consumption rate remains stable, we obtain for each interest rate for the corresponding inflation rate a value on a parabola, which shows at its minimum the critical inflation rate $\hat{\pi}_t$ for the critical

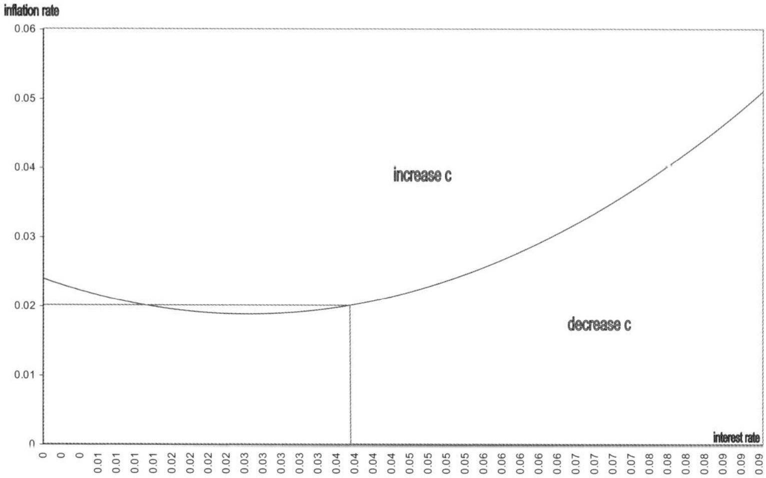

Fig. 15.1. Inflation versus interest.

interest rate ϕ_t. On this parabola the optimal inflation rate $\tilde{\pi}_t$, see (15.30), is found for the choice of the optimal interest rate $\tilde{i}_t$, see (15.25). Above that parabola the consumption rate c_t is growing, whereas below the parabola it is falling. We note that for $\pi < \hat{\pi}$ there is no optimal wealth evolution possible with a stable consumption rate.

When applying the optimal interest rate (15.25) in equation (15.39), it follows that

$$e_t = \pi_t - \hat{\pi}_t - \frac{\gamma_t \, \sigma_t^2}{2} \left(1 - \frac{\bar{\gamma}_t}{\gamma_t} \, \alpha_t \right)^2 . \tag{15.42}$$

Consequently, if one assumes a stable consumption rate, which appears to be reasonable for a sustainable monetary policy, together with the optimal interest rate $\tilde{i}_t$, then the corresponding optimal inflation rate $\tilde{\pi}_t$ is given by the *inflation rate rule*

$$\pi_t - \hat{\pi}_t + \frac{\gamma_t \, \sigma_t^2}{2} \left(1 - \frac{\bar{\gamma}_t}{\gamma_t} \, \alpha_t \right)^2 . \tag{15.43}$$

We will show below that equation (15.43) suggests an inflation rate level for which a monetary policy, based on setting an optimal interest rate, is likely to work.

15.7. Economic regimes

We discuss in the following several economic regimes:

(1) **Subcritical inflation:**
In this case the inflation rate is subcritical, that is below the critical inflation rate with $\pi_t < \hat{\pi}_t$ such that equation (15.43) is violated. In this case it follows from (15.39) that the consumption rate decreases, no matter where the interest rate may be set. If the interest rate is close to the critical interest rate, then the decline is minimized. In this situation there is a downward trend in the consumption rate. Due to the resulting decrease in the consumption rate, the expected return a_t of the equity index of the real economy is likely to decrease. A recession and eventually a depression could result if nothing else changes. This is in line with observations in [2] who emphasize that investors' net worth moves procyclically and can magnify economic disturbances.
We emphasize that the case of subcritical inflation can be dangerous for the economy. In such a low inflation regime there are not many possibilities to counteract the decrease in the consumption rate. Setting the interest rate will not work, as we have seen. The investors will ignore the level of the interest rate and reduce further their consumption rate. An optimal wealth evolution with stable consumption rate is impossible. The fact that subcritical inflation does not allow a sustainable optimal wealth evolution is an important insight. It questions the popular policy of low inflation targeting. When targeting a too low subcritical inflation rate level in a given economic situation such monetary policy can have a long lasting negative impact on the consumption rate and, thus, on the real economy. The price that the economy may have to pay to get out of such an economic trap is a period of deliberate higher inflation. A government can create an increase in the inflation rate by "printing" money. A problem may be later to get inflation back near to its minimal possible level.

(2) **Supercritical inflation:**
Let us now study the case when the inflation rate is supercritical, $\pi_t > \hat{\pi}_t$. We discuss first the case without sufficient credit clearing, which means that $\frac{\bar{\gamma}_t}{\gamma_t}\alpha_t$ is not near the value one:

(a) **Supercritical inflation without sufficient credit clearing:**
In the following supercritical inflation rate regime we assume that $\pi_t \geq \hat{\pi}_t$ and the monetary authority implements the optimal interest rate (15.25) but does not have sufficient credit clearing such that instead of equality (15.43) the following inequality still remains:

$$\pi_t - \hat{\pi}_t < \frac{\gamma_t \, \sigma_t^2}{2} \left(1 - \frac{\bar{\gamma}_t}{\gamma_t} \, \alpha_t \right)^2 . \tag{15.44}$$

This means that the inflation rate is here lower than the optimal one as given in (15.43). This yields by (15.42) and (15.44) a decreasing consumption rate c_t. The resulting downward trend in the consumption rate is not as strong as in the previous case under subcritical inflation. A reversal can be achieved, assuming an optimal interest rate (15.25), by letting the inflation rate become sufficiently high or by balancing better the economy's borrowing and lending. Credit clearing with $\frac{\bar{\gamma}_t}{\gamma_t} \, \alpha_t = 1$, will safely increase the consumption rate under supercritical inflation. It is important to point out that increased borrowing, when one has already $\frac{\bar{\gamma}_t}{\gamma_t} \, \alpha_t > 1$, will further decrease the consumption rate and, thus, may not stop a subsequent downturn in the real economy. These findings should give a warning to monetary authorities of economies that have foreign debt and where the inflation rate is too low. Increasing outside debt without any other change will not stop a consumption rate falling. The inflation rate has to increase to move into another economic regime.

(b) **Supercritical inflation with sufficient credit clearing:**
In the case of a supercritical inflation rate $\pi_t \geq \hat{\pi}_t$ and sufficient credit clearing one can make (15.44) an equality by choosing the inflation rate optimal, see (15.43). Once this optimal balance is achieved, the monetary policy should focus on maintaining such a situation with approximately optimal interest and inflation rate. For this purpose let us assume that the monetary authority no longer targets exactly the optimal interest rate (15.25). Instead let the monetary policy deliberately deviate from it by setting the interest rate slightly above or below the optimal interest rate so that the consumption

rate moves according to relation (15.39) in one or the other direction. Now, the aim is to identify a control mechanism that keeps the consumption rate stable when both inflation rate and interest rate are approximately optimal.

Under supercritical inflation we know from (15.39) that in the case

$$(i_t - \phi_t)^2 \leq 2\gamma_t \sigma_t^2 (\pi_t - \hat{\pi}_t) \tag{15.45}$$

the consumption rate is increasing. If nothing else changes the economy may drift in such a situation to a high consumption rate and the economy may even overheat. However, by increasing the interest rate i_t above the level

$$\bar{i}_t = \phi_t + \sqrt{2\,\gamma_t\,\sigma_t^2(\pi_t - \hat{\pi}_t)} \tag{15.46}$$

the monetary authority can stop the increase of the consumption rate. If it moves afterwards the interest rate back to a level below $\bar{i}_t$, then the consumption rate should start to increase again. This provides a convenient mechanism for controlling the consumption rate via the setting of the interest rate above or below some target level $\bar{i}_t$. For an economy with $\alpha_t < \frac{\gamma_t}{\bar{\gamma}_t}$, representing outside lending, this mechanism works well and can deliver approximately optimal interest and inflation rates.

When for $\alpha_t < \frac{\gamma_t}{\bar{\gamma}_t}$ the inflation rate π_t is such that the inflation rate rule (15.43) holds, then the interest rate $\bar{i}_t$ becomes the optimal interest rate, see (15.25), and we have practically an optimal wealth evolution when following the above described control mechanism.

Note by (15.43) and (15.35) that for the case of credit clearing $\alpha_t = \frac{\gamma_t}{\bar{\gamma}_t}$ one obtains the minimal possible optimal inflation rate and maximal consumption rate. In this case the setting of the interest rate above or below $\bar{i}_t = \phi$ will only decrease the consumption rate and the control mechanism breaks down. Still, the monetary authority could increase the inflation rate beyond $\hat{\pi}_t$ to increase the consumption rate, see Fig. 15.1. They could increase the interest rate beyond the critical interest rate $\bar{i}_t = \phi_t$ to decrease the consumption rate. This is another mechanism which allows to obtain an approximately optimal wealth evolution.

For the case $\alpha_t > \frac{\gamma_t}{\bar{\gamma}_t}$, when the economy substantially borrows from abroad, the interest rate $\bar{i}_t$, given in (15.46) is not the optimal one. The control mechanism still works similar as in the case above. However, we emphasize that in this case the total wealth evolution is *not* optimal. The monetary authority could in this case try to keep the inflation rate rather low because of the otherwise even less optimal wealth evolution. For an economy with fraction $\alpha_t > \frac{\gamma_t}{\bar{\gamma}_t}$ the wealth evolution remains suboptimal when using the control mechanism that is setting the interest rate above and below $\bar{i}_t$. Alternatively, the monetary authority could target the optimal interest rate $\tilde{i}_t$ which is below the critical interest rate and can become easily negative. One could achieve with lower inflation a decrease in the consumption rate and vice versa. However, due to inflation persistence, a decline in inflation is often difficult to achieve by the monetary authority. There is the danger that the optimal interest rate may become negative under such a strategy. This demonstrates that an economy that borrows has an optimal interest rate below the critical interest rate and its monetary policy is either not optimal or faces a delicate task.

15.8. Evaluation of current monetary policy

The currently popular monetary policy rule, the Taylor rule, is supposed to respond to the output and inflation gap. These are in our notation given by $a_t - \bar{a}$ and $\pi_t - \bar{\pi}$, respectively, where $\bar{a}$ denotes the long run average expected return and $\bar{\pi}$ the long run average inflation rate. This may work reasonably well in normal circumstances, since it reflects roughly what our findings suggest in the case $\alpha_t > \frac{\gamma_t}{\bar{\gamma}_t}$ and $\tilde{i}_t > 0$. However, this may not be adequate for extreme situations as typical after a financial crisis or in a deflationary period. These times require unconventional monetary policies.

When compared with the Taylor rule our results are more differentiated and clearly different. Current monetary policy focuses primarily on holding the inflation rate low. For example, it has been set at a fixed target level of about 2% in several countries. Holding the inflation rate low is still reasonable as long as the inflation rate is above the critical inflation rate $\hat{\pi}_t$

given in (15.40). Holding it dogmatically at a fixed 2% level may potentially bring it below the critical inflation rate level and can in this way create long lasting negative effects to the economy. As discussed above, for subcritical inflation the consumption rate will move uncontrolled downward and may drive the economy into a recession.

Probably most important is our insight that it is necessary to make inflation increase to a level where monetary policy works. This questions the low inflation rate targeting policies that have been implemented, supported by many politicians and economists.

The interest rate rule (15.25) links the productivity of the economy, reflected by the expected return a_t of the equity index, with other economic and financial quantities. Under this rule higher expected return on equity trigger higher interest rates, which is in line with common perception. The interest rate rule (15.25) puts this perception into a clear quantitative relationship involving also the fraction borrowed from abroad, the risk aversion and the variance of the returns on equity.

To implement a successful monetary policy, the monetary authority has to create an economic environment where the optimal interest rate $\tilde{i}_t$ is safely located above zero. This means by (15.25) one needs to have

$$a_t > \bar{\gamma}_t \, \alpha_t \, \sigma_t^2. \tag{15.47}$$

For this to happen, the expected return a_t of the equity index has to be large enough. Extremely high risk aversion but also excessive foreign debt can reverse the relationship (15.47). Governmental stimulus can increase economic activity and, thus, the expected return on equity. However, increased debt may offset this.

The fraction α_t of wealth invested in the risky asset index plays in the interest rate rule (15.25) an important role, which has been probably not much perceived among central bankers and economists but resonates well with common sense. An economy that has significant foreign debt, that is $\alpha_t > \frac{\gamma_t}{\bar{\gamma}_t}$, is more susceptible with its interest rate policy to the variance of its equity returns than is one which runs a surplus. Furthermore, we saw in our previous discussion that for an economy with significant foreign debt it is more difficult to create and maintain an optimal wealth evolution by setting the interest rate than for an economy which is lending. On the other hand, an economy with significant foreign lending, that is $\alpha_t < \frac{\gamma_t}{\bar{\gamma}_t}$, faces potentially the danger of a deflation, as can be seen in (15.30).

The above discussion shows that the monetary authority should always have the long term prosperity of an economy in mind and needs to avoid an economic trap of zero interest rate and subcritical inflation in the first place. Since politicians concentrate, in general, on rather short time horizons due to the election cycle, this requires some autonomy of monetary policy. Exceptions may arise in times of a financial meltdown where the government should help with fiscal policy to bring the economy back on track so that monetary policy can work again. It is particularly dangerous to let credit become too easily available. Cheap credit first pushes equity prices up into a potential equity bubble and the necessary market correction makes it likely that negative optimal interest rates occur and subcritical inflation emerges leading to an economic trap. Furthermore, debt will be accumulated by the economy, which is a burden and makes monetary policy difficult in critical times, in particular after a crisis.

One notes from the inflation rate rule (15.25) that the inflation rate is minimal in the case when the credit market clears. In an economy with balanced lending and borrowing all wealth is invested. In this way the economy uses all its own wealth to support its productive activities. This also maximizes the consumption rate. A reliance on credit for securing economic output increases in the long run the optimal inflation rate. On the other hand, when lending to other economies, the optimal inflation may be lower and the danger of deflation may emerge in certain situations. In such a situation the potential of the economy is not fully utilized.

15.9. Some empirical facts

It makes sense to discuss the theoretically derived interest rate rule by looking at long term historical data when assuming the case of power utility, that is, $\gamma = \bar{\gamma}_t = \gamma_t$. Since our results are independent of the particular market dynamics they should be reasonably robust.

Let us assume that the expected equity index return, the interest rate, the inflation rate, the equity index volatility and the consumption rate are stationary processes. In this case one can obtain rough estimates for their averages. At least in the long run, when averaging over various markets, the presented equations should be reasonably reflected. In Table 15.1 we show long run estimates over the last century for the equity index return a, the T-bill return i, the inflation rate π, the equity index volatility σ, together with the equity index dividend rate d, obtained for sixteen economics by [10].

The long run estimate for the market price of risk $\theta \approx \frac{a-i}{\sigma}$, see (15.4), is calculated in the table. It reaches on average 23.6%, which is not far from a typical equity index volatility, which shows an average of 22.7%.

Table 15.1. Estimates from sixteen markets.

Country	a	i	π	σ	d	θ	$\gamma\alpha$	c	π_{min}
Australia	0.119	0.045	0.041	0.177	0.048	0.418	2.362	0.091	0.034
Belgium	0.082	0.052	0.055	0.228	0.028	0.132	0.577	0.0056	0.039
Canada	0.097	0.049	0.031	0.168	0.041	0.286	1.701	0.059	0.032
Denmark	0.089	0.07	0.041	0.201	0.044	0.095	0.47	0.0334	0.035
France	0.121	0.043	0.079	0.231	0.036	0.338	1.462	0.021	0.046
Germany	0.097	0.046	0.051	0.323	0.036	0.158	0.489	0.0075	0.035
Ireland	0.095	0.058	0.045	0.222	0.033	0.167	0.751	0.027	0.044
Italy	0.12	0.047	0.091	0.294	0.038	0.248	0.845	-0.013	0.045
Japan	0.125	0.054	0.076	0.303	0.057	0.234	0.773	0.0055	0.033
Netherlands	0.09	0.037	0.03	0.21	0.049	0.252	1.202	0.039	0.015
S. Africa	0.12	0.057	0.048	0.228	0.025	0.276	1.212	0.047	0.064
Spain	0.1	0.065	0.061	0.22	0.047	0.159	0.723	0.030	0.036
Sweden	0.116	0.058	0.037	0.228	0.035	0.254	1.116	0.053	0.052
Switzerland	0.076	0.033	0.022	0.204	0.029	0.211	1.033	0.033	0.026
UK	0.101	0.051	0.041	0.2	0.038	0.25	1.25	0.041	0.038
US	0.101	0.041	0.032	0.202	0.041	0.297	1.47	0.053	0.030
Average	0.103	0.05	0.049	0.227	0.039	0.236	1.09	0.033	0.038

Of particular interest is the product $\gamma\alpha$ of the risk aversion parameter with the fraction of wealth held in the equity index. It follows by (15.23) under optimal wealth evolution that

$$\gamma\,\alpha \approx \frac{a-i}{\sigma^2} = \frac{\theta}{\sigma}.$$

This quantity turns out to be rather different for the sixteen economies. Surprisingly, it yields on average a level of about one. The risk aversion times the fraction invested in the equity index, $\gamma\alpha$, seems to be higher for Anglo-Saxon economies, which use a well developed equity market for capitalizing their industries and have in most cases foreign debt. Countries which used during the last century the credit system to provide mostly the capital for their companies or which have been lending significantly to other economies, seem to give lower estimates for $\gamma\alpha$. Since borrowing and lending have to offset each other in the global economy, it is probably reasonable to expect α to be on average one. Since $\gamma\alpha$ is in our informal discussion on average close to one, the risk aversion parameter γ could on average be close to one. This would suggest that logarithmic utility may reflect on average in a global economy the preferences of the investors. From (15.15) and (15.16) it follows in this case that $\tilde{W}$ is not only a martingale but also a constant. When using logarithmic utility, this means that the real wealth consumed per unit of time is growing with the maximum possible rate, see [19]. For the economy this is a very reasonable objective. It can be shown, see [19], that the trajectory of real consumed wealth per unit of time is for logarithmic utility after sufficiently long time almost surely above that for any other utility.

Now, let us estimate the long run average consumption rate c for each country based on relation (15.39), where we set

$$c \approx i - \pi + \frac{(a-i)^2}{2\,\gamma\,\sigma^2}.$$

The resulting estimate of this value average over all countries amounts to a consumption rate of 3.3%, which appears to be reasonable. Note that this is also close to the average dividend rate d of about 3.9%.

Finally, we show in Table 15.1 also the minimal inflation rate

$$\pi_{\min} = \hat{\pi} \approx a - c - \frac{\gamma\,\sigma^2}{2},$$

see (15.40), which is also the critical inflation rate. The resulting average yields a critical inflation rate of 3.8%, which is roughly about 1% below the average inflation rate of 4.9% obtained in Table 15.1. This makes good sense, since the inflation rate should be slightly above the minimal possible inflation rate to make the mechanics of monetary policy working.

So far, the relationships that we revealed appear empirically to be plausible. To obtain more precise evidence about their validity in a dynamical sense further research has to be performed. However, already now it seems clear that the relationships described make good economic sense.

15.10. Conclusions

Monetary policy was challenged by the recent financial crisis. This led to a visible shift of monetary policy from inflation targeting to intervention into the financial market. This has called for a type of monetary policy model that is unconventional and does allow to consider investor's risk perception, inflation, return on equity, variance of equity returns, consumption rate and potentially other economic variables. We suggested here a dynamic portfolio approach, which allows one to study the interaction between investors' strategic behavior and monetary policy. We derived policy rules that explain how monetary authorities could set the interest rates in relation to inflation rates, expected return on equity, risk aversion, equity index volatility and consumption rate. We also discussed the dangers of prolonged deflation, subcritical inflation and zero interest rates. As we showed, the inflation rate needs to reach a certain minimal level to allow the interest rate to be a viable monetary control instrument. If the proposed monetary policy rules are applied properly, then the consumption rate can be controlled to remain stable if desired. The inflation rate can be kept at a minimal possible level, and the interest rate and consumption rate can be maximal. As the monetary policy focuses currently under the Taylor rule on the inflation and output gap, the paper suggests that it should also have an eye on the necessary minimum level of the inflation rate and should target the corresponding optimal interest rate reflecting financial market conditions. Our discovered relationships also indicate how to act in a financial meltdown and crisis to bring the economy back on track.

We pursued a nonlinear approach when we spelled out the new inflation rate rule and interest rate rule in the context of dynamic portfolio optimization for investors. As we have shown, there can arise important knife edge problems in the interrelation of economic and financial variables. Our aim was to contribute to a better understanding about the impact of monetary policy also under extreme financial and economic circumstances, which are currently characteristic for many economies.

References

[1] Beaudry, P. & F. Portier (2004). An exploration into Pigou's theory of cycles. *J. Monetary Economy* **51**(6), 1183–1216.

[2] Bernanke, B., M. Gertler, & S. Gilchrist (1999). The financial accelerator in a quantitative business cycle framework. In J. Taylor and M. Woodford (Eds.), *Handbook of Macroeconomics*, pp. 1341–1393. Amsterdam, North-Holland.

[3] Bernanke, B., V. R. Reinhard, & B. P. Sack (2004). Monetary policy alternatives at the zero bound: An empirical assessment. Federal Reserve Board, Washington, D.C. no 48.

[4] Campbell, J. & L. Viceira (2002). *Strategic Asset Allocation, Portfolio Choice for Long Term Investors.* Oxford, Oxford University Press.

[5] Cecchetti, S., H. Genberg, J. Lipsky, & S. Wadhwani (2000). Asset prices and central bank policy. Geneva Reports on the World Economy 2.

[6] Christiano, L., R. Motto, & M. Rostagno (2005). Monetary policy and a stock market boom-bust cycle. Manuscript, Northwestern University and the European Central Bank.

[7] Cochrane, J. H. (2001). *Asset Pricing.* Princeton University Press.

[8] Coenen, G. & V. Wieland (2004). Exchange-rate policy and the zero bound on nominal interest rates. Working Paper Series 350, European Central Bank.

[9] Cox, J. C. & C. F. Huang (1989). Optimum consumption and portfolio policies when asset prices follow a diffusion process. *J. Economic Theory* **49**, 33–83.

[10] Dimson, E., P. Marsh, & M. Staunton (2002). *Triumph of the Optimists: 101 Years of Global Investment Returns.* Princeton University Press.

[11] Dupor, B. I. (2005). Stabilizing non-fundamental asset price movements under discretion and limited information. *J. Monetary Economy* **52**, 727–747.

[12] Eggertsson, G. B. & M. Woodford (2003). Optimal monetary policy in a liquidity trap. NBER Working Papers 9968, National Bureau of Economic Research.

[13] Frankel, J. (1995). *Financial Market and Monetary Policy.* Cambridge: MIT Press.

[14] Gertler, M. & P. Karadi (2009). A model of unconventional monetary policy. Dept. of Economics, New York University (working paper).

[15] Merton, R. C. (1973). An intertemporal capital asset pricing model. *Econometrica* **41**, 867–888.

[16] Merton, R. C. (1992). *Continuous-Time Finance.* Blackwell, Oxford.

[17] Platen, E. (1999). A short term interest rate model. *Finance Stoch.* **3**(2), 215–225.

[18] Platen, E. (2009). Real world pricing of long term contracts. Technical report, University of Technology, Sydney. QFRC Research Paper 262.

[19] Platen, E. & D. Heath (2010). *A Benchmark Approach to Quantitative Finance.* Springer Finance. Springer.

[20] Svensson, L. E. O. (1997). Inflation forecast targeting: Implementation and monitoring inflation targets. *European Economic Review* **41**, 1111–1146.

[21] Svensson, L. E. O. (1999). Inflation targeting: Some extensions. Stockholm School of Economics, manuscript.

[22] Taylor, J. B. (1993). Discretion versus policyrules in practice. *Carnegie-Rochester Conference Series on Public Policy* **39**, 195–214.

[23] Tobin, J. (1969). A general equilibirum approach to monetary theory. *J. of Money, Credit and Banking* **1**(1), 15–29.

[24] Tobin, J. (1980). *Asset Accumulation and Economic Activity.* Oxford: Basic Blackwell.

[25] Woodford, M. (2003). *Interest and Prices: Foundation of a Theory of Monetary Policy.* NJ, Princeton University Press.

Chapter 16

Mean-variance portfolio selection under regime-switching diffusion asset models: A two-time-scale limit

G. Yin and Y. Talafha

Department of Mathematics,
Wayne State University,
Detroit, MI 48202, USA
gyin@math.wayne.edu, ee7542@wayne.edu

Dedicated to Professor Robert J. Elliott on the occasion of his 70th Birthday

This work is concerned with a mean-variance optimal portfolio selection problem. The asset model under consideration is a switching diffusion where the switching is represented by a continuous-time Markov chain. Our consideration focuses on the case where the Markov chain has a fast changing part and a slowly varying part. We show that under rather mild conditions, the wealth process converges weakly to a limit process, which can be considered as an average with respect to the stationary distribution corresponding to the fast varying part of the generator of the Markov chain.

Contents

16.1. Introduction

This work is devoted to mean-variance portfolio selection problems in continuous time. The asset model includes regime switching modeled by a continuous-time Markov chain with a finite state space. The main motivation of the study stems from our recent work [Ref. [40]]. Our new contributions are to provide a model in which the Markov chain has fast varying part and slowly changing part. This leads to a two-time-scale limit. The model is motivated by the following consideration. Suppose that the market parameters depend on the market mode that switches among a finite number of states, which may reflect the state of the underlying economy and the mood of investors in the market etc. For instance, the market can be roughly divided to "bullish" and "bearish" states. The appreciation and volatility rates under bullish and bearish states can be quite different. Normally there are more intermediate states between the bullish and bearish market. The situation that we are interested in is where the switching among these states is driven by two types of motions, one fast switching and one slowly changing. Mathematically, this can be done by introducing a small parameter $\varepsilon > 0$ into the model. Using our recent work [Ref. [40]], it is possible to get the solution of the mean-variance portfolio selections. If the discrete event process (the switching process) has a large state space, we have to solve a system with a large number of equations, hence the amount of computation can be intensive. In this paper, we introduce the idea of averaging into the problem. We show that, associated with the original model, there is a limit system that has a much simpler structure. Using the limit problem as a guide, we can construct near-optimal portfolio selection strategies.

In the literature, Markowitz proposed the mean-variance model in [Ref. [21]] for portfolio construction in a single period. The salient feature of the model is that it enables an investor to seek the highest return after specifying his/her acceptable risk level, as quantified by the variance of the return. In fact, the mean-variance approach has become the foundation of modern finance theory and has inspired numerous subsequent works. One natural generalization is to investigate dynamic mean-variance models. Along this line, multi-period mean-variance portfolio selection was studied in, for example, Mossin [Ref. [22]], Samuelson [Ref. [25]], Hakansson [Ref. [14]], Grauer and Hakansson [Ref. [12]], and Pliska [Ref. [24]] among others. Another extension is to the consideration of continuous-time mean-variance hedging problems. In fact, such problems were dealt with by Duffie and Richardson [Ref. [7]] and Schweizer [Ref. [26]] where optimal dynamic strategies were derived, based on the projection theorem, to hedge contingent claims in incomplete markets. In [Ref. [7]], this result was derived under the assumption that all the coefficients (interest rate, volatility rate, etc.) are deterministic and time-invariant constants. The model treated in [Ref. [26]] is mathematically general; however the solution is based on a martingale measure and is thus not easy to use in practice. Therefore, our work concentrates on a more concrete example.

Our work focuses on regime-switching models. Early work on regime-switching models for option pricing can be found in [Ref. [1]]. Subsequent work in economic time series [Ref. [15]] extends the traditional ARMA model setting and allows the models under consideration to involve regime changes. Recently, there have been growing interests in using regime-switching diffusions to delineate the financial market, where the switching processes are used to describe stochastic volatility resulting from market modes, interest rates, as well as other economic factors; see for example, [Refs. [3, 5, 6, 8, 13, 23, 30, 38, 40]] and references therein. A host of models, including geometric Brownian motion and mean-reversion models have been extended to include regime switching. Applications have also been found in using switching models in insurance and risk models for calculating ruin probability and possibility of default etc. (see, for example, [Refs. [28, 31]]). In fact, regime-switching models have gained much popularity. Regime-switching models have also been used in such areas as filtering, stability, and ergodic control etc. (see [Refs. [2, 4, 11, 16, 18, 29, 32, 34, 35, 37]]) among others.

Using the stochastic linear-quadratic (LQ) control framework (see [Ref. [36]]), Zhou and Li [Ref. [39]] studied the mean-variance problem for a continuous-time model from another angle. By embedding the original (not readily solvable) problem into a tractable auxiliary problem, following a similar embedding technique introduced in Li and Ng [Ref. [19]] for the multi-period model, it was shown that this auxiliary problem in fact is a stochastic optimal LQ problem and can be solved explicitly by the LQ theory. Such an approach establishes a natural connection between portfolio selection problems and standard stochastic control models.

The rest of the paper is arranged as follows. The precise formulation is given next. Section 16.3 develops a limit theorem that shows that the assets under consideration have a limit that is an average with respect to the associated measure of the switching process. Section 16.4 makes further remarks to conclude the paper.

16.2. Problem formulation

We adopt the notation in [Ref. [40]]. Let $(\Omega, \mathcal{F}, P)$ be a fixed complete probability space on which is defined a standard d-dimensional Brownian motion $W(t) \equiv (W^1(t), \ldots, W^d(t))'$ and a continuous-time stationary Markov chain $\alpha^\varepsilon(t)$ taking values in a finite state space $\mathcal{M} = \{1, 2, \ldots, l\}$, such that $W(t)$ and $\alpha^\varepsilon(t)$ are independent. Let $\varepsilon > 0$ be a small parameter. The Markov chain has a generator $Q^\varepsilon = (q_{ij}^\varepsilon)_{l\times l}$ satisfying $q_{ij}^\varepsilon \geq 0$ for $i \neq j$ and $\sum_{i=1}^{l} q_{ij}^\varepsilon = 0$ for each $i \in \mathcal{M}$. Note that the generator depends on a small parameter $\varepsilon > 0$. In this paper, we consider the case that the generator is given by

$$Q^\varepsilon = \frac{Q}{\varepsilon} + Q_0, \tag{16.1}$$

where both Q and Q_0 are generators of some appropriate Markov chains and Q is irreducible. The rationale is that the Markov chain is subject to both fast and slow motion. The Q/ε represents the fast changing part, whereas Q_0 reflects the slowly varying part. Effectively, there are two time scales involved. One of them is the running time t and the other one is the stretched time or fast time $\tau = t/\varepsilon$.

Define $\mathcal{F}_t = \sigma\{W(s), \alpha^\varepsilon(s) : 0 \leq s \leq t\}$. and denote by $L^2_{\mathcal{F}}(0, T; \mathbb{R}^\iota)$ the set of all $\mathbb{R}^\iota$-valued, measurable stochastic processes $f(t)$ adapted to $\{\mathcal{F}_t\}_{t\geq 0}$, such that $E\int_0^T |f(t)|^2 dt < +\infty$. Consider a market in which $d+1$ assets are traded continuously. One of the assets is a bank account whose price $P_0(t)$ is the solution of the ordinary differential equation with random switching

$$\begin{aligned} &dP_0^\varepsilon(t) = r(t, \alpha^\varepsilon(t))P_0^\varepsilon(t)dt, \quad t \in [0, T], \\ &P_0^\varepsilon(0) = p_0 > 0, \end{aligned} \tag{16.2}$$

where $r(t, i) \geq 0$, for $i \in \mathcal{M}$, are the interest rates corresponding to different market modes. The other d assets are stocks with prices $P_m^\varepsilon(t)$, $m = 1, 2, \ldots, d$, satisfying the following system of stochastic differential equations with random switching

$$\begin{aligned} &dP_m^\varepsilon(t) = P_m^\varepsilon(t)\left[b_m(t, \alpha^\varepsilon(t))dt + \sum_{n=1}^{d} \sigma_{mn}(t, \alpha^\varepsilon(t))dW_n(t)\right], \quad t \in [0, T], \\ &P_m^\varepsilon(0) = p_m > 0, \end{aligned} \tag{16.3}$$

where for each $i \in \mathcal{M}$, $b_m(t, i)$ is the appreciation rate process and $\sigma_m(t, i) := (\sigma_{m1}(t, i), \ldots, \sigma_{md}(t, i))$ is the volatility or the dispersion rate of the mth stock, associated with regime $\alpha^\varepsilon(t) = i$ with $i \in \mathcal{M}$. Define the volatility matrix by

$$\sigma(t,i) = (\sigma_{mn}(t,i))_{d\times d}, \quad \text{for each } i \in \mathcal{M}. \tag{16.4}$$

Assumption (A). *In this paper, we assume that for some $\delta > 0$,*

$$\sigma(t,i)\sigma'(t,i) \geq \delta I, \ \forall (t,i) \in [0,T] \times \mathcal{M}. \tag{16.5}$$

We also assume that all the functions $r(t,i), b_m(t,i), \sigma_{mn}(t,i)$ are measurable and uniformly bounded in t.

Condition (16.5) is simply a non-degeneracy requirement. Suppose that the initial market mode $\alpha^\varepsilon(0) = i$. Consider an agent with an initial wealth $x_0^\varepsilon > 0$. Denote by $x^\varepsilon(t)$ the total wealth at time $t \geq 0$. Assuming that the trading of shares takes place continuously and there is no transaction cost and consumptions, then by virtue of [Ref. [36, p. 57]]

$$\begin{aligned} dx^\varepsilon(t) = \Big[r(t,\alpha^\varepsilon(t))x^\varepsilon(t) + \sum_{m=1}^{d} \big[b_m(t,\alpha^\varepsilon(t)) - r(t,\alpha^\varepsilon(t)) \big] u_m(t) \Big] dt \\ + \sum_{n=1}^{d}\sum_{m=1}^{d} \sigma_{mn}(t,\alpha^\varepsilon(t)) u_m(t) dW_n(t), \\ x^\varepsilon(0) = x > 0, \ \ \alpha^\varepsilon(0) = i, \end{aligned} \tag{16.6}$$

where $u_m(t)$ is the total market value of the agent's wealth in the mth asset, $m = 0, 1, \ldots, d$, at time t. We call $u(\cdot) = (u_1(\cdot), \ldots, u_d(\cdot))'$ a portfolio of the agent. Note that once $u(\cdot)$ is determined, $u_0(\cdot)$, the asset in the bank account is completed specified since

$$u_0(t) = x^\varepsilon(t) - \sum_{i=1}^{d} u_i(t).$$

Thus, in the following analysis, only $u(\cdot)$ is considered. Let

$$B(t,i) := (b_1(t,i) - r(t,i), \ldots, b_d(t,i) - r(t,i)), \ i \in \mathcal{M}, \tag{16.7}$$

we can rewrite the wealth equation (16.6) as

$$\begin{aligned} &dx^\varepsilon(t) = [r(t,\alpha^\varepsilon(t))x^\varepsilon(t) + B(t,\alpha^\varepsilon(t))u(t)]dt + u'(t)\sigma(t)dW(t), \\ &x^\varepsilon(0) = x, \ \ \alpha^\varepsilon(0) = i. \end{aligned} \tag{16.8}$$

We say that a portfolio $u(\cdot)$ is admissible if $u(\cdot) \in L^2_{\mathcal{F}}(0,T;\mathbb{R}^d)$ and the SDE (16.8) has a unique solution $x(\cdot)$ corresponding to $u(\cdot)$. In this case, we refer to $(x(\cdot), u(\cdot))$ as an admissible (wealth, portfolio) pair.

The agent's objective is to find an admissible portfolio $u(\cdot)$ among all the admissible portfolios whose expected terminal wealth is $Ex^\varepsilon(T) = z$ for a given $z \in \mathbb{R}^1$, so that the risk in-terms of the variance of the terminal wealth

$$\text{Var } x^\varepsilon(T) \equiv E[x^\varepsilon(T) - Ex^\varepsilon(T)]^2 = E[x^\varepsilon(T) - z]^2 \tag{16.9}$$

is minimized. The problem of finding such a portfolio $u(\cdot)$ is referred to as the mean-variance portfolio selection problem.

The mean-variance portfolio selection is a constrained stochastic optimization problem, parameterized by $z \in \mathbb{R}^1$:

$$\begin{aligned} &\text{minimize} \quad J_{\text{MV}}(x, i, u(\cdot)) := E[x^\varepsilon(T) - z]^2, \\ &\text{subject to} \begin{cases} Ex^\varepsilon(T) = z, \\ (x^\varepsilon(\cdot), u(\cdot)) \ \text{admissible}. \end{cases} \end{aligned} \tag{16.10}$$

The problem is said to be feasible if there is at least one portfolio satisfying all the constraints. The problem is finite if it is feasible and the infimum of $J_{\text{MV}}(x, i, u(\cdot))$ is finite. If an optimal portfolio to the above problem exists, it is said to be an efficient portfolio corresponding to z, and the corresponding $(\text{Var } x^\varepsilon(T), z) \in \mathbb{R} \times \mathbb{R}$ and $(\sigma_{x^\varepsilon(T)}, z) \in \mathbb{R} \times \mathbb{R}$ are interchangeably called an efficient point, where $\sigma_{x^\varepsilon(T)}$ denotes the standard deviation of $x^\varepsilon(T)$. The set consisting of all the efficient points is called the efficient frontier.

16.3. Limit results

16.3.1. *Preliminary results*

Using the results of [Ref. [40]], we can have the feasibility and optimality result below. First define

$$\rho(t, i) := B(t, i)[\sigma(t, i)\sigma'(t, i)]^{-1} B'(t, i), \ i \in \mathcal{M}. \tag{16.11}$$

Consider the following systems of ordinary differential equations:

$$\begin{aligned} &\dot{P}^\varepsilon(t, i) = [\rho(t, i) - 2r(t, i)]\, P^\varepsilon(t, i) - \sum_{j=1}^{l} q_{ij}^\varepsilon P^\varepsilon(t, j), \\ &P^\varepsilon(T, i) = 1, \quad i \in \mathcal{M}, \end{aligned} \tag{16.12}$$

and

$$\dot{H}^\varepsilon(t,i) = r(t,i)H^\varepsilon(t,i) - \frac{1}{P^\varepsilon(t,i)}\sum_{j=1}^{l} q_{ij}^\varepsilon P^\varepsilon(t,j)[H^\varepsilon(t,j) - H^\varepsilon(t,i)],$$
$$H^\varepsilon(T,i) = 1, \quad i \in \mathcal{M}. \tag{16.13}$$

There exist unique solutions to the above two systems (16.12) and (16.13) equations since they are linear with uniformly bounded coefficients.

Theorem 16.1. *The following results hold:*

- *The mean-variance problem* (16.10) *is feasible for every* $z \in \mathbb{R}$ *if and only if*

$$E\int_0^T |B(t,\alpha^\varepsilon(t))|^2 dt > 0. \tag{16.14}$$

- *An optimal feedback control is given by*

$$u^{\varepsilon,*}(t,x,i) = -[\sigma(t,i)\sigma'(t,i)]^{-1}B'(t,i)[x+(\lambda-z)H^\varepsilon(t,i)]. \tag{16.15}$$

- *The corresponding optimal value is given by*

$$\begin{aligned}\inf J(x,i,u(\cdot),\lambda) = {} & [P^\varepsilon(0,i)H^\varepsilon(0,i)^2 + \theta - 1](\lambda - z)^2 \\ & + 2[P^\varepsilon(0,i)H^\varepsilon(0,i)x - z](\lambda - z) \\ & + P^\varepsilon(0,i)x^2 - z^2,\end{aligned} \tag{16.16}$$

where inf *is taken over those* $u(\cdot)$ *that are admissible, and*

$$\begin{aligned}\theta := {} & E\int_0^T \sum_{j=1}^{l} q_{\alpha^\varepsilon(t)j}^\varepsilon P^\varepsilon(t,j)[H^\varepsilon(t,j) - H^\varepsilon(t,\alpha^\varepsilon(t))]^2 dt \\ = {} & \sum_{\ell=1}^{l}\sum_{j=1}^{l}\int_0^T P^\varepsilon(t,j)p_{i\ell}^\varepsilon(t)q_{\ell j}^\varepsilon[H^\varepsilon(t,j) - H^\varepsilon(t,\ell)]^2 dt \\ \geq {} & 0,\end{aligned} \tag{16.17}$$

with the transition probabilities $p_{i\ell}^\varepsilon(t)$ *of the Markov chain* $\alpha^\varepsilon(t)$ *associated with the generator* Q.

16.3.2. *Limit results*

Note that since Q is assumed to be irreducible, there is an invariant measure associated with the generator. That is, there is a $\nu = (\nu_1, \ldots, \nu_l) \in \mathbb{R}^{1\times l}$

such that $\nu_i > 0$ for each $i \in \mathcal{M}$ and ν is the unique solution of the system of equations

$$\nu Q = 0, \quad \nu \mathbb{1} = 1, \quad \text{where } \mathbb{1} = (1, \ldots, 1)' \in \mathbb{R}^l.$$

The ν is known as stationary distribution associated with the generator Q. With Theorem 16.1 at our hands, we proceed to obtain limit results for the system as $\varepsilon \to 0$. The result is presented in the following theorem. For notational simplicity, we write

$$\begin{aligned} u^{\varepsilon,*}(t,x,i) &= G(t,x,i)x + h(t,i) \\ G(t,x,i) &= -[\sigma(t,i)\sigma'(t,i)]^{-1}B'(t,i) \in \mathbb{R}^d \\ h(t,x,i) &= G(t,x,i)(\lambda - z)H^{\varepsilon}(t,i) \in \mathbb{R}^d. \end{aligned} \tag{16.18}$$

Theorem 16.2. *Consider system* (16.8). *With the feedback control given in* (16.18), *the pair of wealth and control process* $(x^\varepsilon(\cdot), u^{\varepsilon,*}(\cdot))$ *converges weakly to* $(x(\cdot), u(\cdot))$ *such that* $x(t)$ *is given by*

$$dx = [\overline{r}(t)x(t) + \overline{B}(t)u(t,x(t))]dt + u'(t,x(t))\overline{\sigma}(t)dW(t), \tag{16.19}$$

where

$$\begin{aligned} \overline{g}(t) &= \sum_{i=1}^{l} \nu_i g(t,i) \quad \textit{for } g(t,i) = r(t,i),\ g(t,i) = B(t,i), \\ \overline{\sigma}(t)\overline{\sigma}'(t) &= \sum_{i=1}^{l} \nu_i \sigma(t,i)\sigma'(t,i). \end{aligned}$$

The limit control is given by

$$\begin{aligned} u(t,x(t)) &= \overline{G}(t,x(t)) + \overline{h}(t,x(t)) \qquad \textit{where} \\ \overline{G}(t,x) &= \sum_{i=1}^{l} \nu_i G(x,t,i), \\ \overline{h}(t,x,i) &= \sum_{i=1}^{l} \nu_i h(t,x,i). \end{aligned} \tag{16.20}$$

The proof will be carried out by establishing a number of lemmas. First, we obtain a second moment bound. Throughout the rest of the paper, we use K to denote a generic positive constant, whose values may change from line to line.

Lemma 16.1. *Under assumption (A),*

$$\sup_{t\in[0,T]} |x^\varepsilon(t)|^2 \le K$$

for some $K > 0$.

Proof. For notational simplicity, we will write $u^\varepsilon(t,x,i)$ in lieu of $u^{\varepsilon,*}(t,x,i)$ in what follows. Using the feedback control (16.18) in (16.8), we obtain

$$\begin{aligned} dx^\varepsilon(t) &= [r(t,\alpha^\varepsilon(t))x^\varepsilon(t) + B(t,\alpha^\varepsilon(t))u^\varepsilon(t,x^\varepsilon(t),\alpha^\varepsilon(t))]dt \\ &\qquad\qquad + u'(t,x^\varepsilon(t),\alpha^\varepsilon(t))\sigma(t)dW(t), \\ x^\varepsilon(0) &= x, \quad \alpha^\varepsilon(0) = i. \end{aligned} \tag{16.21}$$

By virtue of assumption (A), $P(t,i)$ and $H(t,i)$ are all bounded uniformly in $t \in [0,T]$. By an elementary inequality, we have

$$\begin{aligned} E|x^\varepsilon(t)|^2 &\le KE|x^\varepsilon(0)|^2 \\ &\quad + KE\left|\int_0^t B(v,\alpha^\varepsilon(v))[G(v,x^\varepsilon(v),\alpha^\varepsilon(v))x^\varepsilon(v) + h(v,\alpha^\varepsilon(v))]dv\right|^2 \\ &\quad + KE\left|\int_0^t u'(v,x^\varepsilon(v),\alpha^\varepsilon(v))\sigma(v)dW(v)\right|^2 \\ &\le K + K\int_0^t E|x^\varepsilon(v)|^2 dv. \end{aligned}$$

An application of the Gronwall inequality leads to

$$E|x^\varepsilon(t)|^2 \le K\exp(KT) < \infty. \tag{16.22}$$

Moreover, taking sup over $[0,T]$ in (16.22), the desired result follows. □

Next we prove a lemma that gives us the desired compactness. This is also stated as a lemma.

Lemma 16.2. *The* $(x^\varepsilon(\cdot), u^\varepsilon(\cdot))$ *is tight in* $D([0,T];\mathbb{R}\times\mathbb{R}^d)$ *for any* $T > 0$, *where* $D([0,T];\mathbb{R}\times\mathbb{R}^d)$ *denotes the space of* $\mathbb{R}\times\mathbb{R}^d$*-valued functions that are right continuous, have left limits, endowed with the Skorohod topology.*

Proof. For any $\eta > 0$ and any $t, s > 0$ satisfying $s \le \eta$, using the boundedness of $r(\cdot,i)$, $b_m(\cdot,i)$, $P(\cdot,i)$, and $H(\cdot,i)$, we have

$$\begin{aligned} &E|x^\varepsilon(t+s) - x^\varepsilon(t)|^2 \\ &\le KE\left|\int_t^{t+s} r(v,\alpha^\varepsilon(v))x^\varepsilon(v)dv\right|^2 \\ &\quad + KE\left|\int_t^{t+s} B(t,\alpha^\varepsilon(t))[G(v,x^\varepsilon(v),\alpha^\varepsilon(v))x^\varepsilon(v) + h(v,\alpha^\varepsilon(v))]dv\right|^2 \\ &\quad + KE\left|\int_t^{t+s} u'(v,x^\varepsilon(v),\alpha^\varepsilon(v))\sigma(v)dW(v)\right|^2, \end{aligned}$$

here and hereafter K is a generic positive constant whose values may be different for different appearances. By means of Hölder's inequality,

$$E\left|\int_t^{t+s} r(v,\alpha^\varepsilon(v))x^\varepsilon(v)dv\right|^2 \leq Ks\int_t^{t+s} E|x^\varepsilon(v)|^2 dv. \tag{16.23}$$

Similarly,

$$\begin{aligned}&E\left|\int_t^{t+s} B(t,\alpha^\varepsilon(t))[G(v,x^\varepsilon(v),\alpha^\varepsilon(v))x^\varepsilon(v)+h(v,\alpha^\varepsilon(v))]dv\right|^2\\&\quad\leq Ks\int_t^{t+s} E|x^\varepsilon(v)|dv + Ks^2.\end{aligned} \tag{16.24}$$

By using the elementary properties of stochastic integral with respect to the Brownian motion, we obtain

$$\begin{aligned}&E\left|\int_t^{t+s} u'(v,x^\varepsilon(v),\alpha^\varepsilon(v))\sigma(v)dW(v)\right|^2\\&\quad\leq K\int_t^{t+s} E|G(v,x^\varepsilon(v),\alpha^\varepsilon(v))x^\varepsilon(v)+h(v,\alpha^\varepsilon(v))|^2 dv\\&\quad\leq K\int_t^{t+s} E|x^\varepsilon(v)|^2 dv + Ks.\end{aligned} \tag{16.25}$$

Combining (16.23)–(16.25), and using Lemma 16.1, we have

$$E|x^\varepsilon(t+s)-x^\varepsilon(t)|^2 \leq O(s).$$

It then follows that

$$\lim_{\eta\to 0}\limsup_{\varepsilon\to 0} E|x^\varepsilon(t+s)-x^\varepsilon(t)|^2 = 0.$$

Finally noting the linear dependence of $u^\varepsilon(\cdot)$ on x, $u^\varepsilon(\cdot)$ is also tight. Thus, the desired tightness is obtained. □

To proceed, we define an operator

$$\begin{aligned}\mathcal{L}f(t,x)=&\ f_t(t,x)+f_x(t,x)[\overline{r}(t)x+\overline{B}(t)(\overline{G}(t,x)x+\overline{h}(t,x))]\\&+\frac{1}{2}\mathrm{tr}[f_{xx}(t,x)(\overline{G}(t,x)x+\overline{h}(t,x))(\overline{G}(t,x)x+\overline{h}(t,x))'],\end{aligned} \tag{16.26}$$

where $f(\cdot,\cdot)\in C_0^{1,2}$ ($C^{1,2}$ functions with compact support). We shall characterize the limit $x(\cdot)$ by showing that it is the solution of the martingale problem with operator $\mathcal{L}$. We first state a lemma.

Lemma 16.3. *The martingale problem with operator $\mathcal{L}$ has a unique solution.*

In fact, we need only show that the solution is unique in distribution. This can easily be observed from the linear dependence on x in the operator $\mathcal{L}$ or (16.21). We thus omit the details.

By virtue of Lemma 16.2, $(x^\varepsilon(\cdot), u^\varepsilon(\cdot))$ is tight. Thus, by the well-known Prohorov's theorem, we can extract a convergent subsequence. We denote this subsequence using the index ε for notational simplicity. Denote the limit by $(x(\cdot), u(\cdot))$. By Skorohod's representation, with a slight abuse of notation, we may assume that $(x^\varepsilon(\cdot), u^\varepsilon(\cdot))$ converges to $(x(\cdot), u(\cdot))$ w.p.1. Next, we derive another lemma to characterize the limit process.

Lemma 16.4. *The limit $x(\cdot)$ is a solution of the martingale problem with operator $\mathcal{L}$.*

Proof. We begin by examining

$$\begin{aligned} \mathcal{L}^\varepsilon f(t,x,i) &= f_t(t,x) + f_x(t,x)[r(t)x + B(t)(G(t,x,i)x + h(t,x,i))] \\ &\quad + \frac{1}{2}\mathrm{tr}[f_{xx}(t,x,i)(G(t,x,i)x + h(t,x,i))(G(t,x,i)x + h(t,x,i))'] \\ &\quad + \sum_{j=1}^{l} q_{ij}^\varepsilon f(t,x), \end{aligned} \tag{16.27}$$

where $f(\cdot,\cdot) \in C_0^{1,2}$. We aim to show that

$$f(t,x(t)) - f(0,x) - \int_0^t \mathcal{L}f(v,x(v))dv \quad \text{is a martingale.}$$

We need only show that for any bounded and continuous function $\varphi(\cdot,\cdot)$, any $0 \le t, s < T$ satisfying $t+s \le T$, any positive integer i_0, and any $t_i \le t$, for $i \le i_0$

$$\begin{aligned} E\Big[\varphi(t_i, x(t_i) : i \le i_0)\Big[f(t+s, x(t+s)) - f(t,x(t)) \\ \int_t^{t+s} \mathcal{L}f(v,x(v))dv\Big]\Big] = 0. \end{aligned} \tag{16.28}$$

Note that since $(x^\varepsilon(\cdot), \alpha^\varepsilon(\cdot))$ is a Markov process,

$$\begin{aligned} E\Big[\varphi(t_i, x^\varepsilon(t_i) : i \le i_0)\Big[f(t+s, x^\varepsilon(t+s)) - f(t,x^\varepsilon(t)) \\ - \int_t^{t+s} \mathcal{L}f(v,x^\varepsilon(v))dv\Big]\Big] = 0. \end{aligned} \tag{16.29}$$

By the weak convergence of $x^\varepsilon(\cdot)$ to $x(\cdot)$ together with the use of Skorohod representation,

$$\begin{aligned} &E\Big[\varphi(t_i, x^\varepsilon(t_i) : i \le i_0)[f(t+s, x^\varepsilon(t+s)) - f(t, x^\varepsilon(t))]\Big] \\ &\quad \to E\Big[\varphi(t_i, x(t_i) : i \le i_0)[f(t+s, x(t+s)) - f(t, x(t))]\Big] \end{aligned} \tag{16.30}$$

as $\varepsilon \to 0$.

On the other hand, we have

$$\begin{aligned} &\int_t^{t+s} \mathcal{L}^\varepsilon f(v, x^\varepsilon(v))dv \\ &= \int_t^{t+s} f_v(v, x^\varepsilon(v))dv + \sum_{i=1}^{l} \int_t^{t+s} f_x(v, x^\varepsilon(v)) r(v,i) x^\varepsilon(v) I_{\{\alpha^\varepsilon(v)=i\}} dv \\ &\quad + \sum_{i=1}^{l} \int_t^{t+s} f_x(v, x^\varepsilon(v)) B(v)(G(v, x^\varepsilon(v), i) x^\varepsilon(v) \\ &\qquad\qquad + h(v, x^\varepsilon(v), i)) I_{\{\alpha^\varepsilon(v)=i\}} dv \\ &\quad + \frac{1}{2} \sum_{i=1}^{l} \int_t^{t+s} \text{tr}[f_{xx}(v, x^\varepsilon(v)) \Sigma(v, x^\varepsilon(v), i) I_{\{\alpha^\varepsilon(v)=i\}} dv \\ &\quad + \sum_{i=1}^{l} \int_t^{t+s} Q^\varepsilon f(v, x^\varepsilon(v))(i) I_{\{\alpha^\varepsilon(v)=i\}} dv, \end{aligned} \tag{16.31}$$

where

$$\begin{aligned} \Sigma(t,x,i) &= [G(t,x,i)x + h(t,x,i)][G(t,x,i)x + h(t,x,i)]', \\ Q^\varepsilon f(t,x)(i) &= \sum_{j=1}^{l} q_{ij}^\varepsilon f(t,x), \end{aligned}$$

for $i \in \mathcal{M}$. Since $f(\cdot)$ is a function independent of i, the last term in (16.31) is 0 identically.

By weak convergence and the Skorohod representation,

$$\begin{aligned} &E\Big[\varphi(t_i, x^\varepsilon(t_i) : i \le i_0) \int_t^{t+s} f_v(v, x^\varepsilon(v))dv\Big] \\ &\quad \to E\Big[\varphi(t_i, x(t_i) : i \le i_0) \int_t^{t+s} f_v(v, x(v))dv\Big]. \end{aligned} \tag{16.32}$$

Next, it is readily seen that for a fixed $i \in \mathcal{M}$,

$$\begin{aligned}&\int_t^{t+s} f_x(v, x^\varepsilon(v)) r(v,i) x^\varepsilon(v) I_{\{\alpha^\varepsilon(v)=i\}} dv \\ &\quad = \int_t^{t+s} f_x(v, x^\varepsilon(v)) r(v,i) x^\varepsilon(v) [I_{\{\alpha^\varepsilon(v)=i\}} - \nu_i] dv \\ &\quad\quad + \int_t^{t+s} f_x(v, x^\varepsilon(v)) r(v,i) x^\varepsilon(v) \nu_i dv.\end{aligned}$$

Using [Ref. [33, Lemma 7.2]], it can be shown that

$$E\left|\int_t^{t+s} f_x(v, x^\varepsilon(v)) r(v,i) x^\varepsilon(v) [I_{\{\alpha^\varepsilon(v)=i\}} - \nu_i] dv\right|^2 \to 0 \tag{16.33}$$

as $\varepsilon \to 0$. Moreover, it can be seen that

$$\begin{aligned}&E\Big[\varphi(t_i, x^\varepsilon(t_i) : i \le i_0) \int_t^{t+s} f_x(v, x^\varepsilon(v)) r(v,i) x^\varepsilon(v) \nu_i dv\Big] \\ &\quad \to E\Big[\varphi(t_i, x(t_i) : i \le i_0) \int_t^{t+s} f_x(v, x(v)) r(v,i) x(v) \nu_i dv\Big].\end{aligned} \tag{16.34}$$

Thus,

$$\begin{aligned}&E\bigg[\varphi(t_i, x^\varepsilon(t_i) : i \le i_0)\Big[\sum_{i=1}^{l} \int_t^{t+s} f_x(v, x^\varepsilon(v)) r(v,i) x^\varepsilon(v) I_{\{\alpha^\varepsilon(v)=i\}} dv\Big]\bigg] \\ &\quad \to E\bigg[\varphi(t_i, x(t_i) : i \le i_0) \int_t^{t+s} f_x(v, x(v)) \overline{r}(v) x(v) dv\bigg] \quad \text{as} \quad \varepsilon \to 0.\end{aligned} \tag{16.35}$$

Likewise, we can show that as $\varepsilon \to 0$,

$$\begin{aligned}&E\bigg[\varphi(t_i, x^\varepsilon(t_i) : i \le i_0)\Big[\sum_{i=1}^{l} \int_t^{t+s} f_x(v, x^\varepsilon(v)) B(v) [G(v, x^\varepsilon(v), i) x^\varepsilon(v) \\ &\qquad + h(v, x^\varepsilon(v), i)] I_{\{\alpha^\varepsilon(v)=i\}} dv\Big]\bigg] \\ &\quad \to E\bigg[\varphi(t_i, x(t_i) : i < i_0)\Big[\sum_{i=1}^{l} \int_t^{t+s} f_x(v, x(v)) B(v) [G(v, x(v), i) x(v) \\ &\qquad + h(v, x(v), i)] \nu_i dv\Big]\bigg]\end{aligned} \tag{16.36}$$

and

$$E\bigg[\varphi(t_i, x^\varepsilon(t_i) : i \le i_0)\Big[\sum_{i=1}^{l}\int_t^{t+s} \mathrm{tr}[f_{xx}(v, x^\varepsilon(v))\Sigma(v, x^\varepsilon(v), i)]I_{\{\alpha^\varepsilon(v)=i\}}\Big]\bigg]$$
$$\to E\bigg[\varphi(t_i, x^\varepsilon(t_i) : i \le i_0)\Big[\sum_{i=1}^{l}\int_t^{t+s} \mathrm{tr}[f_{xx}(v, x^\varepsilon(v))\Sigma(v, x^\varepsilon(v), i)]\nu_i dv\bigg]. \tag{16.37}$$

Combining the estimates obtained so far, we obtain the desired martingale limit. Hence the lemma is proved. □

Putting together the results proved thus far, the proof of the theorem is completed. Thus the desired limit system is obtained.

16.4. Further remarks

This paper has focused on two-time-scale mean-variance portfolio selection problems for regime-switching diffusion price models. We showed that system under consideration converges weakly to a limit system. The rationale is that the limit system is much simpler and easier to deal with. Thus, one can construct strategies leading to asymptotic optimality. Using the results obtained in this paper, we can further study the mutual fund theorem.

The results obtained accommodate mean-variance problems for regime-switching models and provide further insight. When we focus on the associated minimal variance, we would need a condition on $z_{\min}$. This was discussed in [Ref. [40, Section 5]]. In a special case, when the interest rate is unaffected by the regime (i.e., independent of the Markov chain $\alpha^\varepsilon(t)$), $z_{\min} = e^{\int_0^T r(t)dt}x$, which was discussed in [Ref. [40, Section 6]].

Acknowledgments

This research was supported in part by the National Science Foundation under grant DMS-0907753, in part by the Air Force Office of Scientific Research under grant FA9550-10-1-0210, and in part by National Natural Science Foundation of China under grant #70871055.

References

[1] G. Barone-Adesi and R. Whaley, Efficient analytic approximation of American option values, *J. Finance*, **42** (1987), 301–320.

[2] G.K. Basak, A. Bisi, and M.K. Ghosh, Stability of Degenerate Diffusions with State-Dependent Switching, *J. Math. Anal. Appl.*, **240** (1999), 219-248.

[3] F. Biagini and P. Guasoni, Mean-variance hedging for stochastic volatility models, *Math. Fin.* **10** (2000), 109-123.

[4] T. Björk, Finite dimensional optimal filters for a class of Itô-processes with jumping parameters, *Stochastics* **4** (1980), 167-183.

[5] J. Buffington and R. Elliott, American options with regime switching, preprint, (2001).

[6] G.B. Di Masi, Y.M. Kabanov and W.J. Runggaldier, Mean variance hedging of options on stocks with Markov volatility, *Theory Prob. Appl.* **39** (1994), 173-181.

[7] D. Duffie and H. Richardson, Mean-variance hedging in continuous time, *Ann. Appl. Probab.* **1** (1991), 1-15.

[8] R.J. Elliott, L. Chan, and T.K. Siu, Option pricing and Esscher transform under regime switching, *Annals of Finance*, **1** (2005), 423–432.

[9] A. Friedman, *Stochastic Differential Equations and Applications*, Vols. I and II, Academic Press, New York, 1975.

[10] J. P. Fouque, G. Papanicolaou, and K.R. Sircar, *Derivatives in Financial Markets with Stochastic Volatility*, Cambridge University Press, 2000.

[11] M.K. Ghosh, A. Arapostathis, and S.I. Marcus, Ergodic control of switching diffusions, *SIAM J. Control Optim.*, **35** (1997), 1952-1988.

[12] R.R. Grauer and N.H. Hakansson, On the use of mean-variance and quadratic approximations in implementing dynamic investment strategies: a comparison of returns and investment policies, *Management Sci.* **39** (1993), 856-871.

[13] X. Guo, *Insider Information and Stock Fluctuations*, PhD Thesis, Rutgers University, (1999).

[14] N.H. Hakansson, Multi-period mean-variance analysis : Toward a general theory of portfolio choice, *J. Fin.* **26** (1971), 857-884.

[15] J.D. Hamilton and R. Susmel, Autoregressive conditional heteroskedasticity and changes in regime. *J. Econometrics* **64**, (1994), 307-333.

[16] A.M. Il'in, R.Z. Khasminskii, and G. Yin, Asymptotic expansions of solutions of integro-differential equations for transition densities of singularly perturbed switching diffusions, *J. Math. Anal. Appl.*, **238** (1999), 516-539.

[17] R.Z. Khasminskii, *Stochastic Stability of Differential Equations*, Sijthoff and Noordhoff, Alphen aan den Rijn, Netherlands, 1980.

[18] R.Z. Khasminskii, C. Zhu, and G. Yin, Stability of regime-switching diffusions, *Stochastic Proc. Appl.*, **117** (2007), 1037-1051.

[19] D. Li and W.L. Ng, Optimal dynamic portfolio selection: Multi-period mean-variance formulation, *Math. Fin.* **10** (2000), 387-406.

[20] D.G. Luenberger, *Investment Science*, Oxford University Press, New York, 1998.

[21] H. Markowitz, Portfolio selection, *J. Fin.* **7** (1952), 77-91

[22] J. Mossin, Optimal multiperiod portfolio policies, *J. Business* **41** (1968), 215-229.

[23] M. Pemy, Q. Zhang, and G. Yin, Liquidation of a large block of stock with regime switching, *Math. Finance*, **18** (2008), 629-648.
[24] S.R. Pliska, *Introduction to Mathematical Finance*, Basil Blackwell, Malden, 1997.
[25] P.A. Samuelson, Lifetime portfolio selection by dynamic stochastic programming, *Rev. Econ. Statist.* **51** (1969), 239-246.
[26] M. Schweizer, Approximation pricing and the variance-optimal martingale measure, *Ann. Probab.* **24** (1996), 206-236.
[27] A.V. Skorohod, *Asymptotic Methods in the Theory of Stochastic Differential Equations*, Amer. Math. Soc., Providence, RI, 1989.
[28] H.L. Yang and G. Yin, Ruin probability for a model under Markovian switching regime, in *Probability, Finance and Insurance*, World Sci., 206-217, T.L. Lai, H. Yang, and S.P. Yung Eds., 2004.
[29] G. Yin, V. Krishnamurthy, and C. Ion, Regime switching stochastic approximation algorithms with application to adaptive discrete stochastic optimization, *SIAM J. Optim.*, **14** (2004), 1187-1215.
[30] G. Yin, R.H. Liu, and Q. Zhang, Recursive algorithms for stock Liquidation: A stochastic optimization approach, *SIAM J. Optim.*, **13** (2002), 240–263.
[31] G. Yin, Y.J. Liu, and H.L. Yang, Bounds of ruin probability for regime-switching models using time scale separation, *Scandinavian Actuarial Journal* **2006** (2006), 111–127.
[32] G. Yin, X. Mao, C. Yuan, and D. Cao, Approximatio methods for hybrid diffusion systems with state-dependent switching diffusion processes: Numerical alogorithms and existence and uniqueness of solutions, *SIAM J. Math. Anal.*, **41** (2010), 2335–2352.
[33] G. Yin and Q. Zhang, *Continuous-time Markov Chains and Applications: A Singular Perturbations Approach*, Springer-Verlag, New York, NY, 1998.
[34] G. Yin and C. Zhu, *Hybrid Switching Diffusions: Properties and Applications*, Springer, New York, 2010.
[35] G. Yin and C. Zhu, Properties of solutions of stochastic differential equations with continuous-state-dependent switching, *J. Differential Eqs.*, **249** (2010), 2409–2439.
[36] J. Yong and X.Y. Zhou, *Stochastic Controls: Hamiltonian Systems and HJB Equations,* Springer, New York, 1999.
[37] C. Yuan and X. Mao, Asymptotic stability in distribution of stochastic differential equations with Markovian switching, *Stochastic Proc. Appl.*, **103** (2003), 277-291.
[38] Q. Zhang, Stock trading: An optimal selling rule, *SIAM J. Control Optim.*, **40** (2001), 64–87.
[39] X.Y. Zhou and D. Li, Continuous-time mean-variance portfolio selection: A stochastic LQ framework, *Appl. Math. Optim.* **42** (2000), 19–33.
[40] X.Y. Zhou and G. Yin, Markowitz's mean-variance portfolio selection with regime switching: a continuous-time model, *SIAM J. Control Optim.*, **42** (2003), 1466–1482.

PART 4

Filtering and Control

Chapter 17

Existence and uniqueness of solutions for a partially observed stochastic control problem

Alain Bensoussan[1,2,*], Metin Çakanyıldırım[1], Meng Li[1] and Suresh P. Sethi[1]

[1]*School of Management, University of Texas at Dallas, Richardson, TX, USA*
{*alain.bensoussan, metin, meng.li, sethi*}*@utdallas.edu*

[2]*School of Business, The Hong Kong Polytechnic University, Hong Kong, and Ajou University, Department of Financial Engineering, Suwon, South Korea*

Dedicated to Professor Robert Elliott

We develop a general methodology for a partially observed stochastic control problem. The dynamics are governed by a discrete-time Markov process. We describe an application to an inventory system with the possibility of shrinkage, and introduce unnormalized conditional probabilities to transform the nonlinear state evolution into a linear one. We then prove the existence and uniqueness of the solution for the Bellman equation between the floor and the ceiling functions in the case of unbounded costs.

Contents

*This research was supported by the WCU (World Class University) program through the National Research Foundation of Korea funded by the Ministry of Education, Science and Technology (R31 - 20007).

17.1. Introduction

The existence and uniqueness of the solution for the optimal control of Markov processes in discrete time is well known in the case of fully observed systems. For a given discrete-time stochastic process defined by its transition probability function, a set of controls, and an initial state, we can use Kolmogorov's equations to obtain the probability measure of the process being in a certain state in a future period. These state probabilities can be used to write the Bellman equation, and thereby, to obtain the existence and uniqueness of its solution, in a given class of functions. The analysis becomes quite challenging if the state of the system is partially observed.

The mathematical treatment consists then in introducing a filtering process for the partially observed state [1]. The filtering process is defined by the conditional probability law of the state given the observed history. We can then reformulate the problem as a Markov decision problem with complete observation, but for an infinite dimensional stochastic system. Besides the difficulties associated with the infinite dimensional state space, the equation defining the filtering process is highly nonlinear (Kushner equation). However, by introducing the unnormalized probability, we can transform this equation into a linear one (Zakai equation).

The unnormalized probability simplifies the analysis, and we can derive an existence and uniqueness result for the solution of the corresponding Bellman equation. In the case of bounded one-period costs, the Bellman equation operator exhibits nice contractive properties which allow us to proceed easily. However, the contraction mapping theorem does not apply to a problem with unbounded one-period cost function, so we need to consider *floor and ceiling functions.* These functions, in turn, will define the appropriate functional space where the solution to the equation exists and is unique. Moreover, this solution corresponds to the value function. The existence and uniqueness of the solution of the Bellman equation for stochastic control problems has been studied in various contexts. In this paper, we improve the results of [2, 3]. In particular, we remove an assumption of smallness of the discount factor.

The remainder of this paper is organized as follows. We introduce the model in Section 17.2. In Section 17.3, we address the issues of the existence and uniqueness of the solution to the Bellman equation and the existence of an optimal feedback control. Section 17.4 concludes this paper. The preliminary results are relegated to the Appendix.

17.2. The model

We study a partially observed inventory model with shrinkage. We use the following notation for period t: I_t is the beginning state (inventory), q_t is the control (amount ordered), D_t is the demand, and S_t is the shrinkage. The state is described by

$$I_{t+1} = (I_t + q_t - D_t - S_t)^+ = \max\{I_t + q_t - D_t - S_t, 0\}. \tag{17.1}$$

We assume D_t (resp. S_t), $t = 1, 2, \ldots$, is independent and identically distributed (i.i.d.) and the the sequences $\{D_t\}$ and $\{S_t\}$ are independent. We use the notations f and g to denote the densities, F and G to denote the cumulative distributions of the demand and the shrinkage, and $\bar{F}(x) = 1 - F(x) = \mathrm{P}(D_t \geq x)$ and $\bar{G}(x) = 1 - G(x) = \mathrm{P}(S_t \geq x)$ to denote the complementary cumulative distribution functions respectively. Let $\bar{D}$ and $\bar{S}$ denote the means of the demand and shrinkage, respectively, i.e., $\bar{D} = \mathrm{E}(D_t)$ and $\bar{S} = \mathrm{E}(S_t)$.

The observed variable (the sales) is $z_{t+1} = \min\{I_t + q_t, D_t\}$ for $t \geq 1$. The information available at the beginning of period t is denoted by the σ-algebra generated by the sales

$$\mathcal{Z}_t := \sigma(z_2, z_3, \ldots, z_t), \quad t \geq 2.$$

We define $\mathcal{Z}_1$ to be the minimum σ-algebra $\{\emptyset, \Omega\}$. An admissible control q_t is adapted to the sigma algebra $\mathcal{Z}_t$.

One-period cost is captured by the cost function $c(I_t, q_t)$ that depends on the state I_t and the control q_t. Specifically, $c(I_t, q_t)$ has lower and upper bound estimates

$$c_2 I_t + c_3 q_t \leq c(I_t, q_t) \leq c_0 + c_1 q_t + c_2 I_t, \quad \text{for } I_t \geq 0, \tag{17.2}$$

where $c_0 \geq 0$ and c_1, c_2 c_3 are positive constants. Given a sequence of order quantities $\tilde{q} = \{q_1, q_2, \ldots\}$, an initial state's probability law $\pi(\cdot)$, and a discount factor $\alpha \in (0, 1)$, the total discounted cost can be written as

$$J(\pi, \tilde{q}) = \mathrm{E} \sum_{t=1}^{\infty} \alpha^{t-1} c(I_t, q_t).$$

Define the value function as $V(\pi) := \inf_{\tilde{q}} J(\pi, \tilde{q})$. We look for q_t, $t \geq 1$, to minimize $J(\pi, \tilde{q})$. For the never ordering policy, $\tilde{q}_0 = \{0, 0, \ldots\}$, the inventory evolves as $I_{t+1} = (I_t - S_t - D_t)^+$. Therefore, $I_t \leq I_1$ for $t > 1$. From (17.2)

$$J(\pi, \tilde{q}_0) \leq \mathrm{E} \sum_{t=1}^{\infty} \alpha^{t-1}(c_0 + c_2 I_t) \leq \frac{c_0 + c_2 \mathrm{E} I_1}{1 - \alpha}. \tag{17.3}$$

Note that the total discounted function can be rewritten as

$$J(\pi, \tilde{q}) = \mathrm{E} \sum_{t=1}^{\infty} \alpha^{t-1} \mathrm{E}[c(I_t, q_t)|\mathcal{Z}_t] = \mathrm{E}[c(I_1, q_1)] + \mathrm{E} \sum_{t=2}^{\infty} \alpha^{t-1} \mathrm{E}[c(I_t, q_t)|\mathcal{Z}_t]. \tag{17.4}$$

The admissible controls are limited within an a priori compact set as follows.

Lemma 17.1. *The admissible controls are bounded, i.e.,*

$$q_t \in \left[0, \frac{c_0}{c_3(1-\alpha)} + \frac{\alpha c_2(\bar{D} + \bar{S})}{c_3(1-\alpha)^2}\right] := \mathcal{Q}, \quad t \geq 1. \tag{17.5}$$

Proof. From (17.1), $I_{t+1} = (I_t + q_t - D_t - S_t)^+ \geq I_t - D_t - S_t$, for $t \geq 1$. Therefore, $I_{t+1} \geq I_1 - \sum_{i=1}^{t}(D_i + S_i)$. Then, from (17.2), the total discounted cost for any policy $\tilde{q}$ satisfies

$$\begin{aligned}
J(\pi, \tilde{q}) &\geq c_2 \sum_{t=1}^{\infty} \alpha^{t-1} \mathrm{E} I_t \geq \frac{c_2}{1-\alpha} \mathrm{E} I_1 - c_2 \mathrm{E} \sum_{t=2}^{\infty} \sum_{i=1}^{t-1} \alpha^{t-1}(D_i + S_i) \\
&= \frac{c_2}{1-\alpha} \mathrm{E} I_1 - c_2 \sum_{t=2}^{\infty} \alpha^{t-1} \mathrm{E}\Big[\sum_{i=1}^{t-1}(D_i + S_i)\Big] \\
&= \frac{c_2}{1-\alpha} \mathrm{E} I_1 - c_2(\bar{D} + \bar{S}) \sum_{t=2}^{\infty} \left[(t-1)\alpha^{t-1}\right] \\
&= \frac{c_2}{1-\alpha} \mathrm{E} I_1 - \frac{\alpha c_2(\bar{D} + \bar{S})}{(1-\alpha)^2},
\end{aligned}$$

where the last equality is from $\sum_{t=2}^{\infty}(t-1)\alpha^{t-1} = \alpha[d(\sum_{t=0}^{\infty} \alpha^t)/d\alpha] = \alpha/(1-\alpha)^2$. Since $\tilde{q}$ is arbitrary,

$$V(\pi) \geq \frac{c_2}{1-\alpha} \mathrm{E} I_1 - \frac{\alpha c_2(\bar{D} + \bar{S})}{(1-\alpha)^2}.$$

From (17.3) and (17.4),

$$\mathrm{E}c(I_1, q_1) + \mathrm{E}\sum_{t=2}^{\infty} \alpha^{t-1}\mathrm{E}[c(I_t, q_t)|\mathcal{Z}_t] \le \frac{c_2}{1-\alpha}\mathrm{E}I_1 + \frac{c_0}{1-\alpha} \tag{17.6}$$

holds to ensure the controls admissible. Following a similar analysis in the previous part of the proof, we have

$$\mathrm{E}\sum_{t=2}^{\infty} \alpha^{t-1}\mathrm{E}[c(I_t, q_t)|\mathcal{Z}_t] = \sum_{t=2}^{\infty} \alpha^{t-1}\mathrm{E}c(I_t, q_t) \ge \frac{\alpha c_2}{1-\alpha}\mathrm{E}I_2 - \frac{\alpha^2 c_2(\bar{D}+\bar{S})}{(1-\alpha)^2},$$

which, combined (17.2) and (17.6), yields

$$c_2\mathrm{E}I_1 + c_3 q_1 + \frac{\alpha c_2}{1-\alpha}\mathrm{E}I_2 - \frac{\alpha^2 c_2(\bar{D}+\bar{S})}{(1-\alpha)^2} \le \frac{c_2}{1-\alpha}\mathrm{E}I_1 + \frac{c_0}{1-\alpha}.$$

Since $\mathrm{E}I_2 = \mathrm{E}(I_1 + q_1 - D_1 - S_1)^+ \ge \mathrm{E}(I_1 - D_1 - S_1) = \mathrm{E}I_1 - \bar{D} - \bar{S}$,

$$\begin{aligned} c_2\mathrm{E}I_1 + c_3 q_1 + \frac{\alpha c_2}{1-\alpha}\mathrm{E}I_1 - \frac{\alpha c_2(\bar{D}+\bar{S})}{1-\alpha} - \frac{\alpha^2 c_2(\bar{D}+\bar{S})}{(1-\alpha)^2} \\ \le \frac{c_2}{1-\alpha}\mathrm{E}I_1 + \frac{c_0}{1-\alpha}. \end{aligned}$$

Canceling the terms out, we conclude

$$q_1 \le \frac{c_0}{c_3(1-\alpha)} + \frac{\alpha c_2(\bar{D}+\bar{S})}{c_3(1-\alpha)^2}.$$

Therefore, $q_1 \in \mathcal{Q}$. Note that $\mathcal{Q}$ does not depend on the initial inventory I_1. Similarly, $q_t \in \mathcal{Q}$, for $t > 1$. □

We then study the Bellman equation with this prior limitation, which facilitates the proof of the uniqueness of the solution for the Bellman equation. Finally, this prior limitation is justified by the verification theorem.

17.2.1. *Conditional probability evolution*

We now develop the probability law $\pi_t(\cdot)$ of I_t given $\mathcal{Z}_t$ and $I_t \ge 0$. This probably law of I_t can be expressed as

$$\pi_t(\Gamma) = \mathrm{E}[\mathbb{1}_\Gamma(I_t)|\mathcal{Z}_t] = A_t\mathbb{1}_\Gamma(0) + \int B_t(\eta)\mathbb{1}_\Gamma(\eta)d\eta,\ \Gamma \subseteq [0,\infty), \tag{17.7}$$

where A_t is a number, B_t is a probability density, and $\mathbb{1}$ is the indicator function. As a convention, any integral without the limit is understood to be from 0 to ∞. Since π_t is a probability law, we must have $A_t + \int B_t(\eta)d\eta = 1$. Using the Dirac measure $\delta_0(\cdot)$, we can formally write $\pi_t(d\eta) = A_t\delta_0(\eta) + B_t(\eta)d\eta$. We derive recursion equations for A_t and B_t as follows.

Theorem 17.1. *We have*

$$
\begin{aligned}
A_{t+1} = {} & \mathbb{1}_{z_{t+1}=q_t}\mathbb{1}_{A_t\neq 0} \\
& + \mathbb{1}_{z_{t+1}<q_t}\left[A_t\bar{G}(q_t - z_{t+1}) + \int B_t(\lambda)\bar{G}(\lambda + q_t - z_{t+1})d\lambda\right] \\
& + \mathbb{1}_{z_{t+1}>q_t}\left(\frac{\bar{F}(z_{t+1})B_t(z_{t+1}-q_t)}{\bar{F}(z_{t+1})B_t(z_{t+1}-q_t) + f(z_{t+1})\int_{z_{t+1}-q_t}^{\infty} B_t(\lambda)d\lambda}\right. \\
& \left. + \frac{f(z_{t+1})\int_{z_{t+1}-q_t}^{\infty} B_t(\lambda)\bar{G}(\lambda + q_t - z_{t+1})d\lambda}{\bar{F}(z_{t+1})B_t(z_{t+1}-q_t) + f(z_{t+1})\int_{z_{t+1}-q_t}^{\infty} B_t(\lambda)d\lambda}\right),
\end{aligned} \tag{17.8}
$$

$$
\begin{aligned}
B_{t+1}(\eta) = {} & \mathbb{1}_{z_{t+1}<q_t}\left[\mathbb{1}_{\eta<q_t-z_{t+1}}g(q_t - z_{t+1} - \eta)A_t\right. \\
& \left. + \int_{\eta+z_{t+1}-q_t}^{\infty} B_t(\lambda)g(\lambda + q_t - z_{t+1} - \eta)d\lambda\right] \\
& + \mathbb{1}_{z_{t+1}>q_t}\frac{f(z_{t+1})\int_{\eta+z_{t+1}-q_t}^{\infty} B_t(\lambda)g(\lambda + q_t - z_{t+1} - \eta)d\lambda}{\bar{F}(z_{t+1})B_t(z_{t+1}-q_t) + f(z_{t+1})\int_{z_{t+1}-q_t}^{\infty} B_t(\lambda)d\lambda},
\end{aligned} \tag{17.9}
$$

for $\eta > 0$.

From Theorem 17.1, π_t evolves according to a highly nonlinear equation akin to the Kushner equation [4]. The Zakai equation [5] provides a linear recursion of the probability law of the inventory. Analogous to (17.7), the unnormalized probability $\rho_t(\cdot)$ can be expressed as $\rho_t(\Gamma) = \phi_t\mathbb{1}_\Gamma(0) + \int \psi_t(\eta)\mathbb{1}_\Gamma(\eta)d\eta$ by using the number ϕ_t and the function ψ_t, which can be obtained recursively as

$$
\begin{aligned}
\phi_{t+1} = {} & \mathbb{1}_{z_{t+1}=q_t}\mathbb{1}_{\phi_t\neq 0} \\
& + \mathbb{1}_{z_{t+1}<q_t}\left[\phi_t\bar{G}(q_t - z_{t+1}) + \int \psi_t(\lambda)\bar{G}(\lambda + q_t - z_{t+1})d\lambda\right] \\
& + \mathbb{1}_{z_{t+1}>q_t}\left[\bar{F}(z_{t+1})\psi_t(z_{t+1}-q_t)\right. \\
& \left. + f(z_{t+1})\int_{z_{t+1}-q_t}^{\infty} \psi_t(\lambda)\bar{G}(\lambda + q_t - z_{t+1})d\lambda\right]
\end{aligned} \tag{17.10}
$$

and

$$\psi_{t+1}(\eta) = \mathbb{1}_{z_{t+1}<q_t}\Big[\mathbb{1}_{0\leq\eta<q_t-z_{t+1}} g(q_t - z_{t+1} - \eta)\phi_t \\ + \int_{\eta+z_{t+1}-q_t}^{\infty} \psi_t(\lambda) g(\lambda + q_t - z_{t+1} - \eta) d\lambda\Big] \\ + \mathbb{1}_{z_{t+1}>q_t} f(z_{t+1}) \int_{\eta+z_{t+1}-q_t}^{\infty} \psi_t(\lambda) g(\lambda + q_t - z_{t+1} - \eta) d\lambda, \tag{17.11}$$

starting with the initial conditions $\phi_1 = A_1$ and $\psi_1(\cdot) = B_1(\cdot)$. We have also the relation $\pi_t(x) = \rho_t(x)/[\phi_t + \int \psi_t(x)dx]$, as justified in [6]. Since π_t integrates to 1 while ρ_t does not, ρ_t is called the unnormalized probability.

Equations (17.8), (17.9), (17.10) and (17.11) can be written in the operator form in the functional space $\Re^+ \times L^1(\Re^+)$. We define the space

$$\Pi := \Big\{(A, B) \in \Re^+ \times L^1(\Re^+) : |A| + \int x|B(x)|dx < \infty\Big\},$$

and $L^1(\Re^+)$ is the space of integrable functions whose domain is the set of nonnegative real numbers. We also define the space

$$\mathcal{H} := \Big\{(\phi, \psi) \in \Re^+ \times L^1(\Re^+) : |\phi| + \int_0^\infty x|\psi(x)|dx < \infty\Big\}.$$

We are ultimately interested in unnormalized probabilities, which are nonnegative. For them, we will specify an appropriate subset of $\mathcal{H}$ as

$$\mathcal{H}^+ = \{\rho \in \mathcal{H} | \rho(x) \geq 0, x \in \Re^+\}.$$

Observe that $\mathcal{H}^+$ is not a subspace of $\mathcal{H}$, for it does not include $-\rho$ for some ρ. We identify unnormalized probabilities in $\mathcal{H}^+$ with the natural norm

$$||\rho|| = |\phi| + \int_0^\infty |\psi(x)|dx + \int_0^\infty x|\psi(x)|dx.$$

For a sequence $\{\rho_n\}$ in $\mathcal{H}^+$ and $\rho \in \mathcal{H}^+$, "$\rho_n \to \rho$" means $||\rho_n - \rho|| \to 0$, which is equivalent to

$$|\phi_n - \phi| \to 0, \quad \int_0^\infty |\psi_n(x) - \psi_n(x)|dx \to 0, \\ \int_0^\infty x|\psi_n(x) - \psi_n(x)|dx \to 0. \tag{17.12}$$

We let $\mathcal{L}(\mathcal{H};\mathcal{H})$ denote the space of bounded linear maps form $\mathcal{H}$ to $\mathcal{H}$. For $\rho = (\phi, \psi(\cdot))$, we define the operator $\ominus\rho \in \mathcal{L}(\mathcal{H},\mathcal{H})$, and the family of operators $T_y\rho \in \mathcal{L}(\mathcal{H};\mathcal{H})$ and $\Theta_{y,x}\rho \in \mathcal{L}(\mathcal{H};\mathcal{H})$. Namely, $\ominus\rho = (\ominus^1\rho; \ominus^2\rho)$, where $\ominus^1\rho = \mathbb{1}_{\phi\neq 0}$ and $\ominus^2\rho = 0$. The family $T_y\rho = (T_y^1\rho; T_y^2\rho(\cdot))$ is indexed by $y \geq 0$, where

$$T_y^1\rho = \phi\bar{G}(y) + \int \psi(\lambda)\bar{G}(\lambda + y)d\lambda,$$

$$T_y^2\rho(\cdot) = \mathbb{1}_{0\leq\cdot<y}g(y - \cdot)\phi + \int_{\cdot-y}^{\infty} \psi(\lambda)g(\lambda + y - \cdot)d\lambda.$$

The family $\Theta_{y,x}\rho = (\Theta_{y,x}^1\rho; \Theta_{y,x}^2\rho(\cdot))$ is indexed by $y, x > 0$, where

$$\Theta_{y,x}^1\rho = \psi(y)\bar{F}(x) + f(x)\int_y^{\infty} \psi(\lambda)\bar{G}(\lambda - y)d\lambda,$$

$$\Theta_{y,x}^2\rho(\cdot) = f(x)\int_{\cdot+y}^{\infty} \psi(\lambda)g(\lambda - y - \cdot)d\lambda.$$

We can define the nonlinear operator $\tilde{\Theta}_{y,x} = (\tilde{\Theta}_{y,x}^1\rho; \tilde{\Theta}_{y,x}^2\rho(\cdot))$ corresponding to $\Theta_{y,x}$ as follows:

$$\tilde{\Theta}_{y,x}^1\rho = \frac{\psi(y)\bar{F}(x) + f(x)\int_y^{\infty} \psi(\lambda)\bar{G}(\lambda - y)d\lambda}{\psi(y)\bar{F}(x) + f(x)\int_y^{\infty} \psi(\lambda)d\lambda},$$

$$\tilde{\Theta}_{y,x}^2\rho(\cdot) = \frac{f(x)\int_{\cdot+y}^{\infty} \psi(\lambda)g(\lambda - y - \cdot)d\lambda}{\psi(y)\bar{F}(x) + f(z)\int_y^{\infty} \psi(\lambda)d\lambda}.$$

We can now write the recursions of π_t and ρ_t succinctly as follows:

$$\begin{aligned}\pi_{t+1}(\cdot) = \mathbb{1}_{z_{t+1}=q_t} \ominus \pi_t(\cdot) + \mathbb{1}_{z_{t+1}<q_t}T_{q_t-z_{t+1}}\pi_t(\cdot) \\ + \mathbb{1}_{z_{t+1}>q_t}\tilde{\Theta}_{z_{t+1}-q_t,z_{t+1}}\pi_t(\cdot),\end{aligned} \tag{17.13}$$

$$\begin{aligned}\rho_{t+1}(\cdot) = \mathbb{1}_{z_{t+1}=q_t} \ominus \rho_t(\cdot) + \mathbb{1}_{z_{t+1}<q_t}T_{q_t-z_{t+1}}\rho_t(\cdot) \\ + \mathbb{1}_{z_{t+1}>q_t}\Theta_{z_{t+1}-q_t,z_{t+1}}\rho_t(\cdot).\end{aligned} \tag{17.14}$$

There is a one-to-one correspondence between the unnormalized probability ρ_t and the probability π_t. We mainly work with the unnormalized probability ρ_t in the following sections. Using ρ_t, the Bellman equation can be written in a relatively simpler form.

17.2.2. *The Bellman equation*

The Bellman equation is [6]

$$V(\pi) = \inf_{q\in\mathcal{Q}} \Big\{ Ac(0,q) + \int B(\eta)c(\eta,q)d\eta + \alpha A\bar{F}(q)V(\delta_0) \\ + \alpha\int_0^q f(z)V(T_{q-z}\pi)dz \\ + \alpha\int_q^\infty \Big[\bar{F}(z)B(z-q) + f(z)\int_{z-q}^\infty B(\eta)d\eta\Big] V(\tilde{\Theta}_{z-q,z}\pi)dz\Big\}, \tag{17.15}$$

for $\pi \in \Pi$. Since π evolves according to the nonlinear operator $\tilde{\Theta}$, a direct study of the above system is not easy. If we use the unnormalized probability $\rho_1 = \rho$ and the associated $(\phi, \psi(\cdot))$ as the system state instead of $\pi = \pi_1$, then the analysis becomes much easier, because ρ evolves with the linear operator Θ. To make the ideas concrete, we define

$$W(\rho) := \left(\phi + \int \psi(x)dx\right) V\left(\frac{\rho}{\phi + \int \psi(x)dx}\right). \tag{17.16}$$

Using this definition, the Bellman Equation (17.15), and some algebra, we can show that W satisfies the Bellman equation

$$W(\rho) = \inf_{q\in\mathcal{Q}} \Big\{ \phi c(0,q) + \int \psi(\eta)c(\eta,q)d\eta + \alpha\phi\bar{F}(q)W(\delta_0) \\ + \alpha\int_0^q f(z)W\left(T_{q-z}\rho\right)dz + \alpha\int_q^\infty W(\Theta_{z-q,z}\rho)dz\Big\}, \tag{17.17}$$

for $\rho \in \mathcal{H}^+$. Note that the optimality equation has similarities but also significant differences with that of [2, 3].

We define the functional space

$$\mathcal{B} := \left\{\phi(\rho) : \mathcal{H} \to \Re : \sup_{\rho\in\mathcal{H}} \frac{|\phi(\rho)|}{||\rho||} < \infty\right\}$$

with the norm

$$||\phi|| := \sup_{\rho\in\mathcal{H}} \frac{|\phi(\rho)|}{||\rho||}.$$

From this definition we can see that $\mathcal{B}$ is a Banach space. For any $\phi \in \mathcal{B}$, we must have $\phi(0) = 0$. We define the space

$$\mathcal{B}^+ := \left\{\phi(\rho) : \mathcal{H}^+ \to \Re : \sup_{\rho\in\mathcal{H}^+} \frac{|\phi(\rho)|}{||\rho||} < \infty\right\} \subseteq \mathcal{B}$$

and denote by $\mathcal{B}^{\Pi}$ the subset of $\mathcal{B}^+$ restricted to Π. That is,

$$\mathcal{B}^{\Pi} := \Big\{\phi(\pi) : \Pi \to \Re : \sup_{\pi\in\Pi} \frac{|\phi(\pi)|}{||\pi||} < \infty\Big\}.$$

We consider the ceiling function $v^0 \in \mathcal{B}^+$ defined as

$$v^0(\rho) := \frac{c_0}{1-\alpha}\Big(|\phi| + \int |\psi(\eta)|d\eta\Big) + \frac{c_2}{1-\alpha}\int \eta|\psi(\eta)|d\eta.$$

Using v^0, we define $\mathcal{G}$ as

$$\mathcal{G} := \{\phi : \mathcal{H}^+ \to \Re : \phi(\rho) \le v^0(\rho)\},$$

and $\tilde{\mathcal{G}}$ as

$$\tilde{\mathcal{G}} := \{\tilde{\phi} : \Pi \to \Re : \tilde{\phi}(\pi) \le v^0(\pi)\}.$$

For $W \in \mathcal{G}$, we define the operator $\mathcal{T}_q$ as follows:

$$\mathcal{T}_q(W) := \phi c(0,q) + \int \psi(\eta)c(\eta,q)d\eta + \alpha\phi\bar{F}(q)W(\delta_0)$$
$$+ \alpha\int_0^q f(z)W\left(T_{q-z}\rho\right)dz + \alpha\int_q^\infty W(\Theta_{z-q,z}\rho)dz.$$

Let

$$\mathcal{T}W(\rho) := \inf_{q\in\mathcal{Q}} \mathcal{T}_qW(\rho), \quad \rho \in \mathcal{H}^+. \tag{17.18}$$

Lemmas 17.2 and 17.3 show two important properties of the operator $\mathcal{T}$.

Lemma 17.2. *The operator $\mathcal{T}$ maps $\mathcal{G}$ into itself, i.e., if $W(\rho) \le v^0(\rho)$, then $\mathcal{T}W(\rho) \le v^0(\rho)$, for $\rho \in \mathcal{H}^+$.*

Proof. First, for the one-period cost,

$$\phi c(0,0) + \int c(\eta,0)|\psi(\eta)|d\eta \le c_0\phi + \int |\psi(\eta)|(c_0 + c_2\eta)d\eta = (1-\alpha)v^0(\rho).$$

We also have

$$|\Theta^1_{y,x}\rho| \le \bar{F}(x)|\psi(y)| + f(x)\int_y^\infty |\psi(\lambda)|\bar{G}(\lambda - y)d\lambda.$$

Second, by changing the order of integration, we obtain

$$\begin{aligned}\int |\Theta^2_{y,x}\rho(\eta)|d\eta &\leq f(x)\int\int_{\eta+y}^{\infty}|\psi(\lambda)|g(\lambda-y-\eta)d\lambda d\eta\\ &= f(x)\int_y^{\infty}|\psi(\lambda)|\int_0^{\lambda-y} g(\lambda-y-\eta)d\eta d\lambda\\ &= f(x)\int_y^{\infty}|\psi(\lambda)|G(\lambda-y)d\lambda,\\ \int \eta|\Theta^2_{y,x}\rho(\eta)| &\leq f(x)\int\eta\int_{\eta+y}^{\infty}|\psi(\lambda)|g(\lambda-y-\eta)d\lambda d\eta\\ &= f(x)\int_y^{\infty}|\psi(\lambda)|\int_0^{\lambda-y}\eta g(\lambda-y-\eta)d\eta d\lambda\\ &\leq f(x)\int_y^{\infty}(\lambda-y)|\psi(\lambda)|G(\lambda-y)d\lambda.\end{aligned}$$

Therefore,

$$\begin{aligned}&c_0|\Theta^1_{y,x}\rho| + c_0\int|\Theta^2_{y,x}\rho(\eta)|d\eta + c_2\int\eta|\Theta^2_{y,x}\rho(\eta)|d\eta\\ &\leq c_0\bar{F}(x)|\psi(y)| + c_0 f(x)\int_y^{\infty}|\psi(\lambda)|d\lambda\\ &\quad + c_2 f(x)\int_y^{\infty}(\lambda-y)|\psi(\lambda)|G(\lambda-y)d\lambda.\end{aligned}$$

Furthermore,

$$\begin{aligned}&\int\Big[c_0|\Theta^1_{z,z}\rho| + c_0\int|\Theta^2_{z,z}\rho(\eta)|d\eta + c_2\int\eta|\Theta^2_{z,z}\rho(\eta)|d\eta\Big]dz\\ &\leq c_0\int\bar{F}(z)|\psi(z)|dz + c_0\int f(z)\int_z^{\infty}|\psi(\lambda)|d\lambda dz\\ &\quad + c_2\int f(z)\int_z^{\infty}(\lambda-z)|\psi(\lambda)|\bar{G}(\lambda-z)d\lambda dz\\ &\leq c_0\int\bar{F}(z)|\psi(z)|dz + c_0\int F(z)|\psi(z)|dz\\ &\quad + c_2\int f(z)\int_z^{\infty}(\lambda-z)|\psi(\lambda)|\bar{G}(\lambda-z)d\lambda dz\\ &\leq c_0\int|\psi(z)|dz + c_2\int f(z)\int\lambda|\psi(\lambda)|d\lambda dz\\ &= c_0\int|\psi(z)|dz + c_2\int\lambda|\psi(\lambda)|d\lambda.\end{aligned}$$

Therefore, for $W \leq v^0$,

$$\begin{aligned}
\mathcal{T}(W) = \inf_q \mathcal{T}_q(W) \leq \mathcal{T}_0(W) &= \phi c(0,0) + \int c(\eta,0)\psi(\eta)d\eta \\
&\quad + \alpha\phi\bar{F}(0)W(\delta_0) + \alpha\int W(\Theta_{z,z}\rho)dz \\
&\leq (1-\alpha)v^0(\rho) + \alpha\phi v^0(\delta_0) + \alpha\int v^0(\Theta_{z,z}\rho)dz \\
&= (1-\alpha)v^0(\rho) + \frac{c_0\alpha\phi}{1-\alpha} \\
&\quad + \frac{\alpha}{1-\alpha}\int\left[c_0\Theta^1_{z,z}\rho + c_0\int\Theta^2_{z,z}\rho(\eta)d\eta + c_2\int\eta\Theta^2_{z,z}\rho(\eta)d\eta\right]dz \\
&\leq (1-\alpha)v^0(\rho) + \frac{c_0\alpha\phi}{1-\alpha} + \frac{\alpha}{1-\alpha}\left[c_0\int|\psi(z)|dz + c_2\int\lambda|\psi(\lambda)|d\lambda\right] \\
&= (1-\alpha)v^0(\rho) + \alpha v^0(\rho) = v^0(\rho),
\end{aligned}$$

which implies that $\mathcal{T}(W) \leq v^0(\rho)$ if $W(\rho) \leq v^0(\rho)$. □

Lemma 17.3. *For each continuous function $W \in \mathcal{G}$, $\mathcal{T}W(\rho)$ is continuous in $\rho \in \mathcal{H}^+$.*

Proof. Let $\{\rho_n\}$ be a sequence in $\mathcal{H}$ such that $\rho_n \to \rho \in \mathcal{H}^+$ and $q_n = g_W(\rho_n) \in \mathcal{Q}$, satisfying $\mathcal{T}W(\rho_n) = \mathcal{T}_{q_n}W(\rho_n)$ from the selection theorem and Lemma 17.1. Since q_n remains in a compact set, we can extract a subsequence $\{(\rho_{n_k}, q_{n_k})\}$ of $\{(\rho_n, q_n)\}$ such that $(\rho_{n_k}, q_{n_k}) \to (\rho, q')$ for some $q' \in \mathcal{Q}$. Hence, using the continuity of $\mathcal{T}_qW(\rho)$ (see Remark 17.2), we have $\lim_{k\to\infty}\mathcal{T}_{q_{n_k}}W(\rho_{n_k}) = \mathcal{T}_{q'}W(\rho)$. Therefore,

$$\mathcal{T}W(\rho) = \inf_q \mathcal{T}_qW(\rho) \leq \mathcal{T}_{q'}W(\rho) = \liminf_{n\to\infty}\mathcal{T}W(\rho_n). \tag{17.19}$$

On the other hand, by the continuity of $\mathcal{T}_qW(\rho)$, $\limsup_{n\to\infty}\mathcal{T}W(\rho_n) \leq \limsup_{n\to\infty}\mathcal{T}_qW(\rho_n) = \mathcal{T}_qW(\rho)$, which yields

$$\limsup_{n\to\infty}\mathcal{T}W(\rho_n) \leq \mathcal{T}W(\rho). \tag{17.20}$$

Therefore, by combining (17.19) and (17.20), we obtain the desired result. □

Remark 17.1. If $W \in \mathcal{G}$ is only l.s.c., we can follow similar arguments as in the proof of Lemma 17.3 (see also Remark 17.2) to show that the functions $\mathcal{T}_qW(\rho)$ and $\mathcal{T}W(\rho)$ are l.s.c. in (q,ρ) and ρ respectively. Hence, the selection theorem ensures the existence of a function $g_W : \mathcal{H}^+ \to \Re^+$ satisfying $\mathcal{T}W(\rho) = \mathcal{T}_{g_W}W(\rho)$, for $\rho \in \mathcal{H}^+$.

For a function $V(\pi) \in \tilde{\mathcal{G}}$, $\pi(\cdot) = (A, B(\cdot))$, define

$$\begin{aligned}\tilde{\mathcal{T}}_q V(\pi) := \inf_{q \in \mathcal{Q}} \Big\{ & Ac(0,q) + \int B(\eta)c(\eta,q)d\eta \\ & + \alpha A\bar{F}(q)V(\delta_0) + \alpha \int_0^q f(z)V(T_{q-z}\pi)dz \\ & + \alpha \int_q^\infty \Big[\bar{F}(z)B(z-q) + f(z)\int_{z-q}^\infty B(\eta)d\eta\Big] V(\tilde{\Theta}_{z-q,z}\pi)dz \Big\}.\end{aligned}$$

Let

$$\tilde{T}V(\pi) := \inf_{q \in \mathcal{Q}} \tilde{T}_q V(\pi), \quad \pi \in \Pi. \tag{17.21}$$

This equation corresponds to (17.15).

17.3. Main results

In this section, we will show the details of the methodology we use to prove the existence and uniqueness of the solution W in an interval of functions. Moreover, the unique solution of the Bellman equation is identical to the value function.

17.3.1. *Existence of a solution to the Bellman equation*

We define a value iteration procedure as follows. Let $\{W_n\}$ be sequences of functions defined by $W_0 = 0$, and for $n \geq 1$,

$$W_{n+1}(\rho) = \mathcal{T}(W_n)(\rho). \tag{17.22}$$

Note that from Lemma 17.3, the value iteration functions defined in (17.22) are continuous on $\mathcal{H}^+$. From the definition of the operator $\mathcal{T}$, $\{W_n\}$ is an increasing sequence. This can be established by induction. Clearly, $W_0 \leq W_1$. Let us assume that $W_n \leq W_{n+1}$. Then,

$$\begin{aligned}W_{n+1} = & \inf_{q \in \mathcal{Q}} \Big\{ \phi c(0,q) + \int \psi(\eta)c(\eta,q)d\eta + \alpha\phi\bar{F}(q)W_n(\delta_0) \\ & + \alpha \int_0^q f(z)W_n\left(T_{q-z}\rho\right)dz + \alpha \int_q^\infty W_n(\Theta_{z-q,z}\rho)dz \Big\} \\ \leq & \inf_{q \in \mathcal{Q}} \Big\{ \phi c(0,q) + \int \psi(\eta)c(\eta,q)d\eta + \alpha\phi\bar{F}(q)W_{n+1}(\delta_0) \\ & + \alpha \int_0^q f(z)W_{n+1}\left(T_{q-z}\rho\right)dz + \alpha \int_q^\infty W_{n+1}(\Theta_{z-q,z}\rho)dz \Big\} = W_{n+2}.\end{aligned}$$

Theorem 17.2. *(a) There exists a lower semicontinuous (l.s.c.) function* $\underline{W} \leq v^0$ *such that* $W_n \nearrow \underline{W}$*, and* $\underline{W} = \mathcal{T}\underline{W}$*.*

(b) For each $\rho \in \mathcal{H}^+$*, there exists a measurable function* $g_{\underline{W}} : \mathcal{H}^+ \to \Re^+$ *such that* $\mathcal{T}\underline{W}(\rho) = \mathcal{T}_{g_{\underline{W}}}\underline{W}(\rho)$*.*

Proof. (a) Observe that since $W_0 = 0$, we have $W_1 \leq v^0$. Since $\{W_n\}$ is an increasing sequence of continuous functions such that $W_n \leq v^0$ for $n \geq 1$, there is a l.s.c function $\underline{W} \leq v^0$ such that

$$W_n \nearrow \underline{W} \text{ as } n \to \infty. \tag{17.23}$$

Using the fact that the operator $\mathcal{T}$ is monotone, we have $W_{n+1} = \mathcal{T}W_n \leq T\underline{W}$. Thus, from (17.23),

$$\underline{W}(\rho) \leq \mathcal{T}\underline{W}(\rho), \quad \rho \in \mathcal{H}^+. \tag{17.24}$$

To obtain the reverse inequality, let $\acute{q}_n$ be the feedback control such that $W_{n+1}(\rho) = \mathcal{T}W_n(\rho) = \mathcal{T}_{\acute{q}_n}W(\rho)$. Thus, observe that for any N, $W_{n+1}(\rho) \geq \mathcal{T}_{\acute{q}_n}W_N(\rho)$, for all $n \geq N$. Then, from (17.23),

$$\underline{W}(\rho) \geq \mathcal{T}_{\acute{q}_n}W_N(\rho). \tag{17.25}$$

In addition, we can extract a subsequence $\{\acute{q}_{n_k}\}$ of $\{\acute{q}_n\}$ such that $\acute{q}_{n_k} \to \acute{q}$ as $k \to \infty$. Then by the continuity of the function $q \to \mathcal{T}_q W_N(\rho)$ (see Lemma 17.5), we have that $\mathcal{T}_{\acute{q}_{n_k}}W_N(\rho) \to \mathcal{T}_{\acute{q}}W_N(\rho)$ as $k \to \infty$. Therefore, from (17.25), we have that $\underline{W}(\rho) \geq \mathcal{T}_{\acute{q}}W_N(\rho)$. Letting $N \to \infty$, we obtain $\underline{W}(\rho) \geq \mathcal{T}_{\acute{q}}\underline{W}(\rho) \geq \mathcal{T}\underline{W}(\rho)$, which combined with (17.24), yields $\underline{W}(\rho) = \mathcal{T}\underline{W}(\rho)$ for each ρ.

(b) From (a), Lemma 17.1 and the selection theorem, for each $\rho \in \mathcal{H}^+$, there exists a measurable function $g_{\underline{W}} : \mathcal{H}^+ \to \Re^+$ such that $\mathcal{T}\underline{W}(\rho) = \mathcal{T}_{g_{\underline{W}}}\underline{W}(\rho)$. □

From Theorem 17.2, $\underline{W} \in \mathcal{H}^+$. So far we have analyzed the unnormalized system (17.17) and we show that there exists a solution of this system. Next, we will show that there is a solution of system (17.15). This result immediately follows from Theorem 17.2. Let $\{V_n\}$ be the normalized value iteration functions corresponding to $\{W_n\}$. That is V_n is a function on Π satisfying (see (17.16))

$$W_n(\rho) = V_n\left(\frac{\rho}{\phi + \int \psi(\lambda)d\lambda}\right)\left(\phi + \int \psi(\lambda)d\lambda\right). \tag{17.26}$$

Then, $W_0 = V_0 = 0$. Moreover,

$$V_{n+1}(\pi) = \tilde{T}V_n(\pi). \tag{17.27}$$

We know that $W_n(\pi) = V_n(\pi)$ for all $\pi \in \Pi$ from (17.26). Theorem 17.2 (a) yields the existence of an l.s.c. function $\underline{V}(\pi) \leq v^0(\pi)$, such that $V_n \nearrow \underline{V}$ as $n \to \infty$ and $\underline{V}(\pi) = \tilde{\mathcal{T}}\underline{V}(\pi)$. Hence,

$$\begin{aligned}\underline{V}(\pi) := \inf_{q\in\mathcal{Q}} \Big\{ & Ac(0,q) + \int B(\eta)c(\eta,q)d\eta \\ & + \alpha A\bar{F}(q)\underline{V}(\delta_0) + \alpha \int_0^q f(z)\underline{V}(T_{q-z}\pi)dz \\ & + \alpha \int_q^\infty \Big[\bar{F}(z)B(z-q) + f(z)\int_{z-q}^\infty B(\eta)d\eta\Big]\underline{V}(\tilde{\Theta}_{z-q,z}\pi)dz\Big\}. \end{aligned} \tag{17.28}$$

Furthermore, from Theorem 17.2 (b), there exists a map $\underline{g} : \Re^+ \times \Pi \to \Re^+$ such that

$$\underline{V}(\pi) = \tilde{\mathcal{T}}\underline{V}(\pi) = \tilde{\mathcal{T}}_{\underline{g}(\pi)}\underline{V}(\pi), \quad \pi \in \Pi. \tag{17.29}$$

So we see that there is a solution of (17.15).

We have shown that, starting from V_0, W_0, sequences $\{V_n\}$ and $\{W_n\}$ are increasing and converge to $\underline{V}$ and $\underline{W}$, respectively, and $\underline{V}$ and $\underline{W}$ are the solutions of systems (17.15) and (17.17). Furthermore, it can be seen easily that $\underline{V}$ and $\underline{W}$ are respectively the minimum solutions of (17.15) and (17.17) in the interval $[0, v^0]$.

We compare the minimal solution of the Bellman equation, $\underline{V}$, and the value function. Let $\grave{q}_t$ be the feedback control such that $\underline{V}(\pi_t) = A_t c(0,\grave{q}_t) + \int B_t(\eta)c(\eta,\grave{q}_t)d\eta + \alpha \mathrm{E}[\underline{V}(\pi_{t+1}|\mathcal{Z}_{t+1})]$. Hence, $\mathrm{E}[\alpha^{t-1}\underline{V}(\pi_t)] - \mathrm{E}[\alpha^t\underline{V}(\pi_{t+1})] = \alpha^{t-1}[A_t c(0,\grave{q}_t) + \int B_t(\eta)c(\eta,\grave{q}_t)d\eta]$. Summing up for $t = 1, 2, \cdots, M$ yields

$$\begin{aligned}\underline{V}(\pi) &= \sum_{t=1}^{M} \alpha^{t-1}\Big[A_t c(0,\grave{q}_t) + \int B_t(\eta)c(\eta,\grave{q}_t)d\eta\Big] + \alpha^M \mathrm{E}[\underline{V}(\pi_{M+1})] \\ &\geq \sum_{t=1}^{M} \alpha^{t-1}\Big[A_t c(0,\grave{q}_t) + \int B_t(\eta)c(\eta,\grave{q}_t)d\eta\Big]. \end{aligned} \tag{17.30}$$

Letting $M \to \infty$, we have $\underline{V}(\pi) \geq V(\pi)$, for any $\pi \in \Pi$. Next, we will show that the solutions of the systems (17.15) and (17.17) are unique.

17.3.2. *Uniqueness and continuity of the solution of the Bellman equation*

We consider the unnormalized value iteration procedure, but starting with the function v^0. That is, for $\rho \in \mathcal{H}^+$, $W^0(\rho) = v^0(\rho)$ and $W^{n+1}(\rho) =$

$\mathcal{T}W^n(\rho)$. Let V^n defined as (17.27) be the corresponding normalized value iteration function. Using induction arguments, we know that $\{W^n\}$ is nonnegative decreasing sequences. We have $W^1 \le W^0 = v^0(\rho)$, then assume that $W^n \le W^{n-1}$. Therefore,

$$\begin{aligned}
W^{n+1} &= \inf_{q\in\mathcal{Q}}\Big\{\phi c(0,q) + \int \psi(\eta)c(\eta,q)d\eta + \alpha\phi\bar{F}(q)W^n(\delta_0) \\
&\quad + \alpha\int_0^q f(z)W^n\left(T_{q-z}\rho\right)dz + \alpha\int_q^\infty W^n(\Theta_{z-q,z}\rho)dz\Big\} \\
&\le \inf_{q\in\mathcal{Q}}\Big\{\phi c(0,q) + \int \psi(\eta)c(\eta,q)d\eta + \alpha\phi\bar{F}(q)W^{n+1}(\delta_0) \\
&\quad + \alpha\int_0^q f(z)W^{n+1}\left(T_{q-z}\rho\right)dz + \alpha\int_q^\infty W^{n+1}(\Theta_{z-q,z}\rho)dz\Big\} \\
&= W^{n+2}.
\end{aligned}$$

Similarly, we can show $\{V^n\}$ is a nonnegative decreasing sequence. Since $\{W^n\}$ is nonnegative decreasing and has a lower bound 0, then there exists a function $\overline{W} \le W^0$ such that $W^n \searrow \overline{W}$ as $n \to \infty$. Since W^n is a decreasing sequence, we have $W^n \ge \overline{W}$ for any finite n. Apply $\mathcal{T}$ on both sides of $W^n \ge \overline{W}$, we obtain $W^{n+1} = \mathcal{T}W^n \ge \mathcal{T}\overline{W}$. Apply $\mathcal{T}$ infinitely many times on the left hand side of this inequality, we obtain

$$\overline{W} \ge \mathcal{T}\overline{W}. \tag{17.31}$$

Moreover, from the decreasing property of the sequence $\{W^n\}$ we can show that for all $q \in \mathcal{Q}$, $\mathcal{T}_q W^n \ge \overline{W}$. Consequently, $\overline{W} \le \mathcal{T}_q\overline{W}$, for all $q \in \mathcal{Q}$, which implies $\overline{W} \le \mathcal{T}\overline{W}$. This inequality combined with (17.31) yields $\overline{W}(\rho) = \mathcal{T}\overline{W}(\rho)$ for any $\rho \in \mathcal{H}^+$. Similarly, we can show $\overline{V}(\pi) = \tilde{\mathcal{T}}\overline{V}(\pi)$ for any $\pi \in \Pi$. The next theorem states that the solutions of (17.15) and (17.17) are unique.

Theorem 17.3. (*a*) *For each $\pi \in \Pi$, $V(\pi) = \overline{V}$. Moreover, $\overline{V}$ is the maximal solution in $\tilde{\mathcal{G}}$ of the optimal equation* (17.15). *Hence, $W(\rho) = \overline{W}(\rho)$ for each $\rho \in \mathcal{H}^+$, and $\overline{W}$ is the maximal solution of* (17.17) *in $\mathcal{G}$. Moreover, the solutions of* (17.15) *and* (17.17) *are unique.*

(*b*) *The value functions $V(\cdot)$ and $W(\cdot)$ are continuous on Π and $\mathcal{H}^+$, respectively.*

Proof. (a) First, we show that any solution of the Bellman equation $V(\pi) \le \overline{V}(\pi)$. Because $V \in \tilde{\mathcal{G}}$ from (17.15), we have $V(\pi) \le v^0(\pi)$. Then as $V(\pi)$ is a fixed point of the operator $\tilde{\mathcal{T}}$, we obtain $V(\pi) \le V^n(\pi)$ for all

n, which implies from the decreasing property of the sequence $\{V^n\}$ that $V(\pi) \leq \overline{V}(\pi)$ for any $\pi \in \Pi$.

Next, we compare the value function V with $\overline{V}$, the maximal solution of the Bellman equation. Proceeding as in the previous analysis, we can establish the relation

$$\overline{V}(\pi) = \sum_{t=1}^{N} \alpha^{t-1} \mathrm{E} c(I_t, q_t) + \alpha^N \mathrm{E}[\overline{V}(\pi_{N+1})]. \tag{17.32}$$

Now, as $\overline{V} \in \tilde{\mathcal{G}}$, it follows that

$$\overline{V}(\pi_{N+1}) \leq v^0(\pi_{N+1}). \tag{17.33}$$

In addition,

$$\begin{aligned} v^0(\pi_{N+1}) &= c_0 A_{N+1} + c_0 \int B_{N+1}(x)dx + c_2 \int x B_{N+1}(x)dx \\ &= c_0 + c_2 \mathrm{E}(I_{N+1} | \mathcal{Z}_{N+1}). \end{aligned} \tag{17.34}$$

Then, $\mathrm{E}v^0(\pi_{N+1}) = c_0 + c_2 \mathrm{E} I_{N+1}$. Since inventory $\{I_t\}$ is obtained through the sequence $I_{t+1} = (I_t + q_t - D_t - S_t)^+$ and we assume the bound is m on q_t from (17.5), then $I_{t+1} \leq I_t + m$, and so $I_N \leq I_1 + m(N-1)$, and

$$\alpha^N \mathrm{E} I_N \leq \alpha^N [\mathrm{E} I_1 + m(N-1)] \to 0, \text{ as } N \to \infty. \tag{17.35}$$

Therefore, combining (17.33), (17.34) and (17.35), we have $\alpha^N \mathrm{E}[\overline{V}(\pi_{N+1})] \to 0$ as $N \to \infty$. Let $N \to \infty$ in (17.32), we get $\overline{V} \leq J(\pi, \tilde{q})$, and as $\tilde{q} \in \mathcal{Q}$ is arbitrary, we have $\overline{V}(\pi) \leq V(\pi)$.

Then, we see that the minimum solution $\underline{V}$ and the maximum solution $\overline{V}$ of the Bellman equation (17.15) coincide and are equal to the value function.

(b) According to part (a), we see that the value iteration functions W^n converge decreasingly to the value function W. On the other hand, since $v^0(\rho)$ is a continuous function, we have from lemma that $W^n(\rho)$ is continuous on $\mathcal{H}^+$ for all $n \geq 1$. Therefore, from the deceasing property of sequence $\{W^n\}$, we can ensure that W is u.s.c.. Hence, from the lower semicontinuity of W, we can conclude that W is continuous on $\mathcal{H}^+$. This result also yields the continuity of V on Π. This completes the proof. □

From Theorems 3 and 4, we have shown that the Bellman equation (17.15) has a unique solution V. From the verification theorem, we can then show that V is the value function. That is, the unique solution of the Bellman equation is identical to the value function. Moreover, the prior bound on the control is justified.

17.4. Concluding remarks

We have studied an infinite-horizon partially observed stochastic control problem in the inventory context. In this system, partially observed states eventually lead to a dynamic program in the space of probability laws. This dynamic program is highly nonlinear. We apply the methodology of unnormalized probability to linearize the dynamic programming equations, facilitating the analysis.

For our problem with unbounded one-period cost function, we first restrain the admissible controls in a compact set. Then, in Theorem 17.2, using the floor and ceiling functions, we prove the existence of the solution for the Bellman equation, as well as the existence of an optimal feedback control. Furthermore, from Theorem 17.3, the solution of the optimality equation is unique and continuous. Finally, from the verification theorem, the bound on the admissible control is justified, and the unique solution of the Bellman equation is identical to the value function.

We conclude the paper by describing an extension of the model. For example, one might consider a model with Markov demand. It is expected that the Zakai type evolution, the floor and ceiling functions will encourage other related existence and uniqueness proofs for the partially observed discrete-time optimal control problem.

Appendix: Preliminaries

Lemma 17.4. *If $q_n \to q$ in the Euclidean norm and $\phi_n \to \psi_n$ in the norm $||\cdot||$ as $n \to \infty$, then $\int c(\eta, q_n)\psi_n(\eta)d\eta \to \int c(\eta, q)\psi(\eta)d\eta$.*

Proof. Since $c(\eta, q) \geq 0$, and $c(\cdot, q)$ is of linear growth,

$$\lim_{n\to\infty}\left|\int c(\eta, q_n)\psi_n(\eta)d\eta - \int c(\eta, q)\psi(\eta)d\eta\right|$$

$$= \lim_{n\to\infty}\left|\int c(\eta, q_n)\psi_n(\eta)d\eta - \int c(\eta, q)\psi_n(\eta)d\eta\right.$$

$$\left. + \int c(\eta, q)\psi_n(\eta)d\eta - \int c(\eta, q)\psi(\eta)d\eta\right|$$

$$\leq \lim_{n\to\infty}\left|\int c(\eta,q_n)\psi_n(\eta)d\eta - \int c(\eta,q)\psi_n(\eta)d\eta\right|$$
$$+\lim_{n\to\infty}\left|\int c(\eta,q)\psi_n(\eta)d\eta - \int c(\eta,q)\psi(\eta)d\eta\right|$$
$$\leq \lim_{n\to\infty}\left|\int [c(\eta,q_n)-c(\eta,q)]\psi_n(\eta)d\eta\right| + \lim_{n\to\infty}\left|\int c(\eta,q)[\psi_n(\eta)-\psi(\eta)]d\eta\right|$$
$$\leq \lim_{n\to\infty}\left|\int [c(\eta,q_n)-c(\eta,q)]\psi_n(\eta)d\eta\right|$$
$$+\lim_{n\to\infty}\left|\int (c_0+c_1q+c_2\eta)[\psi_n(\eta)-\psi(\eta)]d\eta\right|,$$

where both limits approach zero, respectively, on account of continuity of $c(\eta, q)$ in q and (17.12). □

From Lemma 17.4, we can see that the function

$$(q,\psi(\cdot)) \to \int c(\eta,q)\psi(\eta)d\eta \tag{17.36}$$

is continuous.

Lemma 17.5. *For each $z \geq 0$ and the bounded function $\rho : \Re^+ \to \Re^+$, the functions $(q,\rho) \to T_{q-z}\rho$ and $(q,\rho) \to \Theta_{z-q,z}\rho$ are continues, where $(q,\rho) = (q,\phi,\psi(\cdot)) \in \Re^+ \times \mathcal{H}^+$.*

Proof. Let $\{(q_n,\rho_n)\}$ be a sequence in $\Re^+ \times \mathcal{H}^+$ converging to $(q,\rho) \in \Re^+ \times \mathcal{H}^+$. Then, by adding and subtracting the term $\phi\bar{G}(q_n - z) + \int \psi(\lambda + q_n - z)d\lambda$, we have

$$\begin{aligned}\left|T^1_{q_n-z}\rho_n - T^1_{q-z}\rho\right| &= \left|\phi_n\bar{G}(q_n-z) + \int \psi_n(\lambda)\bar{G}(\lambda+q_n-z)d\lambda - \phi\bar{G}(q-z)\right.\\ &\quad \left. - \int \psi(\lambda)\bar{G}(\lambda+q-z)d\lambda\right|\\ &\leq |\psi_n-\psi|G(q_n-z) + \left|\bar{G}(q_n-z)-\bar{G}(q-z)\right|\phi\\ &\quad + \int |\psi_n(\lambda)-\psi(\lambda)|\bar{G}(\lambda+q_n-z)d\lambda\\ &\quad + \int \phi(\lambda)|\bar{G}(\lambda+q_n-z)-\bar{G}(\lambda+q-z)|d\lambda,\end{aligned}$$

which, by the Dominated Convergence Theorem and the continuity of $G(\cdot)$, converges to zero as $n \to \infty$. This proves the continuity of the $(q,\rho) \to T_{q-z}\rho$. Similarly, we can show that $(q,\rho) \to \Theta_{z-q,z}\rho$ is continuous. □

Lemma 17.6. *For each nonnegative measurable function* $W : \mathcal{H}^+ \to \Re^+$*, the functions* $(q,\rho) \to \int_0^q f(z)\, W(T_{q-z}\rho)dz$ *and* $(q,\rho) \to \int_q^\infty W(\Theta_{z-q,z}\rho)dz$ *are continuous.*

Proof. From Lemma 17.5 and dominated convenience theorem, for every bounded function $W : \rho \to \Re^+$, the functions $(q,\rho) \to \int f(z)W(T_{q-z}\rho)dz$ and $(q,\rho) \to \int W(\Theta_{z-q,z}\rho)dz$ are continuous.

Let $\{q_n, \rho_n\}$ be a sequence in $\Re^+ \times \mathcal{H}^+$ converging to $(q,\rho) \in \Re^+ \times \mathcal{H}$ and let W be a measurable function on $\mathcal{H}^+$. Then, there exists a sequence $\{W_k\}$ of bounded functions on $\mathcal{H}$ such that $W_k \nearrow W$ as $k \to \infty$.

Therefore, for each $k \in \mathbb{N}$, we have

$$\liminf_{n\to\infty} \int f(z)W_k(T_{q_n - z}\rho_n)dz = \int f(z)W_k(T_{q-z}\rho)dz.$$

Thus,

$$\begin{aligned}\liminf_{n\to\infty} \int f(z)W(T_{q_n - z}\rho_n)dz &\geq \liminf_{n\to\infty} \int f(z)W_k(T_{q_n - z}\rho_n)dz \\ &= \int f(z)W_k(T_{q-z}\rho)dz.\end{aligned}$$

Letting $k \to \infty$ and using Fatou's Lemma, we obtain

$$\lim_{k\to\infty} \int f(z)W_k(T_{q-z}\rho)dz \geq \int f(z)W(T_{q-z}\rho)dz.$$

Thus, $\liminf_{n\to\infty} \int f(z)W(T_{q_n-z}\rho_n)dz \geq \int f(z)W(T_{q-z}\rho)dz$. That is, the function $(q,\rho) \to \int f(z)W(T_{q-z}\rho)dz$ is l.s.c.. Applying the same arguments to the function $-W$, we have that $\int f(z)W(T_{q-z}\rho)dz$ is continuous with respect to (q,ρ). Similarly, we show that $(q,\rho) \to \int W(\Theta_{z-q,z}\rho)dz$ is continuous. □

Remark 17.2. Let $W \in \mathcal{G}$ be a continuous function on $\mathcal{H}^+$. Then in view of (17.36), Lemmas 17.5 and 17.6, we have $\mathcal{T}_q W(\rho)$ is continuous in $(q,\rho) \in \Re^+ \times \mathcal{H}^+$.

References

[1] A. Bensoussan, M. Çakanyildirim, and S. Sethi, Filtering for discrete-Time Markov processes and applications to inventory control with incomplete information, *Oxford Handbook of Nonlinear Filtering.* (2010).

[2] A. Bensoussan, M. Çakanyildirim, and S. Sethi, Partially observed inventory systems: The case of zero-balance walk, *SIAM Journal on Control & Optimization.* **46**(1), 176–209, (2007).

[3] A. Bensoussan, M. Çakanyildirim, J. Minjárez-Sosa, S. Sethi, and R. Shi, Partially observed inventory systems: The case of rain checks, *SIAM Journal on Control and Optimization*. **47**, 2490, (2008).
[4] H. Kushner, Dynamical equations for optimal nonlinear filtering, *Journal of Differential Equations*. **3**, 179–190, (1967). ISSN 0022-0396.
[5] M. Zakai, On the optimal filtering of diffusion processes, *Probability Theory and Related Fields*. **11**(3), 230–243, (1969).
[6] A. Bensoussan, M. Çakanyildirim, M. Li, and S. Sethi, Inventory control with a cash register: Sales recorded but not demand or shrinkage (Unabridged). (2011). Working Paper, The University of Texas at Dallas.

Chapter 18

Continuous control of piecewise deterministic Markov processes with long run average cost

O.L.V. Costa[1,*] and F. Dufour[2,†]

[1] *Departamento de Engenharia de Telecomunicações e Controle*
Escola Politécnica da Universidade de São Paulo
CEP: 05508 900-São Paulo, Brazil
oswaldo@lac.usp.br

[2] *Université de Bordeaux*
IMB, Institut Mathématiques de Bordeaux
INRIA Bordeaux Sud Ouest, Team: CQFD
351 cours de la Liberation
33405 Talence Cedex, France
dufour@math.u-bordeaux1.fr

The main goal of this chapter is to derive sufficient conditions for the existence of an optimal control strategy for the long run average continuous control problem of piecewise deterministic Markov processes (PDMP's) taking values in a general Borel space and with compact action space depending on the state variable. In order to do that we apply the so-called vanishing discount approach to obtain a solution to an average cost optimality inequality (ACOI) associated to the long run average cost problem. Our main assumptions are written in terms of some integro-differential inequalities related to the so-called expected growth condition, and geometric convergence of the post-jump location kernel associated to the PDMP.

Contents

*This author received financial support from CNPq (Brazilian National Research Council), grant 301067/2009-0.
†This author was supported by ARPEGE program of the French National Agency of Research (ANR), project "FAUTOCOES", number ANR-09-SEGI-004.

18.1. Introduction

A general family of non-diffusion stochastic models suitable for formulating optimization problems in several areas of operations research, namely piecewise deterministic Markov processes (PDMP's), was introduced in [1, 2]. These processes are determined by three local characteristics; the flow ϕ, the jump rate λ, and the transition measure Q. Starting from x the motion of the process follows the flow $\phi(x,t)$ until the first jump time T_1 which occurs either spontaneously in a Poisson-like fashion with rate λ or when the flow $\phi(x,t)$ hits the boundary of the state-space. In either case the location of the process at the jump time T_1 is selected by the transition measure $Q(\phi(x,T_1),.)$ and the motion restarts from this new point as before. A suitable choice of the state space and the local characteristics ϕ, λ, and Q provide stochastic models covering a great number of problems in operations research, see M.H.A. Davis [2].

There exist basically two types of control for PDMP's: *continuous control* and *impulse control*. This terminology has been introduced by M.H.A. Davis in [2, page 134] where continuous control is used to describe situations in which the control variable acts at all times on the process through the characteristics (ϕ, λ, Q) by influencing the deterministic motion and the probability of the jumps. On the other hand the terminology impulse control refers to a control that intervenes on the process by moving it to a new point in the state space at some times specifed by the controller.

In [3] we studied the long run average continuous control problem of PDMP's taking values in a general Borel space. At each point x of the state space a control variable is chosen from a compact action set $\mathbb{U}(x)$ and it is applied on the jump parameter λ and transition measure Q. The goal was to minimize the long run average cost, which is composed of a running cost and a boundary cost (which is added each time the PDMP touches the boundary). Both costs are assumed to be positive but not necessarily bounded. As far as the authors are aware, this was the first time that this kind of problem was considered in the literature. Indeed, results are available for the long run average cost problem but for impulse control see Costa [4], Gatarek [5] and the book by M.H.A. Davis [2] (see the references therein). On the other hand, the continuous control problem has been studied only for discounted costs by A. Almudevar [6], M.H.A. Davis [2, 7], M.A.H. Dempster and J.J. Ye [8, 9], Forwick, Schäl, and Schmitz [10], M. Schäl [11], A.A. Yushkevich [12, 13].

The aim of the present work, which is based on [3] and [14], is to provide tractable conditions that are directly related to the primitive data (ϕ, λ, Q) of the PDMP to ensure the existence of an optimal control. The problem of finding sufficient conditions based on the primitive data of the process is a traditional subject in the literature of the long run average control problem for Markov Decision Processes. Without attempting to present an exhaustive panorama on this topic, the interested reader may consult [15–19] and the references therein for detailed discussions on this problem. More precisely, we present in this paper some assumptions based on integro-differential inequalities related to a positive test function g and $\overline{r}$ (see Assumption 18.6), and on the geometric convergence of the post-jump location kernel associated with the PDMP (see Assumption 18.7), so that, under these hypotheses, we can show that $\alpha \mathcal{J}_{\mathcal{D}}^{\alpha}(x_0)$ (where $\mathcal{J}_{\mathcal{D}}^{\alpha}(x_0)$ is the optimal value function for the discounted cost problem starting from x_0 and with discount factor $\alpha > 0$) is bounded in a neighborhood of $\alpha = 0$ and that the relative difference of the α-discount value function $h_\alpha(\cdot) = \mathcal{J}_{\mathcal{D}}^{\alpha}(\cdot) - \mathcal{J}_{\mathcal{D}}^{\alpha}(x_0)$ belongs to a weighted-norm space of functions, labeled $\mathbb{B}_g(E)$. Another important point is that in the present work h_α is not necessarily bounded below by a constant.

The paper is organized in the following way. Some notation, basic assumptions, the problem formulation and the one-stage optimization operators that will be associated to the ACOI are introduced in section 18.2. Section 18.3 deals with assumptions related to the continuity of the parameters, the expected growth condition and geometric convergence of the post-jump location of the PDMP. The aim of section 18.4 is to derive the optimality equation for the discounted optimal control problem. The main results are presented in section 18.5, which provides sufficient conditions for the existence of a solution to the ACOI and an optimal control strategy for the long run average continuous control problem of a PDMP in terms of the parameters defining the process. The proofs of the main results are presented in the appendix.

18.2. Notation, basic assumptions, and problem formulation

18.2.1. *Preliminaries*

In this section we first present some standard notation and some basic definitions related to the motion of a controlled PDMP $\{X(t)\}$. Then we discuss the control problem and the one-stage optimization operators associated to the ACOI (see (18.12)) that we will consider throughout the paper. The reader is referred to [2] for further details and properties for the definition of a PDMP.

The set of admissible control strategies $\mathcal{U}$, the associated motion of the controlled process and the long run average and discounted optimal control problems are presented in sub-section 18.2.2, together with some basic notation and assumptions. Sub-section 18.2.3 deals with the sets of the discrete-time relaxed and ordinary controls, which will be used in the definition of the one-stage optimization operators. The basic idea is that, by relaxed control we mean control functions that are probability measures on the control action space, while by ordinary control we mean control functions that are deterministic functions on the control action space. The advantage of considering relaxed controls is that it yields a compact set, which implies many important properties. On the other hand we are interested in obtaining, as an optimal solution, an ordinary control. Having that in mind, we present in sub-section 18.2.4 (Definition 18.1) two one-stage optimization operators, namely $\mathcal{R}_\alpha$ and $\mathcal{T}_\alpha$, the first one that minimizes over the set of relaxed controls, while the second one minimizes over the set of ordinary controls. It follows from Theorem 18.7 in the appendix that, under the assumptions of section 18.3, the two operators have the same value and, moreover, the optimal control can be written in a feedback form. This leads to the desired result, that is, the optimal control depends in a deterministic way only on the state-variable x.

18.2.2. *Problem formulation and some definitions*

Let us start by presenting some notation that will be used throughout the paper: $\mathbb{N}$ denotes the set of natural numbers, $\mathbb{R}$ the set of real numbers, $\mathbb{R}_+$ the set of positive real numbers and $\mathbb{R}^d$ the d-dimensional euclidian space. We write η as the Lebesgue measure on $\mathbb{R}$. For X a metric space $\mathcal{B}(X)$ represents the σ-algebra generated by the open sets of X. $\mathcal{M}(X)$

(respectively, $\mathcal{P}(X)$) denotes the set of all finite (respectively probability) measures on $(X, \mathcal{B}(X))$. Let X and Y be metric spaces. The set of all Borel measurable (respectively bounded) functions from X into Y is denoted by $\mathbb{M}(X;Y)$ (respectively $\mathbb{B}(X;Y)$). Moreover, for notational simplicity $\mathbb{M}(X)$ (respectively $\mathbb{B}(X)$, $\mathbb{M}(X)^+$, $\mathbb{B}(X)^+$) denotes $\mathbb{M}(X;\mathbb{R})$ (respectively $\mathbb{B}(X;\mathbb{R})$, $\mathbb{M}(X;\mathbb{R}_+)$, $\mathbb{B}(X;\mathbb{R}_+)$). For $g \in \mathbb{M}(X)$ with $g(x) > 0$ for all $x \in X$, $\mathbb{B}_g(X)$ is the set of functions $v \in \mathbb{M}(X)$ such that

$$||v(x)||_g = \sup_{x\in X} \frac{|v(x)|}{g(x)} < +\infty.$$

$\mathbb{C}(X)$ denotes the set of continuous functions from X to $\mathbb{R}$. For $h \in \mathbb{M}(E)$, h^+ (respectively h^-) denotes the positive (respectively, negative) part of h.

Next we define the motion of a controlled PDMP, and the long run average and discounted optimal control problems. Let E be a Borel subset of $\mathbb{R}^n$, ∂E its boundary, and $\overline{E}$ its closure. A controlled PDMP is determined by its local characteristics (ϕ, λ, Q), as presented in the sequel. The flow $\phi(x,t)$ is a function $\phi : \mathbb{R}^n \times \mathbb{R}_+ \longrightarrow \mathbb{R}^n$, continuous in (x,t) and such that $\phi(x, t+s) = \phi(\phi(x,t), s)$.

For each $x \in E$ the time the flow takes to first reach the boundary starting from x is defined as $t_*(x) \doteq \inf\{t > 0 : \phi(x,t) \in \partial E\}$. For $x \in E$ such that $t_*(x) = \infty$ (that is, the flow starting from x never touches the boundary), we set $\phi(x, t_*(x)) = \Delta$, where Δ is a fixed point in ∂E. The space of functions absolutely continuous along the flow with limit towards the boundary is defined as follows:

$$\begin{aligned}\mathbb{M}^{ac}(E) = \big\{g \in \mathbb{M}(E) \ : \ & g(\phi(x,t)) : [0, t_*(x)) \mapsto \mathbb{R} \\ & \text{is absolutely continuous for each } x \in E \\ & \text{and whenever } t_*(x) < \infty \text{ the limit } \lim_{t\to t_*(x)} g(\phi(x,t)) \text{ exists}\big\}.\end{aligned}$$

For $g \in \mathbb{M}^{ac}(E)$ and $z \in \partial E$ for which there exists $x \in E$ such that $z = \phi(x, t_*(x))$ where $t_*(x) < \infty$ we define $g(z) = \lim_{t\to t_*(x)} g(\phi(x,t))$ (note that the limit exists by assumption). As shown in Lemma 2 in [20], for $g \in \mathbb{M}^{ac}(E)$ there exists a function $\mathcal{X}g \in \mathbb{M}(E)$ such that for all $x \in E$ and $t \in [0, t_*(x))$

$$g(\phi(x,t)) - g(x) = \int_0^t \mathcal{X}g(\phi(x,s))ds.$$

The local characteristics λ and Q depend on a control action $u \in \mathbb{U}$ where $\mathbb{U}$ is a metric space defined in the following way: $\lambda \in \mathbb{M}(\overline{E} \times \mathbb{U})^+$

and Q is a stochastic kernel on E given $\overline{E} \times \mathbb{U}$. For each $x \in \overline{E}$ we define the subsets $\mathbb{U}(x)$ of $\mathbb{U}$ as the set of feasible control actions that can be taken when the state process is in $x \in \overline{E}$, that is, the control action that will be applied to λ and Q must belong to $\mathbb{U}(x)$. The following assumptions will be made throughout the paper:

Assumption 18.1. For all $x \in \overline{E}$, $\mathbb{U}(x)$ is a closed subspace of the compact metric space $\mathbb{U}$.

Assumption 18.2. The set $K = \big\{(x,a) : x \in \overline{E}, a \in \mathbb{U}(x)\big\}$ is a Borel subset of $\overline{E} \times \mathbb{U}$.

A control policy U is a pair of functions $(u, u_\partial) \in \mathbb{M}(\mathbb{N} \times E \times \mathbb{R}_+; \mathbb{U}) \times \mathbb{M}(\mathbb{N} \times E; \mathbb{U})$ satisfying $u(n,x,t) \in \mathbb{U}(\phi(x,t))$, and $u_\partial(n,x) \in \mathbb{U}(\phi(x,t_*(x)))$ for all $(n,x,t) \in \mathbb{N} \times E \times \mathbb{R}_+$. The class of admissible control strategies will be denoted by $\mathcal{U}$. Consider the state space $\widehat{E} = E \times E \times \mathbb{R}_+ \times \mathbb{N}$. For a control policy $U = (u, u_\partial)$ let us introduce the following parameters for $\hat{x} = (x,z,s,n) \in \widehat{E}$: the flow $\widehat{\phi}(\hat{x},t) = (\phi(x,t), z, s+t, n)$, the jump rate $\widehat{\lambda}^U(\hat{x}) = \lambda(x, u(n,z,s))$, and the transition measure

$$\widehat{Q}^U(\hat{x}, A \times B \times \{0\} \times \{n+1\}) = \begin{cases} Q(x, u(n,z,s)); A \cap B) & \text{if } x \in E, \\ Q(x, u_\partial(n,z); A \cap B) & \text{if } x \in \partial E, \end{cases}$$

for A and B in $\mathcal{B}(E)$. From [2, section 25], it can be shown that for any control strategy $U = (u, u_\partial) \in \mathcal{U}$ there exists a filtered probability space $(\Omega, \mathcal{F}, \{\mathcal{F}_t\}, \{P^U_{\hat{x}}\}_{\hat{x} \in \widehat{E}})$ such that the piecewise deterministic Markov process $\{\widehat{X}^U(t)\}$ with local characteristics $(\widehat{\phi}, \widehat{\lambda}^U, \widehat{Q}^U)$ may be constructed as follows. For notational simplicity the probability $P^U_{\hat{x}_0}$ will be denoted by $P^U_{(x,k)}$ for $\hat{x}_0 = (x,x,0,k) \in \widehat{E}$. Take a random variable T_1 such that

$$P^U_{(x,k)}(T_1 > t) \doteq \begin{cases} e^{-\Lambda^U(x,k,t)} & \text{for } t < t_*(x) \\ 0 & \text{for } t \geq t_*(x) \end{cases}$$

where for $x \in E$ and $t \in [0, t_*(x)[$, $\Lambda^U(x,k,t) \doteq \int_0^t \lambda(\phi(x,s), u(k,x,s))ds$. If T_1 is equal to infinity, then for $t \in \mathbb{R}_+$, $\widehat{X}^U(t) = \big(\phi(x,t), x, t, k\big)$. Otherwise select independently an $\widehat{E}$-valued random variable (labelled $\widehat{X}^U_1$) having distribution

$$\begin{aligned} &P^U_{(x,k)}(\widehat{X}^U_1 \in A \times B \times \{0\} \times \{k+1\} | \sigma\{T_1\}) \\ &\quad = \begin{cases} Q(\phi(x,T_1), u(k,x,T_1)); A \cap B) & \text{if } \phi(x,T_1) \in E, \\ Q(\phi(x,T_1), u_\partial(k,x); A \cap B) & \text{if } \phi(x,T_1) \in \partial E. \end{cases} \end{aligned}$$

The trajectory of $\{\widehat{X}^U(t)\}$ starting from $(x, x, 0, k)$, for $t \leq T_1$, is given by

$$\widehat{X}^U(t) \doteq \begin{cases} (\phi(x,t), x, t, k) & \text{for } t < T_1, \\ \widehat{X}_1^U & \text{for } t = T_1. \end{cases}$$

In a similar way we select the next inter-jump time $T_2 - T_1$ and post-jump location $\widehat{X}^U(T_2) = \widehat{X}_2^U$, starting from $\widehat{X}^U(T_1) = \widehat{X}_1^U$. Let us define the components of the PDMP $\{\widehat{X}^U(t)\}$ by $\widehat{X}^U(t) = \big(X(t), Z(t), \tau(t), N(t)\big)$. It is easy to see from the previous construction that $X(t)$ corresponds to the trajectory of the system, $Z(t)$ is the value of $X(t)$ at the last jump time before t, $\tau(t)$ is the time elapsed between the last jump and time t, and $N(t)$ is the number of jumps of the process $\{X(t)\}$ at time t. As in Davis [2], we consider the following assumption to avoid any accumulation point of the jump times:

Assumption 18.3. For any $x \in E$, $U = (u, u_\partial) \in \mathcal{U}$, and $t \geq 0$, we have

$$E_{(x,0)}^U \left[\sum_{i=1}^{\infty} I_{\{T_i \leq t\}} \right] < \infty.$$

We suppose that the costs of our control problem contain two terms, a running cost f and a boundary cost r, satisfying the following properties:

Assumption 18.4. $f \in \mathbb{M}(\overline{E} \times \mathbb{U})^+$, and $r \in \mathbb{M}(\partial E \times \mathbb{U})^+$.

Define, for $\alpha \geq 0$, $t \in \mathbb{R}_+$, and $U \subset \mathcal{U}$,

$$\begin{aligned} \mathbf{J}^\alpha(U,t) = & \int_0^t e^{-\alpha s} f\big(X(s), u(N(s), Z(s), \tau(s))\big) ds \\ & + \int_0^t e^{-\alpha s} r\big(X(s-), u_\partial(N(s-), Z(s-))\big) dp^*(s), \end{aligned}$$

where $p^*(t) = \sum_{i=1}^{\infty} I_{\{T_i \leq t\}} I_{\{X(T_i-) \in \partial E\}}$ counts the number of times the process hits the boundary up to time t. For notational simplicity, set $\mathbf{J}(U,t) = \mathbf{J}^0(U,t)$. We want to minimize over $\mathcal{U}$ the long-run average cost given by

$$\mathcal{A}(U,x) = \overline{\lim_{t \to +\infty}} \frac{1}{t} E_{(x,0)}^U [\mathbf{J}(U,t)]$$

and we set

$$\mathcal{J}_{\mathcal{A}}(x) = \inf_{U \in \mathcal{U}} \mathcal{A}(U,x).$$

For the α discounted case, with $\alpha > 0$, we want to minimize the cost given by

$$\mathcal{D}^{\alpha}(U,x) = E^{U}_{(x,0)}[\mathbf{J}^{\alpha}(U,\infty)]$$

and we set

$$\mathcal{J}^{\alpha}_{\mathcal{D}}(x) = \inf_{U\in\mathcal{U}} \mathcal{D}^{\alpha}(U,x).$$

18.2.3. *Relaxed and ordinary discrete-time controls*

The goal of this sub-section is to present the definitions of the discrete-time ordinary and relaxed control sets used in the formulation of the one-stage optimization operators, which will be introduced in sub-section 18.2.4. In order to do that we consider $\mathbb{C}(\mathbb{U})$ equipped with the topology of uniform convergence and $\mathcal{M}(\mathbb{U})$ equipped with the weak* topology $\sigma(\mathcal{M}(\mathbb{U}),\mathbb{C}(\mathbb{U}))$. For $x \in E$, define $\mathcal{P}_x(\mathbb{U})$ as the set of measures $\mu \in \mathcal{P}(\mathbb{U})$ satisfying $\mu(\mathbb{U}(\phi(x,t_*(x)))) = 1$. $\mathcal{P}(\mathbb{U})$ and $\mathcal{P}_x(\mathbb{U})$ for $x \in E$ are subsets of $\mathcal{M}(\mathbb{U})$ and are equipped with the relative topology.

We define $\mathcal{V}^r$ (respectively $\mathcal{V}^r(x)$ for $x \in E$) as the set of all η-measurable functions μ on $\mathbb{R}_+$ taking values in $\mathcal{P}(\mathbb{U})$ such that $\mu(t,\mathbb{U}) = 1$ η-a.e. (respectively $\mu(t,\mathbb{U}(\phi(x,t))) = 1$ η-a.e.). As shown in sub-section 3.1 in [3], $\mathcal{V}^r(x)$ is a compact set of the metric space $\mathcal{V}^r$. It means that a sequence $(\mu_n)_{n\in\mathbb{N}}$ in $\mathcal{V}^r(x)$ converges to μ if and only if for all $g \in L^1(\mathbb{R}_+;\mathbb{C}(\mathbb{U}))$

$$\lim_{n\to\infty} \int_{\mathbb{R}_+} \int_{\mathbb{U}(\phi(x,t))} g(t,u)\mu_n(t,du)dt = \int_{\mathbb{R}_+} \int_{\mathbb{U}(\phi(x,t))} g(t,u)\mu(t,du)dt.$$

We define the sets of relaxed controls as follows: $\mathbb{V}^r(x) = \mathcal{V}^r(x) \times \mathcal{P}_x(\mathbb{U})$, for $x \in E$ and $\mathbb{V}^r = \mathcal{V}^r \times \mathcal{P}(\mathbb{U})$.

The set of ordinary controls, denoted by $\mathbb{V}$ (respectively $\mathbb{V}(x)$ for $x \in E$), is defined as above, except that it is composed of deterministic functions instead of probability measures. In other words, we have $\mathcal{V}(x) = \{\nu \in \mathbb{M}(\mathbb{R}_+,\mathbb{U}) : (\forall t \in \mathbb{R}_+), \nu(t) \in \mathbb{U}(\phi(x,t))\}$, $\mathbb{V}(x) = \mathcal{V}(x) \times \mathbb{U}(\phi(x,t_*(x)))$, $\mathbb{V} = \mathbb{M}(\mathbb{R}_+,\mathbb{U}) \times \mathbb{U}$. It follow that the set of ordinary controls is a subset of the set of relaxed controls $\mathbb{V}^r$ (respectively $\mathbb{V}^r(x)$ for $x \in E$) by identifying any control action $u \in \mathbb{U}$ with the Dirac measure concentrated on u, so that $\mathbb{V} \subset \mathbb{V}^r$ (respectively $\mathbb{V}(x) \subset \mathbb{V}^r(x)$ for $x \in E$). From now on we will consider that $\mathbb{V}$ (respectively $\mathbb{V}(x)$ for $x \in E$) will be endowed with the topology generated by $\mathbb{V}^r$.

Notice that in general there does not exist a topology for which $\mathbb{V}$ and $\mathbb{V}(x)$ are compact sets. However from the previous construction, it follows that $\mathbb{V}^r$ and $\mathbb{V}^r(x)$ are compact sets, justifying the introduction of the class of relaxed control $\mathbb{V}^r$.

We need the following assumption to guarantee that set of feasible state/relaxed-control pairs is a measurable subset of $\mathcal{B}(E) \times \mathcal{B}(\mathbb{V}^r)$ (see also [17], page 14).

Assumption 18.5. $\mathcal{K} \doteq \big\{(x, \Theta) : \Theta \in \mathbb{V}^r(x), x \in E\big\} \in \mathcal{B}(E) \times \mathcal{B}(\mathbb{V}^r)$.

18.2.4. *One-stage optimization operators*

Some important operators associated to the ACOI (see (18.12)) will be introduced in this sub-section. These operators will be called one-stage optimization operators (see Definition 18.1 below) and associated to them we define the set of measurable selectors. One of special interest will be named the ordinary feedback measurable selectors.

We first introduce the following notation: for $x \in \overline{E}$, $\mu \in \mathcal{P}(\mathbb{U})$, $h \in \mathbb{M}(E)^+$ and $w \in \mathbb{M}(\overline{E} \times \mathbb{U})^+$, set

$$\begin{aligned} w(x,\mu) &\doteq \int_{\mathbb{U}} w(x,u)\mu(du), \\ Qh(x,\mu) &\doteq \int_{\mathbb{U}}\int_{E} h(z)Q(x,u;dz)\mu(du), \text{ and} \\ \lambda Qh(x,\mu) &\doteq \int_{\mathbb{U}} \lambda(x,u)\int_{E} h(z)Q(x,u;dz)\mu(du). \end{aligned}$$

Fix $\Theta = \big(\mu, \mu_\partial\big) \in \mathbb{V}^r$, $(x, A) \in E \times \mathcal{B}(E)$, $\alpha \in \mathbb{R}$, and define

$$\begin{aligned} \Lambda^\mu(x,t) &\doteq \int_0^t \lambda(\phi(x,s),\mu(s))ds \\ G_\alpha(x,\Theta;A) &\doteq \int_0^{t_*(x)} e^{-\alpha s-\Lambda^\mu(x,s)}\lambda Q I_A(\phi(x,s),\mu(s))ds \\ &\quad + e^{-\alpha t_*(x)-\Lambda^\mu(x,t_*(x))}Q(\phi(x,t_*(x)),\mu_\partial;A). \qquad (18.1) \end{aligned}$$

For $h \in \mathbb{M}(E)^+$, we define $G_\alpha h(x,\Theta) \doteq \int_E h(y)G_\alpha(x,\Theta;dy)$ and for $x \in E$, $\Theta = \big(\mu, \mu_\partial\big) \in \mathbb{V}^r$, $v \in \mathbb{M}(E \times \mathbb{U})^+$, $w \in \mathbb{M}(\partial E \times \mathbb{U})^+$ and $\alpha \in \mathbb{R}$, set

$$L_\alpha v(x,\Theta) \doteq \int_0^{t_*(x)} e^{-\alpha s-\Lambda^\mu(x,s)}v(\phi(x,s),\mu(s))ds, \qquad (18.2)$$

$$H_\alpha w(x,\Theta) \doteq e^{-\alpha t_*(x)-\Lambda^\mu(x,t_*(x))}w(\phi(x,t_*(x)),\mu_\partial). \qquad (18.3)$$

For $h \in \mathbb{M}(E)$ (respectively, $v \in \mathbb{M}(E \times \mathbb{U})$), $G_\alpha h(x,\Theta) = G_\alpha h^+(x,\Theta) - G_\alpha h^-(x,\Theta)$ (respectively, $L_\alpha v(x,\Theta) = L_\alpha v^+(x,\Theta) - L_\alpha v^-(x,\Theta)$) provided the difference has a meaning. It will be useful in the sequel to define the function $\mathcal{L}_\alpha(x,\Theta)$ as follows: $\mathcal{L}_\alpha(x,\Theta) \doteq L_\alpha I_{E\times\mathbb{U}}(x,\Theta)$. In particular for $\alpha = 0$ we write for simplicity $G_0 = G$, $L_0 = L$, $H_0 = H$, $\mathcal{L}_0 = \mathcal{L}$.

We now present measurability properties of the operators G_α, L_α, and H_α.

Lemma 18.1. *Let $h \in \mathbb{M}(\mathbb{R}_+ \times E \times \mathcal{V}^r \times \mathbb{R}_+ \times \mathbb{U})^+$. Then the $\mathbb{R}_+$-valued mapping defined on $\mathbb{R}_+ \times E \times \mathcal{V}^r$ by*

$$(t,x,\mu) \to \int_{\mathbb{R}_+} \int_{\mathbb{U}} h(t,x,\mu,s,u)\mu(s,du)ds$$

is $\mathcal{B}(\mathbb{R}_+ \times E \times \mathcal{V}^r)$-measurable.

Proof. Let $\mathcal{H}$ be the class of functions $h \in \mathbb{B}(\mathbb{R}_+ \times E \times \mathcal{V}^r \times \mathbb{R}_+ \times \mathbb{U})$ such that the real-valued mapping defined on $\mathbb{R}_+ \times E \times \mathcal{V}^r$ by

$$(t,x,\mu) \to \int_{\mathbb{R}_+} e^{-s} \int_{\mathbb{U}} h(t,x,\mu,s,u)\mu(s,du)ds$$

is $\mathcal{B}(\mathbb{R}_+ \times E \times \mathcal{V}^r)$-measurable. This set is closed relative to addition, multiplication by constants and bounded pointwise passage to the limit. Consider the class $\mathcal{C}$ of functions h such that $h = h_1 h_2$ with $h_1 = \mathbb{B}(\mathbb{R}_+ \times E \times \mathcal{V}^r)$ and $h_2 = \mathbb{B}(\mathbb{R}_+ \times \mathbb{U})$. Then from Lemma A.3 in [10], $\mathcal{C} \subset \mathcal{H}$. Moreover, $\mathcal{C}$ is closed relative to multiplication. Applying the monotone class Theorem, it follows that $\mathcal{H}$ contains $\mathbb{B}(\mathbb{R}_+ \times E \times \mathcal{V}^r \times \mathbb{R}_+ \times \mathbb{U})$. Consider $g \in \mathbb{M}(\mathbb{R}_+ \times E \times \mathcal{V}^r \times \mathbb{R}_+ \times \mathbb{U})^+$ and define $g_k(t,x,\mu,s,u) = \big(e^s g(t,x,\mu,s,u)\big) \wedge k$. Therefore the mapping

$$(t,x,\mu) \to \int_{\mathbb{R}_+} e^{-s} \int_{\mathbb{U}} g_k(t,x,\mu,s,u)\mu(s,du)ds$$

is $\mathcal{B}(\mathbb{R}_+ \times E \times \mathcal{V}^r)$-measurable. By using the monotone convergence theorem, the result follows. □

Proposition 18.1. Let $\alpha \in \mathbb{R}_+$, $g \in \mathbb{M}(E)$ be bounded from below, $w_1 \in \mathbb{M}(E \times \mathbb{U})$ be bounded from below, and $w_2 \in \mathbb{M}(\partial E \times \mathbb{U})^+$. Then the mappings $G_\alpha g(x,\Theta)$, $L_\alpha w_1(x,\Theta)$, and $H_\alpha w_2(x,\Theta)$ defined on $E \times \mathbb{V}^r$ with values in $\mathbb{R}$ are $\mathcal{B}(E \times \mathbb{V}^r)$-measurable.

Proof. Applying Lemma 18.1 to $h(t,x,\mu,s,u) = I_{\{s \le t\}} \lambda(\phi(x,s),u)$ implies that the mapping $\Lambda^\mu(x,t)$ defined on $\mathbb{R}_+ \times E \times \mathcal{V}^r$ with value in $\mathbb{R}$ is measurable with respect to $\mathcal{B}(\mathbb{R}_+ \times E \times \mathcal{V}^r)$. There is no loss of generality in assuming that g is positive. Clearly, $Qg(.,.)$ is measurable with respect to $\mathcal{B}(\overline{E} \times \mathbb{U})$. From Lemma 18.1, we obtain that the $\mathbb{R}_+$-valued mapping defined on $E \times \mathcal{V}^r$ by

$$(x, \mu) \to \int_0^{t_*(x)} e^{-\alpha s - \Lambda^{\mu}(x,s)} \lambda Qg(\phi(x,s), \mu(s)) ds$$

is $\mathcal{B}(E \times \mathcal{V}^r)$-measurable. Moreover, the $\mathbb{R}_+$-valued function defined on $E \times \mathcal{V}^r \times \mathbb{U}$ by $(x, \mu, u) \to e^{-\alpha t_*(x) - \Lambda^{\mu}(x, t_*(x))} Qg(\phi(x, t_*(x)), u)$ is clearly $\mathcal{B}(E \times \mathcal{V}^r \times \mathbb{U})$-measurable. Consequently, it follows that $G_\alpha g(x, \Theta)$ defined on $E \times \mathbb{V}^r$ with value in $\mathbb{R}$ is $\mathcal{B}(E \times \mathbb{V}^r)$-measurable. By using the same arguments, the same property can be shown for the mappings $L_\alpha w_1(x, \Theta)$, and $H_\alpha w_2(x, \Theta)$. □

We now present the definitions of the one-stage optimization operators.

Definition 18.1. Let $\alpha \in \mathbb{R}_+$, $\rho \in \mathbb{R}$, and $h \in \mathbb{M}(E)$. In what follows assume that, for any $x \in E$ and $\Upsilon \in \mathbb{V}(x)$, $-\rho \mathcal{L}_\alpha(x, \Upsilon) + L_\alpha f(x, \Upsilon) + H_\alpha r(x, \Upsilon) + G_\alpha h(x, \Upsilon)$ is well defined. The (ordinary) one-stage optimization operator is defined by

$$\mathcal{T}_\alpha(\rho, h)(x) = \inf_{\Upsilon \in \mathbb{V}(x)} \Big\{ -\rho \mathcal{L}_\alpha(x, \Upsilon) + L_\alpha f(x, \Upsilon) + H_\alpha r(x, \Upsilon) + G_\alpha h(x, \Upsilon) \Big\}.$$

Assume that for any $x \in E$ and $\Theta \in \mathbb{V}^r(x)$, $-\rho \mathcal{L}_\alpha(x, \Theta) + L_\alpha f(x, \Theta) + H_\alpha r(x, \Theta) + G_\alpha h(x, \Theta)$ is well defined. The relaxed one-stage optimization operator is defined by

$$\mathcal{R}_\alpha(\rho, h)(x) = \inf_{\Theta \in \mathbb{V}^r(x)} \Big\{ -\rho \mathcal{L}_\alpha(x, \Theta) + L_\alpha f(x, \Theta) + H_\alpha r(x, \Theta) + G_\alpha h(x, \Theta) \Big\}.$$

In particular for $\alpha = 0$ we write for simplicity $\mathcal{T}_0 = \mathcal{T}$, and $\mathcal{R}_0 = \mathcal{R}$.

The sets of measurable selectors associated to $\big(\mathbb{U}(x)\big)_{x \in E}$, $\big(\mathbb{V}(x)\big)_{x \in E}$ and $\big(\mathbb{V}^r(x)\big)_{x \in E}$ are defined by

$$\mathcal{S}_{\mathbb{U}} = \Big\{ u \in \mathbb{M}(\overline{E}, \mathbb{U}) : (\forall x \in \overline{E}), u(x) \in \mathbb{U}(x) \Big\},$$

$$\mathcal{S}_{\mathbb{V}} = \Big\{ (\nu, \nu_\partial) \in \mathbb{M}(E, \mathbb{V}) : (\forall x \in E), \big(\nu(x),\ \nu_\partial(x)\big) \in \mathbb{V}(x) \Big\},$$

$$\mathcal{S}_{\mathbb{V}^r} = \Big\{ (\mu, \mu_\partial) \in \mathbb{M}(E, \mathbb{V}^r) : (\forall x \in E), \big(\mu(x),\ \mu_\partial(x)\big) \in \mathbb{V}^r(x) \Big\}.$$

For $\alpha \in \mathbb{R}_+$, $\rho \in \mathbb{R}$, and $v \in \mathbb{M}(E)$, the one-stage optimization problem associated to the operator $\mathcal{T}_\alpha(\rho, v)$, respectively $\mathcal{R}_\alpha(\rho, v)$, consists of finding a measurable selector $\Upsilon \in \mathcal{S}_{\mathbb{V}}$, respectively $\Theta \in \mathcal{S}_{\mathbb{V}^r}$ such that for all $x \in E$,

$\mathcal{T}_\alpha(\rho,v)(x) = -\rho\mathcal{L}_\alpha(x,\Upsilon) + L_\alpha f(x,\Upsilon) + H_\alpha r(x,\Upsilon) + G_\alpha v(x,\Upsilon)$ and respectively $\mathcal{R}_\alpha(\rho,v)(x) = -\rho\mathcal{L}_\alpha(x,\Theta) + L_\alpha f(x,\Theta) + H_\alpha r(x,\Theta) + G_\alpha v(x,\Theta)$.

We notice that there exist two natural mappings from $\mathcal{S}_{\mathbb{U}}$ to $\mathcal{S}_{\mathbb{V}}$ and from $\mathcal{S}_{\mathbb{U}}$ to $\mathcal{U}$ (see Propositions 3.8 and 3.10 in [3] for measurability properties), as presented in the next definitions.

Definition 18.2. For $u \in \mathcal{S}_{\mathbb{U}}$, we define the mapping u_ϕ of the space E into $\mathbb{V}$ by $u_\phi : x \to \big(u(\phi(x,.)), u(\phi(x,t_*(x)))\big)$.

Definition 18.3. For $u \in \mathcal{S}_{\mathbb{U}}$, we define the mapping U_{u_ϕ} of the space $\mathbb{N} \times E \times \mathbb{R}_+$ into $\mathbb{U} \times \mathbb{U}$ by $U_{u_\phi} : (n,x,t) \to \big(u(\phi(x,t)), u(\phi(x,t_*(x)))\big)$.

We have the following proposition.

Proposition 18.2. If $u \in \mathcal{S}_{\mathbb{U}}$ then $u_\phi \in \mathcal{S}_{\mathbb{V}}$. If $u \in \mathcal{S}_{\mathbb{U}}$ then $U_{u_\phi} \in \mathcal{U}$.

Proof. From Lemma A.3 in [10] and item (i) of Lemma 3 in [21] it follows that the mapping $x \to u(\phi(x,.))$ of the space E into $\mathbb{M}(\mathbb{R}_+,\mathbb{U})$ is measurable. Moreover, for all $(x,t) \in E \times \mathbb{R}_+$, $u(\phi(x,t)) \in \mathbb{U}(\phi(x,t))$. Therefore, u_ϕ belongs to $\mathcal{S}_{\mathbb{V}}$. The second affirmative is a straightforward consequence of the measurability properties of u and ϕ and the fact that $u(x) \in \mathbb{U}(x)$. □

Remark 18.1. The measurable selectors of the kind u_ϕ as in Definition 18.2 will be called ordinary feedback measurable selectors in the class $\mathcal{S}_{\mathbb{V}} \subset \mathcal{S}_{\mathbb{V}^r}$ and the control strategies of the kind U_{u_ϕ} as in Definition 18.3 will be called ordinary feedback control strategies in the class $\mathcal{U}$.

18.3. Assumptions on the parameters of the PDMP

The purpose of this section is to introduce several tractable hypotheses directly on the parameters of the PDMP in order to obtain sufficient conditions for the existence of a solution to ACOI (see (18.12)) and from this derive the existence of an optimal control for the long run average control problem of a PDMP posed in section 18.2.

The following assumption will be used, among other things, to guarantee the uniform boundedness of $\alpha\mathcal{J}_D^\alpha(x)$ with respect to α (see Theorem 18.2), and is related to the so-called expected growth condition (see, for instance, Assumption 3.1 in [16] for the discrete-time case, or Assumption A in [15] for the continuous-time case).

Assumption 18.6. There exist $b \geq 0$, $c > 0$, $\delta > 0$, $M \geq 0$ and $g \in \mathbb{M}^{ac}(E)$, $g \geq 1$, and $\overline{r} \in \mathbb{M}(\partial E)$, $\overline{r}(z) \geq 0$, satisfying for all $x \in E$

$$\sup_{a \in \mathbb{U}(x)} \Big\{ \mathcal{X}g(x) + cg(x) - \lambda(x,a)\left[g(x) - Qg(x,a)\right] \Big\} \leq b, \tag{18.4}$$

$$\sup_{a \in \mathbb{U}(x)} \Big\{ f(x,a) \Big\} \leq Mg(x), \tag{18.5}$$

and for all $x \in E$ with $t_*(x) < \infty$

$$\sup_{a \in \mathbb{U}(\phi(x,t_*(x)))} \{ \overline{r}(\phi(x,t_*(x))) + Qg(\phi(x,t_*(x)),a) \} \leq g(\phi(x,t_*(x))), \tag{18.6}$$

$$\sup_{a \in \mathbb{U}(\phi(x,t_*(x)))} \Big\{ r(\phi(x,t_*(x)),a) \Big\} \leq \frac{M}{c+\delta} \overline{r}(\phi(x,t_*(x))). \tag{18.7}$$

The next assumption is related to geometric ergodic properties of the operator G (see for example the comments on page 122 in [18] or Lemma 3.3 in [16] for more details on this kind of assumption). Notice that for any $u \in \mathcal{S}_{\mathbb{U}}$, $G(.,u_\phi;.)$ can be seen as the stochastic kernel associated to the post-jump location of a PDMP. It will be used in particular to ensure that the relative difference of the α-discount value functions, h_α, defined by $h_\alpha(x) = \mathcal{J}^\alpha_{\mathcal{D}}(x) - \mathcal{J}^\alpha_{\mathcal{D}}(x_0)$, belongs to the weighted-norm space of functions $\mathbb{B}_g(E)$ (see Theorem 18.3).

Assumption 18.7. For each $u \in \mathcal{S}_{\mathbb{U}}$ there exists a probability measure ν_u, such that $\nu_u(g) < +\infty$ and

$$\Big| G^k h(x,u_\phi) - \nu_u(h) \Big| \leq a\|h\|_g \kappa^k g(x), \tag{18.8}$$

for all $h \in \mathbb{B}_g(E)$ and $k \in \mathbb{N}$, with $a > 0$ and $0 < \kappa < 1$ independent of u.

Assumptions 18.8, 18.9, 18.10 and 18.11, to be presented in the sequel, are needed to guarantee some convergence and semi-continuity properties of the one-stage optimization operators, the equality between the operators $\mathcal{T}_\alpha$ and $\mathcal{R}_\alpha$, and the existence of an ordinary feedback measurable selector (see sub-section A.3 and in particular Theorem 18.7).

Assumption 18.8. We assume that:

i) for each $x \in E$, the restriction of $\lambda(x,.)$ to $\mathbb{U}(x)$ is continuous.
ii) there exists $\overline{\lambda} \in \mathbb{M}(\overline{E})^+$ such that for all $x \in E$, $a \in \mathbb{U}(x)$, $\lambda(x,a) \leq \overline{\lambda}(x)$ and for $t \in [0, t_*(x))$, $\int_0^t \overline{\lambda}(\phi(x,s))\,ds < \infty$ and if $t_*(x) < \infty$ then $\int_0^{t_*(x)} \overline{\lambda}(\phi(x,s))\,ds < \infty$.

Assumption 18.9. There exists a sequence of measurable functions $(f_j)_{j\in\mathbb{N}}$ in $\mathbb{M}(\overline{E}\times\mathbb{U})^+$ such that for all $y\in\overline{E}$, $f_j(y,.)\uparrow f(y,.)$ as $j\to\infty$ and the restriction of $f_j(y,.)$ to $\mathbb{U}(y)$ is continuous. There exists a sequence of measurable functions $(r_j)_{j\in\mathbb{N}}$ in $\mathbb{M}(\partial E\times\mathbb{U})^+$ such that for all $z\in\partial E$, $r_j(z,.)\uparrow r(z,.)$ as $j\to\infty$ and the restriction of $r_j(z,.)$ to $\mathbb{U}(z)$ is continuous.

Assumption 18.10. For all $x\in\overline{E}$ and $h\in\mathbb{B}(E)$, the restriction of $Qh(x,.)$ to $\mathbb{U}(x)$ is continuous, and the restriction of $Qg(x,.)$ to $\mathbb{U}(x)$ is continuous.

The final assumption is:

Assumption 18.11. There exist $\underline{\lambda}\in\mathbb{M}(E)^+$, $\overline{f}\in\mathbb{M}(E)^+$, $K_\lambda\in\mathbb{R}_+$ such that

a) $\lambda(y,a)\geq\underline{\lambda}(y)$ and $f(y,a)\leq\overline{f}(y)$ for all $y\in E$ and $a\in\mathbb{U}(y)$,
b) $\int_0^{t_*(x)} e^{ct-\int_0^t\underline{\lambda}(\phi(x,s))ds}dt\leq K_\lambda$, for all $x\in E$,
c) $\lim_{t\to+\infty} e^{ct-\int_0^t\underline{\lambda}(\phi(x,s))ds}=0$, for all $x\in E$ with $t_*(x)=+\infty$,
d) $\lim_{t\to+\infty} e^{-\int_0^t\underline{\lambda}(\phi(x,s))ds}g(\phi(x,t))=0$, for all $x\in E$ with $t_*(x)=\infty$,
e) $\int_0^{t_*(x)} e^{-\int_0^t\underline{\lambda}(\phi(x,s))ds}\overline{f}(\phi(x,t))dt<\infty$.

The next definition will be important to characterize the optimal feedback measurable selector and optimal control strategy for our problem.

Definition 18.4. Consider $w\in\mathbb{M}(E)$ and $h\in\mathbb{B}_g(E)$. We define

D1) $\widehat{u}(w,h)\in\mathcal{S}_{\mathbb{U}}$ as the measurable selector satisfying

$$\inf_{a\in\mathbb{U}(x)}\Big\{f(x,a)-\lambda(x,a)\Big[w(x)-Qh(x,a)\Big]\Big\}$$
$$=f(x,\widehat{u}(w,h)(x))-\lambda(x,\widehat{u}(w,h)(x))\Big[w(x)-Qh(x,\widehat{u}(w,h)(x))\Big],$$

and

$$\inf_{a\in\mathbb{U}(z)}\{r(z,a)+Qh(z,a)\}=r(z,\widehat{u}(w,h)(z))+Qh(z,\widehat{u}(w,h)(z)).$$

D2) $\widehat{u}_\phi(w,h)\in\mathcal{S}_{\mathbb{V}}$ as the measurable selector derived from $\widehat{u}(w,h)$ through the Definition 18.2.

Remark 18.2. Notice that from Assumption 18.10 and Lemma 8.3.7 in [18] we have that $Qh(x,.)$ is continuous in $\mathbb{U}(x)$. From this and Assumptions 18.8 i) and 18.9, we see that the existence of $\widehat{u}(w,h)$ in D1) follows from Proposition D.5 in [17].

Remark 18.3. Notice the following consequences of Assumption 18.11:

i) Assumption 18.11 c) implies that

$$G_\alpha(x,\Theta;A) = \int_0^{t_*(x)} e^{-\alpha s - \Lambda^\mu(x,s)} \lambda Q I_A(\phi(x,s),\mu(s))ds,$$

and $H_\alpha w(x,\Theta) = 0$, for any $x \in E$ with $t_*(x) = +\infty$, $A \in \mathcal{B}(E)$, $\alpha \geq -c$, $\Theta = (\mu, \mu_\partial) \in \mathbb{V}^r(x)$, $w \in \mathbb{M}(\partial E \times \mathbb{U})$.

ii) Assumptions 18.11 a) and b) imply that $\mathcal{L}_\alpha(x,\Theta) \leq K_\lambda$ for any $\alpha \geq -c$, $x \in E$, $\Theta \in \mathbb{V}^r(x)$.

18.4. Optimality equation for the discounted case

The aim of this section is to derive the optimality equation for the discounted optimal control problem. As usual in this kind of problem we characterize first the optimality equation for the truncated on the jump times T_m problems and then take the limit as $m \to \infty$. Throughout this section we consider $\alpha > 0$ fixed. For any $g \in \mathbb{M}(E)^+$, we set $\mathcal{W}g$ as the function on E defined as $\mathcal{W}g(x) = \mathcal{R}_\alpha(0,g)(x)$, for $x \in E$. The following proposition is an immediate consequence of the results derived in the previous section.

Proposition 18.3. For $g \in \mathbb{M}(E)^+$ consider $w = \mathcal{W}g$ and suppose that for all $x \in E$, $w(x) \in \mathbb{R}$. Under these conditions, $w \in \mathbb{M}(E)^+$ and $\widehat{u}_\phi(w,g) \in \mathcal{S}_\mathbb{V}$ (see Definition 18.4). Moreover, w satisfies

$$w(x) = L_\alpha f(x,\widehat{u}_\phi(w,g)(x)) + H_\alpha r(x,\widehat{u}_\phi(w,g)(x)) + G_\alpha g(x,\widehat{u}_\phi(w,g)(x)). \tag{18.9}$$

Proof. From Theorem 6.4 in [3], we obtain the first equality and that $\widehat{u}_\phi(w,g)$ satisfies equation (18.9). Now applying Proposition 3.3 in [3] and by using the fact that $f \in \mathbb{M}(\overline{E} \times \mathbb{U})^+$ and $r \in \mathbb{M}(\partial E \times \mathbb{U})^+$, we obtain that $\mathcal{W}g \in \mathbb{M}(E)^+$ □

Define the sequence of functions $(v_m)_{m\in\mathbb{N}}$ as $v_{m+1} = \mathcal{W}v_m$, $v_0 = 0$. We have the following proposition.

Proposition 18.4. For all $x \in E$ and $m \in \mathbb{N}$ we have that $v_m(x) = \inf_{U\in\mathcal{U}} \mathcal{D}_m^\alpha(U,x)$.

Proof. This follows from the same lines as the proof of the Propositions 4.3 and 4.4 and in [3]. □

Since $v_m(x) = \inf_{U \in \mathcal{U}} \mathcal{D}_m^\alpha(U,x) \leq \mathcal{J}_{\mathcal{D}}^\alpha(x)$, the functions $v_m \in \mathbb{M}(E)^+$ and are non-decreasing. Consequently, there exists $v \in \mathbb{M}(E)^+$ such that $v_m \uparrow v$, and it follows that $v \leq \mathcal{J}_{\mathcal{D}}^\alpha$. We need the following propositions:

Proposition 18.5. If $h \in \mathbb{M}(E)^+$ is such that $h(x) \geq \mathcal{W}h(x)$ then $h(x) \geq \mathcal{J}_{\mathcal{D}}^\alpha(x)$.

Proof. By using Theorem 6.4 in [3] with $\rho = 0$, we obtain that there exists $\widehat{u}_\phi \in \mathcal{S}_{\mathbb{V}}$ such that

$$h(x) \geq \mathcal{T}_\alpha(0,g)(x) = L_\alpha f(x,\widehat{u}_\phi(x)) + H_\alpha r(x,\widehat{u}_\phi(x)) + G_\alpha h(x,\widehat{u}_\phi(x)).$$

Define $w(x) = L_\alpha f(x,\widehat{u}_\phi(x)) + H_\alpha r(x,\widehat{u}_\phi(x)) + G_\alpha h(x,\widehat{u}_\phi(x))$. Clearly $w(x) \geq 0$. Moreover, the hypotheses of Proposition 4.4 in [3] are satisfied with $\rho = 0$. Consequently, it follows that there exists $\widehat{U} \in \mathcal{U}$ such that for all $m \in \mathbb{N}$, $(t,x,k) \in \mathbb{R}_+ \times E \times \mathbb{N}$, $E_{(x,k)}^{\widehat{U}}\Big[\mathbf{J}^\alpha(\widehat{U}, t \wedge T_m)\Big] \leq h(x)$. From assumption 18.3 (which implies that $T_m \to \infty$ $P^{\widehat{U}}$ a.s.), and taking the limit as $t \to \infty$ we obtain that $h(x) \geq \mathcal{J}_{\mathcal{D}}^\alpha(x)$. □

Proposition 18.6. We have that $v(x) = \mathcal{W}v(x)$.

Proof. Let us show first that $v(x) \leq \mathcal{W}v(x)$. By using the definition of $\mathcal{W}$ we have for any $\Upsilon \in \mathbb{V}^r(x)$ that

$$v_{m+1}(x) \leq L_\alpha f(x,\Upsilon) + H_\alpha r(x,\Upsilon) + G_\alpha v_m(x,\Upsilon).$$

Taking the limit as $m \uparrow \infty$ and from the monotone convergence theorem we obtain that

$$\begin{aligned} v(x) \leq & L_\alpha f(x,\Upsilon) + H_\alpha r(x,\Upsilon) + \lim_{m\to\infty} G_\alpha v_m(x,\Upsilon) \\ & = L_\alpha f(x,\Upsilon) + H_\alpha r(x,\Upsilon) + G_\alpha v(x,\Upsilon) \end{aligned}$$

showing that $v(x) \leq \mathcal{W}v(x)$. From Proposition 18.3, there exists for any $m \in \mathbb{N}$, $u_\phi^m \in \mathcal{S}_{\mathbb{V}}$ such that

$$\mathcal{W}v_m(x) = L_\alpha f(x,u_\phi^m(x)) + H_\alpha r(x,u_\phi^m(x)) + G_\alpha v_m(x,u_\phi^m(x)). \tag{18.10}$$

Fix $x \in E$. Since $u_\phi^m(x) \in \mathbb{V}(x) \subset \mathbb{V}^r(x)$ and $\mathbb{V}^r(x)$ is compact we can find a further subsequence, still written as $u_\phi^m(x)$ for notational simplicity, such that $u_\phi^m(x) \to \hat{\Theta} \in \mathbb{V}^r(x)$. From Proposition 5.7 in [3],

$$\begin{aligned} v(x) &= \lim_{m\to\infty} v_{m+1}(x) \\ &= \varliminf_{m\to\infty} \Big\{L_\alpha f(x,u_\phi^m(x)) + H_\alpha r(x,u_\phi^m(x)) + G_\alpha v_m(x,u_\phi^m(x))\Big\}, \end{aligned}$$

and so

$$v(x) \geq L_\alpha f(x, \hat{\Theta}) + H_\alpha r(x, \hat{\Theta}) + G_\alpha v(x, \hat{\Theta}) \geq \mathcal{R}_\alpha(0, g)(x) = \mathcal{W}v(x), \tag{18.11}$$

giving the result. □

Finally we have the following theorem characterizing the optimality equation for the discounted optimal control problem and showing the convergence of the truncated problems.

Theorem 18.1. *We have that* $v_n \uparrow \mathcal{J}_{\mathcal{D}}^{\alpha}$ *and* $\mathcal{J}_{\mathcal{D}}^{\alpha}(x) = \mathcal{W}\mathcal{J}_{\mathcal{D}}^{\alpha}(x)$.

Proof. All we need to show is that $\mathcal{J}_{\mathcal{D}}^{\alpha}(x) \leq v(x)$. But this is immediate from Propositions 18.6 and 18.5. □

18.5. Main results

We present in this section the main results of the paper, supposing that the assumptions on the parameters of the PDMP presented in section 18.3 hold. Theorem 18.2 shows that for a fixed state x_0 we have that $\alpha\mathcal{J}_{\mathcal{D}}^{\alpha}(x_0)$ is bounded in a neighborhood of $\alpha = 0$. Theorem 18.3 states that the relative difference $h_\alpha(x) = \mathcal{J}_{\mathcal{D}}^{\alpha}(x) - \mathcal{J}_{\mathcal{D}}^{\alpha}(x_0)$ of the α-discount value function $\mathcal{J}_{\mathcal{D}}^{\alpha}$ belongs to $\mathbb{B}_g(E)$. As a consequence of Theorems 18.2 and 18.3, it is shown in Theorem 18.4 that there exists a pair (ρ, h) satisfying the ACOI,

$$h \geq \mathcal{T}(\rho, h), \tag{18.12}$$

where $\rho \in \mathbb{R}_+$ and $h \in \mathbb{B}_g(E)$. Note that in the present case, we only have that $h(x)$ is bounded below by $-Cg(x)$ for some $C > 0$ and, consequently, the approach developed previously by the authors in [3] cannot be used. The idea to overcome this difficulty is to show in Lemma 18.3 that under the assumptions presented in section 18.3 one can get that for $\widehat{u} \subset \mathcal{S}_{\mathbb{U}}$,

$$\overline{\lim_{t \to +\infty}} \frac{1}{t} \overline{\lim_{m \to \infty}} E_{(x,0)}^{U_{\widehat{u}_\phi}} \Big[\mathcal{T}_{\alpha}(\rho, h)\big(X(t \wedge T_m)\big) \Big] \geq 0$$

This technical result will lead in Theorem 18.5 to the main result of the paper, which is the existence of an optimal ordinary feedback control strategy for the long run average-cost problem of a PDMP, derived from the pair (ρ, h) satisfying the ACOI (18.12). For the sake of clarity in the exposition, the proofs of some intermediate results and of Theorems 18.2 and

18.3 are presented in the appendix. The following theorem states that for any discount factor α and for any state $x \in E$, $\alpha \mathcal{J}_D^{\alpha}(\cdot)$ is bounded.

Theorem 18.2. *For any $\alpha > 0$ and $x \in E$,*

$$\mathcal{J}_D^{\alpha}(x) \leq \frac{M}{c+\alpha} g(x) + \frac{Mb}{c\alpha}. \tag{18.13}$$

Proof. The proof of this result can be found in the appendix, see section A.1. □

It is shown next that for any state y fixed in E, the difference $\mathcal{J}_D^{\alpha}(\cdot)$-$\mathcal{J}_D^{\alpha}(y)$ belongs to $\mathbb{B}_g(E)$.

Theorem 18.3. *For any $\alpha > 0$ and $x \in E$, $y \in E$,*

$$\left| \mathcal{J}_D^{\alpha}(x) - \mathcal{J}_D^{\alpha}(y) \right| \leq \frac{aM'}{1-\kappa}(1+g(y))g(x). \tag{18.14}$$

Proof. The proof of this result can be found in the appendix, see section A.2. □

By combining the two previous theorems we show next that there exists a pair (ρ, h) in $\mathbb{R}_+ \times \mathbb{B}_g(E)$ satisfying (18.12). In order to show this we also need Theorem 18.7 (proved in the appendix, section A.3), which states that for any function $h \in \mathbb{B}_g(E)$, the one-stage optimization operators $\mathcal{R}_\alpha(\rho, h)(x)$ and $\mathcal{T}_\alpha(\rho, h)(x)$ are equal, and that there exists an ordinary feedback measurable selector for the one-stage optimization problems.

Theorem 18.4. *Set $\rho_\alpha = \alpha \mathcal{J}_D^{\alpha}(x_0)$ and $h_\alpha(\cdot) = \mathcal{J}_D^{\alpha}(\cdot) - \mathcal{J}_D^{\alpha}(x_0)$ for a fixed state $x_0 \in E$. Then,*
i) there exists a decreasing sequence of positive numbers $\alpha_k \downarrow 0$ such that $\rho_{\alpha_k} \to \rho$ and for all $x \in E$, $\lim_{k\to\infty} \alpha_k \mathcal{J}_D^{\alpha_k}(x) = \rho$.
ii) h defined by $h = \varliminf_{k\to\infty} h_{\alpha_k}$ belongs to $\mathbb{B}_g(E)$ and satisfies for any $x \in E$

$$\begin{aligned} h(x) &\geq \mathcal{T}(\rho, h)(x) \\ &= -\rho \mathcal{L}(x, \widehat{u}_\phi(x)) + Lf(x, \widehat{u}_\phi(x)) + Hr(x, \widehat{u}_\phi(x)) + Gh(x, \widehat{u}_\phi(x)), \end{aligned}$$

where $\widehat{u} = \widehat{u}(\mathcal{T}(\rho, h), h)$ and $\widehat{u}_\phi = \widehat{u}_\phi(\mathcal{T}(\rho, h), h)$ (see Definition 18.4).

Proof. Item *i)* is a straightforward consequence of Theorem 18.2.
Now, from Theorem 18.3, one obtains that there exists a constant C independent of k and x such that $|h_{\alpha_k}(x)| \leq Cg(x)$. This implies that $h = \varliminf_{k\to\infty} h_{\alpha_k}$ belongs to $\mathbb{B}_g(E)$. Following the same arguments as in the

proof of Proposition 8.4 in [3], it can be shown that $h(x) \geq \mathcal{R}(\rho, h)(x)$. Applying Theorem 18.7 it follows that

$$\begin{aligned} h(x) &\geq \mathcal{T}(\rho, h)(x) \\ &= -\rho\mathcal{L}(x, \widehat{u}_\phi(x)) + Lf(x, \widehat{u}_\phi(x)) + Hr(x, \widehat{u}_\phi(x)) + Gh(x, \widehat{u}_\phi(x)), \end{aligned}$$

where $\widehat{u}$ is defined by $\widehat{u}(\mathcal{T}(\rho, h), h)$. This shows that item *ii*) holds, and thereby gives the result. □

We wish to obtain now from $\widehat{u}_\phi$ as defined in Theorem 18.4 an optimal control strategy for the long run average continuous control problem. First we need some technical results, stated in the next two lemmas.

Lemma 18.2. *Consider an arbitrary $u \in \mathcal{S}_{\mathbb{U}}$ and let u_ϕ and U_{u_ϕ} be as in Definitions 18.2 and 18.3 respectively. For all $x \in E$ define $\widehat{g}(x) = -b\mathcal{L}_{-c}(x, u_\phi(x)) + G_{-c}g(x, u_\phi(x))$. Then $\widehat{g} \in \mathbb{B}_g(E)$ and U_{u_ϕ} satisfies*

$$\begin{aligned} E^{U_{u_\phi}}_{(x,0)}\Big[\widehat{g}(X(t \wedge T_m))\Big] \leq & e^{-ct}g(x) + \frac{b}{c}[1 - e^{-ct}] + u\|\widehat{g}\|_g g(x)\kappa^m \\ & + \|\widehat{g}\|_g \nu_u(g) + bK_\lambda. \end{aligned} \quad (18.15)$$

Proof. From Corollary 18.1 with $\alpha = -c$ and recalling that $\overline{r}(z) \geq 0$ we obtain that $-b\mathcal{L}_{-c}(x, u_\phi(x)) + G_{-c}g(x, u_\phi(x)) \leq g(x)$. Clearly, $\widehat{g} \in \mathbb{M}(E)$ is bounded from below by $-bK_\lambda$ from Assumption 18.11 b) and thus $\widehat{g} \in \mathbb{B}_g(E)$. Since $\widehat{g} \in \mathbb{M}(E)$ is bounded from below, it is easy to show that

$$bE^{U_{u_\phi}}_{(x,0)}\Big[\int_0^{t\wedge T_m} e^{cs} ds\Big] + E^{U_{u_\phi}}_{(x,0)}\Big[e^{c(t\wedge T_m)}\widehat{g}(X(t \wedge T_m))\Big] \leq g(x),$$

by using the same arguments as in the proof of Proposition 4.4 in [3]. Combining Fatou's Lemma and Assumption 18.3 we obtain that

$$E^{U_{u_\phi}}_{(x,0)}\Big[\widehat{g}(X(t))\Big] \leq e^{-ct}g(x) + \frac{b}{c}[1 - e^{-ct}]. \quad (18.16)$$

Clearly, we have

$$E^{U_{u_\phi}}_{(x,0)}\Big[\widehat{g}(X(t\wedge T_m))\Big] = E^{U_{u_\phi}}_{(x,0)}\Big[I_{\{t<T_m\}}\widehat{g}(X(t))\Big] + E^{U_{u_\phi}}_{(x,0)}\Big[I_{\{t\geq T_m\}}\widehat{g}(X(T_m))\Big].$$

Consequently, we see

$$E^{U_{u_\phi}}_{(x,0)}\Big[\widehat{g}(X(t \wedge T_m))\Big] \leq E^{U_{u_\phi}}_{(x,0)}\Big[\widehat{g}(X(t))\Big] + G^m\widehat{g}(x, u_\phi(x)) + bK_\lambda,$$

by recalling that $\widehat{g}$ is bounded from below by $-bK_\lambda$. The result follows by using Assumption 18.7 and equation (18.16). □

As a consequence of Lemma 18.2, we get the next result:

Lemma 18.3. *For all* $x \in E$, $E_{(x,0)}^{U_{\widehat{u}_\phi}}\Big[\mathcal{T}(\rho,h)\big(X(t\wedge T_m)\big)\Big]$ *is well defined and satisfies*

$$\overline{\lim_{t\to+\infty}}\,\frac{1}{t}\,\overline{\lim_{m\to\infty}}\,E_{(x,0)}^{U_{\widehat{u}_\phi}}\Big[\mathcal{T}(\rho,h)\big(X(t\wedge T_m)\big)\Big]\geq 0. \tag{18.17}$$

Proof. By definition, we have that $\mathcal{T}(\rho,h)(x)\geq -\rho\mathcal{L}(x,\widehat{u}_\phi(x))+Gh(x,\widehat{u}_\phi(x))$. Therefore, using the definition of $\widehat{g}$ in Lemma 18.2 with $u=\widehat{u}$ we obtain that

$$\mathcal{T}(\rho,h)(x)\geq -(\rho+b\|h\|_g)K_\lambda-\|h\|_g\widehat{g}(x). \tag{18.18}$$

Consequently, combining Lemma 18.2 and equation (18.18) we obtain the result. □

Finally we have the main result of this paper, which shows that the ordinary feedback control $U_{\widehat{u}_\phi}$ (obtained from the ACOI (18.12) and $\widehat{u}_\phi$ as in Theorem 18.4) is an optimal strategy for the long run average-cost problem of a PDMP.

Theorem 18.5. *For all* $x\in E$,

$$\rho=\mathcal{J}_{\mathcal{A}}(x)=\mathcal{A}(U_{\widehat{u}_\phi},x).$$

Proof. Define

$$\begin{aligned}J_m^{U_{\widehat{u}_\phi}}(t,x)=&E_{(x,0)}^{U_{\widehat{u}_\phi}}\Bigg[\int_0^{t\wedge T_m}\Big[f\big(X(s),\hat{u}(X(s))\big)-\rho\Big]ds\\&+\int_0^{t\wedge T_m}r\big(X(s-),\hat{u}_\partial(X(s-))\big)dp^*(s)+\mathcal{T}(\rho,h)\big(X(t\wedge T_m)\big)\Bigg].\end{aligned}$$

From Lemma 18.3 we have that $E_{(x,0)}^{U_{\widehat{u}_\phi}}\Big[\mathcal{T}(\rho,h)\big(X(t\wedge T_m)\big)\Big]$ is well defined. Consequently, by using Theorem 18.4, we can show that $J_m^{U_{\widehat{u}_\phi}}(t,x)\leq h(x)$ for all $m\in\mathbb{N}$, $(t,x)\in\mathbb{R}_+\times E$. Therefore,

$$\begin{aligned}&E_{(x,0)}^{U_{\widehat{u}_\phi}}\Bigg[\int_0^{t\wedge T_m}\Big[f\big(X(s),\hat{u}(X(s))\big)\Big]ds+\int_0^{t\wedge T_m}r\big(X(s-),\hat{u}_\partial(X(s-))\big)dp^*(s)\Bigg]\\&\quad+E_{(x,0)}^{U_{\widehat{u}_\phi}}\Big[\mathcal{T}(\rho,h)\big(X(t\wedge T_m)\big)\Big]\leq\rho\, t+h(x).\end{aligned}$$

Now by using Lemma 18.3, it follows that

$$\varlimsup_{t\to+\infty}\frac{1}{t}\varlimsup_{m\to\infty}E^{U_{\widehat{u}_\phi}}_{(x,0)}\Big[\int_0^{t\wedge T_m}\Big[f\big(X(s),\hat{u}(X(s))\big)\Big]ds + \int_0^{t\wedge T_m} r\big(X(s-),\hat{u}_\partial(X(s-))\big)dp^*(s)\Big] \leq \rho\varlimsup_{t\to+\infty}\frac{1}{t}\varlimsup_{m\to\infty}\big[t\wedge T_m\big].$$

Combining Assumption 18.3 and the monotone convergence theorem, we have

$$\varlimsup_{t\to\infty}\frac{1}{t}E^{U_{\widehat{u}_\phi}}_{(x,0)}\Big[\int_0^t\Big[f\big(X(s),\hat{u}(X(s))\big)\Big]ds + \int_0^t r\big(X(s-),\hat{u}_\partial(X(s-))\big)dp^*(s)\Big] \leq \rho$$

showing that $\mathcal{J}_{\mathcal{A}}(x) \leq \mathcal{A}(U_{\widehat{u}_\phi},x) \leq \rho$. From Theorem 1 in [22, chapter 5] we easily obtain the reverse inequality, completing the proof. □

Appendix A. Proofs of the main results

The proofs of the main results of the paper are presented in this appendix.

A.1. Proof of Theorem 18.2

The next two propositions will be crucial for the boundedness results on $\mathcal{J}^{\alpha}_{D}(\cdot)$ to be developed in the sequel. They establish a connection between a general integro-differential inequality (respectively equality) related to the local characteristics of the PDMP and an inequality (respectively equality) related to the operators G_α, L_α and H_α.

Proposition 18.7. Suppose that there exist $v \in \mathbb{M}^{ac}(E,\mathbb{R}_+)$, $\ell \in \mathbb{M}(E)^+$, $k \in \mathbb{M}(E)^+$, $p \in \mathbb{M}(\partial E)^+$, $\Theta = (\mu,\mu_\partial) \subset \mathcal{S}_{\mathbb{V}^r}$, $d \geq 0$, and $\alpha \geq -c$ satisfying

$$\mathcal{X}v(\phi(x,t)) - \big[\alpha + \lambda(\phi(x,t),\mu(x,t))\big]v(\phi(x,t)) + \ell(\phi(x,t)) + \lambda(\phi(x,t),\mu(x,t))Qk(\phi(x,t),\mu(x,t)) \leq d, \tag{A.1}$$

for all $x \in E$, $t \in [0,t_*(x))$ and

$$v(\phi(x,t_*(x))) \geq p(\phi(x,t_*(x))) + Qk(\phi(x,t_*(x)),\mu_\partial(\phi(x,t_*(x)))), \tag{A.2}$$

for all $x \in E$ with $t_*(x) < \infty$.
Then

$$v(x) \geq -d\mathcal{L}_\alpha(x, \Theta(x)) + L_\alpha \ell(x, \Theta(x)) + H_\alpha p(x, \Theta(x)) + G_\alpha k(x, \Theta(x)). \tag{A.3}$$

Proof. Multiplying both sides of equation (A.1) by $e^{-\alpha t - \Lambda^{\mu(x)}(x,t)}$ and integrating over $[0, s]$ for $s \in [0, t_*(x))$ we get that

$$\begin{aligned} d \int_0^s e^{-\alpha t - \Lambda^{\mu(x)}(x,t)} dt \geq\ & e^{-\alpha s - \Lambda^{\mu(x)}(x,s)} v(\phi(x, s)) - v(x) \\ & + \int_0^s e^{-\alpha t - \Lambda^{\mu(x)}(x,t)} \big[\ell(\phi(x,t)) + \lambda(\phi(x,t), \mu(x,t)) Qk(\phi(x,t), \mu(x,t))\big]\, dt. \end{aligned} \tag{A.4}$$

Consider the case in which $t_*(x) < \infty$. By using the fact that $v \in \mathbb{M}^{ac}(E)$, we obtain from Remark 18.3 *ii*) and equation (A.4) that

$$\begin{aligned} v(x) \geq & - d\mathcal{L}_\alpha(x, \Theta(x)) + L_\alpha \ell(x, \Theta(x)) \\ & + e^{-\alpha t_*(x) - \Lambda^{\mu(x)}(x, t_*(x))} v(\phi(x, t_*(x))) \\ & + \int_0^{t_*(x)} e^{-\alpha t - \Lambda^{\mu(x)}(x,t)} \lambda(\phi(x,t), \mu(x,t)) Qk(\phi(x,t), \mu(x,t)) dt. \end{aligned} \tag{A.5}$$

However, from equation (A.2), it follows that

$$v(x) \geq - d\mathcal{L}_\alpha(x, \Theta(x)) + L_\alpha \ell(x, \Theta(x)) + H_\alpha p(x, \Theta(x)) + G_\alpha k(x, \Theta(x)).$$

Now consider the case in which $t_*(x) = +\infty$. From equation (A.4) (and recalling that v is positive), we have that

$$\begin{aligned} d \int_0^s e^{-\alpha t - \Lambda^{\mu(x)}(x,t)} dt \geq & - v(x) + \int_0^s e^{-\alpha t - \Lambda^{\mu(x)}(x,t)} \big[\ell(\phi(x,t)) \\ & + \lambda(\phi(x,t), \mu(x,t)) Qk(\phi(x,t), \mu(x,t))\big]\, dt, \end{aligned}$$

and so, by taking the limit as s tends to infinity in the previous equation, it yields

$$\begin{aligned} v(x) \geq & -d\mathcal{L}_\alpha(x, \Theta(x)) + L_\alpha \ell(x, \Theta(x)) \\ & + \int_0^{t_*(x)} e^{-\alpha t - \Lambda^{\mu(x)}(x,t)} \lambda(\phi(x,t), \mu(x,t)) Qk(\phi(x,t), \mu(x,t)) dt. \end{aligned}$$

However, by using the fact that $t_*(x) = +\infty$ and Remark 18.3 *i*), we have that $H_\alpha p(x,\Theta(x)) = 0$ and $G_\alpha k(x,\Theta(x)) = \int_0^{t_*(x)} e^{-\alpha t - \Lambda^{\mu(x)}(x,t)} \lambda(\phi(x,t),$ $\mu(x,t))Qk(\phi(x,t),\mu(x,t))dt$, showing the result. □

Notice that if the inequalities in (A.1) and (A.2) are replaced by equalities then the hypotheses of Proposition 18.7 must be restricted to $\alpha \geq 0$ to show that the inequality in (A.3) becomes an equality. More specifically, we have the following result:

Proposition 18.8. Suppose that there exist $v \in \mathbb{M}_w^{ac}(E,\mathbb{R}_+)$, $\ell \in \mathbb{M}(E)^+$, $k \in \mathbb{M}(E)^+$, $p \in \mathbb{M}(\partial E)^+$, $\Theta = (\mu, \mu_\partial) \in \mathcal{S}_{\mathbb{V}^r}$, $d \geq 0$, and $\alpha \geq 0$ satisfying

$$\begin{aligned}\mathcal{X}v(\phi(x,t)) - &[\alpha + \lambda(\phi(x,t),\mu(x,t))]\, v(\phi(x,t)) + \ell(\phi(x,t)) \\ &+ \lambda(\phi(x,t),\mu(x,t))Qk(\phi(x,t),\mu(x,t)) = d, \qquad \text{(A.6)}\end{aligned}$$

for all $x \in E$, $t \in [0, t_*(x))$ and

$$v(\phi(x,t_*(x))) = p(\phi(x,t_*(x))) + Qk(\phi(x,t_*(x)), \mu_\partial(\phi(x,t_*(x)))), \qquad \text{(A.7)}$$

for all $x \in E$ with $t_*(x) < \infty$.

Then

$$\begin{aligned}v(x) = &-d\mathcal{L}_\alpha(x,\Theta(x)) + L_\alpha \ell(x,\Theta(x)) + H_\alpha p(x,\Theta(x)) \\ &+ G_\alpha k(x,\Theta(x)). \qquad \text{(A.8)}\end{aligned}$$

Proof. By following the same steps as in the first part of the proof of Proposition 18.7 we have that for all $s \in [0, t_*(x))$,

$$\begin{aligned}d\int_0^s e^{-\alpha t - \Lambda^{\mu(x)}(x,t)} dt = &\; e^{-\alpha s - \Lambda^{\mu(x)}(x,s)} v(\phi(x,s)) - v(x) \\ &+ \int_0^s e^{-\alpha t - \Lambda^{\mu(x)}(x,t)} \big[\ell(\phi(x,t)) \\ &+ \lambda(\phi(x,t),\mu(x,t))Qk(\phi(x,t),\mu(x,t))\big]\, dt \qquad \text{(A.9)}\end{aligned}$$

The case in which $t_*(x) < \infty$ can be treated in the same manner as in the proof of Proposition 18.7. However, the case in which $t_*(x) = +\infty$ is different. By using Assumption 18.11 d) and the fact that $0 \leq v \leq \|v\|_g g$, we have that for any $\alpha \geq 0$,

$$\begin{aligned}\lim_{s\to+\infty} e^{-\alpha s - \Lambda^{\mu(x)}(x,s)} v(\phi(x,s)) &\leq \|v\|_g \lim_{s\to+\infty} e^{-\int_0^{t_*(x)} \underline{\lambda}(\phi(x,t))dt} g(\phi(x,s)) \\ &= 0.\end{aligned}$$

Therefore, taking the limit as s tends to infinity in equation (A.9), we have that

$$d\mathcal{L}_\alpha(x,\Theta(x)) = -\, v(x) + L_\alpha \ell(x,\Theta(x)) + \int_0^s e^{-\alpha t - \Lambda^{\mu(x)}(x,t)} \lambda(\phi(x,t),\mu(x,t)) Qk(\phi(x,t),\mu(x,t)) dt,$$

and this shows equation (A.8) by using Remark 18.3 *i*). □

We obtain the following corollary by applying Proposition 18.7 to the inequalities (18.4) and (18.6):

Corollary 18.1. *For any $u \in \mathcal{S}_{\mathbb{U}}$, $\alpha \geq -c$, and $x \in E$*

$$g(x) \geq -b\mathcal{L}_\alpha(x,u_\phi(x)) + (c+\alpha)L_\alpha g(x,u_\phi(x)) + H_\alpha \overline{r}(x,u_\phi(x)) + G_\alpha g(x,u_\phi(x)), \tag{A.10}$$

and for all $\Theta \in \mathcal{S}_{\mathbb{V}^r}$

$$(c+\alpha)L_\alpha g(x,\Theta(x)) + H_\alpha \overline{r}(x,\Theta(x)) + G_\alpha g(x,\Theta(x)) \leq bK_\lambda + g(x). \tag{A.11}$$

Proof. Clearly from Proposition 18.2 and Remark 18.1, it follows that $u_\phi \in \mathcal{S}_{\mathbb{V}^r}$. Consequently, setting $d = b$, $v = g$, $\ell = (c+\alpha)g$, $p = \overline{r}$, $k = g$, and $\Theta = u_\phi$ in Proposition 18.7 we get equation (A.10). Similarly, from Remark 18.3 *ii*), the inequality (A.11) is a straightforward consequence of the inequality (A.3). □

The next theorem provides bounds for a sequence of functions defined by a general recursive equation and for the functions Lf, Hr and Lg, in function of α and g.

Theorem 18.6. *Define the sequence $(q_m(x))_{m\in\mathbb{N}}$ by*

$$q_0^\alpha(x) = 0,$$
$$q_{m+1}^\alpha(x) = L_\alpha f(x,u_\phi^{m+1}(x)) + H_\alpha r(x,u_\phi^{m+1}(x)) + G_\alpha q_m^\alpha(x,u_\phi^{m+1}(x)), \tag{A.12}$$

where $x \in E$, $(u^m)_{m\in\mathbb{N}} \in \mathcal{S}_{\mathbb{U}}$ and $\alpha > 0$.

Then the following assertions hold:

i) for any $x \in E$, $m \in \mathbb{N}$ and $\alpha \in [0,\delta)$, we have that

$$q_m^\alpha(x) \leq \frac{M}{c+\alpha} g(x) + \frac{Mb}{c\alpha}. \tag{A.13}$$

ii) for any $x \in E$, $u \in \mathcal{S}_{\mathbb{U}}$,

$$0 \leq Lf(x, u_\phi(x)) + Hr(x, u_\phi(x)) \leq \frac{M(1 + bK_\lambda)}{c} g(x), \tag{A.14}$$

$$0 \leq Lg(x, u_\phi(x)) \leq \frac{(1 + bK_\lambda)}{c} g(x). \tag{A.15}$$

Proof. Let us show (A.13) by induction. For $m = 0$ it is immediate since $q_0^\alpha = 0$. Suppose it holds for m. Combining (A.12) and (A.13) we have

$$\begin{aligned} q_{m+1}^\alpha(x) &\leq L_\alpha f(x, u_\phi^m(x)) + H_\alpha r(x, u_\phi^m(x)) \\ &\quad + \frac{M}{c + \alpha} G_\alpha g(x, u_\phi^m(x)) + \frac{Mb}{c\alpha} G_\alpha 1(x, u_\phi^m(x)). \end{aligned} \tag{A.16}$$

Moreover, from equations (A.10) and (A.11), we obtain that

$$\begin{aligned} G_\alpha g(x, u_\phi^m(x)) &\leq g(x) + b\mathcal{L}_\alpha(x, u_\phi^m(x)) - (c + \alpha) L_\alpha g(x, u_\phi^m(x)) \\ &\quad - H_\alpha \overline{r}(x, u_\phi^m(x)). \end{aligned} \tag{A.17}$$

Replacing (A.17) into (A.16) and using (18.5) and (18.7), we get

$$\begin{aligned} q_{m+1}^\alpha(x) &\leq L_\alpha (f - Mg)(x, u_\phi^m(x)) + H_\alpha \big(r - \frac{M}{c + \alpha} \overline{r}\big)(x, u_\phi^m(x)) \\ &\quad + \frac{M}{c + \alpha} g(x) + Mb\Big(\frac{1}{c\alpha} G_\alpha 1(x, u_\phi^m(x)) + \frac{1}{c + \alpha} \mathcal{L}_\alpha(x, u_\phi^m(x))\Big) \\ &\leq \frac{M}{c + \alpha} g(x) + \frac{Mb}{c\alpha}\Big(G_\alpha 1(x, u_\phi^m(x)) + \alpha \mathcal{L}_\alpha(x, u_\phi^m(x))\Big) \\ &\leq \frac{M}{c + \alpha} g(x) + \frac{Mb}{c\alpha} \end{aligned} \tag{A.18}$$

since that $G_\alpha 1(x, u_\phi^m(x)) + \alpha \mathcal{L}_\alpha(x, u_\phi^m(x)) = 1$.

Let us show now (A.14) and (A.15). For $\alpha = 0$ it follows from Remark 18.3 *ii)* and equation (A.10)

$$\begin{aligned} g(x) + bK_\lambda &\geq g(x) + b\mathcal{L}(x, u_\phi(x)) \\ &\geq cLg(x, u_\phi(x)) + H\overline{r}(x, u_\phi(x)) + Gg(x, u_\phi(x)), \end{aligned} \tag{A.19}$$

showing equation (A.15) since $g \geq 1$ and $\overline{r} \geq 0$. Now, combining equations (18.5), (18.7) and (A.19) we get (A.14), which yields the final part of the result. □

Proof. [Proof of Theorem 18.2] By using Proposition 18.3 and Theorem 18.1, it can be shown that there exists $u_\phi^m \in \mathcal{S}_{\mathbb{V}}$ such that the sequence $\big(v_m^\alpha(x)\big)_{m \in \mathbb{N}}$ defined by $v_{m+1}^\alpha(x) = L_\alpha f(x, u_\phi^m(x)) + H_\alpha r(x, u_\phi^m(x)) + G_\alpha v_m^\alpha(x, u_\phi^m(x))$ and $v_0^\alpha(x) = 0$ satisfies $v_{m+1}^\alpha \uparrow \mathcal{J}_D^\alpha(x)$ as $m \uparrow \infty$. Therefore, considering $q_m^\alpha = v_m^\alpha$ in Theorem 18.6 and taking the limit as $m \uparrow \infty$ we get (18.13). □

A.2. Proof of Theorem 18.3

The next lemma shows that $\mathcal{J}_D^\alpha(x)$ can be written as an infinite sum of iterates of the stochastic kernel G_α. Using this result $\mathcal{J}_D^\alpha(x)$ can be characterized in terms of the Markov kernel G in Proposition 18.9. From classical hypotheses on G, such as the geometric ergodic condition in Assumption 18.7, it will be shown in Theorem 18.3 that the mapping defined by $\mathcal{J}_D^\alpha(\cdot)$-$\mathcal{J}_D^\alpha(y)$ for y fixed in E belongs to $\mathbb{B}_g(E)$.

Lemma 18.4. *For each $\alpha > 0$ there exists $u^\alpha \in \mathcal{S}_{\mathbb{U}}$ such that*

$$\mathcal{J}_D^\alpha(x) = \sum_{k=0}^{\infty} G_\alpha^k (L_\alpha f + H_\alpha r)(x, u_\phi^\alpha(x)). \tag{A.20}$$

Proof. From Theorem 18.1, $\mathcal{J}_D^\alpha \in \mathbb{M}(\mathbb{E})$ and $\mathcal{J}_D^\alpha(x) = \mathcal{R}_\alpha(0, \mathcal{J}_D^\alpha)(x)$. Moreover, from Theorem 6.4 in [3], there exists $u^\alpha \in \mathcal{S}_{\mathbb{U}}$ such that the ordinary feedback measurable selector $u_\phi^\alpha \in \mathcal{S}_{\mathbb{V}}$ satisfies

$$\begin{aligned} \mathcal{J}_D^\alpha(x) &= \mathcal{R}_\alpha(0, \mathcal{J}_D^\alpha)(x) = \mathcal{T}_\alpha(0, \mathcal{J}_D^\alpha)(x) \\ &= L_\alpha f(x, u_\phi^\alpha)(x) + H_\alpha r(x, u_\phi^\alpha) + G_\alpha \mathcal{J}_D^\alpha(x, u_\phi^\alpha). \end{aligned} \tag{A.21}$$

Iterating (A.21) and recalling that $\mathcal{J}_D^\alpha(y) \geq 0$ for every y, yields for every $m \in \mathbb{N}$ that,

$$\begin{aligned} \mathcal{J}_D^\alpha(x) &= \sum_{k=0}^{m-1} G_\alpha^k (L_\alpha f + H_\alpha r)(x, u_\phi^\alpha(x)) + G_\alpha^m \mathcal{J}_D^\alpha(x, u_\phi^\alpha(x)) \\ &\geq \sum_{k=0}^{m-1} G_\alpha^k (L_\alpha f + H_\alpha r)(x, u_\phi^\alpha(x)). \end{aligned} \tag{A.22}$$

For the control $U_{u_\phi^\alpha} \in \mathcal{U}$ (see Definition 18.3), it is easy to show that

$$\begin{aligned} &\sum_{k=0}^{m-1} G_\alpha^k (L_\alpha f + H_\alpha r)(x, u_\phi^\alpha(x)) \\ &\quad = E_{(x,0)}^{U_{u_\phi^\alpha}} \bigg[\int_0^{T_m} e^{-\alpha s} f\big(X(s), u(N(s), Z(s), \tau(s))\big) ds \\ &\qquad + \int_0^{T_m} e^{-\alpha s} r\big(X(s-), u_\partial(N(s-), Z(s-))\big) dp^*(s) \bigg], \end{aligned} \tag{A.23}$$

where $U_{u_\phi^\alpha} = \big(u, u_\partial\big)$ From Assumption 18.3, $T_m \to \infty$, $P^{U_{u_\phi^\alpha}}$ a.s. Therefore from the monotone convergence theorem, equation (A.23) implies that

$\sum_{k=0}^{\infty} G_\alpha^k (L_\alpha f + H_\alpha r)(x, u_\phi^\alpha(x)) = \mathcal{D}^\alpha(U_{u_\phi^\alpha}, x)$, and from equation (A.22)

$$\mathcal{J}_D^\alpha(x) \geq \sum_{k=0}^{\infty} G_\alpha^k (L_\alpha f + H_\alpha r)(x, u_\phi^\alpha(x)) = \mathcal{D}^\alpha(U_{u_\phi^\alpha}, x). \tag{A.24}$$

But since $U_{u_\phi^\alpha} \in \mathcal{U}$ and $\mathcal{J}_D^\alpha(x) = \inf_{U \in \mathcal{U}} \mathcal{D}^\alpha(U, x)$ it is clear that $\mathcal{D}^\alpha(U_{u_\phi^\alpha}, x) \geq \mathcal{J}_D^\alpha(x)$, so that (A.24) yields (A.21). □

The next proposition gives a characterization of $\mathcal{J}_D^\alpha(x)$ in terms of G.

Proposition 18.9. For $\alpha > 0$ and u_ϕ^α as in Lemma 18.4, define the sequence $\left(s_m^\alpha(x)\right)_{m\in\mathbb{N}}$ for $x \in E$ by $s_0^\alpha(x) = 0$ and $s_{m+1}^\alpha(x) = L_\alpha f(x, u_\phi^\alpha(x)) + H_\alpha r(x, u_\phi^\alpha(x)) + G_\alpha s_m^\alpha(x, u_\phi^\alpha(x))$. Then

$$\mathcal{J}_D^\alpha(x) = \lim_{m\to\infty} \sum_{k=0}^{m} G^k(L(f - \alpha s_{m+1-k}^\alpha) + Hr)(x, u_\phi^\alpha(x)). \tag{A.25}$$

Proof. By definition for all $m \in \mathbb{N}$, $s_m^\alpha \in \mathbb{M}(E)$ and $s_{m+1}^\alpha(x) = \sum_{k=0}^{m} G_\alpha^k(L_\alpha f + H_\alpha r)(x, u_\phi^\alpha(x))$ and clearly from Lemma 18.4, we have that $s_m^\alpha \uparrow \mathcal{J}_D^\alpha$ as $m \uparrow \infty$. Applying Lemma A.2 in [3], it can be shown that $s_m^\alpha \in \mathbb{M}^{ac}(E)$ and for all $x \in E$, and $t \in [0, t_*(x))$,

$$\begin{aligned} s_{m+1}^\alpha(x) = \int_0^t & e^{-\alpha s - \int_0^s \lambda(\phi(x,\theta), u^\alpha(\phi(x,\theta)))d\theta} \Big[f(\phi(x,s), u^\alpha(\phi(x,s))) \\ & + \lambda(\phi(x,s), u^\alpha(\phi(x,s))) Q s_m^\alpha(\phi(x,s), u^\alpha(\phi(x,s))) \Big] ds \\ & + e^{-\alpha t - \int_0^t \lambda(\phi(x,s), u^\alpha(\phi(x,s)))ds} s_{m+1}^\alpha(\phi(x,t)), \end{aligned}$$

implying that

$$\begin{aligned} \mathcal{X} s_{m+1}^\alpha(x) - [\alpha + \lambda(x, u^\alpha(x))]\, s_{m+1}^\alpha(x) + f(x, u^\alpha(x)) & \\ + \lambda(x, u^\alpha(x)) Q s_m^\alpha(x, u^\alpha(x)) = 0. & \end{aligned} \tag{A.26}$$

Consider the case in which $t_*(x) < \infty$. Since $s_{m+1}^\alpha \in \mathbb{M}^{ac}(E)$, this yields that

$$\begin{aligned} & s_{m+1}^\alpha(x) \\ & = L_\alpha f(x, u_\phi^\alpha(x)) + e^{-\alpha t_*(x) - \int_0^{t_*(x)} \lambda(\phi(x,s), u^\alpha(\phi(x,s)))ds} s_{m+1}^\alpha(\phi(x, t_*(x))) \\ & \quad + \int_0^{t_*(x)} e^{-\alpha s - \int_0^s \lambda(\phi(x,\theta), u^\alpha(\phi(x,\theta)))d\theta} \lambda(\phi(x,s), u^\alpha(\phi(x,s))) \\ & \qquad \cdot Q s_m^\alpha(\phi(x,s), u^\alpha(\phi(x,s))) ds. \end{aligned} \tag{A.27}$$

From Assumption 18.8, we have that $e^{-\int_0^{t_*(x)} \lambda(\phi(x,s),u^\alpha(\phi(x,s)))ds} > 0$. Therefore, combining the definition of $s_m^\alpha(x)$ and equation (A.27), gives

$$s_{m+1}^\alpha(\phi(x,t_*(x))) = Qs_m^\alpha(\phi(x,t_*(x)),u(\phi(x,t_*(x)))) + r(\phi(x,t_*(x)),u(\phi(x,t_*(x)))). \quad \text{(A.28)}$$

Using Proposition 18.8 we get from (A.26), (A.28) that

$$s_{m+1}^\alpha(x) = L(f-\alpha s_{m+1}^\alpha)(x,u_\phi^\alpha(x)) + Hr(x,u_\phi^\alpha(x)) + Gs_m^\alpha(x,u_\phi^\alpha(x)). \quad \text{(A.29)}$$

Iterations of (A.29) over m yields (A.25). □

Before showing that the mapping defined by $\mathcal{J}_D^\alpha(\cdot)$-$\mathcal{J}_D^\alpha(y)$ for y fixed in E belongs to $\mathbb{B}_g(E)$, we need to prove that the mapping $L(f-\alpha s_{m+1}^\alpha)(.,u_\phi^\alpha(.)) + Hr(.,u_\phi^\alpha(.))$ belongs to $\mathbb{B}_g(E)$.

Lemma 18.5. *Define* $M' = \frac{M(1+\frac{b}{c})(1+bK_\lambda)}{c}$. *For* $\alpha > 0$, u_ϕ^α *as in Lemma 18.4,* s_m^α *as in Lemma 18.9, and* $x \in E$, *we have that*

$$\left|L(f-\alpha s_{m+1}^\alpha)(x,u_\phi^\alpha(x)) + Hr(x,u_\phi^\alpha(x))\right| \le M'g(x). \quad \text{(A.30)}$$

Proof. Notice that

$$-\alpha L s_{m+1}^\alpha(x,u_\phi^\alpha(x)) \le L(f-\alpha s_{m+1}^\alpha)(x,u_\phi^\alpha(x)) + Hr(x,u_\phi^\alpha(x)) \le Lf(x,u_\phi^\alpha(x)) + Hr(x,u_\phi^\alpha(x)). \quad \text{(A.31)}$$

Considering $q_m^\alpha = s_m^\alpha$ in Theorem 18.6 and recalling that $g \ge 1$ we get from equation (A.13) that

$$s_m^\alpha(x) \le \frac{M}{c+\alpha}g(x) + \frac{Mb}{c\alpha} \le \frac{M(1+\frac{b}{c})}{\alpha}g(x). \quad \text{(A.32)}$$

Therefore from (A.32) we have that $\alpha s_m^\alpha \le M(1+\frac{b}{c})g$ and thus, from (A.15),

$$\alpha L s_{m+1}^\alpha(x,u_\phi^\alpha(x)) \le \frac{M(1+\frac{b}{c})(1+bK_\lambda)}{c}g(x). \quad \text{(A.33)}$$

By combining equations (A.14), (A.31) and (A.33) the result follows. □

Proof. [Proof of Theorem 18.3] From Assumption 18.7 and Lemma 18.5 we get that for all $x \in E$,

$$\Big|G^k(L(f-\alpha s_{m+1-k}^\alpha) + Hr)(x,u_\phi^\alpha(x)) - \pi_{u^\alpha}\big(L(f-\alpha s_{m+1-k}^\alpha) + Hr\big)\Big| \le aM'\kappa^k g(x).$$

Consequently,

$$\left|\sum_{k=0}^{m} G^k(L(f-\alpha s^\alpha_{m+1-k})+Hr)(x,u^\alpha_\phi(x)) \right.$$
$$\left. - G^k(L(f-\alpha s^\alpha_{m+1-k})+Hr)(y,u^\alpha_\phi(y))\right| \leq aM'(g(x)+g(y))\frac{1-\kappa^{m+1}}{1-\kappa}.$$

Taking the limit as $m \uparrow \infty$ in the previous equation and recalling that $g \geq 1$ we get the desired result from Proposition 18.9. □

A.3. Existence of an ordinary feedback measurable selector

The main purpose of this sub-section is to show that for any function $h \in \mathbb{B}_g(E)$, the one-stage optimization operators $\mathcal{R}_\alpha(\rho,h)(x)$ and $\mathcal{T}_\alpha(\rho,h)(x)$ are equal and that there exists an ordinary feedback measurable selector for the one-stage optimization problems associated to these operators (see Theorem 18.7). First we need two technical lemmas.

Lemma 18.6. *Let $\alpha \geq 0$, $\rho \in \mathbb{R}_+$, $h \in \mathbb{B}_g(E)$ and set $w = \mathcal{R}_\alpha(\rho,h)$. Then there exists $\hat{\Theta} \in \mathcal{S}_{\mathbb{V}^r}$ such that*

$$w(x) = -\rho\mathcal{L}_\alpha(x,\hat{\Theta}(x)) + L_\alpha f(x,\hat{\Theta}(x)) + H_\alpha r(x,\hat{\Theta}(x)) + G_\alpha h(x,\hat{\Theta}(x)). \tag{A.34}$$

Moreover, $w \in \mathbb{M}^{ac}(E)$, and satisfies for all $x \in E$ and $t \in [0,t_*(x))$,

$$w(x) = \inf_{\mu\in\mathcal{V}^r(x)} \left\{ \int_0^t e^{-\alpha s-\Lambda^\mu(x,s)}\Big[-\rho + f(\phi(x,s),\mu(s)) + \lambda Qh(\phi(x,s),\mu(s))\Big]ds + e^{-\alpha t-\Lambda^\mu(x,t)}w(\phi(x,t))\right\} \tag{A.35}$$

$$= \int_0^t e^{-\alpha s-\Lambda^{\hat{\mu}(x)}(x,s)}\Big[-\rho + f(\phi(x,s),\hat{\mu}(x,s)) + \lambda Qh(\phi(x,s),\hat{\mu}(x,s))\Big]ds + e^{-\alpha t-\Lambda^{\hat{\mu}(x)}(x,t)}w(\phi(x,t)), \tag{A.36}$$

where $\hat{\Theta}(x) = (\hat{\mu}(x),\hat{\mu}_\partial(x))$.

Proof. From Corollary 18.1, it follows that the mapping V defined on $\mathcal{K}$ by

$$V(x,\Theta) = -\rho\mathcal{L}_\alpha(x,\Theta) + L_\alpha f(x,\Theta) + H_\alpha r(x,\Theta) + G_\alpha h(x,\Theta)$$

takes values in $\mathbb{R}$. Moreover, from Assumption 18.5 and Proposition 18.1, it follows that V is measurable. Furthermore, by using Corollary 5.7 in [3] it follows that for all $x \in E$, $V(x,.)$ is lower semicontinuous on $\mathbb{V}^r(x)$. Recalling that $\mathbb{V}^r(x)$ is a compact subset of $\mathbb{V}^r$ and by using Proposition D.5 in [17], we obtain that there exists $\hat{\Theta} \in \mathcal{S}_{\mathbb{V}^r}$ such that equation (A.34) is satisfied. The rest of the proof is similar to the proof of Proposition 4.2 in [3] and it is therefore omitted. □

Lemma 18.7. *Let $\alpha \geq 0$, $\rho \in \mathbb{R}_+$ and $h \in \mathbb{B}_g(E)$. Then, for all $x \in E$*

$$\mathcal{R}_\alpha(\rho,h)(x) \geq -(\rho + b\|h\|_g)K_\lambda - \|h\|_g g(x), \tag{A.37}$$

and for all $x \in E$ such that $t_*(x) = \infty$ and $\Theta = (\mu, \mu_\partial) \in \mathbb{V}^r(x)$

$$\begin{aligned} -\rho\mathcal{L}_\alpha(x,\Theta) + L_\alpha f(x,\Theta) + H_\alpha r(x,\Theta) + G_\alpha h(x,\Theta) \\ = \lim_{t\to+\infty} \int_0^t e^{-\alpha s - \Lambda^\mu(x,s)} \Big[-\rho + f(\phi(x,s),\mu(s)) \\ + \lambda Q h(\phi(x,s),\mu(s)) \Big] ds. \end{aligned} \tag{A.38}$$

Proof. From equation (A.11) we have

$$G_\alpha g(x,\Theta) \leq bK_\lambda + g(x), \tag{A.39}$$

for all $x \in E$ and $\Theta \in \mathbb{V}^r$. Consequently, by using equation (A.34) and the fact that $f \geq 0$ and $r \geq 0$ it follows that $\mathcal{R}_\alpha(\rho,h)(x) \geq -\rho\mathcal{L}_\alpha(x,\hat{\Theta}(x)) + G_\alpha h(x,\hat{\Theta}(x)) \geq -(\rho + b\|h\|_g)K_\lambda - \|h\|_g g(x)$, showing the first part of the result.

From Assumptions 18.11 a), b) and e), we have that $\lim_{t\to+\infty} \int_0^t e^{-\alpha s - \Lambda^\mu(x,s)} \big[-\rho + f(\phi(x,s),\mu(s)) \big] ds$ exists in $\mathbb{R}$, and from equation (A.39), $\lim_{t\to+\infty} \int_0^t e^{-\alpha s - \Lambda^\mu(x,s)} \lambda Q g(\phi(x,s),\mu(s)) ds$ exists in $\mathbb{R}$. By using the fact that $h \in \mathbb{B}_g(E)$, it follows that the limit on the right hand side of equation (A.38) exists. Finally, from Remark 18.3 *i)* we get the last part of the result. □

We show next the main result of this sub-section.

Theorem 18.7. *Let $\alpha \geq 0$, $\rho \in \mathbb{R}_+$, $h \in \mathbb{B}_g(E)$ and set $w = \mathcal{R}_\alpha(\rho,h)$. Then $w \in \mathbb{M}^{ac}(E)$ and the ordinary feedback measurable selector $\widehat{u}_\phi(w,h) \in \mathcal{S}_{\mathbb{V}}$ (see item D2) of Definition 18.4) satisfies the following one-stage opti-*

mization problems:

$$\begin{aligned}\mathcal{R}_\alpha(\rho,h)(x) &= \mathcal{T}_\alpha(\rho,h)(x)\\ &= -\rho\mathcal{L}_\alpha(x,\widehat{u}_\phi(w,h)(x)) + L_\alpha f(x,\widehat{u}_\phi(w,h)(x))\\ &\quad + H_\alpha r(x,\widehat{u}_\phi(w,h)(x)) + G_\alpha h(x,\widehat{u}_\phi(w,h)(x)). \quad (A.40)\end{aligned}$$

Proof. According to Lemma 18.6, there exists $\hat{\Theta} \in \mathcal{S}_{\mathbb{V}^r}$ such that for all $x \in E$ and $t \in [0, t_*(x))$ we have

$$e^{-\alpha t-\Lambda^{\hat{\mu}(x)}(x,t)}w(\phi(x,t)) - w(x) = \int_0^t e^{-\alpha s-\Lambda^{\hat{\mu}(x)}(x,s)}\Big[\rho - f(\phi(x,s),\hat{\mu}(x,s)) - \lambda Qh(\phi(x,s),\hat{\mu}(x,s))\Big]ds, \quad (A.41)$$

where $\hat{\Theta}(x) = (\hat{\mu}(x), \hat{\mu}_\partial(x))$. Since $w \in \mathbb{M}^{ac}(E)$, we obtain from equation (A.41) that

$$\begin{aligned}&-\mathcal{X}w(\phi(x,t)) + \alpha w(\phi(x,t)) + \rho\\ &\geq \inf_{\mu\in\mathcal{P}\left(\mathbb{U}(\phi(x,t))\right)}\Big\{f(\phi(x,t),\mu) - \lambda(\phi(x,t),\mu)w(\phi(x,t)) + \lambda Qh(\phi(x,t),\mu)\Big\}.\end{aligned}$$

However, notice that

$$\begin{aligned}&\inf_{\mu\in\mathcal{P}\left(\mathbb{U}(\phi(x,t))\right)}\Big\{f(\phi(x,t),\mu) - \lambda(\phi(x,t),\mu)w(\phi(x,t)) + \lambda Qh(\phi(x,t),\mu)\Big\} - \rho\\ &= \inf_{a\in\mathbb{U}(\phi(x,t))}\Big\{f(\phi(x,t),a) - \lambda(\phi(x,t),a)\big[w(\phi(x,t)) - Qh(\phi(x,t),a)\big]\Big\} - \rho.\end{aligned}$$

Consequently, by considering the measurable selector $\overline{u} \in \mathcal{S}_{\mathbb{U}}$ given by $\overline{u} = \widehat{u}(w,h)$ (see Definition 18.4, D1)), we have that

$$\begin{aligned}-\mathcal{X}w(\phi(x,t)) + \alpha w(\phi(x,t)) &= -\rho + f(\phi(x,t),\overline{u}(\phi(x,t)))\\ &\quad - \lambda(\phi(x,t),\overline{u}(\phi(x,t)))\big[w(\phi(x,t)) - Qh(\phi(x,t),\overline{u}(\phi(x,t)))\big],\end{aligned}$$

$\eta - a.s.$ on $[0, t_*(x))$, otherwise this would lead to a contradiction with equation (A.35). Consequently, for all $t \in [0, t_*(x))$ it follows that

$$\begin{aligned}w(x) - e^{-(\alpha t+\overline{\Lambda}(x,t))}w(\phi(x,t)) + \int_0^t e^{-(\alpha s+\overline{\Lambda}(x,s))}\Big[f(\phi(x,s),u(\phi(x,s)))\\ + \lambda(\phi(x,s),\overline{u}(\phi(x,s)))Qh(\phi(x,s),\overline{u}(\phi(x,s))) - \rho\Big]ds,\end{aligned} \quad (A.42)$$

where we set $\overline{\Lambda}(x,t) = \int_0^t \lambda(\phi(x,s),\overline{u}(\phi(x,s)))ds$.

First consider the case in which $t_*(x) < \infty$. We obtain, by taking the limit as t tends to $t_*(x)$ in the previous equation, that the ordinary feedback measurable selector $\widehat{u}_\phi(w,h) \in \mathcal{S}_{\mathbb{V}}$ (see item D2) of Definition 18.4) satisfies:

$$\begin{aligned} w(x) = {} & e^{-(\alpha t_*(x)+\overline{\Lambda}(x,t_*(x)))} w(\phi(x,t_*(x))) - \rho\mathcal{L}_\alpha(x,\widehat{u}_\phi(w,h)(x)) \\ & + L_\alpha f(x,\widehat{u}_\phi(w,h)(x)) \\ & + \int_0^{t_*(x)} e^{-(\alpha s+\overline{\Lambda}(x,s))}\lambda(\phi(x,s),\overline{u}(\phi(x,s)))Qh(\phi(x,s),\overline{u}(\phi(x,s)))ds. \end{aligned} \tag{A.43}$$

Define the control $\Theta(x)$ by $(\hat{\mu}(x),\mu)$ for $\mu \in \mathcal{P}\big(\mathbb{U}(\phi(x,t_*(x)))\big)$. Therefore, we have that

$$\begin{aligned} w(x) \le & -\rho\mathcal{L}_\alpha(x,\hat{\Theta}(x)) + L_\alpha f(x,\hat{\Theta}(x)) \\ & + \int_0^{t_*(x)} e^{-\alpha s-\Lambda^{\hat{\mu}(x)}(x,s)}\lambda Qh(\phi(x,s),\hat{\mu}(x,s))ds \\ & + e^{-\alpha t_*(x)-\Lambda^{\hat{\mu}(x)}(x,t_*(x))}\big[Qh(\phi(x,t_*(x)),\mu) + r(\phi(x,t_*(x)),\mu)\big]. \end{aligned} \tag{A.44}$$

From equation (A.36) and since $w \in \mathbb{M}^{ac}(E)$, we have that

$$\begin{aligned} w(x) = & -\rho\mathcal{L}_\alpha(x,\hat{\Theta}(x)) + L_\alpha f(x,\hat{\Theta}(x)) \\ & + \int_0^{t_*(x)} e^{-\alpha s-\Lambda^{\hat{\mu}(x)}(x,s)}\lambda Qh(\phi(x,s),\hat{\mu}(x,s))ds \\ & + e^{-\alpha t_*(x)-\Lambda^{\hat{\mu}(x)}(x,t_*(x))} w(\phi(x,t_*(x))). \end{aligned} \tag{A.45}$$

By Assumption 18.8, we have that $e^{-\Lambda^{\hat{\mu}(x)}(x,t_*(x))} > 0$. Therefore, combining equations (A.44) and (A.45), gives that for all $x \in E$ and $\mu \in \mathcal{P}\big(\mathbb{U}(\phi(x,t_*(x)))\big)$,

$$w(\phi(x,t_*(x))) \le Qh(\phi(x,t_*(x)),\mu) + r(\phi(x,t_*(x)),\mu).$$

Clearly, by using equation (A.34), it can be claimed that the previous inequality becomes an equality for $\mu = \hat{\mu}_\partial(x)$, implying that

$$\begin{aligned} w(\phi(x,t_*(x))) & = \inf_{\mu\in\mathcal{P}(\mathbb{U}(\phi(x,t_*(x))))} \{r(\phi(x,t_*(x)),\mu) + Qh(\phi(x,t_*(x)),\mu)\} \\ & = \inf_{a\in\mathbb{U}(\phi(x,t_*(x)))} \{r(\phi(x,t_*(x)),a) + Qh(\phi(x,t_*(x)),a)\}. \end{aligned}$$

Consequently, we have that

$$\begin{aligned} w(\phi(x,t_*(x))) = {} & r(\phi(x,t_*(x)),\overline{u}(\phi(x,t_*(x)))) \\ & + Qh(\phi(x,t_*(x)),\overline{u}(\phi(x,t_*(x)))). \end{aligned} \tag{A.46}$$

Combining equations (A.43) and (A.46), it follows that

$$\begin{aligned} w(x) = &- \rho\mathcal{L}_\alpha(x, \widehat{u}_\phi(w,h)(x)) + L_\alpha f(x, \widehat{u}_\phi(w,h)(x)) + H_\alpha r(x, \widehat{u}_\phi(w,h)(x)) \\ &+ G_\alpha h(x, \widehat{u}_\phi(w,h)(x)). \end{aligned}$$

Consider now the case in which $t_*(x) = \infty$. By using equation (A.42) and (A.37) we obtain that

$$\begin{aligned} w(x) \geq &- e^{-(\alpha t + \overline{\Lambda}(x,t))}\big[(\rho + b\|h\|_g)K_\lambda + \|h\|_g g(\phi(x,t))\big] \\ &+ \int_0^t e^{-(\alpha s + \overline{\Lambda}(x,s))}\Big[f(\phi(x,s), \overline{u}(\phi(x,s))) \\ &+ \lambda(\phi(x,s), \overline{u}(\phi(x,s)))Qh(\phi(x,s), \overline{u}(\phi(x,s))) - \rho\Big]ds. \end{aligned} \quad \text{(A.47)}$$

However, from Assumptions 18.11 a) and d) we obtain that

$$\lim_{t\to+\infty} e^{-(\alpha t + \overline{\Lambda}(x,t))}\big[(\rho + b\|h\|_g)K_\lambda + \|h\|_g g(\phi(x,t))\big] = 0. \quad \text{(A.48)}$$

Consequently, combining equations (A.38), (A.47) and (A.48), the ordinary feedback measurable selector $\widehat{u}_\phi(w,h) \in \mathcal{S}_{\mathbb{V}}$ satisfies:

$$\begin{aligned} w(x) \geq &- \rho + \mathcal{L}_\alpha(x, \widehat{u}_\phi(w,h)(x)) + L_\alpha f(x, \widehat{u}_\phi(w,h)(x)) \\ &+ H_\alpha r(x, \widehat{u}_\phi(w,h)(x)) + G_\alpha h(x, \widehat{u}_\phi(w,h)(x)). \end{aligned}$$

By using equation (A.35) it follows that the inequality in the previous equation is in fact an equality.

In conclusion, since $\mathbb{V}(x) \subset \mathbb{V}^r(x)$ it follows that $\mathcal{R}_\alpha(\rho,h)(x) \leq \mathcal{T}_\alpha(\rho,h)(x)$. However, we have shown that $\widehat{u}_\phi(w,h) \in \mathcal{S}_{\mathbb{V}}$ satisfies

$$\begin{aligned} \mathcal{R}_\alpha(\rho,h)(x) = &-\rho\mathcal{L}_\alpha(x, \widehat{u}_\phi(w,h)(x)) + L_\alpha f(x, \widehat{u}_\phi(w,h)(x)) \\ &+H_\alpha r(x, \widehat{u}_\phi(w,h)(x)) + G_\alpha h(x, \widehat{u}_\phi(w,h)(x)), \end{aligned}$$

which is the desired result. □

References

[1] M. Davis, Piecewise-deterministic Markov processes: A general class of non-diffusion stochastic models, *J.Royal Statistical Soc. (B)*. **46**, 353–388, (1984).

[2] M. Davis, *Markov Models and Optimization*. (Chapman and Hall, London, 1993).

[3] O. Costa and F. Dufour, Average continuous control of piecewise deterministic Markov processes, *SIAM J. Control and Optim.* **48**, 4262–4291, (2010).

[4] O. Costa, Average impulse control of piecewise deterministic processes, *IMA J. Math. Control Inform.* **6**(4), 375–397, (1989). ISSN 0265-0754.

[5] D. Gatarek, Impulsive control of piecewise-deterministic processes with long run average cost, *Stochastics Stochastics Rep.* **45**(3-4), 127–143, (1993). ISSN 1045-1129.

[6] A. Almudevar, A dynamic programming algorithm for the optimal control of piecdewise deterministic Markov processes, *SIAM J. of Control and Optim.* **40**(2), 525–539, (2001).

[7] M. Davis. Control of piecewise-deterministic processes via discrete-time dynamic programming. In *Stochastic differential systems (Bad Honnef, 1985)*, vol. 78, *Lecture Notes in Control and Inform. Sci.*, pp. 140–150. Springer, Berlin, (1986).

[8] M. Dempster and J. Ye, Necessary and sufficient optimality conditions for control of piecewise deterministic processes, *Stochastic and Stochastics Reports.* **40**, 125–145, (1992).

[9] M. Dempster and J. Ye, Generalized Bellman-Hamilton-Jacob optimality conditions for a control problem with boundary conditions, *Appl. Math. Optimization.* **33**, 211–225, (1996).

[10] L. Forwick, M. Schäl, and M. Schmitz, Piecewise deterministic Markov control processes with feedback controls and unbounded costs, *Acta Appl. Math.* **82**(3), 239–267, (2004). ISSN 0167-8019.

[11] M. Schäl, On piecewise deterministic Markov control processes: control of jumps and of risk processes in insurance, *Insurance Math. Econom.* **22**(1), 75–91, (1998). ISSN 0167-6687.

[12] A. Yushkevich, Bellman inequalities in Markov decision deterministic drift processes, *Stochastics.* **23**, 235–274, (1987).

[13] A. Yushkevich, Verification theorems for Markov decision processes with controlled deterministic drift and gradual and impulsive controls., *Theory Probab. Appl.* **34**(3), 474–496, (1989).

[14] O. Costa and F. Dufour, The vanishing discount approach for the average continuous control of piecewise-deterministic Markov processes, *Journal of Applied Probability.* **46**, 1157–1183, (2009).

[15] X. Guo and U. Rieder, Average optimality for continuous-time Markov decision processes in polish spaces, *The Annals of Applied Probability.* **16**, 730–756, (2006).

[16] X. Guo and Q. Zhu, Average optimality for Markov decision processes in Borel spaces: A new condition and approach, *Journal of Applied Probability.* **43**, 318–334, (2006).

[17] O. Hernández-Lerma and J. Lasserre, *Discrete-time Markov control processes.* vol. 30, *Applications of Mathematics*, (Springer-Verlag, New York, 1996). ISBN 0-387-94579-2. Basic optimality criteria.

[18] O. Hernández-Lerma and J. Lasserre, *Further topics on discrete-time Markov control processes.* vol. 42, *Applications of Mathematics*, (Springer-Verlag, New York, 1999). ISBN 0-387-98694-4.

[19] Q. Zhu, Average optimality for continuous-time Markov decision processes with a policy iteration approach, *Journal of Mathematical Analysis and Applications.* **339**, 691–704, (2008).

[20] O. Costa and F. Dufour. Relaxed long run average continuous control of

piecewise deterministic markov processes. In *Proceedings of the European Control Conference*, pp. 5052–5059, Kos, Greece, (July, 2007).

[21] A. Yushkevich, On reducing a jump controllable Markov model to a model with discrete time., *Theory Probab. Appl.* **25**, 58–69, (1980).

[22] D. Widder, *The Laplace Transform.* Princeton Mathematical Series, v. 6, (Princeton University Press, Princeton, N. J., 1941).

Chapter 19

Stochastic linear-quadratic control revisited

Tyrone E. Duncan*

Department of Mathematics,
University of Kansas,
Lawrence, KS 66045, USA

The control of a linear stochastic system with a Brownian motion and a quadratic cost functional in the state and the control is probably the most well known explicitly solvable stochastic control problem in continuous time. Some results are given in this chapter where this control problem is generalized by replacing Brownian motion by other stochastic (noise) processes such as the family of fractional Brownian motions and explicit solutions for an optimal control and the optimal cost can be given. The problem formulations include both finite and infinite time horizons, and completely and partially observed systems.

Contents

19.1. Introduction

Optimal control for deterministic control systems experienced a significant development in the late 1950s with Pontryagin's maximum principle [21] which provided necessary conditions for optimality generalizing the results for the calculus of variations and with Kalman's solution ([17]) of the control of a linear deterministic system with a quadratic cost functional using the solution of a Hamilton-Jacobi equation. This latter solution was gen-

*Research supported by NSF grant DMS 0808138, AFOSR grant FA9550-09-1-0554 and ARO grant W911NF-10-1-0248.

eralized to the control of a linear system with a Brownian motion in the 1960s (e.g. Kushner [17] and Wonham [23]). These finite dimensional system results have been extended to controlled linear equations in an infinite dimensional Hilbert space (e.g. [19]) that provide models for some linear partial differential equations and likewise the corresponding stochastic linear equations in a Hilbert space that model some linear stochastic partial differential equations (e.g. [7, 9]).

For the linear-quadratic stochastic control problem it is natural to question why the optimal control is the same as for the deterministic problem especially because the equation for the solution of the stochastic problem, a Hamilton-Jacobi-Bellman equation, is a second order nonlinear PDE whereas for the deterministic problem the equation, a Hamilton-Jacobi equation, is a first order nonlinear PDE. While both solutions are quadratic functions of the state, this fact does not explain intrinsically why the two solutions should be essentially the same, differing only by a function of time.

The above problems are finite time horizon, but these problems can be extended to some infinite time horizon problems. For the deterministic problem, then the cost is an integral on the positive half line and the solution follows by a passage to the limit verifying that the steady state system describes an optimal system that is stable in the Lyapunov sense, that is, all of the eigenvalues of this system have strictly negative real parts. A natural corresponding infinite time horizon stochastic problem has an ergodic (long run average) quadratic cost. In this case the optimal control is the same as the aforementioned infinite time horizon deterministic control problem. For the stochastic problems there are also the partially observed problems where the admissible controls are required to be adapted to the noisy partial observations of the state.

A fundamental question in the above problems for completely observed systems is, 'Why are the solutions of the deterministic and the stochastic problems so similar?' in that they have the same optimal controls and optimal costs that differ only by a function of time. If a clear explanation is provided for this question, then there should be a more fundamental understanding of the effects of adding a term to a linear system to obtain an affine system, so that a Brownian motion is only a particular affine term that may be replaced by other stochastic (noise) processes. Thus the models with Markov processes or martingales should be extended to other stochastic processes, such as the family of fractional Brownian motions, whose statistics can be empirically justified for many physical phenomena.

However more generally the assumption of Gaussian processes should also be able to be relaxed.

A brief outline of the paper is given now. In Section 19.2 initially a finite time horizon control problem is formulated with a linear system driven by a fractional Brownian motion and a quadratic cost functional. Subsequently this basic model is modified by considering a partially observed control problem as well as an ergodic control problem. Some comments are made on the methods of proof and the basic ideas that provide the results. In all of these control problems a fractional Brownian motion for the linear system can be replaced by a suitable stochastic process that is square integrable and has continuous sample paths. In Section 19.3 some concluding remarks are made.

This paper is dedicated to Robert J. Elliott in celebration of his seventieth birthday.

19.2. Some linear stochastic control problems

Consider the following controlled stochastic system

$$\begin{aligned} dX(t) &= (AX(t) + CU(t))dt + dB(t) \\ X(0) &= X_0 \end{aligned} \tag{19.1}$$

where X_0 is an $\mathbb{R}^n$-valued random variable with zero mean and covariance Σ, $X(t) \in \mathbb{R}^n, U(t) \in \mathbb{R}^m, A \in \mathcal{L}(\mathbb{R}^n, \mathbb{R}^n), C \in \mathcal{L}(\mathbb{R}^m, \mathbb{R}^n)$, $(B(t), t \in [0,T])$ is an $\mathbb{R}^n$-valued standard Brownian motion,

$$\begin{aligned} J^0(U) = &\frac{1}{2}\int_0^T \Big(\langle QX(s), X(s)\rangle + \langle RU(s), U(s)\rangle\Big) ds \\ &+\frac{1}{2}\langle MX(T), X(T)\rangle \end{aligned} \tag{19.2}$$

and the cost functional is

$$J(U) = \mathbb{E}J^0(U) \tag{19.3}$$

where $Q \in \mathcal{L}(\mathbb{R}^n, \mathbb{R}^n), R \in \mathcal{L}(\mathbb{R}^m, \mathcal{R}^m), M \in \mathcal{L}(\mathbb{R}^n, \mathbb{R}^n)$, $Q > 0$, $R > 0$ and $M \geq 0$ are symmetric linear transformations, $T > 0$ is fixed, $<, >$ is the standard inner product on the appropriate Euclidean space, and the family of admissible controls U is specified subsequently. It is assumed that X_0 and $(B(t), t \in [0,T])$ are independent.

The system (19.1) has an obvious formal relation with the following deterministic affine controlled system

$$\frac{dx}{dt} = Ax(t) + Cu(t) + b(t) \tag{19.4}$$

where $x(t) \in \mathbb{R}^n$, A and C are as in (19.1), $(b(t), t \geq 0)$ is an $\mathbb{R}^n$-valued (piecewise continuous) curve and the cost functional is $J^0(u)$. An elementary method to solve this deterministic control problem uses the classical method of completion of squares (e.g. [24]). This method uses the natural multidimensional extension of the well known method of expressing a quadratic polynomial as the sum of the square of a linear polynomial and a constant term. In this method, for a linear-quadratic control problem, there is a squared term that contains a linear term in the control, u, and a "constant" term that does not contain the control so that the minimization is clearly obtained by choosing u to make the squared term zero.

This method can be extended to stochastic linear systems, though now the appropriate measurability of the control must be introduced as an additional requirement. This approach can demonstrate why the stochastic problem with Brownian motion has an optimal control that is identical to the deterministic control problem. More importantly, this approach can be generalized to other processes that are not martingales. To extend the deterministic affine system result to a stochastic affine system, a sequence of piecewise continuous processes are constructed that converges uniformly to the given process with continuous sample paths. For the stochastic system, an optimal control is initially obtained that lacks the property of being adapted to the solution of the state system. Passing to the limit of the sequence of piecewise continuous processes which requires some elementary results from stochastic analysis gives an optimal nonadapted control. A suitable conditioning and a verification of optimality provide an optimal adapted control. For some other problems where the stochastic analysis is insufficient, another strategy can be used to verify an optimal control by using the uniform approximation of the sequence of processes that converges to the noise process.

Since an important component to define an adapted control is obtained by an appropriate conditional expectation, some more general control problems can be solved in a similar way. One important example is a partially observed linear-quadratic problem. Furthermore, other partially observed problems with some more general observation equations can also be solved and an infinite time horizon control problem with an ergodic (long run average) quadratic cost functional can be solved.

To be specific about some stochastic processes, initially the family of standard fractional Brownian motions is defined. This family of processes was introduced by Kolmogorov [17] and the first computation of a statistic of these processes from data was made by Hurst [12].

Definition 19.1. Let $(\Omega, \mathcal{F}, \mathbb{P})$ be a complete probability space. For each $H \in (0,1)$, a real-valued standard fractional Brownian motion $(B(t), t \geq 0)$ on $(\Omega, \mathcal{F}, \mathbb{P})$ is a Gaussian process with continuous sample paths that satisfies

$$\begin{aligned} \mathbb{E}[B(t)] &= 0 \\ \mathbb{E}[B(s)B(t)] &= (1/2)(t^{2H} + s^{2H} - |t-s|^{2H}) \end{aligned} \tag{19.5}$$

for all $s, t \in \mathbb{R}_+$.

An $\mathbb{R}^n$-valued standard fractional Brownian motion with Hurst parameter H is an n-vector of n independent real-valued standard fractional Brownian motions each with the Hurst parameter H.

Associated with each Gaussian process is a "natural" Hilbert space. For a fractional Brownian motion, let L^2_H be the Hilbert space whose inner product $\langle \cdot, \cdot \rangle_H$ is given by

$$\langle f, g \rangle_H = \rho(H) \int_0^T u^2_{\frac{1}{2}-H}(r)(I^{H-\frac{1}{2}}_{T-} u_{H-\frac{1}{2}} f)(r)(I^{H-\frac{1}{2}}_{T-} u_{H-\frac{1}{2}} g)(r)dr$$

where $\rho(H) = \frac{2H\Gamma(H+\frac{1}{2})\Gamma(\frac{3}{2}-H)}{\Gamma(2-2H)}$ and $u_a(r) = r^a$ for $r \geq 0$ and $a \in \mathbb{R}$. The term $I^{H-\frac{1}{2}}_{T-}$ is a fractional integral for $H \in (\frac{1}{2}, 1)$ and a fractional derivative for $H \in (0, \frac{1}{2})$. For $\alpha \in (-1, 0)$ the fractional derivative is usually denoted $D^{-\alpha}$. These are defined in [22] as

$$\left(I^\alpha_{T-}\varphi\right)(t) = \frac{1}{\Gamma(\alpha)} \int_t^T (s-t)^{\alpha-1} \varphi(s)\, ds,$$

$$\left(D^\alpha_{T-}\psi\right)(t) = \frac{1}{\Gamma(1-\alpha)} \left(\frac{\psi(t)}{(T-t)^\alpha} + \alpha \int_t^T \frac{\psi(s) - \psi(t)}{(s-t)^{\alpha+1}}\, ds \right).$$

If $(B(t), t \geq 0)$ is a real-valued standard fractional Brownian motion with Hurst parameter $H \in (0,1)$ fixed, then

$$\begin{aligned} &\mathbb{E}[B(s)B(t)] \\ &= \langle 1_{[0,s]}, 1_{[0,t]} \rangle_H \\ &= \rho(H) \int_0^T u^2_{\frac{1}{2}-H}(r)(I^{H-\frac{1}{2}}_{T-} u_{H-\frac{1}{2}} 1_{[0,s]})(r)(I^{H-\frac{1}{2}}_{T-} u_{H-\frac{1}{2}} 1_{[0,t]})(r)dr. \end{aligned} \tag{19.6}$$

This covariance factorization in the Lebesgue space L^2 demonstrates the relevance of the Hilbert space L^2_H to the fractional Brownian motion. It can be shown that the covariance function $\mathbb{E}[B(s)B(t)]$ is an analytic function of H.

Let $(B(t), t \in [0,T])$ be an $\mathbb{R}^n$-valued standard fractional Brownian motion. Let $(\mathcal{F}(t), t \in [0,T])$ be the filtration of $(B(t), t \in [0,T])$. The family of adapted, admissible controls, $\mathcal{U}_a$, is

$$\mathcal{U}_a = \Big\{U : U \text{ is an } \mathbb{R}^m\text{-valued process adapted to } (\mathcal{F}(t), t \in [0,T]) \\ \text{such that } U \in L^2([0,T]) \text{ a.s.}\Big\}.$$

The following theorem provides an optimal control for the control problem (19.1) and (19.3).

Theorem 19.1. *For the optimal control problem* (19.1) *and* (19.3) *where* $(B(t), t \in [0,T])$ *is an* $\mathbb{R}^n$*-valued standard fractional Brownian motion with the Hurst parameter* $H \in (0,1)$ *and the family of admissible, adapted controls is* $\mathcal{U}_a$*, there is an optimal control* $\hat{U}^*$ *that can be expressed as*

$$\hat{U}^*(t) = -R^{-1}C^T(P(t)X(t) + V(t)), \tag{19.7}$$

where $(P(t), t \in [0,T])$ *is the unique symmetric positive definite solution of the Riccati equation*

$$\frac{dP}{dt} = -PA - A^T P + PCR^{-1}C^T P - Q, \tag{19.8}$$

$$P(T) = M, \tag{19.9}$$

and $(V(t), t \in [0,T])$ *is the process given by*

$$V(t) = \mathbb{E}[\int_t^T \Phi_P(s,t)P(s)dB(s)|\mathcal{F}(t)], \tag{19.10}$$

$$= \int_0^t u_{1/2-H} I_{t-}^{1/2-H}(I_{T-}^{H-1/2} u_{H-1/2} 1_{[t,T]} \Phi_P(.,t)P)dB_t$$

and Φ_P *is the fundamental solution of the matrix equation*

$$\frac{d\Phi_P(s,t)}{dt} = -(A^T - P(t)CR^{-1}C^T)\Phi_P(s,t), \tag{19.11}$$

$$\Phi_P(s,s) = I. \tag{19.12}$$

While a proof is given in [4] that is valid for an arbitrary standard fractional Brownian motion, and some special cases are given in [3, 14], a more elementary proof than in [4] is sketched here that is applicable to

$H \in (\frac{1}{2}, 1)$ that uses some stochastic analysis for these processes and is motivated by the proof in [8] for the linear-quadratic control of stochastic systems in a Hilbert space with a fractional Brownian motion.

Proof. (Sketch). Let $H \in (\frac{1}{2}, 1)$ be fixed. Let $(B_n(t), t \in [0,T]), n \in \mathbb{N})$ be a sequence of processes with piecewise linear sample paths such that B_n is obtained from B by linear interpolation at the dyadic points of $[0,T]$, $(\frac{kT}{2^n}, k = 0,1,2,\ldots)$. Let $(X_n(t), t \in [0,T])$ be the solution of

$$\begin{aligned} dX_n(t) &= (AX_n(t) + CU(t))dt + dB_n(t) \\ X_n(0) &= X_0\,. \end{aligned} \tag{19.13}$$

Fix $n \in \mathbb{N}$. The cost functional $J_n(U)$ for the system (19.13) can be expressed in the following form by integrating the differentials of $(\langle \Gamma(t)X(t), X(t)\rangle, t \geq 0)$ and $(\langle \phi(t), X(t)\rangle, t \geq 0)$ and using a completion of squares

$$\begin{aligned} J_n^0(U) &= \frac{1}{2}\langle \Gamma(0)X_0, X_0\rangle + \langle \phi_n(0), X_0\rangle \\ &\quad + \frac{1}{2}\int_0^T [(|R^{-1/2}[RU + C^T PX_n + C^T\phi_n]|^2 - |R^{-1/2}C^T\phi_n|^2)dt \\ &\qquad + 2\langle \phi_n, dB_n\rangle], \end{aligned} \tag{19.14}$$

where

$$\begin{aligned} \frac{d\phi_n}{dt} &= -[(A^T - P(t)CR^{-1}C^T)\phi_n(t) + P(t)\frac{dB_n}{dt}], \\ \phi_n(T) &= 0, \\ \phi_n(t) &= \int_t^T \Phi_P(s,t)P(s)dB_n(s), \end{aligned} \tag{19.15}$$

and

$$J_n^0(U) = \frac{1}{2}\int_0^T (\langle QX_n, X_n\rangle + \langle RU, U\rangle)dt + \langle MX_n(T), X_n(T)\rangle. \tag{19.16}$$

From (19.14) an optimal control is clearly

$$U_n^*(t) = -R^{-1}(C^T P(t)X_n(t) + C^T\phi_n(t))\,. \tag{19.17}$$

It follows from the relation between stochastic integrals of Itô-type and Statonovich-type for a fractional Brownian motion [6] that

$$\lim_{n\to\infty} \mathbb{E}\left[\int_0^T \phi_n dB_n\right] = \int_0^T\int_0^s \mathrm{tr}(P(s)\Phi_P^T(s,r))\phi_{\bar{H}}(s-r)drds \tag{19.18}$$

where

$$\phi_{\bar{H}}(s) = \bar{H}(2\bar{H}-1)|s|^{2\bar{H}-2}. \tag{19.19}$$

Let Π be the orthogonal projection given by

$$\Pi : L^2([0,T]\times\Omega,\mathbb{R}^m) \to L^2_{\mathcal{F}}([0,T]\times\Omega,\mathbb{R}^m), \tag{19.20}$$

where $L^2_{\mathcal{F}}([0,T]\times\Omega,\mathbb{R}^m)$ is the family of $\mathbb{R}^m$-valued $(\mathcal{F}(t), t\in[0,T])$ adapted (predictable) processes and let $\Pi^{\perp} = I - \Pi$ be the complementary orthogonal projection. By the result for nonadapted controls, it follows that

$$\begin{aligned} J(U) - \frac{1}{2}\mathbb{E}\langle P(0)X_0, X_0\rangle = \frac{1}{2}\mathbb{E}\int_0^T &\Big[|R^{-\frac{1}{2}}(RU + \Pi C^T PX + \Pi C^T\phi|^2 \\ &+ |\Pi^{\perp}(R^{-1}C^TPX + R^{-1}C^T\phi)|^2\Big]dt \\ &+\int_0^T\int_0^s \operatorname{tr}(P(s)\Phi_P(s,r))\phi_{\bar{H}}(s-r)drds. \end{aligned} \tag{19.21}$$

Clearly the right side of (19.21) is minimized in $L^2_{\mathcal{F}}([0,T]\times\Omega,\mathbb{R}^m)$ by choosing the control

$$\begin{aligned} U^*(t) &= -R^{-1}C^TP(t)\Pi X(t) - R^{-1}C^T\Pi\phi(t) \\ &= -R^{-1}C^TP(t)X(t) - R^{-1}C^TV(t), \end{aligned} \tag{19.22}$$

because the family of orthogonal projections is conditional expectation with respect to $(\mathcal{F}(t), t\in[0,T])$. The explicit expression for the conditional expectation in (19.10) follows from a result in [1]. The optimal cost follows directly from (19.22) as

$$\begin{aligned} J(U^*) = \frac{1}{2}\operatorname{tr}(P(0)\Sigma) + \frac{1}{2}\mathbb{E}\int_0^T (|\Pi^{\perp}R^{-1}C^T\phi(t)|^2)dt \\ +\int_0^T\int_0^s \operatorname{tr}(P(s)\Phi_P^T(s,r))\phi_H(s-r)drds. \end{aligned} \tag{19.23}$$

□

Since the previous sketch of a proof relies basically on some notions from projections in Hilbert space (completion of squares and conditional expectation) and some elementary stochastic analysis, it can be extended to other stochastic control models, in particular, partially observed systems. A linear observation model is given as follows

$$dY(t) = FX(t)dt + d\widetilde{B}(t) \tag{19.24}$$

where $F \in \mathbb{L}(\mathbb{R}^n, \mathbb{R}^p)$ and $(\widetilde{B}(t), t \geq 0)$ is an $\mathbb{R}^p$-valued standard fractional Brownian motion with $H \in (0,1)$ and independent of the fractional Brownian motion B and X_0. The family $(\mathcal{G}(t), t \in [0,T])$ is the filtration for $(Y(t), t \in [0,T])$ for $U \equiv 0$. This family of σ-algebras is equal to the one with the optimal control (cf. [11], [20]).

The family of admissible controls $\mathcal{U}$ is defined as

$$\mathcal{U}_a = \Big\{ U : U \text{ is an } \mathbb{R}^m\text{-valued process adapted to } (\mathcal{G}(t), t \in [0,T]) \\ \text{such that } U \in L^2([0,T]) \text{ a.s.} \Big\}.$$

The following result gives an optimal control. This result is discussed in [5]. The case of a scalar system is given in [16].

Theorem 19.2. *Let $H \in (\frac{1}{2}, 1)$ and $(B(t), t \in [0,T])$ be the corresponding $\mathbb{R}^n$-valued standard fractional Brownian motion. For the control problem described by* (19.1), (19.24), *and* (19.3), *there is an optimal control $(U^*(t), t \in [0,T])$ in $\mathcal{U}$ that is given by*

$$U^*(t) = -R^{-1}C^T(P(t)\widehat{X}(t|t) + \widehat{\phi}(t|t)), \tag{19.25}$$

where $(P(t), t \in [0,T])$ is the unique, symmetric, positive definite solution of the Riccati equation

$$\frac{dP}{dt} = -PA - A^TP + PCR^{-1}C^TP - Q, \tag{19.26}$$
$$P(T) = M,$$

and

$$\widehat{X}(t|t) = \mathbb{E}[X(t)|\mathcal{G}(t)] \tag{19.27}$$

$$\phi(t) = \int_t^T \Phi_P(s,t)P(s)dB(s), \tag{19.28}$$

$$\widehat{\phi}(t|t) = \mathbb{E}[\phi(t)|\mathcal{G}(t)], \tag{19.29}$$

and Φ_P satisfies the following equation

$$\frac{d\Phi_P(s,t)}{dt} = -(A^T - P(t)CR^{-1}C^T)\Phi_P(s,t), \tag{19.30}$$

$$\Phi_P(s,s) = I. \tag{19.31}$$

The optimal cost is

$$J(U^*) = \frac{1}{2}\mathbb{E}\int_0^T |R^{-1}C^T(P(t)\widetilde{X}(t|t) + \widetilde{\phi}(t|t))|^2 dt \tag{19.32}$$
$$+ \int_0^T \int_0^s \operatorname{tr}(P(s)\Phi_P^T(s,r))\phi_{\bar{H}}(s-r)drds,$$

where $\phi_{\bar{H}}(s) = \bar{H}(2\bar{H}-1)|s|^{2\bar{H}-2}$, $\widetilde{X}(t|t) = X(t) - \widehat{X}(t|t)$, *and* $\widetilde{\phi}(t|t) = \phi(t) - \widehat{\phi}(t|t)$.

The methods for verifying this theorem are a natural generalization of the methods for the previous theorem.

The finite time horizon optimal control result in Theorem 19.2 can be extended to an infinite time horizon control problem with an ergodic (long run average) quadratic cost. The cost functional J_∞ is an ergodic quadratic functional of X and U that is given by

$$J_\infty(U) = \limsup_{T\to\infty} \frac{1}{T} J_T(U), \tag{19.33}$$

where

$$J_T(U) = \frac{1}{2}\int_0^T (\langle QX(s), X(s)\rangle + \langle RU(s), U(s)\rangle)ds \tag{19.34}$$

and $Q \in \mathcal{L}(\mathbb{R}^n, \mathbb{R}^n)$, $R \in \mathcal{L}(\mathbb{R}^m, \mathcal{R}^m)$, $M \in \mathcal{L}(\mathbb{R}^n, \mathbb{R}^n)$, $Q > 0$, $R > 0$ and $M \geq 0$ are symmetric linear transformations. For simplicity of discussion, only a problem with a scalar system is given. This scalar problem is also considered in [15]. A more general result for a multidimensional system and an arbitrary fractional Brownian motion is given in [2].

Theorem 19.3. *Let* $n = 1$ *in* (19.1), $(B(t), t \geq 0)$ *be a real-valued standard fractional Brownian motion with* $H \in (\frac{1}{2}, 1)$ *and let* $A = a$ *and* $C = c$ *in* (19.1) *and* $R = r$ *and* $Q = q$ *in* (19.3) *where* $r > 0$, $q > 0$ *and* $\{a,c\}$ *are nonzero real numbers. An optimal ergodic control for the control problem* (19.1) *and* (19.33) *is*

$$U^*(t) = -\frac{c}{r}(\rho X(t) + V(t)), \tag{19.35}$$

where ρ *is the positive solution of the quadratic polynomial*

$$0 = -2a\rho + \frac{c^2}{r}\rho^2 - q \tag{19.36}$$

and

$$V(t) = \mathbb{E}\Big[\int_t^\infty e^{\widetilde{a}(s-t)}\rho dB(s)|\mathcal{F}(t)\Big], \tag{19.37}$$

$$\widetilde{a} = a - \frac{c\rho}{r} < 0. \tag{19.38}$$

The optimal cost is

$$J(U^*) = \lim_{T\to\infty} \frac{1}{T} J_T(U^*) = \frac{\Gamma(2H+1)}{4|\tilde{a}|^{2H}}\Big(q + \frac{\rho^2 c^2}{r}\sin(\pi H)\Big) \quad a.s. \tag{19.39}$$

where

$$J_T(U) = \frac{1}{2}\int_0^T (qX^2(t) + rU^2(t))dt. \tag{19.40}$$

To verify the optimal control (19.35) a suitable passage to the limit as $T \to \infty$ of a family of controls obtained from (19.7) is made. A basic computation for the optimal cost is given now.

$$\begin{aligned}
&\lim_{T\to\infty} \frac{1}{T}\mathbb{E}\Big[\frac{1}{2}\Big(\int_0^T e^{\tilde{a}(T-s)}dB(s)\Big)^2\Big] \\
&= \frac{1}{2}\int_0^\infty \int_0^\infty e^{\tilde{a}p}e^{\tilde{a}q}H(2H-1)|p-q|^{2H-2}dpdq \\
&= \frac{1}{2}\frac{\Gamma(2H+1)}{2|\tilde{a}|^{2H}}.
\end{aligned}$$

19.3. Concluding remarks

The results in the previous theorems demonstrate how the optimal control results for linear systems with Brownian motions can be generalized to the family of fractional Brownian motions. Furthermore, the methods to verify the optimal controls in the theorems do not depend on the process being a fractional Brownian motion. Other Gaussian and non-Gaussian processes may be used to obtain analogous results. For example, Theorem 19.1 can be generalized by considering linear stochastic systems with stochastic (noise) processes that are square integrable and have continuous sample paths and for Theorem 19.3, the linear stochastic systems can be modified by replacing the fractional Brownian motion with some other processes that are square integrable and have stationary increments. The methods should more naturally relate the areas of deterministic and stochastic linear control.

References

[1] T. E. Duncan, Prediction for some processes related to a fractional Brownian motion, *Stat.Prob.Lett.* 76 (2006), 128-134
[2] T. E. Duncan, Ergodic control of linear systems with fractional Brownian motions, preprint.
[3] T. E. Duncan and B. Pasik-Duncan, Control of some linear stochastic systems with a fractional Brownian motion, *Proc. 48th IEEE Conference Decision and Control*, Shanghai 2009, 8518-8522
[4] T. E. Duncan and B. Pasik-Duncan, Linear-quadratic fractional Gaussian Control, preprint.

[5] T. E. Duncan and B. Pasik-Duncan, Control of some partially observed linear stochastic systems with fractional Brownian motions, *Proc. 50th IEEE Conference on Decision and Control*, Orlando, 5035-5040, 2011.
[6] T. E. Duncan, J. Jakubowski, and B. Pasik-Duncan, Stochastic integration for fractional Brownian motion in a Hilbert space, *Stochastics Dynamics* 6 (2006), 53-75.
[7] T. E. Duncan, B. Maslowski, and B. Pasik-Duncan, Adapted boundary and point control of linear stochastic distributed systems, *SIAM J. Control Optim.* 32 (1994), 648-672.
[8] T. E. Duncan, B. Maslowski, and B. Pasik-Duncan, Linear-quadratic control for stochastic equations in a Hilbert space with fractional Brownian motions, *SIAM J. Control Optim.*, to appear..
[9] F. Flandoli, Direct solutions of a Riccati equation arising in a stochastic control problem with control and observation on the boundary, *Appl. Math. Optim.*, 14 (1986), 107-129.
[10] W. H. Fleming and R. W. Rishel, *Deterministic and Stochastic Optimal Control*, Springer, New York, 1975.
[11] T. T. Georgiou and A. Lindquist, The separation principle in stochastic control, redux, preprint.
[12] H. E. Hurst, Long-term storage capacity in reservoirs, *Trans. Amer. Soc. Civil Eng.*, 116 (1951), 400-410.
[13] R. E. Kalman, Contributions to the theory of optimal control, *Bol. Soc. Mat. Mexicana*, 5 (1960), 102-1119.
[14] M. L. Kleptsyna, A. Le Breton, and M. Viot, About the linear quadratic regulator problem under a fractional Brownian perturbation, *ESAIM Probab. Stat.* 9 (2003), 161-170.
[15] M. L. Kleptsyna, A. Le Breton, and M. Viot, On the infinite time horizon linear-quadratic regulator problem under a fractional Brownian perturbation, *ESAIM Probab. Stat.* 9 (2005), 185-205.
[16] M. L Kleptsyna, A. Le Breton, and M. Viot, Separation principle in the fractional Gaussian linear-quadratic regulator problme with partial observation, *ESAIM Probab. Stat.*, 17 (2008), 94-124.
[17] A. N. Kolmogorov, Wienersche spiralen und einige andere interessante kurven in Hilbertschen Raum, *C.R. (Doklady) Acad. USSS (N.S.)*, 26 (1940), 115-118.
[18] H. J. Kushner, Optimal stochastic control, *IRE Trans Auto. Control*, AC-7 (1962), 120-122.
[19] I. Lasiecka and R. Triggiani, The regulator problem for parabolic equations with Dirichlet boundary control I, *Appl. Math. Optim.*, 16 (1987), 147-168.
[20] A. Lindquist, On feedback control of linear stochastic systems, *SIAM J. Control*, 11 (1973), 323-343,
[21] L. S. Pontryagin, Optimal processes of regulation, *Uspekhi Mat. Nauk*, 14 (1959), 3-20 (in Russian).
[22] S. G. Samko, A. A. Kilbas, and O. I. Marichev, *Fractional Integrals and Derviatives*, Gordon and Breach, Yverdon, 1993.

[23] W. M. Wonham, On a matrix Riccati equation of stochastic control, *SIAM J. Conntrol*, 6 (1968), 681-697.

[24] J. Yong and X. Y. Zhou, *Stochastic Controls*, Springer-Verlag, New York, 1999.

Chapter 20

Optimization of stochastic uncertain systems: Entropy rate functionals, minimax games and robustness

Farzad Rezaei[1], Charalambos D. Charalambous[2] and N.U. Ahmed[3]

[1]*frezaei@alumni.uottawa.ca*

[2]*Department of Electrical and Computer Engineering, University of Cyprus, Nicosia, 1678, Cyprus*
chadcha@ucy.ac.cy

[3]*School of Information Technology and Engineering, University of Ottawa, Ottawa, K1N 6N5, Canada*
ahmed@site.uottawa.ca

In large deviations theory Legendre–Fenchel transform is employed to identify the rate functional or action functional which is used to compute probabilities of rare events. On the other hand, minimax games are introduced to address robustness issues in control and estimation problems of uncertain systems, and to derive dissipation inequalities. This chapter aims at establishing explicit connections between the Legendre–Fenchel transform, robustness of uncertain systems, and risk-seeking and risk-averse optimization problems. The connections are obtained by introducing new classes of stochastic uncertain systems in which the uncertainty is described by a family of probability measures, which are absolutely continuous with respect to an a priori nominal measure. With respect to this formulation, minimization of the relative entropy over the set of uncertain measures, subject to certain fidelity criteria is shown to be equivalent to the Legendre–Fenchel transform. Subsequently, Legendre–Fenchel transform is employed to relate information theoretic concepts and robustness of uncertain systems, and to derive several properties associated with minimax games arising in robustness problems. These include a characterization of the solution to the disturbance attenuation problem, connections to dissipation inequalities, and connections to risk-sensitive problems, in which the strategies are risk-seeking, emphasizing optimistic scenarios; and risk-averse emphasizing pessimistic scenarios. The results are applied to uncertain nonlinear partially observable systems, in which the nominal measure is modelled through conditional distributions, and existence of optimal minimax strategies is proved using wide-sense controls.

Contents

20.1. Introduction

The main objective of this paper is to identify and explore specific relations between robustness and information theoretic concepts. In particular, to determine the exact relations between the Legendre–Fenchel transform [1–4], which is often employed to identify entropy rate functionals or action functionals in Large Deviations theory, dissipation [5–7], which is used to address robustness of uncertain systems [8, 9], and risk-seeking and risk-averse optimization problems [10–13]. Further, to explore these relations in the context of robustness of uncertain nonlinear partially observable systems [14–19]. Identification of the entropy rate functional associated with stochastic differential equations driven by Brownian Motion which is scaled by a small parameter can be found in [1, 18, 20, 21], while related material are also found in [22–24]. On the other hand, robustness of stochastic systems subject to relative entropy uncertainty can be found in [25–29].

The main contribution of this paper is the derivation of several properties associated with robustness of uncertain systems, in an abstract setting, which are then shown to be applicable to uncertain stochastic nonlinear partially observed control systems. The connections discussed above are established by introducing a new class of stochastic uncertain systems, in which the pay-off is described by the relative entropy or Kullback-Leibler distance between an unknown probability measure and a known (fixed) probability measure [22], subject to a fidelity constraint expressed in terms of the unknown probability measure.

The abstract results are derived in the first part of the paper. The second part of the paper is devoted to the application of the results to nonlinear partially observed uncertain stochastic systems described by stochastic differential equations. Here, an example is presented in which the dynamics and observations are linear subject to additive Brownian motion processes, while the constraint is a quadratic functional of the state and control.

The abstract formulation of the new class of stochastic systems is introduced next.

20.1.1. *Abstract formulation*

Let $\left(\Sigma, d\right)$ denote a complete separable metric space (a Polish space), and $(\Sigma, \mathcal{B}(\Sigma))$ the corresponding measurable space, in which $\mathcal{B}(\Sigma)$ is the σ-algebra generated by open sets in Σ. Let $\mathcal{M}(\Sigma)$ denote the set of probability measures on $(\Sigma, \mathcal{B}(\Sigma))$. Let $\mathcal{U}_{ad}$ denote the set of admissible controls. By choosing a control policy $u \in \mathcal{U}_{ad}$ for the nominal system (system perfectly known and assumed to be free from unknown external disturbances), Player-1 can induce a probability measure $\mu^u \in \mathcal{M}(\Sigma)$ in order to meet its performance objectives. Player-2, representing an adversary, tries to thwart the objective of Player-1 by trying to gain knowledge of the measure μ^u through jamming signals perturbing the system. Such information may be used by Player-2 to diminish the effectiveness of control actions of the Player-1.

For a given $u \in \mathcal{U}_{ad}$, let $M(u)(\neq \emptyset) \subset \mathcal{M}(\Sigma)$ denote the set of probability measures induced by the perturbed system while Player-1 exercises control u. In other words, $M(\cdot)$ is a multivalued map given by

$$M : \mathcal{U}_{ad} \longrightarrow 2^{\mathcal{M}(\Sigma)} \setminus \emptyset,$$

where $2^{\mathcal{M}(\Sigma)}$ denotes the power set of $\mathcal{M}(\Sigma)$. Let $\ell^u : \Sigma \rightarrow \Re$ be a real-valued measurable function bounded from below; and $H(\nu|\mu)$ denote the entropy of ν relative to the measure μ both from $M(u)$ (see Definition 20.1). The relative entropy $H(\nu|\mu)$ is also called the Kulback-Leibler distance between the two measures. In fact, it is often used in information theory as a measure of divergence between the two probability measures. Often the dependence of various measures on u appearing as superscripts will be often omitted.

Let $\mu \in M(u)$ denote the measure induced by the nominal system. In order to gain knowledge of the measure $\mu(\equiv \mu^u)$, Player-2 tries to minimize the relative entropy $H(\nu|\mu)$ by choosing a measure $\nu \in M(u)$ subject to some additional constraints. Let $M_c(u) \subset M(u)$ denote such a constraint. Then clearly the objective of Player-2 is to find a $\nu \in M_c(u)$ that matches as closely as possible with μ. This leads to the following minimization problem:

$$\inf_{\nu \in M_c(u)} H(\nu|\mu),$$

for every control $u \in \mathcal{U}_{ad}$ exercised by the Player-1. Since such knowledge gained by Player-2 may be damaging to the Player-1, it tries to choose a control policy to maximize this. This gives rise to the maxmin problem

$$\sup_{u \in \mathcal{U}_{ad}} \inf_{\nu \in M_c(u)} H(\nu|\mu).$$

This is the general problem considered in this paper for different constraints $M_c(u)$.

Another situation that leads to similar maxmin problem can be described as follows: Let μ^0 denote the measure induced by an uncontrolled stochastic system. Suppose this system without control is intrinsically unstable in the sense that there exists an $\varepsilon > 0$ such that

$$\mu^0\{x \in \Sigma : d(x, o) > r\} > \varepsilon$$

for every finite positive number r. The objective of Player-1 is to stabilize the system (stay away from μ^0) by choosing an appropriate control policy while Player-2 tries to destabilize the process using its limited resources (energy limitations). This leads to the maxmin problem $\sup_{u \in \mathcal{U}_{ad}} \inf_{\nu \in M_c(u)} H(\nu|\mu^0)$.

Note that another equivalent formulation is the minimax problem $\inf_{u \in \mathcal{U}_{ad}} \sup_{\nu \in M_c(u)} -H(\nu|\mu)$. Here it is assumed that nature assumes a worst case strategy by maximizing $-H(\nu|\mu)$ for a fixed nominal measure $\mu \in M(u)$ over $\nu \in M_c(u)$ while the control policy $u \in \mathcal{U}_{ad}$ is chosen to minimize it. This formulation is shown in Section 20.5.2 to be equivalent to minimax robust control methods.

There are two possible physically meaningful choices of the constraint set M_c. These are

$$M_c(u) = M_o(u) \equiv \{\nu \in M(u) : \int \ell^u d\nu \leq \gamma\}$$

for some $\gamma < \int \ell^u d\mu \equiv m$, and

$$M_c(u) = M_p(u) \equiv \{\nu \in M(u) : \int \ell^u d\nu \geq \gamma\}$$

for $\gamma > m$.

Next, we discuss the primal and dual problems and their relations to robustness of stochastic systems.

20.1.1.1. *Primal minimizing problems*

(PP1): For a given control law $u \in \mathcal{U}_{ad}$, and a fixed nominal measure $\mu \in M(u)$, define the functional

$$J_o(u) \equiv \inf\{H(\nu|\mu) : \nu \in M_o(u)\}. \tag{20.1}$$

The ultimate objective is to find a control policy that maximizes this functional over the set $\mathcal{U}_{ad}$. As an intermediate step, we prove the existence of a $\nu^* \in M_o(u)$ at which the infimum indicated above is attained. Then we present an explicit characterization of this measure.

(PP2): Similarly, for a given control law $u \in \mathcal{U}_{ad}$, and a fixed measure $\mu \in M(u)$, define the functional

$$J_p(u) \equiv \inf\{H(\nu|\mu) : \nu \in M_p(u)\}. \tag{20.2}$$

Again we derive similar results for this problem as in (PP1.)

Clearly, the minimizing measures in (20.1), (20.2) will depend on the nominal measure $\mu \in M(u)$ and the energy function ℓ^u. Further, it is interesting to point out that the generalization of (PP1), (PP2) to multiple energy constraints $\int_\Sigma \ell_j^u d\nu \leq \gamma_j$, $\int_\Sigma \ell_j^u d\nu \geq \gamma_j$, respectively, $1 \leq j \leq M$ is a natural extension, and can be dealt with without any extra effort leading to similar conclusions.

Note also that problem PP1 (and similarly PP2) can be viewed as a generalization of Jayne's [30, 31] maximum entropy density estimation described by $\sup_{f \in A(u)} - \int_{\Re^n} \log(f(x))f(x)dx$, where $A(u) = \{f; \int_{\Re^n} \ell^u(x)f(x)dx \leq \gamma\}$, f is the probability density of the system defined on a finite dimensional space which is unknown.

Now we proceed to re-cast problems (PP1) and (PP2) in terms of maxmin games, in which it is assumed that the objectives of the control and uncertainty are conflicting (as it is usually done in robust control problems).

20.1.1.2. *Primal maxmin problems*

(PMP1): For a given $\mu \in M(u)$, find a control law $u^* \in \mathcal{U}_{ad}$ so that

$$J_o(u^*) = \sup_{u \in \mathcal{U}_{ad}} J_o(u). \tag{20.3}$$

(PMP2): For a given $\mu \in M(u)$, find a control law $u^* \in \mathcal{U}_{ad}$ so that

$$J_p(u^*) = \sup_{u \in \mathcal{U}_{ad}} J_p(u). \tag{20.4}$$

In subsequent sections (specifically, Lemma 20.1 and Theorem 20.4) it is shown that the Primal Problems are explicitly related to information theoretic concepts of large deviations, dissipation inequalities of robustness, and risk-sensitive optimization. These relations are summarized below.

20.1.2. *Summary of the main relations*

20.1.2.1. *Disturbance attenuation in robustness (see Section 20.5.2)*

For a given $u \in \mathcal{U}_{ad}$ let

$$L_2(\nu; \mathcal{H}) \triangleq \Big\{ \phi^u : \Sigma \to \mathcal{H} : \mathcal{H} \quad \text{a Hilbert Space}, \phi^u \text{ a random variable such that } \int_\Sigma \|\phi\|^2_{\mathcal{H}} d\nu < \infty \Big\}$$

denote the Hilbert Space of random variables. Let $L_2(\nu; \mathcal{Z})$ and $L_2(\nu; \mathcal{D})$ denote the Hilbert Spaces of tracking signals and disturbance signals, respectively. For a given $u \in \mathcal{U}_{ad}$, let $T^u : \mathcal{D} \to \mathcal{Z}$ be a bounded linear operator with induced norm defined by

$$J(u) \triangleq \|T^u\| = \sup_{\|d\|_{L_2(\nu;\mathcal{D})} \neq 0} \frac{\|z\|^2_{L_2(\nu;\mathcal{Z})}}{\|d\|^2_{L_2(\nu;\mathcal{D})}}, \qquad z = T^u d, \quad d \in \mathcal{D}. \tag{20.5}$$

The robust optimal control problem is to find a $u^* \in \mathcal{U}_{ad}$ which minimizes the worst case induced norm, i.e.,

$$J(u^*) = \inf_{u \in \mathcal{U}_{ad}} \|T^u\|.$$

The sub-optimal disturbance attenuation is to ensure that for all $u \in \mathcal{U}_{ad}$ that $J(u) \leq \frac{1}{2s}$, $s > 0$, which is equivalent to

$$\begin{aligned} J^s(u) &= \sup_{d \in L_2(\nu;\mathcal{D})} \Big\{ s \int \|z\|^2_{\mathcal{Z}} d\nu^u - \frac{1}{2} \int \|d\|^2_{\mathcal{D}} d\nu \Big\} \\ &= - \inf_{d \in L_2(\nu;\mathcal{D})} \Big\{ \frac{1}{2} \int \|d\|^2_{\mathcal{D}} d\nu - s \int \|z\|^2_{\mathcal{Z}} d\nu \Big\}, \qquad (20.6) \end{aligned}$$

and ensuring that the pay-off is non-positive.

Consider next, the primal problem (PP2); then, the dual functional associated with converting the primal problem (PP2) into an equivalent unconstrained optimization in which $s \in \Re$ is the Lagrange multiplier given by

$$J^{s,\gamma}(u, \nu^*) - \inf_{\nu \in \mathcal{M}(\Sigma)} \Big\{ H(\nu|\mu) - s\Big(E_\nu\Big(\ell^u\Big) - \gamma\Big)\Big\}. \qquad (20.7)$$

If $H(\nu|\mu) < \infty$, then ν is absolutely continuous with respect to μ, and in many problems it can be shown that $H(\nu|\mu) = \frac{1}{2}\int \|d\|^2_{\mathcal{D}} d\nu$ (see the case of Nonlinear partially observable controlled problem of Section 20.5.2). Thus, by substituting $H(\nu|\mu) = \frac{1}{2}\int \|d\|^2_{\mathcal{D}} d\nu$, $\ell^u = \|z\|^2_{\mathcal{Z}}$ into (20.7) we recover (20.6). Hence, the sub-optimal disturbance attenuation problem (20.6) is equivalent to the the primal problem (PP2), in which s is the Lagrange multiplier. Moreover, larger values of s imply higher attenuation and hence higher dissipation. An application of the above results to general nonlinear partially observable systems is discussed in Section 20.5.2.

20.1.2.2. *Legendre–Fenchel or Cramer Transform (see Lemma 20.1)*

The dual functionals associated with converting the primal problems (PP1) and (PP2) into equivalent unconstrained optimization problems are equal to the Legendre–Fenchel transforms of ℓ^u defined by

$$\begin{aligned} \Psi^\star_\mu(\gamma) &\triangleq \sup_{s \in \Re} \Big\{ s\gamma - \log E_\mu \Big\{ e^{s\ell^u} \Big\} \Big\} \\ &= \sup_{s \in \Re} \inf_{\nu \in \mathcal{M}(\Sigma)} \Big\{ H(\nu|\mu) - s\Big(E_\nu\Big(\ell^u\Big) - \gamma\Big)\Big\}. \qquad (20.8) \end{aligned}$$

The Legendre–Fenchel transform of ℓ^u is employed in Large Deviations Theory [1–3] to identify the entropy rate functional $\Psi^\star_\mu(\gamma)$ associated with rare events.

In particular, for (PP1) the supremum is taken over $s \leq 0$ and for (PP2) the supremum is taken over $s \geq 0$. Moreover, the non-trivial solutions (e.g., $\nu \neq \mu$) corresponds for (PP1) to $m \stackrel{\triangle}{=} E_\mu\Big\{\ell^u\Big\} > \gamma$, and for (PP2) to $m \stackrel{\triangle}{=} E_\mu\Big\{\ell^u\Big\} < \gamma$. The Legendre–Fenchel transform of ℓ^u enjoys several monotonicity properties, one of which is a condition for searching for the value of s which maximizes the pay-off $J^s(u)$. The identification of the entropy rate functional associated with stochastic differential equations driven by Brownian Motion which is scaled by a small parameter can be found in [1]. Related material are also found in [22, 23].

20.1.2.3. *Large deviations and Chernoff bounds*

Consider a controlled random process, $\ell^u \stackrel{\triangle}{=} \{\ell^u(t)\}_{t\geq 0}$, $\ell^u : [0,T]\times\Omega \to \Re$, in which $u \stackrel{\triangle}{=} \{u(t)\}_{t\geq 0}$ is the controlled process, belonging to some admissible set $\mathcal{U}_{ad}$, all defined on a complete probability space with filtration $\Big(\Omega, \mathcal{F}, P^u; \{\mathcal{F}_t\}_{t\geq 0}\Big)$.

We are interested in the following optimization problems.

$$J_{min}(u^*) = \inf_{u\in\mathcal{U}_{ad}} \frac{1}{T}\log P^u\Big(\frac{\ell^u(T)}{T} \leq \gamma\Big) \tag{20.9}$$

$$J_{max}(u^*) = \max_{u\in\mathcal{U}_{ad}} \frac{1}{T}\log P^u\Big(\frac{\ell^u(T)}{T} \geq \gamma\Big). \tag{20.10}$$

Such problems are often encountered in various disciplines, and they are very difficult to solve. However, by using the Chernoff bound one deduces the inequalities

$$J_{min}(u^*) \leq \inf_{u\in\mathcal{U}_{ad}} \Big\{ -s\gamma + \frac{1}{T}\log E_{P^u}\Big\{e^{s\ell^u(T)}\Big\}\Big\}, \quad s<0, \quad \forall\gamma\in\Re$$

$$J_{max}(u^*) \leq \inf_{u\in\mathcal{U}_{ad}} \Big\{ -s\gamma + \frac{1}{T}\log E_{P^u}\Big\{e^{s\ell^u(T)}\Big\}\Big\}, \quad s>0, \quad \forall\gamma\in\Re.$$

Define $\bar{\ell}^u(T) \stackrel{\triangle}{=} \frac{\ell^u(T)}{T}$, and suppose $\ell^u \stackrel{\triangle}{=} \{\ell^u(t)\}_{t\geq 0}$ satisfies the law of large numbers which states that $\bar{\ell}^u(T)$ converges almost surely as $T\to\infty$. Then under the condition that an entropy rate functional exists (the large deviations principle holds) [1] it is expected that the Chernoff bounds are exact, that is,

$$P^u\Big(\frac{\ell^u(T)}{T} \leq \gamma\Big) \simeq e^{-T\Psi^\star_{P^u}(\gamma)}, \qquad P^u\Big(\frac{\ell^u(T)}{T} \geq \gamma\Big) \simeq e^{-T\Psi^\star_{P^u}(\gamma)},$$

as $T\to\infty$, where $\Psi^\star_{P^u}(\gamma)$ is the entropy rate functional given by

$$\Psi^\star_{P^u}(\gamma) = \sup_{s\in\Re}\Big\{s\gamma - \Psi_{P^u}(s)\Big\}, \qquad \Psi_{P^u}(s) \stackrel{\triangle}{=} \lim_{T\to\infty}\frac{1}{T}\log E_{P^u}\Big\{e^{s\ell^u(T)}\Big\}. \tag{20.11}$$

Specifically,

$$\lim_{T\to\infty} \frac{1}{T} \log P^u\Big(\frac{\ell^u(T)}{T} \leq \gamma\Big) = -\sup_{s\leq 0}\Big\{s\gamma - \Psi_{P^u}(\gamma)\Big\}, \tag{20.12}$$

for $\gamma < \lim_{T\to\infty} \frac{1}{T} E_{P^u}(\ell^u(T))$, and

$$\lim_{T\to\infty} \frac{1}{T} \log P^u\Big(\frac{\ell^u(T)}{T} \geq \gamma\Big) = -\sup_{s\geq 0}\Big\{s\gamma - \Psi_{P^u}(\gamma)\Big\}, \tag{20.13}$$

for $\gamma > \lim_{T\to\infty} \frac{1}{T} E_{P^u}(\ell^u(T))$. Clearly, if the relative entropy and average energy constraints in (PP1) and (PP2) are defined per unit time, in the limit as, $T \to \infty$, the dual functional associated with these problems are equivalent to the computation of the entropy rate functional as given in the right side of (20.12), (20.13), respectively.

20.1.2.4. *Risk-seeking versus risk-averse optimization (see Theorem 20.4)*

In the context of robust disturbance attenuation of uncertain systems, the measure $\mu \in \mathcal{M}(\Sigma)$ corresponds to the nominal measure, $\nu \in \mathcal{M}(\Sigma)$ corresponds to the uncertain measure, and the fidelity constraints $E_\nu\{\ell^u\} \leq \gamma$ and $E_\nu\{\ell^u\} \geq \gamma$ represent average energy constraints.

For (PP1), it is shown in Theorem 20.4 that the average energy constraint with respect to all uncertain measures $\nu \ll \mu$, $E_\nu\{\ell^u\} \leq \gamma$, is below the average energy of the nominal model, $E_\mu\{\ell^u\} > \gamma$; hence it represents a risk-seeking or an optimistic scenario. For (PP2), it is shown in Theorem 20.4 that the average energy constraint with respect to all uncertain measures $\nu \ll \mu$, $E_\nu\{\ell^u\} \geq \gamma$ is above the average energy of the nominal model $E_\mu\{\ell^u\} < \gamma$; hence it represents a risk-averse or a pessimistic scenario. The parameter $s \in \Re$ is the Lagrange multiplier associated with the dual functional of the primal problems (PP1) and (PP2). In particular, $s \leq 0$ corresponds to PP1) while $s \geq 0$ corresponds to (PP2).

Let ν^* denote the measure obtained by carrying the infimum in (20.8) over all measures ν, such that $H(\nu|\mu) < \infty$. Since entropy rate functionals are nonnegative [1–3], that is, $\Psi^\star_\mu(\gamma) \geq 0, \forall \gamma \in \Re$, then (20.8) implies the inequality

$$\Big\{E_{\nu^*}\{\ell^u\} - \frac{1}{s} H(\nu^*|\mu)\Big\}\Big|_{s=s^*} \leq \gamma \tag{20.14}$$

where s^* is obtained by finding the supremum in (20.8). Since $s \geq 0$ is associated with the dual functional of (PP2), then condition (20.14) is a dissipation inequality associated with the optimal solution of (PP2) (this connection is established in Section 20.5.2). In addition, s^* is obviously inversely proportional to the optimal attenuation level (see Section 20.5.2).

The above statements which are derived in subsequent sections provide a brief summary of the main relations between the Legendre–Fenchel transform, robustness, and risk-seeking and risk-averse optimization. Additional connections are established through the Legendre–Fenchel transform in Theorem 20.1. This theorem includes monotonicity properties of solutions to (PP1) and (PP2), existence, uniqueness and characterization of the value of the Lagrange multiplier "s" associated with PP1) and PP2), and several properties associated with solutions to stochastic minimax games.

The paper is organized as follows. In Section 20.2, the basic definitions and duality relations are introduced between relative entropy, cumulant generating functions and Legendre–Fenchel transforms. In Section 20.3, existence of the minimizing measures associated with (PP1), (PP2) is proved under the topology of weak convergence. In Section 20.4, the equivalence between the constrained and unconstrained problems is derived, and the basic results leading to the relations between information theoretic concepts of large deviations, dissipation inequalities of robustness, and risk-sensitive optimization summarized above are derived (Lemma 20.1 and Theorem 20.4). In Section 20.5, the abstract primal problems (PP1), (PP2) are applied to nonlinear continuous time stochastic partially observable control systems. Moreover, their relations with the sub-optimal disturbance attenuation problems are discussed. Existence of optimal control laws and worst case measure are proved by reformulating the problem using conditional distributions, which satisfy certain Partial Differential Equations (PDE's). Finally the paper is concluded with an example.

20.2. Duality relations and tilted measures

In this section basic definitions and duality relations between relative entropy, cumulant generating functions, and Legendre–Fenchel transforms are introduced.

The concepts introduced here are used extensively in subsequent sections to characterize the optimal strategies, to establish the various relations discussed earlier in an abstract setting, and to apply the abstract setting to nonlinear stochastic control of partially observable systems.

The following definition is standard; it will be used throughout the paper.

Definition 20.1. Let $\nu, \mu \in \mathcal{M}(\Sigma)$ and $\ell : \Sigma \to \Re$ a measurable function.

1) The moment generating function of ℓ with respect to μ is defined by

$$M_\mu(s) \stackrel{\triangle}{=} E_\mu\left(e^{s\ell}\right) = \int_\Sigma e^{s\ell} d\mu, \quad s \in \Re. \tag{20.15}$$

2) The cumulant generating function of ℓ with respect to μ is defined by

$$\Psi_\mu(s) \stackrel{\triangle}{=} \log M_\mu(s) = \log \int_\Sigma e^{s\ell} d\mu, \quad s \in \Re. \tag{20.16}$$

3) The Legendre–Fenchel Transform of $\Psi_\mu(s)$ is defined by

$$\Psi^\star_\mu(x) \stackrel{\triangle}{=} \sup_{s \in \Re} \left\{ sx - \Psi_\mu(s) \right\}, \quad x \in \Re, \tag{20.17}$$

and this is also called the Entropy Rate Functional of ℓ.

4) The relative entropy of ν with respect to μ is defined by

$$H(\nu|\mu) \stackrel{\triangle}{=} \begin{cases} \int_\Sigma \log(\frac{d\nu}{d\mu}) d\nu & \text{if } \nu << \mu \\ \\ +\infty & \text{otherwise.} \end{cases} \tag{20.18}$$

It can be shown that $\Psi_\mu(s)$ as a function of ℓ is convex, $H(\nu|\mu)$ as a function of $\nu, \mu \in \mathcal{M}(\Sigma)$ is convex in both the arguments, $H(\nu|\mu) \geq 0$, and $H(\nu|\mu) = 0$, if and only if $\nu = \mu$. Thus, $H(\nu|\mu)$ can be considered as a measure of discrepancy between the two measures.

The moment generating function (20.15) and cumulant generating function (20.16), when employed in the context of stochastic control and filtering, represent the so-called risk-sensitive pay-off [10, 14, 16, 32]. For linear quadratic problems, the solution of risk-sensitive problems is equivalent to the solution of the minimax game formulation of the disturbance attenuation problem [10, 16]. Similarly, connections between risk-sensitive pay-off functionals and deterministic and stochastic minimax games with square integrable disturbances are also established in [13, 18, 21, 25]. The Legendre–Fenchel Transform (20.17) is employed in Large Deviations Theory to identify the entropy rate functional, which describes the exponential rate of convergence to zero of rare events. Finally, it is noted that the cumulant generating function and the relative entropy are in duality with respect to a Legendre–Fenchel transform, and the following result is a variant of a theorem found in [25].

Theorem 20.1. *For a given $s \in \Re$, and $\ell : \Sigma \to \Re$ a measurable function such that $s\ell$ is bounded below,*

$$-\Psi_\mu(s) = -\log E_\mu(e^{s\ell}) = \inf_{\nu \in \mathcal{M}(\Sigma)} \Big\{ H(\nu|\mu) - \int_\Sigma s\ell d\nu \Big\}. \quad (20.19)$$

Moreover, if $\ell e^{s\ell} \in L_1(\mu)$, then the infimum in (20.19) is attained by the tilted probability measure ν^* given by

$$d\nu^* = \frac{e^{s\ell} d\mu}{\int_\Sigma e^{s\ell} d\mu}. \quad (20.20)$$

Proof. The proof is given in [25]. □

20.3. Existence of minimizing measures

This section is concerned with establishing the existence of minimizing measures associated with the problems (20.1) and (20.2). For simplicity of notations, the explicit dependence of measures and functions on control u is suppressed.

Let Σ be a complete separable metric space with metric d and sigma-algebra $\mathcal{B}(\Sigma)$ generated by open sets in Σ. Let $BC(\Sigma)$ denote the space of bounded continuous real valued functions defined on Σ. Furnished with the sup norm topology, this is a Banach space. Let $\mathcal{M}(\Sigma)$ denote the space of all countably additive Borel probability measures on $\mathcal{B}(\Sigma)$. For any sequence $\{\nu_n\}, n \geq 1$ in $\mathcal{M}(\Sigma)$ weak convergence to $\nu \in \mathcal{M}(\Sigma)$ is defined by $\int_\Sigma f d\nu_n \to \int_\Sigma f d\nu$, for every $f \in BC(\Sigma)$. The space of probability measures $\mathcal{M}(\Sigma)$ can be metrized in such a way that $\mathcal{M}(\Sigma)$ itself becomes a complete separable metric space (Polish space), and that weak convergence is equivalent to convergence in this metric.

We are interested in the following minimization problems associated with the problems (20.1) and (20.2) as stated before.

Problem 20.1. (Basic Extremal Problems) *Let $\mathcal{M}_o(\Sigma)$ be a weakly compact subset of $\mathcal{M}(\Sigma)$ and $\ell \in BC(\Sigma)$, $\mu \in \mathcal{M}_o(\Sigma)$ a fixed (nominal) measure, $m \stackrel{\triangle}{=} E_\mu(\ell) = \int_\Sigma \ell d\mu$ and $\gamma \in \Re$.*

Define the functional $J(\nu) \stackrel{\triangle}{=} H(\nu|\mu)$, and the (constraint) sets $M_o = \{\nu \in \mathcal{M}_o(\Sigma);\ \int_\Sigma \ell d\nu \leq \gamma\}$ and $M_p = \{\nu \in \mathcal{M}_o(\Sigma);\ \int_\Sigma \ell d\nu \geq \gamma\}$. Then

P1) for $\gamma < m$, find $\nu^* \in M_o$ so that

$$J(\nu^*) = \inf_{\nu \in M_o} J(\nu) \stackrel{\triangle}{=} J_o(\nu^*). \quad (20.21)$$

P2) for $\gamma > m$, find $\nu^* \in M_p$ so that

$$J(\nu^*) = \inf_{\nu \in M_p} J(\nu) \stackrel{\triangle}{=} J_p(\nu^*). \tag{20.22}$$

Since $J : \mathcal{M}(\Sigma) \to [0, \infty]$ is a convex functional and the sets M_o and M_p are convex, the problems as stated above are convex optimization problems. In the following theorem we establish the existence of a minimizing measure.

Theorem 20.2.

(1) Consider the set $\mathcal{K} \stackrel{\triangle}{=} \{(\nu, \mu) \in \mathcal{M}(\Sigma) \times \mathcal{M}(\Sigma) : \nu << \mu\}$. The relative entropy functional $(\nu, \mu) \to H(\nu|\mu)$ from $\mathcal{K} \longrightarrow [0, \infty]$ is (jointly) weakly lower semi continuous.
(2) For each $\ell \in BC(\Sigma)$ and $\gamma \in \Re$, the sets M_o and M_p are compact in the weak topology. Moreover, for P1 (resp. P2) the same result holds if ℓ is continuous and bounded from below (resp. bounded above).
(3) For each problem of 20.1, there exists a minimizing measure $\nu^* \in M_o$ and $\nu^* \in M_p$. Moreover, for P1 (resp. P2) the same result holds if ℓ is continuous and bounded from below (resp. bounded above).

Proof. (1): An elegant proof of this statement, based on the duality between relative entropy and free energy, can be found in [4].

(2): For $\ell \in BC(\Sigma)$, it is easy to verify that the sets M_o and M_p are weakly closed. Being weakly closed subsets of the weakly compact set $\mathcal{M}_o(\Sigma)$, they are weakly compact. Next we consider the case of ℓ continuous and bounded from below. The proof is similar for M_0 and M_p. We consider M_0 and suppose we have a net $\{\nu_\alpha\}$ in M_0, where $\alpha \in (\mathcal{D}, \leq)$ (a directed set). Since $\mathcal{M}_o(\Sigma)$ is weakly compact and $M_o \subset \mathcal{M}_o(\Sigma)$, then $\{\nu_\alpha\}$ is a net in $\mathcal{M}_o(\Sigma)$, and by assumption there exists a subnet $\{\nu_{\alpha_k}\}$ and a measure $\nu \in \mathcal{M}_o(\Sigma)$ such that $\nu_{\alpha_k} \stackrel{w}{\to} \nu$.

We will show that $\nu \in M_0$. Since ℓ is bounded below, there exists $L \in \Re$ such that $g(x) \stackrel{\triangle}{=} \ell(x) - L \geq 0$, where L denotes the lower bound. Then, for any $m \subset N$, we have

$$\overline{\lim} \int_\Sigma g(x) \wedge m \ \ d\nu_\alpha(x) = \int_\Sigma g(x) \wedge m \ \ d\nu(x) \leq \int_\Sigma g(x) d\nu(x).$$

Hence there exists $\alpha_0 \in \mathcal{D}$ such that

$$\begin{aligned} &\sup_{\alpha \geq \alpha_0} \int_\Sigma g(x) \wedge m \ \ d\nu_\alpha(x) \leq \int_\Sigma g(x) d\nu(x), \\ \text{and} \qquad &\sup_m \sup_{\alpha \geq \alpha_0} \int_\Sigma g(x) \wedge m \ \ d\nu_\alpha(x) \leq \int_\Sigma g(x) d\nu(x). \end{aligned} \tag{20.23}$$

Clearly,

$$\begin{aligned}\overline{\lim}\int_\Sigma g(x)d\nu_\alpha(x) &= \overline{\lim}\sup_{m\in N}\int_\Sigma g(x)\wedge m\ \ d\nu_\alpha(x)\\ &= \inf_{\alpha_0}\sup_{\alpha\geq\alpha_0}\sup_{m}\int_\Sigma g(x)\wedge m\ \ d\nu_\alpha(x)\\ &= \inf_{\alpha_0}\sup_{m}\sup_{\alpha\geq\alpha_0}\int_\Sigma g(x)\wedge m\ \ d\nu_\alpha(x)\\ &\leq \sup_{m}\sup_{\alpha\geq\alpha_0}\int_\Sigma g(x)\wedge m\ \ d\nu_\alpha(x).\end{aligned}$$

Combining the above with (20.23) we obtain

$$\overline{\lim}\int_\Sigma g(x)d\nu_\alpha(x)\leq\int_\Sigma g(x)d\nu(x). \tag{20.24}$$

For the reverse inequality, notice that for every $m\in N$ we have

$$\underline{\lim}\int_\Sigma g(x)d\nu_\alpha(x)\geq\underline{\lim}\int_\Sigma g_m(x)d\nu_\alpha(x)=\int_\Sigma g_m(x)d\nu(x),\quad\forall m\in N,$$

where $g_m = g\wedge m$. The equality in the above holds because g_m is bounded and continuous. Letting $m\to\infty$ and noting that $g_m\uparrow g$, it follows from Lebesgue monotone convergence theorem that

$$\underline{\lim}\int_\Sigma g(x)d\nu_\alpha(x)\geq\int_\Sigma g(x)d\nu(x). \tag{20.25}$$

By combining (20.24) and (20.25), we have $\lim_\alpha\int_\Sigma g(x)d\nu_\alpha(x) = \int_\Sigma g(x)d\nu(x)$. Hence $\nu\in M_0$, proving the weak compactness of the set M_0. The proof for M_p is similar.

(3): Existence of a minimizing measure follows from (1), (2) and Weierstrass theorem.

For uniqueness, note that, for fixed μ, the real valued function $\nu\longrightarrow H(\nu|\mu)$ defined on the set $\{\nu\in\mathcal{M}(\Sigma); H(\nu|\mu)<\infty\}$, is strictly convex* and therefore the minimizing measure is unique. □

Remark 20.1. In general the identification of a weak compact set is not easy when considering abstract spaces. However, for complete separable spaces, using Prokhorov's theorem [4], the following sufficient conditions for a family of probability of measures on Σ are derived.

Let $\ell:\Sigma:\to\Re\cup\{\infty\}$ be a measurable function satisfying

*We have $\frac{d^2}{dx^2}\Big(x\log x\Big)>0$ for $x>0$, hence $x\log x$ is strictly convex on $(0,\infty)$. Also $f=\frac{d\nu}{d\mu}\geq 0$. Since $0.\log 0=0$, then one can redefine $H(\nu|\mu)=\int_{\{x\in\Sigma;f(x)>0\}}f\log f d\mu$, which is strictly convex.

i) $\inf_{x\in\Sigma} \ell(x) > -\infty$,

ii) The level set $\mathcal{Z}_\ell(M) \triangleq \{x \in \Sigma : \ell(x) \leq M\}$ is relatively compact subset of Σ, for each $M < \infty$.

Such a function is called a tightness function on Σ.

Define a functional $G : \mathcal{M}(\Sigma) \to \Re \cup \{\infty\}$ by $G(\nu) \triangleq \int_\Sigma \ell(x)d\nu(x)$, assuming ℓ is a tightness function on Σ. Then it can be shown [4] that the set $\mathcal{Z}_G(M) \triangleq \{\nu \in \mathcal{M}(\Sigma) : G(\nu) \leq M\}$ is tight for each $M < \infty$. Using the the tightness of $\mathcal{Z}_G(M)$ and Prokhorov's theorem it follows that $\mathcal{Z}_G(M)$ is a relatively compact subset of $\mathcal{M}(\Sigma)$.

Consider a non-negative Borel measurable function $\ell : \Sigma \to \Re^m$ such that the set

$$\mathcal{Z}_\ell(M) \triangleq \Big\{x \in \Sigma : \ell_i(x) \leq M_i, i = 1, 2, \ldots, m\Big\},$$

is compact for every $M \triangleq (M_1, \ldots, M_m)^{tr} \in \Re^m$. Then the family

$$\mathcal{Z}_G(M) \triangleq \Big\{\nu \in \mathcal{M}(\Sigma) : \int \ell_i(x)d\nu(x) \leq M_i, i = 1, 2, \ldots, m\Big\},$$

is relatively compact for each $M \in \Re^m$.

20.4. Equivalence of constrained and unconstrained optimization and Legendre–Fenchel transform

In this section we reformulate the constrained optimization problem 20.1, as an unconstrained one using the theory of Lagrange multipliers, and then show the equivalence of the two problems. Here we also demonstrate interesting relations between the rate functional associated with Large Deviations theory, Legendre–Fenchel transform and the problem of disturbance attenuation. Through the control version of Legendre–Fenchel transform, several properties of the minimizing measure are established including the range of values of the Lagrange multiplier for which the constrained and unconstrained problems are equivalent.

In addition, relations with optimization of stochastic systems in which the pay-off is an exponential functional, known as risk-sensitive problems, are also obtained. In particular, it is shown that problem $J_o(\nu^*)$ in (20.21) corresponds to the optimistic scenario (emphasizing the best cases) in which strategies are risk-seeking, while problem $J_p(\nu^*)$ in (20.22) corresponds to the pessimistic scenario (emphasizing the worst cases) in which strategies are risk-averse.

The constrained problems 20.1 are converted into unconstrained ones by introducing the Lagrangian functionals and the dual functionals. In Theorem 20.4 the equivalence of this dual functional with the two problems in 20.1 is established by specifying the range of values of the Lagrange multiplier $s \in \Re$.

Recall that our uncertain system is modeled as a map

$$M : \mathcal{U}_{ad} \longrightarrow 2^{\mathcal{M}(\Sigma)} \setminus \emptyset.$$

This is too general a class. For applications, we restrict to the class of (perturbed) systems that induce a compact family of measures. Let $wk(\mathcal{M}(\Sigma)) \subset 2^{\mathcal{M}(\Sigma)} \setminus \emptyset$ denote the class of nonempty weakly compact subsets of the set $\mathcal{M}(\Sigma)$. Our system is then given by the map

$$M : \mathcal{U}_{ad} \longrightarrow wk(\mathcal{M}(\Sigma)),$$

which assigns to each admissible control $u \in \mathcal{U}_{ad}$ a weakly compact set $M(u)$ of $\mathcal{M}(\Sigma)$. Recall the constrained sets $\{M_o(u), M_p(u)\} \subset M(u)$ for any $u \in \mathcal{U}_{ad}$. For a fixed $\mu \in M(u)$ and each arbitrary but fixed $\gamma \in R$ and every $s \in \Re$, define the Lagrangian

$$J^{s,\gamma}(u,\nu) \triangleq H(\nu|\mu) - s\Big(E_\nu(\ell^u) - \gamma\Big). \tag{20.26}$$

Then introduce the unconstrained problem,

$$J^{\gamma}(u,\nu^*) \equiv \sup_{s \in \Re} \inf_{\nu \in M(u)} J^{s,\gamma}(u,\nu),$$

and the constrained problems as follows:

$$J_o^{\gamma}(u,\nu^*) \equiv \inf_{\nu \in M_o(u)} H(\nu|\mu) \tag{20.27}$$

$$J_p^{\gamma}(u,\nu^*) \equiv \inf_{\nu \in M_p(u)} H(\nu|\mu). \tag{20.28}$$

Note that these are the problems P1) and P2), respectively, of Problem 20.1. The following theorem establishes the equivalence of the constrained and unconstrained optimization problems.

Theorem 20.3. *Consider the problems P1) and P2) (see Problem 20.1) corresponding to a given control policy $u \in \mathcal{U}_{ad}$. Suppose they are well defined in the sense described below*[†]*. Then*

(1) if $\inf_{x \in \Sigma} \ell^u(x) < \gamma$, we have

$$J_o^{\gamma}(u,\nu^*) = J^{\gamma}(u,\nu^*). \tag{20.29}$$

[†] i.e., there exists at least one measure $\nu_o \in M_o(u)$ and one measure $\nu_p \in M_p(u)$ for which $H(\nu_o|\mu) < \infty$ and $H(\nu_p|\mu) < \infty$.

(2) if $\sup_{x\in\Sigma} \ell^u(x) > \gamma$, *we have*

$$J_p^\gamma(u, \nu^*) = J^\gamma(u, \nu^*). \tag{20.30}$$

Moreover, for (1) (resp. (2)) the same conclusions hold if ℓ^u *is continuous and bounded below (resp. above).*

Proof. Let X denote the vector space of countably additive regular signed measures $\nu : \mathcal{B}(\Sigma) \to \Re$ having finite total variation and

$$\Omega \equiv \Big\{\nu \in X;\ \nu \text{ is a regular countably additive probability measure}\Big\}.$$

Also, Ω is a convex set, as if $\nu_1, \nu_2 \in \Omega$ we have $\nu(\Sigma) = \lambda\nu_1(\Sigma) + (1 - \lambda)\nu_2(\Sigma) = 1$ so $\nu = \lambda\nu_1 + (1-\lambda)\nu_2 \in \Omega$ for any $\lambda \in [0, 1]$. For an arbitrary but fixed $u \in \mathcal{U}_{ad}$, define the mapping

$$\nu \longrightarrow G(\nu) \triangleq \int_\Sigma \ell^u d\nu - \gamma.$$

Clearly this is a linear map from X into the ordered vector space $(\Re, \prec)$ and so convex. Also it follows from the properties of relative entropy that $\nu \to H(\nu|\mu)$ is a convex functional. Thus the functional

$$\nu \longrightarrow H(\nu|\mu) - sG(\nu) \equiv J^{s,\gamma}(u, \nu),$$

is a convex map from X into the ordered vector space $(\Re, \prec)$. Therefore, the equivalence of constrained and unconstrained problems follows from the Lagrange Duality theorem in [33, pp. 224-225] provided Ω has interior points. However, this follows trivially from the following argument.

Consider the first problem. By our assumption, the set $\Gamma = \{x \in \Sigma : \ell^u(x) < \gamma\}$ is nonempty. Since the elements of $M(u)$ are regular measures, we can choose a measure $\nu_1 \in M(u)$ with support Γ, that is, $\nu_1(\Gamma) = 1$ and $\nu_1(\Gamma^c) = 0$. Then it is clear that $\int_\Sigma \ell^u d\nu_1 < \gamma$. The inequality is strict since ν_1 is a probability measure. This shows that Ω has nonempty interior. For the second problem the argument is similar.

Since the existence of the solution to P1) (resp. P2)) also holds for ℓ^u which are continuous and bounded from below (resp. above), the equivalence of the constrained and unconstrained problems also holds for this class of functions. □

The following result (see Lemma 20.1) presents several properties of the unconstrained problems, including monotonicity properties of the dual functional with respect to γ and conditions for finding the Lagrange multiplier.

Lemma 20.1. *For a given $u \in \mathcal{U}_{ad}$, assume $\ell^u : \Sigma \to \Re$ is a measurable function which is bounded below. Define the Legendre–Fenchel transform of $\Psi_\mu(s)$, corresponding to $\ell = \ell^u$, by*

$$\Psi_\mu^*(\gamma) = \sup_{s \in \Re} \Big\{ s\gamma - \Psi_\mu(s) \Big\}.$$

Then the following statements hold.

(1) $\inf_{\nu \in M(u)} J^{s,\gamma}(u,\nu) = s\gamma - \Psi_\mu(s), \quad \gamma \in \Re, \quad s \in \Re.$

(2) *Letting $\nu^{*,s}$ denote the minimizer in (1), the functional $J^{s,\gamma}(u,\nu^{*,s})$ is concave in $s \in \Re$.*

(3) *$\Psi_\mu^\star(\gamma)$ is a convex function of $\gamma \in \Re$.*

(4) *$\Psi_\mu^\star(\gamma) \geq 0, \forall \gamma \in \Re$.*

(5) *If $\ell^u \in L_1(\mu)$ and $E_\mu(\ell^u) = m^u$ then*

   (a) *$\Psi_\mu^\star(\gamma) = 0$ for $\gamma = m^u$.*

   (b) *$\Psi_\mu^\star(\gamma)$ is non-decreasing for $\gamma \in [m^u, \infty)$.*

   (c) *$\Psi_\mu^\star(\gamma)$ is non-increasing for $\gamma \in (-\infty, m^u]$.*

   (d) *Suppose there exists an open connected set $I \subset \Re$ such that $\ell^u e^{s\ell^u} \in L_1(\mu)$ for all $s \in I$, then $J^{s,\gamma}(u,\nu^{*,s})$ is continuously differentiable with respect to s with the derivative given by*

$$\frac{d}{ds} J^{s,\gamma}(u,\nu^{*,s}) = \gamma - \frac{E_\mu\Big(\ell^u e^{s\ell^u}\Big)}{E_\mu\Big(e^{s\ell^u}\Big)} = \gamma - E_{\nu^{*,s}}\Big(\ell^u\Big), \qquad (20.31)$$

   where

$$d\nu^{*,s} = \frac{e^{s\ell^u} d\mu}{\int_\Sigma e^{s\ell^u} d\mu}. \qquad (20.32)$$

   In addition, if $(\ell^u)^2 e^{s\ell^u} \in L_1(\mu)$ for all $s \in I$, then $J^{s,\gamma}(u,\nu^{*,s})$ is twice continuously differentiable with respect to s satisfying

$$\frac{d^2}{ds^2} J^{s,\gamma}(u,\nu^{*,s}) = -\Big\{ E_{\nu^{*,s}}\Big((\ell^u)^2\Big) - \Big(E_{\nu^{*,s}}(\ell^u)\Big)^2 \Big\} < 0. \quad (20.33)$$

   (e) *If the first assumption of (d) holds then the supremum of the function $s \to J^{s,\gamma}(u,\nu^{*,s})$ is attained on the boundary of the constraint sets M_o, M_p for some $s^* \in I$, i.e.,*

$$E_{\nu^{*,s}}\Big(\ell^u\Big)|_{s=s^*} = \gamma,$$

   where $\nu^{*,s}$ is given by (20.32) and it is unique if the second assumption holds.

Proof. The proof for (1)-(5a-5c) is standard [1, 3]. For the proof of (5d) and (5e) see the Appendix. □

In the following theorem we identify the regions over which $J^{s,\gamma}(u,\nu^{*,s})$ is maximized, and present conditions for finding the Lagrange multipliers for the problems P1 and P2 (see Problem 20.1).

Theorem 20.4. *Recall the constrained Problems of (20.27), (20.28), and suppose $J_o^\gamma(u,\nu^*)$ and $J_p^\gamma(u,\nu^*)$ are finite. Then we have the following statements for Risk-seeking and Risk-Averse Scenarios.*

1) Risk-Seeking Scenario: When $\gamma < m^u \triangleq E_\mu(\ell^u)$, there exists a minimizing measure $\nu^* \in M(u)$ which satisfies

$$J_o^\gamma(u,\nu^*) = \sup_{s\le 0}\Big\{J_o^{s,\gamma}(u,\nu^*)\Big\} = \sup_{s\le 0}\Big\{s\gamma - \Psi_\mu(s)\Big\} \tag{20.34}$$

$$= \Psi_\mu^\star(\gamma) = \inf_{\nu\in M_o(u)} H(\nu|\mu), \tag{20.35}$$

and if $\ell^u e^{s\ell^u} \in L_1(\mu)$ for all $s \in I_-$, where $I_- \subset (-\infty,0]$ is an open connected set, then $\nu^* \in M(u)$ is given by

$$d\nu^* = \frac{e^{s^*\ell^u}d\mu}{\int_\Sigma e^{s^*\ell^u}d\mu}, \qquad \text{for certain } s^* \in I_-. \tag{20.36}$$

Moreover, if both $\ell^u e^{s\ell^u}$ and $(\ell^u)^2 e^{s\ell^u} \in L_1(\mu)$ for all $s \subset I_-$, then the supremum over $s \le 0$ in (20.35) is attained at $s^* \in I_-$ and

$$\gamma = E_{\nu^{*,s}}\Big\{\ell^u\Big\}|_{s=s^*} \le E_{\nu^{*,s}}\Big\{\ell^u\Big\} < E_\mu\Big\{\ell^u\Big\} = m^u, \quad \forall s\in[s^*,0]. \tag{20.37}$$

2) Risk-Averse Scenario: When $\gamma > m^u \triangleq E_\mu(\ell^u)$, there exists a minimizing measure $\nu^* \in M(u)$ which satisfies

$$J_p^\gamma(u,\nu^*) = \sup_{s\ge 0}\Big\{J_p^{s,\gamma}(u,\nu^{*,s})\Big\} = \sup_{s\ge 0}\Big\{s\gamma - \Psi_\mu(s)\Big\} \stackrel{\nabla}{=} \Psi_\mu^\star(\gamma) \tag{20.38}$$

$$= \inf_{\nu\in M_P(u)} H(\nu|\mu) \tag{20.39}$$

and if $\ell^u e^{s\ell^u} \in L_1(\mu)$, for all $s \in I_+$, where $I_+ \subset [0,\infty)$ is an open connected set, then $\nu^* \in M(u)$ is given by

$$d\nu^* = \frac{e^{s^*\ell^u}d\mu}{\int_\Sigma e^{s^*\ell^u}d\mu}, \quad \text{for certain} \quad s^* \in I_+. \tag{20.40}$$

Moreover, if both $\ell^u e^{s\ell^u}$ *and* $(\ell^u)^2 e^{s\ell^u} \in L_1(\mu)$ *for all* $s \subset I_+$*, then the supremum over* $s \geq 0$ *in (20.38) is attained at some* $s^* \in I_+$ *given by*

$$\gamma = E_{\nu^*}\Big\{\ell^u\Big\}|_{s=s^*} \geq E_{\nu^*}\Big\{\ell^u\Big\} > E_{\mu}\Big\{\ell^u\Big\} = m^u, \quad \forall s \in [0, s^*]. \quad (20.41)$$

Further, the above statements also hold for those ℓ^u *which are continuous and bounded from below.*

Proof. See Appendix. □

Remark 20.2. In view of part 2 of Lemma 20.1 the following inequality holds.

$$\Big\{E_{\nu^*}\Big\{\ell^u\Big\} - \frac{1}{s}H(\nu^*|\mu)\Big\}|_{s=s^*} \leq \gamma. \quad (20.42)$$

For $s^* > 0$ corresponding to the risk-averse scenario (e.g., P2), inequality (20.42), when applied to dynamical systems, is equivalent to a dissipation inequality. Thus, the risk-averse scenario leads to strategies of the type which arise in optimal H^∞ disturbance attenuation problems [12]. This connection to disturbance attenuation of dynamical systems is further discussed in Section 20.5.2.

20.5. Applications to partially observed stochastic uncertain systems

In this section, the abstract formulation of Problem 20.1 leading to Lemma 20.1 and Theorem 20.4 is applied to partially observable problems. However, for these results to be applicable, certain transformations of the partially observable problems are required in order to describe the uncertainty and the nominal model through conditional distributions, rather than a priori distribution. Once this is done, existence of the minimax strategies (from the class of wide sense controls [34]), and several properties of the optimal solution are presented, which are exactly equivalent to those stated under Lemma 20.1 and Theorem 20.4.

20.5.1. *Partially observed control of uncertain stochastic systems*

Let $\{x(t)\}_{t\geq 0}$ denote the state process subject to control, $\{y(t)\}_{t\geq 0}$ the observation process, and $\{u(t)\}_{t\geq 0}$ the control process, all defined for a fixed and finite time interval $I \triangleq [0, T]$.

For each $u \in \mathcal{U}_{ad}$ (set of admissible controls to be defined shortly) the nominal state and the observation process, giving rise to a nominal probability measure P, are governed by the following system of stochastic differential equations:

$$\left(\Sigma, \mathcal{B}(\Sigma), P\right) : \begin{cases} dx(t) = f(x(t), u(t))dt + \sigma(x(t))dw(t), & x(0) = \xi \\ dy(t) = h(x(t))dt + Ndv(t), & y(0) = 0. \end{cases} \tag{20.43}$$

Here $x(t) \in \Re^n, y(t) \in \Re^d, u(t) \in \mathcal{U} \subset \Re^k$, $\{w(t)\}_{t\geq 0}$ and $\{v(t)\}_{t\geq 0}$ are independent Brownian motions taking values in $\Re^m, \Re^d$, respectively, which are also independent of the initial state $x(0) = \xi$. The following assumptions are introduced.

Assumption 20.1. The nominal system satisfies the following assumptions:

1) The control $\{u(t); t \in I\}$ is non anticipative and takes values in $\mathcal{U} \subset \Re^k$ which is compact and convex.
2) $f : \Re^n \times \mathcal{U} \to \Re^n, \sigma : \Re^n \to \mathcal{L}(\Re^m; \Re^n), f(x, u) = f_0(x) + f_1(x)u$, and f_0, f_1, σ are bounded and Lipschitz continuous.
3) $h : \Re^n \to \Re^d, h \in BC(\Re^n, \Re^d)$ and twice continuously differentiable.
4) $N \in \mathcal{L}(\Re^d; \Re^d)$ and $\exists \beta > 0$ such that $NN' \geq \beta I_d$.
5) The initial state ξ has distribution P_0.
6) $\lambda : \Re^n \times \mathcal{U} \to \Re, \kappa : \Re^n \to \Re$, are continuous, bounded and $u \longrightarrow \lambda(x, u)$ is convex for all $x \in \Re^n$.

Next, the problem statement is made precise by introducing the precise definition of admissible controls and identifying the spaces on which the nominal system is defined. Consider the sample space

$$\Omega = \Omega^w \times \Omega^x \times \Omega^y \times \Omega^u,$$

where

$$\Omega^w = C(I; \Re^m), \quad \Omega^x = C(I; \Re^n), \quad \Omega^y = C(I, \Re^d), \quad \Omega^u = L_2(I; \mathcal{U}),$$

and $y(0) = 0$ is assumed throughout. Here, $\Omega^w, \Omega^x, \Omega^y$ are endowed with the usual sup-norm topology, while Ω^u is endowed with the weak topology (which is metrizable and separable). A typical element of Ω is

$$\omega(t) = \Big(w(t, \omega), x(t, \omega), y(t, \omega), u(t, \omega)\Big), \qquad 0 \leq t \leq T.$$

Let $\Omega^{w,x} = \Omega^w \times \Omega^x, \Omega^{y,u} = \Omega^y \times \Omega^u$. Then Ω is provided with a filtration $\{\mathcal{F}_t; t \in [0, T]\}$ which is defined as follows.

Let $\mathcal{F}_t^w = \sigma\{w(s); 0 \leq s \leq t\}$, $\mathcal{F}_t^x = \sigma\{x(s); 0 \leq s \leq t\}$, $\mathcal{F}_t^y = \sigma\{y(s); 0 \leq s \leq t\}$, which may be regarded as the Borel $\sigma-$algebras on $C(I; \Re^q), q = m, n, d$, respectively, and $\mathcal{F}_t^u = \sigma\{\int_0^s u(\tau)d\tau; 0 \leq s \leq t\}$, which is the Borel $\sigma-$algebra on Ω^u.

Introduce the product $\sigma-$algebras

$$\mathcal{F}_t \stackrel{\triangle}{=} \mathcal{F}_t^{w,x} \times \mathcal{F}_t^{y,u}, \quad \mathcal{F}_t^{w,x} \stackrel{\triangle}{=} \mathcal{F}_t^w \times \mathcal{F}_t^x, \quad \mathcal{F}_t^{y,u} \stackrel{\triangle}{=} \mathcal{F}_t^y \times \mathcal{F}_t^u,$$

where $\mathcal{F}_t^{w,x}$, $\mathcal{F}_t^{y,u}$, $\mathcal{F}_t$ are the Borel $\sigma-$algebras on $\Omega^{w,x}, \Omega^{y,u}, \Omega$, respectively.

Fix a sample path for the observation and control process $\{y(\cdot,\omega), u(\cdot,\omega)\}$. Given the initial data $x(0) = \xi, y(0) = 0, w(0) = 0$, Assumption 20.1 implies existence of a unique probability measure $\bar{P}_\xi^{y,u}$ on $\left(\Omega^{w,x}, \mathcal{F}_T^{w,x}\right)$, which coincides with the distribution measure of $\{x(t), w(t); t \in [0,T]\}$ given $\{(y(\cdot,\omega), u(\cdot,\omega)\}$, such that $\{w(t); t \in [0,T]\}$ is a Brownian motion process and

$$\left(\Omega^{w,x}, \mathcal{F}_T^{w,x}, \bar{P}_\xi^{y,u}\right) : x(t) = \xi + \int_0^t f(x(s), u(s))ds + \int_0^t \sigma(x(s))dw(s). \tag{20.44}$$

In addition, Assumption 20.1 also guarantees that the map $(\xi, u) \longrightarrow \bar{P}_\xi^{y,u}$ is continuous in the weak sense, that is, for each $\varphi \in BC(\Omega^{x,w})$, $(\xi, u) \longrightarrow \int_{\Omega^{x,w}} \varphi(x,w)d\bar{P}_\xi^{y,u}(x,w)$ is continuous on $\Re^n \times \Omega^u$.

Definition 20.2. The set of admissible controls denoted by $\mathcal{U}_{ad}$ consists of probability measures π on $\left(\Omega^{y,u}, \mathcal{F}_T^{y,u}\right)$, that is, $\pi \in \mathcal{M}(\Omega^{y,u})$ such that $\{y(t); t \in [0,T]\}$ is $\mathcal{F}_T^{y,u} - \pi$-a.s. Brownian motion (see [34], p. 264). These are called wide sense (relaxed) controls.

The projection $\{y(\cdot,\omega), u(\cdot,\omega)\} \mapsto y(\cdot,\omega)$ maps $\pi \in \mathcal{M}(\Omega^{y,u})$ onto a Wiener measure, and for all $t \in [0,T]$, $u(t)$ and $\sigma\{y(r) - y(t); 0 \leq t \leq r \leq T\}$ are independent under π. Given the measure $\Pi_0 \in \mathcal{M}(\Re^n)$ corresponding to the initial state $x(0) = \xi$, by Bayes rule, for each $A \in \mathcal{F}_T^{w,x}$, we have

$$\bar{P}^{y,u}(A) = \int_{\Re^n} \bar{P}_\xi^{y,u}(A) d\Pi_0(\xi), \tag{20.45}$$

which is the unique joint (conditional) distribution measure of $\{x(t), w(t); t \in [0,T]\}$ given $\{y(\cdot,\omega), u(\cdot,\omega)\}$.

For each $\pi \in \mathcal{U}_{ad}$, define the joint distribution measure $\tilde{P}^\pi$ on $\left(\Omega, \mathcal{F}_T\right)$ by

$$\tilde{P}^\pi(dw, dx, du, dy) \stackrel{\triangle}{=} \bar{P}^{y,u}(dw, dx) \times \pi(dy, du) \in \mathcal{M}(\Omega). \tag{20.46}$$

Notice that the projection

$$\{w(\cdot,\omega), x(\cdot,\omega), y(\cdot,\omega), u(\cdot,\omega)\} \mapsto \{y(\cdot,\omega), u(\cdot,\omega)\}$$

under $\tilde{P}^\pi$ is $\pi \in \mathcal{M}(\Omega^{y,u})$.

Finally, define the nominal measure P^π as follows: Let $\{y(t) - Nv(t); t \in [0,T]\}$, be an $(\mathcal{F}_t, \tilde{P}^\pi)$-Brownian motion process with covariance $NN't$, and introduce the $(\mathcal{F}_t, \tilde{P}^\pi)$-adapted martingale process

$$\begin{aligned}\Lambda^u(t) = \exp\Big\{ \int_0^t & h'(x(s))(NN')^{-1}dy(s) \\ & - \frac{1}{2}\int_0^t h'(x(s))\Big(NN'\Big)^{-1} h(x(s))ds\Big\}.\end{aligned} \tag{20.47}$$

Define the nominal measure through the Radon-Nikodym derivative

$$\frac{dP^\pi(w,x,y,u)}{d\tilde{P}^\pi(w,x,y,u)}|_{\mathcal{F}_T} \triangleq \Lambda^u(T). \tag{20.48}$$

Then, Assumption 20.1 implies that $P^\pi(\Omega) = E_{\tilde{P}^\pi}\{\Lambda^u(t)\} = 1, \forall t \in [0,T]$, and thus $P^\pi \in \mathcal{M}(\Omega)$. Moreover, by Girsanov's theorem,

$$Nv^\pi(t) \triangleq y(t) - \int_0^t h(x(s))ds$$

is a Brownian motion process with covariance $NN't$ under P^π, and the distribution of $\{w(t), x(t); t \in [0,T]\}$ is invariant under the measure change of (20.48). Thus, under the measure P^π, $\{v^\pi(t); t \in [0,T]\}$ and $\{w(t); t \in [0,T]\}$ are independent Wiener processes. Hence, for each $\pi \in \mathcal{U}_{ad}$, there exists a unique nominal measure $P^\pi \in \mathcal{M}(\Omega)$ under which the state $\{x(t); t \in [0,T]\}$ and observation process $\{y(t); t \in [0,T]\}$ satisfy (20.43) in the sense of weak (martingale) solutions.

The following result is given in [34].

Lemma 20.2. *The set of admissible controls $\mathcal{U}_{ad}$ is compact under weak sequential convergence (or equivalently with respect to the Prohorov metric).*

Proof. This follows from the compactness and convexity assumptions for $\mathcal{U}$ and Assumption 5.1(3). (See [34], Lemma 2.3, p. 265.) □

We can now state the problem precisely, as follows.

Problem 20.2. *Suppose the nominal system given by (20.43) is perturbed by an unknown (measurable) process $\{\gamma, \delta\}$ taking values from the set $\mathcal{H} =$*

$L_2(I;\Re^m) \times L_2(I;\Re^d)$ *(of square integrable functions with respect to* $Q^{\gamma,\delta}$*). This gives rise to the uncertain system described by*

$$\left(\Omega, \mathcal{F}_t, Q^{\gamma,\delta}\right) : \begin{cases} dx(t) = f(x(t), u(t))dt + \sigma(x(t))\gamma(t)dt + \sigma(x(t))dw^u(t), \\ dy(t) = h(x(t))dt + N\delta(t)dt + N dv^u(t), \\ x(0) = \xi, \qquad y(0) = 0, \end{cases} \tag{20.49}$$

where $Q^{\gamma,\delta}$ *denotes the probability measure on* Ω *induced by the system (20.49) corresponding to a given realization* $\{\gamma,\delta\}$ *and a given control* $\pi \in \mathcal{U}_{ad}$*. For a given* $\pi \in \mathcal{U}_{ad}$*, let*

$$M(\pi) \equiv \{Q^{\gamma,\delta} \in \mathcal{M}(\Omega) : \{\gamma,\delta\} \in \mathcal{H}\}$$

denote the family of probability measures induced by the uncertain system. Clearly the graph of this multi-measure $M(\pi)$ *is given by*

$$\Xi \equiv \{(\pi, Q) \in \mathcal{U}_{ad} \times \mathcal{M}(\Omega) : Q \in M(\pi)\}.$$

Given the nominal measure $P^\pi \in \mathcal{M}(\Omega)$*, define* $J(\pi,Q) = H(Q|P^\pi)$ *on* Ξ*. Then, find a pair* $(\pi^*, Q^*) \in \Xi$ *which solves the following constrained optimization problem:*

$$J(\pi^*, Q^*) = \sup_{\pi \in \mathcal{U}_{ad}} \inf_{Q \in M(\pi)} J(\pi, Q) \tag{20.50}$$

subject to fidelity

$$E_Q\Big\{\int_I \lambda(x(t), u(t))dt + \kappa(x(T))\Big\} \leq \alpha, \quad \alpha \in \Re \tag{20.51}$$

or subject to fidelity

$$E_Q\Big\{\int_I \lambda(x(t), u(t))dt + \kappa(x(T))\Big\} \geq \alpha, \quad \alpha \in \Re. \tag{20.52}$$

Following a similar procedure as in Section 20.4, Problem 20.2 can be reformulated using the dual functional as follows:

For every $s \in \Re$, define the Lagrangian,

$$\begin{aligned} J^{s,\alpha}(\pi, Q) &\triangleq H(Q|P^\pi) - s\Big(E_Q\Big\{\int_I \lambda(x(t), u(t))dt + \kappa(x(T))\Big\} - \alpha\Big) \\ &= H(Q|P^\pi) - s\Big(E_Q(\ell^u) - \alpha\Big), \end{aligned} \tag{20.53}$$

where $\ell^u(x) \equiv \int_I \lambda(x(t), u(t))dt + k(x(T))$. The associated dual functional is given by

$$J^{s,\alpha}(\pi, Q^*) = \inf_{Q \in M(\pi)} J^{s,\alpha}(\pi, Q). \tag{20.54}$$

In addition, define

$$\Psi_\pi^*(\alpha) \stackrel{\triangle}{=} \sup_{s\in\Re} J^{s,\alpha}(\pi, Q^*), \tag{20.55}$$

which may take values from the extended real number system. The statements of the next Corollary follow directly from Lemma 20.1 and Theorem 20.4.

Corollary 20.1. *For an arbitrary but fixed $\pi \in \mathcal{U}_{ad}$, the results stated in Lemma 20.1 and Theorem 20.4 hold for the functionals (20.54) and (20.55).*

Proof. Identifying,

$$\Sigma \mapsto \Omega; \quad u \mapsto \pi \in \mathcal{M}(\Omega^{y,u}); \quad \mu \mapsto P^\pi; \quad \nu \mapsto Q \in M(\pi);$$
$$\ell^u \mapsto \int_I \lambda(x(t), u(t))dt + \kappa(x(T)),$$

the proof is identical to that of Lemma 20.1 and Theorem 20.4. □

20.5.2. *Robust control and minimax games*

Here, the exact relation between Problem 20.2 and the usual formulation of robust control is established. The robust control problem is defined as follows [35]. Consider an uncertain partially observable system given by

$$\left(\Omega, \mathcal{F}_t, Q^{\gamma,\delta}\right) : \begin{cases} dx(t) = f(x(t), u(t))dt + \sigma(x(t))\gamma(t)dt + \sigma(x(t))dw^u(t), \\ dy(t) = h(x(t))dt + N\delta(t)dt + Ndv^u(t), \\ x(0) = \xi, \qquad y(0) = 0, \end{cases} \tag{20.56}$$

where $Q^{\gamma,\delta} \in M(\pi)$, and $\{\gamma, \delta\}$ taking values from $\mathcal{H} = L_2(I; \Re^m) \times L_2(I; \Re^d)$. Let $B_r = \{(\gamma, \delta) \in \mathcal{H}; \frac{1}{2}\|(\gamma, \delta)\|_{\mathcal{H}} \leq r\}$. Define the pay-off functional $J : \mathcal{U}_{ad} \to \Re$ by

$$J(\pi) = \sup_{(\gamma,\delta)\in B_r} E_Q\left\{ \int_I \lambda(x(t), u(t))dt + \kappa(x(T))\right\}, \tag{20.57}$$

where $r > 0$ is a fixed constant and $Q = Q^{\gamma,\delta}$. The optimal robust control problem is to find a control law $\pi^* \in \mathcal{U}_{ad}$ such that

$$J(\pi^*) = \inf_{\pi\in\mathcal{U}_{ad}} J(\pi). \tag{20.58}$$

It is clear from the above description that the objective of the controller is to minimize the cost functional, while the uncertainty tries to maximize it.

This problem can be reformulated in an unconstrained form as follows

$$J^{\beta}(\pi) = \sup_{(\gamma,\delta)\in\mathcal{H}} \Big\{E_Q\Big[\int_I \lambda(x(t),u(t))dt + \kappa(x(T))\Big] - \beta\, E_Q\Big[\frac{1}{2}\int_I \Big(\|\gamma(t)\|^2 + \|\delta(t)\|^2\Big)dt\Big]\Big\}, \tag{20.59}$$

where $\beta > 0$ is the Lagrange multiplier associated with the constraint. Therefore, the actual robust control problem which is often solved is to find a $\pi^* \in \mathcal{U}_{ad}$ so that

$$J^{\beta}(\pi^*) = \inf_{\pi\in\mathcal{U}_{ad}} J^{\beta}(\pi). \tag{20.60}$$

By use of Girsanov's theorem, in particular the density

$$\Upsilon \equiv \exp\Big(\int_I \gamma'(s)dw(s) - \frac{1}{2}\int_I \|\gamma(s)\|^2 ds\Big) \times \exp\Big(\int_I \delta'(s)dv(s) - \frac{1}{2}\int_I \|\delta(s)\|^2 ds\Big),$$

giving $dQ = \Upsilon dP^{\pi}$, one can verify (see [22]) that the entropy of Q relative to P^{π} is given by

$$H(Q|P^{\pi}) = \frac{1}{2}E_Q\Big\{\int_I \Big(\|\gamma(t)\|^2 + \|\delta(t)\|^2\Big)dt\Big\}. \tag{20.61}$$

However, letting $s > 0$ denote the Lagrange multiplier, the unconstrained form of problem (20.50), (20.52) is

$$\begin{aligned}
&\sup_{\pi\in\mathcal{U}_{ad}} \inf_{Q\in M(\pi)} \Big\{H(Q|P^{\pi}) - sE_Q\Big(\int_I \lambda(x(t),u(t))dt + \kappa(x(T))\Big)\Big\} \\
&= -s \inf_{\pi\in\mathcal{U}_{ad}} \sup_{(\gamma,\delta)\in\mathcal{H}} \Big\{E_Q\Big(\int_I \lambda(x(t),u(t))dt + \kappa(x(T))\Big) \\
&\quad -\frac{1}{2s}E_Q\Big[\frac{1}{2}\int_I \Big(\|\gamma(t)\|^2 + \|\delta(t)\|^2\Big)dt\Big]\Big\} \\
&= -sJ^{\frac{1}{s}}(\pi^*).
\end{aligned}$$

Hence, (20.60) is equivalent to the unconstraint problem (20.50). Thus we conclude that the robust control problem is equivalent to min-max games problem which has been our main theme.

20.5.3. *Duality of separated uncertain systems with wide sense controls*

In this section, separated strategies are introduced by describing the nominal systems using conditional distributions rather than a priori distributions. In the theory of partially observable systems such strategies are called separated strategies. They are important in answering questions on existence of optimal policies and in reformulating partially observable problems as fully observable ones in which the role of a state variable is played by conditional distributions.

Introduce the functional χ^u given by

$$\chi^u(t) \stackrel{\triangle}{=} \exp\left(s\int_0^t \lambda(x(\tau), u(\tau))d\tau\right), t \geq 0, s \in R, \tag{20.62}$$

and, for each sample path $\left\{u(\cdot,\omega), y(\cdot,\omega)\right\}$, define the measure-valued process $\{M_t^{y,u}\}_{t\geq 0}$ so that for each $\varphi \in BC(\Re^n)$,

$$M_t^{y,u}(\phi) \stackrel{\triangle}{=} E_{\bar{P}^{y,u}}\left\{\phi(x(t))\chi^u(t)\Lambda^u(t)\right\}, \tag{20.63}$$

is $\mathcal{F}_t^{y,u}$ adapted where $\Lambda^u(t)$ is given by (20.47). Denote the transition kernel associated with the conditional probability measure $M_t^{y,u}$ by $(y,u) \mapsto M_t(dx|y,u) \stackrel{\nabla}{=} M_t^{y,u}(dx)$. Then

$$M_t^{y,u}(\phi) = \int_{\Re^n} \phi(z) M_t^{y,u}(dz), \qquad \phi \in BC(\Re^n), \tag{20.64}$$

and moreover, the measure-valued process $\{M_t^{y,u}\}_{t\geq 0}$ satisfies the following stochastic partial differential equation written in the weak form

$$\begin{aligned} dM_t^{y,u}(\phi) &= M_t^{y,u}(L_s(u)\phi)dt + \langle M_t^{y,u}(h'(NN')^{-1}\phi), dy(t)\rangle \\ M_0(\phi) &= P_0(\phi), \end{aligned} \tag{20.65}$$

where for any given control policy u, $L_s(u)$ is the second order partial differential operator given by

$$L_s(u) = \frac{1}{2}\sum_{i,j=1}^n (\sigma\sigma')_{i,j}(x)\frac{\partial^2}{\partial x_i \partial x_j} + \sum_{i=1}^n f_i(x, u(t))\frac{\partial}{\partial x_i} + s\lambda(x, u(t)).$$

Note that L_0 denotes the backward Kolmogorov operator associated with the state process of the nominal system. Introduce the normalized probability measure

$$d\mu_T^{\pi}(x,y,u) \stackrel{\triangle}{=} \frac{d\mu_{T,un}^{\pi}(x,y,u)}{\int_{\Re^n \times \Omega^{y,u}} d\mu_{T,un}^{\pi}(x,y,u)} = \frac{d\mu_{T,un}^{\pi}(x,y,u)}{\mu_{T,un}^{\pi}(\Re^n \times \Omega^{y,u})}, \tag{20.66}$$

where

$$d\mu^{\pi}_{T,un}(x,y,u) = dM^{y,u}_{T}(x)d\pi(y,u),$$

and $\pi \in \mathcal{M}(\Omega^{y,u})$ is a probability measure.

Remark 20.3. For an arbitrary but fixed $\pi \in \mathcal{U}_{ad}$, the dual functional, $J^{s,\alpha}(\pi, Q^*)$, given in (5.42), can be expressed in terms of the measure $M^{y,u}_{T}$ and π as follows.

$$J^{s,\alpha}(\pi, Q^*) = s\alpha - \log \int_{\Omega^{y,u}} M^{y,u}_{T}(e^{s\kappa})d\pi(y,u) = s\alpha - \Psi_{P^{\pi}}(s),$$

where κ is as defined in Assumption 5.1.

Corollary 20.2. *The dual functional $J^{s,\alpha}(\pi, Q^*)$ is the result of an optimization (minimization) problem involving measures on $\Re^n$ as follows.*

$$\begin{aligned}
&J^{s,\alpha}(\pi, Q^*)\\
&= s\alpha - \log\Big(\int_{\Re^n\times\Omega^{y,u}} e^{s\kappa(z)}d\mu^{\pi}_{T,un}(z,y,u)\Big)\\
&= s\alpha - \log\Big(\int_{\Re^n\times\Omega^{y,u}} e^{s\kappa(z)}d\mu^{\pi}_{T}(z,y,u)\Big) - \log\Big(\mu^{\pi}_{T,un}(\Re^n\times\Omega^{y,u})\Big)\\
&= \inf_{\nu\in\mathcal{M}(\Re^n\times\Omega^{y,u})}\Big\{H(\nu|\mu^{\pi}_{T}) - s\Big(\int_{\Re^n\times\Omega^{y,u}}\kappa(z)d\nu(z,y,u)-\alpha\Big)\Big\}\\
&\quad - \log\Big(\mu^{\pi}_{T,un}(\Re^n\times\Omega^{y,u})\Big),
\end{aligned}$$

where the infimum is attained by $\nu^{s,*} \in \mathcal{M}(\Re^n \times \Omega^{y,u})$ given by

$$d\nu^{s,*}(x,y,u) = \frac{e^{s\kappa(x)}d\mu^{\pi}_{T}(x,y,u)}{\int_{\Re^n\times\Omega^{y,u}} e^{s\kappa(z)}d\mu^{\pi}_{T}(z,y,u)}. \tag{20.67}$$

Here the parameter s is the same as that in equation (20.65).

Proof. Identifying

$$\begin{aligned}
&\Sigma \mapsto \Re^n \times \Omega^{y,u}; \qquad u \mapsto \pi \in \mathcal{M}(\Omega^{y,u});\\
&\mu \mapsto \mu^{\pi} \in \mathcal{M}(\Re^n\times\Omega^{y,u}); \qquad \ell^u \mapsto \kappa; \qquad \lambda = 0,
\end{aligned} \tag{20.68}$$

the proof is identical to that of Lemma 20.1. □

Existence of the optimal control policy $\pi^* \in \mathcal{U}_{ad}$ was proved in [34, Theorem 4.1, p. 269; Theorem 7.2, p. 279]. Note that in (20.67) we have suppressed the dependence of $\nu^{s,*}$ on the control measure π. For a fixed $s \in \Re$, for optimization with respect to π, we suppress the variable s and emphasize the dependence on π and write $\nu^{\pi,*}$.

Theorem 20.5. *For any admissible* s,

(a) $\pi \longrightarrow J^{s,\alpha}(\pi, \nu^{\pi,*})$ *is weakly upper-semi continuous on* $\mathcal{U}_{ad}$,
(b) *There exists a* $\pi^* \in \mathcal{U}_{ad}$ *such that* $J^{s,\alpha}(\pi^*, \nu^{\pi^*,*}) \geq J^{s,\alpha}(\pi, \nu^{\pi,*}), \forall \pi \in \mathcal{U}_{ad}$.

Proof. The proof is similar to that given in [34]. □

Remark 20.4. There is a smaller class of control laws (strict sense controls) denoted by $\mathcal{U}_{ad}^s$ which is the subset of those $\pi \in \mathcal{U}_{ad}$ in which the measure π can be written as

$$\pi(dy, du) = \delta_{\underline{u}(y)}(du) \times W_y(dy), \tag{20.69}$$

for some mapping $\underline{u} : \Omega^y \to \Omega^u$, which is $(\mathcal{F}_t^y, \mathcal{F}_t^u)$-measurable, where $\delta_{\underline{u}(y)}(\cdot)$ denotes the Dirac measure concentrated at $\underline{u}(y)$. It is well known that for non convex control problems optimal controls in the strict sense may not exist and this is the reason why one considers wide sense (relaxed) controls. Nevertheless, any expression derived for $\pi \in \mathcal{U}_{ad}$ applies to the class of strict-sense controls $\mathcal{U}_{ad}^s$ provided it contains the optimal policy. Clearly in this case we have $\int_{\Omega^{y,u}} \psi(y,u) d\pi(y,u) \mapsto \int_{\Omega^y} \psi(y, u(y)) dW_y(dy)$ for any $\psi \in BC(\Omega^{y,u})$.

Assumption 20.2. Some additional conditions are sufficient to guarantee that the measure-valued process $\{M_t^{y,u}, t \geq 0\}$ has a density $q^{y,u}$ given by $\{q^{y,u}(t,\cdot) \equiv e^{y \cdot h(\cdot)} q^{y,u}(t,\cdot), t \geq 0\}$. They are as follows:

7) $n = m$, $a(x) \triangleq \sigma(x)\sigma'(x) \geq \alpha I_n$, for some $\alpha > 0$, for all $x \in \Re^n$, and $\frac{\partial}{\partial x_j} a_{i,j} \in L^\infty(\Re^n)$, for all i, j.
8) P_0 has a density $p_0(x)$ and $p_0 \in L_2(\Re^n)$. Then,

$$M_t^{y,u}(\phi) = \int_{\Re^n} \phi(z) e^{y'(t)h(z)} q^{y,u}(t,z) dz, \qquad \phi \in BC(\Re^n). \tag{20.70}$$

Moreover, $q^{y,u}(\cdot, z)$ is the solution of the following partial differential equation

$$\begin{aligned} \frac{\partial}{\partial t} q(t,x) &= A^*(y(t), u(t)) q(t,x) + e(x, y(t), u(t)) q(t,x) \\ &\quad + s\lambda(x, u(t)) q(t,x), \qquad (t,x) \in (0,T] \times \Re^n, \end{aligned} \tag{20.71}$$

$$q(0,x) = p_0(x), \qquad x \in \Re^n, \tag{20.72}$$

where $A^*(y,u)$ is the formal adjoint of the operator $A(y,u)$ given by

$$A(y,u) = \frac{1}{2} \sum_{i,j=1}^n a_{i,j}(x) \frac{\partial^2}{\partial x_i \partial x_j} + \sum_{i=1}^n b_i(x,u) \frac{\partial}{\partial x_i} - \sum_{i=1}^n (ay \cdot \nabla h)_i \frac{\partial}{\partial x_i},$$

$$e(x, y, u) = \frac{1}{2}(ay \cdot \nabla h, y \cdot \nabla h) - \hat{f} \cdot (y \cdot \nabla h) - \|h\|^2_{\Re^d},$$
$$\hat{f}_i \equiv f_i - (1/2) \sum_{j \geq 1} \frac{\partial}{\partial x_j} a_{i,j},$$

in which $\cdot$, (,) are the dot products in $\Re^d$, $\Re^n$, respectively.

Using the density of the measure valued process the pay-off functional can be expressed as

$$J^{s,\alpha}(\pi, \nu^{\pi,*}) = s\alpha - \log \int_{\Omega^{y,u}} \Big\{ \int_{\Re^n} e^{s\kappa(z)} e^{y'(T)h(z)} q^{y,u}(T, z) dz \Big\} \times d\pi(y, u). \tag{20.73}$$

Existence of the optimal control policy $\pi^* \in \mathcal{U}_{ad}$ follows from Theorem 20.5.

20.5.3.1. *An example*

For the purpose of illustrating the concepts presented earlier, we consider the following partially observed linear LQG problem.

The system dynamics and observations of of the nominal system (20.43) are restricted to the following linear structure

$$\Big(\Omega, \mathcal{F}_t, P^\pi\Big) : \begin{cases} dx(t) = Fx(t)dt + Bu(t)dt + Gdw(t), & x(0) = \xi, \\ dy(t) = Hx(t)dt + Ndv(t), & y(0) = 0. \end{cases} \tag{20.74}$$

The density of the initial state $x(0)$ is

$$p_0(x) = \frac{\exp(-\frac{1}{2}\|P_0^{-\frac{1}{2}}(x - \xi)\|^2)}{(2\pi)^{\frac{n}{2}}|P_0|^{\frac{1}{2}}}, \quad P_0 = P_0' > 0.$$

Given the nominal measure $P^\pi \in \mathcal{M}(\Omega)$ find a pair (π^*, Q^*) which solves the following constrained optimization problem:

$$J(\pi^*, Q^*) = \sup_{\pi \in \mathcal{U}_{ad}} \inf_{Q \in \mathcal{M}(\Omega)} H(Q|P^\pi)$$

subject to fidelity

$$\frac{1}{2} E_Q \Big\{ \int_0^T \Big(x'(t)Qx(t) + u'(t)Ru(t) \Big) dt + x'(T)Mx(T) \Big\} \geq \alpha, \quad \alpha \in \Re,$$

where $Q = Q' \geq 0, M = M' \geq 0, R = R' > 0$.

Next, we employ the results of Section 20.5.3, in which the resulting dual functional is expressed in terms of the conditional measure $M_t^{y,u}$ (see

Remark 20.3 and Corollary 20.2). It can be shown that $M_t^{y,u}$ has a density $m^{y,u}(t,x)$ given by

$$\frac{dM^{y,u}(t,x)}{dx} = m^{y,u}(t,x)$$
$$= \nu_{0,t}^{u} \times \frac{\exp\left(-\frac{1}{2}\|P(t)^{-\frac{1}{2}}(x-r(t))\|^2\right)}{(2\pi)^{\frac{n}{2}}|P(t)|^{\frac{1}{2}}} \times \exp\frac{s}{2}\left(\mathcal{C}_{0,t}^{u}+\mathcal{I}_{0,t}\right),$$

where

$$\nu_{0,t}^{u} = \exp\left\{\int_0^t (Hr(s))'\left(NN'\right)^{-1}dy - \frac{1}{2}\int_0^t \|N^{-1}Hr(s)\|^2 ds\right\},$$
$$\mathcal{C}_{0,t}^{u} \triangleq \int_0^t \{r'Qr + u'(s)Ru(s)\}\,ds, \qquad \mathcal{I}_{0,t} \triangleq \int_0^t Tr(PQ)ds,$$

and $P:[0,T]\to\mathcal{L}(\mathcal{R}^n,\mathcal{R}^n)$, $P=P'\geq 0$, $r:[0,T]\times\Omega\to\mathcal{R}^n$, are given by

$$\begin{cases} \dot{P} = FP + PF' + sPQP + GG' - PH'\left(NN'\right)^{-1}HP, \\ dr = (F+sPQ)\,rdt + Budt + PH'\left(NN'\right)^{-1}\left(dy - Hrdt\right), \\ r(0)=\xi, \qquad P(0)=P_0, \\ y(\cdot) \text{ is an } \{\mathcal{F}_t^{y,u}\}_{t\in T}\text{-adapted Wiener process with correlation } NN'. \end{cases} \tag{20.75}$$

Denote by $\tilde{\rho}(AB)$ the spectral radius of AB (A,B are matrix-valued functions), and define

$$\tilde{s}^* \triangleq \sup\{s : P(t)\geq 0, \quad \tilde{\rho}(P(t)M) < \frac{1}{s} \quad \forall t\in[0,T]\},$$

where P is given by the solution of (20.75). Then

$$\int_{\Re^n} e^{\frac{s}{2}x'Mx}dM_T^{y,u}(x)$$
$$= \frac{1}{|I-sP(T)M|^{\frac{1}{2}}} \times \exp\frac{s}{2}\left\{r'(T)(I-sP(T)M)^{-1}Mr(T)\right\} \tag{20.76}$$
$$\times \exp\frac{s}{2}\left(\mathcal{C}_{0,T}^{u}+\mathcal{I}_{0,T}\right)\times\nu_{0,T}^{u}, \qquad s\in[0,\tilde{s}^*].$$

Therefore, by Corollary 20.2, the resulting pay-off which should be minimized with respect to $\pi \in \mathcal{U}_{ad}$ is given by

$$\begin{aligned} J^{s,\alpha}(\pi,\nu^{\pi,*}) &= s\alpha + \frac{1}{2}\log|I - sP(T)M| \\ &\quad - \log \int_{\Omega^{y,u}} \Big\{ \exp\frac{s}{2}\Big(r'(T)(I - sP(T)M)^{-1}Mr(T)\Big) \\ &\qquad \times \exp\frac{s}{2}\left(\mathcal{C}^u_{0,T} + \mathcal{I}_{0,T}\right) \Big\} d\pi(y,u), \qquad s \in [0, \tilde{s}^*]. \end{aligned} \tag{20.77}$$

If we further restrict the class of control laws to strict-sense then the optimal control law can be found, and it is given in [16, 19, 36].

20.6. Conclusion

This paper is concerned with a new class of stochastic control problems, in which the pay-off is described by the relative entropy between the nominal measure and the uncertain measure, while the uncertain measures satisfy certain energy inequality constraints. The controller seeks to maximize the pay-off while the uncertainty seeks to minimize it. Throughout, it is shown that this problem is equivalent to max-min games in which the strategies observe the conditional distribution associated with the nominal model. Moreover, this problem incorporates both pessimistic and optimistic strategies, depending on the constraint chosen. Certain monotonicity properties of the optimal solution are discussed, while relations to the well-known Legendre–Fenchel transform are introduced. In addition, connections to minimax games of partially observable stochastic systems and to risk-sensitive control problems are investigated.

Appendix

Proof. [Proof of Lemma 20.1.] It is sufficient to prove the statements over those ν for which $H(\nu|\mu) < \infty$. To avoid confusion, we suppress the dependence of ν on u.

(5d): Since by our assumption, $\ell^u e^{s\ell^u} \in L_1(\mu)$ for $s \in I$, an open connected subset of $\Re$, we can interchange the order of integration and differentiation. Hence on I we have

$$\frac{d}{ds}J^{s,\gamma}(u,\nu^{s,*}) = \gamma - \frac{d}{ds}\Big(\log\int_{\Sigma} e^{s\ell^u}\,d\mu\Big) = \gamma - \frac{\int_\Sigma \ell^u e^{s\ell^u}\,d\mu}{\int_\Sigma e^{s\ell^u}\,d\mu}$$
$$= \gamma - \frac{E_\mu\Big(\ell^u e^{s\ell^u}\Big)}{E_\mu\Big(e^{s\ell^u}\Big)} = \gamma - E_{\nu^{s,*}}\Big(\ell^u\Big),$$

where

$$d\nu^{s,*} = \frac{e^{s\ell^u}\,d\mu}{\int_\Sigma e^{s\ell^u}\,d\mu}.$$

In addition, if $(\ell^u)^2 e^{s\ell^u} \in L_1(\mu)$ for $s \in I$, $E_\mu\Big(\ell^u e^{s\ell^u}\Big)$ is differentiable on I, and so $J^{s,\gamma}(u,\nu^{s,*})$ is twice differentiable on I and

$$\frac{d^2}{ds^2}J^{s,\gamma}(u,\nu^{s,*}) = -\Big\{E_{\nu^{s,*}}\Big((\ell^u)^2\Big) - \Big(E_{\nu^{s,*}}(\ell^u)\Big)^2\Big\} \le 0.$$

(5e): The first part is obvious. If $s \to J^{s,\gamma}(u,\nu^{s,*})$ attains its supremum on I, it follows from (4.20) that for some $s^* \in I$ we have

$$\frac{d}{ds}J^{s,\gamma}(u,\nu^{s,*})\mid_{s=s^*} = \gamma - E_{\nu^{s^*,*}}\Big(\ell^u\Big) = 0, \tag{20.78}$$

where

$$d\nu^{s^*,*} = \frac{e^{s^*\ell^u}\,d\mu}{\int_\Sigma e^{s^*\ell^u}\,d\mu}.$$

Hence we have

$$E_{\nu^{s,*}}\Big(\ell^u\Big)\mid_{s=s^*} = \gamma. \tag{20.79}$$

Now if $\ell^u e^{s\ell^u} \in L_1(\mu)$, then $s \to J^{s,\gamma}(u,\nu^{s,*})$ is differentiable on I, and since it is concave,(20.78) gives the maximum point, and therefore the supremum is attained at $s = s^*$, and it occurs on the boundary by (20.79). Also it $(\ell^u)^2 e^{s\ell^u} \in L_1(\mu)$, then $J^{s,\gamma}(u,\nu^{u,*})$ is twice differentiable and the supremum is uniquely attained at $s = s^*$. □

Proof. [Proof of Theorem 20.4.]

1) Risk-Seeking Scenario

The existence of a minimizing measure was proved in Theorem 20.2. Also it follows from Theorems 20.1, 20.3 and part 1) of Lemma 20.1 that the constrained and the unconstrained problems are equivalent. Then using "s" as the Lagrange multiplier, we have

$$\begin{aligned}\inf_{\nu\in M_o(u)} H(\nu \mid \mu) &= \sup_{s\in\Re}\inf_{\nu\in M(u)}\Big(H(\nu\mid\mu) - s(E_\nu(\ell^u)-\gamma)\Big)\\ &= \sup_{s\in\Re}\inf_{\nu\in M(u)} J^{s,\gamma}(u,\nu) = \sup_{s\in\Re} J^{s,\gamma}(u,\nu^{s,*})\\ &= \sup_{s\in\Re}\Big(s\gamma - \log\int_\Sigma e^{s\ell^u}\,d\mu\Big) = \Psi_\mu^*(\gamma). \qquad (20.80)\end{aligned}$$

Assume $\gamma < m^u = E_\mu\{\ell^u\}$ ‡. The supremum in right hand side of (20.80), can be split into two parts, as follows

$$\sup_{s\in\Re} J^{s,\gamma}(u,\nu^{s,*}) = \max\{\sup_{s>0} J^{s,\gamma}(u,\nu^{s,*})\ ,\ \sup_{s\le 0} J^{s,\gamma}(u,\nu^{s,*})\}. \qquad (20.81)$$

If $s>0$, then $s\gamma < sm^u$. Also Jensen inequality implies that

$$\log\int_\Sigma e^{s\ell^u}\,d\mu^u \ge s\int_\Sigma \ell^u d\mu = sm^u.$$

Hence we have

$$J^{s,\gamma}(u,\nu^{s,*}) = s\gamma - \log\int_\Sigma e^{s\ell^u}\,d\mu \le 0 \qquad \forall s\in(0,\infty). \qquad (20.82)$$

Also, by Lemma 20.1, part 5), $J^{s,\gamma}(u,\nu^{s,*})$ is always nonnegative for all $s\in\Re$. Hence

$$\sup_{s>0} J^{s,\gamma}(u,\nu^{s,*}) = 0.$$

Then from (20.81) and the nonnegativity of $J^{s,\gamma}(u,\nu^{u,*})$ we obtain

$$\sup_{s\in\Re} J^{s,\gamma}(u,\nu^{u,*}) = \sup_{s\le 0} J^{s,\gamma}(u,\nu^{u,*}) \qquad \text{if } \gamma < m^u.$$

This establishes equality in (20.34). Combine this result with (20.80), to get equation (20.35). By Theorem 20.1 and the above conclusion, the minimizing measure $\nu^{s,*}$ is given by

$$d\nu^{s,*} = \frac{e^{s\ell^u}\,d\mu}{\int_\Sigma e^{s\ell^u}\,d\mu}, \qquad s\le 0,$$

which establishes (20.36).

If $\ell^u e^{s\ell^u}\in L_1(\mu)$ and $(\ell^u)^2 e^{s\ell^u}\in L_1(\mu)$, for all $s\in I_-$ then by Lemma 20.1, $J^{s,\gamma}(u,\nu^{s,*})$ is twice differentiable, and $\frac{d^2}{ds^2}J^{s,\gamma}(u,\nu^{s,*})\le 0$, hence $\frac{d}{ds}J^{s,\gamma}(u,\nu^{s,*}) = \gamma - E_{\nu^{s,*}}(\ell^u)$ is a non increasing function of s. Therefore

$$\frac{d}{ds}J^{s,\gamma}(u,\nu^{s,*})|_{s=0} \le \frac{d}{ds}J^{s,\gamma}(u,\nu^{s,*}) \le \frac{d}{ds}J^{s,\gamma}(u,\nu^{s,*})|_{s=s^*}, \quad \forall s\in[s^*,0].$$

‡The other case, i.e. $m^u \le \gamma$ is trivial.

By Lemma 20.1, part e), $\frac{d}{ds}J^{s,\gamma}(u,\nu^{s,*})|_{s=s^*}=0$, and, for $s=0$, $\nu^{0,*}=\mu$. Thus,

$$\gamma - E_\mu(\ell^u) \leq \gamma - E_{\nu^{s,*}}(\ell^u) \leq 0, \qquad \forall s \in [s^*,0].$$

By re-arranging the inequalities, we obtain

$$\gamma = E_{\nu^{s,*}}(\ell^u)\mid_{s=s^*} \leq E_{\nu^{s,*}}(\ell^u) \leq E_\mu(\ell^u), \qquad \forall s \in [s^*,0],$$

and since by assumption $\gamma < m^u$,

$$\gamma \leq E_{\nu^{s,*}}(\ell^u) < m^u, \qquad \forall s \in [s^*,0].$$

This establishes equation (20.37).

2) Risk-Averse Scenario

The existence of minimizing measure was shown in Theorem 20.2. Also by virtue of the result of Theorems 20.1, 20.3 and part 1) of Lemma 20.1, constrained problem and unconstrained problem are equivalent. Then using "s" as the Lagrange multiplier, we have

$$\begin{aligned}\inf_{\nu\in M_p(u)} H(\nu\mid\mu) &= \sup_{s\in\Re}\inf_{\nu\in M(u)}\Big(H(\nu\mid\mu) - s(E_\nu(\ell^u)-\gamma)\Big)\\ &= \sup_{s\in\Re}\inf_{\nu\in M(u)} J^{s,\gamma}(u,\nu) = \sup_{s\in\Re} J^{s,\gamma}(u,\nu^{s,*})\\ &= \sup_{s\in\Re}\Big(s\gamma - \log\int_\Sigma e^{s\ell^u}d\mu\Big). \end{aligned}\tag{20.83}$$

Note that for the last equality it is sufficient to consider only those ν which satisfy $H(\nu|\mu)<\infty$. Assume $\gamma > m^u = E_\mu\{\ell^u\}$.[§]
The supremum in right hand side of (20.83), can be split into two parts, as follows:

$$\sup_{s\in\Re} J^{s,\gamma}(u,\nu^{s,*}) = \max\{\sup_{s\geq 0} J^{s,\gamma}(u,\nu^{s,*})\ ,\ \sup_{s<0} J^{s,\gamma}(u,\nu^{s,*})\}. \tag{20.84}$$

If $s<0$, then $s\gamma < sm^u$. Also, by Jensen inequality

$$\log\int_\Sigma e^{s\ell^u}d\mu \geq s\int_\Sigma \ell^u d\mu = sm^u.$$

Combining the above inequalities, we obtain

$$J^{s,\gamma}(u,\nu^{s,*}) = s\gamma - \log\int_\Sigma e^{s\ell^u}d\mu \leq 0, \qquad \forall s\in(-\infty,0) \tag{20.85}$$

[§]The other case, i.e. $m^u \geq \gamma$ is trivial.

Also, By Lemma 20.1, part 5), $J^{s,\gamma}(u,\nu^{s,*})$ is always nonnegative for all $s \in \Re$. Hence

$$\sup_{s<0} J^{s,\gamma}(u,\nu^{s,*}) = 0.$$

Then it follows from (20.84) and nonnegativity of $J^{s,\gamma}(u,\nu^{s,*})$ that

$$\sup_{s\in\Re} J^{s,\gamma}(u,\nu^{s,*}) = \sup_{s\geq 0} J^{s,\gamma}(u,\nu^{s,*}), \qquad \text{for } \gamma > m^u.$$

This establishes equality in (20.38). Combine this result with (20.83), to get equation (20.39). By Theorem 20.1, the minimizing measure $\nu^{s,*}$ is given by

$$d\nu^{s,*} = \frac{e^{s\ell^u} d\mu}{\int_\Sigma e^{s\ell^u} d\mu}, \qquad (s \geq 0),$$

and this establishes (20.40).

If $\ell^u e^{s\ell^u} \in L_1(\mu)$ and $(\ell^u)^2 e^{s\ell^u} \in L_1(\mu)$, for all $s \in I_+$ then by Lemma 20.1, $J^{s,\gamma}(u,\nu^{s,*})$ is twice differentiable, and $\frac{d^2}{ds^2} J^{s,\gamma}(u,\nu^{s,*}) \leq 0$, hence $\frac{d}{ds} J^{s,\gamma}(u,\nu^{s,*}) = \gamma - E_{\nu^{s,*}}(\ell^u)$ is a non increasing function of s. Therefore,

$$\frac{d}{ds} J^{s,\gamma}(u,\nu^{s,*})|_{s=s^*} \leq \frac{d}{ds} J^{s,\gamma}(u,\nu^{s,*}) \leq \frac{d}{ds} J^{s,\gamma}(u,\nu^{s,*})|_{s=0}, \qquad \forall s \in [0,s^*].$$

By Lemma 20.1, part e), $\frac{d}{ds} J^{s,\gamma}(u,\nu^{s,*})|_{s=s^*} = 0$, and for $s = 0$, $\nu^{s,*} = \mu$. Thus,

$$0 \leq \gamma - E_{\nu^{s,*}}(\ell^u) \leq \gamma - E_\mu(\ell^u), \qquad \forall s \in [0,s^*].$$

By re-arranging the inequalities we obtain,

$$E_\mu(\ell^u) \leq E_{\nu^{s,*}}(\ell^u) \leq \gamma = E_{\nu^{s,*}}(\ell^u)\,|_{s=s^*}, \qquad \forall s \in [0,s^*],$$

and since by assumption $\gamma > m^u$, we have

$$m^u < E_{\nu^{s,*}}(\ell^u) \leq \gamma, \qquad \forall s \in [0,s^*],$$

and this establishes equation (20.41). □

References

[1] J.-D. Deuschel and D. Stroock, *Large Deviations.* (Academic Press Inc., Boston, 1989).
[2] S. Varadhan, *Large Deviations and Applications.* (Society for Industrial and Applied Mathematics, Philadelphia, 1984).
[3] D. Stroock, *An Introduction to the Theory of Large Deviations.* (Springer-Verlag, Berlin, 1984).

[4] P. Dupuis and R. Ellis, *A Weak Convergence Approach to the Theory of Large Deviations.* (Wiley-Interscience, New York, 1997).
[5] J. Willems, Dissipative dynamical systems part I: General theory, *Arch. Rational Mech. Anal.* **45**(5), 321–351, (1972).
[6] A. J. van der Schaft, L^2-gain analysis of nonlinear systems and nonlinear state feedback H_∞ control, *IEEE Trans. Automat. Contr.* **37**(6), 770–784, (1992).
[7] A. Isidori and A. Astolfi, Disturbance attenuation and $\mathcal{H}_\infty-$control via measurement feedback in nonlinear systems, *IEEE Trans. Automat. Contr.* **37** (9), 1283–1293, (1992).
[8] G. Zames, Feedback and optimal sensitivity: Model reference transformations, multiplicative seminorms, and approximate inverses, *IEEE Trans. Automat. Contr.* **26**(2), 301–320, (1981).
[9] K. Glover and J. Doyle, State-space formulae for all stabilizing controllers that satisfy an H_∞-norm bound and relations to risk-sensitivity, *Systems & Control Letters.* **11**, (1988).
[10] D. Jacobson, Optimal stochastic linear systems with exponential performance criteria and their relation to deterministic differential games, *IEEE Trans. Automat. Contr.* **18**(2), 124–131, (1973).
[11] D. Jacobson, *Extensions of Linear-Quadratic Control, Optimization and Matrix Theory.* (Academic Press Inc., Boston, 1977).
[12] P.Whittle, *Risk-Sensitive Optimal Control.* (Wiley-Interscience, New York, 1990).
[13] P. Whittle, A risk-sensitive maximum principle, *Systems and Control Letters.* **15**, (1990).
[14] J. Speyer, An adaptive terminal guidance scheme based on an exponential cost criterion with applications to homing missile guidance, *IEEE Trans. Automat. Contr.* **21**(3), 371–375, (1976).
[15] P. Kumar and J. van Schuppen, On the optimal control of stochastic systems with an exponential-of-integral performance index, *Journal of mathematical analysis and applications.* **80**, 312–332, (1981).
[16] A. Bensoussan and J. H. van Schuppen, Optimal control of partially observable stochastic systems with an exponential-of-integral performance index, *SIAM Journal on Control and Optimization.* **23**(4), 599–613, (1985).
[17] P. Whittle, A risk-sensitive maximum principle: The case of imperfect state observations, *IEEE Trans. Automat. Contr.* **36**(7), 793–801, (1991).
[18] M. James, J. Baras, and R. Elliott, Risk-sensitive control and dynamic games for partially observed discrete time nonlinear systems, *IEEE Trans. Automat. Contr.* **39**(4), 780–792, (1994).
[19] C. Charalambous and J. Hibey, Minimum principle for partially observable nonlinear risk-sensitive control problems using measure-valued decompositions, *Stochastics and Stochastics Reports.* **57**(3-4), 247–288, (1996).
[20] C. Charalambous, The role of informations state and adjoint in relating nonlinear output feedback risk-sensitive control and dynamic games, *IEEE Trans. Automat. Contr.* **42**(8), 1163–1170, (1997).
[21] W. H. Fleming and W. M. McEneaney. Risk sensitive optimal control and

differential games. In *Stochastic theory and adaptive control*, pp. 185–197, Lawrence, Kansas (Sept., 1991).

[22] O. Hijab, *Stabilization of Control Systems.* (Springer-Verlag, Berlin, 1987).

[23] P. Dupuis and W. M. McEneaney, Risk-sensitive and robust escape criteria, *SIAM Journal on Control and Optimization.* **21**(3), 2021–2049, (1997).

[24] C. D. Charalambous and F. Rezaei. Optimization of stochastic uncertain systems: Large deviations and robustness of partially observable diffusions. In *American Control Conference*, pp. 3152–3157, Boston, U.S., (2004).

[25] P. D. Pra, L. Meneghini, and W. Runggaldier, Some connections between stochastic control and dynamic games, *Mathematics of Control, Signals, and Systems.* **9**(4), 303–326, (1996).

[26] V. A. Ugrinovskii and I. R. Petersen, Finite horizon minimax optimal control of stochastic partially observed time varying uncertain systems, *Mathematics of Control, Signals and Systems.* **12**(1), 1–23, (1999).

[27] I. R. Petersen, M. R. James, and P. Dupuis, Minimax optimal control of stochastic uncertain systems with relative entropy constraints, *IEEE Transactions on Automatic Control.* **45**, 398–412, (2000).

[28] V. A. Ugrinovskii and I. R. Petersen, Minimax LQG control of stochastic partially observed uncertain systems, *SIAM J. Control Optim.* **40**, 1189–1226, (2001).

[29] C. Charalambous and F. Rezaei, Stochastic uncertain systems subject to relative entropy constraints: Induced norms and monotonicity properties of minimax games, *IEEE Trans. Automat. Contr.* **52**(4), 647–663, (2007).

[30] E. T. Jaynes, Information theory and statistical mechanics, *Physics Review.* **106**, 620–630, (1957).

[31] E. T. Jaynes, Information theory and statistical mechanics ii, *Physics Review.* **108**, 171–190, (1957).

[32] P. Whittle, Risk-sensitive Linear/Quadratic/Gaussian control, *Advances in Applied Probability.* **13**, (1981).

[33] D. Luenberger, *Optimization by Vector Space Methods.* (John Wiley, New York, 1969).

[34] W. H. Fleming and E. Pardoux, Optimal control of partially observed diffusions, *SIAM J. Control Optim.* **20**, 261–285, (1982).

[35] M. Green and D. J. N. Limebeer, *Linear Robust Control.* (Prentice-Hall, Englewood Cliffs, NJ, 1995).

[36] C. D. Charalambous and R. J. Elliott, Classes of nonlinear partially observable stochastic optimal control problems with explicit optimal control laws, *SIAM J. Control Optim.* **36**, 542–578, (1998).

Chapter 21

Gradient based policy optimization of constrained Markov decision processes

Vikram Krishnamurthy[1] and Felisa J. Vázquez Abad

[1]*Department of Electrical and Computer Engineering, University of British Columbia, Vancouver, V6T 1Z4, Canada*
vikramk@ece.ubc.ca

[2]*Department of Computer Science, Hunter College, City University of New York, NY 10065, USA*
felisav@hunter.cuny.edu

We present stochastic approximation algorithms for computing the locally optimal policy of a constrained, average cost, finite state Markov Decision Process. The stochastic approximation algorithms require estimation of the gradient of the cost function with respect to the parameter that characterizes the randomized policy. We propose a spherical coordinate parameterization and implement a simulation based gradient estimation scheme using potential realization factors. In this paper we analyse boundedness of moments of the estimators and provide new analysis of the computational requirements of the estimators. We present a stochastic version of a primal-dual (augmented) Lagrange multiplier method for the constrained algorithm. We show that the "large sample size" limit of the procedure is unbiased, but for small sample size (which is relevant for on-line real time operation) it is biased. We give an explicit expression for the asymptotic bias and discuss several possibilities for bias reduction. Numerical examples are given to illustrate the performance of the algorithms. In particular we present a case study in transmission control in wireless communications where the optimal policy has a simplified threshold structure and apply our stochastic approximation method to this example.

Contents

21.1. Introduction

Let S denote an arbitrary finite set called the *state space*. Let $\mathcal{U}_i$, $i \in S$ denote an arbitrary collection of finite sets called *action sets*. A Markov Decision Process [1] (MDP) $\{X_n\}$ with finite state space S evolves as follows. When the system is in state $i \in S$, a finite number of possible actions from the finite set $\mathcal{U}_i$ can be taken. Let u_n denote the action taken by the decision maker at time n and let $d(i)+1$ denote the cardinality of the action set $\mathcal{U}_i$. The evolution of the system is Markovian with a transition probability matrix $A(u)$ that depends on the action $u \in \mathcal{U}_i$, that is for $i, j \in S$,

$$A_{ij}(u) \stackrel{\triangle}{=} \mathbb{P}[X_{n+1} = j | X_n = i, u_n = u], \quad u \in \mathcal{U}_i, \quad n = 0, 1, \ldots \tag{21.1}$$

Denote by $\mathfrak{F}_n, n \geq 1$ the σ-algebra generated by the observed *system trajectory* $(X_0, \ldots, X_n, u_0, \ldots, u_{n-1})$ and set $\mathfrak{F}_0$ as the σ-algebra generated by X_0. The filtration of the process is the increasing sequence of σ-algebras $\{\mathfrak{F}_n, n \geq 0\}$. Define the set of **admissible** policies $\mathcal{D} = \{\mathbf{u} = \{u_n\} : u_n \text{ is measurable w.r.t. } \mathfrak{F}_n, \ \forall n \in \mathbb{N}\}$. To avoid being distracted by technicalities, except in Sec. 21.6, we assume that all elements of $A(u)$ for each

u are non-zero, so that the resulting process is ergodic. In Sec. 21.6, a wireless telecommunications example is presented and sufficient conditions are given such that every feasible policy induces a recurrent Markov chain.

The cost incurred at stage n is a known bounded function $c(X_n, u_n) \geq 0$ where $c : S \times \mathcal{U} \to \mathbb{R}$. For any admissible policy $\mathbf{u} \in \mathcal{D}$, let $\mathbb{E}_{\mathbf{u}}$ denote the corresponding expectation and define the infinite horizon average cost

$$J_{x_0}(\mathbf{u}) = \limsup_{N \to \infty} \frac{1}{N} \mathbb{E}_{\mathbf{u}} \left[\sum_{n=1}^{N} c(X_n, u_n) \mid X_0 = x_0 \right]. \tag{21.2}$$

Motivated by several problems in telecommunication network optimization such as admission control in wireless networks [2–4], we consider the cost (21.2), subject to L sample path constraints

$$\mathbb{P}_{\mathbf{u}} \left[\limsup_{N \to \infty} \frac{1}{N} \sum_{n=1}^{N} \beta_l(X_n, u_n) \leq \gamma_l \right] = 1, \quad l = 1, 2, \ldots, L,$$

where $\beta_l : S \times \mathcal{U} \to \mathbb{R}$ are known bounded functions and γ_l are known constants. These are used, for example, in admission control of telecommunication networks to depict quality of service (QoS) constraints, see [2, 3]. For ergodic MDPs, these sample path constraints are equivalent to the average constraints [5]:

$$\limsup_{N \to \infty} \frac{1}{N} \mathbb{E}_{\mathbf{u}} \left[\sum_{n=1}^{N} \beta_l(X_n, u_n) \right] \leq \gamma_l, \quad l = 1, \ldots, L. \tag{21.3}$$

The aim is to compute an optimal policy $\mathbf{u}^* \in \mathcal{D}$ that satisfies

$$J_{x_0}(\mathbf{u}^*) = \inf_{\mathbf{u} \in \mathcal{D}} J_{x_0}(\mathbf{u}) \quad \forall x_0 \in S, \tag{21.4}$$

that is, $\mathbf{u}^*$ has the minimum cost for all initial states $x_0 \in S$ subject to the constraints (21.3). It is well known [6] that if $L > 0$ then the optimal policy $\mathbf{u}^*$ is *randomized* for at most L of the states. If the transition probabilities $A_{ij}(u)$ (21.1) are known, then the optimal policy $\mathbf{u}^*$ for the constrained MDP (21.2, 21.3) is straightforwardly computed as the solution of a linear programming problem.

Objectives. Problem (21.2, 21.3) is an adaptive constrained Markov decision process problem. This paper presents stochastic approximation algorithms for adaptively computing the optimal policy $\mathbf{u}^*$ of the above constrained MDP (21.2, 21.3).

There are two methodologies that are used in the literature for solving stochastic adaptive control problems: ***direct methods***, where the unknown transition probabilities $A_{ij}(u)$ are estimated *simultaneously* while updating the control policy, and ***implicit methods*** – such as simulation based methods, where the transition probabilities are not directly estimated in order to compute the control policy. In this paper, we focus on implicit simulation-based algorithms for solving the MDP (21.2, 21.3). By simulation based we mean that although the transition probabilities $A(u)$, $u \in \mathcal{U}_i$, $i \in S$ are unknown, the decision maker can observe the system trajectory under any choice of control actions $\mathbf{u} = \{u_n\}$. In other words, the adaptive control algorithms we present are adapted to the filtration $\{\mathfrak{F}_n; n \geq 0\}$ defined above. Moreover, these algorithms can deal with slowly time varying transition probabilities $A(u)$.

Neurodynamic programming methods [7] such as Q-learning and temporal difference methods are also examples of simulation based implicit methods that have been widely used to solve unconstrained MDPs where the optimal policy $\mathbf{u}^*$ is a pure policy, that is, u_n^* is a deterministic function of X_n. However, for the constrained MDP (21.2, 21.3), as the optimal policy is randomized, there seems to be no obvious way of modifying such methods.

Summary of main results In this paper we present a stochastic version of gradient-based optimisation methods. In Section 21.4, we derive a consistent weak derivative (WD) "phantom" estimator. This gradient estimator does not require explicit knowledge of the transition probabilities $A(u)$ – so for brevity we call it a "parameter free" gradient estimator.

Section 21.4.2 presents the implementation of our WD estimators over short batch lengths to facilitate real time control of the MDP.The key advantage is that they are amenable to adaptive control of MDPs with time varying parameters. Moreover, these gradient estimators yield a reduction in variance of several orders of magnitude compared to the score function method in [8, 9]. However, as we discuss in Section 21.4.2, a non negligible asymptotic bias may occur. We derive the statistical properties (consistency, bias for small batch size, and efficiency) of the gradient estimates. We give stochastic bounds for the number of parallel phantoms, which allows us to approximate the moments of the coupling time as well as the computational complexity of the algorithm. To our knowledge, these results, which are applicable to realisation perturbation factors, are new.

Putting the parameter free gradient estimation algorithm into a stochastic gradient algorithm results in a new simulation based algorithm for the adaptive control of a constrained MDP. The control problem Eq. (21.4) under constraints Eq. (21.3) can be reformulated as an optimization problem using a parametrization, as will be stated in Section 21.2. The main challenge that we address in this paper is as follows: Given N observations of the process under a given policy parametrized by the control parameter $\alpha(n)$, say $Z_{(n-1)N}, \ldots, Z_{nN}$, and without explicit knowledge of the transition probabilities $P_{ij}(u)$, adjust the control variable $\alpha(n+1)$ to approximate the optimal control. We assume that the actual instantaneous cost and constraints values are observed exactly. In the stationary case where the transition probabilities, cost and constraint functions do not change, one usually constructs algorithms such that $\|\alpha(n) - \alpha^*\|$ is "small" in some sense, usually with probability one. For tracking, however, we focus on an algorithm with adaptive capabilities that will approach the current optimum, while at the same time reacting to changes in the environment. There are two important results that we present in this context.

First, because the constraints in the MDP are in the form of long term averages, that are not known to the decision maker, it is not possible to use stochastic approximation algorithms with the usual gradient projection methods, as in [10], for example. We present primal-dual based stochastic approximation algorithms with a penalty function, and augmented Lagrangian (multiplier) algorithms, for solving the time varying constrained MDP. Weak convergence of the action probability estimates to the optimal action probabilties is established when the sample size for the estimation is large. However, for a short batch length implementation, we show that the algorithm is asymptotically suboptimal and we characterise the asymptotic bias. Sec. 21.5.3 illustrates the performance of the algorithms on a constrained MDP with time varying transition probabilities.

Second, because the action probabilities must always add up to one and be nonnegative, there is a strong motivation to develop a parametrization that ensures feasibility of the estimates generated by the stochastic approximation algorithm at every step. To do this, we parameterize the action probabilities of the constrained MDP using *spherical coordinates*. This parametrization is particularly well suited to the MDP problem and has a superior convergence rate, see discussion in Sec. 21.2.4. In [8, 9], a different parameterization is used which is similar to the generalized gradient approach of [11].

Sec. 21.6 gives a case study of how the results in the paper can be used to estimate a monotone scheduling policy that arises in transmission scheduling over fading wireless communication channels. We prove that the optimal scheduling policy is a randomized mixture of two monotone (threshold) policies – this proof is of independent interest as it generalizes the well known results of the classical paper [12] (see also [13]) to the case of a correlated Markov chain channel and with delay constraints.

21.2. Problem formulation and discussion

In this section we formulate the constrained MDP (21.2, 21.3) as a stochastic optimization problem, in terms of parameterized action probabilities. Then, for convenience, a summary of the key algorithms in this paper is given. Finally we briefly summarize how our approach differs from two other approaches in the literature.

21.2.1. *Spherically parameterized randomized policies*

The randomized optimal policy for the constrained MDP can be defined in terms of the action probabilities θ, parameterized by ψ, as

$$\mathbb{P}[u_n = a | X_n = i] = \theta_{ia}(\psi), \quad a \in \mathcal{U}_i,\ i \in S \tag{21.5}$$

where $\theta_{ia}(\psi) \geq 0$, $\sum_{a \in \mathcal{U}_i} \theta_{ia}(\psi) = 1$ for every state $i \in S$. Here $\psi \in \Psi$ is a finite dimensional vector which parameterizes the action probabilities θ, and Ψ is some suitably defined compact subset of the Euclidean space. The unconstrained problem with pure optimal policy is a degenerate case, where, for each $i \in S$, $\theta_{ia} = 1$ for some $a \in \mathcal{U}_i$.

The most obvious parameterization ψ for θ, which we shall call *canonical coordinates*, is to choose $\psi = \theta$. Thus $\psi = \{\psi_i\} = \{\theta_i\}$, $i \in S$ is the set of action probability vectors $(\psi_{ia}; a \in \mathcal{U}_i)$ satisfying Eq. (21.5). As discussed in Sec. 21.2.4, this parameterization has several disadvantages.

In this paper we use a more convenient *spherical coordinate* parameterization $\psi = \alpha$ that automatically ensures the feasibility of $\theta(\alpha)$ (that is, the constraints in (21.5) hold) without imposing a hard constraint on α. Adapted to our MDP problem, it reads as follows:

Fix the control agent $i \in S$. Suppose without loss of generality that $\mathcal{U}_i = \{0, \ldots, d(i)\}$. To each value $\theta_{ia}, a \in \mathcal{U}_i$, associate the values $\lambda_{ia} = \sqrt{\theta_{ia}}$. Then Eq. (21.5) yields $\sum_{a \in \mathcal{U}_i} \lambda_{ia}^2 = 1$, and λ_{ia} can be interpreted as the coordinates of a vector that lies on the surface of the unit sphere in $\mathbb{R}^{d(i)+1}$, where $d(i)+1$ is the size of $\mathcal{U}_i$ (that is, the number of actions). In spherical coordinates, the angles are $\alpha_{ip}, p = 1, \ldots, d(i)$, and the radius is always of size unity. For $d(i) \geq 1$, the spherical coordinates parameterization α satisfies:

$$\theta_{ia}(\alpha) = \lambda_{ia}^2,$$
$$\lambda_{ia} = \begin{cases} \cos(\alpha_{i,1}), & a = 0, \\ \cos(\alpha_{i,(a+1)}) \prod_{k=1}^{a} \sin(\alpha_{i,k}), & 1 \leq a \leq d(i) - 1, \\ \sin(\alpha_{i,d(i)}) \prod_{k=1}^{d(i)-1} \sin(\alpha_{i,k}), & a = d(i). \end{cases} \tag{21.6}$$

Note that $\theta_{ia}(\alpha) = \lambda_{ia}^2$ is an analytic function of α, that is, infinitely differentiable in α. It is clear that under this α parameterization, the control variables $\alpha_{ip}, p \in \{1, \ldots, d(i)\}$, do not need to satisfy any constraints in order for $\theta(\alpha)$ to be feasible. Furthermore, since $\theta_{ia}(\alpha) = \lambda_{ia}^2$ involves even powers of $\sin(\alpha_{i,p})$ and $\cos(\alpha_{i,p})$, it suffices to consider $\alpha_{ip} \in \boldsymbol{\alpha}$ where $\boldsymbol{\alpha}$ denotes the compact set

$$\boldsymbol{\alpha} = \{\alpha_{ip} \in [0, \pi/2];\ i \in S,\ p \in \{1, \ldots, d(i)\}\}, \tag{21.7}$$

that is, for any $\alpha \in \mathbb{R}^{d(i)}$, there is a unique $\alpha \in \boldsymbol{\alpha}$ which yields the same value of $\theta(\alpha)$. Let $\boldsymbol{\alpha}^o$ denote the interior of the set $\boldsymbol{\alpha}$. Finally, define the compact set $\boldsymbol{\alpha}^\mu \subset \boldsymbol{\alpha}^o$ for user defined small parameter $\mu > 0$ as

$$\boldsymbol{\alpha}^\mu = \{\alpha_{ip} \in [\mu, \pi/2 - \mu];\ i \in S,\ p \in \{1, \ldots, d(i)\}\}. \tag{21.8}$$

Note that $\boldsymbol{\alpha}^\mu$ is the set obtained by removing small intervals at 0 and $\pi/2$ from $\boldsymbol{\alpha}$. As discussed in detail after the statement of Proposition 21.2 in Sec. 21.5, excluding 0 and $\pi/2$ is necessary to prove convergence of the algorithms we propose – however, since μ can be chosen arbitrarily small, it is not important in the actual algorithmic implementation.

21.2.2. *Parameterized constrained MDP formulation*

We now formulate the above MDP problem as a stochastic optimization problem where the instantaneous random cost is independent of α but the expectation is with respect to a measure parameterized by α. Such a "parameterized integrator" formulation is common in gradient estimation, see [14], and will be subsequently used to derive our gradient estimators.

Consider the augmented (homogeneous) Markov chain $Z_n \triangleq (X_n, u_n)$ with state space $\mathcal{Z} = S \times \mathcal{U}$ and transition probabilities parameterized by

α given by

$$\begin{aligned} P_{i,a,j,a'}(\alpha) &\triangleq \mathbb{P}(X_{n+1} = j, u_{n+1} = a' \mid X_n = i, u_n = a) \\ &= \theta_{j,a'}(\alpha)\, A_{ij}(a), \end{aligned} \tag{21.9}$$

for $i, j \in S, a \in \mathcal{U}_i, a' \in \mathcal{U}_j$. It follows that for any $\alpha \in \boldsymbol{\alpha}^o$ (interior of set $\boldsymbol{\alpha}$), the chain $\{Z_n\}$ is ergodic, and it possesses a unique invariant probability measure $\pi_{i,a}(\alpha)$, $i \in S$, $a \in \mathcal{U}_i$. Let $\mathbb{E}_{\pi(\alpha)}$ denote the expectation w.r.t the measure $\pi(\alpha)$ parameterized by α. From (21.2) we have $J_{x_0}(\mathbf{u}) = C(\theta(\alpha))$, where

$$C(\theta(\alpha)) \triangleq \mathbb{E}_{\pi(\alpha)}[c(Z)] = \sum_{i\in S} \sum_{a\in\mathcal{U}_i} \pi_{i,a}(\alpha) c(i,a). \tag{21.10}$$

We subsequently denote $C(\theta(\alpha))$ as $C(\alpha)$, whenever it is convenient. The L constraints (21.3) can be expressed as

$$B_l(\alpha) \triangleq \mathbb{E}_{\pi(\alpha)}[\beta_l(Z)] - \gamma_l = \sum_{i\in S} \sum_{a\in\mathcal{U}_i} \pi_{i,a}(\alpha)\beta_l(i,a) - \gamma_l \le 0, \tag{21.11}$$

for $l = 1, \ldots, L$. Define $B = (B_1, \ldots, B_l)$. Thus the optimization problem (21.4) with constraints (21.3) can be written as

$$\textbf{Problem S1:} \qquad \min_{\alpha\in\boldsymbol{\alpha}^\mu} C(\alpha) \tag{21.12}$$

$$\text{subject to} \quad B_l(\alpha) \le 0, \quad l = 1, \ldots, L. \tag{21.13}$$

The constraints $B_l(\alpha)$ in (21.13) will be called **MDP constraints**. As should be evident from this formulation, the optimal control problem (21.12, 21.13) depends uniquely on the *invariant* distribution $\pi(\alpha)$ of the chain, rather than the (unknown) transition probabilities $A_{ij}(u)$.

It is important to note that in this paper, the expected cost $C(\alpha)$ and expected constraints $B_l(\alpha) \le 0$ are not known to the decision maker (since the transition probabilities are not known). Our aim is to devise a recursive (on-line) stochastic approximation algorithm to solve Problem S1 without explicit knowledge of the transition probabilities $A(a)$. Such an algorithm operates recursively on the observed system trajectory $(X_0, \ldots, X_n, u_0, \ldots, u_{n-1})$ to yield a sequence of estimates $\{\alpha(n)\}$ of the optimal solution. If the unknown dynamics $A(a)$ of the MDP are constant, then the proposed algorithm ensures that the estimates $\alpha(n)$ approach the optimal solution. On the other hand, if the unknown underlying dynamics $A(a)$ evolve slowly with time or jump changes infrequently, then the proposed algorithm will track the optimal trajectory in a sense to be made clear later. Stochastic approximation algorithms for tracking slowly evolving parameters has been widely studied in the literature, see [15, 16]. The tracking of parameters that jump infrequently has been analysed more recently in [17].

Assumption 21.1. The minima α^* of Problem S1 (that is, (21.12, 21.13)) are regular, that is, $\nabla_\alpha B_l(\alpha^*)$, $l = 1, \ldots, L$ are linearly independent. Also, α^* belongs to the set of Kuhn Tucker points

$$\begin{aligned} \mathrm{KT} = \Big\{ \alpha^* \in \boldsymbol{\alpha}^\mu : \ \exists\, \mu_l \geq 0,\ l = 1, \ldots, L \\ \text{such that} \quad \nabla_\alpha C + \nabla_\alpha B \mu = 0, \quad B' \mu = 0 \Big\} \end{aligned} \tag{21.14}$$

where $\mu = (\mu_1 \ldots, \mu_L)'$. Moreover α^* satisfies the second order sufficiency condition ${\nabla_\alpha}^2 C(\alpha^*) + {\nabla_\alpha}^2 B(\alpha^*)\mu > 0$ (positive definiteness) on the subspace

$$\Big\{ y \in \mathbb{R}^L : \nabla_\alpha B_l(\alpha^*) y = 0 \text{ for all } l \text{ s.t. } B_l(\alpha^*) = 0, \mu_l > 0 \Big\}.$$

21.2.3. *User's guide*

A summary of the key equations for implementing the learning based algorithm proposed in this paper in spherical coordinates for constrained MDP Problem S1 (21.12, 21.13) is as follows:

Input Parameters:
Cost matrix $(c(i,a))$, constraint matrix $(\beta(i,a))$, batch size N.

Step 0, Initialize: Set $n = 0$, initialize $\alpha(n) \in \boldsymbol{\alpha}^o$ and vector $\lambda(n) \in \mathbb{R}_+^L$. *

Step 1, System Trajectory Observation: Observe MDP over batch $I_n \triangleq \{k \in [nN, (n+1)N - 1]\}$ using randomized policy $\theta(\alpha(n))$ of (21.6) and compute estimate $\hat{B}(n)$ of the constraints, (cf. Eq. (21.49) or (21.51) of Sec. 21.5.3).

Step 2, Gradient Estimation without explicit knowledge of parameters: Compute gradient estimates $\widehat{\nabla_\alpha C}(n)$, $\widehat{\nabla_\alpha B}(n)$ over the batch I_n using the WD phantom estimates $\hat{\hat{G}}_n$ given in Eq. (21.40).

*More precisely, $\alpha(0)$ needs to be initialized in $\boldsymbol{\alpha}^\mu$, where $\boldsymbol{\alpha}^\mu \subset \boldsymbol{\alpha}^o$ excludes a small μ size ball around the boundary of $\boldsymbol{\alpha}$, see (21.8) and Sec. 21.5.1.

Step 3, Update Policy $\theta(\alpha(n))$ using constrained stochastic gradient algorithm: Use a penalty function primal-dual based stochastic approximation algorithm to update α as

$$\alpha(n+1) = \alpha(n) - \epsilon\Big(\widehat{\nabla_\alpha C}(n) + \widehat{\nabla_\alpha B}(n) \max\big[0, \lambda(n) + \rho\widehat{B}(n)\big]\Big) \tag{21.15}$$

$$\lambda(n+1) = \max\left[\left(1 - \frac{\epsilon}{\rho}\right)\lambda(n), \lambda(n) + \epsilon\widehat{B}(n)\right]. \tag{21.16}$$

The "penalization" ρ is a suitably large positive constant and $\max[\cdot,\cdot]$ above is taken element wise, see (21.22, 21.23).

Step 4, Iterate: Set $n = n + 1$ and go to Step 1.

Remark 21.1.

1. For large batch size N, the bias of the estimates $\hat{B}$ and $\widehat{\nabla_\alpha B}$ are $O(1/N)$ and the algorithm (21.15, 21.16) is asymptotically optimal. However, for fast tracking of time-varying MDPs it is necessary to choose N small. In this case, the main source of bias in the estimation of α^* using (21.15, 21.16) is the covariance of $\widehat{\nabla_\alpha B}$ with $\hat{B}$. Several approaches to deal with this bias are discussed in Sec. 21.5.3.
2. Another alternative to (21.16) is to update λ via a multiplier (augmented Lagrangian) algorithm (see Sec. 21.3.2). A third alternative is to fix λ. For sufficiently large ρ, $\alpha(n)$ will converge to $\alpha(\infty)$ which is in a pre-specified ball around a local minimum α^*.
3. If the true parameters of the MDP jump change at infrequent intervals, then iterate averaging [18] (as long as the minimal window of averaging is smaller than the jump change time) and adaptive step size algorithms can be implemented in the above stochastic approximation algorithms to improve efficiency and tracking capabilities.
4. For the special case of two actions and a monotone policy (see Section 21.6), $\alpha(n)$ can be further parameterized and the dimension of the optimization Problem S1 can be substantially reduced.

21.2.4. *Discussion of other approaches in literature*

Before presenting the details of the algorithm proposed in this paper, we briefly summarize two works in the literature that also use stochastic approximation methods to solve MDPs. It is well known [1] that Problem S1 formulated in terms of the invariant measure π is the following linear program:

$$\min_{\pi} \sum_{i \in S} \sum_{a \in \mathcal{U}_i} \pi_{ia} c(i,a) \tag{21.17}$$

$$\text{subject to} \begin{cases} \sum_{i \in S} \sum_{a \in \mathcal{U}_i} \pi_{ia} \beta_l(i,a) < \gamma_l \\ \sum_{i \in S, a \in \mathcal{U}_j} \pi_{ia} = 1, \\ \sum_{a \in \mathcal{U}_j} \pi_{ja} = \sum_{i \in S} \sum_{a \in \mathcal{U}_j} \pi_{ia} A_{ij}(a), \\ 0 \leq \pi_{ia} \leq 1, \quad i \in S, a \in \mathcal{U}_j. \end{cases}$$

With θ^* and π^* denoting the optimal solutions of (21.12) and (21.17), respectively, it is straightforward to show that

$$\theta^*_{ia} = \frac{\pi^*_{ia}}{\sum_{u \in \mathcal{U}_i} \pi^*_{iu}}. \tag{21.18}$$

The closest approach to our paper is that presented in [8, 9]. The MDP in [8, 9] is without constraints and uses the parameterization

$$\theta_{ia}(\psi) = \frac{e^{\psi_{ia}}}{\sum_{u \in \mathcal{U}_i} e^{\psi_{iu}}}, \quad \psi_{ia} \in \mathbb{R}, i \in S, \ a \in \mathcal{U}_i.$$

This exponential parameterization satisfies

$$\frac{\partial \theta_{iu}}{\partial \psi_{ia}} = \begin{cases} \theta_{iu}(1-\theta_{iu}) & u = a \\ -\theta_{iu}\theta_{ia} & u \neq a. \end{cases}$$

Using the chain rule of differentiation on the cost function $C(\theta(\psi))$ (defined similarly as in (21.10))

$$\begin{aligned} \frac{\partial}{\partial \psi_{ia}} C[\theta(\psi)] &= \sum_{u \in \mathcal{U}_i} \frac{\partial}{\partial \theta_{ia}} C(\theta) \left(\frac{\partial \theta_{iu}}{\partial \psi_{ia}} \right) \\ &= \theta_{ia} \left(\frac{\partial}{\partial \theta_{ia}} C(\theta) - \sum_{u \in \mathcal{U}_i} \theta_{iu} \frac{\partial}{\partial \theta_{ia}} C(\theta) \right), \end{aligned} \tag{21.19}$$

which is identical to the *Generalized Gradient* of [11]. In our report [19], we explain why this formulation yields the appropriate descent directional derivative for canonical coordinates, as defined in Sec. 21.2.1. See also [11] and references therein. However, gradient algorithms based on this parameterization can exhibit slow convergence, particularly when the optimal probability vector θ^* is degenerate. When a component of θ_{ia} is zero, (21.19) is zero and hence a gradient based algorithm remains at this point. Because the drift of the update is proportional to the size of the updated component, as a component approaches zero, the magnitude of future updates decreases (to prevent crossing outside the feasible set). This mechanism slows down convergence of the gradient algorithm using canonical coordinates $\psi = \theta$, particularly close to the optimal solution when this has components representing pure strategies, as is often the case. We refer the reader to [19] for numerical examples that demonstrate that the parameterization involving spherical coordinates has superior convergence properties compared to canonical coordinates.

In addition, the approach for derivative estimation in [8, 9] is via the Score Function method, which usually suffers from unbounded variance for infinite horizon costs. To alleviate this problem, the authors use a forgetting factor that introduces a bias in the derivative estimation. Our derivative estimators are more efficient and consistent, with provably bounded variance over infinite horizon, in $\boldsymbol{\alpha}^o$. In Sec. 21.7.1, numerical examples show that the variance of the score function method is several orders of magnitude larger than that of the measured valued derivative estimator.

The above methods all use a "simulation optimization" approach, where almost sure convergence to the true optimal value can be shown under an appropriate choice of parameters of the algorithms. In particular, all stochastic approximations involved in the above mentioned methodologies use a decreasing step size. One of the motivations of the present work is to implement a stochastic approximation procedure with constant step size, in order for the controlled Markov chain to be able to deal with tracking slowly varying external conditions, which result in slowly varying $A(u)$.

Remark 21.2. Our MDP setting assumes perfect observation of the process $\{X_n\}$. The paper [8, 9] considers a partially observed MDP (POMDP), but assumes that the observations Y_n of the process X_n belong to a finite set. In that work they consider suboptimal strategies of the form $\theta_{ia} = \mathbb{P}\{u_n = a \,|\, Y_n = i\}$. Such a policy is clearly not optimal for a POMDP, as the optimal policy is a measurable function of the history $(Y_1, \ldots, Y_k, u_1, \ldots, u_{k-1})$, which is summarized by a continuous-valued information state. Such suboptimal POMDP models are a special case of the problem considered here and our method can be applied in a straightforward manner. We refer the reader to [20–22] for structural results in POMDPs.

21.3. Ordinary differential equations for solving constrained MDPs

As mentioned above, to find the optimal value α^* defined in (21.14) (or equivalently, θ^* defined in (21.18)), our plan is to use a stochastic approximation algorithm of the form (21.15, 21.16). A key result in stochastic approximation theory (averaging theory), see, for example, [18], states that under suitable regularity and stability conditions, the behavior of the stochastic approximation algorithm is captured by a *deterministic* ordinary differential equation (ODE) (or, more generally, inclusion) as the step size ϵ goes to zero. Thus, to design stochastic approximation algorithms and give insight into their performance, we will first focus on designing ODEs whose solutions converge to the solution of the constrained MDP (21.12, 21.13).In other words, we will construct suitable ODEs whose stable points will be Kuhn-Tucker points of the optimization problem (21.12, 21.13).

Once such ordinary differential equations have been designed, the corresponding stochastic approximation algorithms follow naturally by replacing the gradient with the gradient estimate (which is computed from the sample path of the Markov chain), i.e, by replacing $\nabla_\alpha C$, $\nabla_\alpha B$, B in the deterministic algorithms presented below with the estimators $\widehat{\nabla_\alpha C}$, $\widehat{\nabla_\alpha B}$, $\widehat{B}$. These estimates are computed using the parameter free gradient estimation algorithms given in Sec. 21.4.2. The proofs of convergence of the resulting stochastic approximation algorithms are given in Sec. 21.5.

In this section, we present a primal-dual and an augmented Lagrangian algorithm. Our technical report [19] also presents a primal algorithm based on gradient projection which requires higher computational complexity.

21.3.1. *First-order primal-dual algorithm for constrained MDP*

A widely used deterministic optimization method (with extension to stochastic approximation in [10, p. 180]) for handling constraints is based on the Lagrange multipliers and uses a first-order primal-dual algorithm [23, p. 446]. First, convert the inequality constraints (21.13) in Problem S1 to equality constraints, by introducing the variables $z = (z_1, \ldots, z_L) \in \mathbb{R}^L$, so that $B_l(\alpha) + z_l^2 = 0$, $l = 1, \ldots, L$. Define the Lagrangian for the constrained MDP as

$$\mathcal{L}(\alpha, z, \lambda) \triangleq C(\alpha) + \sum_{l=1}^{L} \lambda_l (B_l(\alpha) + z_l^2). \qquad (21.20)$$

Here, $\lambda_l \in \mathbb{R}$, $l = 1, \ldots, L$ are Lagrange multipliers for the constraint $B_l(\alpha) + z_l^2 = 0$. In order to converge, a primal-dual algorithm requires the Lagrangian to be locally convex at the the optimum, that is, the Hessian must be positive definite at the optimum (which is much more restrictive than the second order sufficiency condition of Assumption 1 in Sec. 21.2.2). Numerical examples show that this positive definiteness condition on the Hessian, which Luenberger [24, p. 397] terms "local convexity", seldom holds in the MDP case. We can "convexify" Problem S1 by adding a penalty term to the objective function (21.12). The resulting problem is:

$$\min_{\alpha \in \boldsymbol{\alpha}^\mu, z \in \mathbb{R}^L} C(\alpha) + \frac{\rho}{2} \sum_{l=1}^{L} \left(B_l(\alpha) + z_l^2\right)^2,$$

subject to (21.13). Here ρ denotes a large positive constant. The optimum of the above problem [24, p. 429] is identical to that of (21.12, 21.13). Define the augmented Lagrangian,

$$\mathcal{L}_\rho(\alpha, z, \lambda) \triangleq C(\alpha) + \sum_{l=1}^{L} \lambda_l (B_l(\alpha) + z_l^2) + \frac{\rho}{2} \sum_{l=1}^{L} \left(B_l(\alpha) + z_l^2\right)^2. \quad (21.21)$$

Note that the original Lagrangian may not be convex near the solution (and hence the primal-dual algorithm may not work). However, for sufficiently large ρ, [23] shows that the augmented Lagrangian is locally convex. After some further calculations, detailed in [23, pp. 396 and 397], the primal-dual algorithm operating on $\mathcal{L}_\rho(\alpha(n), z(n), \lambda(n))$ reads

$$\begin{aligned} \alpha^\epsilon(n+1) = \alpha^\epsilon(n) - \epsilon\Big(& \nabla_\alpha C(\alpha^\epsilon(n)) + \nabla_\alpha B(\alpha^\epsilon(n)) \\ & \cdot \max\Big[0, \lambda^\epsilon(n) + \rho B(\alpha^\epsilon(n))\Big]\Big) \end{aligned} \quad (21.22)$$

$$\lambda^\epsilon(n+1) = \max\left[\left(1 - \frac{\epsilon}{\rho}\right)\lambda^\epsilon(n), \lambda^\epsilon(n) + \epsilon B(\alpha^\epsilon(n))\right] \quad (21.23)$$

where $\epsilon > 0$ denotes the step size and the notation $z = \max[x, y]$ for any two equal dimensional vectors x, y denotes the vector z with components $z_i = \max[x_i, y_i]$.

Lemma 21.1. *Under Assumption 21.1, for sufficiently large $\rho > 0$, there exists $\bar{\epsilon} > 0$, such that for all $\epsilon \in (0, \bar{\epsilon}]$, the sequence $\{\alpha^\epsilon(n), \lambda^\epsilon(n)\}$ generated by the primal-dual algorithm (21.22) is attracted to a local KT pair (α^*, λ^*) of Problem S1.*

Proof. Since $\mathcal{L}_\rho$ is locally convex for sufficiently large $\rho > 0$ [24], the proof follows from Proposition 4.4.2 in [23]. □

Let $T \in \mathbb{R}^+$ denote a fixed constant and $t \in [0, T]$ denote continuous time. Define the piecewise constant interpolated continuous-time process

$$\alpha^\epsilon(t) = \alpha^\epsilon(n) \quad t \in [n\epsilon, (n+1)\epsilon) \tag{21.24}$$

$$\lambda^\epsilon(t) = \lambda^\epsilon(n) \quad t \in [n\epsilon, (n+1)\epsilon). \tag{21.25}$$

Lemma 21.1 implies that $\{\lambda^\epsilon(n)\}$ lies in a compact set Λ for all n. Recall from (21.8), that $\alpha^\epsilon(n) \in \boldsymbol{\alpha}^\mu$ where $\boldsymbol{\alpha}^\mu$ is compact. Then the following result follows directly from the above lemma and [10] where the convergence of the stochastic version is proved.

Theorem 21.1. *Let $\{\alpha(t), \lambda(t)\}$ satisfy thc ODEs*

$$\begin{aligned}
\frac{d}{dt}\alpha(t) &= -\nabla_\alpha C(\alpha(t)) - \nabla_\alpha B(\alpha(t)) \max\Big[0, \lambda(t) + \rho B(\alpha(t))\Big], \\
\frac{d}{dt}\lambda_l(t) &= \begin{cases} B_l(\alpha(t)) & \text{if } \lambda_l(t) + \rho B_l(\alpha(t)) \geq 0 \\ \lambda_l(t)/\rho & \text{if } \lambda_l(t) + \rho B_l(\alpha(t)) < 0, \end{cases} \\
&\quad \text{for } \alpha(0) \in \boldsymbol{\alpha}^\mu,\ \lambda_l(0) \in \mathbb{R}, \quad l = 1, \ldots, L.
\end{aligned} \tag{21.26}$$

Then, under Assumption 21.1, for sufficiently large ρ (see Lemma 21.1), the interpolated process $\{\alpha^\epsilon(t), \lambda^\epsilon(t)\}$ defined in (21.24, 21.25) converges uniformly as $\epsilon \to 0$ to the process $\{\alpha(t), \lambda(t)\}$, that is,

$$\lim_{\epsilon \downarrow 0} \sup_{0<t\leq T} |\alpha^\epsilon(t) - \alpha(t)| = 0, \qquad \lim_{\epsilon \downarrow 0} \sup_{0<t\leq T} |\lambda^\epsilon(t) - \lambda(t)| = 0, \tag{21.27}$$

The attraction point of Eq. (21.26) is the local KT pair (α^*, λ^*) of Problem S1.

21.3.2. *Augmented Lagrangian (multiplier) algorithms for constrained MDP*

We outline two augmented Lagrangian (multiplier) deterministic algorithms for optimizing the constrained MDP assuming full knowledge of the objective $C(\alpha)$, constraints $B(\alpha)$ and their derivatives.

21.3.2.1. *Inexact primal minimization multiplier algorithm*

The augmented Lagrangian approach (also known as a multiplier method) consists of the following coupled ODE and difference equation:

$$\begin{aligned}\frac{d\alpha^{(n+1)}(t)}{dt} &= -\nabla_\alpha C(\alpha^{(n+1)}(t)) \\ &\quad - \nabla_\alpha B(\alpha^{(n+1)}(t)) \max\left[0, \lambda(n) + \rho B(\alpha^{(n+1)}(t)\right]\end{aligned} \tag{21.28}$$

$$\lambda_l(n+1) = \max\left[0,\ \lambda_l(n) + \rho B_l(\alpha^{(n+1)}(\infty))\right], \quad l = 1, \ldots, L, \tag{21.29}$$

where $\alpha^{(n+1)}(\infty)$ denotes the stable point of the ODE (21.28). Iteration (21.29) is a first order update for the multiplier, while (21.28) represents an ODE which is attracted to the minimum of the augmented Lagrangian $\mathcal{L}_\rho$. The max in (21.29) arises in dealing with the inequality constraints, see [23, p. 396]. Proposition 4.2.3 in [23] shows that if $(\alpha^{(0)}(0), \lambda_0(0))$ lies in the domain of attraction of a local KT pair (α^*, λ^*), then (21.28, 21.29) converges to this KT pair.

A practical alternative to the above exact primal minimization is first order *inexact minimization* of the primal. The iterative version of the algorithm reads [23, p. 406]: At time $n+1$ set $\alpha^{(0)}(n+1) = \alpha(n)$. Then run $j = 0, \ldots, J-1$ iterations of the following gradient minimization of the primal

$$\begin{aligned}\alpha^{(j+1)}(n+1) = \alpha^{(j)}(n+1) - \epsilon\Big(&\nabla_\alpha C(\alpha^{(j)}(n)) \\ &+ \nabla_\alpha B(\alpha^{(j)}(n)) \max\left[0, \lambda(n) + \rho B(\alpha^{(j)}(n))\right]\Big),\end{aligned} \tag{21.30}$$

yielding $\alpha(n+1) = \alpha^{(J)}(n+1)$, followed by a first order multiplier step

$$\lambda_l(n+1) = \max\left[0,\ \lambda_l(n) + \rho B_l(\alpha(n+1))\right], \quad l = 1, \ldots, L. \tag{21.31}$$

Iteration (21.30) represents a first order fixed step size inexact minimization of the augmented Lagrangian $\mathcal{L}_\rho$ (*inexact* because (21.30) is terminated after a finite number of steps J). It is shown, see [23] and references therein, that as long as the inexact minimization of the primal is such that the error tolerances are decreasing with n but summable, then the algorithm converges to a Kuhn Tucker point.

21.3.2.2. *Fixed multiplier*

A trivial case of the multiplier algorithm is to fix $\lambda(n) = \bar{\lambda}$ for all time n and only update α according to (21.30) with $I = 1$ iteration at each time instant. This is clearly equivalent to the primal update (21.22) with fixed $\lambda(n) = \bar{\lambda}$. From Theorem 21.1, the interpolated trajectory of this algorithm converges uniformly as $\epsilon \to 0$ to the trajectory of the ODE

$$\frac{d}{dt}\alpha(t) = -\nabla_\alpha C(\alpha(t)) - \nabla_\alpha B(\alpha(t)) \max\left[0, \bar{\lambda} + \rho B(\alpha(t))\right], \quad \alpha(0) \in \boldsymbol{\alpha}^\mu. \tag{21.32}$$

The following result in [23] shows that the attraction point of this ODE is close to α^* for sufficiently large ρ, resulting in a near optimal solution. First convert the L inequality constraints to equality constraints as outlined in Sec. 21.3.1. Let α^*, λ^* denote the corresponding KT pair.

Proposition 21.1. [23, Proposition 4.2.3]. Let $\bar{\rho} > 0$ be scalar such that ${\nabla_\alpha}^2\mathcal{L}_\rho(\alpha^*, \lambda^*) > 0$. Then there exist positive scalars δ and K such that, for $(\bar{\lambda}, \rho) \in D \in \mathbb{R}^{L+1}$ defined by

$$D = \{(\lambda, \rho) : \|\lambda - \lambda^*\| < \delta\rho,\ \rho \geq \bar{\rho}\},$$

the attraction point $\alpha^{\lambda,\rho}$ of the ODE (21.32) is unique. Moreover, $\|\alpha^{\lambda,\rho} - \alpha^*\| \leq K(\|\bar{\lambda} - \lambda^*\|)/\rho$.

21.4. Gradient estimation algorithms for constrained MDP

In the previous section, we presented deterministic algorithms for optimizing the constrained MDP Problem S1 that assume complete knowledge of the gradient of the expected cost and constraints. This section deals with how to estimate the gradient of the cost and constraint by simulation. As mentioned earlier, this gradient estimation step is a key step in the stochastic approximation algorithms to optimize the constrained MDP.

Throughout this section, we focus on the process with a fixed value of the control parameter α. We use $\mathbb{E}_\alpha$ to denote expectation w.r.t the underlying probability measure of $\{Z_n, n = 1, 2, \ldots\}$. We will also use $\mathcal{E}_i$ to denote expectation w.r.t. to the distribution of the random action, given that the state is $X_n = i$. Finally, Z_n will be refereed to as the "*nominal process.*"

21.4.1. *Infinite-horizon measure-valued derivative estimation*

Although we are interested in estimating the gradient given short data lengths (so as to deal with time varying transition probabilities), to fix ideas, we first focus on the infinite horizon case, that is, we build gradient estimators of

$$\nabla_\alpha \mathbb{E}_{\pi(\alpha)}[F(Z)] = \nabla_\alpha \left(\sum_{i\in S} \sum_{u\in\mathcal{U}_i} F(i,u)\pi_{i,u}(\alpha) \right). \tag{21.33}$$

Here F indicates a function of the state-action pair $Z = (i,u)$, such as the cost c or the constraint values b_l, and $\pi_{i,u}(\alpha)$ is the stationary measure of the chain $\{Z_n\}$ under fixed parameter value α. The resulting estimators $\widehat{\nabla_\alpha C}$ and $\widehat{\nabla_\alpha B}$ are parameter free (learning) gradient estimators.

In the case of spherical coordinates, it follows from the relationship Eq. (21.6) that the action u_{k+1} (given $i_{k+1} = i$) has a distribution

$$u_{k+1} = \begin{cases} 0 & \text{w.p. } \cos^2(\alpha_{i1}) \\ Y_1 & \text{w.p. } \sin^2(\alpha_{i1}) \end{cases} \tag{21.34}$$

$$Y_1 = \begin{cases} 1 & \text{w.p. } \cos^2(\alpha_{i2}) \\ Y_2 & \text{w.p. } \sin^2(\alpha_{i2}) \end{cases}$$

$$\vdots$$

$$Y_{d(i)-1} = \begin{cases} d(i)-1 & \text{w.p. } \cos^2(\alpha_{i,d(i)}) \\ d(i) & \text{w.p. } \sin^2(\alpha_{i,d(i)}). \end{cases}$$

Let $Y_{d(i)} = d(i)$. The value of α_{ip}, $i \in S$, $p \in \{1, 2, \ldots d(i)\}$, does not affect the distribution of u_{k+1} if $i_{k+1} \neq i$, therefore the gradient of the one-step transition expectation is non-null only when $i_{k+1} = i$. In this case we have

$$\begin{aligned} &\frac{\partial}{\partial \alpha_{ip}} \mathbb{E}[f(i, u_{k+1})] \\ &= \frac{\partial}{\partial \alpha_{ip}} \mathbb{E}\left[f(i,p-1)\cos^2(\alpha_{ip}) + f(i,Y_p)\sin^2(\alpha_{ip}) \right] \prod_{k=1}^{p-1} \sin^2(\alpha_{ik}) \\ &= -2\sin(\alpha_{ip})\cos(\alpha_{ip}) \prod_{k=1}^{p-1} \sin^2(\alpha_{ik}) \mathbb{E}[f(i,p-1) - f(i,Y_p)], \end{aligned}$$

as the terms $f(i,k), k < p-1$ have weights which are independent of α_{ip}. The random variable Y_p is called the "phantom action" and it has a distribution concentrated on $\{p, \ldots, d(i)\}$ corresponding to

$$\mathbb{P}(Y_p = a) = \frac{\theta_{ia}(\alpha)}{\prod_{m=1}^{p-1} \sin^2(\alpha_{im})}, \quad a \geq p, \ p \in \{1, 2, \ldots d(i)\}. \tag{21.35}$$

Note that by construction, for $p = d(i)$ the random variable $Y_{d(i)} = d(i)$ is degenerate.

The above analysis for the one-step transition expectation is an example of what is called a "weak derivative" [14]. To obtain the gradient of an infinite horizon expectation (21.33), a measure-valued approach for weak derivatives is used in [19]. In the proof of Theorem 21.2, we will use the approach of the perturbation realization factors. The equivalence between the two approaches has been shown in [25].

First introduce the following notation. Let $\{\tilde{u}_k\}$ denote a sequence of iid random variables (independent of $\mathfrak{F}_n$) as follows

$$\tilde{u}_k = \begin{cases} Y_p, & p < d(i), \\ d(i)\mathbf{1}_{\{u_k = d(i)-1\}} + (d(i)-1)\mathbf{1}_{\{u_k = d(i)\}} & p = d(i). \end{cases} \tag{21.36}$$

We now provide the basis for the interpretation of $\tilde{u}_k$. For $p - 1 < d(i)$,

$$\mathbb{P}(u_{k+1} = p - 1 \mid X_{k+1} = i) = \cos^2(\alpha_{ip}) \prod_{k=1}^{p-1} \sin^2(\alpha_{ik}),$$

so we can rewrite the derivative expression as

$$\begin{aligned} \frac{\partial}{\partial \alpha_{ip}} \mathbb{E}[f(i, u_{k+1})] &= -2\tan(\alpha_{ip})\mathbb{P}(u_{k+1} = p - 1 \mid X_{k+1} = i) \\ &\quad \cdot \mathbb{E}[f(i, u_{k+1}) - f(i, \tilde{u}_{k+1})] \mid u_{k+1} = p - 1]. \end{aligned}$$

This is the basis for using an "on-line" estimator, as we will shortly explain in detail. When the state is $(X_{k+1}, u_{k+1} = (i, p-1)$, we calculate the difference between the observed trajectory and that of the phantom process, where the phantom action is now u_k. Because the case $p = d(i)$ has a degenerate distribution for Y_p, the estimation can be made both when $u_{k+1} = d(i)$ (in which case the phantom action is $d(i) - 1$), or when it is $d(i) - 1$ (in which case the phantom action is $d(i)$). The complete details of this reasoning appears in [19].

Define the hitting time

$$\nu(k) = \min\{n > 0 : Z_{k+n} = (i, \tilde{u}_k)\}. \tag{21.37}$$

Theorem 21.2. *Consider the MDP $\{Z_n\}$ governed by (21.5, 21.9) with $\alpha \in \boldsymbol{\alpha}^\mu$. Fix (i,p) and let $a = p-1$. Let $\hat{\Gamma}$ denote any estimator that converges as $N \to \infty$ to $C(\alpha)$ a.s. (e.g., $\hat{\Gamma} = \frac{1}{N}\sum_{n=1}^{N} c(Z_n)$) and $K_{ip}(\alpha,k)$ defined by*

$$K_{ip}(\alpha,k) = \begin{cases} -\tan(\alpha_{ip})\,\delta_{i,p}(k), & p = 1,\ldots,d(i)-1 \\ -\cos(\alpha_{i,d(i)})\sin(\alpha_{i,d(i)})[\delta_{i,d(i)}(k) - \delta_{i,d(i)+1}(k)], & p = d(i). \end{cases} \tag{21.38}$$

Here $\delta_{ip}(k) = \mathbf{1}_{\{Z_k=(i,p-1)\}}$.

Then for fixed $\alpha \in \boldsymbol{\alpha}^o$, a parameter free consistent estimator for the gradient in (21.33) is given by

$$\hat{G}_N(i,p) = \frac{2}{N}\sum_{k=1}^{N} K_{ip}(\alpha,k)\left([c(i,u_k) - c(i,\tilde{u}_k)] + \sum_{n=k+1}^{k+\nu(k)} c(Z_n) - \nu(k)\,\hat{\Gamma}\right), \tag{21.39}$$

for $p = 1,\ldots,d(i)$.

The proof of the above theorem is in the appendix. The algorithm computes the differences between the *nominal* (observed) process trajectory and the "parallel" ones that start at time k with action $\tilde{u}_k$. Such parallel processes are called "phantom" processes, and we call our algorithm the *WD phantom estimator.*

21.4.2. *Fast WD for tracking time-varying MDPs*

In Theorem 21.2, consistency of the WD phantom estimator for large sample size N was established. However, for adaptive control of constrained MDPs with time varying transition probabilities, it is necessary to implement the estimation over small batch sizes of the observed system trajectory, so that the iterates of the stochastic gradient algorithm are performed more frequently, to track the optimal time varying α^*. The aim of this section is to present an implementation of the WD phantom estimators over short batch sizes N and to show that the resulting gradient estimate is still consistent. We call these "fast WD phantoms".

The implementation of the fast WD phantoms over short batch sizes proceeds as follows. Suppose that the gradient $\nabla_\alpha C(\alpha)$ is to be estimated using the observed MDP trajectory over the nth batch $I_n \equiv$

$\{nN+1,\ldots,(n+1)N\}$. As in Theorem 21.2, let $\widehat{\Gamma}_n$ denote an estimator of $C(\alpha)$ using the observed trajectory of the MDP in I_n. The fast WD phantom estimators for the components (i,p), $i \in S$, $p \in \{1,2,\ldots,d(i)\}$, of the gradient are (compare with (21.39)):

$$\widehat{\widehat{G}}_n(i,p) = \frac{2}{N}\left[\sum_{k=nN+1}^{(n+1)N} K_{ip}(\alpha,k)\left[c(i,u_k) - c(i,\tilde{u}_k) + \sum_{j=k+1}^{\mu(k)} c(Z_j) - \nu(k)\widehat{\Gamma}_n \mathcal{D}_n(k)\right] + \sum_{k\in\mathcal{L}(nN)} K_{ip}(\alpha,k)\left(\sum_{j=nN+1}^{\mu(k)} c(Z_j) - \nu(k)\widehat{\Gamma}_n\mathcal{D}_n(k)\right)\right] \tag{21.40}$$

where $\mu(k) = \min\{k+\nu(k),(n+1)N\}$, $\mathcal{D}_n(k) = \mathbf{1}_{\{k+\nu(k)\in I_n\}}$ (phantom k dies in I_n), and $\mathcal{L}(j) = \{k \le j \colon \nu(k)+k > j\}$ is the list of living phantoms at stage j. A similar estimator holds for the gradient of the constraints.

The interpretation of (21.40) is as follows. At each step k, the state $Z_k = (i,a)$ is observed and a new phantom system (labelled by k) is started, generating the phantom decision $\tilde{u}_k$ as described above. The k-th phantom system "dies" at the hitting time $j = k+\nu(k)$, otherwise it is contained in the set of "living" phantoms $\mathcal{L}(j)$. In a computer program, this corresponds to a list. This phantom will be used to estimate the partial derivative of all the functions (cost and constraints) with respect to $\alpha_{i,a+1}$ if $a < d(i)-1$. If $a = d(i)-1$ or $a = d(i)$, the corresponding phantom system contributes (with opposite signs) to the estimation of the gradient w.r.t. $\alpha_{i,d(i)}$. The difference in costs inside the brackets in (21.40) contains the initial contribution of a phantom system. Afterwards, while a phantom system k is alive, it contributes to (21.40) the term $c(Z_j)$ at each step, and when it dies ($\mathcal{D}_n(k) = 1$) it contributes the term $\nu(k)\widehat{\Gamma}_n$ (if death occurs within the interval I_n). The above equation takes only observations of the trajectory within the *current* interval, thus a final term appears considering the contributions of all the living phantoms at the start of the interval, because it is possible that phantom systems may survive several estimation intervals. For complete details on the implementation program code and other variations, see [26].

Theorem 21.3. *Assume that for any* $\alpha \in \boldsymbol{\alpha}^o$, $\mathbb{E}_{\pi(\alpha)}\widehat{\Gamma}_n = C(\alpha)$. *Then the bias of the fast WD phantom gradient estimator* $\widehat{\widehat{G}}_n(i,p)$ *of (21.40) per* *batch of size N is given by (with $K_{ip}(\alpha,k)$ defined in (21.38))*

$$\begin{aligned}&\mathbb{E}_{\pi(\alpha)}[\widehat{\widehat{G}}_n(i,p)] - \nabla_\alpha C(\alpha)\\&= \frac{2}{N}\sum_{\{k:\ k+\nu(k)\in I_n\}} K_{ip}(\alpha,k)\,\mathrm{Cov}_{\pi(\alpha)}(\nu(k),\widehat{\Gamma}_n) + o\Big(\frac{1}{N}\Big).\end{aligned} \tag{21.41}$$

Proof. Let $\widehat{\widehat{G}}{}^o_n(i,p)$ denote the gradient estimate (21.40) when $\hat{\Gamma}_N$ is replaced by $C(\alpha)$. We first show that the gradient estimate $\widehat{\widehat{G}}{}^o_n(i,p)$ is unbiased under the invariant measure $\pi(\alpha)$, that is, $\mathbb{E}_{\pi(\alpha)}[\widehat{\widehat{G}}{}^o_n(i,p)] = \nabla_\alpha C(\alpha)$.

Note that, under $\pi(\alpha)$, consecutive estimators have the same distribution (although they are not independent), because the invariant distribution of the number of living phantoms at the start of the interval is independent of n. From the ergodicity of the underlying MDP, $\mathbb{E}_{\pi(\alpha)}[\widehat{\widehat{G}}{}^o_n(i,p)] = \lim_{n\to\infty}(1/n)\sum_{m=0}^{n-1}\widehat{\widehat{G}}{}^o_m(i,p)$. Because the estimation by batches considers breaking up the partial sums of the estimation, then the difference

$$\begin{aligned}\frac{1}{n}\sum_{m=0}^{(n-1)}\widehat{\widehat{G}}{}^o_m(i,p) - \frac{2}{nN}\sum_{k=1}^{nN} K_{ip}(\alpha,k)\Bigg[c(i,a) - c(i,\tilde{u}_k)\\ + \sum_{j=(k+1)}^{k+\nu(k)} c(Z_j) - \nu(k)C(\alpha)\Bigg]\end{aligned}$$

tends to zero in absolute value, a.s. Using Theorem 21.2, the $\frac{2}{nN}\sum_k$ term in the above equation converges a.s. to $\nabla_\alpha C(\alpha)$ as $n\to\infty$, for any fixed value of the batch size N. This establishes the claim that the gradient estimate is unbiased. Next, using the above expression, consider $\widehat{\widehat{G}}{}^o_n - \widehat{\widehat{G}}_n$. Using the fact that the consecutive estimators are additive, we obtain

$$\frac{1}{n}\sum_{m=0}^{n-1}(\widehat{\widehat{G}}{}^o_n - \widehat{\widehat{G}}_n) = \frac{1}{n}\sum_{m=0}^{n-1}\left[\frac{2}{N}\sum_{k=1}^{nN}\nu(k)(\hat{\Gamma}_m - C(\alpha))\mathbf{1}_{\{k+\nu(k)\in I_m\}}\right] + e(nN).$$

The error term $e(nN)$ comes from the phantom systems $k\colon k+\nu(k)\le nN$, and their contribution tends to zero a.s. as $n\to\infty$. Under the invariant measure, we have

$$\begin{aligned}&\mathbb{E}_{\pi(\alpha)}\left(\nu(k)(\hat{\Gamma}_m - C(\alpha))\mathbf{1}_{\{k+\nu(k)\in I_m\}}\right)\\&= \mathrm{Cov}_{\pi(\alpha)}(\nu(k)\mathbf{1}_{\{k+\nu(k)\in I_m\}},\hat{\Gamma}_m) + C(\alpha)\mathbb{E}_{\pi(\alpha)}(\nu(k)\mathbf{1}_{\{k+\nu(k)\in I_m\}}).\end{aligned}$$

□

21.4.3. *Computational cost and memory requirement*

Consider the fast WD phantom algorithm. Let $\alpha \in \boldsymbol{\alpha}^{\mu}$. We will bound stochastically the number of living phantoms in terms of Binomial random variables.

Consider the process $\{Z_n\}$ in stationary operation, and call $\mathcal{L}_{nN}(i,a)$ the number of living phantoms in Eq. (21.40) that are waiting to hit state (i,a). All these phantoms were created at some earlier time instant when the chain hit the state $Z_k = (i,p)$ and the phantom decision was $\tilde{u}_k = a$. Clearly, the maximum number of phantoms in $\mathcal{L}_{nN}(i,a)$ satisfies

$$\|\mathcal{L}_{nN}(i,a)\| = \sum_{a>p} \sum_{k=t(i,a)}^{nN} \mathbf{1}_{\{X_k=i;u_k=p,\tilde{u}_k=a\}},$$

where $t(i,a) = \max(j \le nN \colon X_j = 1, u_j = a)$, because, if the state (i,a) is visited at time j, at that time all living phantoms that were in $\mathcal{L}_{j-1}(i,a)$ die and $\mathcal{L}_j(i,a)$ is empty. Call $\tau(i,a)$ the return time to state (i,a) and let $n(i)$ be the number of visits to state i ($X_k = i$) within two consecutive visits to state (i,a). Then the number of phantom systems in $\mathcal{L}_{nN}(a)$ is bounded by a Binomial$(\mathbf{p}(a), n(i))$, where

$$\mathbf{p}(a) = \frac{\theta_{ia}}{\sum_{m=1}^{p-1} \sin^2(\alpha_{im})},$$

according to the creation of the phantom systems. Clearly considering the maximum value of all such probabilities, we can bound the number of phantoms on each list for every value of $\alpha \in \boldsymbol{\alpha}^{\mu}$.

In Eq. (21.40), the estimator $\widehat{\widehat{G}}{}^{o}_{n}(i,p)$ is composed of bounded quantities (the state space is finite) plus a contribution of N terms of the order of $\nu(k)$ each, plus a contribution which is proportional to the random variable:

$$\sum_{a>p} \sum_{k \in \mathcal{L}_{nN}(i,a)} \nu(k) \le \sum_{a>p} \sum_{i=1}^{n(i)} \nu(k) \le \sum_{a>p} \sum_{k=1}^{\tau(i,a)} \nu(k).$$

Note that, as in the proof of Theorem 21.2, $\nu(k) < \max(\tau(i,a) : a > p) = \tau$ a.s. As α^{μ} is a compact set with ergodic states, the return times are all finite a.s. and $\mathbb{E}_{\alpha}[\tau(i,a)] = 1/\pi_{ia}$. Therefore, the boundedness of the m th moment of $\widehat{\widehat{G}}^{o}_{n}(i,p)$ now follows from the boundedness of the $2m$-th moment of τ.

21.5. Stochastic gradient algorithms for constrained MDP

In this section we present the stochastic gradient algorithms that use the parameter free gradient estimators (fast WD phantoms) of Sec. 21.4.2 to optimize the constrained MDP given in Problem S1 (21.12, 21.13). Also weak convergence proofs of these algorithms are presented. The stochastic algorithms presented are stochastic versions of the two deterministic algorithms of Sec. 21.3.1 and Sec. 21.3.2. Using the simulation-based fast WD phantom gradient estimator (21.40), with local sample averages for the estimation of the constraint functions, may lead to a noticeable bias for small batch size N, thus making the algorithm suboptimal.

For notational convenience we consider equality constraints here (as mentioned in Sec. 21.3.1 the inequality constraints can be handled with minor modifications) so that the constrained MDP problem (21.12, 21.13) reads

$$\min_{\alpha \in \boldsymbol{\alpha}^{\mu}} C(\alpha), \quad \text{s.t. } B_l(\alpha) = 0, \quad l = 1, \ldots, L.$$

A control "agent" is associated with each of the possible visited states $i \in S$. The control parameter for this agent is the vector $(\alpha_{ip}, p \in \{1, \ldots, d(i)\})$, plus an agent for the (artificial control) variable representing the Lagrange multiplier λ. The scheme works by observing the process over a batch size N during which the value of the control parameter does not change. Over this batch, the constraint is estimated as $\hat{B}(n)$, see Sec. 21.5.3, and the gradients are estimated as $\widehat{\nabla_\alpha C}(n)$, $\widehat{\nabla_\alpha B}(n)$ using Eq. (21.40). We assume that $\mathbb{E}_{\pi(\alpha)}[\hat{B}(n)] = B(\alpha)$. Let

$$h_0(\alpha) = \mathbb{E}_{\pi(\alpha)}[\widehat{\nabla_\alpha C}(n)], \quad h_l(\alpha) = \mathbb{E}_{\pi(\alpha)}[\widehat{\nabla_\alpha B_l}(n)], \quad l = 1, \ldots, L$$

be the invariant averages of the batch estimation (refer to Theorem 21.3).

21.5.1. *Stochastic first-order primal-dual algorithm*

Consider the stochastic approximation where the control parameter (α, λ) is updated as (c.f. (21.22, 21.23))

$$\alpha^\epsilon(n+1) = \alpha^\epsilon(n) - \epsilon\Bigg(\widehat{\nabla_\alpha C}(n) + \sum_{l=1}^{L}\Big(\lambda_l^\epsilon(n) + \rho\hat{B}_l(n)\Big)\,\widehat{\nabla_\alpha B_l}(n) + Z^\mu(n)\Bigg), \quad \alpha^\epsilon(0) \in \boldsymbol{\alpha}^\mu \tag{21.42}$$

$$\lambda^\epsilon(n+1) = \lambda^\epsilon(n) + \epsilon\,\hat{B}(n) \tag{21.43}$$

where the gradient estimators are given by the fast WD phantoms in (21.40) and $Z^\mu(n)$ is a projection that ensures that $\{\alpha^\epsilon(n)\} \in \boldsymbol{\alpha}^\mu$. The above truncation is a mathematical artifice to prove convergence. In practical implementation, that is, when $\epsilon > 0$, truncation is not important since one can choose $\mu \ll \epsilon$, e.g., close to the numerical resolution of the computer, see remark below.

Proposition 21.2. For $\alpha^\epsilon(n), \lambda^\epsilon(n)$ given by (21.42, 21.43), define the interpolated process $\{\alpha^\epsilon(t), \lambda^\epsilon(t)\}$ as in (21.24, 21.25). Then as $\epsilon \to 0$, $\{a^\epsilon(t), \lambda^\epsilon(t)\}$ converges in distribution to the solution of the ODE (c.f. (21.26))

$$\begin{aligned}\frac{d}{dt}\alpha(t) &= -\left[h_0[\alpha(t)] + \sum_{l=1}^{L}(\lambda_l(t) + \rho\, B_l[\alpha(t)])\, h_l[\alpha(t)] + \kappa[\alpha(t)]\right]_\mu, \\ \frac{d}{dt}\lambda(t) &= B[\alpha(t)],\end{aligned} \tag{21.44}$$

with $\alpha(0) \in \boldsymbol{\alpha}^\mu$, where the notation $[\cdot]_\mu$ refers to the truncated ODE onto $\boldsymbol{\alpha}^\mu$ and the added drift is defined as

$$\kappa(\alpha) = \rho \lim_{n\to\infty} \sum_{l=1}^{L} \mathrm{Cov}_{\pi(\alpha)}[\hat{B}_l(n), \widehat{\nabla_\alpha B_l}(n)]. \tag{21.45}$$

Remark 21.3. A weak convergence proof of the stochastic gradient algorithm (21.42, 21.43) requires uniform integrability of the gradient estimates. Without the above truncation, if $\alpha_{ip} = \pi/2$ or $\alpha_{ip} = 0$, (equivalently one or more action probabilities $\theta_{iu} = 0$), then the hitting time $\nu(k)$ in (21.37) of the phantom system k with initial state (i, u) is not uniformly bounded and the gradient estimator (21.40) is not well defined. So in the weak convergence proof below we place a μ size ball around the boundary of $\boldsymbol{\alpha}$ (recall $\boldsymbol{\alpha}^\mu$ above is defined as $\boldsymbol{\alpha}$ minus this ball) and the estimates $\alpha^\epsilon(n)$

are truncated to $\boldsymbol{\alpha}^\mu$. However, this truncation is very different to standard truncations in the stochastic approximation literature:

1. Recall from Sec. 21.2.1 that the boundary of the set $\boldsymbol{\alpha}$ is a fictitious boundary and the action probabilities $\theta(\alpha)$ are symmetric about this fictitious boundary (since they are functions of $\sin^2(\alpha)$ and $\cos^2(\alpha)$). For example, suppose that $\alpha_{ip}^\epsilon(n) < \pi/2 - \mu$ and that (21.42) generates the estimate $\alpha_{ip}^\epsilon(n+1) = \pi/2 + \mu + K$ for some small constant $K > 0$. Then truncation is not required since, by symmetry, $\alpha_{ip}^\epsilon(n+1) = \pi/2 - (\mu + K) \in \boldsymbol{\alpha}^\mu$. Thus in practical implementation, that is, when $\epsilon > 0$, truncation is not important since we can choose $\mu \ll \epsilon$. The probability that an update lies precisely in a ball of radius μ is negligibly small – since if the estimate overshoots or undershoots this ball, it is automatically in $\boldsymbol{\alpha}^\mu$.
2. Suppose that the untruncated version of the ODE (21.44) has a stable point on the boundary of $\boldsymbol{\alpha}$, ie., a pure policy. Then clearly, the truncated ODE will have a stable point within $O(\mu)$ of the stable point of the untruncated ODE. Hence the truncation is very different to standard ODE truncations such as (21.26).

Proof. We first show that the term in parenthesis on the RHS of (21.42) is uniformly integrable. Note that all the terms in the gradient estimate (21.40) apart from $\mathcal{L}(\cdot)$ and $\nu(\cdot)$ are uniformly bounded for $\alpha \in \boldsymbol{\alpha}^\mu$ since the MDP is finite state and the batch size N is fixed and finite. Hence a sufficient condition for uniform integrability is to show that $\sum_{k\in\mathcal{L}(nN)} \nu(k)$ in (21.40) has finite variance.

Fix (i,p) for the estimator in (21.40). We focus on the phantoms that have an initial decision $\tilde{u}_k = a$ for a fixed a, and we call these a-phantoms. By definition, all a-phantoms die simultaneously at time n_1 when the process $Z_{n_1} = (i,a)$. Let n_0 denote the previous time instant at which $\{Z_n\}$ visited (i,a), that is, $Z_{n_0} = (i,a)$. The longest hitting time $\nu(k)$ of all these phantoms is bounded by the time between successive returns $n_1 - n_0$. Consider now the number of living a-phantoms in $\mathcal{L}(n)$ at any time $n \in [n_0, n_1]$. Each of these a-phantoms must have been created when the process hits the state (i,p) and the phantom decision chosen is a, which happens with probability $\frac{\theta_{ia}}{\sum_{m\geq a}\theta_{im}}$. Therefore an almost sure upper bound for the cardinality of $\mathcal{L}(n)$ is $n_1 - n_0$. Note that $n_1 - n_0$ is the return time of a ergodic MDP on a finite state and therefore has all moments bounded.

Having established uniform integrability of the updates, the result follows by direct application of Theorem 5.2.1 in [10]. The continuity of the invariant expectations follows from the fact that the transition kernel of Z_n is analytic in α. To characterize the drift functions, use:

$$\mathbb{E}_{\pi(\alpha)}\left[\hat{B}_l(n)\widehat{\nabla_\alpha B}_l(n)\right] = \mathbb{E}_{\pi(\alpha)}[\hat{B}_l(n)]\,\mathbb{E}_{\pi(\alpha)}[\widehat{\nabla_\alpha B}_l(n)] + \mathrm{Cov}_{\pi(\alpha)}[\hat{B}_l(n), \widehat{\nabla_\alpha B_l}(n)],$$

which establishes the result. □

21.5.2. *Stochastic augmented Lagrangian multiplier algorithm*

Consider the following stochastic approximation version of the multiplier algorithm Eq. (21.31),

$$\alpha^\epsilon(n+1) = \alpha^\epsilon(n) - \epsilon\left(\widehat{\nabla_\alpha C}(n) + \sum_{l=1}^{L}\left(\lambda_l^\epsilon(n) + \rho\hat{B}(n)\right)\widehat{\nabla_\alpha B_l}(n) + Z^\mu(n)\right), \tag{21.46}$$

with $\alpha^\epsilon(0) \in \boldsymbol{\alpha}^\mu$, where $Z^\mu(n)$ is a projection that ensures that $\{\alpha^\epsilon(n)\} \in \boldsymbol{\alpha}^\mu$ and $\{\lambda^\epsilon(n), \epsilon > 0, n \in \mathbb{N}\}$ is any tight sequence. A trivial example is when $\lambda^\epsilon(n)$ is a bounded constant (a.s.).

The following result regarding the weak convergence of (21.46) is proved in the appendix.

Proposition 21.3. Assume that $\{\lambda^\epsilon(n), \epsilon > 0, n \in \mathbb{N}\}$ is tight. Define the interpolated process $\alpha^\epsilon(t)$ of (21.46) as in Eq. (21.24). Then as $\epsilon \to 0$, the interpolated process $\alpha^\epsilon(t)$ converges in distribution to the solution of the ODE

$$\frac{d\alpha(t)}{dt} = -\left[h_0[\alpha(t)] + \sum_{l=1}^{L}(\bar{\lambda}_l + \rho\, B_l[\alpha(t)])\, h_l[\alpha(t)] + \kappa[\alpha(t)]\right]_{\mu}, \tag{21.47}$$

with $\alpha(0) \in \boldsymbol{\alpha}^\mu$, where $\bar{\lambda}_l$ is an accumulation point of the sequence $\{\bar{\lambda}(\epsilon), \epsilon > 0\}$ of (convergent) Cesaro sums,

$$\bar{\lambda}(\epsilon) \equiv \lim_{N\to\infty} \frac{1}{N}\sum_{n=1}^{N} \lambda^\epsilon(n),$$

and $\kappa(\alpha)$ is defined in (21.45).

Remark 21.4. If the bias in $h_l(\alpha)$ and $\kappa(\alpha)$ are negligible, then, under no truncation, the ODE (21.47) reduces to (21.32) – which is the ODE for the deterministic fixed multiplier algorithm. Proposition 21.1 of Sec. 21.3.2 implies that the estimates converge weakly to a near optimal point, provided that the pair $(\bar{\lambda}, \rho)$ is well chosen. The bias in $h_l(\alpha)$ and $\kappa(\alpha)$ is of order $O(1/N)$.

Consider the following update of the multiplier λ in (21.46). Define $\mathcal{I} = \lfloor 1/\epsilon \rfloor$ and consider the recursion

$$\lambda(n+1) = \lambda(n) + \bar{B}(n/\mathcal{I})\mathbf{1}_{\{\frac{n}{\mathcal{I}} \in \mathbb{N}\}}, \text{ where } \bar{B}(n/\mathcal{I}) = \frac{1}{\mathcal{I}} \sum_{j=(n-1)\mathcal{I}+1}^{n\mathcal{I}} \hat{B}(j), \tag{21.48}$$

together with (21.46). Thus the multiplier is updated once every $\mathcal{I}$ time points. If the bias in $h_l(\alpha)$ and $\kappa(\alpha)$ is negligible, then as $\epsilon \to 0$, the algorithm (21.46) – (21.48) converges weakly to the deterministic system (21.28, 21.29) which is the exact multiplier algorithm. As mentioned in Sec. 21.3.2, this in turn converges to a local KT point. In a practical implementation, one would choose $\mathcal{I}$ as a large positive integer. In our numerical examples, see [19], even a choice of $\mathcal{I} = 10$ resulted in convergence to a KT point.

21.5.3. *Tradeoff between bias and tracking ability*

The three sources of bias in the stochastic gradient algorithm (21.42, 21.43) are the bias in the estimates $\widehat{\nabla_\alpha B}$, $\widehat{\nabla_\alpha C}$ and $\hat{B}$. A quick mathematical artifice for eliminating the bias is to use batch sizes $N(\epsilon) \to \infty$ as $\epsilon \to 0$. Then the ODE (21.44) becomes identical to (21.26). In the numerical examples of [19] we chose $N = 1000$, as for finite N the bias is $O(1/N)$. Although choosing $N(\epsilon) \to \infty$ is theoretically appealing, it is of no practical use since the stochastic gradient algorithm will not respond quickly to changes in the optimal policy caused by time variations in the parameters of the MDP. In our conference paper [26] we use batch sizes $N = 5, 10$ to update the parameter $\alpha(n)$ frequently, and indeed the stochastic approximation algorithm can be implemented even for $N = 1$. The bias in the gradient estimator of Theorem 21.3 can be controlled using averaging of the estimation of the cost function, or other smoothing statistical techniques. However, what we really are interested in is the resulting bias of the stable point of the limiting ODE.

The results of extensive numerical studies indicate that the bias in $h_l(\alpha)$, $l = 0, 1, \ldots, L$, has negligible effect on the behaviour of the stochastic gradient algorithm even for small batch sizes of $N = 5$. For example, we performed comparisons using the local sample average $\hat{\Gamma}_n = \sum_{k \in I_n} c(Z_k)/N$ and then using the actual theoretical value $\hat{\Gamma}_n = C(\alpha)$ in $\widehat{\widehat{G}}_n$ of (21.40) as well as for the constraints with no remarkable difference in the stochastic gradient algorithm.

The main source of bias for small batch sizes is that introduced by $\kappa(\alpha)$, in Propositions 21.2 and 21.3. Using the local sample average over the n-th batch

$$\hat{B}_l^\epsilon(n) = \widehat{\widehat{B}}_l(n) \triangleq \frac{1}{N} \sum_{m \in I_n} \beta_l(Z_m) \tag{21.49}$$

yields a noticeable asymptotic bias $\kappa(\alpha)$. A better alternative is to use the Cesaro sum $\hat{B}_l^\epsilon(n) = \frac{1}{n} \sum_{k=1}^{n} \widehat{\widehat{B}}_l(k)$. Since $\hat{B}_l^\epsilon(n) \to B_l(\alpha)$ a.s. for all $\alpha \in \boldsymbol{\alpha}^\phi$ as $n \to \infty$, this Cesaro sum estimator would correct the asymptotic bias of the stochastic gradient algorithm. However, running averages do not respond to changes in the underlying parameters (e.g. transition probabilities) of the MDP, since they are decreasing step size algorithms. Hence they cannot be used for tracking time varying optimal policies. To handle this tracking case, we use in [26] an exponential smoothing

$$\alpha^\epsilon(n+1) = \alpha^\epsilon(n) + \epsilon \left(\widehat{\nabla_\alpha C}(n) - \sum_{l=1}^{L} \left(\lambda_l^\epsilon(n) + \rho \hat{B}_l^\epsilon(n) \right) \widehat{\nabla_\alpha B_l}(n) \right) \tag{21.50}$$

$$\lambda_l^\epsilon(n+1) = \lambda^\epsilon(n) + \epsilon \, \hat{B}_l^\epsilon(n), \quad \hat{B}_l^\epsilon(n+1) = \hat{B}_l^\epsilon(n) + \delta \left(\widehat{\widehat{B}}_l(n) - \hat{B}_l^\epsilon(n) \right), \tag{21.51}$$

for $l = 1, \ldots, L$, where $\delta > 0$ and $\widehat{\widehat{B}}_l(n)$ is the local sample average in (21.49). Using a two time scale stochastic approximation argument, it can be shown that if $\epsilon/\delta \to 0$, e.g., if $\delta = \sqrt{\epsilon}$, and $\epsilon \to 0$, then the asymptotic limit points of the corresponding ODE are unbiased. In practical implementation, for non zero δ, the estimates are biased. While the asymptotic bias can be controlled, the exponential smoothing delays the reaction time of the stochastic gradient algorithm, since a faster time scale has been introduced, as illustrated in Section 21.7.2.

21.6. Case study: Monotone policies for packet transmission scheduling over correlated wireless fading channels

In this section we consider a special case of Problem S1 that arises in transmission scheduling in wireless telecommunication systems. The constrained MDP we consider, models a transmission scheduling problem in a wireless telecommunication network. The action set is $\mathcal{U} = \{0, 1\}$ corresponding to *transmit* and *do not transmit*, respectively. By using a Lagrangian formulation for dynamic programming, we show that the optimal policy is a randomized mixture between two deterministic monotone (threshold) policies; such a policy is a two-step staircase function as plotted in Fig. 21.1. So the weak derivative based stochastic approximation algorithms presented above can be used to estimate this structured optimal policy. Because of the threshold structure of the optimal policy, the algorithm implementation is very efficient.

Consider the following transmission scheduling problem over a correlated fading wireless channel. At each time slot, a user has to decide whether to transmit a packet, unless the packet storage buffer is empty. The objective is to minimize the infinite horizon average transmission cost subject to a constraint on the average delay penalty cost. As in [27, 28], we model the correlated fading wireless channel by a finite state Markov chain (FSMC). That is, we assume the channel state evolves according to a FSMC, and the channel state realization is known at every time. At time $n = 0, 1, \ldots$, the system state is the 3-tuple $X_n = [X_n^b, X_n^c, X_n^y]$, where:

(i) $X_n^b \in \mathcal{B} = \{0, 1, \ldots\}$ is the buffer occupancy state
(ii) X_n^c is the state of the correlated wireless communication channel. Assume $X_n^c \in \mathcal{C} = \{\mathcal{C}_1, \mathcal{C}_2, \ldots, \mathcal{C}_K\}$, where $\mathcal{C}$ is the finite channel state space and $\mathcal{C}_i$ corresponds to a better channel state than $\mathcal{C}_j$ for all $i > j$; X_n^c evolves as a Markov chain with the transition probabilities $A^c(X_{n+1}^c = \mathcal{C}_j | X_n^c = \mathcal{C}_i) = a_{ij}^c$.
(iii) X_n^y is the number of new packets arriving at the buffer. For simplicity, assume an i.i.d binary packet arrival process, that is at any time n $X_n^y \in \mathcal{Y} = \{0, 1\}$, which denotes either the arrival of no packet or one packet, with the probability mass function $P(X_n^y = 1) = \delta$ and $P(X_n^y = 0) = 1 - \delta$.

The system state space is then the countable set $\mathcal{S} = \mathcal{B} \times \mathcal{C} \times \mathcal{Y}$.

Let the action sets be $\mathcal{U} = \{0, 1\}$ for every buffer occupancy state $X^b > 0$, and $\mathcal{U}_0 = \{0\}$ for the buffer occupancy state $X^b = 0$, where 0 and 1 correspond to the action of not transmitting and transmitting respectively. We now define the costs and constraints of constrained MDP in Problem S1.

- The transmission cost function $c(\cdot,\cdot) : \mathcal{C} \times \mathcal{U} \to \mathbb{R}_+$ is a function of the channel state. Assume that when a transmission is not attempted, no transmission cost is incurred, that is $c(\cdot, 0) = 0$.
- The constraint is specified by a delay penalty cost $\beta(\cdot,\cdot) : \mathcal{C} \times \mathcal{U} \to \mathbb{R}_+$, which is applicable only for buffer occupancy state $x^b > 0$. Assume that when a transmission is attempted, there is no delay penalty cost, that is $\beta(\cdot, 1) = 0$.
  For channel utilization enhancement, it is assumed that $c(\cdot, u)$ is decreasing and $\beta(\cdot, u)$ is increasing in the channel state, that is the transmission cost is lower and the delay penalty cost is higher for better channel states.

If the transmission of a packet over the channel is attempted, that is, action $u = 1$ is selected, a packet will be successfully received (and hence removed from the buffer) with probability given by the function $f : \mathcal{C} \to [0, 1]$. Here, $f(\cdot)$ is a user-defined increasing function, that is, a higher (better) channel state has a higher success probability.

The transition probabilities of the constrained MDP are then given by

$$\begin{aligned}
\mathbb{P}(X_{n+1}|X_n, u = 0) &= \mathbb{P}(X^c_{n+1}|X^c_n)\mathbb{P}(X^y_{n+1})\mathbf{I}(X^b_{n+1} = X^b_n + X^y_n) \\
\mathbb{P}(X_{n+1}|X_n, u = 1) &= \mathbb{P}(X^c_{n+1}|X^c_n)\mathbb{P}(X^y_{n+1})f(X^c_n)\mathbf{I}(X^b_{n+1} = X^b_n + X^y_n - 1) \\
&\quad + \mathbb{P}(X^c_{n+1}|X^c_n)\mathbb{P}(X^y_{n+1})(1 - f(X^c_n))\mathbf{I}(X^b_{n+1} = X^b_n + X^y_n),
\end{aligned}$$

where $\mathbf{I}(\cdot)$ is the indicator function. The corresponding constrained MDP is then given by (21.2), (21.3) with only one constraint, that is, $L = 1$,

$$\limsup_{N\to\infty} \frac{1}{N}\mathbb{E}_u\left[\sum_{n=1}^{N} \beta(X^c_n, u_n)\right] \leq \gamma. \tag{21.52}$$

In the remainder of the section we will outline the steps involved in proving the threshold structure of the optimal policy of the above constrained MDP and describe how the threshold structure can be exploited in the proposed weak derivative based stochastic approximation algorithm.

The steps involved in proving the structural result for the considered countable state, infinite horizon average cost constrained MDP include

- Derive a condition for which all constrained policies induce a stable buffer and recurrent Markov chains.

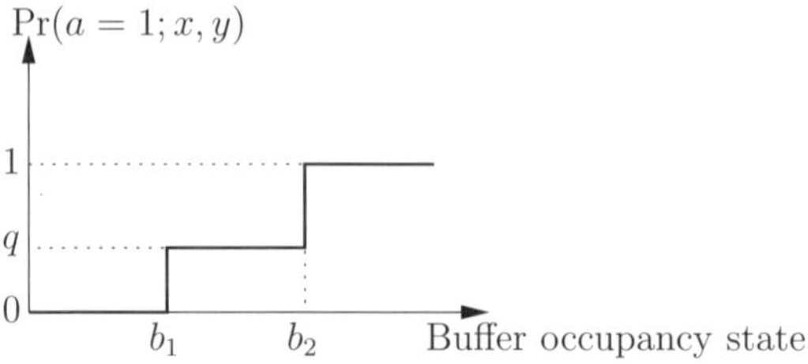

Fig. 21.1. The constrained average cost optimal policy $\mathbf{u}^*([x^b, x^c, x^y])$ is a randomized mixture between two policies that are deterministic and monotonically increasing in the buffer occupancy state x^b.

- Use the Lagrange multiplier formulation and prove the existence of an unconstrained optimal policy under the condition for buffer stability.
- Prove the threshold structure of the unconstrained (Lagrangian cost) optimal policies by using the supermodularity concept. The structure of the constrained optimal policy then follows due to a well-known result that relates unconstrained and constrained optimal policies [6, 29].

The condition for buffer stability and recurrence of the Markov chains is as follows, see [3] for proof.

Lemma 21.2. *Denote*

$$\min_{x^c \in \mathcal{C}} \{f(x^c)\} = \underline{f}, \qquad \min_{x^c \in \mathcal{C}} \beta(x^c, 0) = \underline{\beta}.$$

If $\frac{\delta}{\underline{f}} < 1 - \frac{\gamma}{\underline{\beta}}$, *then every policy* $\mathbf{u}$ *satisfying the constraint (21.52) induces a stable buffer, and a recurrent Markov chain.*

21.6.0.1. *Lagrange formulation and existence of an optimal policy*

We convert the constrained MDP to an unconstrained MDP using the Lagrange multiplier method. In particular, for a Lagrange multiplier λ, the instantaneous Lagrangian cost at time n is $c(X_n, u; \lambda) = c(X_n^c, u) + \lambda\beta(X_n^c, u)$. The Lagrangian average cost for a policy $\mathbf{u}$ is then given by

$$J_{x_0}(\mathbf{u}; \lambda) = \limsup_{N \to \infty} \frac{1}{N} \mathbb{E}_{\mathbf{u}} \left[\sum_{n=1}^{N} c(X_n, u_n; \lambda) | X_0 = x_0 \right], \tag{21.53}$$

and the corresponding unconstrained MDP is to minimize the above Lagrangian average cost.

The existence and threshold structure of an unconstrained stationary average Lagrangian cost optimal policy are established by viewing the average cost MDP model as a limit of discounted cost MDPs with discount factors approaching 1. In particular, [13, 30] provide the theory for relating average cost optimal policies to discounted cost optimal policies. Define the discounted cost by

$$J_{x_0}^{\nu}(\mathbf{u};\lambda) = \lim_{N\to\infty} \mathbb{E}_{\mathbf{u}}\left[\sum_{n=1}^{N} \nu^{n-1} c(X_n, u_n; \lambda) | X_0 = x_0\right], \tag{21.54}$$

where $0 \leq \nu \leq 1$ is the discount factor. Define the optimal discounted cost by $V^{\nu}(x_0) = \inf_{u\in\mathcal{D}} J_{x_0}^{\nu}(\mathbf{u};\lambda)$ (for notational convenience we omit the Lagrange multiplier λ in $V^{\nu}(x_0)$).

In [13, 30], the authors proved that the average cost optimal policy exists and inherits the structure of the discounted cost optimal policies under the following conditions:

A1. For each state x and discount factor ν, the optimal discounted cost $V^{\nu}(x)$ is finite.

A2. Assume a reference state 0. There exists a nonnegative N such that $-N \leq h_{\nu}(x) \overset{\triangle}{=} V^{\nu}(x) - V^{\nu}(0)$ for all $x \in \mathcal{S}$ and $\nu \in (0,1)$.

A3. There exists nonnegative M_x, such that $h_{\nu}(x) \leq M_x$ for every $x \in \mathcal{S}$ and ν. For every x there exists an action $u(x)$ such that $\sum_{x'} \mathbb{P}(x'|x, u(x)) M_{x'} < \infty$.

Define the reference state by $X^0 = [x^b = 0, x^c = \mathcal{C}_K, x^y = 0]$. In light of Lemma 21.2, it is clear that the policy of always transmitting whenever the buffer is not empty will induce a stable buffer, and hence finite expected time and cost for first passage to the reference state. As a result, due to Propositions 5(i) and 4(ii) in [30], A1 and A3 hold. Furthermore, as all instantaneous costs are bounded, the following value iteration converges to the optimal discounted cost $V^{\nu}(\cdot)$ for all discount factors ν

$$V_{n+1}^{\nu}(x) = \min_{u\in\mathcal{U}} Q_{n+1}^{\nu}(x,u) \tag{21.55}$$

$$Q_{n+1}^{\nu}(x,u) = c(x,u;\lambda) + \nu \sum_{x'\in\mathcal{S}} \mathbb{P}(x'|x,u) V_n^{\nu}(x',a). \tag{21.56}$$

Using the recursion (21.55, 21.56) it can be shown by induction that $V^{\nu}([x^b, x^c, x^y])$ is increasing in x^b and x^y [31], which implies that A2 holds.

21.6.0.2. *Threshold structure of discounted/average cost optimal policies*

Due to convergence of (21.55, 21.56) for all initial conditions, in order to show that the discounted cost optimal policy is monotonically increasing in the buffer state b it suffices to show $Q^\infty([x^b, x^c, x^y], u)$ is submodular in (x^b, u) for all $x^b \geq 1$, for some initial condition, see [32]. This can be done via mathematical induction as in the theorem below, see [3] for proof.

Theorem 21.4. *The discounted (Lagrangian) cost optimal policy is a threshold policy of the form*

$$u_\nu^*([x^b, x^c, x^y]) = \begin{cases} 0 & \textit{if } 0 \leq x^b < b(x^c, x^y) \\ 1 & \textit{if } b(x^c, x^y) \leq x^b, \end{cases} \tag{21.57}$$

where $b(x^c, x^y) : \mathcal{C} \times \mathcal{Y} \to \mathcal{B}$ defines the threshold for the pair of channel state and packet arrival event (x^c, x^y).

Therefore, the (unconstrained) Lagrangian average cost optimal policy, which inherits the threshold structure of some sequence of discounted cost optimal policies, is of the form (21.57). Due to Theorem 4.3 in [29], the constrained optimal policy for the constrained MDP is a randomized mixture of two threshold policies

$$\begin{aligned} \mathbf{u}^* &= q(x^c, x^y)\mathbf{u}_1^* + (1 - q(x^c, x^y))\mathbf{u}_2^* \\ &= \begin{cases} 0, & \text{if } 0 \leq x^b < b_1(x^c, x^y), \\ q(x^c, x^y), & \text{if } b_1(x^c, x^y) < x^b < b_2(x^c, x^y), \\ 1, & \text{if } x^b > b_2(x^c, x^y). \end{cases} \end{aligned} \tag{21.58}$$

Here for each channel state $x^c \in \mathcal{C}$ and packet arrival state $x^y \in \mathcal{Y}$, $q(x^c, x^y) \in [0, 1]$ denotes the mixture probability and $\mathbf{u}_1^*, \mathbf{u}_2^*$ are monotone policies in the buffer state x^b of the form (21.57), with threshold states $b_1(x^c, x^y)$ and $b_2(x^c, x^y)$ respectively. Therefore, the optimal policy $\mathbf{u}$ has a simple threshold structure.

21.7. Numerical examples

21.7.1. *Comparison of efficiency of gradient estimators*

Since efficient gradient estimation of the costs and constraints is an essential ingredient of the adaptive control algorithms proposed in this paper, in this subsection we compare the gradient estimators with other gradient estimators.

Score Function Estimator of Bartlett & Baxter [8, 9]: The Score Function gradient estimator of (21.19) can be derived using the measure-valued approach of [33]. Let an arbitrary finite-valued random variable Z take value j with probability $p_\theta(j)$. Then

$$\begin{aligned}\frac{\partial}{\partial\theta_{ia}}F(Z) &= \frac{\partial}{\partial\theta_{ia}}\sum_j F(j)\,p_\theta(j)\\ &= \sum_j\left(\frac{\partial}{\partial\theta_{ia}}\ln[p_\theta(j)]\right)F(j)p_\theta(j)\\ &= \mathbb{E}_\alpha[F(Z)S(\theta_{ia},Z)]\end{aligned}$$

where $S(\theta_{ia},Z)\triangleq\frac{\partial}{\partial\theta_{ia}}\ln[p_\theta(Z)]$ is known as the Score Function. Returning to the Markov process $\{Z_n\}$, using $P(Z_{k+1}=(i,a)|Z_k=(j,u))=A_{ji}(u)\theta_{ia}$ yields $S(\theta_{ia},Z_{k+1})=\frac{1}{\theta_{ia}}\mathbf{1}_{\{Z_{k+1}=(i,a)\}}$, which is not uniformly bounded in θ: when one or more components of the control parameter tend to zero (which they do when a policy is pure instead of randomized) the estimator blows up. To overcome this problem a Score Function estimator is used in [8, 9] with the exponential parameterization. When inserted in the formula (21.19) for the chain rule, the Score Function estimator for the Markov Chain is of the form $\sum_n S(\theta_{ia},Z_n)$ and it is well known that the variance increases with time. Numerous variance reduction techniques have been proposed in the literature, for example, [14] includes regenerative estimation, finite horizon approximations and more recently [8, 9] propose to use a forgetting factor for the derivative estimator. Their method suffers therefore from a variance/bias trade-off, while our estimation method is consistent and has uniformly bounded variance (in N).

System Parameters: We simulated the following MDP: $S=\{1,2\}$ (2 states), $d(i)+1=3$ (3 actions),

$$A(0)=\begin{pmatrix}0.9 & 0.1\\ 0.2 & 0.8\end{pmatrix},\quad A(1)=\begin{pmatrix}0.3 & 0.7\\ 0.6 & 0.4\end{pmatrix},\quad A(2)=\begin{pmatrix}0.5 & 0.5\\ 0.1 & 0.9\end{pmatrix}.$$

The action probability matrix $(\theta(i,a))$ and cost matrix $(c(i,a))$ were chosen as:

$$(\theta(i,a))=\begin{bmatrix}0.2 & 0.6 & 0.2\\ 0.4 & 0.4 & 0.2\end{bmatrix},\quad (c(i,a))=-\begin{bmatrix}50.0 & 200.0 & 10.0\\ 3.0 & 500.0 & 0.0\end{bmatrix}$$

Gradient Estimates in Spherical Coordinates: The theoretical values of the gradients are $\nabla_\alpha C_N(\alpha) = \begin{pmatrix} 45.05 & -55.07 \\ 187.58 & -159.91 \end{pmatrix}$. The gradient estimates Eq. (21.39) for batch sizes $N = 100, 1000$ are:

$$\widehat{G}_{100} = \begin{pmatrix} 43.333 \pm 4.791 & -54.191 \pm 2.096 \\ 196.956 \pm 8.951 & -161.200 \pm 4.918 \end{pmatrix}$$

$$\widehat{G}_{1000} = \begin{pmatrix} 44.322 \pm 1.323 & -53.656 \pm 0.712 \\ 189.549 \pm 2.829 & -164.329 \pm 1.231 \end{pmatrix}.$$

In the above expression, the numbers following the $\pm$ sign are confidence intervals which were estimated at level 0.05 using a normal approximation with 100 batches.

WD Phantoms versus Score Function Method: As mentioned in Sec. 21.2.4 the closest approach to the algorithms in this paper is that in [8, 9], which uses a Score Function method to estimate the gradients. Here we compare our gradient estimator with the score function gradient estimator of [8, 9]. Since the score function gradient estimator in [8, 9] uses canonical coordinates θ, to make a fair comparison in this example we work with canonical coordinates. A detailed development of the appropriate modifications to the corresponding factors multiplying the realisation perturbations is in [19]. The theoretical values of the generalized gradient (21.19) for the above MDP are

$$\nabla_\psi[C(\theta(\psi))] = \begin{pmatrix} -9.010 & 18.680 & -9.670 \\ -45.947 & 68.323 & -22.377 \end{pmatrix}. \tag{21.59}$$

We simulated the WD phantom and score function (SF) estimators for (21.19) in canonical coordinates, see [19] for implementation details. For batch sizes $N = 100$ and 1000, the WD phantom gradient estimates are

$$BF_{100} = \begin{pmatrix} -7.851 \pm 0.618 & 17.275 \pm 0.664 & -9.425 \pm 0.594 \\ -44.586 \pm 1.661 & 66.751 \pm 1.657 & -22.164 \pm 1.732 \end{pmatrix},$$

$$BF_{1000} = \begin{pmatrix} -8.361 \pm 0.215 & 17.928 \pm 0.240 & -9.566 \pm 0.211 \\ -46.164 \pm 0.468 & 68.969 \pm 0.472 & -22.805 \pm 0.539 \end{pmatrix}.$$

Again the numbers after $\pm$ above, denote the confidence intervals at level 0.05 with 100 batches. Note that for some of the elements of BF above, the confidence interval estimates do not contain the theoretical value (21.59). The variance of the WD phantom estimator is shown in Table 21.1, together with CPU time.

Table 21.1. Variance of WD phantom estimator in canonical coordinates.

$N = 1000$	Var$[BF_N]$		
$i = 0$	1.180	1.506	1.159
$i = 1$	5.700	5.800	7.565
CPU		2 secs.	

Table 21.2. Variance of Score Function estimator.

N		Var$[BS_N]$		
$N = 1000$	$i = 0$	89083	135860	89500
	$i = 1$	584012	593443	393015
	CPU		1374 secs.	
$N = 10000$	$i = 0$	876523	1310900	880255
	$i = 1$	5841196	5906325	3882805
	CPU		13492 secs.	

We implemented the score function gradient estimator of [8, 9] with the following parameters: forgetting factor 1 (otherwise the estimates are biased), batch sizes of $N = 1000$ and 10000. In both cases a total number of $10,000$ batches were simulated.

The score function gradient estimates are

$$BS_{10000} = \begin{pmatrix} -3.49 \pm 5.83 & 16.91 \pm 7.17 & -13.42 \pm 5.83 \\ -41.20 \pm 14.96 & 53.24 \pm 15.0 & -12.12 \pm 12.24 \end{pmatrix}$$

$$BS_{1000} = \begin{pmatrix} -6.73 \pm 1.84 & 19.67 \pm 2.26 & -12.93 \pm 1.85 \\ -31.49 \pm 4.77 & 46.05 \pm 4.75 & -14.55 \pm 3.88 \end{pmatrix}.$$

The variance of the score function gradient estimates are given Table 21.2.

Note that even with substantially larger batch sizes and number of batches (and hence computational time), the variance of the score function estimator is orders of magnitude larger than that of the WD phantom estimator.

21.7.2. *Stochastic adaptive control of time varying constrained MDP*

We consider adaptive stochastic control of the following time-varying constrained MDP: For time up to 4000, $S = \{0, 1\}$, $\mathcal{U}_i = \{0, 1, 2\}$, $i \in S$,

$(d(0) = d(1) = 2)$,

$$A(0) = \begin{pmatrix} 0.9 & 0.1 \\ 0.2 & 0.8 \end{pmatrix}, \quad A(1) = \begin{pmatrix} 0.3 & 0.7 \\ 0.6 & 0.4 \end{pmatrix}, \quad A(2) = \begin{pmatrix} 0.5 & 0.5 \\ 0.1 & 0.9 \end{pmatrix}.$$

The cost matrix $(c(i,a))$, two constraints matrices $(\beta_1(i,a))$, $(\beta_2(i,a))$ are

$$(c(i,a)) = -\begin{bmatrix} 50 & 200 & 10 \\ 3 & 500 & 0 \end{bmatrix}, \quad \beta_1 = \begin{bmatrix} 20 & 100 & -8 \\ -3 & 4 & -10 \end{bmatrix}, \; \beta_2 = \begin{bmatrix} 10 & -20 & 22 \\ -19 & 17 & -15 \end{bmatrix}.$$

The optimal control policy incurs a cost of -111.80 (or equivalently a reward of 111.80) and is randomized with probabilities (21.18)

$$\theta^* = \begin{bmatrix} 0 & 0.2 & 0.8 \\ 0 & 0.28 & 0.72 \end{bmatrix}.$$

For time between 4000 and 12000 the transition probabilities are

$$A(0) = \begin{bmatrix} 0.5 & 0.5 \\ 0.5 & 0.5 \end{bmatrix}, \quad A(1) = \begin{bmatrix} 0.9 & 0.1 \\ 0.1 & 0.9 \end{bmatrix}, \quad A(2) = \begin{bmatrix} 0.5 & 0.5 \\ 0.45 & 0.55 \end{bmatrix}.$$

This has an optimal cost of -44.52 (that is reward of 44.52).

The algorithm was initialized with randomized policy

$$\theta(0) = \begin{bmatrix} 0.1 & 0.1 & 0.8 \\ 0.0 & 0.2 & 0.8 \end{bmatrix}.$$

The batch sizes over which the gradients are estimated was chosen as $N = 10$. The parameters used in the primal-dual algorithm are $\rho = 100$, $\epsilon = 2 \times 10^{-7}$ (see (21.42, 21.43)).

As can be seen from Fig. 21.2, it takes only around 100 batches (1000 time points) for the algorithm to rapidly approaches the optimal policy. The algorithm also quickly responds to the change in optimal policy at batch time 400. The choice of the discounting factor δ in (21.51) of the primal-dual method clearly shows the trade off between bias and tracking ability in Fig. 21.2. For $\delta = 1.0$, the algorithm has fast tracking properties but a large bias. For $\delta = 0.5$ and $\delta = 0.1$ the bias gets smaller but the adaptation rate is slower. Our conference paper [26] and report [19] give additional numerical examples.

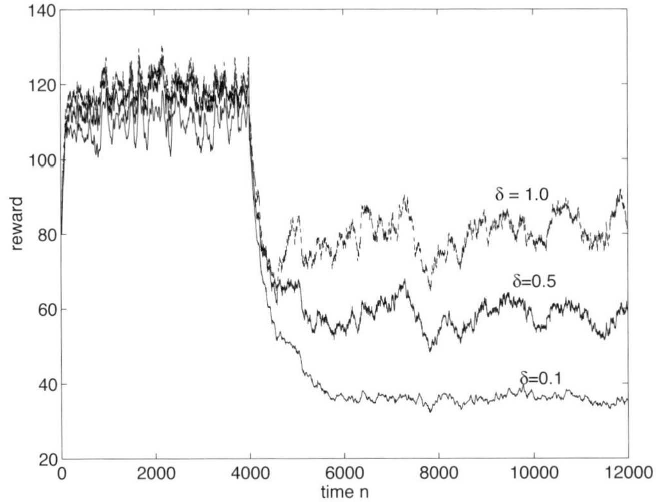

Fig. 21.2. Primal-dual algorithm based stochastic adaptive controller.

21.8. Conclusions

In this paper, simulation based gradient algorithms have been presented for adaptively optimizing a constrained average cost finite state MDP. First, a parameterization of the randomized control policy using spherical coordinates was presented. This was complemented by a novel measure-valued gradient estimator using WD phantoms for short batch sizes. The resulting algorithm for estimating gradients of the cost and constraints does not require explicit knowledge of the transition probabilities of the MDP or any off-line simulations.

As illustrated in Sec. 21.7.1, the resulting gradient estimator has much smaller variance than score function based gradient estimators. The WD phantom gradient estimator was then used in a stochastic gradient algorithm with fixed step size in order to track time varying MDP with unknown transition probability matrices. Primal dual and multiplier based stochastic gradient algorithms were presented for handling the constraints.

Appendix: Proof of Theorem 21.2

Proof. To relate our formula (21.39) to the realization perturbation formulas in [34] use the following argument. The matrix $P(\alpha)$ in Eq. (21.9) denotes the transition probability of $\{Z_n\}$ using a fixed value of the control parameter α. Fix the indices i, p and consider a perturbation of $\alpha_{i,p}$ into $\alpha_{i,p} + \delta$. The new transition matrix is $P(\alpha + \delta) = P(\alpha) + \delta Q(\alpha) + o(\delta^2)$, for $Q(\alpha) = P'(\alpha)$ the matrix with entries equal to the derivatives of the entries of $P(\alpha)$ with respect to $\alpha_{i,p}$. In [34], it is established that

$$\frac{\partial}{\partial \alpha_{ip}} \mathbb{E}_{\pi(\alpha)}[c(Z)] = \pi(Q(\alpha)g), \tag{21.60}$$

where π is the stationary state probability of the chain and the vector g has entries

$$g(i, u) = \lim_{N \to \infty} \mathbb{E}_\alpha \left[\sum_{n=1}^{N} F(Z_n) \mid Z_0 = (i, u) \right] - (N-1)C(\alpha).$$

Let $p = 1, \ldots, d(i) - 1$, then by definition, the matrix Q satisfies

$$\begin{aligned} Q_{(j,u),(i',u')} &= \frac{\partial}{\partial \alpha_{ip}} P_{(j,u)(I',u')}(\alpha) = p_{j,i'}(u) \frac{\partial}{\partial \alpha_{ip}} \theta_{i',u'}(\alpha) \\ &= p_{j,i'}(u) \times \begin{cases} 0, & i' \neq i, \\ 0, & i' = i, u' < p-1, \\ -2\tan(\alpha_{ip}), \theta_{i',p-1} & i' = i, u' = p-1, \\ +2\tan(\alpha_{ip}), \theta_{i',u'} & i' = i, u' \geq p, \end{cases} \end{aligned}$$

where we have used the expression $\theta_{i,u'} = \cos^2(\alpha_{i,u'+1}) \prod_{m=1}^{u'+1} \sin^2(\alpha_{i,u'+1})$.

Using now $\pi P = \pi$, and the expression for Q, Eq. (21.60) becomes:

$$\frac{\partial}{\partial \alpha_{ip}} \mathbb{E}_{\pi(\alpha)}[c(Z)] = -2\tan(\alpha_{ip})\, \pi_{i,u'} \mathcal{E}_i[g(i, p-1) - g(i, Y_p)].$$

As explained in [19], $g(i, p-1) - g(i, Y_p) = \sum_{n=k}^{k+nu(k)} c(Z_n) - \nu(k)C(\alpha)$, with initial condition $Z_k = (i, p-1)$ and $nu(k) = \min\{n > 0 : Z_{k+n} = (i, Y_p)\}$, so the resulting expression is:

$$\begin{aligned} &\frac{\partial}{\partial \alpha_{ip}} \mathbb{E}_{\pi(\alpha)}[c(Z)] \\ &= -2\tan(\alpha_{ip})\theta_{i,p-1}\mathcal{E}_i \left[\sum_{n=k}^{k+nu(k)} \mathbb{E}_\alpha[c(Z_n) - nu(k)C(\alpha) \mid Z_k = (i, p-1)] \right] \end{aligned} \tag{21.61}$$

for $p = 1, \dots, d(i) - 1$. When implementing the above equation on-line, we use the observations of the actual process $\{Z_k\}$ to estimate the various derivatives. The proportion of time that we observe action $p-1$ at state i is, by definition, $\theta_{i,p-1}$, thus the averaging over N observations yields

$$\frac{\partial}{\partial \alpha_{ip}} \mathbb{E}_{\pi(\alpha)}[c(Z)]$$

$$= -2\tan(\alpha_{ip}) \lim_{N\to\infty} \frac{1}{N} \sum_{k=1}^{N} \mathcal{E}_i \left[\delta_{i,p}(k) \sum_{n=k}^{k+\nu(k)} \mathbb{E}_\alpha[c(Z_n) - \nu(k) C(\alpha)\right]. \tag{21.62}$$

The result Eq. (21.39) is established by noticing that $\hat{\Gamma}_N$ converges a.s. to $C(\alpha)$, because for $\alpha \in \boldsymbol{\alpha}^\mu$ $\mathbb{E}_\alpha[\nu(k)] < \infty$, it is a coupling time of a positive recurrent Markov chain. The case $p = d(i)$ consists of a degenerate case and a similar chain rule argument can be shown for the factor $K_{i,d(i)}(\alpha, k)$. A detailed derivation of this formula can be obtained also using measure-valued differentiation [14, 33], and appears in [19]. □

Appendix: Proof of Proposition 21.3

Proof. First, from tightness of $\{\lambda^\epsilon(n)\}$, it follows that the family $\{\bar{\lambda}(\epsilon)\}$ of (deterministic) averages lies in a compact set, thus the accumulation points $\bar{\lambda}$ exist.

From the proof of uniform integrability in Proposition 21.2, and tightness of $\{\lambda^\epsilon(n)\}$, it follows that the sequence $\{(\alpha^\epsilon(n+1) - \alpha^\epsilon(n)/\epsilon)\}$ is uniformly integrable, which implies that $\{\alpha^\epsilon(n)\}$ is tight. Therefore for any sequence $(\alpha^{\epsilon_k}(\cdot), \lambda(\epsilon_k))$ there is at least one (weakly) convergent subsequence with a.s. Lipschitz continuous limit (refer to [18]). For the rest of the proof, until specified, assume that ϵ labels a weakly convergent subsequence (to avoid the cumbersome ϵ_k indexing). We will now identify the limits of such convergent subsequences and show that they all satisfy the same ODE. Also to ease notation in the proof, write $Y_0(n) = \widehat{\nabla_\alpha C}(n)$, $Y_l(n) = \widehat{\nabla_\alpha B_l}(n)$.

From the definition, Eq. (21.24), it follows that:

$$\alpha^\epsilon(t+s) - \alpha^\epsilon(t) = -\epsilon \sum_{n=\lfloor t/\epsilon \rfloor}^{\lfloor (t+s)/\epsilon \rfloor - 1} \left[Y_0(n) + \sum_{l=1}^{L} \lambda_l^\epsilon(n+1)\, Y_l(n) \right].$$

Divide the interval $(t, t+s]$ into subintervals of small size δ_ϵ each containing a number n_ϵ of updates, as shown in Figure 21.3.

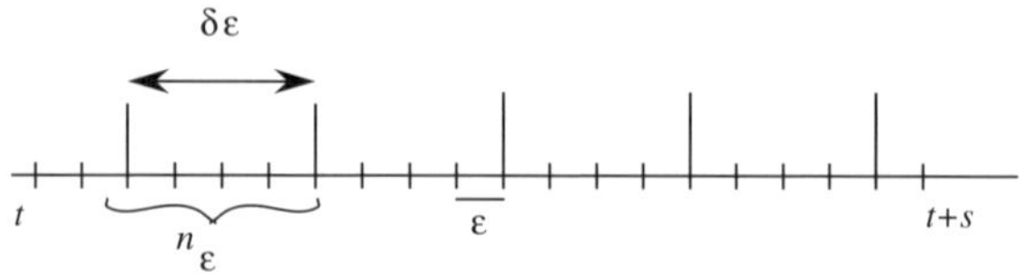

Fig. 21.3. $\delta_\epsilon = \epsilon n_\epsilon$. Condition: $\delta_\epsilon \to 0, n_\epsilon \to \infty$ as $\epsilon \to 0$.

Using the δ_ϵ grouping of subintervals, one obtains the telescopic sum

$$\alpha^\epsilon(t+s) - \alpha^\epsilon(t) = -\sum_{l=\lfloor t/\delta_\epsilon \rfloor}^{\lfloor (t+s)/\delta_\epsilon \rfloor - 1} \delta_\epsilon \times \left(\frac{1}{n_\epsilon} \sum_{j=ln_\epsilon}^{(l+1)n_\epsilon - 1} \left[Y_0(j) + \sum_{l=1}^{L} \lambda_l^\epsilon(j+1)\, Y_l^\epsilon(j) \right] \right).$$

We now show that if $\mathfrak{F}^\epsilon(t)$ denotes the σ-algebra generated by the interpolated process up to time t, then:

$$\begin{aligned} \mathcal{E}_i[\alpha^\epsilon(t+s) - \alpha^\epsilon(t) \mid \mathfrak{F}^\epsilon(t)] \approx \sum_{l=\lfloor t/\delta_\epsilon \rfloor}^{\lfloor (t+s)/\delta_\epsilon \rfloor - 1} \delta_\epsilon \Bigg(h_0[\alpha^\epsilon(ln_\epsilon)] \\ + \sum_{l=1}^{L} (\bar{\lambda}_l(\epsilon) + B_l[\alpha^\epsilon(ln_\epsilon)])\, h_l[\alpha^\epsilon(ln_\epsilon) + \kappa[\alpha^\epsilon(ln_\epsilon)] \Bigg) \end{aligned} \tag{21.63}$$

where the expectation of the absolute error in the end points of the δ_ϵ discretization vanishes as $\epsilon \to 0$. Let $\mathbb{E}_{ln_\epsilon}$ denote the expectation conditioning on the information available up to the start of the current small subinterval of size δ_ϵ. Use conditional expectations to express

$$\mathbb{E}[Y_l(j) \mid \mathfrak{F}^\epsilon(t)] = \mathbb{E}[\mathbb{E}_{ln_\epsilon}[Y_l(j)] \mid \mathfrak{F}^\epsilon(t)], ln_\epsilon \le j < (l+1)n_\epsilon$$

for each term in the telescopic sum $l = 0, \ldots, L$. That is, we use a filter of the terms, focusing on each of the averages within subintervals.

Because one is interested in averages, any version of the process $\{\alpha^\epsilon(n)\}$ can be used to characterize these conditional expectations. In particular, the Skorohod representation establishes that there is a process $\tilde{\alpha}^\epsilon(n)$ for each ϵ in the weakly convergent subsequence, such that $\tilde{\alpha}^\epsilon(n)$ has the same distribution as $\alpha^\epsilon(n)$ and it converges with probability 1 to the same a.s. continuous limit $\alpha(t)$ (see [18]). Because $\alpha(t)$ is Lipschitz continuous w.p.1, $\|\alpha(l\delta_\epsilon + \delta_\epsilon) - \alpha(l\delta_\epsilon)\| = \mathcal{O}(\delta_\epsilon)$ and since $\tilde{\alpha}^\epsilon(n)$ converges w.p. 1, it follows that $\sup_{ln_\epsilon \le j < (l+1)n_\epsilon} \|\tilde{\alpha}^\epsilon(j) - \alpha(l\delta_\epsilon)\| \to 0$ a.s., which implies that the underlying distribution of the batch estimators

$Y_l(j), j = ln_\epsilon, \ldots, (l+1)n_\epsilon - 1$ converges to that of the fixed-α MDP at the parameter value $\alpha = \alpha(l\delta_\epsilon)$. Using the fact that $\alpha^\epsilon(ln_\epsilon)$ converges in distribution to $\alpha(l\delta_\epsilon)$, it follows that

$$\begin{aligned}
&\mathbb{E}_{ln_\epsilon}\left[\frac{1}{n_\epsilon}\sum_{j=ln_\epsilon}^{(l+1)n_\epsilon-1}\left[Y_0(j) + \sum_{l=1}^{L}\lambda_l^\epsilon(j+1)\,Y_l(j)\right]\right] \\
&\quad= \mathbb{E}_{ln_\epsilon}\left[\frac{1}{n_\epsilon}\sum_{j=ln_\epsilon}^{(l+1)n_\epsilon-1}\mathbb{E}_\alpha\left[\widehat{\nabla_\alpha C_0}(j) + \sum_{l=1}^{L}(\lambda_l^\epsilon(j) + \hat{B}(j))\,\widehat{\nabla_\alpha B_l}(j)\right]\right].
\end{aligned}$$

Given the initial state value (with the aggregated information about the living phantoms), and the value of $\lambda_l^\epsilon(j)$, the expectation for the fixed α process of the gradient estimator $\widehat{\nabla_\alpha B_l}(j)$ is independent of $\lambda_l^\epsilon(j)$. As $n_\epsilon \to \infty$, the underlying process $\{Z_n\}$ will have the stationary distribution for the j-th estimation batch, so that

$$\begin{aligned}
&\mathbb{E}_{ln_\epsilon}\left[\frac{1}{n_\epsilon}\sum_{j=ln_\epsilon}^{(l+1)n_\epsilon-1}\left[Y_0(j) + \sum_{l=1}^{L}\lambda_l^\epsilon(j+1)\,Y_l(j)\right]\right] \\
&\quad= h_0[\alpha^\epsilon(ln_\epsilon)] + \sum_{l=1}^{L}\Bigg(h_l[\alpha^\epsilon(ln_\epsilon)]\,\mathbb{E}_{ln_\epsilon}\left[\frac{1}{n_\epsilon}\sum_{j=ln_\epsilon}^{(l+1)n_\epsilon-1}\lambda_l^\epsilon(j)\right] \\
&\qquad + \rho\,\mathbb{E}_{\pi(\alpha)}[\hat{B}_l(n)\,\widehat{\nabla_\alpha B_l}(n)]\Bigg) \\
&\quad\approx h_0[\alpha^\epsilon(ln_\epsilon)] + \sum_{l=1}^{L}(\bar{\lambda}(\epsilon) + \rho B[\alpha^\epsilon(ln_\epsilon)])h_l[\alpha^\epsilon(ln_\epsilon)] + \kappa[\alpha^\epsilon(ln_\epsilon)],
\end{aligned}$$

which establishes Eq. (21.63). Define now a piecewise constant function (on the δ_ϵ-subintervals)

$$\mathcal{G}^\epsilon(\alpha^\epsilon(t), \bar{\lambda}(\epsilon)) = h_0[\alpha^\epsilon(ln_\epsilon)] + \sum_{l=1}^{L}(\lambda_l(\epsilon) + \rho B[\alpha^\epsilon(ln_\epsilon)])h_l[\alpha^\epsilon(ln_\epsilon)] + \kappa[\alpha^\epsilon(ln_\epsilon)],$$

for $\delta_\epsilon < t < (l+1)\delta_\epsilon$, then Eq. (21.63) implies that: $\mathbb{E}[\alpha^\epsilon(t+s) - \alpha^\epsilon(t) \mid \mathfrak{F}^\epsilon(t)] \approx \int_t^{t+s}\mathcal{G}^\epsilon[\alpha^\epsilon(s), \bar{\lambda}(\epsilon)]\,ds$ which implies that the limit process is a martingale with zero quadratic variation. For a detailed presentation of this methodology the reader is referred to [18]. Taking now the limit along the weakly convergent subsequence, $\alpha^\epsilon(t) \to \alpha(t), \lambda(\epsilon) \to \bar{\lambda}$ establishes the limiting ODE for this subsequence. □

References

[1] M. Puterman, *Markov Decision Processes.* (John Wiley, 1994).

[2] F. Yu and V. Krishnamurthy, Optimal joint session admission control in integrated WLAN and CDMA cellular network, *IEEE Transactions Mobile Computing.* **6**(1), 126–139 (Jan., 2007).

[3] M. Ngo and V. Krishnamurthy, Optimality of threshold policies for transmission scheduling in correlated fading channels, *Communications, IEEE Transactions on.* **57**(8), 2474–2483, (2009).

[4] M. Ngo and V. Krishnamurthy, Monotonicity of constrained optimal transmission policies in correlated fading channels with arq, *IEEE Trans. Signal Processing.* **58**(1), 478–451, (2010).

[5] K. Ross and R. Varadarajan, Markov decision processes with sample path constraints: The communication case, *Operations Research.* **37**(5), 780–790 (Sept-Oct, 1989).

[6] E. Altman, *Constrained Markov Decision Processes.* (Chapman and Hall, London, 1999).

[7] D. Bertsekas and J. Tsitsiklis, *Neuro-Dynamic Programming.* (Athena Scientific, Belmont, MA., 1996).

[8] J. Baxter and P. Bartlett. Direct gradient-based reinforcement learning: I. Gradient estimation algorithms. Technical report, Computer Sciences Laboratory, Australian National University, http://discus.anu.edu.au/ml/index.html, (1999).

[9] P. Bartlett and J. Baxter, Estimation and approximation bounds for gradient-based reinforcement learning, *J. Comput. Syst. Sci.* **64**(1), 133–150, (2002).

[10] H. Kushner and D. Clark, *Stochastic Approximation Methods for Constrained and Unconstrained Systems.* (Springer-Verlag, 1978).

[11] F. Vázquez-Abad, Strong points of weak convergence: a study using RPA gradient estimation for automatic learning, *Automatica.* **35**(7), 1255–1274, (1999).

[12] C. Derman, G. Lieberman, and S. Ross, Optimal system allocations with penalty cost, *Management Science.* **23**(4), 399–403 (December, 1976).

[13] S. Ross, *Introduction to Stochastic Dynamic Programming.* (Academic Press, San Diego, California., 1983).

[14] G. Pflug, *Optimization of Stochastic Models: The Interface between Simulation and Optimization.* (Kluwer Academic Publishers, 1996).

[15] A. Benveniste, M. Metivier, and P. Priouret, *Adaptive Algorithms and Stochastic Approximations.* vol. 22, *Applications of Mathematics*, (Springer-Verlag, 1990).

[16] V. Solo and X. Kong, *Adaptive Signal Processing Algorithms – Stability and Performance.* (Prentice Hall, N.J., 1995).

[17] G. Yin, V. Krishnamurthy, and C. Ion, Regime switching stochastic approximation algorithms with application to adaptive discrete stochastic optimization, *SIAM Journal on Optimization.* **14**(4), 117–1215, (2004).

[18] H. Kushner and G. Yin, *Stochastic Approximation Algorithms and Applica-*

tions. (Springer-Verlag, New York, 1997).

[19] F. Vazquez-Abad and V. Krishnamurthy. Self learning control of constrained Markov decision processes– a valued gradient approach. Technical Report G-2003-51, GERAD-HEC Montreal, http://www.gerad.ca/fichiers/cahiers/G-2003-51.pdf (August, 2003).

[20] V. Krishnamurthy, Bayesian sequential detection with phase-distributed change time and nonlinear penalty – a lattice programming pomdp approach, *IEEE Trans. Inform. Theory.* **57**(3) (Oct., 2011). http://arxiv.org/abs/1011.5298.

[21] V. Krishnamurthy and D. Djonin, Optimal threshold policies for multivariate POMDPs in radar resource management, *IEEE Transactions on Signal Processing.* **57**(10), (2009).

[22] V. Krishnamurthy and D. Djonin, Structured threshold policies for dynamic sensor scheduling–a partially observed Markov decision process approach, *IEEE Trans. Signal Proc.* **55**(10), 4938–4957 (Oct., 2007).

[23] D. Bertsekas, *Nonlinear Programming.* (Athena Scientific, Belmont, MA., 2000).

[24] D. Luenberger, *Linear and Nonlinear Programming.* (Addison Wesley, 1984), Second edition.

[25] B. Heidergott and X.-R. Cao, A note on the relation between weak derivatives and perturbation realization, *IEEE Trans. Auto. Control.* **47**(7), 1112–1115, (2002).

[26] V. Krishnamurthy, F. Vazquez-Abad, and K. Martin. Implementation of gradient estimation to a constrained Markov decision problem. In *IEEE Conference on Decision and Control*, Maui, Hawaii, (2003).

[27] H. S. Wang and N. Moayeri, Finite-state Markov channel - A useful model for radio communications channels, *IEEE Trans. Vehicular Tech.* **44**(1), 163–171, (1995).

[28] A. Zhang and S. A. Kassam, Finite-state Markov model for Rayleigh fading channels, *IEEE Trans. Commun.* **47**, 1688–1692 (November, 1999).

[29] F. J. Beutler and K. W. Ross, Optimal Policies for Controlled Markov Chains with a Constraint, *Journal of Mathematical Analysis and Applications.* **112**, 236–252, (1985).

[30] L. I. Sennott, Average Cost Optimal Stationary Policies in Infinite State Markov Decision Processes with Unbounded Costs, *Operations Research.* **37** (4), 626–633 (July-August, 1989).

[31] M. H. Ngo and V. Krishnamurthy. On Optimality of Monotone Channel-aware Transmission Policies: A Constrained Markov Decision Process approach. In *Proceedings of ICASSP'07*, Honolulu, HI (April, 2007).

[32] D. Topkis, *Supermodularity and Complementarity.* (Princeton University Press, 1998).

[33] B. Heidergott and F. Vázquez-Abad, Measure valued differentiation for markov chains, *Journal of Optimization, Theory and Applications.* (2007).

[34] X. R. Cao, The relations amongst potentials, perturbation analysis and Markov decision processes, *J DEDS.* **8**, 71–87, (1998).

Chapter 22

Parameter estimation of a regime-switching model using an inverse Stieltjes moment approach

Xiaojing Xi[1], Marianito R. Rodrigo[2] and Rogemar S. Mamon[3]

[1]*Department of Applied Mathematics, University of Western Ontario, London, Ontario, Canada*

[2]*School of Mathematics and Applied Statistics, University of Wollongong, Wollongong, New South Wales, Australia*

[3]*Department of Statistical and Actuarial Sciences, University of Western Ontario, London, Ontario, Canada*

We address the problem of recovering the time-dependent parameters of the Black–Scholes option pricing model when the underlying stock price dynamics are modelled by a finite-state, continuous-time Markov chain. The coupled system of Dupire-type partial differential equations is derived and formulated as an inverse Stieltjes moment problem. We provide numerical illustration on how to apply our method to simulated financial data. The accuracy of the model parameter estimation is examined and sensitivity analyses are included to study the behaviour of the estimated results when model parameters are varied.

Contents

22.1. Introduction

A fundamental problem in financial mathematics is the recovery of model parameters given observed market prices. Volatility, for instance, is an important but unobservable parameter, whose estimate is necessary when pricing derivatives and enables us to understand price dynamics. Traders calculate implied volatilities from market data for option valuation, as well as using them as a guide to monitor the market's sentiments. In the present work, we focus on recovering the parameters of a regime-switching model from European call option prices.

A number of approaches have been proposed to deal with this type of problem. In a pioneering paper, Dupire [1] verifies empirically that different strikes and maturities lead to different implied volatilities for options on a given asset. Boyle and Thangaraj [2], as well as Andersen and Brotherton-Ratcliffe [3], obtain local implied volatilities by numerically implementing Dupire's equation. Rodrigo and Mamon [4] give a new expression for the volatility by deriving a semi-explicit solution of Dupire's equation. They also provide a different formula in [5], which makes use of the so-called inverse Stieltjes moment approach. Bouchouev and Isakov [6] reduce the identification of the volatility to an inverse parabolic problem with the final observation. Deng et al. [7] employ an optimal control framework with a new terminal condition to solve this kind of inverse problem.

Recently, considerable attention has been given to the use of regime-switching models, or hidden Markov models (HMMs), in finance. In an HMM, the model parameters switch amongst unobservable states of the economy and are governed by a Markov process. A regime-switching volatility is a simple way to incorporate stochastic volatilities. It has the ability to capture long-term and fundamental changes in the economic mechanism that generates the data. Significant empirical evidence from the literature lends support for the appropriateness of regime-switching models. For instance, Chu et al. [8] advocate the use of these models to describe returns and volatility dynamics in the stock market. Turner et al. [9] argue that either the mean or variance, or both, may exhibit differences between two regimes. The investigation of Engel and Hamilton [10], Bekaert and Hodrick [11], and Engel and Hakkio [12] document regime switching in major foreign exchange rates. Dahlquist and Gray [13] and Ang and Bekaert [14] show that various foreign, short-term interest rates are well described by regime-switching models. Some applications of regime-switching models modulated by a hidden Markov chain can be found in the work of Elliott and Mamon [15], as well as in Elliott and Kopp [16].

Regime-switching models have achieved growing importance in various financial problems as they can capture a richer set of empirical and theoretical characteristics of a market. They have also enriched the developments in option pricing theory. For example, Elliott et al. [17] develop a method to price options based on a regime-switching random Esscher transform. In turn, this method was used by Ching et al. [18] to price exotic options under a hidden Markov model with long-range dependence in the states of an economy, which is known as a higher-order HMM. Mamon and Rodrigo [19] present closed-form solutions for European option values when the dynamics of both the short rate and the volatility of the underlying price process are modulated by a continuous-time Markov chain. Boyle and Draviam [20] derive the system of partial differential equations (PDEs) of Black–Scholes type that governs the dynamics of European options in a regime-switching framework and price exotic options by solving the coupled PDEs numerically. Duan et al. [21] develop a family of option pricing models which are based on the GARCH process and the variance-updating schemes also depend on a second factor orthogonal to asset innovations. Other works that feature regime-switching models in other applications include Siu et al. [22] for credit default swaps, Elliott and van der Hoek [23] for asset allocations, and Elliott and Mamon [15] for short-term interest rates.

The above studies in option pricing under a regime-switching framework serve as motivation for investigating the inverse problem of recovering the volatilities when they are governed by HMMs. There is a relatively limited amount of literature on estimating regime-switching parameters using market data. In this paper, we extend the inverse Stieltjes moment approach in [5] by assuming that the volatility of the underlying asset is governed by a continuous-time Markov chain. In this model, the unobservable parameters are the volatilities in each state and the intensity probabilities of the hidden Markov chain. We start with the well-known system of Black–Scholes-type PDEs and derive the coupled system of Dupire-type PDEs that governs the dynamics of European option prices.

The rest of the paper is organised as follows. In Section 22.2, we recall the regime-switching model setup. In Section 22.3, we derive the system of Dupire-type PDEs describing the dynamics of European option prices under this setup. We formulate the inverse Stieltjes moment problem in Section 22.4, and also discuss how our proposed method could determine the model parameters. In Section 22.5, we exhibit an implementation to a set of "theoretical data" which were generated by solving the coupled

Dupire-type PDEs. We conclude with a brief summary in Section 22.6.

22.2. Regime-switching model setup

We wish to value a European option within the standard Black–Scholes market with two basic securities consisting of a riskless asset (a bond whose value is B_t at time $t \geq 0$) and a risky asset (a stock whose price is S_t at time t). Moreover, we assume that the economic state of the world is modelled by a finite-state Markov chain $\mathbf{x}_t$ that evolves in continuous time. This implies that the bank rate process r_t and the stock's volatility σ_t and rate of return μ_t are governed by Markov chain dynamics.

Without loss of generality, we may take the state space of $\mathbf{x}_t$ to be the finite set $\{\mathbf{e}_1, \ldots, \mathbf{e}_N\}$ of canonical vectors in $\mathbb{R}^N$. Assume that $\mathbf{x}_t$ is homogeneous in time and has intensity matrix $\mathbf{A} = (a_{ij})$, i.e.,

$$a_{ji} \geq 0 \quad \text{for } j \neq i, \quad \sum_{i=1}^{N} a_{ij} = 0 \quad \text{for each } j = 1, \ldots, N.$$

If $\mathbf{p}_t = E[\mathbf{x}_t] = (p_t^1, \ldots, p_t^N)^*$ where * is the transpose operator, then $\mathbf{p}_t$ satisfies

$$\frac{\mathrm{d}\mathbf{p}_t}{\mathrm{d}t} = \mathbf{A}\mathbf{p}_t.$$

It can be shown [24] that $\mathbf{x}_t$ has a semi-martingale representation

$$\mathbf{x}_t = \mathbf{x}_0 + \int_0^t \mathbf{A}\mathbf{x}_u \, \mathrm{d}u + \mathbf{M}_t$$

where $\mathbf{M}_t$ is a martingale.

Suppose that $r_t = \langle \mathbf{r}, \mathbf{x}_t \rangle$ for some given vector $\mathbf{r} = (r_1, \ldots, r_N)^*$ in $\mathbb{R}^N$ with $r_1, \ldots, r_N > 0$. Here, $\langle \cdot, \cdot \rangle$ denotes the usual inner product in $\mathbb{R}^N$. Then \$1 invested at time zero becomes

$$B_t = e^{\int_0^t r_u \, \mathrm{d}u} \tag{22.1}$$

at time t. In addition, suppose that the rate of return μ_t and the volatility σ_t depend on the state $\mathbf{x}_t$, i.e., there exist vectors $\boldsymbol{\mu} = (\mu_1, \ldots, \mu_N)^*$ and $\boldsymbol{\sigma} = (\sigma_1, \ldots, \sigma_N)^*$ in $\mathbb{R}^N$ (with $\mu_i, \sigma_i > 0$ for all $i = 1, \ldots, N$) such that $\mu_t = \langle \boldsymbol{\mu}, \mathbf{x}_t \rangle$ and $\sigma_t = \langle \boldsymbol{\sigma}, \mathbf{x}_t \rangle$. Then the dynamics of the stock is described by the stochastic differential equation

$$\mathrm{d}S_t = \mu_t S_t \, \mathrm{d}t + \sigma_t S_t \, \mathrm{d}W_t$$

where W_t is a Brownian motion on a filtered probability space denoted by $(\Omega, \mathcal{F}, P, (\mathcal{F}_t)_{t\geq 0})$ and $(\mathcal{F}_t)_{t\geq 0}$ is taken to be the natural filtration. It can be shown that S_t is expressible as

$$S_t = S_0 e^{\int_0^t (\mu_u - \sigma_u^2/2)\,\mathrm{d}u + \int_0^t \sigma_u \mathrm{d}W_u}. \qquad (22.2)$$

If the bond and stock dynamics are given by Eq. (22.1) and Eq. (22.2), respectively, and if at time $t \in [0, T]$ we have $S_t = S$ and $\mathbf{x}_t = \mathbf{x}$, then the price of a European call option with expiry T and strike price K is

$$c(t, S, T, K, \mathbf{x}) = E^Q \left[e^{-\int_t^T r_u\,\mathrm{d}u} (S_T - K)^+ \mid S_t = S,\ \mathbf{x}_t = \mathbf{x} \right] \qquad (22.3)$$

where $(z)^+ = \max(z, 0)$ and E^Q denotes the expectation evaluated under a risk-neutral measure Q. We remark that regime switching leads to an incomplete market, which can be completed by the introduction of Arrow-Debreu securities [25] related to the cost of switching. Thus, in Eq. (22.3) we are assuming that we are already working under a risk-neutral measure Q. Just like in the classical Black–Scholes case, we assume that $\mu = r$ in the stock price dynamics under Q; hence the rate of return will not appear in Eq. (22.3). We do not rule out the dependence of the market price of risk on the state $\mathbf{x}_t$ at time t. But, irrespective of whether or not we assume a special or functional form for the market price of risk that depends on $\mathbf{x}_t$, or some other more general dependence which we do not know, the information from the market should be implicitly reflected in the parameters that we want to estimate. That is, we do not know what the exact dependence is but what are important to us are the parameter estimates that should encapsulate this information.

Define $c_i(t, S, T, K) = c(t, S, T, K, \mathbf{e}_i)$ for each $i = 1, \ldots, N$. We note that $r_i = \langle \mathbf{r}, \mathbf{e}_i \rangle$ and $\sigma_i = \langle \boldsymbol{\sigma}, \mathbf{e}_i \rangle$. It can be shown [19] that $c_1, \ldots, c_N$ satisfy a system of coupled PDEs of Black–Scholes type in the variables t and S, namely

$$\frac{\partial c_i}{\partial t} + \frac{1}{2}\sigma_i^2 S^2 \frac{\partial^2 c_i}{\partial S^2} + r_i S \frac{\partial c_i}{\partial S} - r_i c_i + \sum_{j=1}^{N} a_{ji} c_j = 0 \quad (i = 1, \ldots, N), \quad (22.4)$$

together with the terminal conditions

$$c_i(T, S, T, K) = (S - K)^+ \quad (i = 1, \ldots, N). \qquad (22.5)$$

Let $\mathbf{c} = (c_1, \ldots, c_N)^*$, $\boldsymbol{\Sigma} = \mathrm{diag}(\sigma_1, \ldots, \sigma_N)$, and $\mathbf{R} = \mathrm{diag}(r_1, \ldots, r_N)$. Then Eq. (22.4), Eq. (22.5) can be recast in matrix form as

$$\frac{\partial \mathbf{c}}{\partial t} + \frac{1}{2} S^2 \boldsymbol{\Sigma}^2 \frac{\partial^2 \mathbf{c}}{\partial S^2} + S\mathbf{R}\frac{\partial \mathbf{c}}{\partial S} - \mathbf{R}\mathbf{c} + \mathbf{A}^*\mathbf{c} = \mathbf{0}, \qquad (22.6)$$

$$\mathbf{c}(T, S, T, K) = (S - K)^+\mathbf{1}, \tag{22.7}$$

respectively, where $\mathbf{0}$ is the N-dimensional zero vector and $\mathbf{1}$ is the N-dimensional vector all of whose components are equal to one.

Our aim here is to solve the inverse problem of recovering the parameters of the underlying model from market data. The inverse problem was first considered by Dupire [1], who showed that if the prices of a European call option were known for all strike prices and maturity dates, then the volatility surface can be recovered from market data. In our case, instead of a local volatility function, we wish to recover the volatility matrix $\mathbf{\Sigma}$, the transition intensity matrix $\mathbf{A}$, and the rate matrix $\mathbf{R}$.

It is important to note that actual market option prices are quoted for varying strikes and times to maturity. Thus, since we want to utilise a PDE-based approach to solve the inverse problem, we must first derive a system of PDEs similar to Eq. (22.6) but with the independent variables being the time to maturity and the strike price. In other words, we wish to obtain the analogue of Dupire's equation for the system of PDEs given in Eq. (22.6), which is the goal of the next section.

22.3. Derivation of a system of Dupire-type PDEs

First, we show that $c_1, \ldots, c_N$ are homogeneous functions of degree one with respect to S and K, i.e.,

$$c_i(t, \lambda S, T, \lambda K) = \lambda c_i(t, S, T, K) \quad (i = 1, \ldots, N) \tag{22.8}$$

for all $\lambda > 0$. To prove Eq. (22.8), we will use a uniqueness argument by showing that $c_i(t, \lambda S, T, \lambda K)$ and $\lambda c_i(t, S, T, K)$ for all $i = 1, \ldots, N$ satisfy the following final-value problem for $v(t, x, u, y)$:

$$\frac{\partial v_i}{\partial t} + \frac{1}{2}\sigma_i^2 x^2 \frac{\partial^2 v_i}{\partial x^2} + r_i x \frac{\partial v_i}{\partial x} - r_i v_i + \sum_{j=1}^{N} a_{ji} v_j = 0 \quad (i = 1, \ldots, N), \tag{22.9}$$

$$v_i(T, x, u, y) = \lambda(x - y)^+ \quad (i = 1, \ldots, N). \tag{22.10}$$

Let $c_1, \ldots, c_N$ be a solution of Eq. (22.4), Eq. (22.5). Take $v_i(t, x, u, y) = \lambda c_i(t, S, T, K)$ where $x = S$, $u = T$, and $y = K$. Then it is easy to see that $v_1, \ldots, v_N$ satisfy Eq. (22.9), Eq. (22.10). Now take $v_i(t, x, u, y) = c_i(t, S, T, K)$ where $x = S/\lambda$, $u = T$, and $y = K/\lambda$. Again, it is straightforward to verify that $v_1, \ldots, v_N$ satisfy Eq. (22.9), Eq. (22.10). Thus, the homogeneity condition Eq. (22.8) follows from the uniqueness of the solution of the final-value problem.

Invoking Euler's theorem on homogeneous functions, we obtain

$$S\frac{\partial c_i}{\partial S} + K\frac{\partial c_i}{\partial K} = c_i \quad (i = 1, \ldots, N).$$

Differentiating the above equation with respect to S and K gives

$$S\frac{\partial^2 c_i}{\partial S^2} = -K\frac{\partial^2 c_i}{\partial S \partial K}, \quad K\frac{\partial^2 c_i}{\partial K^2} = -S\frac{\partial^2 c_i}{\partial K \partial S} \quad (i = 1, \ldots, N),$$

respectively. It follows that

$$S^2\frac{\partial^2 c_i}{\partial S^2} = K^2\frac{\partial^2 c_i}{\partial K^2} \quad (i = 1, \ldots, N)$$

and Eq. (22.4) becomes

$$\frac{\partial c_i}{\partial t} + \frac{1}{2}\sigma_i^2 K^2 \frac{\partial^2 c_i}{\partial K^2} - r_i K \frac{\partial c_i}{\partial K} + \sum_{j=1}^{N} a_{ji} c_j = 0 \quad (i = 1, \ldots, N), \tag{22.11}$$

with the same terminal condition Eq. (22.5). In matrix form we therefore have

$$\frac{\partial \mathbf{c}}{\partial t} + \frac{1}{2}K^2\mathbf{\Sigma}^2\frac{\partial^2 \mathbf{c}}{\partial K^2} - K\mathbf{R}\frac{\partial \mathbf{c}}{\partial K} + \mathbf{A}^*\mathbf{c} = \mathbf{0}, \tag{22.12}$$

$$\mathbf{c}(T, S, T, K) = (S - K)^+\mathbf{1}. \tag{22.13}$$

Finally, letting $\mathbf{c}(t, S, T, K) = \bar{\mathbf{c}}(u, S, K)$ where $u = T - t$ in Eq. (22.12), Eq. (22.13) yields

$$\frac{\partial \bar{\mathbf{c}}}{\partial u} = \frac{1}{2}K^2\mathbf{\Sigma}^2\frac{\partial^2 \bar{\mathbf{c}}}{\partial K^2} - K\mathbf{R}\frac{\partial \bar{\mathbf{c}}}{\partial K} + \mathbf{A}^*\bar{\mathbf{c}}, \tag{22.14}$$

$$\bar{\mathbf{c}}(0, S, K) = (S - K)^+\mathbf{1}, \tag{22.15}$$

respectively. Note that Eq. (22.14), Eq. (22.15) is now an initial-value problem. When $N = 1$, we have

$$\frac{\partial \bar{c}_1}{\partial u} = \frac{1}{2}\sigma_1^2 K^2\frac{\partial^2 \bar{c}_1}{\partial K^2} - r_1 K\frac{\partial \bar{c}_1}{\partial K}, \tag{22.16}$$

which is Dupire's equation with a constant volatility [1]. Hence, Eq. (22.14) is the analogue of Dupire's equation for the regime-switching case.

22.4. Inverse Stieltjes moment problem

Having derived the system of PDEs in the appropriate independent variables u and K, we now proceed to solve the inverse problem of parameter estimation via the inverse Stieltjes moment method first proposed in [5].

To simplify the ensuing notation, we shall write $\mathbf{c}(u, K)$ instead of $\bar{\mathbf{c}}(u, S, K)$. We will also assume for simplicity that $\mathbf{R} = r\mathbf{I}$ for some given $r > 0$, where $\mathbf{I}$ is the identity matrix; the method can be easily extended to the more general case. As in [5], let us define the nth moment of the call price by

$$m_n^{(i)}(u) = \int_0^\infty K^n c_i(u, K)\, \mathrm{d}K \ (i = 1, \ldots, N),$$

where n is a nonnegative integer. Multiplying both sides of Eq. (22.14) by K^n and integrating over $(0, \infty)$, we formally obtain

$$\begin{aligned}\int_0^\infty K^n \frac{\partial \mathbf{c}}{\partial u}\, \mathrm{d}K = \frac{1}{2}\int_0^\infty K^{n+2}\mathbf{\Sigma}^2 \frac{\partial^2 \mathbf{c}}{\partial K^2}\, \mathrm{d}K - \int_0^\infty rK^{n+1}\frac{\partial \mathbf{c}}{\partial K}\, \mathrm{d}K \\ + \int_0^\infty K^n \mathbf{A}^* \mathbf{c}\, \mathrm{d}K.\end{aligned} \tag{22.17}$$

Assuming that the call price decays to zero sufficiently fast as $K \to \infty$, we deduce that

$$\int_0^\infty K^n \frac{\partial \mathbf{c}}{\partial u}\, \mathrm{d}K = \frac{\mathrm{d}\mathbf{m}_n}{\mathrm{d}u},$$

$$\int_0^\infty K^{n+1} \frac{\partial \mathbf{c}}{\partial K}\, \mathrm{d}K = -(n+1)\mathbf{m}_n,$$

$$\int_0^\infty K^{n+2} \frac{\partial^2 \mathbf{c}}{\partial K^2}\, \mathrm{d}K = (n+1)(n+2)\mathbf{m}_n.$$

Thus, Eq. (22.17) simplifies to

$$\frac{\mathrm{d}\mathbf{m}_n}{\mathrm{d}u} - r(n+1)\mathbf{m}_n = \left[\frac{1}{2}(n+1)(n+2)\mathbf{\Sigma}^2 + \mathbf{A}^*\right]\mathbf{m}_n.$$

Considering any N consecutive moments gives

$$\begin{aligned}
\frac{\mathrm{d}\mathbf{m}_n}{\mathrm{d}u} - r(n+1)\mathbf{m}_n &= \left[\frac{(n+1)(n+2)}{2}\boldsymbol{\Sigma}^2 + \mathbf{A}^*\right]\mathbf{m}_n,\\
\frac{\mathrm{d}\mathbf{m}_{n+1}}{\mathrm{d}u} - r(n+2)\mathbf{m}_{n+1} &= \left[\frac{(n+2)(n+3)}{2}\boldsymbol{\Sigma}^2 + \mathbf{A}^*\right]\mathbf{m}_{n+1},\\
&\vdots\\
\frac{\mathrm{d}\mathbf{m}_{n+N-1}}{\mathrm{d}u} - r(n+N)\mathbf{m}_{n+N-1} &= \left[\frac{(n+N)(n+N+1)}{2}\boldsymbol{\Sigma}^2 + \mathbf{A}^*\right]\mathbf{m}_{n+N-1}.
\end{aligned} \tag{22.18}$$

Since the option prices are assumed to be observed, the moments are also known and we wish to estimate $\mathbf{A}$ and $\boldsymbol{\Sigma}$. Given that $\mathbf{A}$ is an intensity matrix, each of its N diagonal entries, say a_{ii}, is expressible as a sum of the entries in the ith column, so there are essentially $N^2 - N$ unknown entries of $\mathbf{A}$. Together with the N unknown diagonal entries of $\boldsymbol{\Sigma}$, we therefore have a total of N^2 parameters to estimate. Note that Eq. (22.18) is a linear system of N^2 equations in N^2 unknowns.

To explain the basic idea, let us take $N = 2$. Then Eq. (22.18) gives

$$\begin{pmatrix}
\frac{(n+1)(n+2)}{2}m_n^{(1)} & 0 & m_n^{(2)} - m_n^{(1)} & 0\\
0 & \frac{(n+1)(n+2)}{2}m_n^{(2)} & 0 & m_n^{(1)} - m_n^{(2)}\\
\frac{(n+2)(n+3)}{2}m_{n+1}^{(1)} & 0 & m_{n+1}^{(2)} - m_{n+1}^{(1)} & 0\\
0 & \frac{(n+2)(n+3)}{2}m_{n+1}^{(2)} & 0 & m_{n+1}^{(1)} - m_{n+1}^{(2)}
\end{pmatrix}$$

$$\times \begin{pmatrix} \sigma_1^2 \\ \sigma_2^2 \\ a_{21} \\ a_{12} \end{pmatrix} = \begin{pmatrix}
\frac{\mathrm{d}m_n^{(1)}}{\mathrm{d}u} - r(n+1)m_n^{(1)}\\
\frac{\mathrm{d}m_n^{(2)}}{\mathrm{d}u} - r(n+1)m_n^{(2)}\\
\frac{\mathrm{d}m_{n+1}^{(1)}}{\mathrm{d}u} - r(n+2)m_{n+1}^{(1)}\\
\frac{\mathrm{d}m_{n+1}^{(2)}}{\mathrm{d}u} - r(n+2)m_{n+1}^{(2)}
\end{pmatrix}. \tag{22.19}$$

Note that from Eq. (22.15) we see that

$$\mathbf{m}_n(0) = \int_0^\infty K^n (S-K)^+ \mathbf{1}\,\mathrm{d}K = \frac{S^{n+2}}{(n+1)(n+2)}\mathbf{1}.$$

Let

$$M_n^{(i)}(u) = \int_0^u m_n^{(i)}(s)\,\mathrm{d}s \quad (i = 1, 2).$$

To incorporate the initial conditions, we integrate Eq. (22.19) over $[0, u]$ with respect to a dummy variable s to get

$$\begin{pmatrix} \frac{(n+1)(n+2)}{2} M_n^{(1)} & 0 & M_n^{(2)} - M_n^{(1)} & 0 \\ 0 & \frac{(n+1)(n+2)}{2} M_n^{(2)} & 0 & M_n^{(1)} - M_n^{(2)} \\ \frac{(n+2)(n+3)}{2} M_{n+1}^{(1)} & 0 & M_{n+1}^{(2)} - M_{n+1}^{(1)} & 0 \\ 0 & \frac{(n+2)(n+3)}{2} M_{n+1}^{(2)} & 0 & M_{n+1}^{(1)} - M_{n+1}^{(2)} \end{pmatrix}$$
$$\times \begin{pmatrix} \sigma_1^2 \\ \sigma_2^2 \\ a_{21} \\ a_{12} \end{pmatrix} = \begin{pmatrix} m_n^{(1)} - \frac{S^{n+2}}{(n+1)(n+2)} - r(n+1)M_n^{(1)} \\ m_n^{(2)} - \frac{S^{n+2}}{(n+1)(n+2)} - r(n+1)M_n^{(2)} \\ m_{n+1}^{(1)} - \frac{S^{n+3}}{(n+2)(n+3)} - r(n+2)M_{n+1}^{(1)} \\ m_{n+1}^{(2)} - \frac{S^{n+3}}{(n+2)(n+3)} - r(n+2)M_{n+1}^{(2)} \end{pmatrix}. \tag{22.20}$$

In summary, given $c_1(u, K)$ and $c_2(u, K)$ where $0 \le u \le T$ and $K \ge 0$, we compute $m_n^{(1)}$, $m_n^{(2)}$, $m_{n+1}^{(1)}$, and $m_{n+1}^{(2)}$ (for a fixed nonnegative integer n), as well as their integrals over $[0, u]$ for some u in $[0, T]$. We then solve the linear system Eq. (22.20) for the unknown parameters σ_1, σ_2, a_{21}, and a_{12}. Note that $a_{11} = -a_{21}$ and $a_{22} = -a_{12}$ by the definition of $\mathbf{A}$. In addition, the choice of u in $[0, T]$ should not matter since in this framework $\mathbf{A}$ and $\mathbf{\Sigma}$ are constant matrices.

22.5. Numerical implementation and results

To test the accuracy of the inverse Stieltes moment method, we need to have "observed" option prices. The "observed" option prices can be taken to be the solution generated by the initial-value problem Eq. (22.14), Eq. (22.15) (after specifying some matrices $\mathbf{A}$ and $\mathbf{\Sigma}$). Then we try to recover $\mathbf{A}$ and $\mathbf{\Sigma}$ using the moment method.

As a trial run, suppose that $\mathbf{A} = \mathbf{0}$, i.e., there is no switching among regimes. Then Eq. (22.14) reduces to a system of uncoupled Dupire equations. Hence, each component of $\mathbf{c}$ solves Dupire's equation, i.e., if $\mathbf{c} = (c_1, \ldots, c_N)^*$, then

$$\begin{aligned} c_i(u, K) &= S\Phi(d_1^{(i)}(u, K)) - Ke^{-ru}\Phi(d_2^{(i)}(u, K)) \quad (i = 1, \ldots, N), \\ d_1^{(i)}(u, K) &= \frac{\log(S/K) + (r + \sigma_i^2/2)u}{\sigma_i\sqrt{u}}, \\ d_2^{(i)}(u, K) &= \frac{\log(S/K) + (r - \sigma_i^2/2)u}{\sigma_i\sqrt{u}}, \end{aligned} \tag{22.21}$$

where Φ denotes the cumulative distribution function of a standard normal variable. Now we assume that $r = 0.02$, $S = 20$, $u = 0$, $T = 1$, $\sigma_1 = 0.1$ and $\sigma_2 = 0.3$. We take 50 values for strike price ranging from 0 to 60 and 100 values for the time to maturity ranging from 0 to 1. Then the moments and their integrals are calculated and momment method in Eq. (22.20) is applied. The estimated parameters are: $a_{21} = 6.67 \times 10^{(-4)}$, $a_{12} = 3.15 \times 10^{(-4)}$, $\sigma_1 = 0.1$ and $\sigma_2 = 0.2957$.

However, for the more realistic case $\mathbf{A} \neq \mathbf{0}$, there is no known explicit solution of Eq. (22.14), Eq. (22.15) in general. So we will have to solve this problem numerically to generate the "observed" option prices. This implies that we have to truncate the interval $(0, \infty)$ to some finite interval $(0, K_{\max})$ where $K_{\max} > 0$, and then impose reasonable boundary conditions for $\mathbf{c}$ at $K = 0$ and $K = K_{\max}$. If the right endpoint $K_{\max}$ is sufficiently large, and recalling that each component of $\mathbf{c}$ tends to zero as $K \to \infty$, then we can assign a positive but small value to each component. However, the boundary condition at the left endpoint $K = 0$ is not clear. When $N = 1$, the Black–Scholes formula evaluated at $K = 0$ gives S for the call price. For $N > 1$, it is not certain whether each component of $\mathbf{c}$ will also have the value S. To get around this problem, we will solve Eq. (22.14), Eq. (22.15) numerically for $K \in [0, K_{\max}]$ and $u \in [0, T]$ by formulating an explicit method with implicit boundary conditions.

Discretise the variables by

$$u \simeq u^i, \quad K \simeq K^j, \quad \mathbf{c} \simeq \mathbf{c}^{i,j} = (c_1^{i,j}, c_2^{i,j})^*$$

where

$$u^i = i\Delta u, \quad \Delta u = \frac{T}{I} \quad (i = 0, \ldots, I)$$

and

$$K^j = j\Delta K, \quad \Delta K = \frac{K_{\max}}{J} \quad (j = 0, \ldots, J).$$

Using an explicit scheme, we discretise Eq. (22.14) to get

$$\begin{aligned}\frac{c_1^{i+1,j} - c_1^{i,j}}{\Delta u} &= \frac{1}{2}\sigma_1^2 (K^j)^2 \frac{c_1^{i,j-1} - 2c_1^{i,j} + c_1^{i,j+1}}{(\Delta K)^2} - rK^j \frac{c_1^{i,j+1} - c_1^{i,j}}{\Delta K} \\ &\quad + a_{11} c_1^{i,j} + a_{21} c_2^{i,j}, \\ \frac{c_2^{i+1,j} - c_2^{i,j}}{\Delta u} &= \frac{1}{2}\sigma_2^2 (K^j)^2 \frac{c_2^{i,j-1} - 2c_2^{i,j} + c_2^{i,j+1}}{(\Delta K)^2} - rK^j \frac{c_2^{i,j+1} - c_2^{i,j}}{\Delta K} \\ &\quad + a_{12} c_1^{i,j} + a_{22} c_2^{i,j}.\end{aligned}$$

This is equivalent to

$$\mathbf{c}^{i+1,j} = \mathbf{c}^{i,j} + \frac{1}{2}\frac{\Delta u}{(\Delta K)^2}(K^j)^2\boldsymbol{\Sigma}^2(\mathbf{c}^{i,j-1} - 2\mathbf{c}^{i,j} + \mathbf{c}^{i,j+1})$$
$$- \frac{r\Delta u}{\Delta K}K^j(\mathbf{c}^{i,j+1} - \mathbf{c}^{i,j}) + \Delta u\mathbf{A}^*\mathbf{c}^{i,j}$$

for all $i = 1,\ldots,I-1$ and $j = 1,\ldots,J-1$. This solves for the option prices at time $i+1$ in the open interval $(0, K_{\max})$ using the option prices calculated at time i over the closed interval $[0, K_{\max}]$. The initial condition is determined by

$$\mathbf{c}^{0,j} = (S - K^j)^+\mathbf{1} \quad (j = 0,\ldots,J),$$

which includes the values at both endpoints.

To determine the boundary values at $K = 0$, we use second-order Taylor expansions at $(u, 0)$ in the continuous variables, i.e.,

$$\mathbf{c}(u, \Delta K) \simeq \mathbf{c}(u, 0) + (\Delta K)\frac{\partial \mathbf{c}}{\partial K}(u, 0) + \frac{1}{2}(\Delta K)^2\frac{\partial^2 \mathbf{c}}{\partial K^2}(u, 0),$$

$$\mathbf{c}(u, 2\Delta K) \simeq \mathbf{c}(u, 0) + (2\Delta K)\frac{\partial \mathbf{c}}{\partial K}(u, 0) + \frac{1}{2}(2\Delta K)^2\frac{\partial^2 \mathbf{c}}{\partial K^2}(u, 0),$$

$$\mathbf{c}(u, 3\Delta K) \simeq \mathbf{c}(u, 0) + (3\Delta K)\frac{\partial \mathbf{c}}{\partial K}(u, 0) + \frac{1}{2}(3\Delta K)^2\frac{\partial^2 \mathbf{c}}{\partial K^2}(u, 0).$$

Define

$$\boldsymbol{\alpha}_L(u) = \frac{\partial \mathbf{c}}{\partial K}(u, 0), \quad \boldsymbol{\beta}_L(u) = \frac{\partial^2 \mathbf{c}}{\partial K^2}(u, 0).$$

Then in discretised variables we get

$$\mathbf{c}^{i,1} = \mathbf{c}^{i,0} + \Delta K\boldsymbol{\alpha}_L(u^i) + \frac{1}{2}(\Delta K)^2\boldsymbol{\beta}_L(u^i),$$

$$\mathbf{c}^{i,2} = \mathbf{c}^{i,0} + 2\Delta K\boldsymbol{\alpha}_L(u^i) + 2(\Delta K)^2\boldsymbol{\beta}_L(u^i),$$

$$\mathbf{c}^{i,3} = \mathbf{c}^{i,0} + 3\Delta K\boldsymbol{\alpha}_L(u^i) + \frac{9}{2}(\Delta K)^2\boldsymbol{\beta}_L(u^i),$$

valid for all $i = 1,\ldots,I$. In matrix form, this is the same as

$$\begin{pmatrix} 1 & \Delta K & \frac{1}{2}(\Delta K)^2 \\ 1 & 2\Delta K & 2(\Delta K)^2 \\ 1 & 3\Delta K & \frac{9}{2}(\Delta K)^2 \end{pmatrix} \begin{pmatrix} c_1^{i,0} \\ \alpha_L^{(1)}(u^i) \\ \beta_L^{(1)}(u^i) \end{pmatrix} = \begin{pmatrix} c_1^{i,1} \\ c_1^{i,2} \\ c_1^{i,3} \end{pmatrix}$$

and

$$\begin{pmatrix} 1 & \Delta K & \frac{1}{2}(\Delta K)^2 \\ 1 & 2\Delta K & 2(\Delta K)^2 \\ 1 & 3\Delta K & \frac{9}{2}(\Delta K)^2 \end{pmatrix} \begin{pmatrix} c_2^{i,0} \\ \alpha_L^{(2)}(u^i) \\ \beta_L^{(2)}(u^i) \end{pmatrix} = \begin{pmatrix} c_2^{i,1} \\ c_2^{i,2} \\ c_2^{i,3} \end{pmatrix}.$$

Note that the vectors $\boldsymbol{\alpha}_L(u^i)$ and $\boldsymbol{\beta}_L(u^i)$ are not known, which is why we need to solve these two linear systems to obtain the really desired quantity $\mathbf{c}^{i,0}$. Moreover, $\mathbf{c}^{i,1}$, $\mathbf{c}^{i,2}$, and $\mathbf{c}^{i,3}$ on the right-hand sides are known if we first solve the PDEs in the interior $(0, K_{\max})$.

Similarly, at the right boundary point $K = K_{\max}$, we expand

$$\begin{aligned} \mathbf{c}(u, K_{\max} - \Delta K) &\simeq \mathbf{c}(u, K_{\max}) + (-\Delta K)\frac{\partial \mathbf{c}}{\partial K}(u, K_{\max}) \\ &\quad + \frac{1}{2}(-\Delta K)^2 \frac{\partial^2 \mathbf{c}}{\partial K^2}(u, K_{\max}), \\ \mathbf{c}(u, K_{\max} - 2\Delta K) &\simeq \mathbf{c}(u, K_{\max}) + (-2\Delta K)\frac{\partial \mathbf{c}}{\partial K}(u, K_{\max}) \\ &\quad + \frac{1}{2}(-2\Delta K)^2 \frac{\partial^2 \mathbf{c}}{\partial K^2}(u, K_{\max}), \\ \mathbf{c}(u, K_{\max} - 3\Delta K) &\simeq \mathbf{c}(u, K_{\max}) + (-3\Delta K)\frac{\partial \mathbf{c}}{\partial K}(u, K_{\max}) \\ &\quad + \frac{1}{2}(-3\Delta K)^2 \frac{\partial^2 \mathbf{c}}{\partial K^2}(u, K_{\max}). \end{aligned}$$

Defining

$$\boldsymbol{\alpha}_R(u) = \frac{\partial \mathbf{c}}{\partial K}(u, K_{\max}), \quad \boldsymbol{\beta}_R(u) = \frac{\partial^2 \mathbf{c}}{\partial K^2}(u, K_{\max}),$$

we obtain

$$\begin{aligned} \mathbf{c}^{i,J-1} &= \mathbf{c}^{i,J} - \Delta K \boldsymbol{\alpha}_R(u^i) + \frac{1}{2}(\Delta K)^2 \boldsymbol{\beta}_R(u^i), \\ \mathbf{c}^{i,J-2} &= \mathbf{c}^{i,J} - 2\Delta K \boldsymbol{\alpha}_R(u^i) + 2(\Delta K)^2 \boldsymbol{\beta}_R(u^i), \\ \mathbf{c}^{i,J-3} &= \mathbf{c}^{i,J} - 3\Delta K \boldsymbol{\alpha}_R(u^i) + \frac{9}{2}(\Delta K)^2 \boldsymbol{\beta}_R(u^i), \end{aligned}$$

or, in matrix form,

$$\begin{pmatrix} 1 & -\Delta K & \frac{1}{2}(\Delta K)^2 \\ 1 & -2\Delta K & 2(\Delta K)^2 \\ 1 & -3\Delta K & \frac{9}{2}(\Delta K)^2 \end{pmatrix} \begin{pmatrix} c_1^{i,J} \\ \alpha_R^{(1)}(u^i) \\ \beta_R^{(1)}(u^i) \end{pmatrix} = \begin{pmatrix} c_1^{i,J-1} \\ c_1^{i,J-2} \\ c_1^{i,J-3} \end{pmatrix}$$

and

$$\begin{pmatrix} 1 & -\Delta K & \frac{1}{2}(\Delta K)^2 \\ 1 & -2\Delta K & 2(\Delta K)^2 \\ 1 & -3\Delta K & \frac{9}{2}(\Delta K)^2 \end{pmatrix} \begin{pmatrix} c_2^{i,J} \\ \alpha_R^{(2)}(u^i) \\ \beta_R^{(2)}(u^i) \end{pmatrix} = \begin{pmatrix} c_2^{i,J-1} \\ c_2^{i,J-2} \\ c_2^{i,J-3} \end{pmatrix}.$$

As before, the vectors $\boldsymbol{\alpha}_R(u^i)$ and $\boldsymbol{\beta}_R(u^i)$ are not known, so we solve these two linear systems to obtain the really desired quantity $\mathbf{c}^{i,J}$. Moreover, $\mathbf{c}^{i,J-1}$, $\mathbf{c}^{i,J-2}$, and $\mathbf{c}^{i,J-3}$ on the right-hand sides are known if we first solve the PDEs in the interior $(0, K_{\max})$.

Summarising, the explicit algorithm incorporating implicit boundary conditions that we propose can be formulated as follows:

(1) Set

$$\mathbf{c}^{0,j} = (S - K^j)^+\mathbf{1} \quad (j = 0, \ldots, J).$$

(2) For all $i = 0, \ldots, I-1$ do

(a) For all $j = 1, \ldots, J-1$ do

$$\mathbf{c}^{i+1,j} = \mathbf{c}^{i,j} + \frac{1}{2}\frac{\Delta u}{(\Delta K)^2}(K^j)^2\boldsymbol{\Sigma}^2(\mathbf{c}^{i,j-1} - 2\mathbf{c}^{i,j} + \mathbf{c}^{i,j+1}) - \frac{r\Delta u}{\Delta K}K^j(\mathbf{c}^{i,j+1} - \mathbf{c}^{i,j}) + \Delta u\mathbf{A}^*\mathbf{c}^{i,j}.$$

(b) Solve

$$\begin{pmatrix} 1 & \Delta K & \frac{1}{2}(\Delta K)^2 \\ 1 & 2\Delta K & 2(\Delta K)^2 \\ 1 & 3\Delta K & \frac{9}{2}(\Delta K)^2 \end{pmatrix} \begin{pmatrix} c_1^{i+1,0} \\ \alpha_L^{(1)}(u^{i+1}) \\ \beta_L^{(1)}(u^{i+1}) \end{pmatrix} = \begin{pmatrix} c_1^{i+1,1} \\ c_1^{i+1,2} \\ c_1^{i+1,3} \end{pmatrix}$$

and

$$\begin{pmatrix} 1 & \Delta K & \frac{1}{2}(\Delta K)^2 \\ 1 & 2\Delta K & 2(\Delta K)^2 \\ 1 & 3\Delta K & \frac{9}{2}(\Delta K)^2 \end{pmatrix} \begin{pmatrix} c_2^{i+1,0} \\ \alpha_L^{(2)}(u^{i+1}) \\ \beta_L^{(2)}(u^{i+1}) \end{pmatrix} = \begin{pmatrix} c_2^{i+1,1} \\ c_2^{i+1,2} \\ c_2^{i+1,3} \end{pmatrix}$$

for $\mathbf{c}^{i+1,0} = (c_1^{i+1,0}, c_2^{i+1,0})^*$.

(c) Solve

$$\begin{pmatrix} 1 & -\Delta K & \frac{1}{2}(\Delta K)^2 \\ 1 & -2\Delta K & 2(\Delta K)^2 \\ 1 & -3\Delta K & \frac{9}{2}(\Delta K)^2 \end{pmatrix} \begin{pmatrix} c_1^{i+1,J} \\ \alpha_R^{(1)}(u^{i+1}) \\ \beta_R^{(1)}(u^{i+1}) \end{pmatrix} = \begin{pmatrix} c_1^{i+1,J-1} \\ c_1^{i+1,J-2} \\ c_1^{i+1,J-3} \end{pmatrix}$$

and

$$\begin{pmatrix} 1 & -\Delta K & \frac{1}{2}(\Delta K)^2 \\ 1 & -2\Delta K & 2(\Delta K)^2 \\ 1 & -3\Delta K & \frac{9}{2}(\Delta K)^2 \end{pmatrix} \begin{pmatrix} c_2^{i+1,J} \\ \alpha_R^{(2)}(u^{i+1}) \\ \beta_R^{(2)}(u^{i+1}) \end{pmatrix} = \begin{pmatrix} c_2^{i+1,J-1} \\ c_2^{i+1,J-2} \\ c_2^{i+1,J-3} \end{pmatrix}$$

for $\mathbf{c}^{i+1,J} = (c_1^{i+1,J}, c_2^{i+1,J})^*$.

A stability criterion for this scheme is

$$\Delta u \leq (\Delta K)^2 \min\left(\frac{1}{\sigma_1^2}, \frac{1}{\sigma_2^2}\right).$$

Although explicit schemes are generally slower than implicit schemes, the programming is straightforward for the former compared to the latter since the linear system to be solved for the implicit scheme is not anymore tridiagonal.

In the following simulations, we take the parameter values to be $r = 0.02$, $S = 20$, $u = 0$, $T = 1$, $K_{\max} = 60$, $\sigma_1 = 0.1$, and $\sigma_2 = 0.3$. We use 1200 nodes to discretize time axes and 120 nodes to discritize strike axes, i.e., $\Delta u = \frac{1}{1200}$ and $\Delta K = 0.5$, in solving the PDEs (22.14), (22.15) numerically. The numerical solution contains a large size of dataset which usually does not exist in practice; in reality there is only a set of small data points corresponding to time and strike nodes. In our example, we pick 13-time and 21-strike nodes from the solutions and these prices are used as market data in the inverse Stieltjes moment approach. Next, in order to calculate the truncated moments and their integrals accurately, we interpolate the call prices from time to maturity- and strike-direction. The size of the dataset increases to 500 by 500 points after interpolation. Here, we use a Matlab built-in function called Piecewise Cubic Hermite Interpolating Polynomial (PCHIP) for the interpolation procedure. Note that in general that the market data is not equally spaced through time to maturity and strike prices. Therefore, the number of nodes interpolated between two prices depends on the differences in time to maturity and strike price of the two prices. Finally we solve the algebraic system (22.20) for the "unknown" parameters.

First, let us suppose that the intensity matrix $\mathbf{A}$ is of the form

$$\mathbf{A} = \begin{pmatrix} -\lambda & \lambda \\ \lambda & -\lambda \end{pmatrix}$$

where $\lambda > 0$. In Table 22.1, we present the estimated parameters for different values of λ and n. Additionally, we evaluate the errors for the estimated

Table 22.1. Example 1: Estimated parameters for different λ and n with $\sigma_1 = 0.1$ and $\sigma_2 = 0.3$.

λ	Estimated parameter	n		
		2	3	4
0.25	(λ_1, λ_2)	(0.2567,0.2240)	(0.2523,0.1995)	(0.2498,0.1477)
	(σ_1, σ_2)	(0.0977,0.3002)	(0.0987,0.2989)	(0.0994,0.2964)
	$(\text{rmse}_1, \text{rmse}_2)$	(0.0021,0.0058)	(0.0013,0.0069)	(7.70×10^{-4},0.0094)
0.5	(λ_1, λ_2)	(0.5073,0.4686)	(0.5016,0.4420)	(0.4982,0.3844)
	(σ_1, σ_2)	(0.0978,0.3001)	(0.0989,0.2989)	(0.0997,0.2964)
	$(\text{rmse}_1, \text{rmse}_2)$	(0.0013,0.0051)	(6.49×10^{-4},0.0060)	(1.7×10^{-4},0.0083)
1	(λ_1, λ_2)	(1.0088,0.9570)	(0.9995,0.9205)	(0.9939,0.8571)
	(σ_1, σ_2)	(0.0981,0.2999)	(0.0994,0.2988)	(0.1002,0.2964)
	$(\text{rmse}_1, \text{rmse}_2)$	(8.4×10^{-4},0.0042)	(7.22×10^{-4},0.0049)	(8.87×10^{-4},0.0065)
2	(λ_1, λ_2)	(2.0094,1.9369)	(1.9900,1.9001)	(1.9782,0.19101
	(σ_1, σ_2)	(0.0987,0.2999)	(0.1004,0.2990)	(0.1015,0.2968)
	$(\text{rmse}_1, \text{rmse}_2)$	(0.0015,0.0034)	(0.0015,0.0019)	(0.0020,0.0047)
5	(λ_1, λ_2)	(4.9470,5.0349)	(4.9522,4.8768)	(4.8405,4.8238)
	(σ_1, σ_2)	(0.1021,0.3017)	(0.1047,0.3007)	(0.1061,0.2992)
	$(\text{rmse}_1, \text{rmse}_2)$	(0.0043,0.0051)	(0.0041,0.0046)	(0.0024,0.0028)
8	(λ_1, λ_2)	(7.7259,8.6577)	(7.5802,8.3862)	(7.5168,8.1685)
	(σ_1, σ_2)	(0.1069,0.3062)	(0.1102,0.3041)	(0.1116,0.3024)
	$(\text{rmse}_1, \text{rmse}_2)$	(0.0162,0.0179)	(0.0143,0.0160)	(0.0120,0.0139)

option prices using Root Mean Square Error (RMSE), which helps us to analyse the sensitivity of λ and n. From Eq. (22.20), the estimated parameters should be independent of n. The error, which occurs by sovling the Dupire PDEs, certainly affects the calculation. To rectify this to some extent, we utilise larger degrees of the moment, i.e., $n \geq 2$, which would put more weight on option prices in the calculation of moments and consequently reduce the impact of the errors in the Dupire PDEs. In this example, the column corresponding to $n = 2$ shows very good agreement between the estimated values and the actual values assigned when $\lambda \leq 2$. However, if the value of n is too large, higher calculation error occurs when the moments are calculated, and it also affects the estimated results. As can be observed, in the case of $\lambda \leq 2$, RMSE is larger when higher n is used. Furthermore, the accuracy of the estimation is also affected by the value of the intensity rate. As the intensity rate increases, the market tends to switch more frequently between the states and thus, it is more difficult to capture the information from the market data reducing the accuracy of the estimation. Again, this problem can be corrected by using larger degrees of the moment. Therefore, when $\lambda \geq 2$, RMSE is smaller when higher n is used.

In our second numerical experiment, we assume different intensity rates for state 1 and state 2, so the intensity matrix is of the form

$$\mathbf{A} = \begin{pmatrix} -\lambda_1 & \lambda_2 \\ \lambda_1 & -\lambda_2 \end{pmatrix}.$$

Other values of parameters remain unchanged. The estimated parameters for different values of (λ_1, λ_2) and n are presented in Tables 22.2 and 22.3. The difference between the intensity parameters, λ_1 and λ_2 is apparently noticeable on the estimated parameters. When the difference is low, the estimated results closely agree to the actual values for all degrees of moments. In cases where the differences between the intensity parameters are quite substantial, for example $\lambda_1 = 0.25$, $\lambda_2 = 5$, the estimated results are inaccurate for low degree of moments. However, by using a higher degree of moment, e.g., $n = 4$, we still obtain the results closer to actual parameters.

After we estimate the unknown parameters, we calculate the call option prices in each states by solving Eq. (22.6) and Eq. (22.7). Figure 22.1 shows the estimated option prices and the actual values using $\lambda_1 = 0.25$, $\lambda_2 = 2$ and $n = 2$. Note that the computed values agree very well with the actual data.

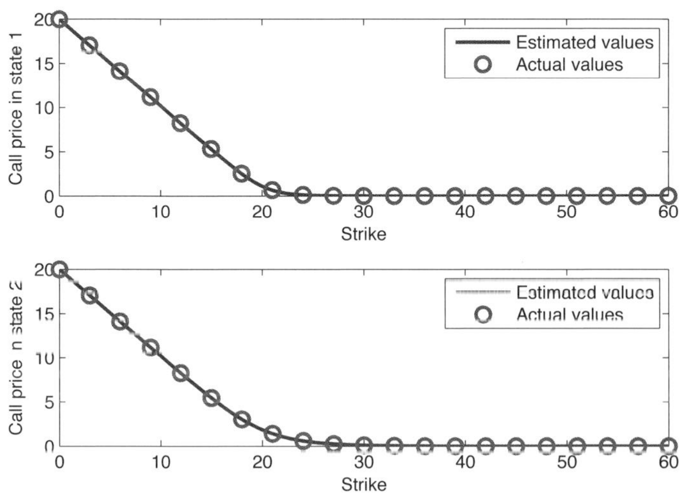

Fig. 22.1. Actual and estimated call prices: $\lambda_1 = 0.25$, $\lambda_2 = 2$ and $n = 2$.

Table 22.2. Example 2: Estimated parameters for different λ and n with $\sigma_1 = 0.1$ and $\sigma_2 = 0.3$.

(λ_1, λ_2)	Estimated parameter	n = 2
(0.25, 0.5)	(λ_1, λ_2)	(0.2578,0.4708)
	(σ_1, σ_2)	(0.0976,0.3001)
	$(\text{rmse}_1, \text{rmse}_2)$	(0.0022,0.0051)
(0.25, 1)	(λ_1, λ_2)	(0.2604,0.9602)
	(σ_1, σ_2)	(0.0974,0.2997)
	$(\text{rmse}_1, \text{rmse}_2)$	(0.0024,0.0039)
(0.25, 2)	(λ_1, λ_2)	(0.2672,1.9039)
	(σ_1, σ_2)	(0.0970,0.2979)
	$(\text{rmse}_1, \text{rmse}_2)$	(0.0027,0.0026)
(0.25, 5)	(λ_1, λ_2)	(0.3044,1.8150)
	(σ_1, σ_2)	(0.0954,0.2253)
	$(\text{rmse}_1, \text{rmse}_2)$	(0.0047,0.0145)
(1, 0.5)	(λ_1, λ_2)	(1.0059,0.4630)
	(σ_1, σ_2)	(0.0983,0.3001)
	$(\text{rmse}_1, \text{rmse}_2)$	(0.0010,0.0053)
(1, 2)	(λ_1, λ_2)	(1.0159,1.9301)
	(σ_1, σ_2)	(0.0977,0.2993)
	$(\text{rmse}_1, \text{rmse}_2)$	$(7.96\times10^{-4}, 0.0029)$
(1, 5)	(λ_1, λ_2)	(1.0518,4.3958)
	(σ_1, σ_2)	(0.0962,0.2887)
	$(\text{rmse}_1, \text{rmse}_2)$	(0.0015,0.0017)
(1, 8)	(λ_1, λ_2)	(1.1127,2.0694)
	(σ_1, σ_2)	(0.0946,0.2077)
	$(\text{rmse}_1, \text{rmse}_2)$	(0.0044,0.0082)
(2, 5)	(λ_1, λ_2)	(2.0416,4.8157)
	(σ_1, σ_2)	(0.0974,0.2980)
	$(\text{rmse}_1, \text{rmse}_2)$	$(5.11\times10^{-4}, 0.0014)$
(2, 8)	(λ_1, λ_2)	(2.0934,7.4342)
	(σ_1, σ_2)	(0.0961,0.2937)
	$(\text{rmse}_1, \text{rmse}_2)$	$(6.35\times10^{-4}, 4.58\times10^{-4})$

Table 22.3. Example 2 (continued): Estimated parameters for different λ and n with $\sigma_1 = 0.1$ and $\sigma_2 = 0.3$.

(λ_1, λ_2)	Estimated parameter	n = 3	n = 4
(0.25, 0.5)	(λ_1, λ_2)	(0.2529,0.4474)	(0.2500,0.3956)
	(σ_1, σ_2)	(0.0987,0.2990)	(0.0993,0.2965)
	$(\text{rmse}_1, \text{rmse}_2)$	(0.0014,0.0060)	$(8.21\times10^{-4}, 0.0084)$
(0.25, 1)	(λ_1, λ_2)	(0.2541,0.9414)	(0.2506,0.8910)
	(σ_1, σ_2)	(0.0985,0.2989)	(0.0993,0.2968)
	$(\text{rmse}_1, \text{rmse}_2)$	(0.0015,0.0046)	$(8.94\times10^{-4}, 0.0064)$
(0.25, 2)	(λ_1, λ_2)	(0.2676,1.9134)	(0.2521,1.8780)
	(σ_1, σ_2)	(0.0983,0.2982)	(0.0091,0.2970)
	$(\text{rmse}_1, \text{rmse}_2)$	(0.0017,0.0028)	(0.0010,0.0037)
(0.25, 5)	(λ_1, λ_2)	(0.2772,3.5618)	(0.2612,4.3568)
	(σ_1, σ_2)	(0.0974,0.2693)	(0.0987,0.2870)
	$(\text{rmse}_1, \text{rmse}_2)$	(0.0024,0.0053)	(0.0014,0.0021)
(1, 0.5)	(λ_1, λ_2)	(0.9982,0.4296)	(0.9934,0.3593)
	(σ_1, σ_2)	(0.0996,0.2987)	(0.1004,0.2960)
	$(\text{rmse}_1, \text{rmse}_2)$	$(8.89\times10^{-4}, 0.0063)$	(0.0011,0.0085)
(1, 2)	(λ_1, λ_2)	(1.0028,1.9159)	(0.9951,1.8564)
	(σ_1, σ_2)	(0.0991,0.2989)	(0.1000,0.2971)
	$(\text{rmse}_1, \text{rmse}_2)$	$(4.447\times10^{-4}, 0.0033)$	$(5.71\times10^{-4}, 0.0042)$
(1, 5)	(λ_1, λ_2)	(1.0194,4.6501)	(1.0007,4.8092)
	(σ_1, σ_2)	(0.0983,0.2984)	(0.0996,0.2973)
	$(\text{rmse}_1, \text{rmse}_2)$	$(7.26\times10^{-4}, 0.0013)$	$(1.84\times10^{-4}, 0.0011)$
(1, 8)	(λ_1, λ_2)	(1.0491,5.1333)	(1.0115,6.8723)
	(σ_1, σ_2)	(0.0974,0.2598)	(0.0992,0.2851)
	$(\text{rmse}_1, \text{rmse}_2)$	(0.0014,0.0026)	$(5.94\times10^{-4}, 8.18\times10^{-4})$
(2, 5)	(λ_1, λ_2)	(2.0013,4.8841)	(1.9788,4.9018)
	(σ_1, σ_2)	(0.0996,0.2991)	(0.1009,0.2994)
	$(\text{rmse}_1, \text{rmse}_2)$	$(6.81\times10^{-4}, 0.0015)$	$(9.45\times10^{-4}, 0.0017)$
(2, 8)	(λ_1, λ_2)	(2.0207,7.7349)	(1.9793,8.0197)
	(σ_1, σ_2)	(0.0990,0.2974)	(0.1006,0.3008)
	$(\text{rmse}_1, \text{rmse}_2)$	$(1.62\times10^{-4}, 5.15\times10^{-4})$	$(3.91\times10^{-4}, 0.29\times10^{-4})$

22.6. Conclusion

In this article, we developed a methodology based on the inverse Stieltjes moment technique to recover the parameters of a regime-switching model from option prices. In particular, the volatility of the asset price, which is the underlying variable of the option, switches over time and modulated by a continuous-time, finite-state Markov chain. The coupled system of Dupire-type PDEs was derived from the well-known coupled system of Black–Scholes PDEs. The inverse Stieltjes moment approach was adopted to formulate the PDEs forming a linear system of equations for the volatilities and the intensity parameters. We demonstrated how to apply this method to "theoretical data", which were obtained by solving the Dupire PDEs. Numerical results were presented to illustrate the accuracy of our method. We also performed various analyses for both cases when the intensity parameters of the intensity matrix $\mathbf{A}$ and the degree of the moment n is varied. Our findings based on the numerical experiments on the two types of data sets indicate the following: (1) for a single intensity rate λ, the higher the intensity rate the lower accuracy of the method, and (2) for two different intensity parameters λ_1 and λ_1, the greater the difference between these intensity rates the less accurate the estimation.

References

[1] B. Dupire, Pricing with a smile, *Risk.* **7**(1), 18–20, (1994).
[2] P. P. Boyle and D. Thangaraj, Volatility estimation from observed option prices, *Decisions in Economics and Finance.* **23**(1), 31–52, (2000).
[3] L. B. G. Andersen and R. Brotherton-Ratcliffe, The equity option volatility smile: An implicit finite-difference approach, *Journal of Computational Finance.* **1**(2), 5–37, (1998).
[4] M. R. Rodrigo and R. S. Mamon, A new representation of the local volatility surface, *International Journal of Theoretical and Applied Finance.* **11**(07), 691–703, (2008).
[5] M. R. Rodrigo and R. S. Mamon, Recovery of time-dependent parameters of a Black-Scholes-type equation: An inverse Stieltjes moment approach, *Journal of Applied Mathematics.* **2007**, (2007). Article ID 62098.
[6] I. Bouchouev and V. Isakov, The inverse problem of option pricing, *Inverse Problems.* **13**(5), L11, (1997).
[7] Z.-C. Deng, J.-N. Yu, and L. Yang, An inverse problem of determining the implied volatility in option pricing, *Journal of Mathematical Analysis and Applications.* **340**(1), 16 – 31, (2008).
[8] C.-S. J. Chu, G. J. Santoni, and T. Liu, Stock market volatility and regime shifts in returns, *Information Sciences.* **94**, 179–190 (October, 1996).

[9] C. M. Turner, R. Startz, and C. R. Nelson, A Markov model of heteroskedasticity, risk, and learning in the stock market, *Journal of Financial Economics.* **25**(1), 3 – 22, (1989).

[10] C. Engel and J. D. Hamilton, Long swings in the dollar: Are they in the data and do markets know it?, *American Economic Review.* **80**(4), 689–713, (1990).

[11] G. Bekaert and R. J. Hodrick, On biases in the measurement of foreign exchange risk premiums, *Journal of International Money and Finance.* **12** (2), 115 – 138, (1993).

[12] C. Engel and C. S. Hakkio, The distribution of exchange rates in the EMS, *International Journal of Finance & Economics.* **1**(1), 55–67, (1996).

[13] M. Dahlquist and S. F. Gray, Regime-switching and interest rates in the European monetary system, *Journal of International Economics.* **50**(2), 399–419, (2000).

[14] A. Ang and G. Bekaert, Regime switches in interest rates, *Journal of Business & Economic Statistics.* **20**(2), 163–82, (2002)

[15] R. J. Elliott and R. S. Mamon, An interest rate model with a Markovian mean reverting level, *Quantitative Finance.* **2**, 454–458(5), (2002).

[16] R. J. Elliott and P. E. Kopp, *Mathematics of Financial Markets.* (Springer, Berlin Heidelberg New York, 1999).

[17] R. J. Elliott, L. Chan, and T. K. Siu, Option pricing and Esscher transform under regime switching, *Annals of Finance.* **1**(4), 423–432, (2005).

[18] W. Ching, K. Siu, and L. Li, Pricing exotic options under a higher-order hidden Markov model, *Journal of Applied Mathematics and Decision Sciences.* pp. 1–15, (2007).

[19] R. S. Mamon and M. R. Rodrigo, Explicit solutions to European options in a regime-switching economy, *Operations Research Letters.* **33**(6), 581–586, (2005).

[20] P. Boyle and T. Draviam, Pricing exotic options under regime switching, *Insurance: Mathematics and Economics.* **40**(2), 267 – 282, (2007).

[21] J.-C. Duan, I. Popova, and P. Ritchken, Option pricing under regime switching, *Quantitative Finance.* **2**, 116–132, (2002).

[22] T. K. Siu, C. Erlwein, and R. S. Mamon, The pricing of credit default swaps under a Markov-modulated Merton's structural model, *North American Actuarial Journal.* **12**(1), 229–238, (2008).

[23] R. J. Elliott and J. van der Hoek, An application of hidden Markov models to asset allocation problems, *Finance and Stochastics.* **1**, 229–238, (1997)

[24] R. J. Elliott, L. Aggoun, and J. B. Moore, *Hidden Markov Models: Estimation and Control.* (Springer, Berlin, 1994).

[25] X. Guo, Information and option pricings, *Quantitative Finance.* **1**, 38–44, (2001).

Chapter 23

An optimal inventory-price coordination policy

Hanqin Zhang[1] and Qing Zhang[2]

[1]*Academy of Mathematics and Systems Sciences,*
Academia Sinica, Beijing, 100190, China, and
School of Business, National University of Singapore,
119245 Singapore

[2]*Department of Mathematics,*
University of Georgia,
Athens, Georgia 30602, USA

This chapter is concerned with an optimal inventory-price coordination strategy under a continuous review inventory model. A simple diffusion process is considered in which the pricing affects the inventory level directly with additive noise. The objective is to choose price over time to maximize a discounted profit function. A closed-form solution is obtained which says the optimal pricing can be given in terms of a hedging point, which can be obtained by solving an algebraic equation.

Contents

23.1. Introduction

This paper studies an optimal inventory-price coordination strategy under a continuous review inventory model. Inventory-price coordination strategies are concerned with dynamically adjusting the price of a product based on real time inventory, so as to maximize an overall profit function. They have attracted much attention from both academia and industry in recent years. Many industrial applications are documented in the literature. For example, industrial applications in the personal computer business can be found in (Byrnes 2003); the retail sale of fashion and seasonal goods is discussed in (Pashigan and Bowen 1991); and the travel and leisure sectors are studied in (Smith et al. 1992).

Academic research along this line has been reported in many papers. For example, Federgruen and Heching (1999) consider a single-product, periodic review inventory model in which demands in consecutive periods are independent and depend on the product's price in accordance with general stochastic demand functions. The ordering cost is a linear replenishment cost and no setup cost is charged. The price implemented in any given period can be set dynamically as a function of the system state. They present the optimality of the base-stock list-price policy. Namely, when the inventory level drops below the base-stock level, it is increased to the base-stock level and the list price is charged; when the inventory level is above the base-stock level, nothing is ordered and a price discount is offered. When a setup cost is charged with each ordering, Chen and Simchi-Levi (2004a, b) prove the optimality of the (s, S, p)-type policy. This policy tells us, at the beginning of each period, if the inventory level is below the reorder point s an order is placed to increase the inventory level up to S, if the inventory level is above the reorder point s no order is placed, and the optimal price is a function of the initial inventory level. Chen and Simchi-Levi (2006), and Feng and Chen (2003) extend this optimality result to the single-product, continuous review inventory model, in which the demands arrive only at the discrete points, and the interarrival times and the demand size of the next arrival depend on the price charged at the latest arrival.

In this paper, we model the cumulative product demand as a price dependent diffusion. The objective is to dynamically adjust the product price so as to maximize a discounted profit function. There are many studies that use diffusion models to characterize cumulative product demand and to investigate the inventory control policy without involving product pricing. For example, Bather (1966) considers a single-product continuous review inventory model in which the cumulative demand is modeled as a Brownian motion with constant drift and constant variance, and the excessive demand is backlogged. Each time an order is made, with a setup cost and a variable cost, the inventory and backlog costs are charged following a general convex function. He finds an optimal policy that minimizes the long-run average expected value of inventory/backlog and ordering costs. Following the Bather model, Richard (1977), and Constantinides and Richard (1978) study the infinite-horizon discounted cost problem with a linear inventory cost. In their model, the inventory level can be adjusted in two directions, namely, the inventory level can be increased and decreased. Both adjustments will be charged by a setup cost and a variable cost. They prove the existence of an optimal policy. Using a quasi-variational inequality (QVI) approach, Sulem (1986) explicitly characterizes the optimal policy. Benkherouf and Bensoussan (2009) and Bensoussan et al. (2005) consider the infinite-horizon discounted cost model, in which the inventory cost is convex and the cumulative demand is modeled as a mixture of Brownian motion and compound Poisson process. After an analytical study of the QVI, they prove the existence of an optimal policy as well as the optimality of an (s, S)-type policy.

With consideration of product pricing, Chen et al. (2010) consider the Brownian demand model in which the holding cost is linear in the quantity held per unit of time, and at any time, an $(0, S)$ inventory policy is implemented. They assume that the price decisions are only dynamically adjusted in the duration between two consecutive replenishment epochs. Specifically, after a replenishment at the beginning of a cycle (that is, the inventory level is just pushed up to S), a price p_1 is charged until the inventory level drops down to level S_1 $(S > S_1)$, then price p_2 is charged until the inventory level drops down to S_2 $(S > S_1 > S_2)$,..., and, finally, when the inventory level drops down to S_N $(S > S_1 > \cdots > S_N)$, the price p_{N+1} is charged until the inventory level drops down to zero, and then another cycle repeats. Given an integer number N and setting $S_i - S_{i+1} = S/(N+1), i = 0, \cdots, N$ with $S_0 = S$, they find an optimal price mechanism $\{p_1, \cdots, p_{N+1}\}$ to minimize the long-run average expected value of inventory/backlog and ordering costs. Note that the price adjustment in Chen et al. (2010) is not done continuously in time over the entire period nevertheless.

In this paper, we consider a price-dependent diffusion model and allow continuous price adjustment. The objective is to dynamically and continuously vary the price over time to maximize the profit function. We study a relatively simple case and show that the optimal control can be given in terms of a hedging point. The hedging point can be obtained by solving an algebraic equation. Such a structural property is desirable from a practical viewpoint because it provides deeper insights into the problem, such as dependence on various parameters, and typically leads to efficient computational methods.

The remainder of the paper is organized as follows. The model and problem are introduced in Section 23.2. The properties of the value function are studied together with the associated HJB equation in Section 23.3. A verification theorem, which is used to characterize the optimal policy, is given in Section 23.4. A numerical example is provided in Section 23.5. We conclude the paper in Section 23.6.

23.2. Problem formulation

In this paper, we model the cumulative demand up to time t by

$$\int_0^t \lambda(p_s)ds - \int_0^t \sigma dW_s,$$

where $\lambda(p) = \lambda_1/(p + \lambda_0)$ for positive constants λ_0 and λ_1, $p \in [a, b]$ for $a > 0$ and $b > 0$, $\sigma > 0$ is the variability measure of demand, and W_s is a standard Brownian motion. We order products with zero leadtime to satisfy the demands. At any time t, according to the current inventory level, the replenishment and price decisions are made. The ordering cost includes a setup cost K and a variable cost c, that is, if the amount ordered is q, then the ordering cost is

$$\begin{cases} 0, & \text{if } q = 0, \\ K + cq, & \text{if } q > 0. \end{cases}$$

We assume the holding cost is linear in the quantity held. Furthermore, backlogging is not allowed, that is, demand must be satisfied immediately. Following the method of Harrison et al. (1983) (see also Yao, 2010), we know the optimal replenishment policy is of $(0, S)$-type. For an $(0, S)$ replenishment policy, let $x_t \in \mathbb{R}$ denote the inventory level at time t. Then

$$x_t = S - \int_0^t \lambda(p_s)ds + \int_0^t \sigma dW_s. \tag{23.1}$$

In view of the optimality of $(0, S)$-type policies, in the following, we assume that the replenishment policy $(0, S)$ is given and only focus on finding the optimally dynamic price p_t over time so as to maximize the discounted expectation of the profit.

We consider admissible p_t in the sense that $p_t \in [a, b]$ for all t and $p = \{p_t,\ t \geq 0\}$ is $\sigma\{W_s :\ 0 \leq s \leq t\}$ progressively measurable. Let τ denote the first time that x_t is reduced to 0, i.e., $\tau = \inf\{t :\ x_t = 0\}$. That is, τ is the length of a replenishment cycle.

Let h be the unit holding cost per unit of time. Given the initial state $x_0 = S$ and an admissible p_t, the corresponding reward function is given as follows:

$$J(S,p) = E\int_0^{\mathcal{T}} e^{-\rho t}\Big[-hx_t + p_t\lambda(p_t)\Big]dt, \tag{23.2}$$

where $\rho > 0$ is the discount factor. The objective of the problem is to choose p over time so as to maximize $J(S,p)$.

The value function is defined by

$$V(S) = \sup_{\text{admissible } p.} J(S,p).$$

Remark 23.1. Here we assume a constant demand variability σ. Chen et al. (2010) consider a more general demand variability case. Namely, the demand variability σ can depend on $\lambda(p)$, with $(\sigma(x))^2/x$ satisfying convexity and being twice continuously differentiable with respect to x. This general demand variability can cover the linear demand variability with $\sigma(x) = \sigma x$ and the square-root demand variability $\sigma(x) = \sigma\sqrt{x}$.

23.3. Properties of the value function

In this section, we study various properties of the value functions, and use the smooth-fit method to solve the corresponding optimal control problem.

Formally, the associated HJB equation should have the form:

$$\rho V(x) = \sup_{p\in[a,b]}\Big\{-\lambda(p)\frac{dV(x)}{dx} + \frac{\sigma^2}{2}\frac{d^2V(x)}{dx^2} + p\lambda(p) - hx\Big\}, \tag{23.3}$$

for $x > 0$, with $V(0) = 0$. The optimal feedback control $p^* = p^*(x)$ is typically given by

$$p^*(x) = \operatorname{argmax}\left\{-\lambda(p)\frac{dV(x)}{dx} + p\lambda(p)\right\}.$$

It is easy to see that

$$\operatorname{argmax}\left\{-\lambda(p)\frac{dV(x)}{dx} + p\lambda(p)\right\} = \operatorname{argmax}\left\{\frac{\lambda_1\left(-\frac{dV(x)}{dx} + p\right)}{p+\lambda_0}\right\}.$$

Note that

$$\frac{\lambda_1\left(-\frac{dV(x)}{dx} + p\right)}{p+\lambda_0} = \lambda_1\left[1 - \frac{\frac{dV(x)}{dx} + \lambda_0}{p+\lambda_0}\right].$$

Therefore, the optimal pricing p^* should have the form:

$$p^*(x) = \begin{cases} b & \text{if } \dfrac{dV(x)}{dx} + \lambda_0 \geq 0, \\ a & \text{if } \dfrac{dV(x)}{dx} + \lambda_0 < 0. \end{cases}$$

In addition, note that the convexity of $\lambda(p)$ and concavity of $p\lambda(p)$. One would expect the concavity of $V(x)$. However, this is not easy to prove directly, as in Sethi and Zhang (1994), due to the presence of the stopping time τ in the reward function. Nevertheless, if we have the concavity of $V(x)$, then there normally exists $x^* > 0$ such that

$$\frac{dV(x)}{dx} + \lambda_0 \geq 0 \text{ for } x \leq x^*,$$
$$\frac{dV(x)}{dx} + \lambda_0 < 0 \text{ for } x < x^*.$$

The optimal pricing p^* can be given in terms of x^*:

$$p^*(x) = \begin{cases} b & \text{if } x \leq x^*, \\ a & \text{if } x > x^*. \end{cases}$$

In this paper, we will show the existence of the hedging point x^*. Using x^*, the state space $[0, \infty)$ can be divided into the high price range (i.e., $p^* = b$) (HPR) $[0, x^*]$ and the low price range (i.e., $p^* = a$) (LPR) (x^*, ∞).

Lemma 23.1. *There exists a constant K such that $|V(x)| \leq K(1 + x)$.*

Proof. First, note that $0 \leq x_t \leq x - K_0 t + \sigma W_t$, for $t \in [0, \tau]$, where $K_0 = \lambda_1/(a + \lambda_0)$. So for all $t \geq 0$, we have $|x_t| \leq |x - K_0 t + \sigma W_t|$. Let $K_1 = b\lambda_1/(a + \lambda_0)$. It follows that, for any admissible p,

$$\begin{aligned} |J(x,p)| &\leq E \int_0^\tau e^{-\rho t}(h|x_t| + K_1)dt \\ &\leq E \int_0^\infty e^{-\rho t}(h|x - K_0 t + \sigma W_t| + K_1)dt \\ &= \int_0^\infty e^{-\rho t}(hE|x - K_0 t + \sigma W_t| + K_1)dt \\ &\leq \int_0^\infty e^{-\rho t}(h(|x| + K_0 t + \sigma E|W_t|) + K_1)dt \\ &\leq \int_0^\infty e^{-\rho t}(h(|x| + K_0 t + \sigma\sqrt{t}) + K_1)dt \\ &\leq K(1 + x), \end{aligned}$$

for some constant K. □

Next we solve the HJB equation. We assume the existence of the hedging point x^* and establish equations that determines its value. Then we obtain the corresponding solution $v(x)$ to the HJB equation (23.3). Finally we prove the solution derived this way is indeed the value function $V(x)$. For convenience, we use $V(x)$ for the value function and $v(x)$ for solution to the HJB equation.

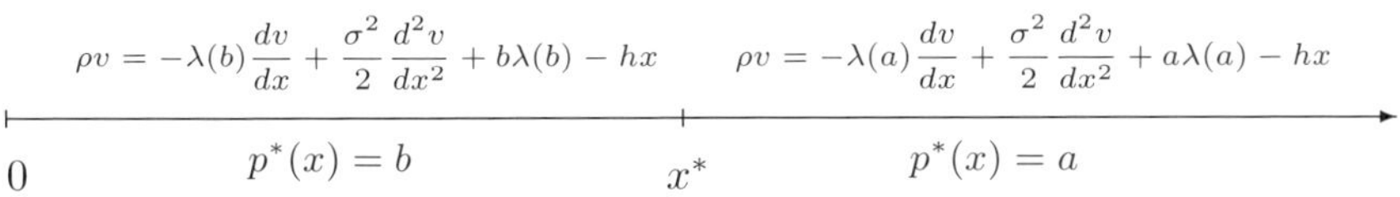

Fig. 23.1. Optimal pricing on different control regions.

Intuitively, if the inventory x_t is high, i.e., $x_t > x^*$, then one should focus on depleting the inventory by lowering the price; on the other hand, it x_t is low, i.e., $x_t < x^*$, there is less pressure from inventory cost, one can raise the price so as to earn more profit. These control policies on different control regions, and the corresponding HJB equations, are marked in Figure 23.1.

Recall that, on the interval $[0, x^*]$, the HJB equation becomes

$$\rho v(x) = -\lambda(b)\frac{dv(x)}{dx} + \frac{\sigma^2}{2}\frac{d^2v(x)}{dx^2} + b\lambda(b) - hx. \tag{23.4}$$

Let β_1 and β_2 denote the eigenvalues of

$$\frac{\sigma^2}{2}\beta^2 - \lambda(b)\beta - \rho = 0.$$

Then

$$\beta_1 = \frac{\lambda(b) - \sqrt{\lambda^2(b) + 2\rho\sigma^2}}{\sigma^2} < 0 \text{ and } \beta_2 = \frac{\lambda(b) + \sqrt{\lambda^2(b) + 2\rho\sigma^2}}{\sigma^2} > 0.$$

A special solution of (23.4) can be given as

$$\phi_1(x) = -\frac{hx}{\rho} + \lambda(b)\left(\frac{h}{\rho^2} + \frac{b}{\rho}\right).$$

Therefore, the general solution is of the form:

$$v(x) = B_1 e^{\beta_1 x} + B_2 e^{\beta_2 x} + \phi_1(x),$$

for constants B_1 and B_2.

We next consider the solution on $[x^*, \infty)$. Similar to the previous case, the HJB equation becomes

$$\rho v(x) = -\lambda(a)\frac{dv(x)}{dx} + \frac{\sigma^2}{2}\frac{d^2v(x)}{dx^2} + a\lambda(a) - hx. \tag{23.5}$$

Let γ_1 and γ_2 denote the eigenvalues of

$$\frac{\sigma^2}{2}\gamma^2 - \lambda(a)\gamma - \rho = 0.$$

Then

$$\gamma_1 = \frac{\lambda(a) - \sqrt{\lambda^2(a) + 2\rho\sigma^2}}{\sigma^2} < 0 \text{ and } \gamma_2 = \frac{\lambda(a) + \sqrt{\lambda^2(a) + 2\rho\sigma^2}}{\sigma^2} > 0.$$

A special solution of (23.5) can be given as

$$\phi_2(x) = -\frac{hx}{\rho} + \lambda(a)\left(\frac{h}{\rho^2} + \frac{a}{\rho}\right).$$

The general solution on this interval is given by

$$v(x) = C_1 e^{\gamma_1 x} + C_2 e^{\gamma_2 x} + \phi_2(x),$$

for constants C_1 and C_2.

In view of Lemma 23.1, it not difficult to see that $C_2 = 0$ because otherwise $|v(x)|$ grows exponentially as $x \to \infty$. Using these two general solutions, we determine the values of the coefficients B_1, B_2, and C_1 based on the following conditions:

$$\begin{cases} v(0) = 0, & \dfrac{dv((x^*)^+)}{dx} + \lambda_0 = 0, \\ \dfrac{dv((x^*)^-)}{dx} + \lambda_0 = 0, & \text{and } v((x^*)^+) = v((x^*)^-). \end{cases}$$

Therefore, we have

$$\begin{cases} B_1 + B_2 + \phi_1(0) = 0, \\ B_1\beta_1 e^{\beta_1 x^*} + B_2\beta_2 e^{\beta_2 x^*} + \dfrac{d\phi_1(x^*)}{dx} = -\lambda_0, \\ C_1\gamma_1 e^{\beta_1 x^*} + \dfrac{d\phi_2(x^*)}{dx} = -\lambda_0, \\ B_1 e^{\beta_1 x^*} + B_2 e^{\beta_2 x^*} + \phi_1(x^*) = C_1 e^{\gamma_1 x^*} + \phi_2(x^*). \end{cases} \tag{23.6}$$

Solving the third equation above for C_1 gives

$$C_1 = \frac{1}{\gamma_1}\left(\frac{h}{\rho} - \lambda_0\right) e^{-\gamma_1 x^*}.$$

Using the second and fourth equations in (23.6), we can write B_1 and B_2 in terms of x^* as follows

$$\begin{pmatrix} B_1 \\ B_2 \end{pmatrix} = \begin{pmatrix} e^{\beta_1 x^*} & e^{\beta_2 x^*} \\ \beta_1 e^{\beta_1 x^*} & \beta_2 e^{\beta_2 x^*} \end{pmatrix}^{-1} \begin{pmatrix} \frac{1}{\gamma_1}\left(\frac{h}{\rho} - \lambda_0\right) + \phi_2(x^*) - \phi_1(x^*) \\ \frac{h}{\rho} - \lambda_0 \end{pmatrix}.$$

Finally, using the first equation in (23.6), we obtain the equation for x^*

$$(1,\,1)\begin{pmatrix} e^{\beta_1 x^*} & e^{\beta_2 x^*} \\ \beta_1 e^{\beta_1 x^*} & \beta_2 e^{\beta_2 x^*} \end{pmatrix}^{-1} \begin{pmatrix} \frac{1}{\gamma_1}\left(\frac{h}{\rho} - \lambda_0\right) + \phi_2(x^*) - \phi_1(x^*) \\ \frac{h}{\rho} - \lambda_0 \end{pmatrix} = -\lambda(b)\left(\frac{h}{\rho^2} + \frac{b}{\rho}\right). \tag{23.7}$$

23.4. A verification theorem

In this section, we give a verification theorem to show that the solution $v(x)$ of equation (23.3) is equal to the value function $V(x)$, and the control $p^*(x)$ is optimal.

Theorem 23.1. *Assume $h \geq \rho\lambda_0$ and $B_1\beta_1^2 + B_2\beta_2^2 \leq 0$. Let x^* be a nonnegative solution to (23.7).*

Let

$$v(x) = \begin{cases} B_1 e^{\beta_1 x} + B_2 e^{\beta_2 x} + \phi_1(x) & \text{if } 0 \leq x < x^*, \\ C_1 e^{\gamma_1 x} + \phi_2(x) & \text{if } x \geq x^*, \end{cases}$$

with

$$\begin{pmatrix} B_1 \\ B_2 \end{pmatrix} = \begin{pmatrix} e^{\beta_1 x^*} & e^{\beta_2 x^*} \\ \beta_1 e^{\beta_1 x^*} & \beta_2 e^{\beta_2 x^*} \end{pmatrix}^{-1} \begin{pmatrix} \frac{1}{\gamma_1}\left(\frac{h}{\rho} - \lambda_0\right) + \phi_2(x^*) - \phi_1(x^*) \\ \frac{h}{\rho} - \lambda_0 \end{pmatrix},$$

$$C_1 = \frac{1}{\gamma_1}\left(\frac{h}{\rho} - \lambda_0\right) e^{-\gamma_1 x^*},$$

$$\phi_1(x) = -\frac{hx}{\rho} + \lambda(b)\left(\frac{h}{\rho^2} + \frac{b}{\rho}\right),$$

$$\textit{and } \phi_2(x) = -\frac{hx}{\rho} + \lambda(a)\left(\frac{h}{\rho^2} + \frac{a}{\rho}\right).$$

Then, $v(x)$ is concave and it is indeed the value function, i.e., $v(x) = V(x)$.

Moreover, let

$$p^*(x) = \begin{cases} b & \text{if } x \le x^*, \\ a & \text{if } x > x^*. \end{cases}$$

Then, $p^*(x)$ is optimal, i.e., $V(x) = J(x, p^*(x.))$.

Proof. The proof is divided into four steps.

Step 1. We show that $v(x)$ is twice continuously differentiable.

Note that $v(x)$ is twice continuously differentiable on $[0, x^*)$ and (x^*, ∞). It suffices to show

$$\frac{d^2 v((x^*)^+)}{dx^2} = \frac{d^2 v((x^*)^-)}{dx^2}. \tag{23.8}$$

Using the continuous differentiability of $v(x)$ and twice continuous differentiability of $v(x)$ on $[0, x^*)$ and (x^*, ∞), we use the HJB equation on these two intervals and take limits as $x \to (x^*)^-$ and $x \to (x^*)^+$, respectively. It follows that

$$\begin{cases} \rho v(x^*) = -\lambda(b)\dfrac{dv(x^*)}{dx} + \dfrac{\sigma^2}{2}\dfrac{d^2 v((x^*)^+)}{dx^2} + b\lambda(b) - hx^*, \\ \rho v(x^*) = -\lambda(a)\dfrac{dv(x^*)}{dx} + \dfrac{\sigma^2}{2}\dfrac{d^2 v((x^*)^-)}{dx^2} + a\lambda(a) - hx^*. \end{cases}$$

Recall that

$$\frac{dv(x^*)}{dx} = -\lambda_0.$$

We have

$$\begin{cases} \rho v(x^*) = \lambda_0\lambda(b) + \dfrac{\sigma^2}{2}\dfrac{d^2 v((x^*)^+)}{dx^2} + b\lambda(b) - hx^*, \\ \rho v(x^*) = \lambda_0\lambda(a) + \dfrac{\sigma^2}{2}\dfrac{d^2 v((x^*)^-)}{dx^2} + a\lambda(a) - hx^*. \end{cases}$$

Eliminating the common terms in these two equations one obtains

$$\lambda_0\lambda(b) + \frac{\sigma^2}{2}\frac{d^2 v((x^*)^+)}{dx^2} + b\lambda(b) = \lambda_0\lambda(a) + \frac{\sigma^2}{2}\frac{d^2 v((x^*)^-)}{dx^2} + a\lambda(a).$$

Recall also that $\lambda(p) = \lambda_1/(p + \lambda_0)$. It is easy to see that

$$\lambda_0\lambda(b) + b\lambda(b) = \lambda_0\lambda(a) + a\lambda(a).$$

This implies (23.8).

Step 2. We show that $dv/dx + \lambda_0 \geq 0$ on $[0, x^*]$ and $dv/dx + \lambda_0 \leq 0$ on $[x^*, \infty)$.

Let $\psi(x) = dv(x)/dx + \lambda_0$. We first consider the interval $[x^*, \infty)$. On this interval, $\psi(x) = C_1\gamma_1 e^{\gamma_1 x} - h/\rho + \lambda_0$. The definition of x^* implies that $\psi(x^*) = 0$. In addition, $\psi(\infty) = \lambda_0 - h/\rho$. Note that $\psi(x)$ is monotone. It follows that $\psi(x) \leq 0$ on $[x^*, \infty)$.

Next, note that, on $[0, x^*]$,

$$\psi(x) = B_1\beta_1 e^{\beta_1 x} + B_2\beta_2 e^{\beta_2 x} - h/\rho + \lambda_0,$$

and $\psi(x^*) = 0$. To see $\psi(x) \leq 0$, it suffices to show $\psi'(x) \leq 0$ on $[0, x^*]$. To this end, note that

$$\psi'(x) = B_1\beta_1^2 e^{\beta_1 x} + B_2\beta_2^2 e^{\beta_2 x} = e^{\beta_1 x}[B_1\beta_1^2 + B_2\beta_2^2 e^{(\beta_2 - \beta_1)x}].$$

Moreover, $\psi'(x^*) = d^2v(x^*)/dx^2 \leq 0$ because of the twice continuous differentiability of $v(x)$ and $d^2v(x^*)/dx^2 \leq 0$ on $[x^*, \infty)$. Furthermore, $\psi'(0) = B_1\beta_1^2 + B_2\beta_2^2 < 0$. These imply $\psi'(x) = e^{\beta_1 x}[B_1\beta_1^2 + B_2\beta_2^2 e^{(\beta_2 - \beta_1)x}] \leq 0$ on $[0, x^*]$ because $[B_1\beta_1^2 + B_2\beta_2^2 e^{(\beta_2 - \beta_1)x}]$ is monotone.

Step 3. We show the convexity of $v(x)$.

Note that the function $v(x)$ is concave on $[0, x^*]$ because $\psi'(x) = d^2v(x)/dx^2 \leq 0$ on this interval. It remains to show $\psi'(x) \leq 0$ on $[x^*, \infty)$. Note that $\psi(x^*) = 0$ implies $C_1 \leq 0$ under the assumption $h > \rho\lambda_0$. Therefore, $\psi'(x) = C_1\gamma_1^2 e^{\gamma_1 x} \leq 0$.

Step 4. Finally, we prove that $v(x) = V(x)$ and $p^*(x)$ is optimal.

Based on Steps 1 and 2, $v(x)$ is a solution to the HJB equation (23.3). Given an admissible p, using Dynkin's formula and $v(0) = 0$, it is easy to see that $Ee^{-\rho\tau}v(x_\tau) = 0$ and

$$\begin{aligned}
v(x) &= E\int_0^\tau e^{-\rho t}[-hx_t + p_t\lambda(p_t)]dt \\
&\quad + E\int_0^\tau e^{-\rho t}\left[\rho v(x_t) + \lambda(p_t)\frac{dv(x)}{dx} - \frac{\sigma^2}{2}\frac{d^v(x)}{dx^2} - p_t\lambda(p_t) + hx_t\right]dt \\
&\geq E\int_0^\tau e^{-\rho t}[-hx_t + p_t\lambda(p_t)]dt \\
&= J(x, p).
\end{aligned}$$

The equality holds if $p_t = p^*(x_t)$. This completes the proof. □

23.5. Numerical examples

In this section, we consider a numerical example with the following specifications:

$$\rho = 0.5,\ h = 1,\ \sigma = 0.3,\ \lambda_0 = 1,\ \lambda_1 = 2,\quad \text{and } [a, b] = [2, 8].$$

Both conditions required in the verification theorem are satisfied. We solve the equation (23.7) which gives $x^* = 1.27$. The corresponding value function $V(x)$ is plotted in Figure 23.2. It is clear from this picture that $V(x)$ is concave.

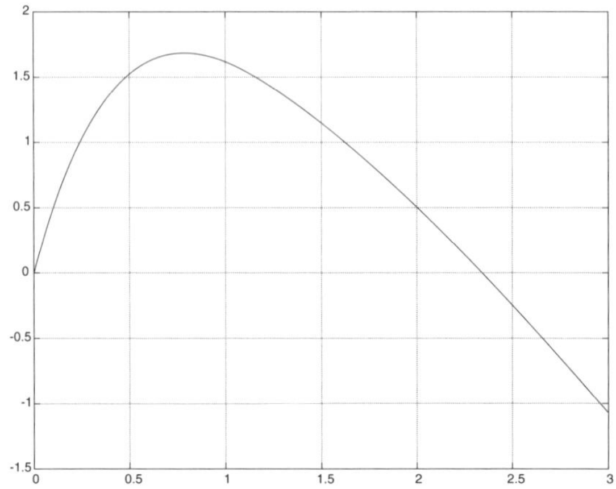

Fig. 23.2. Value function $V(x)$.

We next vary one of the parameters at a time and examine the dependence of x^* on the changes of these parameters.

First we compute the hedging point x^* associated with varying h. Intuitively, larger h would result higher holding cost. Thus the manager would use the price leverage to reduce the inventory level. As a result, the LPR $[x^*, \infty)$ should increase which corresponds decreasing x^*. These are confirmed by the results given in Table 23.1.

Table 23.1. x^* with varying h.

h	0.6	0.8	1	1.2	1.4
x^*	2.18	1.55	1.27	1.09	0.97

Next, we vary σ. A larger σ implies larger variability in the demand. One strategy facing high variability is to raise the price level to maintain higher revenue because it is very likely for x_t to fall below x^* from time to time. This corresponds to larger HPR $[0, x^*]$ and therefore increasing x^*. We see in Table 23.2 that the hedging point x^* is indeed monotonically increasing in σ.

Table 23.2. x^* with varying σ.

σ	0.1	0.2	0.3	0.4	0.5
x^*	1.06	1.16	1.27	1.38	1.47

In Table 23.3, we vary λ_1. Larger λ_1 implies greater capability to reduce inventory when needed, which suggests that there is no need to maintain high level of inventory. The corresponding control strategy should focus on higher revenue and therefore larger HPZ $[0, x^*]$. Table 23.3 shows that the hedging point x^* is increasing in λ_1.

Table 23.3. x^* with varying λ_1.

λ_1	1.2	1.6	2	2.4	2.6
x^*	0.88	1.08	1.27	1.46	1.55

Finally, we select different price ranges $[a, b]$. The resulting values for x^* in Table 23.4 suggest that x^* is decreasing in $[a, b]$. This is because larger $[a, b]$ says less lower range pricing is available. To compensate the lack of low price range, one has to increase LPR $[x^*, \infty)$ which means decreasing x^*.

Table 23.4. x^* with varying $[a, b]$.

$[a, b]$	[0,6]	[1,7]	[2,8]	[3,9]	[4,10]
x^*	1.41	1.33	1.27	1.22	1.18

23.6. Conclusion

A simple pricing model is considered in this paper. It shows that the optimal price can be given in terms of the hedging point x^*, which can be obtained by solving an algebraic equation. It would be interesting to extend the results in this paper to incorporate more general models with possibly jumping demand process. Nevertheless, the closed-form solution obtained in this paper can help provide better understanding in more general cases.

Acknowledgement

This research was inspired by discussions of the first author with Professor Hong Chen at University of British Columbia and Professor David Yao at Columbia University. Part of this work was done when the first author was visiting Department of Mathematics, University of Georgia, and the first author would like to thank the hospitality of the department.

References

[1] J.A. Bather, A continuous inventory model. *Journal of Applied Probability*, **3**(1966), 538-549.

[2] L. Benkherouf and A. Bensoussan, Optimality of an (s, S) policy with compound Poisson and diffusion demands: a quasi-variational inequalities approach. *SIAM Journal of Control and Optimization*, **48**(2009), 756-762.

[3] A. Bensoussan, R.H. Liu and S.P. Sethi, Optimality of an (s, S) policy with compound Poisson and diffusion demands: a quasi-variational inequalities approach. *SIAM Journal of Control and Optimization*, **44**(2005), 1650-1676.

[4] J. Constantinides and S. Richard, Existence of optimal simple policies for discounted-cost inventory and cash management in continuous time. *Operations Research*, **26**(1978), 620-636.

[5] J. Byrnes, Dell manages profitability, not inventory. *Harvard Business School Working Knowledge*, June 2003. Available at http://hbswk.hbs.edu/archive/3497.html.

[6] H. Chen, O. Wu and D. Yao, On the benefit of inventory-based dynamic pricing strategies, *Production and Operations Management*, **19**(2010), 249-260.

[7] X. Chen and D. Simchi-Levi, Coordinating inventory control and pricing strategies: the continuous review model. *Operations Research Letters*, **34**(2006), 323-332.

[8] X. Chen and D. Simchi-Levi, Coordinating inventory control and pricing strategies with random demand and fixed ordering cost: the finite horizon case. *Operations Research*, **52**(2004a), 887-896.

[9] X. Chen and D. Simchi-Levi, Coordinating inventory control and pricing strategies with random demand and fixed ordering cost: the infinite horizon case. *Mathematics of Operations Research*, **29**(2004b), 698-723.

[10] A. Federgruen and A. Heching, Combined pricing and inventory control under uncertainty. *Operations Research*, **47**(1999), 454-475.

[11] Y. Feng and F.Y. Chen, Joint pricing and inventory control with setup costs and demand uncertainty. *Working paper*, Chinese University of Hong Kong, China, 2003.

[12] J.M. Harrison, T.M. Sellke and M.I. Taksar, Impulse control of Brownian motion. *Mathematics of Operations Research*, **8**(1983), 454-466.

[13] P.B. Pashigan and B. Bowen, Why are products sold on sale, *The Quarterly Journal of Economics*, **106**(1991), 1014-1038.

[14] S. Richard, Optimal impulse control of a diffusion process with both fixed and proportional costs of control. *SIAM Journal of Control and Optimization*, **15**(1977), 79-91.

[15] S.P. Sethi and Q. Zhang, *Hierarchical Decision Making in Stochastic Manufacturing Systems*, Birkhäuser, Boston, 1994.

[16] B. Smith, J. Leimkuhler, R. Darrow and J. Samuels, Yield management at American airlines, *Interfaces*, **1**(1992), 8-31.

[17] A. Sulem, A solvable one-dimensional model of a diffusion inventory system. *Mathematics of Operations Research*, **11**(1986), 125-133.

[18] D.C. Yao, *Inventory Control Problem with Brownian Motion Demand.* PhD. Thesis, Institute of Applied Mathematics, Chinese Academy of Sciences, Beijing, China, 2010.

Author Index